# A Flora of Santa Cruz Island

# A Flora of Santa Cruz Island

Steve Junak, Tina Ayers,
Randy Scott, Dieter Wilken,
David Young

Illustrated by
Linda Ann Vorobik

Published by the Santa Barbara Botanic Garden
in collaboration with
the California Native Plant Society

1212 Mission Canyon Road
Santa Barbara, CA 93105

California Native Plant Society
1722 J Street, Suite 17
Sacramento, CA 95814

Text design by Linda Ann Vorobik.
Cover design and Title Page by Beth Hansen.
Cover photographs by Steve Junak.

ISBN 0-943460-23-9

# CONTENTS

## BENEFACTORS

*We gratefully acknowledge the financial support of the institutions who made this book possible.*

Flintridge Foundation
Homeland Foundation
Rogers Family Foundation
Santa Cruz Island Foundation
The Nature Conservancy
The California Native Plant Society

## DEDICATION

*This book is dedicated to the memory of Dr. Carey Stanton (1923-1987), who welcomed scientific exploration of Santa Cruz Island.*

# ACKNOWLEDGMENTS

We thank the late Carey Stanton, the late Henry Duffield, Lyndal Laughrin, Ann Bromfield, Jaret Owens, Thomas Nathanson, Susan Manchester, David and Debbie Welborn, Larry and Nancy Callahan, Keith Herold, and Lauren Melendrez for many favors and the hospitality they extended during visits to the island. Accommodations at the Santa Cruz Island Reserve Field Station and at the island's ranches were much appreciated.

For transportation to and from the island, we thank Aspen Helicopters, Channel Islands Aviation, Island Packers, Lyndal Laughrin, The Nature Conservancy, Charles J. Rennie III, and staff of the U. S. Navy.

For assistance in the field, we thank Carol Bornstein, Stan Davis, Carol Galbraith, Jeanne-Marie Herman, Beth Hobbs, Steve McCabe, Gail Milliken, Antoinette Padgett, Charles J. Rennie III, Karen Rich, Janet Takara, Steve Timbrook, Dirk Van Vuren, and Nancy Vivrette.

For sharing field notes and general observations, we thank Daniel Axelrod, Michael Benedict, Dennis Breedlove, Jim Blakley, Mary Carroll, Marla Daily, Charles Drost, Wayne Ferren, the late Ray Fosberg, Roman Gankin, Keith Herold, the late John T. Howell, Rob Klinger, Lyndal Laughrin, Ralph Philbrick, Peter Raven, Jonathan Sauer, Peter Schuyler, Clif Smith, Robert Thorne, Linda Ann Vorobik, Gary Wallace, Grady Webster, and members of the U.C.S.B. Botanical Society.

For historical research or photographs, we thank Penny Ahlstrand, Bea Beck, Mona Bourell, Marla Daily, Bill Dewey, Rebecca Eldridge, Tibby Gunther, Anita Karg, Terri Sheridan, Clif Smith, Margriet Weatherwax, and staff at the U. S. National Archives. John Johnson and Jan Timbrook provided information on Native Americans. Gloria Maender provided copies of unpublished manuscripts from the Proceedings of the Fourth California Islands Symposium, March 1994. Julie Broughton contributed the glossary. Elizabeth Painter reviewed the text and provided many useful editorial comments.

We thank Eva Hayward-George and Karen Rich for data entry, Jean Hobson for her diligent attention to and preparation of herbarium specimens, and Heidi David and Joann Tanner Rounds for preparing maps of Santa Cruz Island.

For annotations to herbarium specimens, assistance with taxonomically difficult groups, and other helpful comments, we deeply appreciate the help of Kelly Allred, George Argus, Mary Barkworth, Alva Day, Lauramay Dempster, Frederick Essig, J. Robert Haller, Richard Halse, Ron Hartman, Jim Hickman, Meredith Lane, Joy Mastrogiuseppe, Steve McCabe, Reid Moran, Randy Morgan, Robert Ornduff, Ralph Philbrick, Barbara Ralston, Rhonda Riggins, Reed C. Rollins, Andy Sanders, Alan Smith, and Warren L. Wagner. Kevin Nixon and Michael Warnock contributed treatments of *Quercus* and *Delphinium*, respectively.

For access to critical specimens and many favors, we thank the curators and staff of herbaria at the California Academy of Sciences, New York Botanic Garden, Rancho Santa Ana Botanic Garden, San Diego Museum of Natural History, the Smithsonian Institution (U. S. National Herbarium), the University of California at Berkeley, and the University of Notre Dame (Greene Herbarium).

# INTRODUCTION

## The California Channel Islands

On a clear day, several of the eight California Channel Islands can be seen from Los Angeles, one of the largest metropolitan areas in the United States. These islands have long captured the interest and imagination of southern Californians, beginning with the aboriginal inhabitants. Fortunately, the water barrier around the Channel Islands has discouraged casual visitation by mainland residents and the islands remain relatively undeveloped.

The Channel Islands are strikingly different from each other with respect to climate, geology, history, land forms, maximum elevation, size, and vegetation. They are natural areas known for windswept landscapes, rugged coastlines, and unspoiled beaches, often teeming with marine mammals and birds. Their floras, rich in endemic species, have attracted the attention of botanists and horticulturists for more than a century. The eight islands (Figure 1) have been divided into 2 major groups: 1) the Northern Channel Islands of San Miguel, Santa Rosa, Santa Cruz, and Anacapa, and 2) the Southern Channel Islands of Santa Barbara, San Nicolas, Santa Catalina, and San Clemente. The islands represent emergent portions of a complex system of submarine canyons and ridges in a geomorphic province referred to as the California Continental Borderland (Vedder 1976). They are situated between 32° 48' and 34° 05' N latitude and between 118° 21' and 120° 27' W longitude, and range in size from 1 to 96 miles$^2$ (2.6 to 249 km$^2$) (Table 1). Maximum elevations vary between 635 feet (194 m) and 2,470 feet (753 m).

**The Northern Channel Islands**. Geographically, the Northern Channel Islands form a relatively cohesive group of "fringing islands" (Carlquist 1974) about 55 miles (88 km) long. They are a westward extension of the Santa Monica Mountains and are separated from the mainland by the Santa Barbara Channel. Distances from the mainland vary from 13 to 27 miles (20 to 44 km). Channels between individual islands range from 3 to 6 miles (5 to 9 km) in width. All four islands are oriented with their longest axes in an east-west direction. San Miguel, Santa Rosa, and Anacapa islands are part of Channel Islands National Park, although San Miguel is owned by the U. S. Navy. The western 90% of Santa Cruz Island is owned by The Nature Conservancy, an international non-profit conservation organization; the eastern 10% is currently being acquired by the National Park Service from private owners.

**The Southern Channel Islands**. The Southern Channel Islands are more scattered geographically and more isolated from each other and from the mainland than the northern ones. Their longest axes are oriented in a north-south or northwest-southeast direction. They are separated from the mainland by the San Pedro Channel and are located 20 to 61 miles (32 to 98 km) from the mainland. Distances between individual islands range from 21 to 28 miles (34 to 45 km). Santa Barbara Island is part of Channel Islands National Park. San Nicolas and San Clemente islands are owned by the U. S. Navy and are closed to the public. Most of Santa Catalina Island is owned by the Santa Catalina Island Conservancy, a non-profit conservation foundation, and the Santa Catalina Island Company. There also are private holdings at the city of Avalon.

## Regional Geologic and Topographic History

Vedder and Howell (1980) reviewed the geologic and topographic history of the southern California coastal region and the California Channel Islands. They reported that past sea-level fluctuations greatly affected size and shape of the Channel Islands. Large areas on Santa Rosa, Santa Cruz, Santa Catalina, and San Clemente islands apparently have been continuously above sea level for the last 500,000 years. In contrast, San Nicolas and Santa Barbara (and possibly San Miguel and Anacapa) islands were completely submerged during the maximum extent of the ocean, probably during one of the interglacial periods of the Pleistocene Epoch.

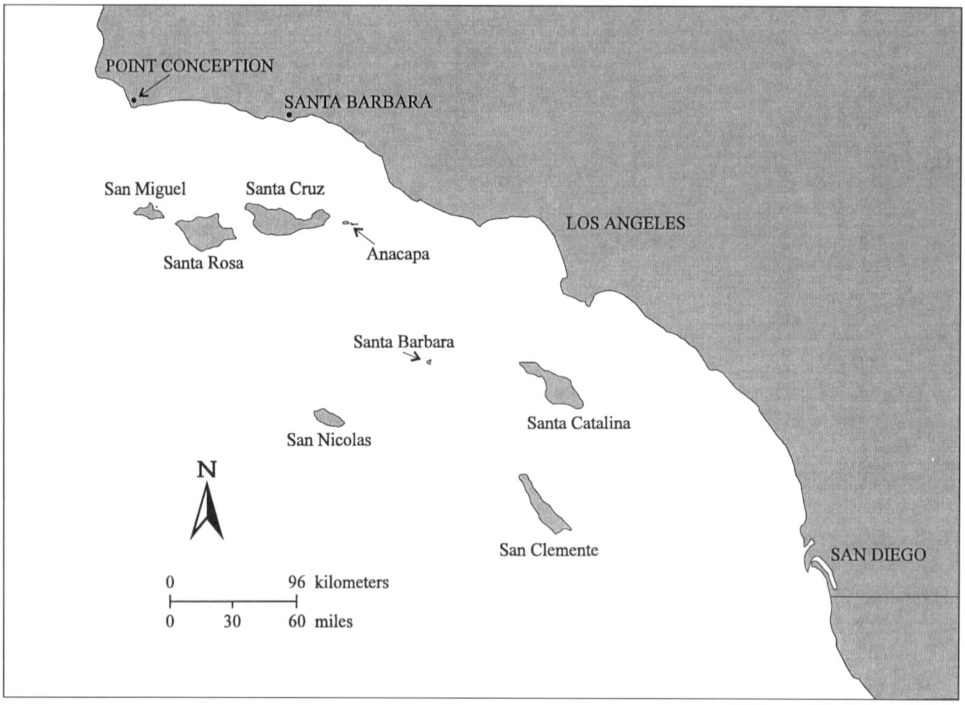


*Figure 1. The California Channel Islands and coastline of southern California.*

During the time of probable minimum sea levels about 17,000 to 18,000 years ago, the four present-day Northern Channel Islands probably were all part of one large landmass, referred to as "Santarosae" (Orr 1968; Vedder and Howell 1980). This landmass has been estimated to have been approximately 724 miles$^2$ (1874 km$^2$) in size. In contrast, the four Southern Channel Islands probably were separated by ocean channels.

Although the four Northern Channel Islands are geographically and geologically aligned with the Santa Monica Mountains, there currently are no geologic or marine geophysical data that suggest the presence of a Pleistocene land bridge between the mainland and the islands. Scientists assumed that a land bridge connected "Santarosae" to the mainland, primarily because mammoth fossils were found on the islands and because it was believed that elephants did not or could not swim. Recent studies on bathymetry of the channel floor have shown, however, that an ocean channel separated "Santarosae" from the mainland during the late Pleistocene (Junger and Johnson 1980, Vedder and Howell 1980). Furthermore, studies of modern elephants (Johnson 1980, Wenner and Johnson 1980) show that they are capable of swimming across channels as wide as the one that existed in the late Pleistocene.

## Physiography

**Topography**. Charles Holder (1910) aptly described the general aspect of Santa Cruz Island:

> *From the sea, Santa Cruz Island is a jumble of lofty hills and mountains, with deep gorges and canons winding in every direction. Hidden away in the very heart of this island is an ideal ranch, with a pronounced foreign atmosphere, in a climate as perfect as that of Avalon on the island to the south.*

Santa Cruz is the largest California Channel Island, with an area of 96 miles$^2$ (249 km$^2$). When visible from the mainland, its rugged outline dominates the horizon. The island is a familiar landmark for residents of the Santa Barbara region. Santa Cruz Island, included in Santa Barbara County, is centered at about 34° N latitude and between 119° 40' and 119° 50' W longitude. The island lies 19 miles (30 km) off the coast of southern California between Santa Rosa Island, 5.5 miles (9 km) to the west, and Anacapa Island, 4.5 miles (7 km) to the east. About 23.5 miles (38 km) in length, Santa Cruz Island is roughly 7.5 miles (12 km) wide at its widest point. The maximum elevation on Santa Cruz Island, 2470 feet (753 m), exceeds that of all other California Channel Islands.

TABLE 1. Physical Characteristics of the California Channel Islands.

| | Area | | Highest Elevation | | Distance to Mainland | |
|---|---|---|---|---|---|---|
| | mi² | (km²) | ft | (m) | mi | (km) |
| | | | Northern Channel Islands | | | |
| San Miguel | 14 | (37) | 830 | (253) | 26 | (42) |
| Santa Rosa | 84 | (217) | 1,589 | (484) | 27 | (44) |
| Santa Cruz | 96 | (249) | 2,470 | (753) | 19 | (30) |
| Anacapa | 1.1 | (2.9) | 930 | (283) | 13 | (20) |
| | | | Southern Channel Islands | | | |
| Santa Barbara | 1 | (2.6) | 635 | (194) | 38 | (61) |
| San Nicolas | 22 | (58) | 910 | (277) | 61 | (98) |
| Santa Catalina | 75 | (194) | 2,125 | (648) | 20 | (32) |
| San Clemente | 56 | (145) | 1,965 | (599) | 49 | (79) |

Santa Cruz is the most rugged and topographically diverse of the Northern Channel Islands. Its landmass is dominated by two longitudinal ridges (the North and South ridges) running in an east-west direction and an intervening valley (the Central Valley), which is 12.5 miles (20 km) long. The western portion of the North Ridge is over 14 miles (22.5 km) long, includes numerous peaks above 2,000 feet (610 m), and is truncated on its east end by Cañada del Puerto. The South Ridge is roughly 13.5 miles (21.7 km) long, with highest peaks at about 1,500 feet (460 m). Another ridge (El Montañon) runs diagonally across the east end of the island, in a northwest-southeast direction. El Montañon is nearly 4 miles (6.4 km) long, reaches a maximum elevation of just over 1,800 feet (550 m), and includes several peaks about 1,500 feet (457 m) high. Although ridge lines are visually dominant, the Central Valley is one of the island's most remarkable landscape features. The size and orientation of the Central Valley are unique features among the Channel Islands. Its presence profoundly affects the microclimate of the island's interior, as well as dividing it geologically and topographically.

The two longitudinal ridges and the Central Valley divide Santa Cruz Island into four ecological zones: 1) the coastal slopes, flats, and canyons on the north side (31.6% of the island); 2) the south-facing interior slopes and canyons above the Central Valley (15.2%); 3) the north-facing interior slopes and canyons on the south side of the same valley (10.9%); and 4) the south-facing coastal slopes, flats, and canyons on the south side (42.3%). A system of unpaved roads provides access to much of the island, with the notable exception of the North Ridge's coastal slopes. The rugged terrain in this area is accessible from the coastline or from the North Ridge.

**Watersheds**. The surface of Santa Cruz Island can be divided into 10 major watersheds, ranging in area from 22.5 miles² (8.7 km²) to 4 miles² (1.5 km²) (California State Water Resources Control Board 1979). On the north side, there are more than 35 major canyons measuring 1.5 to 2.5 miles (2.4-4 km) long. The largest canyons in this region are Cañada del Puerto and Scorpion Canyon, each about 2.5 miles (4 km) long. Many canyons of the North Ridge are steep and rocky, with elevation drops of 1,700 to 2,000 feet (518-610 m).

The Central Valley is actually a series of valleys composed of three separate drainages. Bremner (1932) described events that led to current drainage patterns. The major divide is Centinela, at an elevation of 1,250 feet (380 m). From this point, Cañada Christy drops westward to the ocean, over a distance of 4.6 miles (7.4 km). To the east of Centinela, runoff previously flowed in a relatively straight line to Valley Anchorage, on the south side of the island. As a result of subsequent uplift and stream capture, water in the middle 6.5 mile (10.5 km) section now flows into Cañada del Puerto, on the island's north shore.

The middle section of the Central Valley is a complicated drainage, which includes Cañada de la Portezuela and two branches of Cañada del Medio. Cañada de la Portezuela is now a hanging valley more than 400 feet (122 m) above Cañada del Medio, because eastward stream flow has been captured by Cañada de la Siesta. There apparently was a reversal of stream flow in the eastern branch of Cañada del Medio; the creek now drains westward from El Sobrante into Cañada del Puerto, instead of running east to Valley Anchorage. At the extreme eastern end of the Central Valley, the 1.4 mile (2.25 km) long Cañada Pomona drains eastward from the low divide at El Sobrante into the ocean on the island's south side.

*Aerial view of west end of Santa Cruz Island, looking east with West and Fraser Points in foreground (Pacific Western Aerial Surveys, February 1985, collection of Santa Cruz Island Foundation).*

On the island's south side, there are more than 20 large canyons, with lengths of 1 to 3.5 miles (1.6-5.6 km). The largest of these is Laguna Canyon, about 3.5 miles (5.6 km) long, with an elevation drop of 1250 feet. Other major canyons on the south side include Alamos, Willows, Montañon, and Smugglers' canyons. Alamos and Willows canyons are each about 3 miles (4.8 km) long, while Montañon and Smugglers' are about 2.4 miles (3.9 km) in length. The bottoms of these canyons are generally much wider and have a more gradual slope than those on the north side.

Santa Cruz Island has abundant groundwater, especially in larger watersheds. Numerous springs and seeps flow year-round and permanent groundwater can be found in many larger canyons. Reliable streams can be found on the north side, in most canyons between Valdez Canyon and Cañada del Puerto, and in Cañada del Agua. There also are reliable streams in the Central Valley in Cañada Christy and Cañada de la Siesta, and on the south side in Cañada de Malva Real, Alamos Canyon, and Aguaje Canyon near Yellowbanks Anchorage.

**Climate**. Santa Cruz Island has a Mediterranean climate, with mild wet winters and dry warm summers. However, because of varied topography, the major ecological areas experience significant microclimatic differences in temperature extremes, cold air drainage, precipitation, and the occurrence and persistence of summer fog (Hochberg 1980, James 1982). The prevailing northwest winds drive moisture-laden marine air and summer fogs onto north-facing coastal slopes, while the Central Valley is more isolated from marine influence. Temperature fluctuations are greater in interior sites than at the coast. Freezing temperatures are usually experienced in the Central Valley each winter, but the rest of the island is essentially frost-free (Philbrick and Haller 1977). Average summer temperatures were found to be 25°-28° F (6°-10° C) higher in the Central Valley than at Prisoners' Harbor between 1967 and 1974 (James 1982). Summer fogs and/or low clouds are common along the coastline but less common in the Central Valley. Hochberg (1980) found that the Central Valley was densely overcast only 7% of the time and clear 67% of the time between the months of June and October. Coastal regions, however, were partly or completely overcast as much as 50% of the time.

*Aerial view of northwestern portion of Santa Cruz Island, looking southwest with Santa Rosa and San Miguel Islands in background (S. Junak, February 1987).*

*Aerial view of east end of Central Valley and South Ridge, looking east with Anacapa Island and Santa Monica Mountains in background (S. Junak, January 1988).*

*Coastal bluffs, north side of the island near West Point, looking northeast (S. Junak, June 1980).*

*Coastal bluffs, east of mouth of Willows Canyon, looking east (S. Junak, April 1992).*

*Lower part of Laguna Canyon, looking south (S. Junak).*

TABLE 2. Mean Monthly Temperatures (Stanton Ranch, 1961-1971). Data courtesy of C. Stanton; compiled by R. Yeaton.

| Month | °F | °C |
|---|---|---|
| Jan | 53.4 | 11.9 |
| Feb | 55.2 | 12.9 |
| Mar | 55.7 | 13.2 |
| Apr | 58.2 | 14.6 |
| May | 60.9 | 16.1 |
| Jun | 63.5 | 17.5 |
| Jul | 67.5 | 19.7 |
| Aug | 69.7 | 20.9 |
| Sep | 66.8 | 19.3 |
| Oct | 63.0 | 17.2 |
| Nov | 57.8 | 14.3 |
| Dec | 53.1 | 11.7 |
| Mean: | 60.4 | 15.8 |

TABLE 4. Summary of Mean Monthly Precipitation (Stanton Ranch, 1904-1993 Seasons). Data courtesy of C. Stanton and The Nature Conservancy; compiled by L. Laughrin.

| Month | inches | (s.d.) | mm |
|---|---|---|---|
| Jul | 0.01 | (0.06) | 0.25 |
| Aug | 0.04 | (0.18) | 1.01 |
| Sep | 0.29 | (0.63) | 7.37 |
| Oct | 0.53 | (0.87) | 13.46 |
| Nov | 1.61 | (1.93) | 40.89 |
| Dec | 3.36 | (3.27) | 85.34 |
| Jan | 4.45 | (4.22) | 113.03 |
| Feb | 4.50 | (3.75) | 114.30 |
| Mar | 3.34 | (3.28) | 84.84 |
| Apr | 1.25 | (1.63) | 31.75 |
| May | 0.31 | (0.70) | 7.87 |
| Jun | 0.09 | (0.35) | 2.28 |
| Total: | 19.7 | (8.7) | 500.6 |

TABLE 3. Absolute Maximum and Minimum Temperatures and Dates of Record (Stanton Ranch, 1961-1978). Data courtesy of C. Stanton; compiled by L. Laughrin.

| Year | Absolute Maximum | | | Absolute Minimum | | |
|---|---|---|---|---|---|---|
| | °F | (°C) | Date of Record | °F | (°C) | Date of Record |
| 1961 | 98 | (37) | Oct 14 | 30 | (-1) | Jan 5, Jan 7 |
| 1962 | 94 | (34) | Aug 9 | 25 | (-4) | Dec 28 |
| 1963 | 104 | (40) | Sep 27 | 25 | (-4) | Jan 15 |
| 1964 | 96 | (36) | Sep 23 | 29 | (-2) | Feb 10 |
| 1965 | 96 | (36) | Sep 16 | 30 | (-1) | Dec 20 |
| 1966 | 90 | (32) | Oct 24 | 29 | (-2) | Jan 3 |
| 1967 | 98 | (37) | Aug 30 | 31 | (-1) | Jan 7 |
| 1968 | 96 | (36) | May 26, Sep 25 | 27 | (-3) | Dec 5 |
| 1969 | 90 | (32) | Sep 30 | 25 | (-4) | Dec 31 |
| 1970 | 98 | (37) | May 15, Sep 27 | 22 | (-6) | Jan 30 |
| 1971 | 99 | (37) | Sep 13 | 18 | (-8) | Dec 9 |
| 1972 | 102 | (39) | Aug 22 | 19 | (-7) | Jan 6 |
| 1973 | 101 | (38) | Jun 20 | 18 | (-8) | Jan 6 |
| 1974 | 92 | (33) | Jun 25, Oct 16 | 22 | (-6) | Dec 24 |
| 1975 | 99 | (37) | Sep 23 | 18 | (-8) | Dec 18 |
| 1976 | 99 | (37) | Jun 24 | 21 | (-6) | Jan 26 |
| 1977 | 102 | (39) | Jul 27 | 24 | (-4) | Jan 10 |
| 1978 | 102 | (39) | Jul 15 | 19 | (-7) | Dec 20 |
| Mean: | 98 | (36) | | 24 | (-4) | |

The only long-term temperature data have been collected in the Central Valley at the Stanton Ranch (elevation 60 m). An 11-year record (1961-1971) of temperature data is summarized in Table 2. The data show a mean annual temperature of 60.4° F (15.8° C) and a range in mean monthly temperatures from 53.1°-69.7° F (11.7°-

20.9° C). Absolute maximum and minimum temperatures were recorded at the Stanton Ranch for every year between 1961 and 1978 (Table 3). Absolute minimum temperatures from 18°-31° F (-8° to -1° C) were recorded between December and February. Absolute maximum temperatures from 90°-104° F (32°-40° C) were recorded between May and October.

The only long-term precipitation data also were recorded at the Stanton Ranch. Eighty-eight seasons of rainfall data, measured between July 1904 and June 1993 at the Stanton Ranch, show an annual mean of 19.71 inches (501 mm) (Table 4). During this period, extremes ranged from 6.35 inches (161.3 mm) in 1989-1990 to 56.15 inches (1,426 mm) in 1940-1941. Brumbaugh (1983) produced an isohyet map of estimated mean annual precipitation, based on an analysis of rainfall data from 8 weather stations on the island. According to Brumbaugh, estimated mean annual precipitation ranges from less than 11.8 inches (300 mm) per year in low-lying areas to 23.6 inches (600 mm) per year at highest elevations on Santa Cruz Island. Mean precipitation, based on a 10 year record at four stations (Table 5), ranged from 12.6 inches (321 mm) to 20 inches (508 mm). In most years approximately 80% of annual precipitation falls between November and March, mostly as rainfall. Snowfall is an extremely rare occurrence. The last time that the Central Valley had snow was during the winter of 1949. Summer thundershowers also are rare.

TABLE 5. Summary of Mean Annual Precipitation (Four Stations, 1972-1982), estimated in inches (mm). Data courtesy of C. Stanton; compiled by L. Laughrin.

| Season | Stanton Ranch | | Christy Ranch | | Navy Base | | Prisoners' Harbor | |
|---|---|---|---|---|---|---|---|---|
| 1972-73 | 21.1 | (537) | 11.7 | (297) | 18.2 | (462) | 23.0 | (584) |
| 1973-74 | 16.6 | (423) | 11.0 | (279) | 17.7 | (451) | 13.0 | (330) |
| 1974-75 | 15.0 | (380) | 10.4 | (265) | 12.9 | (327) | 14.9 | (378) |
| 1975-76 | 6.8 | (174) | 4.9 | (124) | 6.2 | (158) | 6.7 | (171) |
| 1976-77 | 12.9 | (328) | 9.8 | (248) | 12.7 | (323) | 11.6 | (295) |
| 1977-78 | 43.8 | (1131) | 28.8 | (732) | 40.7 | (1034) | 30.0 | (763) |
| 1978-79 | 25.4 | (645) | 13.7 | (347) | 22.0 | (559) | 17.6 | (447) |
| 1979-80 | 25.5 | (648) | 13.5 | (342) | 19.5 | (495) | 24.6 | (624) |
| 1980-81 | 16.1 | (408) | 11.3 | (288) | 13.4 | (340) | 13.3 | (337) |
| 1981-82 | 16.8 | (426) | 11.4 | (289) | 14.0 | (355) | 12.0 | (305) |
| Mean | 20.0 | (508) | 12.6 | (321) | 17.7 | (450) | 16.7 | (423) |
| S.D. | 10.1 | (256) | 6.2 | (157) | 9.2 | (234) | 7.1 | (181) |

## Geology and Its Influence on Vegetation

The diverse rock types of Santa Cruz Island, each with different erosional characteristics, physical features, and chemical compositions, have affected development of the island's topography. Easily eroded rock types produce rounded land forms, while rocks more resistant to erosion produce rugged peaks and cliffs. Geology determines many landscape patterns, such as distribution of soil types, groundwater, plant communities, and even individual plant species.

The island's geology has been described by several workers (Bremner 1932, Rand 1933, several papers in Weaver 1969a and Howell 1976, Dibblee 1982). A total of 28 substrates and subunits were mapped by Weaver (1969a), ranging in age from the Jurassic to Quaternary periods. Volcanic rocks, including Santa Cruz Island Volcanics and intrusive rocks, dominate the island's landscape and cover 42.4% of its surface area. Tertiary sedimentary and metamorphic substrates cover 36.3% of the island and include the Monterey Formation (16.1% of its surface area), Blanca Volcaniclastics (12.7%), Cañada Formation (2.2%), Cozy Dell Shale (0.7%), Jolla Vieja Formation (0.5%), Pozo Formation (0.1%), Rincon Member (0.6%), San Onofre Breccia (2.0%), and Vaqueros Sandstone (1.4%). Outcrops of metamorphic and granitic basement rocks cover 13.7%, including Santa Cruz Island Schist (8.6%), Alamos Plutonite (0.4%), Willows Diorite (3.9%), and undifferentiated Santa Cruz Island Schist/Willows Diorite (0.8%). Quaternary deposits are found on 7.6%, including Quaternary alluvium (2.9%), fanglomerates (0.3%), Middle Anchorage alluvium (0.1%), Potato Harbor Formation (0.3%), and terrace gravels (4.0%).

The geographic distribution of major geologic substrates on Santa Cruz Island is shown in Figure 2. The Santa Cruz Island Fault, which runs through the Central Valley, divides the major rock types. On the north side of the fault, Santa Cruz Island Volcanics dominate. This thick sequence of Miocene volcanic and volcaniclastic rocks consists largely of lava flows, flow breccias, and interbedded volcaniclastic rocks (Howell et al. 1976). The composition of these rocks ranges from basalt to dacite or rhyodacite. Near the eastern end of the island, the volcanics are overlain by the Monterey Formation, a Miocene succession of rhythmically bedded, diatomaceous, tuffaceous, and siliceous shale (Weaver and Meyer 1969). The only other rock unit exposed north of the Santa Cruz Island Fault is the Potato Harbor Formation, an eolianite deposit consisting of sandy limestone. The Potato Harbor Formation unconformably overlies volcanic rocks and the Monterey Formation (Weaver and Meyer 1969). South of the Santa Cruz Island Fault, a more diverse geology is exposed. Most high elevation areas are underlain by a west-northwest trending mass of Jurassic basement plutonic and metamorphic rocks. Included in this group are Santa Cruz Island Schist, Willows Diorite, and Alamos Plutonite. The rest of the island is characterized by Eocene to Miocene sedimentary and metamorphic rocks, with minor amounts of Pleistocene alluvial deposits.

In the following discussion, geologic substrates are grouped into four major categories: 1) metamorphic and granitic basement rocks, 2) volcanic substrates, 3) sedimentary substrates, and 4) Quaternary alluvium, terrace deposits, and landslides. We discuss characteristics of these substrates and plant communities usually found on them and similar substrates on the adjacent mainland.

**Metamorphic and Granitic Basement Rocks**. The ancient metamorphic basement rocks of Santa Cruz Island Schist are probably the oldest rocks exposed on any Northern Channel Island (Weaver 1969b). Schist is exposed south of the Santa Cruz Island Fault in a band about 10 miles (16 km) long and 1.5 miles (2.4 km) wide, trending west-northwest. This substrate is intensively faulted and appears relatively homogeneous in texture (Weaver 1969b). There is some disagreement concerning the relationship between Santa Cruz Island Schist and substrates on the adjacent mainland. Although Weaver (1969b) reported that Santa Cruz Island Schist closely resembles Pelona Schist near the east end of the Santa Monica Mountains, Hill (1976) allied Jurassic basement rocks on Santa Cruz Island with the upper part of the ophiolite exposed at Point Sal, on the mainland, both petrologically and chronologically.

Brumbaugh (1980a) found that soils derived from Santa Cruz Island Schist are generally medium-textured loams and silt loams with a medium-acid A horizon (pH 5.7-6.0). Woodring and Bramlette (1950) classified the outcrop at Point Sal as undifferentiated igneous rocks of the Franciscan Formation. Soils derived from volcanic substrates, classified as Lodo Loam (Shipman 1972), are characterized by medium acidity (pH 5.6-7.3), low to very low fertility, and low available water capacity.

Mainland sites support coastal-sage scrub at low elevations and chaparral at higher elevations. Shipman (1972) listed *Arctostaphylos*, *Ceanothus*, and *Quercus dumosa* as dominants. Dibblee (1976) described vegetation on Franciscan and related substrates in Santa Barbara County (including the small outcrop at Point Sal) in very general terms. He reported grasses on soil-covered slopes and scattered oaks and chaparral on rocky knobs. Although closed-cone pines no longer occur at Point Sal, Axelrod (1980) found fossil cones of *Pinus radiata* and *P. remorata* in the vicinity.

On Santa Cruz Island, schist supports a rich mosaic of plant communities, even though most of the outcrop is near the top of an exposed ridge. Plant communities found on this substrate include chaparral, coastal-sage scrub, grasslands, island woodland, pine forest, riparian woodland, riparian scrub, and oak woodland.

Each of the three major pine populations found on Santa Cruz Island occurs on a different substrate. The Christy Pines population, located at the western end of the schist, has the most diverse understory association (Hobbs 1978, Hochberg et al. 1980). The eastern (China Harbor) population is on soils derived from diatomaceous Monterey Shale while the northern (Pelican Bay) population is found on soils derived from weathered andesite and dacite flows of the Santa Cruz Island Volcanics (Weaver 1969b). In addition to the three major populations, minor ones occur around Sierra Blanca ridge on Blanca Volcaniclastics, and a small, scattered population occurs south of the China Harbor population on Monterey Shale.

The best developed and most diverse chaparral stands occur on schist, especially along the north-facing slope south of the Central Valley. Bjorndalen (1978) documented the rich species composition of chaparral on this substrate. The following chaparral dominants are widespread on volcanic and metamorphic substrates but virtually absent on Monterey Shale and other sedimentary rocks on the island: *Arctostaphylos insularis*, *A. tomentosa* subsp. *insulicola*, *Ceanothus arboreus*, *C. megacarpus* subsp. *insularis*, *Cercocarpus betuloides* var. *blancheae*, *Dendromecon rigida* subsp. *harfordii*, and *Rhamnus pirifolia*.

Woodlands are well represented on this substrate and consist primarily of *Lyonothamnus floribundus* subsp. *aspleniifolius* groves associated with chaparral. About 145 of the 1000 groves on the island occur on Santa Cruz

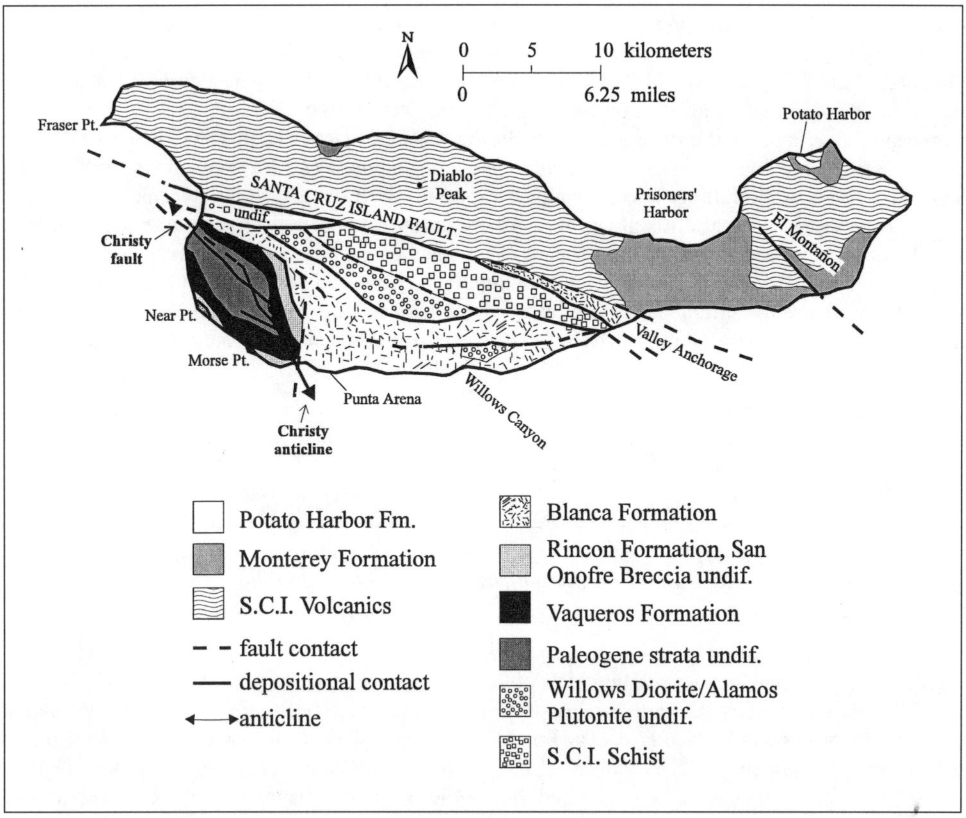


*Figure 2. Simplified geological map of Santa Cruz Island (adapted from Howell et al. 1976).*

Island Schist (Junak 1987). *Quercus tomentella* is rare on schist and *Q. chrysolepis* has not been found on this substrate. *Quercus agrifolia* forms lush woodlands on north-facing slopes and in wetter canyons.

Alamos Tonalite (or Alamos Plutonite) and Willows Diorite are associated with Santa Cruz Island Schist. Alamos Tonalite is a highly quartzose, massive, coarse-grained, somewhat metamorphosed quartz diorite (Weaver 1969b), a substrate thought to be the oldest plutonic rock exposed on the Northern Channel Islands. The main mass of this rock type is rather uniform in color, texture, and mineral composition. The Willows Diorite is another plutonic substrate exposed south of the Santa Cruz Island Schist. Although Alamos Tonalite is markedly homogeneous, Willows Diorite is conspicuously heterogeneous in composition and texture (Rand 1933, Weaver 1969b). Although these rocks are considerably fractured, they are resistant to weathering, so little soil development has taken place. Instead, these rocks are "covered with a thin, gritty rubble of fragments, giving to the surface a yellowish-tan color which is noticeably darker than the light greenish-gray color of the weathered ... Alamos intrusive" (Weaver 1969b).

Vegetation on Alamos Plutonite and Willows Diorite has been heavily grazed by feral sheep, so vegetative cover is sparse (Jones et al. 1993). Plant communities found on these substrates include chaparral, coastal-sage scrub, and grasslands on south-facing slopes. Chaparral, along with small patches of island woodland, pine forest, and oak woodland occur on north-facing slopes. Riparian scrub and riparian woodlands occur in moist canyon bottoms on plutonite and diorite, but are best developed on Quaternary alluvium.

**Volcanic Substrates**. Santa Cruz Island Volcanics are exposed over much of the onshore side of western and central Santa Cruz Island and at the extreme eastern tip of the island. Nolf and Nolf (1969) considered the formation to be lithologically correlated with the Conejo Formation of the western Santa Monica Mountains and also with some older flows of the Topanga Formation farther north and east. Dibblee (1982) mapped Santa Cruz Island Volcanics and San Miguel Volcanics as part of the Conejo Formation.

Santa Cruz Island Volcanics support a variety of tree-dominated plant communities. Because these rocks are exposed on north-facing slopes with abundant ground water, some of the most mesic habitats are found on this substrate. Moist canyons support riparian woodlands dominated by *Acer macrophyllum* and *Salix* spp. *Quercus agrifolia* is common in moist canyons and forms extensive stands on north-facing volcanic slopes. *Quercus chrysolepis* and *Q. tomentella* stands occur on the north sides of the highest peaks. A stand of widely scattered trees of *Pinus muricata* occurs on coastal slopes near Pelican Bay.

Grasslands cover most of the surface area of volcanic substrates (Jones et al. 1993). Chaparral vegetation also is well developed and very diverse. *Pickeringia montana* and most *Garrya* populations are found only on this substrate. The coastal-bluff scrub community is especially well developed on volcanic substrates, which weathers to form nearly vertical cliffs along the northern coast. In contrast, the coastal-bluff community on Monterey Shale is poorly developed and lacks such insular endemics as *Dudleya candelabrum*, *Galium buxifolium*, *Hazardia detonsa*, and *Heuchera maxima*, which are important species on volcanics. The rarest Channel Island endemics are found strictly on igneous substrates or Quaternary terrace gravels. Sixteen plant taxa with extremely limited distributions on Santa Cruz Island are found only on soils derived from igneous and metamorphic rocks:

*Arabis hoffmannii*
*Arbutus menziesii*
*Berberis pinnata* subsp. *insularis*
*Dudleya nesiotica*
*Holodiscus discolor*
*Lepechinia fragrans*
*Malacothamnus fasciculatus* var. *nesioticus*
*Malacothrix indecora*
*Malacothrix squalida*
*Mimulus brandegei*
*Montia fontana*
*Pickeringia montana*
*Populus fremontii*
*Rhamnus californica*
*Sibara filifolia*
*Thysanocarpus conchuliferus*

**Sedimentary Substrates**. Monterey Shale consists of a series of thin, rhythmically bedded, tuffaceous, diatomaceous, and siliceous shales with thin beds of dense limestone near the base (Weaver and Meyer 1969). This mid-Miocene sedimentary substrate is widespread in the coast ranges of California and extends into neighboring provinces (Norris and Webb 1976). The formation occurs in the narrow "isthmus" of Santa Cruz Island. The isthmus results from structural instability of the Monterey Formation, since "it is the propensity of the Monterey Shale to slide into the sea" (Weaver and Meyer 1969). On the island, 38% of the formation's surface area is covered with Quaternary landslides (estimated from map in Weaver 1969a). This instability must also affect soil formation and may inhibit establishment of some plant communities. Medium- to moderate-textured soils predominate on diatomaceous Monterey Shale. Silty loam to silty-clay loams cover slopes under grassland (Brumbaugh 1980b) and grade to loams and fine sandy loams under pine forest (Hobbs 1978).

Norris and Webb (1976) noted that the Monterey Formation usually weathers to a "dark, adobe-like, clay-rich soil that normally supports grass rather than trees." In San Luis Obispo County, Monterey Shale supports an extensive mosaic of grassland with discrete blocs of broad sclerophyll forest and closed-cone pine or chaparral vegetation (Wells 1962). For the Santa Barbara region, Dibblee (1976) reported that the formation is vegetated with "grasses on gentle slopes with residual soil and by coastal-sage scrub on steeper slopes with little soil", while hard porcelaneous to cherty siliceous shale areas support chaparral and oaks. *Pinus torreyana* subsp. *insularis* on Santa Rosa Island grows on a tuffaceous sandstone of the Beechers Bay member of the Monterey Formation (Dibblee 1976, Haller 1986).

Monterey shale is relatively impermeable to water and was described by Worts (1951) as "essentially not water-bearing except for joints or fractures." Surface water is extremely limited on Monterey Shale and results in an almost complete lack of riparian and other mesophytic vegetation, even on north-facing canyons. On Santa Cruz Island, Monterey Shale primarily supports grassland, with scattered stands of chaparral dominated by scrub oaks, coastal-sage scrub, and an extensive Bishop pine forest. Some isolated elements of island woodland (notably *Quercus tomentella*) and riparian woodland (especially *Acer macrophyllum*) are also present but only as isolated trees. As noted above, species composition of plant communities on this substrate differs substantially from that on volcanic substrates. Only one chaparral dominant (*Arctostaphylos viridissima*) is almost completely restricted to Monterey Shale, while at least seven dominants are nearly or entirely restricted to volcanic or metamorphic rocks.

The Blanca Formation is exposed in an area 13 miles (20.9 km) long with a maximum width of 2 miles (3.2 km), in the south-central part of the island. This volcaniclastic substrate is composed of "generally light colored, massive to thick bedded, epiclastic volcanic rocks" (Weaver et al. 1969b). The vegetation on Blanca volcaniclastics, like that on Alamos Plutonite and Willows Diorite, has been subjected to extremely heavy grazing by feral sheep.

Vegetative cover is sparse, with chaparral and small patches of island woodland, pine forest, and oak woodland on north-facing slopes. Grasslands and coastal-sage scrub occur on south-facing slopes. Riparian scrub dominates canyon bottoms, especially those with Quaternary alluvium.

Several similar substrates that occur in the southwestern portion of the island are considered together. They include the Pozo, Cañada, and Jolla Vieja Formations, Cozy Dell and Rincon shales, Vaqueros Sandstone, and San Onofre Breccia. These Tertiary sedimentary rocks were studied intensively by Doerner (1969) and consist of breccias, sandstones, and shales. The soils derived from these substrates are medium- to moderately-textured loams with slightly acid to nearly neutral A horizons (pH 6.0-7.2) (Brumbaugh 1980a).

Dibblee (1976) stated that Rincon Shale in the Santa Ynez Mountains and on Santa Rosa Island supports grassland, with coastal-sage scrub on steep slopes. The clay shales, including Cozy Dell and other Eocene Formations on the Channel Islands, are weathered to residual soil that is covered with grasses (Dibblee 1976). Vaqueros Sandstone supports chaparral on the mainland of Santa Barbara County (Dibblee 1976) but only grassland with scattered shrubs on Santa Cruz Island. On the island, vegetation on these substrates is noticeably less diverse than on surrounding granitic and metamorphic substrates, consisting primarily of grassland and coastal-sage scrub, with scattered stands of chaparral dominated by *Quercus pacifica* and isolated stands of *Pinus muricata*. The coastal-bluff scrub on offshore cliffs is very depauperate. Good examples of other vegetation types are completely lacking on these substrates and the region is subject to heavy erosion.

**Quaternary Alluvium, Terrace Deposits, and Landslides**. These substrates are dominated by grasslands (Jones et al. 1993). Other plant communities occurring on Quaternary alluvium include chaparral, scattered stands of coastal-sage scrub and pine forest, riparian scrub on the south side of the island and in the Central Valley, and riparian woodlands on the north side. Introduced trees, especially *Eucalyptus* and *Cupressus macrocarpa*, have been planted on alluvium in larger canyons and valleys. Quaternary landslides are mostly found on Monterey Shale, where they are dominated by grasslands and coastal-sage scrub. Small stands of pines and riparian scrub also have been reported (Jones et al. 1993). Quaternary terrace gravels support grasslands, coastal-bluff scrub, coastal-sage scrub, small stands of chaparral on the north side of the island, and stands of riparian scrub in moist swales and canyon bottoms. *Dudleya nesiotica*, an endemic to Santa Cruz Island, occurs only on gravels at the west end, where this substrate is derived from Santa Cruz Island Volcanics.

## VEGETATION

Philbrick and Haller (1977) described eight upland plant community types and two wetland community types for Santa Cruz Island. Their upland types include southern beach and dune, coastal bluff, coastal-sage scrub, valley and foothill grassland, island chaparral, southern coastal oak woodland, island woodland, and Bishop pine forest. Wetland types are coastal marsh and southern riparian woodland. Philbrick (1978) provided additional descriptions of vegetation patterns and community structure. Minnich (1980) classified the island's vegetation using six physiognomic categories, combining southern coastal oak woodland with island woodland, and disregarding southern coastal dune, coastal bluff, and coastal marsh communities. Holland (1986) described southern foredune, southern dune scrub, southern coastal-bluff scrub, Venturan coastal-sage scrub, valley needlegrass grassland, non-native grassland, island chaparral, island-oak woodland, southern Bishop-pine forest, coastal and valley freshwater marsh, freshwater seep, southern coast-live-oak riparian forest, and mule-fat scrub communities for the region.

The following classification system for the island's plant communities is primarily based on Philbrick and Haller (1977) and Holland (1986). We have attempted to describe current plant associations; however, habitats are just beginning to recover from decades of overgrazing. This disturbance significantly reduced woody vegetation and, in many areas, artificially maintained grasslands. As recovery of the island's ecosystems continues, plant cover will probably increase and some successional community types may become less prevalent or disappear entirely. The relationships among plant communities on Santa Cruz Island and those on other Channel Islands and the adjacent mainland need further study. Some community types described in the following discussion are widely accepted; others (e.g., some riparian community types) probably will need further refinement.

**Southern Beach and Dune**. A number of sandy beaches are found around the perimeter of the island, especially on the south side. The area above mean high-tide line on these beaches is colonized by low-growing southern foredune dominants, including *Abronia maritima*, *Ambrosia chamissonis*, *Atriplex leucophylla*, *Cakile maritima*, *Camissonia cheiranthifolia* subsp. *cheiranthifolia*, and *Distichlis spicata*. Most plants on these beaches are less than 1 foot (0.3 m) high. Typical southern dune scrub communities (as described by Holland 1986) are rare, because few beaches are expansive enough to allow formation of large dunes. Limited dune systems are found in exposed, windy sites in the southwestern corner of the island, especially between Christy Beach and the Punta Arena area. Characteristic plants in unstabilized dune habitats include *Abronia maritima*, *A. umbellata*, *Ambrosia chamissonis*, *Atriplex californica*, *A. semibaccata*, *Camissonia cheiranthifolia* subsp. *cheiranthifolia*, and *Distichlis spicata*. *Isocoma menziesii* and *Lupinus albifrons* are found in more stabilized dunes. *Astragalus miguelensis* is an important dominant in a small dune area near Forney's Cove. Species with limited distribution include *Calystegia soldanella*, *Cressa truxillensis*, and *Potentilla anserina* subsp. *pacifica*.

**Valley and Foothill Grassland**. Grasslands are widespread and have been disturbed by feral sheep and pigs. Although introduced annual grasses have usually replaced native perennials, *Nassella cernua*, *N. lepida*, and *N. pulchra* occur in open sites throughout the island and can be locally common. In wet years and on richer soils, dominant grasses can be up to 5 feet (1.5 m) tall. Widespread introduced grasses include *Avena* spp., *Bromus* spp., *Gastridium ventricosum*, *Hordeum murinum*, *Lamarckia aurea*, *Lolium multiflorum*, *L. perenne*, and *Phalaris minor*. *Lolium multiflorum* and *L. perenne* dominate grasslands at the west end of the Central Valley, in the Pozo Canyon and Cañada de los Sauces drainages, and on coastal flats east of Forney's Cove. During years with sufficient rainfall, the following native species can be locally common:

*Achillea millefolium*
*Achyrachaena mollis*
*Allium praecox*
*Amsinckia menziesii* var. *intermedia*
*Bloomeria crocea*
*Calandrinia ciliata*
*Calochortus catalinae*
*Dichelostemma capitatum*
*Dodecatheon clevelandii* subsp. *insulare*
*Lasthenia californica*
*Layia platyglossa*
*Lepidium nitidum*
*Lomatium caruifolium*
*Lupinus bicolor*
*Plagiobothrys* spp.
*Ranunculus californicus*
*Sanicula arguta*
*Sisyrinchium bellum*
*Stebbinsoseris heterocarpa*
*Trifolium* spp.
*Viola pedunculata*

Widespread introduced species in grasslands include *Erodium* spp., *Hypochaeris glabra*, *Medicago polymorpha*, *Silene gallica*, and *Sonchus oleraceus*. *Brassica nigra*, *Raphanus raphanistrum*, and *R. sativus* dominate large areas in some wetter years.

Coastal flats near Forney's Cove are especially rich in native species, including:

*Allium praecox*
*Atriplex coulteri*
*Bloomeria crocea*
*Crassula connata*
*Dichelostemma capitatum*
*Dudleya nesiotica*
*Hemizonia fasciculata*
*Hordeum brachyantherum* subsp. *californicum*
*Hordeum intercedens*
*Lasthenia californica*
*Lepidium latipes*
*Lupinus microcarpus*
*Microseris douglasii*
*Nassella pulchra*
*Sidalcea malviflora*
*Spergularia macrotheca*

*Atriplex semibaccata*, *Bromus rubens*, *Erodium cicutarium*, *Hainardia cylindrica*, *Hordeum murinum*, *Lamarckia aurea*, *Lolium multiflorum*, *Medicago polymorpha*, *Mesembryanthemum crystallinum*, *M. nodiflorum*, and *Parapholis incurva* are some common introduced species on these flats.

**Coastal-Bluff Scrub**. This community type is found on seacliffs and coastal slopes around the island's perimeter and on steep canyon walls and rock outcrops in the interior. Dominant plant species vary with slope exposure, geologic substrate, and location. Species richness, especially of endemic taxa, is highest on steep, north-facing, volcanic seacliffs in the central portion of the island. Common species on these cliffs include

*Achillea millefolium*
*Adiantum jordanii*
*Amblyopappus pusillus*
*Antirrhinum nuttallianum*
*Atriplex semibaccata*
*Bromus madritensis* subsp. *rubens*
*Calystegia macrostegia* subsp. *macrostegia*
*Carpobrotus chilensis*
*Coreopsis gigantea*
*Dudleya greenei*
*Epilobium canum*
*Erigeron glaucus*
*Eriogonum arborescens*
*Eriogonum grande* var. *grande*

*Coastal-bluff scrub, dominated by* Dudleya greenei, *mouth of Eagle Canyon (S. Junak, June 1979).*

*Eriophyllum staechadifolium*
*Galium buxifolium*
*Hazardia detonsa*
*Heteromeles arbutifolia*
*Heuchera maxima*
*Lathryus vestitus*
*Leymus condensatus*
*Lotus dendroideus*
*Malacothrix saxatilis* var. *implicata*
*Marah macrocarpus*
*Mimulus flemingii*
*Poa secunda* subsp. *secunda*
*Polypodium californicum*
*Rhus integrifolia*
*Silene laciniata*
*Spergularia macrotheca*
*Stachys bullata*
*Toxicodendron diversilobum*

*Eriogonum grande* var. *rubescens* replaces *E. grande* var. *grande* as a dominant on coastal bluffs in the extreme western portion of the island. Additional taxa common on volcanic seacliffs near the east end of the island include *Artemisia californica*, *Astragalus miguelensis*, *Atriplex californica*, *Hordeum brachyantherum* subsp. *californicum*, and *Opuntia littoralis*. Steep, shaded volcanic canyon walls in the interior support a rich assemblage of additional perennial species, including *Baccharis plummerae*, *Dudleya candelabrum*, *Eriophyllum confertiflorum*, *Galium angustifolium* subsp. *foliosum*, *Melica imperfecta*, *Nassella pulchra*, *Pellaea andromedifolia*, *Sanicula arguta*, *Selaginella bigelovii*, *Solanum clokeyi*, and *Stephanomeria cichoriacea*, as well as most species common on seacliffs. Two rare species, *Arabis hoffmannii* and *Malacothrix indecora*, are restricted to this community type.

On Monterey Shale and other substrates on the south side, common bluff plants include

*Antirrhinum nuttallianum* subsp. *subsessile*
*Aristida adscensionis*
*Artemisia californica*
*Atriplex californica*
*Atriplex semibaccata*
*Brickellia californica*
*Bromus madritensis* subsp. *rubens*
*Calystegia macrostegia* subsp. *macrostegia*
*Castilleja lanata* subsp. *hololeuca*
*Coreopsis gigantea*
*Dudleya greenei*
*Encelia californica*
*Eriogonum arborescens*
*Eriogonum grande* var. *grande*
*Eriophyllum staechadifolium*
*Galium angustifolium* subsp. *foliosum*
*Isocoma menziesii* var. *sedoides*
*Leymus condensatus*
*Lotus dendroideus*
*Malacothrix saxatilis* var. *implicata*
*Mimulus flemingii*
*Mimulus longiflorus*
*Mirabilis californica*
*Muhlenbergia microsperma*
*Opuntia littoralis*
*Opuntia oricola*
*Perityle emoryi*
*Rhus integrifolia*
*Salvia mellifera*
*Spergularia macrotheca*

*Eriogonum grande* var. *rubescens* is an important dominant on coastal bluffs between Forney's Cove and the mouth of Cañada de los Sauces. In years with heavy rainfall, annual plants such as *Camissonia californica*, *Eschscholzia ramosa*, *Hemizonia fasciculata*, *Lupinus hirsutissimus*, and *Phacelia viscida* can be locally abundant between shrubs on south-facing bluffs. *Opuntia prolifera* is an important dominant between Malva Real Anchorage and Horqueta Canyon, where the bluff community is similar to maritime cactus scrub of the Southern Channel Islands and northwestern Baja California. A small population of *Euphorbia misera* occurs with *Opuntia prolifera* on the coastal headland just east of Laguna Canyon.

**Coastal-Sage Scrub**. Coastal-sage scrub communities occur on dry, rocky slopes throughout the island and are most noticeable on the south side and on south-facing slopes in the central and eastern portions of the Central Valley. Westman (1983) sampled coastal-sage scrub and found representative taxa from each of the four major floristic associations of xeric California shrublands. Hobbs (1983) studied factors affecting the distribution of this community in the Central Valley and on the adjacent mainland.

Most stands have been heavily disturbed by feral animals. Some of the least disturbed coastal-sage scrub communities occur on slopes east of Valley Anchorage. Nearly impenetrable thickets can be found there, dominated by *Artemisia californica*, *Castilleja lanata* subsp. *hololeuca*, *Encelia californica*, *Eriogonum arborescens*, *E. grande* var. *grande*, *Hazardia squarrosa*, *Leymus condensatus*, *Malacothrix saxatilis* var. *implicata*, *Opuntia littoralis*, *O. oricola*, *Rhus integrifolia*, and *Salvia mellifera*. Shrub thickets are mostly 3 to 4 feet (ca. 1 m) tall. *Achnatherum diegoense* and *Nassella lepida* are scattered among the shrubs. *Brickellia californica* and *Mirabilis californica* are locally common on south-facing slopes, while *Coreopsis gigantea*, *Eriophyllum confertiflorum*,

*Coastal-sage scrub, dominated by* Artemisia californica *and* Rhus integrifolia, *near Coches Prietos Anchorage, looking north (S. Junak, April 1991).*

and *Helianthemum scoparium* are important dominants on north-facing slopes. The rare *Malacothamnus fasciculatus* var. *nesioticus* is found on south-facing slopes in this community type. Coastal-sage scrub intergrades with coyote-brush scrub and grasslands on gentle slopes with deeper soils and with island chaparral on north-facing slopes.

**Coyote-Brush Scrub**. Shrublands dominated by *Baccharis pilularis* are widespread at elevations below 500 feet (150 m), especially in large canyons on the south side and in the Central Valley. These areas have been heavily disturbed by overgrazing. Many widespread dominant species are introduced grasses, including *Avena barbata*, *A. fatua*, *Bromus diandrus*, *B. hordeaceus*, *Hordeum murinum*, and *Vulpia* spp. In years with sufficient rainfall, large areas are dominated by *Brassica nigra*, *Centaurea melitensis*, and *C. solstitialis*. *Foeniculum vulgare* occurs in large, persistent, and aggressively expanding populations. Coyote-brush scrub occurs primarily on flats with loam to sandy clay loam soils (Clark et al. 1990) and intergrades with coastal-sage scrub on rocky slopes.

**Island Chaparral**. Chaparral is widespread and is most noticeable on north-facing slopes in the Central Valley. Species composition varies greatly with geologic substrate and location, as does size and stature of dominants. Island chaparral intergrades with coastal-sage scrub on south-facing slopes and with island woodland, oak woodland, and Bishop-pine forest on mesic north-facing slopes.

Island chaparral on north-facing interior slopes of the Central Valley and in Islay Canyon is dominated by *Adenostoma fasciculatum* var. *fasciculatum*, *Arctostaphylos insularis*, *Ceanothus arboreus*, *C. megacarpus* subsp. *insularis*, *Cercocarpus betuloides* var. *blancheae*, *Heteromeles arbutifolia*, *Mimulus flemingii*, *M. longiflorus*, *Prunus ilicifolia* subsp. *lyonii*, and *Quercus pacifica*. Bjorndalen (1978) listed species for this portion of the island, where arborescent canopy shrubs are mostly 6 to 12 feet (2-3.5 m) tall.

Island chaparral, dominated by *Adenostoma fasciculatum* var. *prostratum*, *Arctostaphylos viridissima*, *Heteromeles arbutifola*, and *Quercus pacifica*, occurs on Monterey Shale in the "isthmus" region. Associated species include *Castilleja lanata* subsp. *hololeuca*, *Mimulus flemingii*, *M. longiflorus*, and *Quercus parvula* var. *parvula*. *Quercus pacifica* also occurs in large arborescent stands 7 to 12 feet (2-3.5 m) tall. The understory layer beneath pure stands of *Q. pacifica* has been severely disturbed by feral pigs and is primarily composed of introduced annual grasses and herbs.

*Sparse coastal-sage scrub, dominated by* Eriogonum arborescens, *coastal hills near Pozo Beach, looking south (S. Junak, July 1979).*

*Island maritime chaparral, dominated by* Quercus pacifica, *on ridge north of Cañada de los Sauces, looking north (S. Junak, April 1991).*

In exposed sites near the west end of the island, a maritime chaparral is dominated by low shrubs 1 to 4 feet (ca. 0.3-1 m) tall. Thickets of *Quercus pacifica* are common, especially on rocky ridgetops. The low stature of some species in such habitats apparently results from wind-pruning. The prostrate habit in other taxa (e.g., *Adenostoma fasciculatum* var. *prostratum*) appears to be genetically controlled, because representative plants maintain their low stature in cultivation at wind-free sites.

North-facing slopes in the lower portion of Cañada Christy, Cañada Cebada, and Cañada de los Sauces are covered with a rich mixture of species dominating island chaparral and coastal-bluff scrub communities:

*Adenostoma fasciculatum* var. *prostratum*
*Artemisia californica*
*Atragalus trichopodus* var. *lonchus*
*Baccharis pilularis*
*Coreopsis gigantea*
*Eriogonum arborescens*
*Eriophyllum confertiflorum*
*Hazardia squarrosa* var. *grindelioides*
*Isocoma menziesii* var. *vernonioides*
*Heteromeles arbutifolia*
*Lessingia filaginifolia*
*Leymus condensatus*
*Lotus dendroideus*
*Lupinus albifrons*
*Mimulus flemingii*
*Mimulus longiflorus*
*Quercus pacifica*
*Rhus integrifolia*
*Toxicodendron diversilobum*

Typical understory species include *Achillea millefolium*, *Aster radulinus*, *Carex globosa*, *Claytonia perfoliata* subsp. *mexicana*, *Galium nuttallii* subsp. *insulare*, *Jepsonia malvifolia*, *Lathyrus vestitus*, *Marah macrocarpus*, *Melica imperfecta*, *Pentagramma triangularis*, *Poa secunda* subsp. *secunda*, *Sanicula arguta*, *S. hoffmannii*, and *Silene laciniata*. Other associated species grow in rocky openings between shrubs: *Acourtia microcephala*, *Delphinium parishii* subsp. *maritimum*, *Dudleya candelabrum*, *Dodecatheon clevelandii* subsp. *insulare*, *Gnaphalium bicolor*, *G. canescens* subsp. *microcephalum*, *G. purpureum*, *Helianthemum greenei*, *Koeleria macrantha*, *Luzula comosa*, *Nassella pulchra*, *Sisyrinchium bellum*, *Solidago californica*, and *Spergularia macrotheca*.

*Island chaparral on north-facing slopes in Central Valley, looking east (S. Junak, April 1993).*

**Island Woodland**. This community type occurs on relatively deep, moist, somewhat rocky soils of north-facing slopes, ravines, and narrow canyons, especially at higher elevations (Cheatham and Haller 1975, Philbrick and Haller 1977). Island woodland is dominated by a number of endemic trees and shrubs 15-45 feet (4.5-14 m) tall. Island woodland intergrades with island chaparral on drier, rockier slopes and forms savannas with valley and foothill grasslands on fine-textured soils and gentler slopes.

Dominant canopy species include *Heteromeles arbutifolia*, *Lyonothamnus floribundus* subsp. *aspleniifolius*, *Prunus ilicifolia* subsp. *lyonii*, *Q. chrysolepis*, *Q.* ↔*macdonaldii*, and *Q. tomentella*. Associated understory layers are usually not very diverse, especially in populations disturbed by feral sheep and pigs. Some understory species found in *Lyonothamnus* groves (Junak 1987) include

*Agrostis pallens*
*Baccharis pilularis*
*Baccharis plummerae*
*Bromus diandrus*
*Cheilanthes clevelandii*
*Comarostaphylis diversifolia* subsp. *planifolia*
*Galium aparine*
*Galium nuttallii* subsp. *insulare*
*Heteromeles arbutifolia*
*Hieracium argutum*
*Hypochaeris glabra*
*Lathyrus vestitus*
*Marah macrocarpus*
*Mimulus flemingii*
*Mimulus longiflorus*
*Pentagramma triangularis*
*Polypodium californicum*
*Prunus ilicifolia* subsp. *lyonii*
*Quercus pacifica*
*Rhus integrifolia*
*Sanicula arguta*
*Stellaria media*
*Venegasia carpesioides*
*Zigadenus fremontii*

**Southern Coastal Oak Woodland**. This community type occurs throughout the island, primarily on deeper soils on north-facing slopes and in shaded canyons on the north side and in the Central Valley. The dominant species is *Quercus agrifolia*. Common understory plants on slopes include *Heteromeles arbutifolia*, *Stachys bullata*, *Symphoricarpos mollis*, and *Toxicodendron diversilobum*.

In moist canyons on the island's north side, *Quercus agrifolia* dominates many canyon bottoms, including sites similar to those occupied by riparian woodland on the adjacent mainland. Widespread understory plants in this area include *Dryopteris arguta*, *Cardamine californica*, *Eucrypta chysanthemifolia*, *Keckiella cordifolia*, *Lonicera hispidula* var. *vacillans*, *Marah macrocarpus*, *Pholistoma auritum*, *Rubus ursinus*, and *Toxicodendron diversilobum*. *Polystichum munitum* and *Ribes thacherianum* are scattered in the understory layer. The rare *Thysanocarpus conchuliferus* occurs on rocky sites in this community type.

*Small grove of* Lyonothamnus floribundus *subsp.* aspleniifolius, *in canyon south of Lady's Cove (S. Junak, March 1982).*

*Southern coastal oak woodland, island chaparral, and scattered Bishop pines in canyon between Pelican Bay and Prisoners' Harbor, looking south (S. Junak, March 1982).*

*Bishop pine forest on upper slopes, grassy hillsides, and coyote-brush scrub in canyon bottom, upper part of Cañada de los Sauces, looking southeast (S. Junak, May 1991).*

*Aerial view of slopes south of China Harbor, with Bishop pine forest in center, looking south (S. Junak, July 1994).*

**Bishop Pine Forest**. Large populations of Bishop pines occur on north-facing slopes in upper reaches of Cañada Christy, near Pelican Bay, and south of China Harbor. Smaller populations occur in the upper portion of Cañada de los Sauces, on Sierra Blanca ridge, and on the south side near China Harbor. Canopy species include both formas of *Pinus muricata* with occasional stands of *Lyonothamnus floribundus* subsp. *aspleniifolius*, *Quercus agrifolia*, and *Q. tomentella*. The rare *Berberis pinnata* subsp. *insularis* occurs mostly in this community type. Composition of the understory varies significantly between the three large pine populations. The Christy Pines have the richest understory layer (Hobbs 1978, Hochberg et al. 1980). Typical understory species in the Christy population include

*Adenostoma fasciculatum*
*Arctostaphylos insularis*
*Baccharis pilularis*
*Calochortus albus*
*Carex globosa*
*Ceanothus arboreus*
*Cercocarpus betuloides* var. *blancheae*
*Cheilanthes clevelandii*
*Comarostaphylis diversifolia* var. *planifolia*
*Dryopteris arguta*
*Galium californicum* subsp. *flaccidum*
*Galium nuttallii* subsp. *insulare*
*Heteromeles arbutifolia*
*Hieracium argutum*
*Hypochaeris glabra*
*Lotus dendroideus* var. *dendroideus*
*Lilium humboldtii* var. *ocellatum*
*Luzula comosa*
*Mimulus flemingii*
*Mimulus longiflorus*
*Pentagramma triangularis*
*Piperia elongata*
*Polygala californica*
*Quercus pacifica*
*Quercus parvula* var. *parvula*
*Ribes malvaceum*
*Ribes thacherianum*
*Rubus ursinus*
*Sanicula hoffmannii*
*Toxicodendron diversilobum*
*Vaccinium ovatum*
*Venegasia carpesioides*
*Zigadenus fremontii*

**Intertidal and Subtidal Marine Community**. Several species of marine angiosperms occur in intertidal and subtidal zones around the perimeter of the island. *Phyllospadix scouleri* and *P. torreyi* are widespread on rocky sites in the intertidal zone and are often exposed in tidepools at low tides. *Zostera marina* occupies sandy sites in water between mean low tide and 6 feet (0-2 m) deep. *Zostera pacifica* grows in subtidal waters 15-50 feet (4.5-15 m) deep. This community was studied by Apt et al. (1988).

**Coastal Marsh and Estuary**. Small estuaries occur at mouths of several canyons. Each is quite different with respect to dominant plant species, water salinity, and duration of flooding from one year to the next and from season to season. Since the removal of feral sheep from the western 90% of the island, vegetative cover, duration of flooding, and depth of standing water have increased dramatically in estuaries on the south side.

*Distichlis spicata* is a conspicuous dominant found at each site. At Prisoners' Harbor, a small freshwater wetland is dominated by *Scirpus californicus* and *Typha domingensis*. Associated species include *Anemopsis californica*, *Baccharis douglasii*, *B. salicifolia*, *Berula erecta*, *Distichlis spicata*, *Gnaphalium luteo-album*, *Lythrum hyssopifolium*, *Pennisetum clandestinum*, *Plantago major*, *P. lanceolata*, *Pluchea odorata*, *Rumex crispus*, and *Salix lasiolepis*.

At the mouth of Cañada de los Sauces, the area surrounding a small brackish estuary is vegetated with *Atriplex semibaccata*, *A. triangularis*, *Cotula coronopifolia*, *Distichlis spicata*, *Heliotropium curassavicum*, *Jaumea carnosa*, and *Salicornia virginica*. *Scirpus maritimus* has been collected there but plants apparently have not persisted. At the mouth of Cañada de Malva Real, a small estuary is surrounded by *Distichlis spicata*. Associated species include *Atriplex semibaccata*, *Berula erecta*, *Cotula coronopifolia*, *Gnaphalium luteo-album*, *Plantago subnuda*, *Pluchea odorata*, and *Polypogon interruptus*. At mouths of other canyons (e.g., Laguna), there are extensive patches of *Distichlis spicata* in areas that are periodically flooded.

A small, moist alkaline flat that is irregularly flooded occurs at the mouth of Scorpion Canyon. Common species include *Atriplex californica*, *A. coulteri*, *A. lentiformis*, *A. leucophylla*, *Cakile maritima*, *Chenopodium murale*, *Cressa truxillensis*, *Hordeum murinum*, *Parapholis incurva*, *Spergularia bocconii*, *S. macrotheca*, *S. marina*, and *Suaeda taxifolia*. A number of species usually associated with coastal marshes on the mainland (e.g., *Atriplex watsonii*, *Salicornia subterminalis*, and *Suaeda taxifolia*) occur on coastal flats and bluffs of the island.

*Small estuary at mouth of Cañada de Malva Real, looking north (S. Junak, October 1992).*

**Freshwater Seeps and Springs**. Permanent seeps occur throughout the island, especially in canyons on the north side and on coastal bluffs. These seeps are usually located along fault scarps, such as in the second cove east of Fry's Harbor and at Aguaje Escondido near the west end of China Harbor. Several large springs are located in Aguaje Canyon near Yellowbanks Anchorage.

Dominant plant taxa at these seeps and springs vary with location and include many of those found in riparian herbaceous vegetation. Widespread taxa at seeps on the north side of the island include

| | |
|---|---|
| *Adiantum jordanii* | *Mimulus guttatus* |
| *Agrostis viridis* | *Pentagramma triangularis* |
| *Baccharis douglasii* | *Polypogon interruptus* |
| *Baccharis salicifolia* | *Polypogon monspeliensis* |
| *Cotula australis* | *Salix lasiolepis* |
| *Epilobium ciliatum* | *Urtica dioica* subsp. *holosericea* |
| *Gnaphalium luteo-album* | *Verbena robusta* |
| *Mimulus cardinalis* | *Woodwardia fimbriata* |

Dominant species at a seep on south-facing coastal bluffs just west of the mouth of Pozo Canyon are *Baccharis douglasii*, *Distichlis spicata*, *Gnaphalium luteo-album*, *Jaumea carnosa*, and *Polypogon interruptus*. Species with restricted distributions that are found at permanent seeps include *Adiantum aleuticum*, *A. capillus-veneris*, *Andropogon glomeratus* var. *scabriglumis*, *Epipactis gigantea*, *Sagina occidentalis*, and *Samolus parviflorus*.

**Vernal Ponds**. Several vernal ponds occur on the north side between Eagle Canyon and the west end of China Harbor and also on the south side near Coches Prietos Anchorage and Sandstone Point. Another reservoir or artificial pond is located on the headland near Cavern Point. Most of these ponds are located on Monterey Shale and at least some of them were natural depressions. A small "laguna" is shown near Mount Pleasant on a surveyor's map drawn in the 1850s (Greenwell 1856-1857), verifying its presence before widespread agriculture started on the island.

*Freshwater seep, dominated by* Woodwardia fimbriata, *in canyon south of Cueva Valdez, looking south (S. Junak, May 1979).*

Presumably, natural vernal ponds originally supported an interesting native flora, but they have been heavily disturbed by grazing since the late 1800s. Alien plants have invaded disturbed sites and native species are rare in these ponds. Typical species in disturbed ponds near China Harbor include

| | |
|---|---|
| *Amaranthus albus* | *Helianthus annuus* |
| *Amaranthus powellii* | *Hirschfeldia incana* |
| *Atriplex semibaccata* | *Hordeum murinum* |
| *Baccharis pilularis* | *Lactuca serriola* |
| *Bromus carinatus* | *Malva parviflora* |
| *Bromus diandrus* | *Marrubium vulgare* |
| *Bromus hordeaceus* | *Plantago major* |
| *Chenopodium ambrosioides* | *Rumex crispus* |
| *Chenopodium murale* | *Silybum marianum* |
| *Convolvulus arvensis* | *Sisymbrium officinale* |
| *Cynodon dactylon* | *Sonchus oleraceus* |
| *Eremocarpus setigerus* | *Spergularia bocconii* |
| *Foeniculum vulgare* | |

*Gnaphalium luteo-album* and *G. palustre* can be common in some less disturbed vernal ponds. *Beta vulgaris* is common in the reservoir near Cavern Point.

**Riparian Herbaceous Vegetation**. This community type occurs in canyon bottoms where soil moisture is available for most of the year, usually along moist banks and seasonally flooded margins of flowing freshwater creeks. It is characterized by herbaceous species 1 to 3 feet (0.3-1 m) high, with occasional patches of plants up to 8 feet (2.5 m) high. A depauperate version of this community type occurs on Santa Rosa Island (Clark et al. 1990).

Widespread species on volcanic and metamorphic substrates include

| | |
|---|---|
| *Adiantum jordanii* | *Lythrum hyssopifolium* |
| *Agrostis viridis* | *Melilotus indica* |
| *Anagallis arvensis* | *Mimulus cardinalis* |
| *Baccharis douglasii* | *Mimulus guttatus* |
| *Chenopodium ambrosioides* | *Plantago major* |
| *Conyza canadensis* | *Polygonum lapathifolium* |
| *Cotula coronopifolia* | *Polypogon interruptus* |
| *Distichlis spicata* | *Polypogon monspeliensis* |
| *Epilobium ciliatum* | *Rumex crispus* |
| *Equisetum laevigatum* | *Rumex salicifolius* |
| *Gnaphalium luteo-album* | *Typha domingensis* |
| *Juncus bufonius* | *Verbena robusta* |
| *Lotus purshianus* | *Veronica anagallis-aquatica* |

Localized species include *Andropogon glomeratus* var. *scabriglumis*, *Aster subulatus* var. *ligulatus*, *Carex barbarae*, *C. senta*, *Epipactis gigantea*, *Equisetum telmateia* subsp. *braunii*, *Juncus xiphioides*, *Lotus salsuginosus*, *Lythrum californicum*, *Plantago subnuda*, *Trifolium variegatum*, and *Typha latifolia*. Species richness is much reduced on Monterey Shale and other sedimentary substrates.

Submerged freshwater aquatic species are also found in this community type. *Ruppia maritima* occurs in some slow-moving creeks near the west end. Other submerged aquatics (e.g., *Potamogeton pectinatus* and *Zannichellia palustris*) have been collected in Cañada del Puerto. This community intergrades with mule-fat scrub and southern riparian woodland.

**Mule-Fat Scrub**. This community type occurs on nearly level floodplains of large canyons throughout the island, especially in Cañada del Puerto, in the Central Valley, and on the south side. Seasonally flooded alluvial deposits with coarse soils are dominated by *Baccharis salicifolia*, with scattered stands of *Salix lasiolepis* and *Salix exigua* where the water table reaches or approaches the surface.

Associated species include taxa found on gravel floodplains and in coastal-sage scrub, coastal-bluff scrub, and island chaparral communities:

*Astragalus trichopodus* var. *lonchus*
*Calystegia macrostegia* subsp. *macrostegia*
*Ceanothus arboreus*
*Ceanothus megacarpus* subsp. *insularis*
*Cercocarpus betuloides* var. *blancheae*
*Epilobium canum*
*Eriogonum arborescens*
*Eriogonum grande* var. *grande*
*Gnaphalium bicolor*
*Gnaphalium californicum*
*Gnaphalium canescens* subsp. *microcephalum*
*Heterotheca grandiflora*
*Lotus dendroideus* var. *dendroideus*
*Lotus grandiflorus* var. *grandiflorus*
*Lupinus albifrons* var. *douglasii*
*Oenothera elata* subsp. *hirsutissima*
*Rhamnus pirifolia*
*Rhus integrifolia*
*Solanum douglasii*

*Artemisia douglasiana* and *Urtica dioica* subsp. *holosericea* are typical associates in moist sites. *Lepidospartum squamatum* is a conspicuous associate in portions of Cañada del Medio and Cañada del Puerto, while *Lupinus albifrons* is a conspicuous dominant in Cañada Christy. *Tamarix ramosissima* is becoming established in widely scattered localities, especially near mouths of several canyons. On drier margins of floodplains with finer soils, this scrub community intergrades with coyote-brush scrub. Mule-fat scrub also intergrades with riparian herbaceous vegetation and southern riparian woodland in mesic sites.

**Southern Riparian Woodland**. Riparian woodlands occur along permanent streams, especially on the north side between Cueva Valdez and Cañada del Agua and in the Central Valley. On the south side, significant woodlands occur in Alamos Canyon and in the Coches Prietos drainage. Riparian woodlands are dominated by *Populus trichocarpa, Quercus agrifolia*, and *Salix* species, with occasional colonies of *Acer macrophyllum* and *Populus fremontii*. This community type is best developed on Santa Cruz Island Volcanics, especially on the north side. In the wettest canyons of the north side, common understory taxa include *Lonicera hispidula* var. *vacillans*, *Rubus ursinus*, and *Woodwardia fimbriata*, all of which are lacking on Monterey Shale in onshore canyons. *Adiantum aleuticum* and *Athyrium filix-femina* are occasional in wetter canyons and are also restricted to volcanic substrates. Additional understory plants include most species considered widespread in the riparian herbaceous vegetation community. Riparian woodlands intergrade with other riparian communities in canyon bottoms, and with oak and island woodlands on steeper slopes.

Riparian woodlands are noticeably less diverse than similar communities on the adjacent mainland. Three riparian dominants on the mainland, *Alnus rhombifolia*, *Platanus racemosa*, and *Umbellularia californica*, do not occur naturally on the island. About 12 individuals of *Platanus racemosa* were planted in Cañada del Puerto in 1937 or 1938 and have become naturalized (letter from C.Q. Stanton to R.N. Philbrick, 25 December 1975).

## History of Land Use and Biotic Introductions

**Native Americans**. The activities of Native Americans, including harvesting, habitat disturbance, and anthropogenic plant dispersal, influenced the vegetation on Santa Cruz Island for about 7,000-10,000 years. The oldest evidence of human occupation has been dated at about 7,000 years Before Present (BP) (Glassow 1980). Older occupation dates, from 9,000 to over 10,000 years BP, have been reported for San Miguel and Santa Rosa islands (Erlandson 1994). The island's early inhabitants, predecessors of the Chumash Tribe, were hunter-gatherers with an emphasis on fishing and harvesting marine resources. During the early mission period, eleven Chumash villages were reported, with a total population of over 1,100 people (Glassow 1980).

The Chumash affected plants of Santa Cruz Island by 1) selectively harvesting plants for food or other uses, 2) altering habitats in the vicinity of their villages, and, most importantly, 3) transporting plant materials from the mainland to the island and between islands.

The island Chumash apparently harvested large quantities of some plants with underground storage roots, especially *Dichelostemma capitatum* (Timbrook 1993). Disturbance caused by digging, while detrimental to some species, may have promoted propagation of other taxa by aerating heavy soils and scattering bulblets or offshoots, especially if plants were harvested after seeds had been dispersed.

The Chumash traveled between island and mainland villages in wooden canoes or *tomols*. A number of trade items (including plant materials) were transported across the Santa Barbara Channel. Acorns and roots, as well as seeds of "chia," "grasses," and other plants, were reportedly among the items sent to the islands (King 1976). Some oaks may have been introduced from the mainland as a result of the active trade in acorns across the Santa Barbara Channel (Timbrook 1993). The peculiar array of oak species on Santa Cruz Island, including several found mostly at inland sites on the mainland, may be the result of these early introductions.

At least two ceremonial plants, *Datura wrightii* and *Nicotiana clevelandii*, appear to be associated with Chumash shell middens and sites of former villages (Timbrook 1993). The Chumash may have moved these plants from one location to another. However, the observed distributions may be correlated with environmental factors, such as the effects of concentrated shell deposits on soil conditions. Analyses of pollen from archaeological shell middens, such as the preliminary study conducted by Hevly and Hill (1970), will undoubtedly provide valuable data on prehistoric plant usage. Hevly and Hill found significant spatial and temporal variability in pollen samples taken from four localities on the island.

**Europeans**. The history of Europeans on Santa Cruz Island began at a relatively late date, considering the amount of habitat alteration that resulted from their activities. During the last 150 years, the composition and distribution of vegetation on Santa Cruz Island has been greatly altered by large numbers of feral grazing animals, cultivation, disturbance, and the introduction of alien plant taxa (Brumbaugh 1980a, Hochberg et al. 1980, Leishman 1981).

On October 13, 1542, 50 years and one day after Columbus arrived in the Americas, Juan Rodríguez Cabrillo was the first European to sight Santa Cruz Island. He reportedly did not step on land, but passed by as he sailed through the Santa Barbara Channel (Bancroft 1886, Wagner 1966). Later Spanish explorers (including Cermeño in 1595, Vizcaíno in 1602, Portola in 1769, and Vancouver in 1795) also wrote of Santa Cruz Island, but their accounts are vague as to whether any Europeans actually landed. Contrary to some reports, there is no evidence that livestock were placed on Santa Cruz Island by any of these early explorers.

In 1769, all of California, including Santa Cruz Island, became the property of the King of Spain. During Spain's ownership the Chumash continued to live on the island, but population sizes gradually decreased as a result of exposure to diseases introduced by Europeans (Johnson 1982). Between 1814 and 1822, the last of the island Chumash were removed to the Missions at Santa Barbara and San Buenaventura (Glassow 1977, Johnson 1982). Increased activity and exploitation of the island's natural resources by Europeans began in the early 1800s. A report by the President of the California Missions in 1806 mentions the timber, water, fair soil, and ten native rancherias on Santa Cruz Island (Ellison 1937). Captain James Wilcox was reportedly sent to the island in 1815 to get timber for reroofing parts of the Santa Barbara Mission (Stanton 1984). Some wood-cutting also may have taken place in 1817 (Bancroft 1886, Spaulding 1964).

**Early Ranching Years**. In 1821, California became part of the new Republic of Mexico, after Mexico's successful revolt against Spain. In February 1830, a group of more than 100 Mexican convicts was sent to Santa Cruz Island. Don Romualdo Pacheco, then commandante of the Santa Barbara Presidio, provided them with "seed for sowing and some animals such as cattle and a few horses" (Ord 1956). This is the first known reference to livestock on the island. In 1839, the island was granted to Andres Castillero by the Mexican government, to reward him for distinguished military service (Tays 1935). Castillero, the first private owner, held title to the land for 18 years.

Ranching activities had begun by the time Castillero sold the island to William E. Barron in 1857. Under the supervision of Dr. James Barron Shaw, an Englishman residing in Santa Barbara, purebred stock were brought to the island in early 1853, including sheep, cattle, and horses (U.S. District Court 1857). A watercolor by James Madison Alden depicting the Central Valley in 1855 shows buildings, horses, cultivated fields, and fences (Stenzel 1975). A newspaper article in 1857 (*Alta California,* 27 April 1857) reported that there were about 200 acres of land under cultivation at the time.

The Santa Cruz Island ranch soon became one of the best and most admired sheep enterprises in the state. In 1853, Santa Cruz Island was reportedly the chief supplier of mutton to the Los Angeles market (Newmark & Newmark 1930). A newspaper article in 1858 (*Sacramento Union,* 29 September 1858) stated:

> *We have frequently heard of the rare breed of sheep on Santa Cruz Island, but have had no opportunity of seeing a specimen until this week, when we visited the building of Dr. Shaw and were surprised at the sight of a young monster Merino ram - his age being only six months - weight one hundred and two pounds, standing thirty-one inches and covered to the very hoofs with close fine wool - looking to the careless observer as if he had been lately shorn. We are not the best judges of rams, though we are of mutton; but we are convinced we never saw a finer animal, for age, shape, weight, and fineness of wool; he is of the improved French Merino.*

Pigs were apparently introduced during the 1850s as well, although the exact date and circumstances are not known. An 1857 deposition by James Barron Shaw in the U.S. District Court tells of an "intruder" named

*Sheep in corrals, Main Ranch, looking south (Santa Barbara Museum of Natural History, undated).*

James Box, who lived in a shanty and raised pigs as early as 1853. At Shaw's request, the pigs were sold and the intruder left (U.S. District Court 1857). According to Phillips (1927), Captain Martin M. Kimberly "imported a few male and female pigs from San Francisco and turned them loose on the island" sometime around 1854, in order to guarantee a continuous supply of pork. Kimberly reportedly left after three years, gathering up his stock and other belongings, except for the wildest of the pigs (Phillips 1988). The remaining pigs multiplied and eventually became a problem.

In 1869, William Barron sold Santa Cruz Island to ten San Franciscans who formed a corporation called the Santa Cruz Island Company. The number of sheep increased rapidly during the 1860s; by 1870, over 50,000 head were reported (U.S. Bureau of Census 1870). Several drought years in the early 1870s drastically affected sheep operations on San Miguel and San Nicolas islands (Phillips 1927, Johnson 1980) and apparently Santa Cruz Island as well. During the month of April 1872 alone, 20,000 sheep were shipped from Santa Cruz Island to the mainland (Krythe 1958).

Between 1870 and 1885, 20,000 to 30,000 sheep were rounded up annually and shorn (Blake 1887, Symmes and Associates 1922). There were an estimated 60,000 sheep on the island in 1875; 15,000 were apparently killed for hide and tallow in June of that year (Rothrock 1876). During another drought year in 1877, an additional 25,000 sheep had to be slaughtered as a result of insufficient feed (*Santa Barbara Index*, 22 March 1877; Thompson and West 1883). An 1883 account (Thompson and West 1883) described the continuing problem:

> *The island becomes at some times overstocked, and may be said to be in that condition much of the time. The result is that the grasses, being cropped so close, die out, and allow the loosened soil to be removed by wind and rain. Popular opinion is that the amount of fertile land is thus being gradually lessened.*

Other agricultural activities continued and plant introductions apparently increased during the 1870s and 1880s. When Joseph T. Rothrock of the U. S. Coast Survey visited in 1875, he reported that grain was cultivated near Prisoners' Harbor and that "in one or two places pepper-trees [*Schinus*] had been planted and were growing vigorously without care" (Rothrock 1876).

**The Caire Years**. By 1880, through various circumstances Justinian Caire became the sole stockholder of the Santa Cruz Island Company. Under Caire's direction, cattle and sheep ranching were continued, a variety of introduced fruit trees and annual crops were cultivated, grape vineyards were planted, and several additional

*Vineyards surrounding Main Ranch, looking southwest (Santa Barbara Historical Society, undated).*

ranches were developed. Thompson and West (1883) reported that "the pear, apple, cherry, plum, peach, fig, pomegranate, orange, and lemon flourish here." Goodyear (1890) noted that there was a young plantation of fig and olive trees at Smugglers' Cove in May 1889.

*Eucalyptus* apparently was one of the first timber trees to be planted on the island, and several groves were present as early as 1885. Maps (in archives, Santa Cruz Island Foundation) drawn for the Santa Cruz Island Company during the 1880s show groves in Cañada del Puerto (Maps #9 & 13, dated 1885), at the Main Ranch (Map #41, June 1886), and at Prisoners' Harbor (Map #56, July 1886). By 1890, *Eucalyptus* trees also were present at Christy Ranch (then called West Ranch) (Map #44, June 1890). Seeds for additional plantings of *Eucalyptus* were shipped between 1890 and 1894, while *Acacia* seeds were received as early as February 1890 (Santa Cruz Island Company invoices in archives, Santa Cruz Island Foundation). A "proposed forest of Acacia" was mapped in Cañada del Puerto in 1885 (Map #9) and a single large *Acacia* tree is shown on an early map of Rancho del Portezuelo (Map #71, July 1890). In August 1911, a shipment of 300 "cypress plants" (presumably *Cupressus macrocarpa*) was made (Santa Cruz Island Company invoices). Italian stone pines (*Pinus pinea*) also were introduced during the time of Caire ownership.

Vegetable gardens were present as early as 1885 at the Main Ranch, and at the Scorpion and Smugglers' ranches (Santa Cruz Island Company maps). Maps drawn in 1890 show vegetable gardens and corn patches at Christy Ranch and at Rancho del Portezuelo. Garden seeds shipped between 1890 and 1892 included red and white beans, red beet, cabbage, carrot, cauliflower, celery, chervil, onion, parsley, pea, sweet pea, chili pepper, Burbank seed potato, salsify, spinach, and turnip (Santa Cruz Island Company invoices). Two barrels (170 lbs.) of chicory roots were shipped in January 1890 and at least 800 lbs. of alfalfa seeds were purchased between 1890 and 1892.

As early as 1884, a grape vineyard was planted near the Main Ranch (Pinney 1994) and a small vineyard is shown on a map of Smugglers' Ranch drawn in December 1885 (Santa Cruz Island Company Map #28). In 1890, Justinian Caire informed the California State Board of Viticultural Commissioners that he had 60 acres of vineyards, mostly of the Zinfandel variety (Pinney 1994). The vineyards were rapidly expanded in the 1890s and early 1900s and large quantities of wine were shipped by boat to San Francisco, Santa Barbara, and Los Angeles.

By 1890, 50,000 sheep were recorded in the round-up. Symmes and Associates (1922) estimated that about half of the total sheep population were actually shorn, thus bringing the potential total to 100,000. Sheep were allowed to run at will and were rounded up periodically for shearing and slaughter for rendering. Justinian Caire

*Vineyards between Main Ranch and Valley Anchorage with corn field in foreground, looking north (Santa Barbara Museum of Natural History, undated).*

was apparently concerned about the problem of soil erosion and was responsible for ordering construction of the many rock walls and dams placed strategically around the island (Towne and Wentworth 1945).

Justinian Caire died in 1897 and his sons, Arthur and Frederic, continued the agricultural operations. Under their supervision, the vineyards continued to flourish. At the height of wine production in 1910, 83,000 gallons were produced (Pinney 1994). By 1917, the island's vineyard was the largest in Santa Barbara County (Hussey 1961). Grape varieties grown for red wines included Barbera, Burgundy, Cabernet Franc, Cabernet Sauvignon, Carignane, Grenache, Malbec, Mataro, Trousseau, and Zinfandel (Pinney 1994). Varieties grown for white wine included "Bordelais," Burger, Chasselas, Muscat, Riesling, and Sauvignon Vert. The last vintage prior to national Prohibition was in 1918, when 63,000 gallons were produced (Pinney 1994).

Between 1919 and 1932, grapes were shipped to the mainland and no wine was produced. By 1922, the number of varieties had been reduced to six, with Zinfandel being the most extensively grown variety (Symmes and Associates 1922; Pinney 1994). In 1922, about 187 acres were still devoted to vines, with bearing vines on 150 acres (Symmes and Associates 1922). In 1930, 210 tons of grapes were reportedly shipped (Rouse 1980). After the end of Prohibition, wine production began in 1933 and continued until 1936 (Pinney 1994).

By the early 1900s, pig populations were also rapidly growing. A newspaper article in 1904 (*Santa Barbara Morning Press*, 16 March 1904) described the large pig population at that time:

> *The Santa Cruz Island Company, owners of this island, have decided to exterminate the wild boars, which have become so numerous as to be an intolerable nuisance. They are giving their men five cents per head bounty for all they kill.*

In the early 1900s, hay, alfalfa, walnuts, almonds, and vegetables were cultivated (Symmes and Associates 1922). Hay was reportedly cultivated at Christy Ranch (235 acres in 1922), near Scorpion Ranch (309 acres in 1922), and near Prisoners' Harbor (278 acres), with only a small area of unspecified size devoted to hay at the Main Ranch in the Central Valley. Alfalfa cultivation involved a total of about 60 acres (Symmes and Associates 1922).

Vegetable gardens were maintained at the three ranches and at Prisoners' Harbor. Almonds were grown on the south side of the Main Ranch, as were three acres of walnuts. The olive groves cultivated at Smugglers' Cove were abandoned by 1922, but they persisted, along with fig, walnut, and *Eucalyptus* trees. The island remained principally a stock ranch, and sheep were the primary product. Symmes and Associates (1922) reported that, in addition to the sheep, there were nearly 800 head of cattle in 1922, along with almost 100 horses.

A number of fishing camps also were established around the perimeter of the island during the early 1900s. Eaton (1980) described such camps at Forney's Cove and Pelican Bay, at Prisoners', Potato, and China harbors,

*Hay fields near Valley Anchorage, looking southeast (Santa Barbara Museum of Natural History, undated).*

*Rancho Portezuelo with vegetable garden in foreground and chaparral-covered slopes in background, looking south (Santa Barbara Museum of Natural History, undated).*

*Buildings and ornamental plantings at Eaton resort, Pelican Bay, looking south (H. Van Winkle, ca. 1920, collection of E. R. Blakley).*

at Hungryman's Gulch, Smugglers' Cove, and Blue Banks, and at Willows, Johnson, and Pozo beaches. Rouse (1983a) reported another fishing camp at Fry's Harbor and that 15 fishermen lived at various camps on the island in 1933 (Rouse 1983b). Caire (1993) described the fisherman living at Dick's Harbor and the lobster camp at Coches Prietos Anchorage. Gherini (1966) reported camps at Potato and Scorpion harbors, at San Pedro Point, and at Yellow Banks. Residents of some camps introduced plants and/or animals to the island. For example, a vegetable garden planted at the Willows Canyon camp reportedly included "lettuce, carrots, turnips, beets, radishes, young onions, and a patch of garlic" (Eaton 1980).

The fishing camp at Pelican Bay was developed into a resort in 1910 and continued to operate until 1937 (Eaton 1980). The resort was a popular destination for tourists and the Eatons built a number of cabins and tents to accommodate visitors. They also planted a number of alien species around the resort, including a vegetable garden, "white marguerite daisies" [*Chrysanthemum* sp.?], white and black fig trees [*Ficus* sp.], and ice plant [*Carpobrotus* sp.?] (Eaton 1980). Many alien ornamental plants still persist at Pelican Bay, including *Agave americana*, *Carpobrotus edulis*, *Eucalyptus* spp., *Ficus carica*, *Opuntia ficus-indica*, *Pelargonium ↔hortorum*, *Pinus pinea*, and *Vinca major*.

The natural environment of Fry's Harbor was dramatically altered by human activities during the 1920s. The site was used as a movie location in 1919, and numerous plants were introduced for the scenes. According to Eaton (1980), "two large barges were towed from San Pedro, loaded with three dozen big palms, coconut trees, banana trees loaded with fruit, rose bushes, and the like." In the late 1920s, the harbor was heavily disturbed when a quarry was developed to provide rock for construction of a breakwater at Santa Barbara. About 50 to 60 people lived on the site between 1927 and 1930, when the quarry operation was abandoned (Helen 1979).

**The Stanton and Gherini Years**. Because of disagreements among Justinian Caire's heirs, the island was partitioned into seven parcels by court order in 1925. Maria Rossi and Aglae Cappucio (granddaughter and daughter of Justinian Caire, respectively) were awarded two of the seven parcels, encompassing the eastern 10% of the island. Ambrose Gherini, Maria Rossi's husband, took an active interest in sheep ranching and eventually acquired Aglae's interest in the island. After a period of inactivity in the late 1920s and early 1930s, sheep ranching was resumed on the east end. In order to improve the herd through controlled breeding, additional fencing was installed and approximately 400 rams were purchased during the 1930s and early 1940s. Ambrose Gherini's sheep operation continued until 1948, when most of the flock was sold (Warren ca. 1954).

*Site of rock quarry at Fry's Harbor, looking east (September 1930, Santa Cruz Island Foundation).*

*Rock quarry operations at Fry's Harbor, looking north (September 1930, Santa Cruz Island Foundation).*

In 1937, the remaining five parcels, comprising the western 90% of the island, were sold to Edwin Stanton by part of the Caire family. For the first two years following his purchase, Stanton earnestly tried to maintain and improve the sheep flock. About 10,000 domestic sheep were imported in 1938 in an attempt to domesticate the flock (Warren ca. 1954), but the plan proved to be unsuccessful, and Stanton turned his attention to operating a cattle ranch. By the 1930s, years of overgrazing by sheep had apparently facilitated the spread of native prickly pear cacti (*Opuntia littoralis* and *O. oricola*). In 1938, Edwin Stanton estimated that approximately 40% of the rangeland was "rendered useless by dense infestations of prickly pear" and asked researchers from the University of California at Riverside for help in reducing the *Opuntia* populations. In an attempt to control the cacti, several cactus-feeding insects were introduced in the 1940s but were not very effective (Goeden et al. 1967). Eventually, cochineal insects (*Dactylopius* sp.) were released in the early 1950s; these insects proved to be very effective at reducing the *Opuntia* populations. Overall, the biological control program was very effective (Goeden & Ricker 1981). It is believed to be one of the first such programs to successfully control a native weed with introduced insects (Sauer 1988).

In 1939, Edwin Stanton began a systematic roundup of the sheep, shipping all that were caught to the mainland. Warren (ca. 1954) reported

> *Over a short period of years, some 35,000 head were caught and sold, but this number far from depleted the actual supply for many thousands still roamed the less accessible areas. In an attempt to further reduce the size of the sheep population, the island's riders have killed an estimated 20,000 additional head in the years since ...*

Between 1955 and 1962, an additional 27,950 sheep were captured and sent to market. Fencing was installed to restrict sheep primarily to the north side of the island, and an estimated 180,000 sheep were shot during the 1960s and 1970s (Santa Cruz Island Company records). Edwin Stanton died in 1963 and his son Carey assumed management of the Santa Cruz Island Company in 1973 (Daily and Stanton 1983).

**The Nature Conservancy Years**. In 1978, The Nature Conservancy acquired a conservation easement on the western 90% of Santa Cruz Island from the Santa Cruz Island Company. An international non-profit conservation organization, The Nature Conservancy began planning a program of fencing, trapping, and hunting the feral sheep on Stanton's portion of the island soon after the easement was acquired (Schuyler 1993).

Between 1981 and 1983, over 100 miles of existing fences were repaired in order to divide the island into 23 separate pastures, ranging in size from 338 to 11,161 acres (137-4,517 hectares). The Nature Conservancy began hunting feral sheep in December 1981, and by June 1989, a total of 37,171 sheep had been shot by Conservancy hunters or by sport hunters sponsored by the Santa Cruz Island Club (Schuyler 1993). Carey Stanton died in 1987, before the last of the sheep were removed from the western portion of the island. With his death, The Nature Conservancy became sole owner of 90% of Santa Cruz Island.

After taking control, The Nature Conservancy ceased ranching operations and began to remove the cattle. During 1988 and 1989, about 1800 cattle were rounded up and shipped to the mainland. The last cattle were removed from the lower portions of Laguna and Alamos Canyons (Schuyler 1988, R. Philbrick, pers. comm. 1994; P. Schuyler, pers. comm. 1989). By June 1989, feral sheep were completely eliminated from the western 90% of the island, but they remained on the eastern portion. Currently, sheep are reinvading the eastern edge of The Nature Conservancy property. Exclusion of sheep requires continual attention to the fence separating the two properties. Feral pigs are still active throughout most of the island. Several drought years (especially 1989-1990) drastically reduced the numbers of feral animals on the island, but sheep and pig populations are now rapidly recovering. The threat of overgrazing will exist until all introduced herbivores are removed.

In 1980, the four Northern Channel Islands and Santa Barbara Island were designated by Congress as Channel Islands National Park. Neighboring Santa Rosa Island was purchased by the National Park Service in 1986. Currently, the Service has acquired an undivided 75% interest in the Gherini property on Santa Cruz Island. National Park Service acquisition of the island's entire east end will mark the beginning of a new era in resource management.

## Fire History

Abundant charcoal in Pleistocene and Holocene deposits on San Miguel Island (Johnson 1972) and Holocene deposits on Santa Cruz Island (Brumbaugh 1980b) may have resulted from natural prehistoric wildfires. During the last 7,000 years, island Chumash or their predecessors may have purposely set fires for vegetation management (Carroll et al. 1993). Wildfires also could have started accidentally from campfires, which were observed on the Channel Islands in December 1595 (Johnson 1982). During the last 140 years, there were only three documented lightning-caused fires on the California Channel Islands: small burns on Santa Catalina Island in 1967; on Santa Cruz Island in 1987; and on Santa Rosa Island in 1988 (Carroll et al. 1993). There have been a number of other fires on Santa Cruz Island, but most of them were caused by human activities.

Several fires that burned natural vegetation have been documented. A fire reportedly occurred in the 1830s in the Prisoners' Harbor area (Bancroft 1884), shortly after Mexican convicts were exiled to the island in February 1830 (Ord 1956):

> *Those on the island, after being there some time, lost what they had by fire. We saw the flames from here [Santa Barbara]. A long time passed before succor could be sent to them because the schooner which was used to carry supplies to the islands had not arrived. The convicts made some rafts and came here on them.*

Another fire was reported in the early 1870s. Barrett (1935) cited a newspaper article (*Santa Barbara News,* 10 October 1871), which reported that "It is generally believed Santa Cruz Island is on fire. There is great smoke out that way...." The extent of these early fires is unknown.

During the last century, large-scale fires were rare and restricted mostly to the north shore. Between 1917 and 1931, three fires occurred in the vicinity of Pelican Bay, all presumably caused by human activities. McElrath (1967) mentions a wildfire that burned along the north shore from Prisoners' Harbor to Pelican Bay in 1917, presumably the same fire reported by Symmes and Associates (1922). Mason (1927) also described a burn in the pine forest near Pelican Bay, which apparently occurred in August 1927 and consumed understory plants, shrubs and small trees up to one-half mile inland. An additional fire burned approximately 12,000 acres (4850 hectares) along the north shore, west of Prisoners' Harbor in September 1931 (*Santa Barbara Morning Press,* 20-21 September 1931). The abundance and extent of pines between Prisoners' Harbor and the Pelican Bay area was significantly reduced by these fires, because sheep consumed the fire-stimulated seedlings.

Three small fires occurred between 1958 and 1987. Grass fires occurred in the Loma Pelona pasture in August 1958 and on coastal flats near San Pedro Point in August 1979. Another small fire in February 1987 burned an area of 0.32 acres (0.13 hectares) near the north shore of the island between East Twin Harbor and Pelican Bay. This is the first well documented, lightning-caused fire (Carroll et al. 1993). During the 1990s, fires burned additional portions of the island. In June 1990, a fire caused by rescue flares burned about 600 acres (243 hectares) on open slopes above China Harbor. In December 1993, The Nature Conservancy conducted a controlled burn of over 1000 acres (405 hectares) in Pozo Canyon and a smaller one near Campo del Norte. In December 1994, controlled burns were started in Alegria Canyon and near the east end of China Harbor. The China harbor fire escaped and burned most of the China Pines.

## Historical Change in Vegetation

**Pleistocene and Holocene Epochs.** The present Mediterranean climate of coastal southern California results from interaction among several regional factors, primarily the North Pacific subtropical high, and to a lesser extent, the Aleutian low-pressure system, the cool California oceanic current, and the thermal low of the interior valleys and deserts. These factors are integrated with atmospheric and oceanic circulation and were apparently operative throughout late Quaternary time (Johnson 1977). As a result, the Quaternary climate of coastal southern California was semi-arid, even during glacial maxima (Johnson 1977).

The fossil record shows that the climate of the area was somewhat more moist than today. Conifer forests dominated by *Cupressus*, *Pinus*, and *Pseudotsuga* were widespread in the region (Axelrod 1980, 1983). Although fossil evidence is very limited, material recovered from San Miguel, Santa Rosa, and Santa Cruz islands indicates that sclerophyllous shrubs and trees were present on all three islands during the Pleistocene. A sample of late Pleistocene vegetation on Santa Cruz Island was described by Chaney and Mason (1930), who discovered a deposit of macrofossils in the lower portion of Cañada de los Sauces. These fossils, deposited approximately 14,000 years BP (Orr 1968), include nine plant taxa: *Arceuthobium campylopodum*, *Arctostaphylos* sp., *Ceanothus thyrsiflorus, Cornus californica*, *Cupressus goveniana*, *Garrya elliptica*, *Myrica californica*, *Pinus remorata* [*P. muricata* forma *remorata*], and *Pseudotsuga menziesii*. Five of these genera (*Arceuthobium*, *Cornus*, *Cupressus*, *Myrica*, and *Pseudotsuga*) are no longer represented by native taxa on Santa Cruz Island. Chaney and Mason compared the Pleistocene flora of Santa Cruz Island with the California coast near Fort Bragg, about 440 miles (1139 km) north of the island:

> *The rainfall of the Fort Bragg region is considerably higher and the mean temperature is considerably lower than on Santa Cruz Island today, and the conclusion is reached that the Willow Creek flora represents a southward extension of the northern forest which is probably to be correlated with one of the glacial epochs of the Pleistocene.*

Fossils found on San Miguel and Santa Rosa islands confirm that conifers and associated shrubs also were present during the Pleistocene. On San Miguel Island, Johnson (1972, 1977) discovered impressions of *Pinus radiata* and *Ribes* sp. in late Pleistocene and Holocene travertine deposits. He also found fossils of a shrub tentatively identified as *Rhus* sp., which were radiocarbon-dated at approximately 17,000 years BP. On Santa Rosa Island, fossils of *Cupressus goveniana* approximately 16,000 years old were found in Tecolote Canyon (Orr 1968, Johnson 1977). There are no natural populations of *Pinus radiata* or *Cupressus goveniana* on the California Channel Islands at present.

Rapid climatic change in southern California was apparently the most significant factor affecting vegetation change on Santa Cruz Island during the last 12,000 years. Landscape modification by Native Americans also may have affected vegetation patterns during this time period. Evidence for regional changes in climate and vegetation comes from pollen cores taken from the Santa Barbara Channel and neighboring Santa Rosa Island. A preliminary study of pollen from shell middens on Santa Cruz Island (Hevly and Hill 1970) suggests change over time, but samples were not dated.

Heusser (1978) analyzed a 12,000 year record from sediments in the Santa Barbara Basin and concluded that initially wet, cool conditions were gradually replaced by a warmer, drier climate. In general, plant associations were similar to those now present in the Santa Barbara region, but abundance and ranges of dominant plants changed. Heusser inferred that "upland coniferous communities" were dominant in the area during the early part of the period (ca. 12,000 to 7,800 years BP). The conifer communities were replaced by "lowland and cismontane communities" in which *Quercus* and Asteraceae reached optimal development ca. 5,700 years BP. Subsequently, chaparral and coastal-sage scrub associations became increasingly important, particularly since ca. 2,300 years BP. Moist environments, such as river banks and freshwater marshes, gradually became more restricted during the 12,000 year period.

*View of fossil plant locality in Cañada de los Sauces, looking east (S. Junak, February 1979).*

A 5,200 year record, from sediments in a small estuary on Santa Rosa Island and divisible into 3 discrete time zones, also documents climatic and vegetative change. According to Cole and Liu (1994), the lowermost zone in this record (5,200 to 3,250 years BP) represents a time of greater aridity than the present. They found large amounts of *Ambrosia* and Chenopodiaceae pollen and small amounts of Cyperaceae pollen, which suggests an arid climate and little flow of terrestrial fresh water. The next time zone (3,250 years BP to 1800 AD), with large amounts of Asteraceae and Cyperaceae pollen, represents a moister environment. Pollen deposited during the most recent time zone (1800-1990 AD) documents extremely high erosion rates (especially between 1874 and 1920 AD), associated with the introduction of large, non-native grazing animals. Alien plants, especially *Erodium*, spread rapidly on Santa Rosa Island during this period.

**Historical Period, 1850s to Present**. The most significant factors affecting vegetation on Santa Cruz Island during this period include the introduction and subsequent proliferation of feral sheep and pigs, and the arrival and spread of aggressive alien plants. The feral animals, especially sheep, have moderately to severely impacted most island habitats (Hochberg et al. 1980, Van Vuren and Coblentz 1987). In several areas, overgrazing and erosion have reduced large contiguous populations of woody plants to small isolated patches (Brumbaugh 1980b). Evidence for vegetation changes during this period comes from a variety of sources, including historical descriptions and maps, comparisons of successive photographs, grazing exclosures, and herbarium records of alien plant introductions.

Decades of overgrazing by feral sheep have adversely affected most of the island's plant communities by altering their population structure, the natural size and stature of dominant species, as well as species diversity and composition (Hochberg et al. 1980). Selective grazing has reduced the range of many native species (e.g., *Coreopsis gigantea*, *Hazardia detonsa*, *Lupinus albifrons*, and *Mimulus flemingii*) while apparently increasing the range and abundance of other native taxa (e.g., *Eremocarpus setigerus*, *Opuntia littoralis*, *O. oricola*, and *Senecio flaccidus*). Goeden et al. (1967) reported that the range of native *Opuntia* spp. was increased by sheep overgrazing:

> *Infestations vary from scattered, compact clumps, on the better grasslands of the central valleys and on the narrow neck of the island, to dense, nearly unbroken stands that cover many of the southfacing slopes on the north and south ridges. In the past, severe overgrazing by wild sheep greatly facilitated the spread of these cacti. On grasslands denuded by sheep, cactus pads broken from clumps came into immediate contact with the soil, took root, and established many new clumps.*

*Overgrazed slopes and sheep fence in upper part of Laguna Canyon, looking north (S. Junak, December 1979).*

In grassland communities, sheep grazing has reduced herbaceous cover, increased bare ground, altered community structure, decreased litter, and increased erosion (Van Vuren and Coblentz 1987). Browsing by sheep in shrub communities altered growth forms and prevented regeneration in canopy species. Sheep completely defoliated some low-growing shrubs and reduced diversity of understory species (Hochberg et al. 1980, Van Vuren and Coblentz 1987). In the pine forest near Pelican Bay, sheep grazing after fire reduced foliar cover, regeneration, and species richness of the understory (Hobbs 1983). Feral pigs also have adversely affected plant communities, especially by trampling and rooting under oak woodland and chaparral canopies (Baber 1982). Pig activities have inhibited regeneration of native trees and shrubs, caused destruction of the litter layer and upper soil horizons, and promoted accelerated erosion on Santa Cruz and Santa Catalina islands (Baber 1982, Peart et al. 1994).

Aggressive alien plants, especially annual grasses, have negatively impacted the native flora by competing for space, light, soil moisture, nutrients, and pollinators. Habitat disturbance by feral animals has apparently facilitated the spread of alien plants (Hochberg et al. 1980). A number of introduced plant species have spread rapidly after successfully colonizing disturbed sites resulting from overgrazing. Greene (1887) reported only 36 alien plant species, representing 11.2% of the flora known at the time. Now there are 170 alien taxa (26% of the current flora), ranging from tiny annuals to large trees and occurring in a diversity of habitats.

The diversity, range, and abundance of alien annual grasses have increased dramatically since the early 1900s. Greene (1887) reported only 5 alien annual grasses. Brandegee (1888) found 3 additional species during a wetter season. At present, 32 alien annual grass taxa occur on the island and are outcompeting native perennial grasses in many disturbed areas.

**Vegetation Changes on the North Side**. Some of the most noticeable vegetation changes have occurred in chaparral, oak woodland, and pine forest communities on the north side of the island. The volcanic slopes west of Prisoners' Harbor experienced severe overgrazing and gradual loss of long-lived shrubs and trees was observed (Brumbaugh 1980b, Hobbs 1980, Hochberg et al. 1980, Van Vuren and Coblentz 1987). A nearly continuous cover of woody vegetation was reduced to isolated patches, surrounded by eroded slopes, in less than 150 years. The extent and density of the pine forest above Pelican Bay were dramatically reduced by a series of fires, followed by intense grazing, in less than 50 years (Hobbs 1980).

The volcanic slopes on the east end of the island have suffered similar losses of woody vegetation in the chaparral and oak woodland communities. Brumbaugh (1980b) documented vegetation degradation near Coche Point and on High Mount by comparing maps drawn in 1856 with aerial photographs taken in 1929. Observations between the 1850s and 1870s on the north side included descriptions of dense stands of woody vegetation, which were beginning to show the impacts of overgrazing by sheep. U. S. Coast Survey Sub-assistant W. M. Johnson (1858) described the slopes immediately west of Prisoners' Harbor:

> *...so completely overgrown with high brush, that it will be impossible to get through it with the instrument; in putting up our signals the men were forced, in very many places, to move for considerable distances on their hands and knees ... [and] would find themselves at the end of that time, only about one-third of a mile from where they had started.*

About 17 years later, Stehman Forney, of the U. S. Coast Survey, described the vegetation and topography between the west end of the island and Prisoners' Harbor (Forney 1875):

> *The country is rough and almost impassable for man or beast. Deep precipitous canons make down from the main ridge of the island to the sea, the ridges and canons are covered with a dense growth of chaparral, stunted pines, and oak trees ...*

J. T. Rothrock of the Wheeler Expedition spent a week on the island during the same season. He (1876) estimated that there were 60,000 sheep in June 1875 and that they were beginning to have a marked effect on the vegetation and flora:

> *On the grounds most visited by the herds of sheep, all vegetation, save sage-brush,* cactees *[sic] and* erodium *or storksbill, had been entirely swept away. The grass had gone completely, and such plants of the island flora as sheep would eat, it was with difficulty that I could get even a decent botanical specimen of. In fact, pasture had become so thin that the sheep at the time of my visit mere [sic] wandering in very small bands that they might the more readily find food. Even the sage-brush was disappearing, as year after year the sheep had eaten away its leaves and younger shoots, until there was not sufficient of the more green, succulent tissues to elaborate the sap.*

Similar observations on the ecological damage done by sheep were also made by Oscar Loew, who accompanied Rothrock (Loew 1876):

> *The vegetation of Santa Cruz [Island] is scanty; grass is gradually disappearing before the clean sweep made by the sheep herds, whilst cactus, especially* Opuntia, *is spreading more and more. The mountain-sides are here and there covered by small patches of pine, while the low brush attempts in vain to conceal the barren rock, prominent in all the hills. It is said that formerly the entire island was covered with forests; whether true or not, one fact was observed by us, namely, that roots of pine trees exist in the loose soil of the western shore, where no trees in wide circumference occur at present.*

The 1870s were a dry period in southern California, with droughts occurring in 1870-1872 and 1877 (Johnson 1980). As a result, observations made during this period may reflect extreme deviations rather than long-term trends. Nevertheless, Rothrock's and Loew's observations document severe vegetation damage in some parts of the island. When E. L. Greene visited the island in July and August of 1886, the situation may have temporarily improved somewhat, as he seemed to be impressed by the vegetation on the north slope (Greene 1887):

> *The near approach reveals a succession of more or less sharply outlined hills rising one behind another. This lengthwise range of mountains, which forms all that is seen of the island as it is approached from the northern or Santa Barbara side, is intersected at short intervals by deep and narrow gorges which run down to the sea. In most of these running water of good quality is to be found at almost any time in the year, so that the northern slope may be said to be well watered; and the common trees of the nearest mainland mountain districts, such as the large-leaved maple* (Acer macrophyllum), *live oak* (Quercus agrifolia), *poplar* (Populus trichocarpa), *willow* (Salix laevigata), *and many more are found thriving in these canons ... [Here is] a considerable part of the best grazing land of Santa Cruz. It is an open, rolling district, extending back for a half mile or more, everywhere intersected by the narrow canons mentioned, covered with fine grass, dotted*

> *with clumps of scrub oak* (Quercus dumosa), *and some patches of manzanita* (Arctostaphylos), *with here and there a grove of the beautiful fern leaved* Lyonothamnus. *Back of this terrace the land rises more abruptly, breaking into rocky shelves and deep gorges, and the vegetation becomes more arboreal. Here are dense forests of a small pine ... clumps of a large-fruited evergreen cherry-tree ... impenetrable thickets of manzanita, with here and there a group of oaks,* Q.[Quercus] agrifolia *and* Q. chrysolepis.

Despite his seemingly favorable description of the north shore, Greene also noted impacts of sheep grazing on some endemic plants. He found only "two bushes" of *Malvastrum thurberi* var. *laxiflorum* (*Malacothamnus fasciculatus* var. *nesioticus*) "under the protection of large opuntias; perhaps thus kept from the sheep." He described *Syrmatium niveum* (*Lotus argophyllus* var. *niveus*) as being "on the verge of extinction." The range and abundance of a number of native plant taxa have been reduced since Greene's visit, presumably due to disturbance (Hochberg et al. 1980). Some species common on the north side in 1886 are rarely seen there now. Greene (1887) noted that *Habenaria elegans* (*Piperia elongata*) was "frequent on wooded hills at the north" and that *Lilium humboldtii* was "very common in woods everywhere." The abundance of both species has been drastically reduced on the north side of the island; they are rarely seen there now. *Aster radulinus* was found by Greene to be "rather common in open places of the wooded northward side." It is now known from scattered localities in the interior, mostly as an understory species. Greene stated that *Acer macrophyllum* was "common in deep canyons of the north side and very luxuriant." It is occasional now, and common locally only in Diablo Canyon. In his description of *Thysanocarpus conchuliferus*, Greene (1886) wrote that the species was "common on mossy shelves and crevices of the high rocky summits and northward slopes ... ." Currently, this endemic taxon is rarely seen.

Since the 1920s, the pine forest near Pelican Bay has suffered tremendous loss of cover, involving both canopy and understory species. A map drawn in 1922 (Symmes and Associates 1922) shows two large populations of pines ("Pinos Grandes" and "Pinos Chicos") near Pelican Bay. After a series of fires between 1917 and 1931, sheep consumed fire-stimulated seedlings and curtailed establishment of new trees. The size of the pine forest was dramatically reduced by the 1970s. In the population near Pelican Bay, Hobbs (1980) documented reduced foliar cover of woody species, a severe reduction of tree and shrub seedlings, and reduced species diversity in the understory, compared to other groves on the island.

Feral sheep along the north shore were removed during the 1980s (Schuyler 1993) and the pine population near Pelican Bay showed dramatic recovery by 1990. Wehtje (1994) observed an increase in diversity and abundance of woody vegetation in the area and documented successful seedling establishment of native pines and shrubs. Along the north shore, the abundance and range of many native shrubs have increased since sheep removal. Numerous seedlings of *Dendromecon rigida* subsp. *harfordii* have been observed. *Helianthemum greenei* and *Lotus grandiflorus* have become much more common, not only near Pelican Bay but also in scattered locations throughout the western and central areas. *Hazardia detonsa* and *Lotus argophyllus* var. *niveus*, previously restricted to inaccessible cliffs, are becoming common on flats and slopes throughout much of the island.

Vegetation changes on Monterey Shale slopes east of Prisoners' Harbor have not been as dramatic as on volcanic slopes to the west. Historical descriptions of north-facing slopes east of Prisoners' Harbor seem very much like present-day assessments, suggesting that damage from overgrazing was not as severe. Johnson (1858) noted that slopes east of Prisoners' Harbor were "very much cut up by gulches" but had "little or no brush." Forney (1875) wrote that "from Prisoners' harbor eastward the hills are smooth and more regular [compared with those to the west], covered with an abundance of coarse grass."

**Vegetation Changes in the Central Valley**. Vegetation changes in the Central Valley may have been more extreme than those on the north side, especially in oak woodland and coastal-sage scrub communities. Comparing photographs taken in 1869 and the 1970s, Brumbaugh (1980b) documented the dramatic loss of coastal scrub on the south-facing slopes near the main ranch. Delphine Adelaide Caire, the daughter of Justinian Caire, first visited at the age of 24. She described the following changes observed in the Central Valley during her lifetime (Caire 1933):

> *So far as material comforts are concerned, Santa Cruz Island in the year of grace 1880 may in no way be compared to the same property in the year 1933. But its present natural beauty does not come up to that of the past. The bed of the stream that skirts the Main Ranch on its way from Picacho Diablo was much narrower than it is today; mountain slopes were heavily wooded and centuries-old oaks were numerous. In the course of years, rains have accomplished their ruinous*

> *work, carrying off a great amount of topsoil, the innumerable trails cut by sharp sheep trotters having been a contributing factor in such devastation. If wood were needed to heat kitchen ovens, one of those incompetent overseers would send his workmen to the creek with the order to cut wood, without bothering to add any specific directions as to how the job should be done. As a result, many a patriarchal tree fell to the ax of untrained woodcutters. In addition, during the great sheep roundups of the shearing season, in order to keep the sheep from scattering, the vaqueros were accustomed to massing branches as walls to form a corridor for the sheep. In the absence of responsible control, the building of such fences was often at the cost of trees that proper pruning might have saved.*

The abundance of some native species in the Central Valley has been dramatically reduced since the 1880s. Greene noted that *Bloomeria* [*crocea*] was "common on all parts of Santa Cruz [Island]" and that "on grassy knolls in the middle of the island it thrives in such abundance that the umbels touch each other over ... acres ..." *Bloomeria* occurs in much smaller populations today. *Godetia purpurea* (*Clarkia purpurea* subsp. *quadrivulnera*) was "frequent in the interior valley on grassy slopes" when Greene visited the island; it is rarely seen now. Brandegee (1888) wrote that *Heuchera maxima* was "not uncommon throughout Santa Cruz Island." At present this species is occasional, confined primarily to inaccessible sea-cliffs and canyon walls near the coast and to a few scattered inland sites. In contrast to the pine populations near Pelican Bay, the Christy Pines have suffered significant decline since the mid-1980s. Many adult trees have died, apparently due to several years of drought stress and infestations of bark beetles. However, the pine forest is apparently becoming re-established with seedlings and young trees.

**Changes in Alien Plant Populations**. The range and abundance of numerous alien plant species also have changed since the 1880s. Several alien species reported by Greene (1887) as extremely limited in distribution are now common and widespread (e.g., *Chenopodium ambrosioides*, *Convolvulus arvensis*, and *Sisymbrium officinale*). On the other hand, some alien species described as well established by Greene have either disappeared from the island or have much reduced ranges at present (e.g., *Conium maculatum* and *Ricinus communis*). Greene (1887) reported that *Ricinus* was "growing spontaneously along the hills back from the landing, forming small trees;" it has not been seen in recent years.

**Widespread Changes After Sheep and Cattle Removal**. Since the removal of sheep and cattle from the western portion of Santa Cruz Island, populations of several native and alien plant species have spread rapidly. The dramatic expansion of the alien *Foeniculum vulgare* is one of the most conspicuous examples (Beatty and Licari 1992). A number of endemic plant taxa also have expanded their ranges and are becoming more abundant. *Hazardia detonsa*, *Lotus argophyllus* var. *niveus*, and *Mimulus flemingii*, restricted to steep cliffs during the 1970s and 1980s, are rapidly spreading onto open volcanic slopes throughout much of the island. *Lotus argophyllus* var. *niveus* also is becoming widespread and common in dry washes and on gravel floodplains. Vegetation in many parts of the island has begun to recover, especially on the north shore where native pines, shrubs, and bunch grasses are rapidly spreading. In grasslands, Klinger et al. (1994) found that herbaceous cover increased and bare ground decreased after sheep removal. Although recovery of Santa Cruz Island's ecosystem will probably require decades, erosion has been reduced dramatically in many areas and the condition of previously degraded habitats has noticeably improved. Although the pristine vegetation of Santa Cruz Island remains unknown, it appears that at least some damage caused by human activities and introductions is reversible.

## History of Botanical Exploration

Botanists began exploring the Santa Barbara region and describing plants in the 1760s, but the history of botanical discovery on Santa Cruz Island did not begin until the 1870s. By 1900, fewer than 10 naturalists had collected botanical specimens on the island. Botanical exploration intensified during the 1920s and has continued to the present, with a brief hiatus during World War II. Known collectors of vascular plants on Santa Cruz Island, including details of their collecting trips, are listed in Appendix I.

The first botanical collections were made in 1874, when Albert Kellogg and William G. W. Harford discovered four previously undescribed plant taxa: *Coreopsis gigantea*, *Eriogonum arborescens*, *Hazardia detonsa*, and *Jepsonia malvifolia*. Kellogg and Harford often traveled together and were both associated with the California Academy of Sciences. They reached the island with the help of the U. S. Coast Survey and may have been accompanied by Otto Tittmann, who collected *Opuntia littoralis* in January 1874 (Philbrick 1963). Kellogg was a physician and California's first resident botanist. The first new species described by Kellogg was *Lavatera*

*Joseph T. Rothrock, 1839-1922 (Library, Academy of Natural Sciences of Philadelphia).*

*assurgentiflora*, of which the type had been collected on Anacapa Island prior to 1854 (Lenz 1986). Harford was a naturalist especially interested in conchology and served as Director of the Academy from 1876 until 1886 (Jepson 1933, Ewan 1955). Otto Tittmann conducted surveys of the Santa Barbara Channel for the Coast Survey between 1873 and 1876.

Joseph T. Rothrock, physician and professor of botany at the University of Pennsylvania, spent a week in the spring of 1875. He was accompanied by ornithologist and ethnographer Henry W. Henshaw and mineralogist Oscar Loew, as part of the U. S. Geographical Survey under the command of Lieutenant George M. Wheeler. Rothrock and Loew made general observations on the vegetation and flora and Rothrock collected specimens, including the type of *Chorizanthe wheeleri* (Wheeler 1876). Like Kellogg and Harford, they reached the island with the help of the U. S. Coast Survey (Schumacher 1877).

Charles C. Parry, physician, botanist, and geologist, well known for his explorations in the western United States, also collected in 1876. Parry was in the Santa Barbara area in July 1876, apparently for the Centennial Celebration (Smith 1976). Barclay Hazard, a resident of Santa Barbara, visited in the summer of 1885 and collected a tree that was unusual to him (Smith 1976). He sent a specimen to Edward L. Greene of the University of California for identification, which Greene later described as *Lyonothamnus aspleniifolius*. In recognition of Hazard's observations, Greene named the genus *Hazardia* in his honor.

Edward L. Greene, then professor of botany at the University of California (Ewan and Ewan 1981), was a dedicated botanist and Episcopal clergyman who strongly influenced the development of western botany (Reifschneider 1964). Apparently fascinated with island floras, Greene visited Santa Cruz Island in July and August 1886 and increased the known flora from five taxa to 321, of which more than 20 were previously undescribed. Greene was the first botanist to collect extensively on the island, publishing the first annotated checklist (Greene 1887). He stayed at the main ranch in the Central Valley and traversed much of the island on foot, refusing to accept a saddle horse or guide from the Caire family (Jepson 1927). According to Jepson (1927), he botanized until after dark and the Caires often feared that he had lost his way on the unfamiliar terrain. Although most of Greene's collections are represented in the herbarium at the University of Notre Dame, a valuable set of duplicates was sent to the Caire family on Santa Cruz Island but was subsequently lost (Wolf 1935). Two island endemics, *Dudleya greenei* and *Helianthemum greenei*, were named in his honor.

In the spring of 1888, Townshend S. Brandegee, well known botanical explorer of the western United States and Baja California, collected on Santa Cruz Island and added about 80 taxa to Greene's list (Brandegee 1888). Brandegee also published a checklist for five of the California Channel Islands in 1890. *Mimulus brandegei*, a Santa Cruz Island endemic, was named in his honor.

Lorenzo G. Yates and Henry Chapman Ford of the Santa Barbara Society of Natural History collected numerous fern specimens between 1887 and 1893, including the first record of *Polypodium scouleri*. Their trips resulted in three short publications (Ford 1890; Yates 1890a, 1890b). Yates, a dentist and horticulturist, had broad interests in natural history, especially conchology (Camp 1963). Ford was the first professional artist to live in the Santa Barbara area, settling in Carpinteria in 1875. He was very interested in archaeology, botany, and geology, and was one of the founders of the Society of Natural History in 1876.

In 1894, Francesco Franceschi, a Santa Barbara botanist and horticulturist, spent a week on the island and obtained seeds of several taxa then unknown in the horticultural trade, including *Lyonothamnus floribundus* subsp. *aspleniifolius* (Tucker 1945). Franceschi's trip proved to be even more exciting than he had planned (Fenzi 1976):

Edward L. Greene, 1843-1915 (Hunt Institute for Botanical Documentation, Carnegie Mellon University).

Townshend S. Brandegee, 1843-1925 (Library, Academy of Natural Sciences of Philadelphia).

> *In a small sailing boat he [Franceschi] and his two sons set out for the Channel Islands. They landed on Santa Cruz Island and succeeded in digging up a fine specimen which they loaded laboriously into the boat. Their return journey was far from placid. A high wind made the seas rough, and to make matters worse, they were mistaken for outlaws by suspicious coast guards. Shots were fired at them and the tiny vessel began to leak. The sons were obliged to bail out the water while the Doctor, whose one concern was his precious Ironwood tree, navigated the fast sinking vessel to the safety of the harbor at Santa Barbara. In his nursery the tree flourished and in three years' time (1897), sufficient stock had been produced for* Lyonothamnus floribundus *to be listed in his catalogue of general trade.*

Island explorer and field botanist Luella Blanche Trask visited in October 1900. She lived on Santa Catalina Island from 1893 until 1912 (Cantelow and Cantelow 1957) and collected extensively on the Southern Channel Islands during the late 1890s and early 1900s. We have seen only one of her collections from Santa Cruz Island. Her best specimens, deposited at the California Academy of Sciences in San Francisco, were lost during the 1906 earthquake and fire (Millspaugh and Nuttall 1923, Cantelow and Cantelow 1957). Her personal herbarium was destroyed in a large fire at Avalon on Santa Catalina Island in November 1915 (Millspaugh and Nuttall 1923). *Cercocarpus betuloides* var. *blancheae* was named in her honor.

While a student at Stanford University, entomologist Robert E. Snodgrass collected animals on several islands (Mallis 1971). He collected vascular plants and lichens on Santa Cruz Island in July 1901. Joseph Grinnell, zoology professor at the University of California at Berkeley, collected plants in September 1903. Harvey M. Hall of the University of California at Berkeley visited in May 1908. He was apparently accompanied on his trip by Santa Barbara resident Freda Sexauer Niedermuller, who also collected plants on Santa Cruz Island in the 1920s.

Alice Eastwood, of the California Academy of Sciences, collected plants on Santa Cruz Island in July 1917. She published a paper on the island's manzanitas (Eastwood 1933) and a checklist of vascular plants for the California Channel Islands (Eastwood 1941). Unfortunately, the scientific value of Eastwood's checklist is limited, because no voucher specimens have been located for many of her reports and some undocumented records seemingly resulted from transpositions of columns in her table. She made no other trips to the island

*Alice Eastwood (left) and Mrs. Charles E. Miller on Santa Cruz Island (Archives, California Academy of Sciences, July 1917).*

between 1917 and 1941 and her information for the checklist came largely from other sources. Mr. and Mrs. Charles E. Miller collected in August 1916, in July 1917 with Alice Eastwood, and again in May and June 1918. Mrs. Alonson Swain collected plants in 1917 and in 1919, including the type of *Ceanothus megacarpus* subsp. *insularis*. Her husband was Ranch Superintendent on the island during the late 1910s.

Herman Knoche collected specimens in September 1919. A botanist particularly interested in island vegetation, Knoche collected on islands over much of the world. Among his botanical contributions was a four-volume work, *Flora Balearica*, written in French and published at his own expense (Thomas 1959). Other collectors in the 1910s included H. J. Keech (September 1912) and R. Kuhn (1917 and 1920).

During the early 1920s, Mr. and Mrs. Reginald C. Robbins made at least two collecting trips to the island. In 1921 and 1928, Olive Day Thacher, of the Thacher School in Ojai, obtained specimens and sent them to W. L. Jepson. Jepson named *Ribes menziesii* var. *thacherianum* in her honor. Adele Lewis Grant, instructor in botany at Cornell University, visited in July 1923, in conjunction with her study of the genus *Mimulus* (Grant 1924). Marcus E. Jones, botanist, explorer, professional collector, and mining consultant, visited in April 1924, September 1927, and March 1929.

Ralph Hoffmann, Director of the Santa Barbara Museum of Natural History from 1923-1932, began collecting on the island in March 1925. He made at least 32 collecting trips before his untimely death on San Miguel Island on 21 July 1932. His unpublished flora of the Northern Channel Islands and some of his published papers (Hoffmann 1931, 1932a,b) contributed significantly to our current knowledge. He was a meticulous collector, unafraid of steep cliffs, and was able to collect at the site of a 1931 fire near Pelican Bay. Among his many discoveries were *Arabis hoffmannii* and *Sanicula hoffmannii*. During the late 1920s, Willis L. Jepson and Herbert L. Mason, of the University of California at Berkeley, visited the island. Other collectors in this period included May N. Ackley, Anson Blake, Lincoln Ellison, F. Raymond Fosberg, George R. Johnstone, Mabel B. Peirson, and M. Stason.

Botanical activities intensified in the early 1930s. Ralph Hoffmann continued his explorations, often in the company of other botanists such as John T. Howell of the California Academy of Sciences, Carl B. Wolf of Rancho Santa Ana Botanic Garden, and a well known horticulturist, Gertrude Lester Rowntree. Dorothy Irma Daniel Cooke, of the Santa Barbara Museum of Natural History, also accompanied Ralph Hoffmann on some trips and made collections. Philip A. Munz, of Rancho Santa Ana Botanic Garden, and Elizabeth Crow Norland collected during this period, as did Le Roy Abrams and Ira L. Wiggins, both of Stanford University. Howard E. McMinn, of Mills College, collected woody plants. Ira W. Clokey, a mining engineer and later a botanist, made two collecting trips. Subsequently, he published a paper that added 41 plant taxa to previously published checklists for the island (Clokey 1931). *Solanum clokeyi* was later named in his honor. Frank W. Peirson, an amateur botanist and avid collector, and his sister Mabel made numerous collections. Other collectors in this period included Fred Caire, one of the land owners, Margaret Dearing of the Santa Barbara Botanic Garden, G. S. Rand, and zoologist H. H. Sheldon.

In the late 1930s, collectors included Rimo Bacigalupi of the University of California at Berkeley, Norman E. Bilderback, Francis Elmore, Ruth Hartwell of the Santa Barbara Museum of Natural History, Dr. Harvey, O. A. Plunkett, and R. S. Woglum. Maunsell Van Rensselaer, then Director of the Santa Barbara Botanic Garden, made several trips in 1936. Marine biologist and photographer M. Woodbridge Williams collected widely in 1939 and later published a photo essay and a popular article (Williams 1954a, 1954b). Beginning in 1939, Meryl B. Dunkle collected plants on the combined expeditions of the Allan Hancock Foundation and the Los Angeles County Museum of Natural History, which included the island on its itinerary (Dunkle 1950).

*Ralph Hoffmann, 1870-1932 (Santa Barbara Museum of Natural History).*

In 1941, Reid Moran was the botanist on a Los Angeles County Museum trip to the island, and Francis Elmore added about 75 specimens to collections of the Allan Hancock Foundation (Gentry 1949). H. Baer was apparently the last collector on the island before World War II. The onset of World War II dramatically curtailed plant collecting trips. Under war-time regulations, the Channel Islands were closed to yachts and travel across the Santa Barbara Channel was rigidly controlled by defense authorities (Wheeler 1944, Gherini 1966).

After the war, Cornelius Muller, of the University of California at Santa Barbara, made his first trip and Reid Moran returned in 1950. Clifton Smith, of the Santa Barbara Museum of Natural History, made the first of several trips in 1958, accompanied by archaeologist Phil C. Orr. Smith later incorporated data from his collections into a regional flora (Smith 1976). E. R. Blakley began collecting for the Santa Barbara Botanic Garden in 1958, accompanied by E. K. Balls of Rancho Santa Ana Botanic Garden.

The 1960s marked the establishment of a University of California field station, which promoted more field work. The decade also marked the beginning of more frequent botanical trips by staff of the Santa Barbara Botanic Garden. E. R. Blakley made several trips between 1960 and 1962, accompanied by Martin Piehl in 1962. Ralph Philbrick and Michael Benedict began a series of collecting trips in 1964, accompanied on various trips by Richard Broder, Oscar Clarke, Allen Flinck, James McPherson, Donald Ricker, Clifton Smith, Dale Smith, and others. Robert Thorne, of Rancho Santa Ana Botanic Garden, made the first of several collecting trips in 1967, accompanied by Percy Everett. Other collectors during this decade included Dennis Breedlove of the California Academy of Sciences, Roman Gankin, Ed Gedling, Grady Webster and others of the University of California at Davis, Clare Hardham, and Robert L. Hoover of the California Polytechnic University at San Luis Obispo.

Collectors during the 1970s included Frank Almeda and Bruce Bartholomew of the California Academy of Sciences, G. Douglas Barbe of the California Department of Food and Agriculture, Larry Heckard of the Jepson Herbarium, Baki Kasapligil of Mills College, Steven Timbrook of the Santa Barbara Botanic Garden, entomologist David Weissman of the University of California at Irvine, and David Young, then a graduate student at Rancho Santa Ana Botanic Garden. Jonathan D. Sauer and several graduate students of the University of California at Los Angeles collected during this decade. Marla Daily began collecting specimens for a herbarium at the University of California Field Station in 1976, an activity she continued into the 1980s. Ralph Philbrick and Michael Benedict continued collecting for the Santa Barbara Botanic Garden. Steve Junak and Mary Hochberg Carroll began collecting in the late 1970s, primarily in connection with a study conducted for The Nature Conservancy (Hochberg et al. 1980).

In the late 1970s, Wayne Ferren, of the University of California at Santa Barbara, organized a series of annual trips by the UCSB Botanical Society. Tim Ledwith and Ray Collett, of the University of California at

Santa Cruz, collected horticultural materials and some herbarium specimens. Other collectors during this period included Mona Bourell of the California Academy of Sciences, Thomas Elias and Walter Wisura, both of Rancho Santa Ana Botanic Garden, Lyndal Laughrin of the University of California at Santa Barbara, Peter Schuyler of The Nature Conservancy, Sally Unser of the National Park Service, and Julie Vanderwier of the Pacific Missile Test Center.

In March 1989, field work for the Santa Cruz Island Flora project was begun by Tina Ayers, Steven Junak, Randy Scott, and David Young. Other collectors during the early 1990s included Barbara Ertter of the University of California at Berkeley and Steve McCabe of the University of California at Santa Cruz.

In addition to herbarium specimens, botanists and horticulturists also have collected seeds and cuttings from plants on Santa Cruz Island since the early 1900s. Living collections can be seen and studied at several California institutions, including Rancho Santa Ana Botanic Garden, Santa Barbara Botanic Garden, the Regional Parks Botanic Garden in Berkeley, and the University of California (Berkeley) Botanical Garden. Horticultural selections of *Adenostoma*, *Arctostaphylos*, *Ceanothus*, *Epilobium*, and *Garrya* have been introduced into the nursery trade.

## Floristic Analysis

The vascular flora of Santa Cruz Island includes 88 families of plants, 348 genera, 627 species, and 2 common hybrids (*Pelargonium* ×*hortorum*, *Quercus* ×*macdonaldii*). At least 18 species are represented by more than 1 forma, variety, or subspecies; there are 21 such taxa. Consequently 650 taxa at or below the species level occur on Santa Cruz Island. The island has the richest flora of any of the California Channel Islands (Table 6). The 3 largest families are Asteraceae, Poaceae, and Fabaceae; the 3 largest genera are *Lotus*, *Bromus*, and *Atriplex* (Table 7).

**Alien Species**. Eleven of the families and 82 of the genera on Santa Cruz Island are represented exclusively by alien taxa. At least 170 taxa (including 157 species) are considered to be introduced and represent 26% of the island's total flora. Percentages of alien taxa on the other California Channel Islands range from 20% to 47% (Table 6). Most of the alien taxa are native to the Mediterranean region (Europe, n Africa, and w Asia) or Australia. Six of the introduced taxa are native to the California mainland (*Anemopsis californica*, *Cupressus macrocarpa*, *Helianthus annuus*, *Hemizonia fitchii*, *Platanus racemosa*, and *Xanthium strumarium*) but were either intentionally or unintentionally introduced during the last 150 years. Three taxa (*Eriogonum giganteum* subsp. *giganteum*, *Lavatera assurgentiflora* subsp. *assurgentiflora*, and *L. assurgentiflora* subsp. *glabra*) are native to other California Channel islands but were planted on Santa Cruz Island and have subsequently escaped from cultivation. At least 9 alien plant taxa, which have become naturalized, have not yet been reported on the

Table 6. Numbers of Vascular Plant Taxa on the California Channel Islands and Guadalupe Island. Totals shown for native taxa include those endemic to the California Islands. Endemic taxa are those found only on 1 or more of the California Islands. Number in () represents percentage of total taxa in island flora.

| | Taxa | | | |
|---|---|---|---|---|
| | Native | Endemic | Alien | Total |
| | | Northern Channel Islands | | |
| San Miguel | 198 | 18 (7) | 69 (26) | 267 |
| Santa Rosa | 387 | 42 (9) | 98 (20) | 485 |
| Santa Cruz | 480 | 45 (7) | 170 (26) | 650 |
| Anacapa | 190 | 22 (8) | 75 (28) | 265 |
| | | Southern Channel Islands | | |
| Santa Barbara | 88 | 14 (11) | 44 (33) | 132 |
| San Nicolas | 139 | 18 (7) | 131 (48) | 270 |
| Santa Catalina | 421 | 37 (6) | 185 (30) | 604 |
| San Clemente | 272 | 47 (13) | 110 (28) | 382 |
| | | Baja California Islands | | |
| Guadalupe Island | 173 | 53 (24) | 46 (21) | 219 |

other California Channel Islands: *Aegilops cylindrica*, *Albizia lophantha*, *Amaranthus deflexus*, *Bromus stamineus*, *Cichorium intybus*, *Koeleria phleoides*, *Malva nicaeensis*, *Verbascum thapsus*, and *Veronica anagallis-aquatica*.

**Hybridization**. The flora of Santa Cruz Island includes a number of natural hybrids and at least one horticultural introduction of hybrid origin (*Pelargonium* ×*hortorum*). Natural hybridization between species has been observed on the island in the following genera: *Dudleya*, *Eriogonum*, *Grindelia*, *Hazardia*, *Helianthemum*, *Mimulus*, *Opuntia*, *Pentagramma*, *Polypodium*, *Quercus*, *Raphanus*, and *Rhus*.

**Endemism**. The California Channel Islands are known for high levels of endemism (Raven 1967). Some of the characteristics of plant taxa endemic to the California Channel Islands have been discussed by Philbrick (1980) and Thorne (1969). At least 37 of the island's native taxa are island endemics, not found on the mainland but occurring on at least 2 of the California Channel Islands:

*Arabis hoffmannii*
*Arctostaphylos tomentosa* subsp. *insulicola*
*Astragalus miguelensis*
*Berberis pinnata* subsp. *insularis*
*Castilleja lanata* subsp. *hololeuca*
*Ceanothus arboreus*
*Ceanothus megacarpus* subsp. *insularis*
*Chorizanthe wheeleri*
*Calystegia macrostegia* var. *macrostegia*
*Dendromecon rigida* subsp. *harfordii*
*Dudleya candelabrum*
*Dudleya greenei*
*Eriogonum arborescens*
*Eriogonum grande* var. *grande*
*Eriogonum grande* var. *rubescens*
*Erysimum insulare*
*Eschscholzia ramosa*
*Galium angustifolium* subsp. *foliosum*
*Galium buxifolium*
*Galium nuttallii* subsp. *insulare*
*Gilia nevinii*
*Hazardia detonsa*
*Helianthemum greenei*
*Hemizonia clementina*
*Heuchera maxima*
*Jepsonia malvifolia*
*Lotus dendroideus* var. *dendroideus*
*Lyonothamnus floribundus* subsp. *aspleniifolius*
*Malacothrix indecora*
*Malacothrix saxatilis* var. *implicata*
*Malacothrix squalida*
*Mimulus flemingii*
*Quercus pacifica*
*Quercus tomentella*
*Rhamnus pirifolia*
*Sibara filifolia*
*Solanum clokeyi*

Three additional taxa (*Adenostoma fasciculatum* var. *prostratum*, *Marah macrocarpus* var. *major*, and an unrecognized subspecies of *Stachys bullata*) are probably endemic to the California Islands but need further study. Eight taxa are endemic to Santa Cruz Island:

*Arctostaphylos insularis*
*Arctostaphylos viridissima*
*Dudleya nesiotica*
*Lotus argophyllus* var. *niveus*
*Malacothamnus fasciculatus* var. *nesioticus*
*Mimulus brandegei*
*Ribes thacherianum*
*Thysanocarpus conchuliferus*

**Relationships among Islands**. Floristic relationships among the California Channel Islands have been examined and discussed by Raven (1967), Thorne (1969), and Wallace (1985). Santa Cruz Island shares from 17 to 316 of its native taxa (4% to 66%, respectively) with other California and Baja California islands (Table 8). Floristic affinities are related to distance and size of island, with the largest islands (Santa Rosa, Santa Catalina, and San Clemente) having the highest numbers of shared native taxa with Santa Cruz. Among the Baja California islands, Guadalupe shares the greatest number of native taxa with Santa Cruz. Similar patterns exist with respect to island endemics and alien taxa.

**Relationships with the Mainland**. The flora of Santa Cruz Island appears harmonic and balanced when compared to regional floras of comparable size on the adjacent mainland, with a few conspicuous exceptions. Some taxa are widespread on the mainland but have not been found on any of the California Channel Islands, including *Alnus rhombifolia*, *Cordylanthus* spp., *Monardella* spp., *Penstemon* spp., *Platanus racemosa*, *Ribes speciosum*, *Salvia spathacea*, *Umbellularia californica*, and *Yucca whipplei*. The following native taxa are widespread on the mainland but have an insular distribution that only includes Santa Cruz Island:

*Acer macrophyllum*
*Adiantum aleuticum*
*Andropogon glomeratus* var. *scabriglumis*
*Antirrhinum multiflorum*
*Arabis glabra*
*Arbutus menziesii*

*Aristida ternipes* var. *hamulosa*
*Athyrium filix-femina* var. *cyclosorum*
*Blennosperma nanum* var. *nanum*
*Calamagrostis rubescens*
*Calyptridium monandrum*
*Camissonia hirtella*
*Carex barbarae*
*Carex harfordii*
*Carex senta*
*Cheilanthes cooperae*
*Cirsium brevistylum*
*Clematis lasiantha*
*Cucurbita foetidissima*
*Cuscuta subinclusa*
*Cystopteris fragilis*
*Epilobium brachycarpum*
*Epipactis gigantea*
*Equisetum hyemale* subsp. *affine*
*Erodium macrophyllum*
*Galium californicum* subsp. *flaccidum*
*Githopsis diffusa* subsp. *diffusa*
*Holodiscus discolor*
*Lepidium virginicum* var. *robinsonii*
*Lepidospartum squamatum*
*Lonicera interrupta*
*Lotus micranthus*
*Ludwigia peploides* subsp. *peploides*
*Lythrum californicum*
*Marah fabaceus*
*Meconella denticulata*
*Oenothera elata* subsp. *hirsutissima*
*Orobanche uniflora*
*Pectocarya penicellata*
*Phacelia cicutaria* var. *hubbyi*
*Pickeringia montana*
*Piperia michaelii*
*Poa howellii*
*Polygala californica*
*Polygonum lapathifolium*
*Quercus dumosa*
*Quercus kelloggii*
*Quercus parvula* var. *parvula*
*Rhamnus californica* subsp. *californica*
*Salix lucida* subsp. *lasiandra*
*Salvia leucophylla*
*Samolus parviflorus*
*Scirpus maritimus*
*Scutellaria tuberosa*
*Stillingia linearifolia*
*Triteleia hyacinthina*
*Zannichellia palustris*

At least 27 plant taxa apparently reach the northern limits of their ranges on Santa Cruz Island or on the adjacent mainland coast:

*Achnatherum diegoense*
*Antirrhinum nuttallianum* subsp. *subsessile*
*Astragalus trichopodus* var. *lonchus*
*Atriplex pacifica*
*Atriplex watsonii*
*Calandrinia maritima*
*Camissonia robusta*
*Cheilanthes clevelandii*
*Clematis pauciflora*
*Comarostaphylis diversifolia* subsp. *planifolia*
*Dichondra occidentalis*
*Euphorbia misera*
*Hemizonia clementina*
*Lepechinia fragrans*
*Lepidium oblongum* var. *insulare*
*Lycium californicum*
*Malacothrix similis*
*Nicotiana clevelandii*
*Oligomeris linifolia*
*Opuntia littoralis*
*Opuntia oricola*
*Opuntia prolifera*
*Perityle emoryi*
*Pholistoma racemosum*
*Plagiobothrys collinus* var. *gracilis*
*Stephanomeria diegensis*
*Stillingia linearifolia*

At least 14 other plant taxa reach their southern limits on Santa Cruz or neighboring islands:

*Allium lacunosum* var. *lacunosum*
*Aster radulinus*
*Calamagrostis rubescens*
*Centaurium davyi*
*Cirsium brevistylum*
*Erigeron glaucus*
*Erigeron sanctarum*
*Eriophyllum staechadifolium*
*Hemizonia increscens* subsp. *increscens*
*Montia fontana*
*Polypodium scouleri*
*Polygala californica*
*Stephanomeria exigua* subsp. *coronaria*
*Triteleia hyacinthina*

Occurrences of 2 taxa (*Malacothrix coulteri* and *Monolepis nuttalliana*) on Santa Cruz Island appear to represent disjunct populations of plant taxa occurring mostly in the interior of California.

Table 7. Ten Largest Families and Genera of Vascular Plants on Santa Cruz Island. Families and genera are ranked by total number of taxa at or below the level of species.

| | Native (Endemic) | Alien | Total |
|---|---|---|---|
| Family | | | |
| Asteraceae | 82 (5) | 29 | 111 |
| Poaceae | 28 (0) | 42 | 70 |
| Fabaceae | 35 (3) | 11 | 46 |
| Brassicaceae | 18 (4) | 11 | 29 |
| Chenopodiaceae | 16 (0) | 5 | 21 |
| Onagraceae | 17 (0) | 0 | 17 |
| Scrophulariaceae | 15 (3) | 2 | 17 |
| Caryophyllaceae | 11 (0) | 5 | 16 |
| Polygonaceae | 8 (4) | 7 | 15 |
| Boraginaceae | 12 (0) | 0 | 12 |
| Genus | | | |
| *Lotus* | 10 (2) | 1 | 11 |
| *Bromus* | 4 (0) | 7 | 11 |
| *Atriplex* | 9 (0) | 1 | 10 |
| *Quercus* | 9 (2) | 0 | 9 |
| *Trifolium* | 9 (0) | 0 | 9 |
| *Gnaphalium* | 8 (0) | 1 | 9 |
| *Camissonia* | 7 (0) | 0 | 7 |
| *Carex* | 7 (0) | 0 | 7 |
| *Lupinus* | 7 (0) | 0 | 7 |
| *Malacothrix* | 6 (3) | 0 | 6 |

TABLE 8. Floristic Relationships between Santa Cruz and Other Islands of Southern and Baja California. Totals for native taxa include those endemic to the California Islands. Endemic taxa are those found only on 1 or more of the California Islands.

| | Number of Plant Taxa Shared with Santa Cruz Island | | |
|---|---|---|---|
| | Native | Endemic | Alien |
| California Channel Islands | | | |
| San Miguel | 170 | 13 | 65 |
| Santa Rosa | 317 | 31 | 89 |
| Anacapa | 181 | 19 | 64 |
| Santa Barbara | 73 | 3 | 39 |
| San Nicolas | 108 | 5 | 88 |
| Santa Catalina | 306 | 16 | 122 |
| San Clemente | 195 | 12 | 84 |
| Baja California Islands | | | |
| Los Coronados | 75 | 1 | 23 |
| Todos Santos | 75 | 1 | 30 |
| San Martin | 50 | 2 | 16 |
| Guadalupe | 86 | 7 | 37 |
| San Benito | 16 | 1 | 5 |
| Cedros | 78 | 1 | 36 |
| Natividad | 18 | 1 | 5 |

## GUIDE FOR THE READER

**Included Taxa.** All known native and introduced vascular plant taxa growing without cultivation are described. Introduced plants include those originally planted (e.g., *Eucalyptus* spp., *Pinus pinea*) and now naturalized. Taxa planted near buildings or along main roads were not treated unless they appeared to be naturalized. Representative herbarium specimens for taxonomic treatments are cited in Appendix II. Doubtful or unsubstantiated reports are listed in Appendix III.

**Identification Keys.** Dichotomous identification keys are provided for all taxa (families, genera, species, subspecies, and varieties) recognized to occur on Santa Cruz Island. For convenience, a few large keys (families of flowering plants and Poaceae) are separated into two or more "Groups", each of which shares a pattern of morphological variation. Dichotomous keys are composed of a series of consecutively numbered, paired, comparative statements or couplets. Couplets within each key are assigned a unique number. The first phrase of a couplet begins with a number followed by a period; the alternate, second phrase begins with the same number followed by an apostrophe. Phrases comprising couplets were designed to use the most discriminating morphological differences between taxa being compared.

**Names of Taxa.** Scientific names are those considered to be the most correct, based on published studies of taxonomic relationships and current conventions established by the International Code of Botanical Nomenclature (Greuter 1988). Most of the names used in this text are consistent with Hickman (1993). Use of other names for some taxa at the specific, subspecific, or varietal levels are based on more recent literature. Citation and abbreviation of authors follow Brummitt and Powell (1992). Common names are given for genera, species, subspecies, and varieties only when in wide use or otherwise appropriate.

**Descriptions**. Descriptions provide a brief, diagnostic summary of morphological variation within each taxon. Descriptions for families and genera are inclusive for their worldwide distribution. Descriptions of species, subspecies, and varieties are based primarily on variation on Santa Cruz Island, as represented by herbarium specimens deposited at the Santa Barbara Botanic Garden and other institutional herbaria. More complete or supplementary descriptions may be found in Abrams (1940-1944), Abrams and Ferris (1960), Hickman (1993), and Munz and Keck (1968).

Descriptions of all taxa were constructed using the following general sequence of morphological features: habit (including sexuality if not bisexual), roots, stems, leaves, inflorescence, flowers, and fruits. Within these general categories, either semicolons or commas were used for clarity of presentation to separate adjectival statements (e.g., shape, metric length, color, pubescence type) that modify the preceding nouns. Definitions of botanical terms used in the text are listed alphabetically in the Glossary.

Morphological descriptions of families and genera are followed by approximate numbers of included genera and their worldwide distribution. Descriptions of species, subspecies, and varieties are followed by additional information, if pertinent, organized as follows:

**Time of Flowering**. The time during which plants are producing spores, cones, or flowers is given as a range from earliest to latest month. Months are abbreviated to the first 3 letters. Exceptional flowering times are enclosed in parentheses.

**Native versus Introduced**. All species are presumed native to the island unless noted as "Introduced." Such alien species are considered introduced to and naturalized on Santa Cruz Island by human intervention beginning with Spanish colonization in the late 1700s. An estimate of their earliest known occurrence, based on herbarium specimens and publications, is provided for most alien taxa by a note that begins "First collected ..." The geographic origin of alien taxa is given as part of the distribution statement (see below). If native or alien status is in doubt (e.g., *Bromus berterianus*), appropriate comments are appended.

**Endemic Taxa**. Species believed to occur only on the island are noted as "Endemic to Santa Cruz Island." Species known to occur on other islands, but not on the mainland, are noted as "Island endemic."

**Taxa of Special Environmental Concern**. Certain taxa generally recognized as having a limited distribution and declining in numbers are deserving of special attention and protection by the State of California and the United States (Skinner and Pavlik 1994). These are noted as "Taxon (i.e., species, subspecies, or variety) of special environmental concern." Sixty-one such taxa are known to occur on the island.

**Frequency of Distribution**. Frequency of distribution was classified into five categories, each of which provides an estimate of relative abundance. These categories are based on extensive field observations by the senior author and by the known number of localities as determined from herbarium specimens. Localities in this system are considered to be populations separated by distances of at least 0.25 mile (ca. 0.4 kilometer).

Taxa that are exceptionally widespread or conspicuous in many habitat types and known from more than 30 localities are treated as "**Abundant**." Taxa that are widespread or conspicuous within one to several habitat types, regularly encountered throughout the island, and known from 21-30 localities are treated as "**Common**." Taxa that are fairly widespread within one or few particular habitat types, have a scattered distribution, and known from 11-20 localities are treated as "**Occasional**." Taxa somewhat widespread, often locally common, but only known from 4-10 localities are treated as "**Scarce**." Taxa known from fewer than 3, often very restricted localities are treated as "**Rare**." However, in some cases (e.g., *Arbutus menziesii*) such taxa may be common elsewhere and not considered rare for the purposes of legal protection.

**Habitat Types**. Brief descriptive phrases (e.g., disturbed areas, rocky slopes, ridgetops) are given as an estimate of the general physiographic conditions in which each taxon is most likely to occur.

**Plant Communities**. Brief descriptive phrases (e.g., chaparral, oak woodland, pine forest) of major plant communities are provided. The list for each taxon indicates its most common occurrence(s). A more detailed classification of plant communities on Santa Cruz Island is provided in the introductory section entitled "Vegetation."

**Elevation**. The range in elevation above sea level is given in both feet (ft) and meters (m).

**Geographic Distribution**. Geographic distribution, with exception of Santa Cruz Island endemics, is composed of 2 separate statements: the distribution on the island and the distribution elsewhere.

The distribution on the island was divided into 3 major areas corresponding to major physiographic features. General or specific localities, if pertinent, are listed for each of the 3 major areas: the north side of the island, the Central Valley, and the south side of the island. When 2 or 3 major areas are represented, statements for each are separated by semicolons. The "north side" of Santa Cruz Island extends from West Point to San Pedro Point on the east and includes all northern exposures and drainages opening on the north side of the island, including Cañada del Puerto. The "Central Valley" extends from Christy Beach on the west to Valley Anchorage on the east, including the watersheds of Cañada Christy, Cañada de la Portezuela, Cañada del Medio, and Cañada Pomona. The "south side" extends from Fraser Point to San Pedro Point and includes all southern exposures and drainages opening on the south side of the island.

The geographic distribution of taxa occurring outside of Santa Cruz Island is given (if pertinent) for other California Channel Islands, Baja California Islands, the general mainland distribution, and, in the case of "Introduced" taxa, the general geographic region of origin. All distributions are given as a range from north to south, supplemented by eastward distribution beyond California, if appropriate. The sequence for California islands includes San Miguel, Santa Rosa, Anacapa, San Nicolas, Santa Barbara, Santa Catalina, and San Clemente. The sequence for Baja California includes Los Coronados, Todos Santos, San Martin, San Geronimo, Guadalupe, San Benito, Cedros, and Natividad islands. United States Postal Service abbreviations are used for states (e.g., AZ for Arizona, CA for California, WA for Washington). The geographic origin of "Introduced" taxa are given as "native of ..."

**Other Comments**. Other comments include notes pertaining to economic use (especially by Native Americans), toxicity, taxonomic relationships, and first records of occurrence. Native American tribes are not identified in most cases and exact medicinal uses are not given. Before attempting to use plants for food or medicine the reader should consult appropriate references. Primary sources of information on edible and medicinal plants included Bean and Saubel (1972), Clarke (1977), Strike (1994), Tanaka (1976), Timbrook (1984), and Uphof (1968). Information on toxic plants was obtained mostly from Fuller and McClintock (1986).

**Illustrations**. With few exceptions every species, subspecies, and variety known to occur on Santa Cruz Island is illustrated. At least one habit is illustrated for each genus; at least one diagnostic character is illustrated for each species, subspecies, or variety. Taxa not illustrated include those either reported for Santa Cruz Island (but not documented with herbarium specimens) or those observed and documented after final illustration plates were prepared. Illustrations were designed to complement identification keys and morphological descriptions by providing views of diagnostic features. Illustrations are accompanied by vertical or horizontal lines that are drawn to scale and measured in cm or mm. When appropriate, labels were used to identify individual structures. The pertinent scientific name appears immediately below the habit and/or structure(s).

# KEY TO MAJOR TAXONOMIC GROUPS

1. Plants reproducing by means of spores; spores often brown, produced in sporangia; sporangia axillary, on surfaces of leaf blades, or aggregated in a terminal, ± herbaceous cone-like structure; large woody cones, flowers, and seeds not produced ........ **Pteridophytes**

1' Plants reproducing by seeds and pollen; seeds either enclosed in ovaries of flowers or naked and borne on woody, scale-like sporophylls comprising woody cones; pollen often white to yellow, produced either in anthers of flowers or on scale-like sporophylls comprising small cones

  2. Ovules and seeds naked, borne on surfaces of woody, scale-like sporophylls comprising cones; ovary, style, and stigma absent; pollen produced on scale-like sporophylls comprising small cones ........ **Gymnosperms**

  2' Ovules enclosed in an ovary; seeds enclosed in a dry or fleshy fruit; ovary with style(s) and stigma(s) ........ **Angiosperms**

## PTERIDOPHYTES

### Ferns and Fern Allies

Perennial plants reproducing by means of spores; spores produced in sporangia; sporangia axillary, on surfaces of leaf blades, often aggregated into sori, or aggregated in terminal, ± herbaceous cone-like structures.

### Key to Families

1. Stems with hollow internodes, usually green; leaves fused into a toothed sheath at each internode, tan to dark brown; sporangia borne on peltate, scale-like sporophylls comprising a terminal cone ........ **Equisetaceae** (Division Sphenophyta)

1' Stems solid, aerial part often short, brown; leaves green, either scale-like or with well developed blades and petioles, green; sporangia solitary and axillary or borne in sori on lower surfaces of blades

  2. Leaves scale-like, sessile, 1-3.5 mm long; sporangia solitary, axillary, borne at tips of branches ........ **Selaginellaceae** (Division Lycophyta)

  2' Leaves with well developed blades and petioles, most longer than 10 cm, blades usually deeply divided into pinnae; sporangia many, borne in sori on lower surfaces of leaf blades (Division Pterophyta)

    3. Sori borne along blade margin; indusia absent or sori partly covered by reflexed or inrolled leaf blade (false indusium)

      4. Rhizomes long-creeping, with felt-like hairs; petioles deeply grooved ...... **Dennstaedtiaceae**

      4' Rhizomes short, often stout, with linear to ovate scales; petioles terete, not grooved ........ **Pteridaceae**

    3' Sori borne along midvein or away from blade margin; indusia either present or absent

      5. Indusium absent or obscured by mature sorus

        6. Leaf blades pinnatifid; indusia completely absent ........ **Polypodiaceae**

        6' Leaf blades pinnately (1-4 times) compound; indusia present, often deciduous or obscured by mature sori ........ **Dryopteridaceae**

      5' Indusium present

        7. Blade veins fusing to form a network; indusia linear to oblong, opening along side facing midvein ........ **Blechnaceae**

        7' Blade veins mostly free, not forming a network; indusia hemispherical, reniform, or peltate, opening on all sides ........ **Dryopteridaceae**

BLECHNACEAE

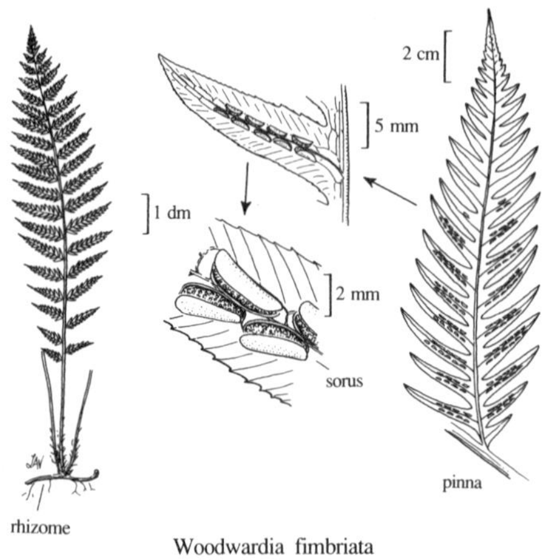


Woodwardia fimbriata

DENNSTAEDTIACEAE

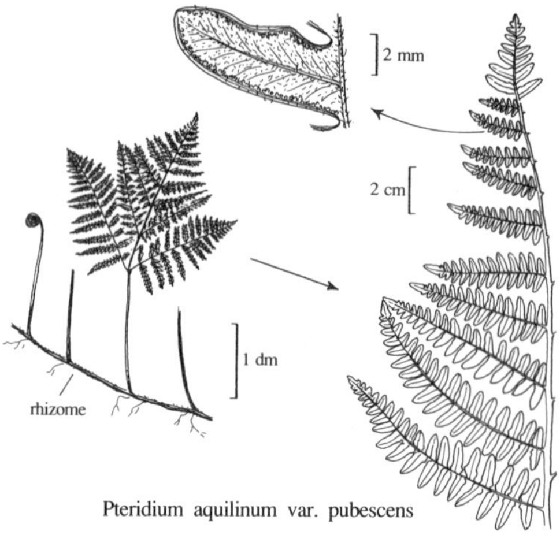


Pteridium aquilinum var. pubescens

# **BLECHNACEAE** C.Presl

## CHAIN FERN FAMILY

Rhizomes erect to often creeping, scaly. Leaf blades deeply pinnatifid to 1-2 times pinnately compound; veins of ultimate segments fused into a network; petioles ± scaly. Sori linear to oblong, parallel to and borne along midvein; indusia present, opening along side facing midvein of blade segment. Ca. 10 genera and 250 species; tropical and temperate, mostly s hemisphere.

### **WOODWARDIA** Smith CHAIN FERN

Rhizome scales brown. Leaves solitary to clustered, blades pinnatifid to once pinnately compound, thick; petioles scaly. Sori ± oblong, in rows, touching or slightly overlapping end-to-end. Ca. 14 species; N. and S. America, Europe.

**W. fimbriata** Smith GIANT CHAIN FERN Rhizomes ascending. Petiole bases with large, reddish brown scales; blades elliptic-lanceolate, 50-150 cm long, 30-50 cm wide, pinnae oblong-lanceolate. Apr-Sep. Occasional; seeps, springs, and stream sides in canyons; grasslands, riparian woodland, oak woodland; 10-1250 ft (3-380 m). North side between Profile Point and Prisoners' Harbor, and in Cañada del Puerto; Central Valley in Cañada de la Siesta and Hooker Canyon; s side in Coches Prietos Anchorage drainage, Aguaje Canyon near Yellowbanks Anchorage. Santa Rosa Island; British Columbia to n Baja CA, e to NV and AZ. Petioles of this species, dyed with red pigments, were used in basketry by Native Americans; rhizomes also were used medicinally.

# **DENNSTAEDTIACEAE** Ching

## BRACKEN FERN FAMILY

Rhizomes short to long, creeping, with felt-like hairs. Leaf blades pinnatifid to 1-4 times pinnately compound. Sori borne near or at blade margin; indusia ± round to linear. Ca. 20 genera and 400 species; mostly tropical, worldwide.

### **PTERIDIUM** Gleditsch ex Scopoli

Rhizomes long-creeping. Leaf blades pinnately (2-4 times) compound; petioles deeply grooved. Sori ± continuous along leaf margin, covered by reflexed margin; indusia obscure, often hidden by sori and reflexed margin. Ca. 6 species; ± worldwide.

**P. aquilinum** L. var. **pubescens** Underw. WESTERN BRACKEN Leaves erect, 3-15 dm tall; ultimate blade segments linear-oblong, lower surfaces densely pubescent, especially along veins, margins entire or sinuate; petioles glabrous. Feb-Sep. Occasional; moist, grassy, n-facing slopes and shaded canyons; grasslands, oak

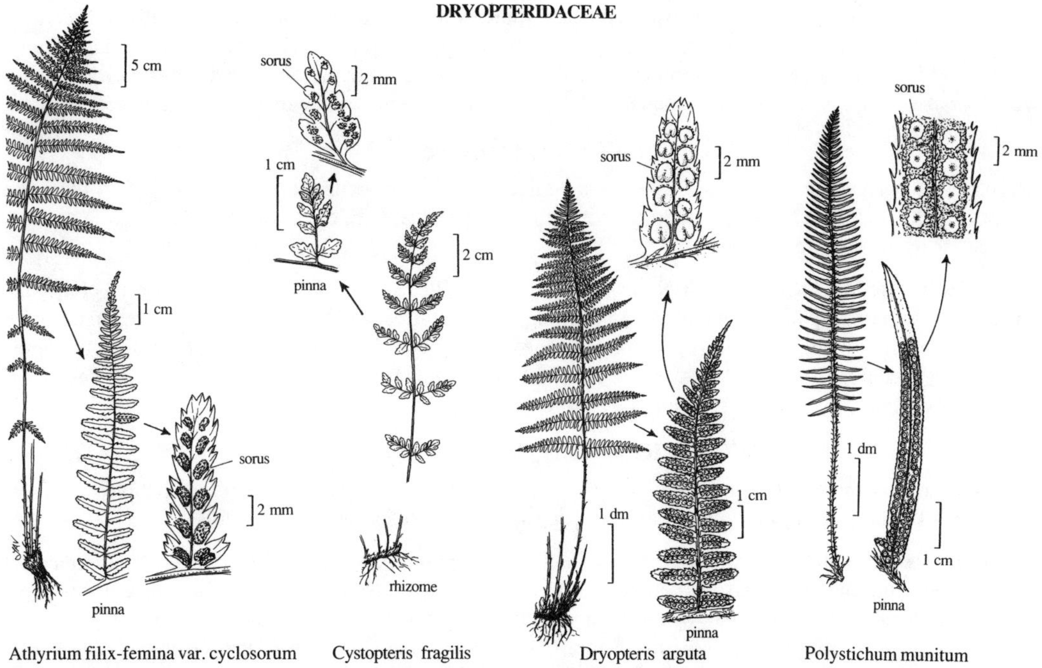


woodland; 200-1700 ft (60-520 m). North side between Cueva Valdez and Prisoners' Harbor (especially at Lagunitas Secas), and on slopes above China Harbor. Santa Rosa and Santa Catalina islands; w N. America, e to MT. Petioles were used in basketry by Native Americans. Although young leaves ("fiddle heads") are reportedly edible, they should be avoided, as they contain shikimic acid, a known carcinogen, which cannot be removed by boiling or leaching.

## **DRYOPTERIDACEAE** Herter
### WOOD FERN FAMILY

Rhizomes erect or short-creeping, scaly. Leaf blades often 2-4 times pinnately compound, often scaly or pubescent; veins of ultimate segments mostly free. Sori usually round, borne on midvein or at tips of veins; indusia reniform, peltate, or hood-like. Ca. 60 genera and 3000 species; worldwide.

1. Leaf blades mostly once compound; indusia peltate .......... **Polystichum**
1' Leaf blades 2-3 times compound; indusia attached at one edge or to one side
  2. Rhizomes slender, short-creeping; reproductive leaves less than 30 cm long; indusia hood-like, arching over sorus .......... **Cystopteris**
  2' Rhizomes thick, erect or creeping; reproductive leaves much longer than 50 cm; indusia variously shaped, but not arching over sorus
    3. Petioles glabrous or nearly so, except at base; indusia oblong to U-shaped, often caducous .......... **Athyrium**
    3' Petioles with scales well above base; indusia round to reniform, persistent .......... **Dryopteris**

### **ATHYRIUM** Roth LADY FERN

Rhizomes ± erect, stout. Leaf blades 1-2 times compound, longer than petioles. Sori along 1 or both sides of veins, away from or near leaf margin; indusia oblong to U-shaped, often caducous. Ca. 100 species; mostly temperate, worldwide.

**A. filix-femina** (L.) Mertens var. **cyclosorum** Rupr. WESTERN LADY FERN Rhizomes erect or ascending, 2-5 cm wide including petiole bases. Blades rhombic, tapering at base, apices acute, pinnae oblong-lanceolate, ultimate segments oblong; petioles 15-35 cm long, black at base, straw-colored above, base scaly, usually persistent.

Sori oblong, circular, or U-shaped; indusia caducous; spores yellow. Jul-Nov. Scarce; waterfalls and seeps; oak woodland; 10-1000 ft (3-30 m). North side in canyons above Lady's and Baby's harbors, and in Tinker's Canyon e of Pelican Bay. AK to s CA, e to SD and CO, disjunct in ne Canada. Small, sterile specimens may be confused with *Cystopteris*. Var. *californicum* Butters, with brown spores, occurs on the mainland at higher elevations.

### CYSTOPTERIS Bernh.

Rhizomes short-creeping. Leaf blades 2-4 times compound, equal to or longer than petioles. Sori along both sides of midvein; indusia hood-like, arching over sorus, often ciliate, thin, often caducous or breaking apart. Ca. 10 species; temperate, worldwide.

**C. fragilis** (L.) Bernh. FRAGILE FERN or BRITTLE FERN Blades 2-3 times compound, narrowly lanceolate, pinnae ovate to narrowly deltate; petioles 2-15 cm long, ca. two-thirds length of blades. Sori round; indusia often obscured at maturity. Jul-Nov. Scarce; seeps and moist canyon bottoms; oak woodland, riparian communities; 50-1500 ft (15-460 m). North side between Cueva Valdez and Pelican Bay (in canyons above Trident, Baby's, Orizaba, and Tinker's coves). Throughout N. America and Eurasia.

### DRYOPTERIS Adans. WOOD or SHIELD FERN

Rhizomes erect to short-creeping, stout. Leaf blades often 1-3 times compound, longer than petioles. Sori along both sides of mid-vein; indusia round-reniform, persistent. Ca. 100 species; worldwide.

**D. arguta** (Kaulf.) Watt. WOOD FERN Rhizomes short-creeping. Blades lanceolate, pinnae oblong-lanceolate, ultimate segments shallowly serrate; petioles 10-45 cm long, scaly. Sori round; indusia with a narrow sinus. Mar-Oct. Common; n-facing slopes, oak woodland; 100-1750 ft (30-535 m). Throughout island on igneous substrates. San Miguel, Santa Rosa, Anacapa, Santa Catalina, and San Clemente islands; British Columbia to n Baja CA, e to AZ. Leaves were used medicinally by Native Americans.

### POLYSTICHUM Roth SWORD or HOLLY FERN

Rhizomes ± erect, stout, scaly. Leaves in a basal rosette, with scales; blades mostly once compound, much longer than petioles, lowest segments sometimes very deeply, pinnately lobed. Sori along both sides of midvein; indusia peltate, persistent to caducous. Ca. 175 species; worldwide.

**P. munitum** (Kaulf.) C.Presl WESTERN SWORD FERN Blades linear-lanceolate, pinnae linear-lanceolate, margins entire to serrate; petioles 6-55 cm long; scales brown to reddish brown. Sori round, submarginal; indusia peltate, ciliate, caducous. Mar-Nov. Scarce; rocky n-facing slopes and shaded ravines, pine forest, oak woodland, riparian woodland; 100-1500 ft (30-460 m). North side between Painted Cave and Punta del Diablo; Central Valley near Christy Pines; s side in upper Laguna Canyon. Santa Rosa Island; Guadalupe Island; British Columbia to central CA, e to SD. Rhizomes were peeled, baked, and eaten by Native Americans.

## EQUISETACEAE Michx. ex DC.
## HORSETAIL FAMILY

With only 1 genus.

### EQUISETUM L. HORSETAIL or SCOURING RUSH

Herbaceous perennials with rhizomes; aerial stems erect to decumbent, jointed, green, impregnated with silica, internodes hollow. Leaves whorled, small, fused into a sheath, not green, tips usually free, tooth-like, deciduous or persistent. Cones terminal, composed of stalked, peltate, scalelike sporophylls. Sporangia on lower surfaces of peltate sporophyll; spores green. Ca. 30 species; worldwide. All species occurring on Santa Cruz Island were used medicinally by Native Americans; the abrasive stems have been used for polishing wooden and metal items.

1. Stems of 2 kinds, vegetative stems green, with many whorled branches; leaf sheaths loose-fitting; cone-bearing stems simple, brown ........ **E. telmateia** subsp. **braunii**

1' Stems all alike, usually simple; leaf sheaths tight-fitting
    2. Leaf sheaths cylindrical, with black bands at both ends; cone apices apiculate ........ **E. hyemale** subsp. **affine**
    2' Leaf sheaths funnelform, with a black band at apex; cone apices rounded ........ **E. laevigatum**

EQUISETACEAE

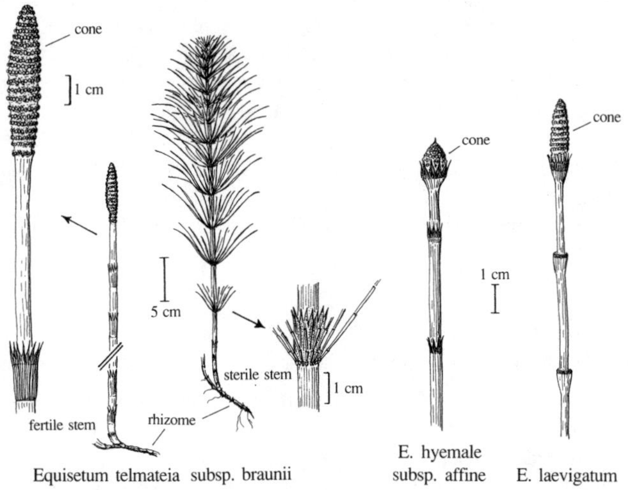


POLYPODIACEAE

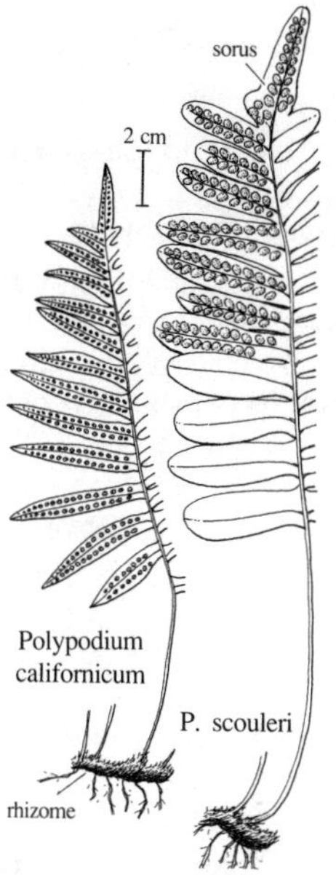


**E. hyemale** L. subsp. **affine** (Engelm.) Calder & R.L.Taylor SCOURING RUSH Aerial stems all green, unbranched, to 1 m tall, 4-12 mm wide. Leaf sheaths cylindrical, slightly longer than wide, tight fitting, with black bands at both ends; teeth 15-50, 1-5 mm long, deciduous. Cone apices apiculate. Apr-Sep. Rare; moist sites; riparian communities; 50-500 ft (15-150 m). North side in Cañada del Agua and at watering hole near China Harbor. Santa Rosa Island; throughout most of N. and C. America.

**E. laevigatum** A.Braun SMOOTH SCOURING RUSH Aerial stems all green, usually unbranched or branched only at bases or when tips damaged, to 1 m tall, 1-5 mm wide. Leaf sheaths narrowly funnel-form, longer than wide, tight fitting, pale green, with narrow black band at apices; teeth 10-25, 1-5 mm long, deciduous. Cone apices rounded. Apr-Jul. Occasional; moist places in canyons; riparian communities; sea level-1250 ft (0-380 m). North side in Hazard's and Diablo canyons, and in Cañada del Puerto; Central Valley in Cañada Christy, Cañada de la Portezuela, and adjoining drainages. Santa Rosa and Santa Catalina islands; British Columbia to n Baja CA, e to central U.S.

**E. telmateia** Ehrh. subsp. **braunii** (J.Milde) Hauke GIANT HORSETAIL Aerial stems of 2 kinds, 25-60 cm tall; vegetative (sterile) stems many-branched, 5-20 mm wide, green; fertile stems unbranched, 10-25 mm wide, brown. Leaf sheaths slightly longer than wide, loose fitting, white-scarious; teeth 15-30, 3-7 mm long, persistent. Cone apices rounded. Apr-Sep. Scarce; moist places in canyon bottoms; riparian communities; 10-100 ft (3-30 m). North side in canyon above Baby's Cove, in Diablo and Fry's canyons, and in Cañada del Agua. Santa Catalina Island; British Columbia to s CA.

## **POLYPODIACEAE** Bercht. and J.Presl
### POLYPODY FAMILY

Rhizomes creeping, scaly. Leaf blades simple to once pinnately compound; veins of ultimate segments free to complexly fused into a network; petioles often glabrous. Sori round to oblong, along veins or scattered over lower surfaces, not near margins; indusia absent. Ca. 40 genera and 500 species; primarily tropical, mostly worldwide.

### **POLYPODIUM** L. POLYPODY

Rhizomes long-creeping; scales brown to black. Leaves solitary to clustered, blades mostly pinnatifid to once pinnately compound, thin to thick; petioles glabrous to scaly. Sori ± round, often in rows parallel to midvein of blade segments. Ca. 100 species; worldwide. Some species grown as ornamentals. Plants from Cueva Valdez and near Twin Harbor, with malformed spores and often difficult to separate clearly, are putative hybrids between *P. californicum* and *P. scouleri*, according to annotations by S. A. Whitmore (SBBG).

1. Most sori 1-3 mm wide; leaf blade segments 5-15 mm wide, margins often serrulate, apices usually acute ........ **P. californicum**
1' Most sori 3-6 mm wide; leaf blade segments 13-20 mm wide, margins usually entire, apices usually entire ........ **P. scouleri**

**P. californicum** Kaulf. CALIFORNIA POLYPODY Rhizomes brown. Leaf blades pinnatifid, deltate to ovate, herbaceous to leathery; segments 5-13(15) mm wide, margins often serrulate, apices acute to obtuse. Sori 1-3(4) mm wide. Dec-Feb. Common; shaded rock outcrops and sea cliffs; chaparral, coastal scrub, oak woodland, pine forest; 10-1250 ft (3-380 m). Throughout island on igneous substrates. All CA Channel islands except San Miguel; Los Coronados, Todos Santos, Guadalupe, and Cedros islands; n CA coast ranges to n Baja CA. Rhizomes were eaten by Native Americans.

**P. scouleri** Hook. & Grev. COAST POLYPODY Rhizomes glaucous. Leaf blades pinnatifid, lanceolate to ovate, stiff, leathery; segments 13-20 mm wide, apices widely obtuse to rounded, margins often entire. Sori (2)3-6 mm wide. Nov-Mar. Rare; moist n-facing sea cliffs; coastal scrub; 10-75 ft (3-23 m). North side near Cueva Valdez and Twin Harbors. Guadalupe Island; British Columbia to n Baja CA.

## PTERIDACEAE Rchb.
### Brake or Maiden Hair Family

Rhizomes erect to long-creeping, with linear to ovate scales. Leaf blades often 2-5 times pinnately compound (palmate-pinnate in some species of *Adiantum*); veins of ultimate segments usually free. Sori borne at or near leaf margins, sometimes along veins, often covered by recurved blade margin (false indusium); indusia absent. Ca. 40 genera and 500 species; temperate and tropical, worldwide. Previously known as Adiantaceae.

1. Lower surfaces of leaf blades pubescent to scaly, or with a white to yellow farina
   2. Lower surfaces of leaf blades pubescent to scaly ........ **Cheilanthes**
   2' Lower surfaces of leaf blades with a white to yellow farina ........ **Pentagramma**
1' Lower surfaces of leaf blades (except for sori) ± glabrous
   3. Leaf blades ± deltate, ultimate segments with acute to mucronate apices ........ **Aspidotis**
   3' Leaf blades not deltate, usually much longer than wide (round in outline in *Adiantum aleuticum*); ultimate segments with rounded to obtuse apices
      4. Ultimate leaf segments thin, membranous, margins lobed to toothed ........ **Adiantum**
      4' Ultimate leaf segments thick, leathery, margins entire ........ **Pellaea**

### ADIANTUM L. MAIDENHAIR FERN

Rhizomes creeping; scales yellow-brown to black. Leaf blades pinnately compound (palmate-pinnately compound in *A. aleuticum*); pinnae usually oblong, petioles firm, dark, glabrous. Sori elongate, appearing marginal, ± covered by reflexed, indusium-like blade margin; sporangia borne along or between ends of the free, forking veins. Ca. 200 species; worldwide, many in tropical America. Some species are cultivated as ornamentals. Species on Santa Cruz Island were used medicinally by Native Americans.

1. Leaf blades palmate-pinnately compound, rachis ± equally forked 2-4 times near apex of petiole ........ **A. aleuticum**
1' Leaf blades pinnately compound, rachis not forked, continuous as 1 axis
   2. False indusia (sori) 3-11 per blade segment, each 1-3 mm long; ultimate blade segments cuneate, acute, or obtuse at base ........ **A. capillus-veneris**
   2' False indusia (sori) 1-3 per blade segment, each 3-10 mm long; ultimate blade segments obtuse to truncate at base ........ **A. jordanii**

**A. aleuticum** (Rupr.) Paris [*A. pedatum* L.] FIVE-FINGER MAIDENHAIR Leaf blades palmate-pinnately compound, rachis forked near base into 2-4 ± equal axes, much wider than long; ultimate segments sessile, oblong, deeply lobed. False indusia distinct, 3-6 per blade segment, 1-4 mm long. Mar-Nov. Scarce; seeps and moist cliffs in canyons; chaparral, oak woodland; 10-1500 ft (3-460 m). North side between Cueva Valdez and Pelican Bay. Alaska to s CA mountains, e to WY and NM, disjunct on serpentine in se Canada and ne U.S.

**A. capillus-veneris** L. SOUTHERN MAIDENHAIR Leaf blades pinnately compound, with only one main axis, longer than wide; ultimate segments stalked, cuneate to reniform, deeply lobed. False indusia distinct, 3-11 per

**PTERIDACEAE**

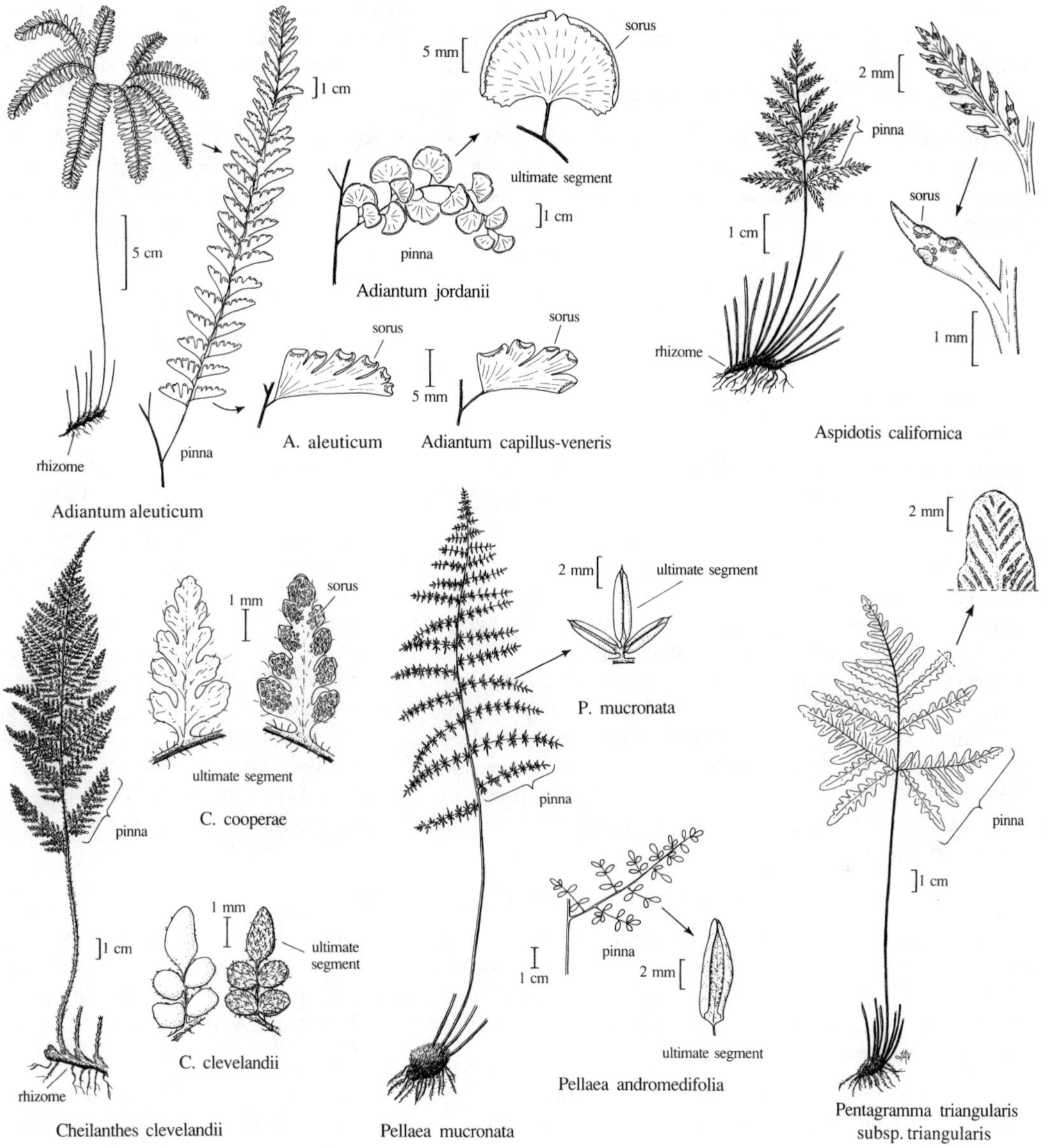


blade segment, 1-3 mm long. Mar-Oct. Scarce; springs and seeps in canyons; many plant communities; 10-1300 ft (3-400 m). North side near Fry's and Platts harbors, in Cañada del Agua, and at Aguaje Escondido near China Harbor; Central Valley between Centinela and Cañada del Puerto; s side near mouth of Laguna Canyon. Santa Rosa, Anacapa, and Santa Catalina islands; Cedros Island; n CA coast Ranges to n Baja CA, e to FL; warm temperate, worldwide. The black petioles of this species were used in basketry by Native Americans.

**A. jordanii** C. Mueller CALIFORNIA MAIDENHAIR Leaf blades pinnately compound, with only one main axis, longer than wide; ultimate segments stalked, reniform, shallowly lobed. False indusia nearly continuous along entire margin. Feb-Jun. Occasional; shaded slopes and ravines; chaparral, oak woodland; 10-1000 ft (3-300 m). Scattered locations throughout island. Santa Rosa, Anacapa, Santa Catalina, San Nicolas, and San Clemente islands; sw OR to n Baja CA.

## **ASPIDOTIS** (Nutt. ex Hook. & Baker) Copel. LACE FERN

Rhizomes creeping; scales dark brown. Leaf blades 3-4 times pinnately compound; petioles firm, dark, glabrous. Sori elongate, appearing marginal, ± covered by reflexed, indusium-like blade margin; sporangia borne along or between ends of free, forking veins. Ca. 200 species; worldwide, many in tropical America. Some species used in folk medicine and cultivated as ornamentals.

**A. californica** (Hook.) Nutt. ex Copel. CALIFORNIA LACE FERN Rhizomes short. Leaf blades deltate, 3-4 times compound, ultimate segments narrowly oblanceolate, glabrous, margins sparsely toothed, apices acute to mucronate. False indusia semi-circular, borne near marginal teeth. Feb-Apr. Scarce; rocky slopes and ledges; chaparral; 100-750 ft (30-230 m). North side between Lady's Harbor and Pelican Bay. Santa Catalina Island; n CA coast ranges to n Baja CA, e to Sierra Nevada foothills.

## **CHEILANTHES** Sw. LIP FERN

Rhizomes short; scales brown, black, or bicolored. Leaf blades pinnatifid to 4 times pinnately compound, lower surfaces densely pubescent to scaly; petioles dark, glabrous to pubescent or scaly. Sporangia continuous or aggregated into sori, either covered by recurved or inrolled, indusium-like blade margin or exposed. Ca. 150 species; worldwide, mostly in N. and S. America.

1. Lower leaf surfaces minutely scaly, scales reddish brown ........................................................ **C. clevelandii**
1' Lower leaf surfaces densely pubescent, trichomes both linear and capitate ................................ **C. cooperae**

**C. clevelandii** D.C.Eaton CLEVELAND'S LIP FERN Rhizomes long-creeping; scales lanceolate, brown with dark brown central stripes. Leaf blades 3 times compound, upper surfaces glabrous, lower surfaces covered with reddish brown scales; ultimate segments ovate to orbicular, margins revolute; petioles brown, scaly. Mar-Sep. Common; rocky slopes; chaparral, pine forest; 100-1800 ft (30-550 m). Throughout island on igneous substrates. Santa Rosa Island; mountains of s CA and n Baja CA.

**C. cooperae** D.C.Eaton MRS. COOPER'S LIP FERN Rhizomes erect to ascending; scales linear-lanceolate, light to dark reddish brown. Leaf blades 3 times compound, both surfaces pubescent with both glandular and nonglandular trichomes; ultimate segments oblong to round-oblong, fertile margins revolute, sterile margins plane; petioles brown, glandular-pubescent. Rare, not seen recently; rock crevices; chaparral; elevation unknown. Known from a single collection made in about 1893. Northern and central CA. Type collection made in Tajiguas Canyon near Santa Barbara; named for Mrs. Ellwood Cooper, early botanist and wife of a well-known *Eucalyptus* grower (Smith 1976).

## **PELLAEA** Link CLIFF-BRAKE

Rhizomes ascending to creeping; scales tan, brown, or bicolored. Leaf blades 1-4 times pinnately compound, upper surfaces usually glabrous; petioles firm, brown to black, glabrous to pubescent. Sporangia ± continuous along margin, ± covered by reflexed or revolute, indusium-like blade margin. Ca. 40 species; worldwide, mostly N. and S. America. Both species occurring on Santa Cruz Island were used by Native Americans in basketry and to make medicinal tea.

1. Petioles straw-colored to gray; ultimate leaf segments ± elliptic, 5-15 mm long .............. **P. andromedifolia**
1' Petioles red-brown to purple-black; ultimate leaf segments ± linear-oblong, 2-6 mm long ...... **P. mucronata**

**P. andromedifolia** (Kaulf.) Fee COFFEE FERN Rhizomes long-creeping, 2-3.5 mm wide. Leaf blades 3 times compound, pinnae ± lanceolate; ultimate segments elliptic, 5-15 mm long, sessile to short-stalked; petioles 8-40 cm long, straw-colored to gray. Feb-Aug. Occasional; dry rocky slopes; coastal scrub, chaparral; sea level-1000 ft (0-300 m). Throughout island on igneous substrates. All CA Channel islands except Santa Barbara; Los Coronados and Cedros islands; sw OR to n Baja CA.

**P. mucronata** (D.C.Eaton) D.C.Eaton var. **mucronata** BIRD'S FOOT FERN Rhizomes short-creeping, 3-5 mm wide. Leaf blades 2-3 times compound, pinnae narrowly oblong; ultimate segments linear-oblong, 2-6 mm long, sessile; petioles 4-20 cm long, red-brown or purple-black. Feb-Jun. Scarce; rocky slopes; chaparral; 200-800 ft (60-245 m). North side of island between Pelican Bay and Prisoners' Harbor, Cañada del Puerto, Smugglers' Canyon; Central Valley in Islay Canyon; s side in Willows Canyon. Santa Rosa, Santa Catalina, and San Clemente islands; Guadalupe Island; n CA coast ranges to n Baja CA, e to NV.

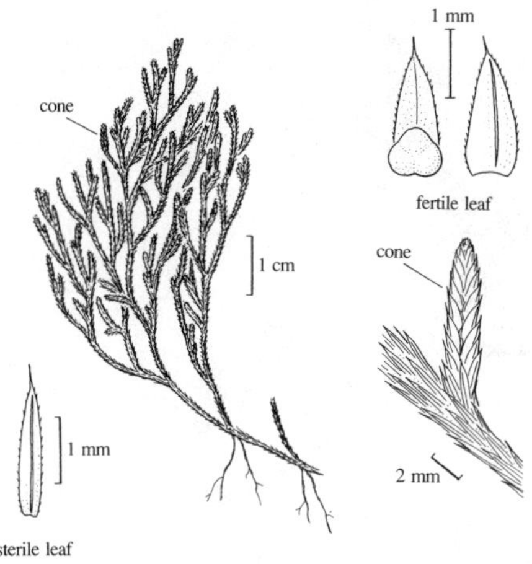

Selaginella bigelovii

**PENTAGRAMMA** Yatsk., Windham, & E.Wollenw.
GOLDBACK or SILVERBACK FERN

Rhizomes short-creeping to erect; scales tan with dark stripe. Leaf blades pinnatifid to 2 times pinnately compound, deltate to pentagonal, lower surfaces covered with white to yellow farina, sometimes glandular; petioles brown to black, glabrous to farinose. Sporangia borne along veins; indusia absent. Ca. 2 species; w N. America. California species were previously included in *Pityrogramma* Link., a tropical genus. Petioles were used by Native Americans for medicinal purposes and in basketry.

**P. triangularis** (Kaulf.) Yatsk., Windham, & E.Wollenw. [*Pityrogramma t.* (Kaulf.) Maxon]. GOLDBACK FERN Leaf blades with upper surfaces glabrous to viscid, lower surfaces yellow-farinose; petioles brown or red-brown. Mar-Jun. Common; rocky slopes; many plant communities; 10-1200 ft (3-365 m). Throughout island on igneous substrates. All CA Channel islands except Santa Barbara; Cedros Island; British Columbia to n Baja CA, e to NV, AZ, and NM. Composed of ca. 4 subspecies. Plants with farina absent and shrunken or malformed spores, previously named var. *viridis* Hoover, are sterile hybrids between the 2 subspecies, according to D. M. Smith (annotations at SBBG).

1. Upper leaf surfaces glabrous to sparsely farinose, not viscid .......... subsp. **triangularis**
1' Upper leaf surfaces viscid .......... subsp. **viscosa**

Subsp. **triangularis** GOLDBACK FERN Leaf blades thin; upper surfaces glabrous to sparsely farinose; lower surfaces light to deep yellow, sometimes white-farinose. Mar-Jun. Common; rocky slopes; many plant communities; 10-1200 ft (3-365 m). Throughout island on igneous substrates. All CA Channel islands except Santa Barbara; Cedros Island; British Columbia to n Baja CA. Plants from Santa Cruz Island, with leaf surfaces white-farinose below, have been called subsp. *semipallida* (J.T.Howell) Yatsk., Windham, & E.Wollenw.

Subsp. **viscosa** (Nutt. ex D.C.Eaton) Yatsk., Windham & E.Wollenw. [*Pityrogramma t.* (Kaulf.) Maxon var. *v.* (Nutt. ex D.C.Eaton) Weath.]. SILVERBACK FERN Leaf blades leathery; upper surfaces viscid; lower surfaces densely white-farinose. Mar-Jun. Common; rocky slopes; many plant communities; 10-1000 ft (3-300 m). Throughout island on igneous substrates. Santa Rosa, Santa Catalina, and San Clemente islands; coastal s CA to n Baja CA.

## **SELAGINELLACEAE** Willk.
### SPIKE-MOSS FAMILY

With only 1 genus.

### **SELAGINELLA** P. Beauv. SPIKE-MOSS

Annuals or perennial herbs with rhizomes; stems prostrate to erect, often dichotomously branched. Leaves all alike, spirally arranged and imbricated or of 2 forms in 4 ranks, scale-like. Cones terminal, usually composed of leaf-like sporophylls subtending sporangia. Sporangia solitary, axillary, of 2 kinds; megasporangia usually with 4 spores (megaspores); microsporangia with many minute spores (microspores). Ca. 700 species; mostly subtropical to tropical, worldwide.

**S. bigelovii** Underw. BUSHY SPIKE-MOSS Plants forming loose clumps, 5-20 cm tall. Sterile leaves narrowly lanceolate, 1.2-1.8 mm long, 0.25-0.35 mm wide, margins ciliate, apices keeled, ferile leaves ovate. Cones oblong, 4-15 mm long. Mar-Jun. Common; rocky slopes and exposed ridge tops, chaparral; 25-1500 ft (7-460 m). Throughout island on igneous substrates. Santa Rosa, Anacapa, Santa Catalina, and San Clemente islands; Cedros Island; n CA to s Baja CA.

CUPRESSACEAE

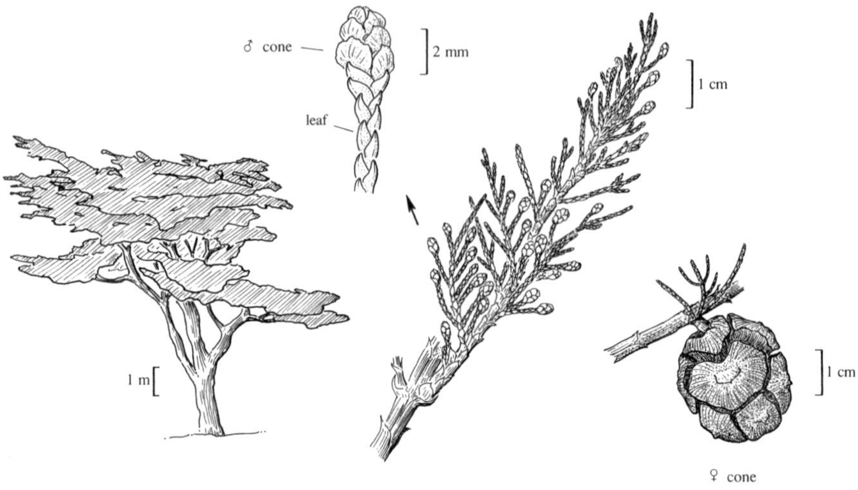


Cupressus macrocarpa

# GYMNOSPERMS

Woody plants, usually with cones or cone-like reproductive structures; reproducing by seeds and pollen; ovules and seeds naked, borne on woody, scale-like sporophylls comprising woody cones; pollen often white to yellow, produced scale-like sporophylls comprising small cones. Generally considered to include 4-5 divisions or phyla, ca. 12 families, ca. 70 genera, and ca. 760 species.

## Key to Families

1. Leaves shorter than 0.5 cm, scale-like, opposite; seed cones with 5-10, peltate scales ........... **Cupressaceae**
1' Leaves longer than 5 cm, needle-like, fascicled; seed cones with many, ± flat scales ...................... **Pinaceae**

## CUPRESSACEAE Bartl.

### CYPRESS FAMILY

Evergreen, monoecious trees. Leaves decussate, scale-like. Pollen produced in sporangia adnate to ± flat scales that are arranged in short, terminal cones. Ovules 1-several, naked, borne on inner surface of a peltate scale; scales comprising persistent cones that mature in 2 years. Seeds 2-many per scale, flat, winged. 17 genera and ± 113 species; worldwide. Many species are cultivated and the wood of some is used commercially.

### CUPRESSUS L. CYPRESS

Shrubs and trees, with rough, brown to grayish bark. Leaves overlapping, olive-green to grayish green. Seed cones woody at maturity, ± globose, ± sessile; scales separating but not falling with age. Ca. 13 species; w N. America and Eurasia. Some species widely cultivated.

**C. macrocarpa** Hartw. MONTEREY CYPRESS Trees to 15 m tall. Twigs cylindrical, 1-2 mm wide. Leaves ovate, clasping at base, 1-2 mm long. Seed cones globose, 2-4 cm long, 2-3 cm wide, with 5-9 scales. Apr-Jul. Introduced. Common; open slopes and canyons; 60-300 ft (20-90 m). Planted on north side at Scorpion Ranch; Central Valley at Christy Ranch, Stanton Ranch, and Rancho del Sur; s side in lower Cañada de los Sauces (where becoming naturalized) and at Smugglers' Cove. Santa Rosa, Anacapa, and Santa Catalina islands; native of the Monterey Peninsula, CA, but widely cultivated throughout CA and elsewhere.

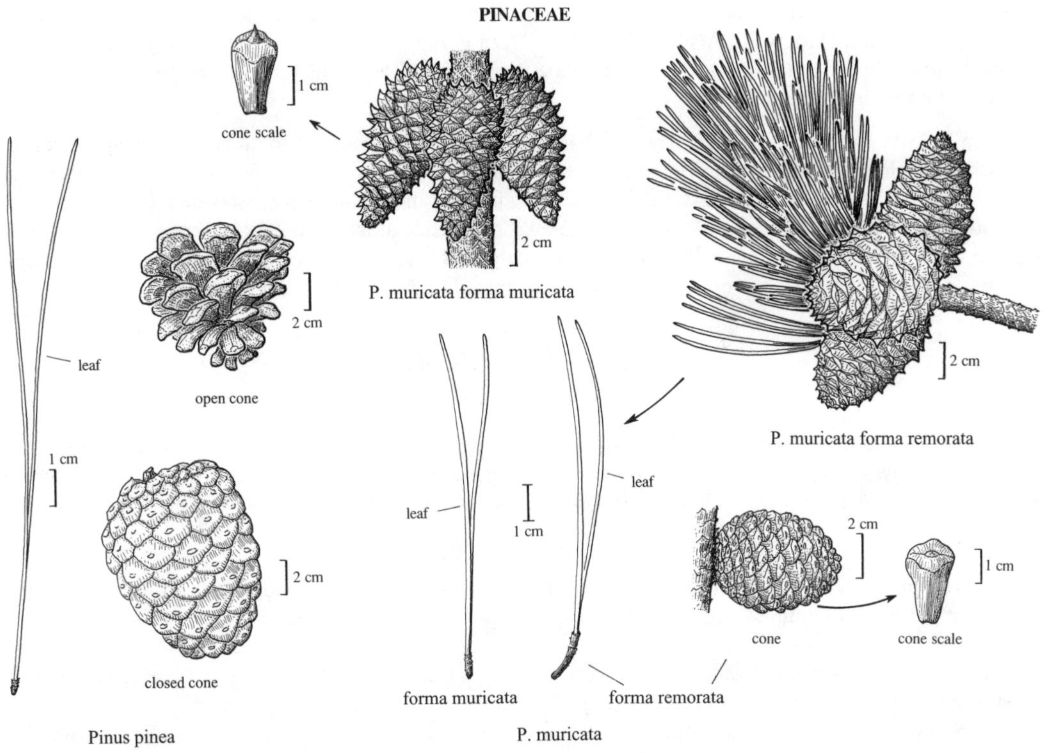


# PINACEAE Lindl.

## PINE FAMILY

Resinous, evergreen, monoecious trees. Leaves needle-like, fascicled, enclosed at base by a membranous sheath. Pollen produced in sporangia adnate to flat scales that are arranged in short, terminal cones. Ovules naked, adnate to upper surface of a ± flat scale; scales arranged in persistent cones that mature in ± 2 years. Seeds 2 per scale, usually winged. Ca. 9 genera and ± 200 species; mostly n temperate.

### PINUS L. PINE

Trees with furrowed or scaly bark; crown of tree slender to umbrella-like. Leaves 1-5 per fascicle, dark green to grayish green. Seed cones woody, ovoid to subglobose, pendulous or spreading; scales strongly appressed or separate from each other at maturity. 93 species; N. and C. America, Eurasia.

1. Seed cones often longer than wide, usually attached in whorls on branches; seed wing more than 10 mm long; crown slender to pyramidal; old bark composed of brown to grayish brown, rounded ridges ........................ **P. muricata**

1' Seed cones wider than long, not whorled on branches; seed wing less than 1 mm wide; crown rounded to umbrella-like; old bark composed of reddish brown, oblong plates ........................ **P. pinea**

**P. muricata** D.Don BISHOP PINE Trees to 20 m tall, crown ± slender to pyramidal; bark composed of rounded ridges. Leaves 5-15 cm long. Seed cones symmetrical to asymmetrical, ± ovoid, usually whorled and either deflexed or borne at right angles to stem; scales appressed. Seeds 6-12 mm long; wing 10-25 mm long. May-Jun. Common; n-facing slopes and ridges; pine forest; 150-1800 ft (45-430 m). North side of island between Orizaba Flats and Pelican Bay, above China Harbor; Central Valley in upper Cañada Christy; s side in Cañada de los Sauces and s of China Harbor. Santa Rosa Island; Humboldt Co. to Santa Barbara Co. and nw Baja CA. Typical *P. muricata* is characterized by ± asymmetrical seed cones that are deflexed relative to subtending stem, cone scales with prolonged, recurved, spiny tips, and dark yellow-green leaves that are usually longer than 8 cm. Plants with ± symmetrical seed cones borne at right angles to stem, cone scales usually with plane to slightly rounded tips, and dark green leaves 5-8 cm long are called *P. muricata* D.Don forma *remorata* (H.Mason) Hoover (*P. remorata* H.Mason) SANTA CRUZ ISLAND PINE. On Santa Cruz Island, intergradation between typical *P. muricata* and forma *remorata* apparently occurs in each of the populations, which differ with respect to the proportion of the 2 taxa.

**P. pinea** L ITALIAN STONE PINE Trees to 10 m tall with a broad, rounded to umbrella-shaped crown; bark composed of reddish brown, irregular, oblong plates. Leaves 10-18 cm long. Seed cones symmetrical, subglobose, ± pendulous; scales spreading. Seeds ca. 15 mm long; wing less than 1 mm long. May-Jun. Introduced. Common; n-facing slopes and canyons; 50-900 ft (15-250 m). North side of island at Pelican Bay and near Prisoners' Harbor; Central Valley near Stanton Ranch. Planted widely in CA; native of Mediterranean Europe. This species was introduced during the Caire ownership and is now spreading into undisturbed vegetation.

# ANGIOSPERMS
## Flowering Plants

Plants with flowers usually composed of sepals, petals, stamens, and pistils; pollen produced in anthers; ovules enclosed in an ovary with style(s) and stigma(s); seeds enclosed in a dry or fleshy fruit. Generally considered to represent 1 division or phylum, including 375 families, ca. 14,000 genera, and ca. 250,000 species.

## Key to Families and Exceptional Groups

1. Leaf venation usually parallel, usually simple (palmately dissected in Potamogetonaceae); perianth parts, if present, usually in multiples of 3 (4 in Potamogetonaceae); vascular bundles in stem cross-section usually scattered ........ **GROUP 1 (MONOCOTS)**
1' Leaf venation pinnate or palmate; perianth parts, if present, usually in multiples of 4 or 5; vascular bundles in stem cross-section usually organized in a ring ........ **(DICOTS)**
  2. Shrubs or Trees ........ **GROUP 2**
  2' Perennial herbs or annuals
    3. Perianth parts apparently absent or only in 1 whorl ........ **GROUP 3**
    3' Perianth parts in 2 or more whorls
      4. Petals free from each other ........ **GROUP 4**
      4' Petals fused together ........ **GROUP 5**

### GROUP 1. MONOCOTS

1. Plants aquatic and often completely submerged
  2. Plants of salt water at or below sea level; inflorescences one-sided, ± flattened ........ **Zosteraceae**
  2' Plants of fresh or brackish water at or above sea level; inflorescence neither flattened nor one-sided
    3. Most leaves alternate; flowers bisexual, in terminal, interrupted spikes or few-flowered umbels ........ **Potamogetonaceae**
    3' Most leaves subopposite, upper ones ± whorled; flowers unisexual, axillary ..... **Zannichelliaceae**
1' Plants terrestrial; if aquatic, then emergent, with roots and bases of stems rooted in mud and most leaves and flowers above water
  4. Perianth either absent or composed of greenish to brown, scale-like to bristly parts
    5. Perianth parts 6, scale-like; fruit a capsule with many seeds ........ **Juncaceae**
    5' Perianth parts absent or, if present, bristle-like; fruit indehiscent, usually with 1 seed
      6. Inflorescence a dense, cylindrical spike 1.5-3.5 cm wide ........ **Typhaceae**
      6' Inflorescences of various types (racemes, spikes, or panicles) but not dense and cylindrical
        7. Stems usually sharply or obtusely 3-angled, usually with solid internodes; leaves in 3 ranks; stamens and pistil usually subtended by 1 bract; fruit lenticular or 3-angled ........ **Cyperaceae**
        7' Stems circular to terete, usually with hollow internodes; leaves in 2 ranks; stamens and pistil usually subtended by 2 bracts; fruit usually elliptic or oblong in outline, not compressed or angled ........ **Poaceae**

4' Perianth present, inner parts white, blue, or yellow, outer ones sometimes also brightly colored
8. Ovary superior
9. Flowers in terminal umbels; umbel subtended by a whorl of 1-few bracts .............. **Alliaceae**
9' Flowers solitary or in racemes, sometimes in ± umbellate clusters; bracts, if present, 1 per flower .......... **Liliaceae**
8' Ovary inferior
10. Leaves fleshy, rigid, spinose, in a large basal rosette .............................. **Agavaceae (*Agave*)**
10' Leaves herbaceous, neither spinose nor in a large basal rosette
11. Flowers bilateral; inflorescence a raceme or spike; ovary with 1 locule .... **Orchidaceae**
11' Flowers radial; inflorescence usually an umbel, flowers sometimes solitary; ovary with 3 locules
12. Flowers usually with a ring-like corona surrounding stamens; perianth parts white, corona white to pale yellow .......... **Amaryllidaceae**
12' Flowers without a ring-like corona; perianth parts blue to bluish purple .......... **Iridaceae**

## GROUP 2. DICOT SHRUBS or TREES

1. Perianth parts apparently absent or only in 1 whorl
2. Flowers and fruits enclosed in a fleshy, ± globose to obovoid receptacle-like structure ......... **Moraceae**
2' Flowers and fruits not enclosed in a fleshy receptacle
3. Leaves ternately or pinnately compound, with 3-15 distinct leaflets ... **Ranunculaceae (*Clematis*)**
3' Leaves simple, entire to deeply lobed, but segments not distinct
4. Leaves opposite at most nodes
5. Leaves scale-like; stems ± succulent, with short, joint-like internode .......... **Chenopodiaceae (*Salicornia*)**
5' Leaves not scale-like; stems not succulent, internodes not joint-like ........... **Garryaceae**
4' Leaves alternate at most nodes
6. Leaf blades palmately veined and palmately lobed; flowers in globose heads .......... **Platanaceae**
6' Leaf blades pinnately veined, margins entire or pinnately toothed to lobed; flowers not in globose heads
7. Pistillate flower stalked, surrounded by few to many staminate flowers, these subtended by an involucre of 5 white-appendaged bracts; sap opaque .......... **Euphorbiaceae (*Euphorbia misera*)**
7' Flowers bisexual or, if unisexual, ovary not stalked, flowers not subtended by an involucre of white-appendaged bracts, and sap usually clear
8. Flowers unisexual; staminate flowers in catkins; pistillate flowers in catkins or small, axillary clusters
9. Plants monoecious; pistillate flowers few, axillary; sepals 5-6; fruit an acorn .......... **Fagaceae**
9' Plants dioecious; pistillate flowers many, in catkins; sepals absent; fruit a capsule .......... **Salicaceae**
8' Flowers bisexual; flowers usually few, in axillary clusters
10. Flower with a cylindrical hypanthium; style longer than ovary, persistent and plumose in fruit ......................................... **Rosaceae (*Cercocarpus*)**
10' Hypanthium either absent or short and cuplike in Myrtaceae; style usually shorter than ovary, neither persistent nor plumose in fruit
11. Hypanthium cuplike; stamens many; leaves strongly aromatic .......... **Myrtaceae**
11' Hypanthium absent; stamens 4-6; leaves not aromatic .......... **Rhamnaceae (*Rhamnus*)**

1' Perianth parts in 2 or more whorls
12. Petals fused, forming a ring or tube, falling from flower as 1 unit
13. Ovary inferior
14. Calyx modified into a pappus or absent; fruit an achene ....................................... **Asteraceae**
14' Calyx usually composed of 4-5 green sepals or lobes; fruit usually fleshy
15. Leaves alternate; stamens usually 10 ......................................... **Ericaceae (*Vaccinium*)**
15' Leaves opposite; stamens 4-5
16. Leaves pinnately compound; style short or absent ................................ **Adoxaceae**
16' Leaves simple; style well developed ................................................ **Caprifoliaceae**
13' Ovary superior
17. Leaves opposite at most nodes; corolla bilateral
18. Leaves aromatic when crushed; ovary deeply 4-lobed .................................. **Lamiaceae**
18' Leaves not aromatic when crushed; ovary globose to ellipsoid, not deeply lobed .......... ............................................................................................................ **Scrophulariaceae**
17' Leaves alternate at most nodes; corolla radial
19. Stamens many, ± fused into a tube surrounding style; leaves palmately veined ............. ............................................................................................................ **Malvaceae**
19' Stamens 2-10, not fused into a tube; leaves usually pinnately veined
20. Stamens 2; leaves with minute, silvery scales .............................. **Oleaceae (*Olea*)**
20' Stamens 5 or 10; leaves glabrous to glandular or tomentose, but not scaly
21. Corolla urceolate; stamens 10, anthers minutely awned .................. **Ericaceae**
21' Corolla rotate, campanulate, or salverform; stamens 5, anthers not awned ..... ............................................................................................ **Solanaceae**
12' Petals free from each other, usually falling separately
22. Ovary inferior (sometimes appearing so in Rhamnaceae)
23. Stems succulent, spiny; leaves apparently absent; sepals and petals many ........... **Cactaceae**
23' Stems not succulent, spines absent; leaves present; sepals and petals each usually 5
24. Hypanthium absent; ovary partly inferior, embedded in a prominent glandular disk ...... ......................................................................................... **Rhamnaceae (*Ceanothus*)**
24' Hypanthium present; ovary completely inferior; glandular disk absent or not prominent
25. Leaves palmately veined and lobed; stamens 5 ............................ **Grossulariaceae**
25' Leaves either pinnately veined or pinnately compound; stamens 10-many ............. ................................................................................................ **Rosaceae**
22' Ovary superior
26. Petals either 3 or 6; sepals and stamens often in multiples of 3
27. Leaves pinnately compound; flowers in racemes or panicles; involucre absent ............. ........................................................................................................ **Berberidaceae**
27' Leaves simple; flowers aggregated into dense clusters each subtended by an involucre, collectively corymbose ....................................................... **Polygonaceae (*Eriogonum*)**
26' Petals 4-5; sepals usually 5 (sometimes 3 in Cistaceae); stamens 5-many, not in multiples of 3
28. Leaves opposite at most nodes
29. Sepals fused into a tube; stamens usually 6; low shrubs on saline soils near beaches ................................................................................................ **Frankeniaceae**
29' Sepals free from each other; stamens usually 5 or 8; trees or shrubs on interior slopes or in canyons away from beach
30. Leaves palmately veined and lobed; petals yellowish to greenish, sepaloid; fruit a samara ........................................................................... **Aceraceae**
30' Leaves pinnately veined, margins entire; petals white; fruit a capsule ........... ............................................................................................ **Rhamnaceae**
28' Leaves alternate at most nodes
31. Leaves pinnately or palmately compound; leaflets 3-many
32. Flowers bilateral (papilionaceous) .................................................. **Fabaceae**
32' Flowers radial

33. Perianth parts and stamens free from each other; stamens usually 4-5 (10 in *Schinus*); stipules absent .......... **Anacardiaceae**
33' Perianth parts and stamens fused into a hypanthium; stamens 10-many; stipules present .......... **Rosaceae**
31' Leaves simple, margins entire to deeply lobed
34. Stamens 4-5; styles or style branches usually 3
35. Leaves scale-like, sessile; ultimate inflorescences cylindrical, dense .......... **Tamaricaceae**
35 Leaves not scale-like, usually petiolate; flowers in ± open racemes, panicles, or umbellate cymes
36. Stipules absent; style branched from near base; fruit drupe-like .......... **Anacardiaceae**
36' Stipules present; style branched ± at middle; fruit a capsule .......... **Rhamnaceae**
34' Stamens 10-many; style 1 or absent, stigma then sessile
37. Low, rounded subshrub less than 0.3 m tall; leaves less than 1 cm wide; outer 2 sepals bract-like .......... **Cistaceae**
37' Shrubs or trees usually more than 1 m tall; most leaves more than 1 cm wide; outer sepals, if any, not bract-like
38. Sepals 2, caducous; petals 4, usually more than 1 cm long; stigmas sessile .......... **Papaveraceae (*Dendromecon*)**
38' Sepals 5, persistent; petals 5, usually less than 1 cm long; style present
39. Inflorescence globose; stamens prominently exserted .......... **Fabaceae (*Acacia*)**
39' Inflorescence a panicle; stamens ± equal to petals and sepals .......... **Rosaceae**

**GROUP 3. DICOT HERBS; flowers with perianth apparently absent or only in 1 whorl**

1. Most cauline leaves alternate
2. Ultimate inflorescence (sometimes all flowers) dense, subtended by an involucre of free to fused bracts
3. Ovary completely inferior or deeply embedded in inflorescence rachis
4. Style branches 2; calyx, if present, modified into a pappus; bracts subtending each flower, if present, not white .......... **Asteraceae**
4' Style branches 3-4; perianth absent; each flower subtended by 1 white bract ... **Saururaceae**
3' Ovary superior
5. Ovary usually stalked, with 3 locules; sap opaque .......... **Euphorbiaceae**
5' Ovary sessile, with 1 locule; sap clear
6. Inflorescence ± terminal; flowers bisexual .......... **Polygonaceae**
6' Inflorescence axillary; flowers unisexual .......... **Urticaceae (*Parietaria*)**
2' Flowers in racemes, in panicles, or solitary, ultimate clusters, if any, not subtended by an involucre
7. Ovary inferior
8. Plants vine-like or twining; stems with tendrils; fruit bristly .......... **Cucurbitaceae (*Marah*)**
8' Plants neither vine-like nor twining; tendrils absent; fruit smooth, except for persistent, horn-like perianth parts .......... **Aizoaceae (*Tetragonia*)**
7' Ovary superior
9. Plants, especially leaves, with grayish, stellate trichomes ... **Euphorbiaceae (*Eremocarpus*)**
9' Plants glabrous to variously pubescent, but trichomes not stellate
10. Stipules present, either free or fused and partly to completely sheathing stem; flowers bisexual
11. Flowers mostly in terminal inflorescences, not concealed by stipules; hypanthium absent; stamens 3-9 .......... **Polygonaceae**
11' Flowers axillary, concealed by sheathing stipules; hypanthium present; stamen 1 .......... **Rosaceae (*Aphanes*)**
10' Stipules apparently absent; flowers usually unisexual
12. Ovary with 3 locules, each with 1 ovule; fruit a capsule .......... **Euphorbiaceae (*Stillingia*)**
12' Ovary with 1 locule with 1 ovule; fruit an achene or utricle

13. Floral bracts scarious, dry in fruit; stems and leaves not succulent, neither farinose nor scurfy ............ .......... **Amaranthaceae**
13' Floral bracts usually not scarious; stems and leaves either succulent or farinose to scurfy or both ........ .......... **Chenopodiaceae**

1' Most cauline leaves opposite or whorled, leaves sometimes all basal
14. Inflorescence an umbel
15. Perianth parts free; leaves lobed to deeply dissected or compound .......... **Apiaceae**
15' Perianth parts fused, tubular; leaves simple, entire .......... **Nyctaginaceae (*Abronia*)**
14' Inflorescence various but not an umbel
16. Ovary (pistillate flower) stalked, apparently surrounded by several stamens (staminate flowers), these subtended by an involucre of 5 white- pink-appendaged petaloid bracts; sap white-opaque .......... **Euphorbiaceae (*Chamaesyce*)**
16' Ovary not stalked; sap not white-opaque
17. Pistils many; stamens many .......... **Ranunculaceae (*Clematis*)**
17' Pistil 1; stamens fewer than 10
18. Ovary inferior; perianth parts fused; fruit usually hispid or prickly
19. Leaves palmately veined and lobed; tendrils present .......... **Cucurbitaceae**
19' Leaves pinnately veined; tendrils absent .......... **Rubiaceae**
18' Ovary superior; perianth parts, if present, free from each other; fruit not neither hispid nor prickly
20. Styles 4-5; ovules 3-many per ovary; fruit a capsule .......... **Caryophyllaceae**
20' Styles 1-3; ovules only 1 per ovary; fruit an achene or utricle
21. Style 1; leaves with stinging trichomes .......... **Urticaceae**
21' Styles 2-3; leaves glabrous to variously pubescent, but trichomes not stinging
22. Stems and leaves either succulent or scurfy or both; fruit ± circular in cross-section; sepals usually 1-5 .......... **Chenopodiaceae**
22' Stems and leaves neither succulent nor scurfy; fruit triangular in cross-section; sepals usually 6 .......... **Polygonaceae**

## GROUP 4. DICOT HERBS; flowers with perianth parts in 2 or more whorls; petals free

1. Flowers bilateral, some petals dissimilar in size or shape, or petals displaced to one side of floral axis
2. Pistils 3 per flower; sepals petaloid, uppermost one obviously spurred .......... **Ranunculaceae (*Delphinium*)**
2' Pistil 1; sepals herbaceous, not obviously spurred (lower petal may be spurred)
3. Leaves compound, leaflets 3-many .......... **Fabaceae**
3' Leaves simple, entire to lobed
4. Annuals; petals 2; inflorescence a terminal spike .......... **Resedaceae**
4' Perennial; petals 5; flowers solitary or few, not arranged in a spike
5. Corolla white to pink; leaves ovate to obovate .......... **Polygalaceae**
5' Corolla yellow to orange; leaves deltate to cordate .......... **Violaceae**

1' Flowers radial, ± equal in size and shape
6. Ovary partly or completely inferior
7. Perianth parts many; stems and leaves, if present, succulent
8. Stems spiny; leaves apparently absent; ovary with 1 locule .......... **Cactaceae**
8' Stems not spiny; leaves present; ovary with 5-20 locules .......... **Aizoaceae**
7' Sepals 2-5; petals 3-5; stems and leaves usually not succulent (except some Portulacaceae)
9. Stamens many
10. Leaves and bracts scabrous, trichomes barbed; stems erect .......... **Loasaceae**
10' Leaves and bracts smooth, glabrous to pubescent, trichomes, if present, not barbed; stems spreading .......... **Portulacaceae (*Portulaca*)**
9' Stamens 1-10
11. Sepals 2 .......... **Portulaceae**
11' Sepals 5

12. Stamens 5
13. Inflorescence an umbel; ovary completely inferior; hypanthium absent .......................... **Apiaceae**
13' Inflorescence a raceme or panicle; ovary partly inferior; hypanthium present ................................ ....................................................................................................... **Saxifragaceae (*Heuchera*)**
12' Stamens 10
14. Stamens in 2 unequal sets; stigma capitate; stems decumbent or creeping .................................. ....................................................................................................... **Onagraceae (*Ludwigia*)**
14' Stamens ± equal; stigma ± filiform; stems erect ...................................................... **Saxifragaceae**
6' Ovary completely superior
15. Stamens many, at least more than twice the number of petals
16. Pistils many
17. Leaves palmately lobed or dissected; stipules absent; hypanthium absent .... **Ranunculaceae**
17' Leaves pinnately dissected; stipules present; hypanthium present ....... **Rosaceae (*Potentilla*)**
16' Pistil 1
18. Leaves palmately veined and lobed; trichomes, if present, stellate ...................... **Malvaceae**
18' Leaves pinnately veined, entire to dissected; plants glabrous to variously pubescent, trichomes, if present, not stellate
19. Sepals 5, outer 2 shorter and narrower than inner 3 ....................................... **Cistaceae**
19' Sepals 2-3, ± equal, free or fused, often caducous .................................. **Papaveraceae**
15' Stamens 1-12, fewer than to no more than twice the number of petals
20. Most leaves compound or nearly so
21. Leaves palmately compound, leaflets 3 ............................................................... **Oxalidaceae**
21' Leaves pinnately compound, leaflets or leaf segments 5 or more
22. Sepals 4; petals 4; stamens 6 ...................................................................... **Brassicaceae**
22' Sepals 5; petals 5; stamens 5 .................................................. **Geraniaceae (*Erodium*)**
20' Most leaves simple, entire to deeply lobed
23. Most leaves alternate and cauline
24. Hypanthium present, cylindrical ................................................................... **Lythraceae**
24' Hypanthium absent
25. Sepals 2; ovary with 1 locule, seeds 2 or more per fruit .................. **Portulacaceae**
25' Sepals 3-5; ovary with 2-5 locules, if with 1 locule, then with 1 seed and triangular in cross-section
26. Ovary with 1 locule and 1 ovule; fruit triangular in cross-section ................... ....................................................................................... **Polygonaceae**
26' Ovary with 2-5 locules; fruit ± circular or flattened in cross-section
27. Sepals 4; petals 4; stamens 6 .............................................. **Brassicaceae**
27' Sepals 5; petals 5; stamens 10 ........................ **Geraniaceae (*Geranium*)**
23' Most leaves opposite, basal, or both
28. Pistils 3-5; stems and leaves succulent ...................................................... **Crassulaceae**
28' Pistil usually 1; stems and leaves not succulent (except some Portulacaceae)
29. Flower with a hypanthium
30. Lower leaves opposite, upper ones alternate; flowers usually axillary at leafy nodes ........................................................................................ **Lythraceae**
30' Leaves mostly basal; flowers on ± bracteate scapes ................ **Saxifragaceae**
29' Flower without a hypanthium, calyx free from corolla
31. Sepals fused into a tube
32. Leaves not decussate; stamens in 1 whorl, ± equal...... **Caryophyllaceae**
32' Leaves decussate; stamens in 2 unequal whorls, outer ones shorter than inner ones ........................................................................ **Frankeniaceae**
31' Sepals free from each other
33. Sepals 2; stipules absent .................................................. **Portulacaceae**
33' Sepals 5; stipules usually present

34. Leaf margins usually entire; ovary not lobed, with 1 locule ....................................... **Caryophyllaceae**
34' Leaves lobed to dissected; ovary with 5 lobes and 5 locules ............................................ **Geraniaceae**

**GROUP 5. DICOT HERBS; flowers with perianth parts in 2 or more whorls; petals connate**

1. Plants without green leaves; stems light yellow, yellow-green, orange, or purplish
   2. Stems twining, vine-like; corolla radial, less than 5 mm long ............................................ **Cuscutaceae**
   2' Stems erect, neither twining nor vine-like; corolla bilateral, 10-30 mm long .............. **Orobanchaceae**

1' Plants with green leaves and stems
   3. Flowers in dense heads subtended by an involucre of bracts; calyx absent or modified into a pappus ........................................ **Asteraceae**
   3' Flowers not in dense heads, if bracteate then not subtended by an involucre; calyx usually herbaceous
      4. Corolla bilateral
         5. Ovary deeply 4-lobed; style arising from notch between lobes .............................. **Lamiaceae**
         5' Ovary not clearly lobed; style at tip of ovary .............................................. **Scrophulariaceae**
      4' Corolla radial
         6. Ovary partly to completely inferior
            7. Cauline leaves opposite
               8. Corolla tube not spurred; calyx lobes erect to spreading, not becoming plumose in fruit ........................................ **Nyctaginaceae**
               8' Corolla tube spurred; calyx lobes coiled, unrolling and becoming plumose in fruit ........................................ **Valerianaceae**
            7' Cauline leaves alternate
               9. Plants twining or vine-like; stems with tendrils; leaves palmately veined ........................................ **Cucurbitaceae**
               9' Plants erect, neither twining nor vine-like; tendrils absent; leaves pinnately veined
                  10. Flowers axillary, subtended by leaves; corolla blue; ovary completely inferior ........................................ **Campanulaceae**
                  10' Flowers in a terminal, naked raceme; corolla white; ovary partly inferior ........................................ **Primulaceae** ***(Samolus)***
         6' Ovary completely superior
            11. Leaves opposite at most nodes
               12. Stamens either 2 or 4
                  13. Stamens 2; some or all main stems decumbent **Scrophulariaceae (*Veronica*)**
                  13' Stamens 4; main stems usually erect
                     14. Ovary deeply 4-lobed, style arising from notch between lobes ........................................ **Lamiaceae**
                     14' Ovary not lobed, ellipsoid, style arising from tip ................ **Verbenaceae**
               12' Stamens 5
                  15. Ovaries 2, separate below, fused at top to a common stigma; sap opaque
                     16. Leaves 2 per node, ovate; anthers free from stigma ............ **Apocynaceae**
                     16' Leaves 3-5 per node, linear-lanceolate; anthers adnate to stigma ........................................ **Asclepiadaceae**
                  15' Ovary 1; sap clear
                     17. Style branches 3; ovary with 3 locules ....... **Polemoniaceae (*Linanthus*)**
                     17' Style either unbranched or with 2 branches; ovary with 1-2 locules
                        18. Stigma unbranched, stigma capitate or fan-shaped
                           19. Stigma fan-shaped; flowers clustered, terminal, erect in fruit ........................................ **Gentianaceae**
                           19' Stigma capitate; flowers solitary, axillary, recurved in fruit ........................................ **Primulaceae (*Anagallis*)**
                        18' Style with 2 branches or lobes
                           20. Leaves entire; ovary with 4 lobes; style arising from notch between lobes ........................................ **Boraginaceae**
                           20' Leaves deeply lobed or dissected; ovary globose to ellipsoid; style arising from tip of ovary ....................... **Hydrophyllaceae**

11' Leaves alternate at most nodes or leaves all basal (plant scapose)
 21. Sepals 2; petals 3; stamens 3 .......... **Portulacaceae (*Calyptridium*)**
 21' Sepals 4-5; petals 4-5; stamens 2, 4, or 5
  22. Sepals 4; petals 4; stamens 2 or 4 .......... **Plantaginaceae**
  22' Sepals 5; petals 5; stamens 5
   23. Inflorescence umbellate; stamens opposite petal lobes .......... **Primulaceae**
   23' Inflorescence a raceme, panicle, or flowers solitary, but not umbellate; stamens alternate petal lobes
    24. Ovary deeply 2- or 4-lobed; fruit dehiscing into 2 or 4 segments
     25. Inflorescence terminal, coiled; leaves pinnately veined .......... **Boraginaceae**
     25' Inflorescence solitary, concealed among leaves; leaves palmately veined .......... **Convolvulaceae (*Dichondra*)**
    24' Ovary globose to ellipsoid, not obviously lobed; fruit a capsule
     26. Style branches 3; ovary with 3 locules .......... **Polemoniaceae**
     26' Style either unbranched or with 2 branches; ovary with 1-2 locules
      27. Style 1, stigma capitate or minutely 2-lobed
       28. Stigma capitate, not evidently lobed .......... **Solanaceae**
       28' Stigma with 2 lobes
        29. Flowers axillary; leaves entire; ovules usually 4 per ovary .......... **Convolvulaceae**
        29' Flowers usually in a terminal raceme; leaves deeply lobed or dissected; ovules 6 or more per ovary .......... **Hydrophyllaceae**
      27' Style evidently 2-branched from near base to above middle
       30. Leaves deeply lobed to dissected .......... **Hydrophyllaceae**
       30' Leaves entire or at most toothed
        31. Flowers solitary in leaf axils; corolla 2-8 mm long, limb white .......... **Convolvulaceae**
        31' Flowers in a coiled raceme or spike; corolla 8-30 mm long, limb violet to blue .......... **Hydrophyllaceae (*Phacelia*)**

# Dicotyledonous Angiosperms

Annuals or perennials. Leaves with pinnate to palmate leaf venation. Vascular bundles in stem cross-section usually organized in a ring. Perianth parts, if present, usually in multiples of 4 or 5; stamens often in multiples of 4 or 5.

## ACERACEAE Juss.

### MAPLE FAMILY

Monoecious, dioecious, or polygamous shrubs and trees. Leaves opposite, simple to compound, often palmately veined or lobed, deciduous; stipules absent. Flowers bisexual or unisexual, radial, in corymbs or racemes. Sepals ± 5, free; petals 4-6, free, sometimes absent; stamens 5-12, often 8; pistil 1, ovary superior, often lobed, locules 2, placentation axile; styles 2, inserted between lobes. Fruit a samara. Two genera and ca. 115 species; temperate n hemisphere and tropical mountains.

### ACER L. MAPLE

Habit, leaves, and flowers as described for family. Fruit with each of 2, single-seeded lobes bearing an elongated, ribbed wing. Ca. 113 species; temperate n hemisphere.

**A. macrophyllum** Pursh BIG LEAF MAPLE Trees, 5-25 m tall; bark grayish. Leaves palmately lobed, 15-30 cm wide, upper surface dark green, shiny, glabrous, lower surface light green, puberulent to glabrous, lobes often 5, margins entire to toothed. Racemes pendulous, dense; flowers many. Fruit densely pubescent; wings 2-2.5 cm long. Feb-Apr. Occasional; streamside in canyons with permanent water; riparian woodland; 10-1500 ft (3-460 m). North side between Lady's Harbor and China Harbor area; locally abundant in Diablo Canyon. Alaska to n Baja CA. Sap is edible and was used by Native Americans to sweeten foods.

ACERACEAE

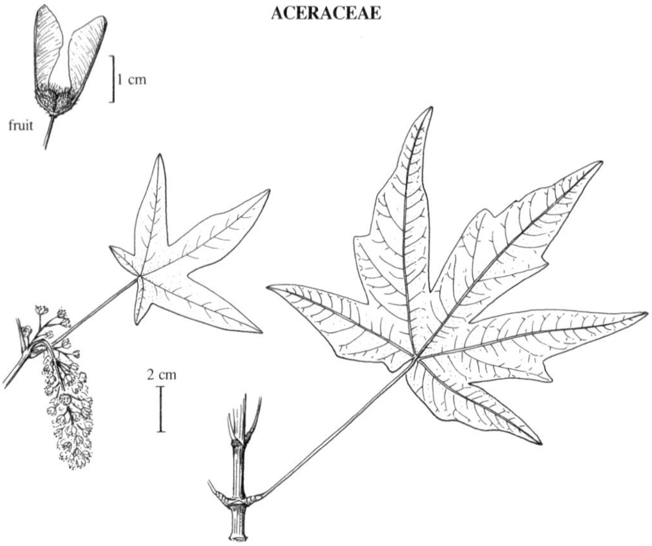


Acer macrophyllum

ADOXACEAE

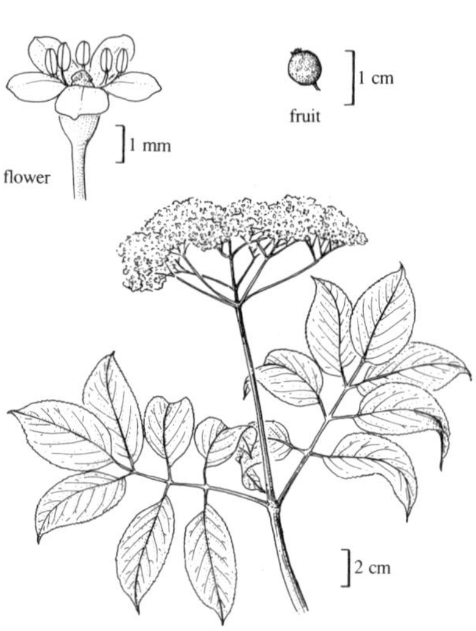


Sambucus mexicana

# **ADOXACEAE** Trautv.

## **ADOXA FAMILY**

Perennials. Leaves opposite, ternately to pinnately compound, deciduous; stipules present or absent. Flowers bisexual, radial, in racemes or panicles. Sepals 2-4, fused; petals 3-5, fused; stamens 5-10, alternate petals; pistil 1, ovary ± inferior, locules 2-5, ovules 1 per locule, pendulous; style branches 3-5. Fruit a berry or drupe. Three genera and ca. 25 species; temperate n hemisphere and subtropical mountains.

### **SAMBUCUS** L. ELDERBERRY

Shrubs or small trees. Leaves pinnately compound; leaflets serrate. Sepals 5; corolla rotate, petals 5; stamens 5; style short or absent. Fruit a berry. Ca. 22 species; temperate n hemisphere and subtropical mountains. Previously treated as a member of the Caprifoliaceae. Often cultivated for fruit used in pies, jellies, wine, and dyes.

**S. mexicana** C.Presl. Shrubs. Leaflets 3-7, elliptic-oblong to ovate, 2-15 cm long. Flowers in flat-topped panicles. Corollas white to cream. Fruit bluish, densely glaucous, dark blue beneath surface. Mar-Jun. Scarce; rocky slopes and canyon bottoms; chaparral, oak woodland; 50-900 ft (15-275 m). North side in ravine above Del Mar Cove, at Valdez Anchorage, Dick's Canyon, and at Prisoners' Harbor; Central Valley in Cañada Christy and Cañada de la Portezuela; s side in upper Pozo Canyon. Santa Rosa, Santa Catalina, and San Clemente islands; British Columbia to Mexico, e to UT and NM. Flowers and fruits are edible after being cooked; uncooked fruits may cause nausea. Although medicinal teas were reportedly made from flowers and roots by Native Americans, foliage and roots can be toxic to humans.

# **AIZOACEAE** Rudolphi

## **FIG-MARIGOLD FAMILY**

Annuals, perennial herbs, or shrubs. Leaves alternate to opposite, simple, often succulent, glabrous; stipules absent. Flowers bisexual, radial, in cymose clusters or solitary. Sepals 3-8; petals often many, free or fused at base, sometimes absent; stamens many; pistil 1, ovary superior to inferior, locules 1-20, placentation axile or parietal, styles 1-20. Fruit a capsule, berry, or nut-like. Ca. 155 genera and 2400 species; tropical and subtropical, mostly s hemisphere, especially s Africa. Many species cultivated as ornamentals.

1. Petals absent; fruit nut-like, sepals adnate to fruit and becoming horn-like .................................. **Tetragonia**

1' Petals present, many; fruit a berry or ± dehiscent capsule, sepals not becoming horn-like in fruit

  2. Perennials; leaves ± triangular in cross-section, glabrous .................................................. **Carpobrotus**

  2' Annuals; leaves cylindrical to flat, with conspicuous, shining, watery papillae .. **Mesembryanthemum**

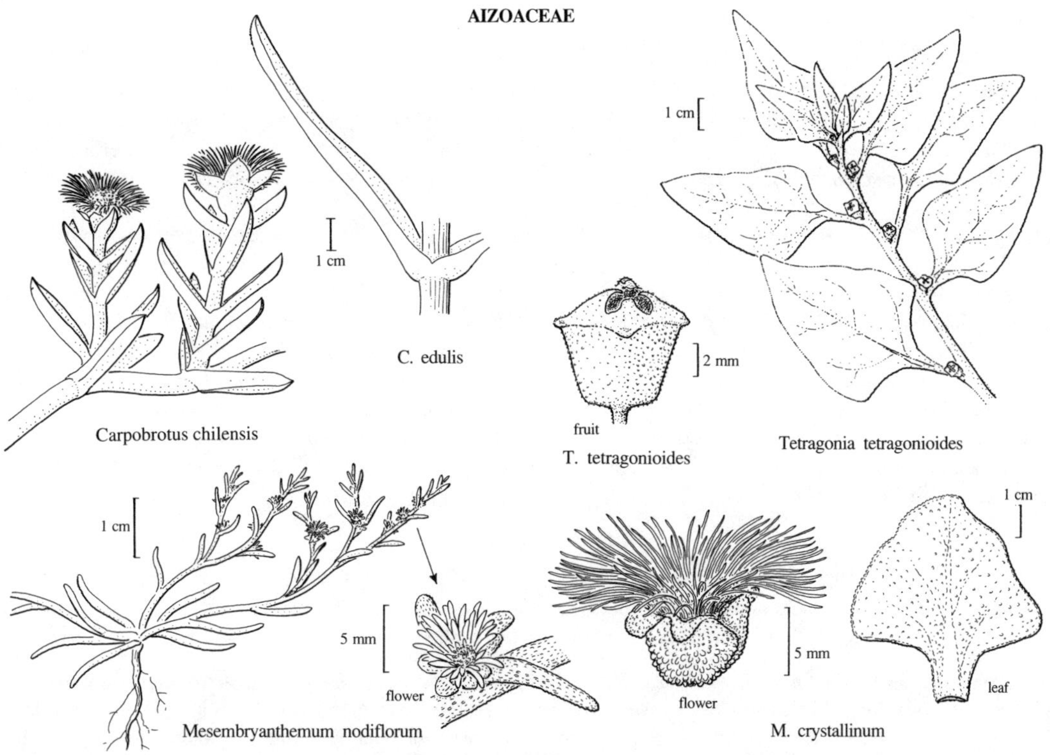


## **CARPOBROTUS** N.E.Br. FIG-MARIGOLD

Subshrubs; stems prostrate, trailing, often forming large mats. Leaves opposite, fused at base, terete to triangular in cross-section. Flowers solitary, terminal. Sepals 5, unequal; petals linear, brightly colored; ovary inferior, locules 8-20, stigmas 8-20, sessile. Fruit berry-like; seeds many. Ca. 30 species; temperate, s Africa, Chile, and Australia.

1. Petals rose-magenta; flower 3-5 cm wide; outer angle of leaf smooth ........................................ **C. chilensis**
1' Petals pink to yellow and aging pink; flower 8-10 cm wide; outer angle of leaf serrate near apex .. **C. edulis**

**C. chilensis** (Molina) N.E.Br. [*C. aequilaterus* (Haw.) N.E.Br. incorrectly applied to this species] SEA FIG Leaves straight, 3-5 cm long, outer angle smooth. Flowers 3-5 cm wide, sessile; sepals 1-2 cm long; petals 1-2 cm long, rose-magenta. Apr.-Sep. Introduced. Occasional; sea cliffs, rocky slopes, and sandy sites; coastal scrub; 10-150 ft (3-45 m). Bluffs on n side between West Point and vicinity of Prisoners' Harbor; s side on coastal flats and bluffs between Forney's Cove and Near Point. San Miguel, Santa Rosa, Anacapa, San Nicolas, Santa Catalina, and San Clemente islands; Los Coronados and Todos Santos islands; OR to Mexico; native of s Africa. Greene (1887) reported that this species was common on the n side of the island in "masses on high rocks overhanging the sea". Leaves and fruits are edible.

**C. edulis** (L.) N.E.Br. HOTTENTOT FIG Leaves ± curved, 7-10 cm long, outer angle serrate. Flowers 8-10 cm wide, with a peduncle; sepals 3-4 cm long; petals 3-4 cm long, yellow or pink. Introduced. Apr-Oct. Rare, persisting where planted; coastal scrub; 10-100 ft (3-30 m). Bluffs at Pelican Bay and around buildings at Scorpion Ranch and Valley Anchorage. Santa Rosa, Anacapa, San Nicolas, and San Clemente islands; OR to Mexico, native of s Africa. First collected at Valley Anchorage in May 1979. Fruits are edible.

## **MESEMBRYANTHEMUM** L. ICEPLANT

Annuals or biennials; stems ascending to prostrate. Leaves alternate or opposite, cylindrical to flat. Flowers in small clusters or axillary and solitary. Sepals 4-6, unequal; petals linear, white; ovary inferior, locules 5-20, styles 5-20, sessile. Fruit a capsule; seeds many. Ca. 70 species; Mediterranean Europe, w Asia, and s Africa. Leaves of some species are cooked and eaten like Spinach.

1. Leaf flat, 2-20 cm long; flower 7-10 mm wide ........ **M. crystallinum**
1' Leaf cylindrical, 1-2 cm long; flower 4-5 mm wide ........ **M. nodiflorum**

**M. crystallinum** L. CRYSTALLINE ICEPLANT Plants prostrate; branches 2-6 dm long. Leaves spatulate to ovate, 2-10 cm long, flat. Flower 7-10 mm wide. Calyx tube campanulate, 8-12 mm long; petals 6-8 mm long, white to reddish. Mar-Oct. Introduced. Common; coastal flats and bluffs, especially in disturbed or eroded sites; coastal strand, coastal scrub, grassland; 10-400 ft (3-135 m). North side between Potato Harbor and San Pedro Point; w and sw portions from Fraser Point to Pozo, Johnson, and Willows Canyon drainages; vicinity of Smugglers Canyon. All CA Channel islands; all Baja CA islands; n CA to Mexico, e to AZ; native of s Africa. Greene (1887) reported that this species was "common at the west end, but not seen elsewhere". Stems, leaves, and fruits are edible.

**M. nodiflorum** L. SLENDER-LEAVED ICEPLANT Plants ± erect to procumbent; branches 0.5-2 dm long. Leaves linear, 1-2 cm long, ± terete. Flower 4-5 mm wide. Calyx tube turbinate, 4-5 mm long; petals ca. 4 mm long, white, aging yellowish. Mar-Nov. Introduced. Occasional; coastal flats and bluffs, especially in disturbed or eroded sites; coastal scrub, grassland; 10-650 ft (3-200 m). North side near West Point peninsula; w and sw portions from Fraser Point to Pozo and Johnson Canyon drainages. All CA Channel islands; all Baja CA islands except San Geronimo; San Francisco Bay region to Mexico, e to AZ; native of s Africa. First collected at Fraser Point in March 1950.

### TETRAGONIA L.

Annuals or perennial herbs. Stems erect to prostrate. Leaves alternate, flat. Flowers axillary, solitary or in clusters. Sepals 3-5, persistent and becoming horn-like in fruit; petals absent; ovary inferior, locules 3-9, stigmas 3-9. Fruit nut-like; seeds few to many. Ca. 50 species; s hemisphere.

**T. tetragonioides** (Pallas) Kuntze NEW ZEALAND SPINACH Annuals, 2-5 dm tall; stems spreading to erect. Leaves 2-5.5 cm long, blades broadly ovate to triangular, bases truncate, entire to undulate, fleshy, papillate. Flowers usually 1-3, 3.5-5 mm long. Apr-Sep. Introduced. Scarce; sandy areas; coastal strand; 10-40 ft (3-12 m). North side at Prisoners' Harbor and mouth of Scorpion Canyon; Central Valley at Christy Beach; s side at mouths of Cañada de los Sauces and Horqueta Canyon. San Miguel, Santa Rosa, Anacapa, and San Nicolas islands; n and s CA; native of s hemisphere. First collected at the mouth of Cañada de los Sauces in October 1958. Fresh or cooked leaves are edible and can be used as a substitute for cultivated spinach.

## AMARANTHACEAE Juss.

### AMARANTH FAMILY

Annuals, perennial herbs, or shrubs. Leaves alternate to opposite, simple, often entire; stipules absent. Flowers bisexual or unisexual, radial, in bracteate spikes or panicles. Sepals often 3-5, free, persistent, sometimes absent, usually scarious; petals absent; stamens 1-5, opposite sepals, sometimes alternating with staminodes; pistil 1, ovary superior, locule 1, placentation basal or pendulous from apex, styles 1-3. Fruit a utricle or circumscissile capsule; seed 1 to several. Ca. 70 genera and 800 species; warm temperate and tropical, worldwide.

### AMARANTHUS L. AMARANTH

Monoecious or dioecious annuals. Inflorescence a panicle or spike-like, dense, axillary or terminal; flowers subtended by scarious, often spinulose bracts. Sepals 3-5, scarious, fused at base, persistent in fruit, glabrous; styles slender, 2-3, papillate. Seed 1 per fruit. Ca. 60 species, worldwide. Some species cultivated as ornamentals or for food (leaves and protein-rich seeds).

1. Fruit a utricle, often ± inflated, ± fleshy ........ **A. deflexus**
1' Fruit a circumscissile capsule, not inflated, dry
    2. Inflorescence an elongated, terminal spike or panicle ........ **A. powellii**
    2' Inflorescence composed of axillary clusters
        3. Plant ± globose, stems ascending to erect; pistillate flowers with 3 sepals ........ **A. albus**
        3' Plant ± depressed, stems prostrate; pistillate flowers with 4-5 sepals ........ **A. blitoides**

**A. albus** L. TUMBLEWEED Plants ± globose; stems erect to ascending, 2-15 dm tall. Leaf blades elliptic to spatulate or oblanceolate, 1-6 cm long, 0.5-3 cm wide. Flowers in axillary clusters; bracts subulate, 2-4 times longer than sepals. Sepals 3, subulate, 1-1.5 mm long. Fruit 1.5-2 mm long, dehiscent; seeds 0.8-1 mm wide,

AMARANTHACEAE

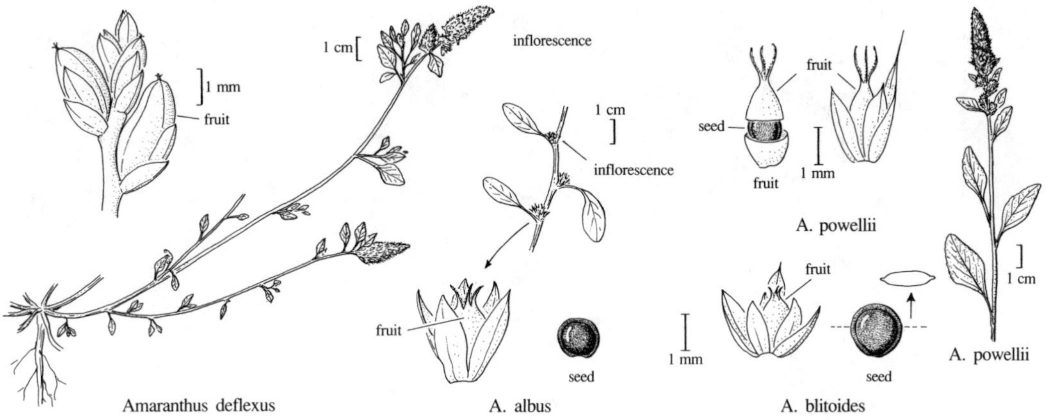


reddish brown. May-Oct. Introduced. Occasional; disturbed areas, gravel floodplains, and margins of seasonal ponds; grassland, riparian communities; 40-800 ft (12-245 m). North side in seasonal ponds near Campo del Norte and China Harbor; Central Valley. Santa Rosa, Anacapa, and Santa Catalina islands; N. America, Europe, and Asia; native of tropical America. First collected at Pelican Bay in June 1930. Leaves of young plants are edible.

**A. blitoides** S.Watson PROSTRATE PIGWEED Plants ± depressed, stems prostrate, 1-5 dm long. Leaf blades spatulate to obovate, 0.8-2.5 cm long, 0.2-1.2 cm wide. Flowers in dense, axillary clusters; bracts ± equal to sepals. Sepals 3-5, oblong, 2-3 mm long. Fruit 2-3 mm long, dehiscent; seeds ca. 1.5 mm wide, black. May-Nov. Scarce; disturbed areas, gravel floodplains, vernal ponds; riparian communities; 5-250 ft (2-75 m). North side in Cañada del Puerto, near Cañada del Agua and China Harbor; Central Valley near Stanton Ranch. Santa Rosa Island; N. America and Europe. Leaves and seeds were used for food by Native Americans.

**A. deflexus** L. LOW AMARANTH Stems ascending to decumbent, 1.5-3 dm long, with many branches. Leaf blades lanceolate to rhombic-ovate, 0.5-3.5 cm long, 0.3-2 cm wide. Flowers in terminal spikes; bracts ovate, ± equal to sepals. Sepals 2-3, oblong, ca. 2 mm long. Fruit 2-3 mm long, indehiscent; seeds ca. 1 mm wide, dark reddish brown. May-Nov. Introduced. Rare; disturbed areas; 100-250 ft (30-75 m). Central Valley at Christy Ranch, Christy Barn, and Stanton Ranch. San Francisco Bay region to s CA; native of Europe. First collected at Stanton Ranch in April 1960.

**A. powellii** S.Watson Stems erect to ascending, 2-10 dm long. Leaf blades ovate to oblong-ovate, 3-10 cm long, 1-5 cm wide. Inflorescence an elongated, terminal spike or panicle; bracts ovate to subulate, 2-3 times as long as sepals. Sepals 4-5, linear-oblong, 2-3.5 mm long. Fruit dehiscent, 2.5-3.5 mm long; seeds 1-1.5 mm wide, black. Jun-Nov. Rare; disturbed areas, margins of seasonal ponds; coastal scrub, grassland; 200-575 ft (60-175 m). North side near head of w fork of Cañada del Agua and near Campo del Norte; Central Valley at Stanton Ranch. Anacapa Island; WA to Mexico and S.America, e to Great Plains. Seeds were eaten by Native Americans.

## ANACARDIACEAE Lindl.

### SUMAC FAMILY

Dioecious or polygamous shrubs and trees. Leaves alternate, simple to compound, deciduous or evergreen; stipules absent. Flowers bisexual or unisexual, radial, in racemes or panicles. Sepals 5, fused at base; petals 5, free; stamens 5 or 10; pistil 1, ovary superior, locule often 1, styles 1-3. Fruit drupe-like, glabrous, glandular, or puberulent, pulp resinous; seed often 1. Ca. 75 genera and 850 species; temperate to tropical, worldwide. Some species used for wood and food (cashew, mango, pistachio). Contact with sap of several species can cause severe dermatitis in humans.

1. Leaves simple ........ **Rhus**

1' Leaves compound, leaflets 3 to many

  2. Leaflets 15-many, lanceolate to linear-lanceolate; stamens 10; fruit pink to red ........ **Schinus**

  2' Leaflets 3-5, oblong to ± orbicular; stamens 5; fruit whitish ........ **Toxicodendron**

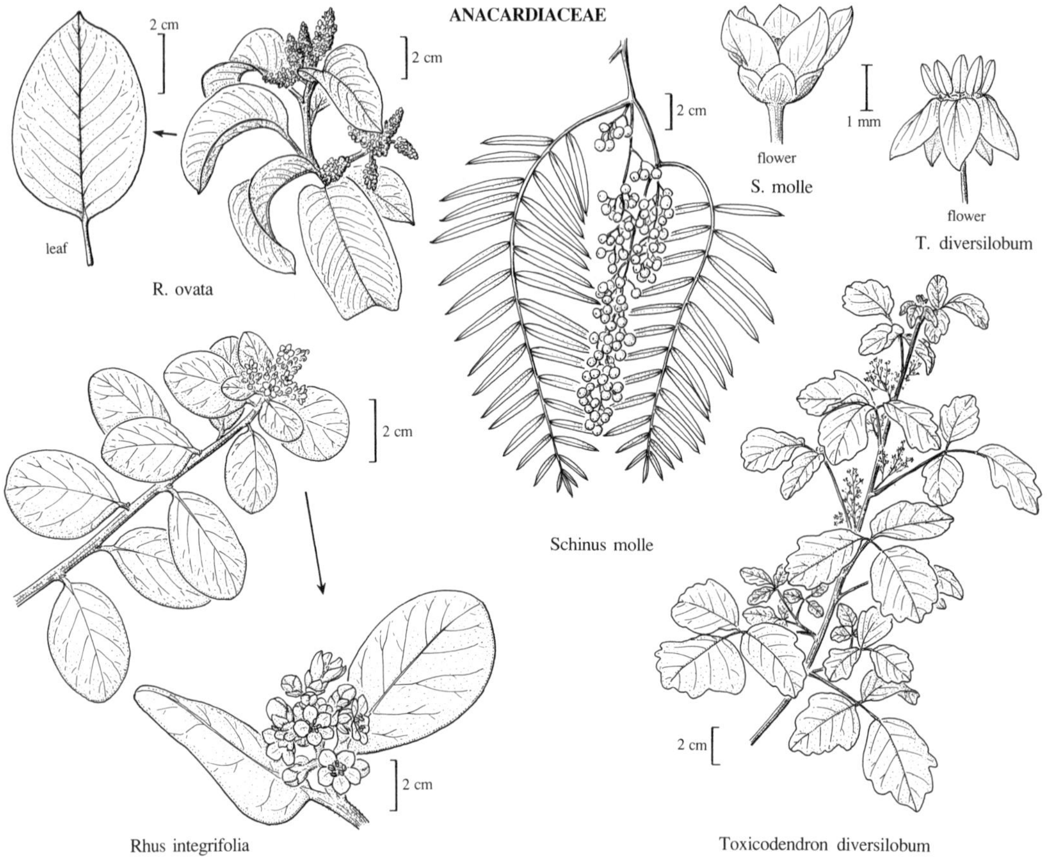


## **RHUS** L. SUMAC

Shrubs or trees, usually dioecious, sometimes polygamous. Leaves simple or compound. Flowers bisexual or unisexual, often sessile, in terminal panicles. Stamens 5; styles 3, free to partly fused. Fruit glandular-puberulent, reddish. Ca. 200 species; temperate, mostly N. and C. America. Some species are natural sources of lacquer, dyes, and tannins; others are used for wood or in basketry.

1. Leaf blades flat, petioles 2-7 mm long; sepals greenish, margins usually glandular-ciliate .... **R. integrifolia**
1' Leaf blades folded along midrib, petioles 10-30 mm long; sepals reddish, margins ciliate but not glandular ........ **R. ovata**

**R. integrifolia** (Nutt. in Torr. & A.Gray) Brewer & S.Watson LEMONADE BERRY Shrubs or small trees. Leaves simple, ovate-elliptic to elliptic-lanceolate, 3-6 cm long, 2-3.5 cm wide, flat, margins entire to serrate, apices and bases rounded-obtuse; petioles 3-7 mm long. Fruit ca. 1 cm wide. Feb-Mar. Common; sea cliffs, rocky slopes, and ridgetops; chaparral, coastal scrub, pine forest; 50-1350 ft (15-410 m). Scattered locations throughout island. All CA Channel islands except San Nicolas and Santa Barbara; Los Coronados, Todos Santos, Guadalupe, and Cedros islands; coastal s CA to n Baja CA. Hybrids between *R. integrifolia* and *R. ovata* have been called *R. i.* var. *traskiae* Barkley, can be recognized by intermediate leaf morphology, and are common on the island. Fruits were used by early settlers and Native Americans to make an acidic drink similar to lemonade.

**R. ovata** S.Watson SUGAR BUSH Shrubs or small trees, 2-10 m tall. Leaves simple, ovate-elliptic to broadly ovate, 4-11 cm long, 2.5-8 cm wide, folded along midrib, bases cuneate to obtuse, margins entire, apices acuminate-acute; petioles 10-30 mm long. Fruit 6-8 mm wide. Mar-Apr. Occasional; rocky slopes and ridgetops; chaparral, coastal scrub; 400-1450 ft (125-440 m). Central Valley between Cañada de la Portezuela and Islay Canyon; s side in Willows and Coches Prietos drainages. Santa Catalina Island; s CA to n Baja CA, e to AZ. Hybridizes with *R. integrifolia.*

## **SCHINUS** L. PEPPER TREE

Shrubs or trees, usually dioecious. Leaves pinnately compound, leaflets often many. Flowers unisexual, in panicles; pedicels short. Stamens 10 in 2 whorls of 5; styles 3, fused at base. Fruit leathery, shiny, reddish, glabrous. Ca. 30 species; tropical and subtropical S. America.

**S. molle** L. PERUVIAN PEPPER-TREE Evergreen trees, 5-15 m tall. Leaves 2-3 dm long, pendulous; leaflets 17 or more, linear-lanceolate, 2.5-6 cm long, usually less than 1 cm wide. Fruits 6-8 mm wide, reddish. Apr-Jun. Introduced. Occasional; s-facing slopes and disturbed sites; chaparral, coastal scrub; 50-700 ft (15-215 m). North side near Campo del Norte and Scorpion Ranch; Central Valley near Christy and Stanton ranches; s side in Laguna Canyon and Coches Prietos drainage. San Nicolas, Santa Catalina, and San Clemente islands; throughout CA; native of Peru. The fruits are toxic to mammals if ingested and are not the source of spice pepper.

## **TOXICODENDRON** Mill. POISON OAK or POISON IVY

Shrubs, trees, or vines, usually dioecious. Leaves ternate to pinnately compound, leaflets 3-9. Flowers unisexual, in axillary racemes or panicles pedicellate. Stamens 5; style 1, with 3 short stigmatic lobes. Fruit becoming membranous or leathery, shiny, white to brownish, glabrous. Ca. 6 species; N. and C. America, e Asia.

**T. diversilobum** (Torr. & A.Gray) Greene POISON OAK Erect shrubs, 1-3 m tall, sometimes viny and climbing. Leaves with 3(-5) leaflets; leaflets ovate to ± orbicular, 2-7 cm long, upper surfaces bright green, lower surfaces pale green, both surfaces becoming reddish in dry season, margins crenulate to lobed. Fruits 4-7 mm wide, whitish. Apr-May. Common; n -facing sea cliffs, slopes, and canyon bottoms; chaparral, coastal scrub, oak woodland, pine forest; 40-1000 ft (12-350 m). Scattered locations throughout island. All CA Channel islands except Santa Barbara; WA to Baja CA. Oils produced by this species can cause severe contact dermatitis in humans. Sap was used medicinally by Native Americans; black dye obtained from sap was also used in basketry.

# **APIACEAE** Lindl. [**Umbelliferae** Juss.]

## CARROT FAMILY

Annuals or perennial herbs, often aromatic; stems usually hollow. Leaves alternate or mostly basal, usually compound, petioles dilated or sheathing at base; stipules minute or absent. Flowers usually bisexual, radial, in simple or compound umbels. Sepals 5 and free, or absent; petals 5, free or absent; stamens 5; pistil 1, ovary inferior, locules 2, placentation axile; styles 2, often borne on an apical projection (stylopodium). Fruit a schizocarp composed of 2 mericarps, each often with 5 ribs; seeds 1 per locule. Ca. 420 genera and 3100 species; temperate and tropical mountains, ± worldwide. Many species are cultivated for food and culinary herbs (e.g., Anise, Caraway, Carrots, Celery, Coriander, Cumin, Dill, Fennel, and Parsley). Some species (e.g., Hemlock) are highly toxic.

1. Leaves simple, palmately lobed; fruit stellate-pubescent; stems decumbent to trailing .................... **Bowlesia**

1' Leaves compound to deeply dissected; fruit glabrous to variously pubescent, but trichomes not stellate; stems usually erect

  2. Ovary and fruit prickly, bristly, or hispid

    3. Stems and leaves glabrous; umbels simple, capitate .................................................... **Sanicula**

    3' Stems and leaves strigose to hispidulous; umbels ± compound, if simple then not capitate

      4. Umbels axillary, not subtended by bracts; ribs of fruit with minute prickles, tuberculate below middle .................................................................................................... **Torilis**

      4' Umbels terminal, subtended by leaf-like bracts; fruit wall minutely prickly or hispid, hispidulous between ribs, not tuberculate

        5. Umbels compound, with many flowers and fruits, peduncles of primary umbels ± equal, flowers ± sessile ............................................................................................ **Daucus**

        5' Umbels ± simple, with 1-10 flowers and fruits, pedicels unequal ........................ **Yabea**

  2' Ovary and fruit glabrous to minutely papillate in *Apiastrum*

    6. Plants ± acaulescent, leaves mostly basal; mericarps dorsally compressed, marginal ribs winged Lomatium

    6' Plants caulescent, leaves mostly cauline; mericarps ± cylindrical to laterally compressed

      7. Plants annual, less than 4.5 dm tall; fruit minutely papillate .............................. **Apiastrum**

      7' Plants perennial, greater than 5 dm tall; fruit glabrous

8. Leaves compound, leaflets lanceolate, oblong, ovate, or ± orbicular; plants of moist or wet habitats
   9. Basal leaves twice pinnately compound, leaflets lanceolate to ± orbicular, irregularly lobed or toothed ........ **Apium**
   9' Basal leaves once compound, leaflets oblong to ovate, regularly toothed ........ **Berula**

8' Leaves pinnately dissected, ultimate segments filiform or oblong in outline; plants of dry habitats
   10. Stems purple-spotted, leaves aromatic but not like anise; fruit ± globose, 2-3 mm long and wide ........ **Conium**
   10' Stems not purple-spotted, leaves with a strong anise odor; fruit oblong-ovoid, mericarps laterally compressed, 3-4.5 mm long ........ **Foeniculum**

## APIASTRUM Nutt.

With only 1 species.

**A. angustifolium** Nutt. WILD CELERY Annuals, 5-30 cm tall; stems erect, often branched from base. Leaves mostly cauline, alternate or opposite, 1-5 cm long, ternately dissected; ultimate leaf segments filiform to linear, 1-6 mm long. Umbels simple, without bracts. Sepals minute or absent; petals white. Fruit orbicular to ± cordate in outline, terete, 1-1.5 mm long, smooth to papillate, ribs obscure. Mar-May. Occasional; mostly s-facing slopes; chaparral, coastal scrub, grassland; 20-1300 ft (6-395 m). Widely scattered locations throughout island. All CA Channel islands; Los Coronados, Todos Santos, and Cedros islands; OR to Baja CA. This species was eaten by Native Americans.

## APIUM L.

Annuals or perennial herbs with rhizomes. Stems erect. Leaves cauline, 1-2 times pinnately compound. Umbel compound. Sepals minute or absent; petals white to greenish. Fruit elliptic to orbicular in outline, glabrous, mericarps ± laterally compressed, ribs subequal, filiform to corky. Ca. 20 species; temperate, ± worldwide.

**A. graveolens** L. CELERY Perennial herbs, 0.5-1.5 m tall. Leaves 5-30 cm long, twice pinnately compound; leaflets lanceolate to ± orbicular, 2-13 cm long, lobed to toothed; petioles 0.5-3 cm long. Fruit 1.5-2 mm long, glabrous, ribs slightly winged. May-Jun. Introduced. Scarce; in marshes and along creeks; 10-50 ft (3-15 m). North side at Prisoners' Harbor; s side at Valley Anchorage. Santa Rosa, San Nicolas, and San Clemente islands; Cedros Island; throughout much of N. America; native of Europe. Petioles are eaten as vegetable and seeds used for flavoring; leaves can accumulate toxic levels of nitrates. First collected in May 1962.

## BERULA Besser ex Koch WATER PARSNIP

With only 1 species.

**B. erecta** (Huds.) Coville Perennial herbs, 0.4-1.3 m tall, from short rhizomes and stolons. Leaves basal and cauline, 10-35 cm long, once pinnately compound, leaflets oblong to ovate, 1-4.5 cm long, toothed. Umbel compound. Sepals minute, persistent; petals white. Fruit 1.5-2 mm long, glabrous, with thick, ± corky walls, mericarps ± laterally compressed, ribs filiform, often inconspicuous. May-Jun. Scarce; in marshes and along creeks; 10-20 ft (3-6 m). North side at Prisoners' Harbor; Central Valley at mouth of Cañada Christy; s side at mouth of Cañada de Malva Real. San Miguel and San Nicolas islands; throughout much of N. America, Europe, and Asia. May be toxic to livestock during some seasons.

## BOWLESIA Ruiz & Pav.

Annuals or perennial herbs. Stems prostrate, decumbent, or erect. Leaves cauline, opposite to alternate, simple to compound. Umbel simple. Sepals absent or minute; petals white to purple. Fruit globose to ovoid, stellate-pubescent to ± glabrous, mericarps ± dorsally compressed, ribs inconspicuous. Ca. 15 species; temperate S. America, with 1 species in N. America.

**B. incana** Ruiz & Pav. Plants 4-50 cm; stems slender, usually decumbent or trailing. Leaves mostly opposite, 2-12 cm long, blade palmately lobed, lobes 3-5, 2-10 mm long, entire. Sepals minute or absent; petals pale yellow. Fruit ± inflated, globose, 1.5-2 mm long, stellate-pubescent, without ribs. Mar-May. Occasional; mostly shaded, grassy, n-facing slopes; 15-2200 ft (5-670 m). North side of island; Central Valley; s side in Smugglers' Canyon. Santa Rosa, Santa Catalina, and San Clemente islands; Cedros Island; coastal northern CA and Sierra Nevada foothills to S. America, e to TX.

APIACEAE: Apiastrum-Bowlesia

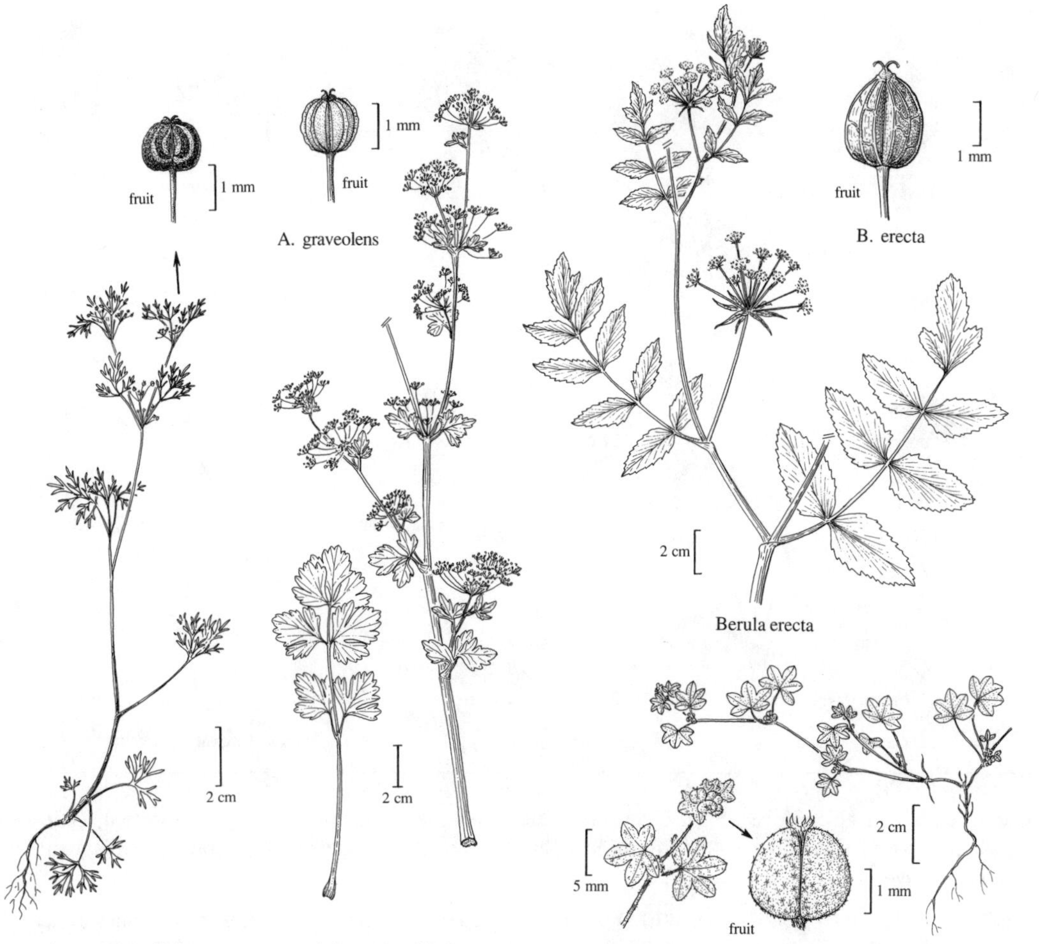


Apiastrum angustifolium Apium graveolens Bowlesia incana

## CONIUM L.

Short-lived perennial herbs. Stems erect. Leaves cauline, 1-3 times compound or pinnately dissected. Umbel compound. Sepals absent; petals white to yellow. Fruit ovate to orbicular in outline, glabrous, mericarps ± cylindrical, slightly compressed laterally, ribs subequal, curved. Ca. 6 species; Mediterranean Europe and s Africa.

**C. maculatum** L. POISON HEMLOCK Plants 1-3 m tall; stems erect, purple-spotted. Leaves 2-3 times pinnately dissected, lower ones to 75 cm long, upper ones 4-7 cm long, ultimate segments oblong to ovate, 2-5 mm long, entire to toothed. Petals white. Fruit ± globose, 2-3 mm long and wide. May-Jun. Introduced. Scarce; floodplains, disturbed flats, grassland, oak woodland; 50-800 ft (15-245 m). North side in Cañada del Puerto near Prisoners' Harbor; Central Valley just w of summit of Portezuela Grade and just w of University of California Field Station. San Nicolas Island; throughout much of North America; native of Europe. This species was observed on a streambank near cultivated ground by E. L. Greene in 1886 and already appeared to be well established (Greene 1887). An extremely toxic species, said to have been used to poison Socrates.

## DAUCUS L.

Annuals or short-lived perennial herbs. Stems often erect. Leaves cauline, pinnately dissected, hispidulous. Umbel compound, inner flowers of umbel radial, outer ones often bilateral. Sepals present or absent; petals

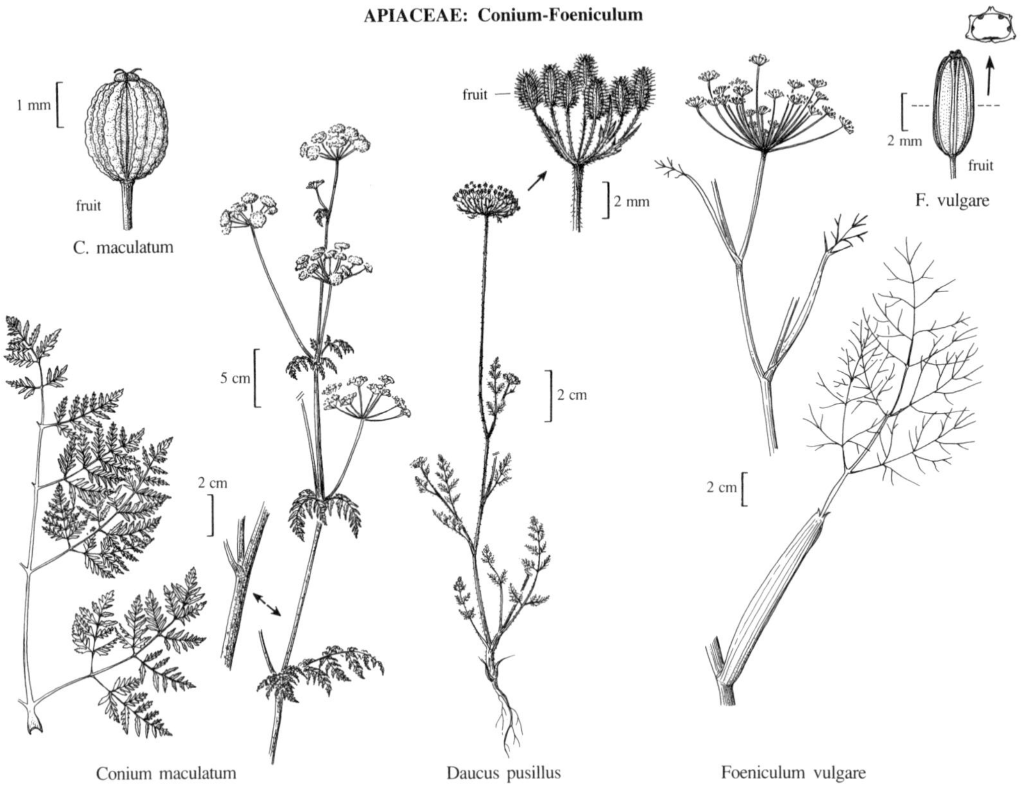


white, often with 2 apical lobes. Fruit oblong to ovate in outline, mericarps dorsally compressed, main ribs winged and bristly, alternating with filiform, hispidulous ribs. Ca. 20 species; temperate n hemisphere and Australia. *Daucus carota* L. is cultivated for its edible roots (carrots).

**D. pusillus** Michx. RATTLESNAKE WEED Annuals, 10-70 cm tall; stems erect. Leaves 1-4 times dissected, 1-15 cm long; ultimate segments filiform to oblong, 1-4 mm long, entire to minutely toothed. Fruit oblong in outline, 2-5 mm long; ribs with prickles. Mar-May. Common; grassy slopes, disturbed flats; coastal scrub, grassland; 20-1700 ft (6-520 m). Throughout island. All CA Channel islands; Los Coronados, Todos Santos, and Guadalupe islands; British Columbia to S. America, e to se U.S. Roots are edible either raw or cooked and were used for food by some Native Americans; this species also was used medicinally.

## FOENICULUM Mill. FENNEL

With only 1 species.

**F. vulgare** Mill. Perennial herbs, 1-2 m tall, stems with strong anise odor. Leaves basal and cauline, 2-4 times dissected, lower ones to 60 cm long, upper ones 10-34 cm; ultimate segments filiform, 5-40 mm long. Umbel compound. Sepals absent; petals yellow. Fruit oblong in outline, 2.5-4.5 mm long, mericarps ± dorsally compressed, ribs thick, prominent. Feb-Jun. Introduced. Abundant; gentle slopes, previously grazed pastures, disturbed sites; 10-1350 ft (3-410 m). Throughout island and rapidly spreading. All CA Channel islands except Anacapa and Santa Barbara; throughout N. and C. America; native of Europe. Greene (1887) reported that this species was "thoroughly established on hillsides near the landing of Prisoners' Harbor" in 1886. A noxious and persistent weed and long cultivated for its essential oils. The leaves, roots, and seeds are edible; seeds are used for seasoning in Italian dishes.

## LOMATIUM Raf.

Perennial herbs, often acaulescent, with vertical, ± slender taproots or rhizomes. Leaves ternately to pinnately dissected. Umbel compound. Sepals usually absent; petals white, yellow, or purple. Fruit linear to obovate in

**APIACEAE: Lomatium-Yabea**

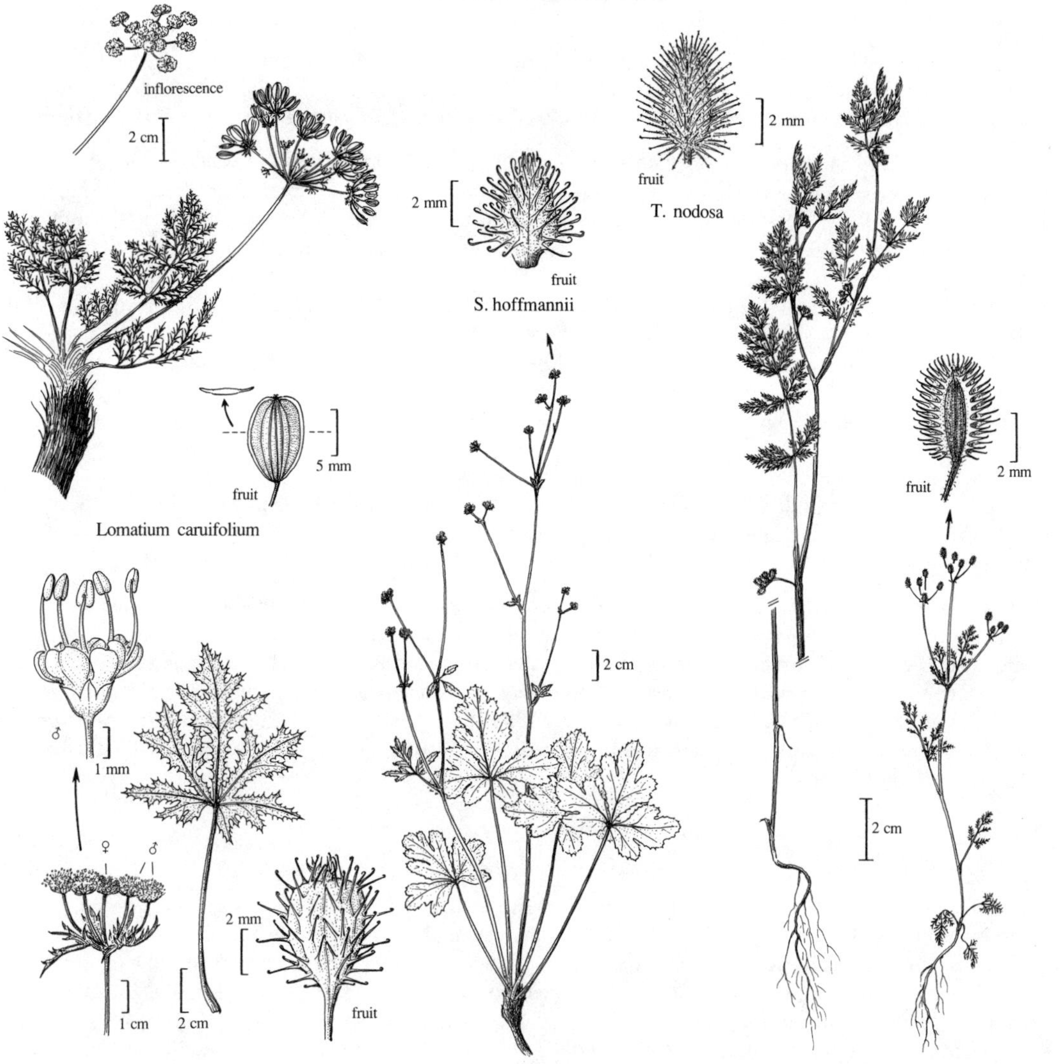


outline, glabrous, dorsally compressed, marginal ribs ± winged, central ribs filiform. Ca. 75 species; temperate N. America. Roots of many species were eaten by Native Americans.

**L. caruifolium** (Hook. & Arn.) J.M.Coult. & Rose CARAWAY-LEAF LOMATIUM Plants acaulescent, 1-5 dm tall. Leaves basal, 2-3 times ternately to pinnately compound, 9-17 cm long, glabrous to scaberulous, ultimate segments linear to oblong, 2-6 mm long. Fruit 5-13 mm long, mericarps oblong to ovate, 5-10 mm wide, wings 2-3 mm wide, ± thick. Mar-May. Common; grassy hillsides and gentle, n-facing slopes; coastal scrub, grassland; 20-500 ft (6-150 m). Widely scattered locations throughout island. San Miguel and Santa Rosa islands; n CA to Santa Barbara Co., e to Sierra Nevada foothills. Some plants have very short stems similar to that of the related species, *L. utriculatum* (Torr. & A.Gray) J.M.Coult. & Rose, but differ in all other diagnostic characters.

## **SANICULA** L. SNAKEROOT or SANICLE

Short- to long-lived perennial herbs, with rhizomes or thick, ± woody taproots, often polygamous (bisexual and staminate flowers). Leaves palmately compound or dissected, ± glabrous. Umbel compound. Sepals prominent, persistent; petals white, yellow, or purplish. Fruit orbicular to oblong in outline, usually prickly, sometimes scaly or tuberculate, mericarps ± cylindrical, slightly compressed laterally, ribs obscure to absent. Ca. 40 species; temperate, worldwide.

1. Leaves palmately to pinnately dissected into ± 7 main segments, segments deeply lobed, ultimate segments sharply toothed .......... **S. arguta**
1' Leaves palmately compound, leaflets mostly 3, often deeply lobed, margins serrate .......... **S. hoffmannii**

**S. arguta** Greene ex J.M.Coult. & Rose Plants 15-50 cm tall. Leaves palmately to ± pinnately compound, 10-35 cm long, ultimate segments sharply toothed. Some flowers with pedicels 1-4 mm long. Petals yellow. Fruit obovoid, 2.5-5 mm long. Feb-May. Common; mostly moist, north-facing slopes; chaparral, coastal scrub, oak woodland, pine forest; 50-1250 ft (15-380 m). Throughout island. All CA Channel islands except Santa Barbara; coastal Monterey Co. to n Baja CA.

**S. hoffmannii** (Munz) C.R.Bell Plants 25-90 cm tall. Leaves palmately compound, leaflets 3, sometimes 5, oval to obovate, 3-7 cm long, ± serrate. Flowers mostly subsessile. Petals greenish yellow. Fruit mostly ovoid, 3-5 mm long. Apr-May. Occasional; moist, shaded n-facing slopes or canyon bottoms; 100-925 ft (30-280 m). North side in Cañada del Puerto; Central Valley from Christy Pines to Stanton Ranch area; s side in Cañada de los Sauces and upper Willows Canyon. Santa Rosa Island; Santa Cruz Co. to Santa Barbara Co.

### **TORILIS** Adans.

Annuals. Leaves pinnately dissected, pubescent. Umbel compound. Sepals minute or absent; petals white to pink. Fruit oblong in outline to ovoid, prickly and often minutely tuberculate, mericarps ± cylindrical, slightly compressed laterally, ribs filiform. Ca. 12 species; Mediterranean Europe, e Asia, and Canary islands.

**T. nodosa** (L.) Gaertn. KNOTTED HEDGE-PARSLEY Plants 15-60 cm tall. Leaves 1-2 times pinnately dissected, 2.5-14 cm long, strigose to hispidulous; ultimate segments linear to lanceolate, 1-4 mm long. Umbels sessile to short-peduncled, usually on opposite side of stem from leaf. Fruits ovoid, 2.5-5 mm long, outer ones bristly, inner ones more tuberculate than prickly. Apr-May. Introduced. Occasional; mostly on disturbed slopes under oaks; grassland, oak woodland, pine forest; 15-1100 ft (5-335 m). Widely scattered locations throughout island. San Miguel, Santa Rosa, and Santa Catalina islands; OR to coastal s CA; native of Europe. First collected at Prisoners' Harbor in June 1930.

### **YABEA** Koso-Pol.

With only 1 species.

**Y. microcarpa** (Hook. & Arn.) Koso-Pol. [*Caucalis m.* Hook. & Arn.] Annuals; stems spreading or trailing, 5-20 cm long. Leaves 3-8 cm long, 1-2 times pinnately dissected, ultimate segments linear, 1-4 mm long, entire hispidulous. Umbels usually compound, with 3-12 flowers. Sepals present; petals white. Fruits oblong-ellipsoid, 2-3.5 mm long, mericarps subcylindrical, ribs ± winged, prickly, surfaces between ribs hispidulous. Mar-Apr. Occasional; mostly on rocky, n-facing slopes; chaparral, coastal scrub, oak woodland; 10-2350 ft (3-715 m). Widely scattered locations throughout w and central portions of island. Santa Catalina and San Clemente islands; WA to Baja CA, e to ID and AZ.

## **APOCYNACEAE** Juss.
### Dogbane Family

Annuals or perennials; sap opaque. Leaves alternate or opposite, simple, entire; stipules absent. Flowers bisexual, radial, in axillary or terminal, cymose clusters. Sepals 5, fused; petals 5, fused; stamens 5, adnate to corolla; pistils 2, sometimes fused at base, ovaries superior, each with 1 locule, styles and stigmas fused into 1 structure. Fruit a follicle, usually 2 per flower. Ca. 215 genera and 2100 species; mostly tropical, worldwide. Some species cultivated as ornamentals (e.g., Oleander, Frangipani); some species, including oleander, highly poisonous.

### **VINCA** L. PERIWINKLE

Perennial herbs with rhizomes; stems erect to trailing. Leaves ± opposite. Flowers solitary, axillary. Calyx lobes imbricated in bud, subequal; corolla funnelform; stamens attached in upper corolla tube; ovaries alternating with 2 prominent nectaries. Ca. 7 species; temperate to subtropical, Europe, c Asia, and Africa.

APOCYNACEAE

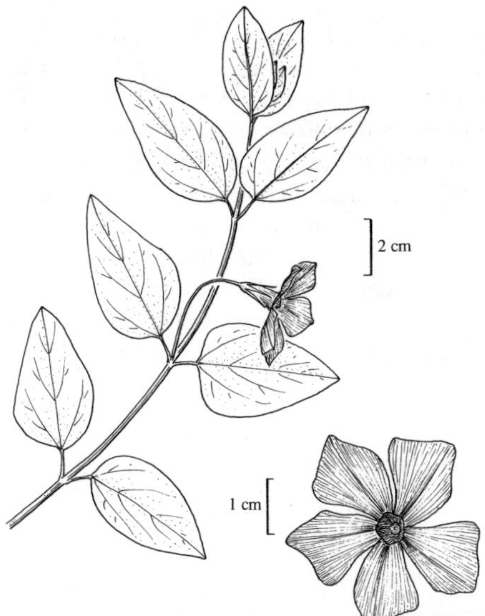


Vinca major

ASCLEPIADACEAE

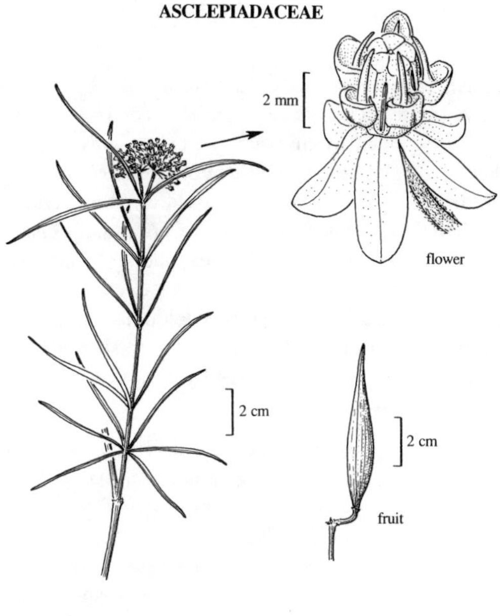


Asclepias fascicularis

**V. major** L. Plants to 3 dm tall. Stems sprawling, often forming large colonies. Leaves widely ovate, 3-8 cm long, dark green, apices acute, petioles 0.5-1 cm long. Pedicels 2.5-5 cm long; corolla blue to violet, tube 1.5-2 cm long, lobes flaring, 2-3.5 cm long. Mar-Jul. Introduced. Scarce; shaded sites, canyon bottoms; Eucalyptus forest, oak woodland; 20-250 ft (6-75 m). North side at Pelican Bay and Cañada del Puerto; Central Valley near Stanton Ranch. WA to s CA, e to se U.S.; native of Europe. First collected in March 1932.

## ASCLEPIADACEAE R.Br.
### MILKWEED FAMILY

Annuals or perennial herbs, sometimes vines; sap milky. Leaves opposite or whorled, simple, entire; stipules absent. Flowers bisexual, radial, in racemes or umbels. Sepals 5, often strongly reflexed; petals 5, fused; stamens 5, anthers adnate to style and stigma to form a columnar structure (gynostegium), pollen of each anther dispersed as 1 unit (pollinium); pistils 2, ovaries superior, each with 1 locule, styles and stigmas fused with anthers above ovaries. Fruit a follicle; seeds comose. Ca. 350 genera and 2850 species; mostly warm temperate and tropical, worldwide.

### ASCLEPIAS L. MILKWEED

Annuals or perennial herbs. Leaf blades linear to cordate. Inflorescence an umbel. Sepals reflexed; corolla tube short, with 5 lower and 5 upper lobes, lower lobes reflexed, often covering sepals, upper lobes erect, often hood-like, often with a horn-like projection, 5 erect lobes forming a corona; gynostegium drum-shaped, apex flat. Follicle usually 1 per flower, other aborted. Ca. 120 species; temperate, mostly N. America.

**A. fascicularis** Decne. in A.DC. NARROW-LEAF MILKWEED Perennial herbs, 0.4-1 m tall. Leaves whorled, 3-6 per node, linear to linear-lanceolate, 3.5-12 cm long. Corolla greenish white to pink, hood-like lobes white or greenish, horn-like projections exserted, incurved, white. Follicles 1-4 per umbel, 5.5-8 cm long, erect. Jun-Oct. Common; rocky slopes and ridges, dry stream beds, and canyon bottoms; 10-2000 ft (3-610 m). Widely scattered locations throughout island. Santa Catalina Island; WA s to Baja CA, e to ID and UT. This species is the food plant for Monarch Butterfly larvae and was an important source of fiber for Native Americans. Although entire plant was reportedly eaten by Native Americans, this species is extremely toxic to livestock.

# ASTERACEAE Dumort. [Compositae Giseke]

## SUNFLOWER FAMILY

Annuals or perennials, bisexual to dioecious. Leaves alternate, opposite, or whorled, simple to compound; stipules absent. Inflorescence a head, composed of 1-many flowers borne on flat to conic, naked, chaffy, or bristly receptacle, surrounded by involucre of bracts (phyllaries). Heads of 4 kinds: bilabiate heads composed entirely of bilabiate flowers; ligulate heads composed entirely of ligulate flowers; discoid heads composed entirely of disk flowers; and radiate heads composed of many central disk flowers and a single series of marginal ray flowers. Bilabiate flowers bisexual; disk flowers usually bisexual with tubular, radial corollas; ligulate flowers bisexual, corolla bilateral, tubular at base, expanding to strap-shaped, often 5-toothed limb; ray flowers often unisexual or sterile, corolla bilateral, tubular at base, expanding to strap-shaped, entire to 3-toothed limb. Calyx absent or highly modified into a pappus, deciduous or persistent in fruit, often composed of scales, bristles, or awns; corolla radial or bilateral, petals 5, fused; stamens 5, anthers fused into a tube surrounding style; pistil 1, ovary inferior, locule 1, ovule 1, basally attached, style branches 2. Fruit an achene. Ca. 1300 genera and 21,000 species; worldwide.

1. Heads with strongly bilabiate corollas, outer ± ligulate ........ **Acourtia**
1' Heads discoid, ligulate, or radiate, corollas radial to bilateral, but not strongly bilabiate
  2. Heads ligulate; plants with milky sap ........ **GROUP 1**
  2' Heads discoid or radiate; plants with clear sap
    3. Heads discoid; corollas usually all tubular and radial
      4. Pappus absent or minute and inconspicuous ........ **GROUP 2**
      4' Pappus present, composed of awns, bristles, or scales ........ **GROUP 3**
    3' Heads radiate; both disk and ray corollas present
      5. Pappus absent or minute and inconspicuous ........ **GROUP 4**
      5' Pappus present, composed of awns, bristles, or scales
        6. Pappus of awns or scales ........ **GROUP 5**
        6' Pappus of bristles ........ **GROUP 6**

### GROUP 1. Heads composed of ligulate flowers; plants with milky sap.

1. Pappus composed of plumose bristles
  2. Flowers subtended by chaffy, scarious bracts; outer pappus bristles much shorter than inner ones ........ **Hypochaeris**
  2' Flowers not subtended by bracts; pappus bristles ± equal, in 1 series
    3. Achene without beak between fruit and pappus ........ **Stephanomeria**
    3' Achene with slender beak between fruit and pappus
      4. Leaves toothed to lobed; phyllaries in 2-3 series, subequal; corollas white ........ **Rafinesquia**
      4' Leaves entire; phyllaries in ± 1 series; corollas purple ........ **Tragopogon**
1' Pappus composed of awns, scales, or capillary bristles, or absent
  5. Pappus composed of scales, these sometimes with awned tips
    6. Corollas blue; pappus in 2-3 series, composed of minute, awnless scales ........ **Cichorium**
    6' Corollas white or yellow; pappus in 1 series, composed of ± 5 awn-tipped scales
      7. Pappus awns usually shorter than scales, glabrous ........ **Uropappus**
      7' Pappus awns equal to or longer than scales, ± barbellate
        8. Pappus scales 5-8 mm, deltate to ovate, tapered to awns; leaves all basal ........ **Microseris**
        8' Pappus scales less than 2 mm long, linear-lanceolate, awned from bifid apices; cauline leaves sometimes present ........ **Stebbinsoseris**
  5' Pappus composed of capillary bristles or absent
    9. Achene truncate at top, pappus attached directly to fruit
      10. Leaves clasping, bases often auriculate; achenes compressed ........ **Sonchus**
      10' Leaves sessile, bases neither auriculate nor clasping; achenes not compressed
        11. Leaves and lower stems tomentose; pappus bristles in 1 series, free to base ........ **Hieracium**
        11' Leaves and lower stems glabrous to pubescent but not tomentose; pappus bristles usually in 2 series, inner ones fused and deciduous as a unit ........ **Malacothrix**

9' Achene with slender beak between fruit and pappus
12. Achenes compressed .......... **Lactuca**
12' Achenes cylindrical to fusiform, not compressed
13. Cauline leaves present, hispid, some trichomes pustulate; pappus plumose **Picris**
13' Leaves mostly basal, glabrous to pubescent, but trichomes neither hispid nor pustulate; pappus of simple, smooth bristles
14. Achenes 10-ribbed or 10-veined, smooth throughout .......... **Agoseris**
14' Achenes 5-ribbed or 5-veined, upper half spinulose .......... **Taraxacum**

**GROUP 2. Heads discoid; corollas ± all tubular and radial; pappus absent or minute and inconspicuous.**

1. Flowers and heads unisexual; pistillate heads with bur-like or spiny involucre; anthers free
2. Staminate heads with phyllaries strongly fused, involucre cup-like; phyllaries of pistillate heads spiny, spines not hooked .......... **Ambrosia**
2' Staminate heads with phyllaries ± free to base; phyllaries of pistillate heads with hooked prickles .......... **Xanthium**
1' Flowers and heads bisexual; involucres not bur-like or spiny; anthers fused into a tube surrounding style
3. Leaves opposite
4. Perennials, glabrous; heads radiate, ± cylindrical to ovoid .......... **Jaumea**
4' Annuals, with woolly pubescence; heads discoid, subglobose .......... **Psilocarphus**
3' Leaves alternate
5. Leaves not lobed or dissected, margins entire
6. Phyllaries 4-6, equal, scarious; style ± lateral .......... **Micropus**
6' Phyllaries absent or 1-5, unequal, minute and scale-like; style apical .......... **Stylocline**
5' Leaves deeply lobed to dissected or margins at least coarsely toothed
7. Heads in racemes or panicles .......... **Artemisia**
7' Heads solitary or in ± flat-topped clusters
8. Receptacle flat to ± convex; outer achenes stalked, winged .......... **Cotula**
8' Receptacle hemispheric to cylndrical; achenes all sessile, wingless .......... **Chamomilla**

**GROUP 3. Heads discoid; corollas ± all tubular and radial, outer ones sometimes minutely ray-like; pappus present, composed of awns, bristles, or scales.**

1. Pappus of scales, these sometimes fused into a minute crown-like ring
2. Pappus a minute, crown-like ring; receptacles conic .......... **Chamomilla**
2' Pappus scales free; receptacles ± flat
3. Annuals; stems glabrous; leaves glandular .......... **Amblyopappus**
3' Shrubs; stems and leaves tomentose .......... **Eriophyllum**
1' Pappus of capillary or plumose bristles
4. Shrubs or subshrubs
5. Corollas white or cream, sometimes pink-tinged
6. Leaves deltate-ovate, base truncate to cordate; phyllaries prominently striate; involucre ± cylindrical .......... **Brickellia**
6' Leaves linear, lanceolate, oblanceolate, or ovate, base tapered; phyllaries not prominently striate; involucre ± hemispheric .......... **Baccharis**
5' Corollas yellow to orange
7. Leaves on flowering stems scale-like, 1-4 mm long .......... **Lepidospartum**
7' Leaves filiform, oblong, or obovate, not scale-like, often longer than 10 mm
8. Phyllaries in ± 2 series, inner ones much longer than outer ones; leaves filiform .......... **Senecio**
8' Phyllaries in many series, inner not much longer than outer ones; leaves oblong to obovate
9. Corollas cylindrical, tube tapering gradually to throat .......... **Hazardia**
9' Corolla tube expanded abruptly to much wider, cylindrical throat .......... **Isocoma**
4' Annuals, herbaceous perennials, or plants vine-like
10. Leaves with spiny teeth or lobes; receptacles bristly (also see *Centaurea*)
11. Corollas yellow; pappus in 2 series, inner series shorter than outer one, ciliate .......... **Cnicus**
11' Corollas white, lavender, or purple; pappus in 1 series, capillary or plumose

12. Pappus of plumose bristles; leaves green throughout ........ **Cirsium**
12' Pappus of barbellate bristles; leaves mottled with white ........ **Silybum**
10' Leaves entire to toothed or lobed, but not spiny; receptacles without bracts subtending disk flowers or with chaffy bracts near margin, bristly only in *Centaurea*
13. Phyllaries prominently spine-tipped ........ **Centaurea**
13' Phyllaries entire to laciniate, but not spine-tipped
14. Receptacle with chaffy bracts subtending or enfolding outer or marginal flowers
15. Outer achenes straight; style apical ........ **Filago**
15' Outer achenes gibbous to outside; style ± lateral ........ **Micropus**
14' Receptacle without bracts subtending disk flowers
16. Phyllaries ± herbaceous throughout, in 1-2 series, inner, if present, much longer than outer
17. Outer flowers pistillate; phyllaries ca. 13 per involucre ........ **Erechtites**
17' Outer flowers bisexual or, if pistillate, then ray-like and in 1 series; phyllaries ca. 8 or 21 per involucre ........ **Senecio**
16' Phyllaries hyaline or scarious to chartaceous, in several overlapping series, ± equal or inner ones slightly smaller than outer ones
18. Corollas purplish; leaves ± ovate ........ **Pluchea**
18' Corollas white; leaves linear, lanceolate, or oblanceolate
19. Stems glabrous, strigose, or hirsute ........ **Conyza**
19' Stems tomentose ........ **Gnaphalium**

**GROUP 4. Heads radiate; both disk and ray corollas present; pappus absent or minute and inconspicuous.**

1. Ray corollas bicolored, upper surface yellow, white, or pinkish, lower surface purple
2. Annuals; disk corollas yellow; phyllaries in 1 series ........ **Blennosperma**
2' Perennials; disk corollas purple; phyllaries in several series ........ **Osteospermum**
1' Ray corollas with same color on both surfaces, white, pink, or yellow
3. Ray corollas white to pink
4. Leaves ± palmately lobed or toothed; plants not strongly aromatic; receptacles without bracts subtending disk flowers ........ **Perityle**
4' Leaves simple to pinnately dissected; plants strongly aromatic; receptacles with ± chaffy bracts
5. Heads clustered in flat-topped panicles; disk corollas white to pink ........ **Achillea**
5' Heads ± solitary; disk corollas yellow ........ **Anthemis**
3' Ray corollas yellow
6. Receptacle without bracts subtending disk flowers
7. Leaves 1-2 times pinnatifid; phyllaries in several overlapping series ........ **Chrysanthemum**
7' Leaves simple, entire to toothed; phyllaries in 2 distinct series ........ **Venegasia**
6' Receptacle with chaffy bracts subtending flowers or in a ring between ray and disk flowers
8. Phyllaries in ± 1 series, enfolding ray achenes ........ **Madia**
8' Phyllaries in 2-3 series, outer ones not enfolding ray achenes
9. Phyllaries in 2 unequal series, outer spreading; disk achenes not enclosed by receptacular bracts; ray flowers fertile ........ **Coreopsis**
9' Phyllaries in 2-3 ± equal, overlapping series; disk achenes enclosed by receptacular bracts; ray flowers sterile ........ **Encelia**

**GROUP 5. Heads radiate; both disk and ray corollas present; pappus of awns or scales.**

1. Receptacle naked, disk flowers not subtended by bracts
2. Ray corollas white; leaves ± palmately veined ........ **Perityle**
2' Ray corollas yellow; leaves pinnately veined
3. Annuals; leaves opposite ........ **Lasthenia**
3' Perennials; leaves alternate
4. Phyllaries in 4-6 overlapping series; pappus of narrow awns, ± V-shaped in cross-section ........ **Grindelia**
4' Phyllaries in ± 1 series; pappus of flat scales ........ **Eriophyllum**

1' Receptacle bristly or with chaffy bracts subtending disk flowers
5. Phyllaries often in 2-3 series; inner phyllaries not enfolding ray achenes; disk flower pappus of 2 deciduous scales .......... **Helianthus**
5' Phyllaries in 1 series, partly or completely enfolding ray achenes; disk flower pappus of 5-many ± persistent scales
6. Phyllaries partly enfolding ray achenes .......... **Hemizonia**
6' Phyllaries completely and often tightly enfolding ray achenes
7. Pappus in 2 series of 5 each, scales conspicuously white, shiny .......... **Achyrachaena**
7' Pappus of 10-many, tan to white, ± linear, basally pubescent scales .......... **Layia**

**GROUP 6. Heads radiate; both disk and ray corollas present; pappus of bristles.**

1. Ray corollas white, pink, purple, or blue
2. Pappus of ray achenes absent or minute and inconspicuous .......... **Lessingia**
2' Pappus or ray achenes well developed, usually conspicuous
3. Phyllaries conspicuously imbricated, ± oblanceolate, each with a tan to whitish, chartaceous base and green apex .......... **Aster**
3' Phyllaries not conspicuously imbricated, linear, oblong, or narrowly lanceolate, with green midrib and ± scarious margins
4. Ray corollas ± equal to disk corollas, inconspicuous; annuals
5. Ray corollas pink to purple .......... **Aster subulatus** var. **ligulatus**
5' Ray corollas white .......... **Conyza**
4' Ray corollas much longer than disk corollas, conspicuous; perennials .......... **Erigeron**
1' Ray corollas yellow
6. Pappus of 1-6 awn-like bristles, ± V-shaped in cross-section; outer phyllaries ± recurved .... **Grindelia**
6' Pappus of many capillary bristles; outer phyllaries not recurved
7. Ray achenes enfolded by phyllaries; ray achenes lacking a pappus .......... **Layia**
7' Ray achenes not enfolded by phyllaries; ray achenes with a well developed pappus
8. Phyllaries in ± 2 series, inner ones much longer than outer ones .......... **Senecio**
8' Phyllaries in 3-6 ± overlapping series
9. Pappus in 2 series, outer of minute scales, much shorter than inner bristles; heads in ± open, flat-topped panicles .......... **Heterotheca**
9' Pappus in 1 series, of bristles; heads in narrow, elongated racemes or panicles .......... **Solidago**

## ACHILLEA L. YARROW

Aromatic perennial herbs. Leaves alternate, entire to pinnately dissected. Heads radiate, in flat-topped clusters; involucres campanulate to ovoid; phyllaries in 3-4 series, unequal, imbricated, receptacles convex, disk flowers subtended by chaffy scales. Ray flowers pistillate; disk flowers bisexual. Achenes compressed, with thick margins; pappus absent. Ca. 85 species; temperate, n hemisphere.

**A. millefolium** L. Plants with rhizomes; stems to 12 dm tall. Leaves lanceolate to linear-lanceolate in outline, 5-20 cm long, finely dissected into linear segments; basal leaves petiolate, cauline ones sessile. Involucres 4-6 mm long; phyllaries green, margins fimbriate, brownish. Rays flowers 3-8, white to pink; disk flowers 15-40. Feb-Jun. Occasional; mostly on grassy, n-facing slopes and coastal bluffs; chaparral, coastal scrub, grasslands, oak woodland, pine forest; 40-1250 ft (12-380 m). Scattered locations throughout island. All CA Channel islands; Cedros Island; N. America, Europe, and Asia. This species is widely used as a medicinal herb and contains essential oils and a bitter alkaloid (achilleine).

## ACHYRACHAENA Schauer BLOW WIVES

With only 1 species.

**A. mollis** Schauer Annuals. Leaves simple, linear, ± entire, sessile, lower ones opposite, upper ones alternate. Heads radiate, solitary; involucres cylindrical to narrowly campanulate; phyllaries 3-8, in ± 1 series, subequal, each tightly enfolding a ray achene; receptacular bracts in 1 series. Ray flowers pistillate, 3-8, yellow to reddish brown; disk flowers bisexual, 5-35, yellow. Achenes black, those of ray flowers smooth, lacking pappus, those of

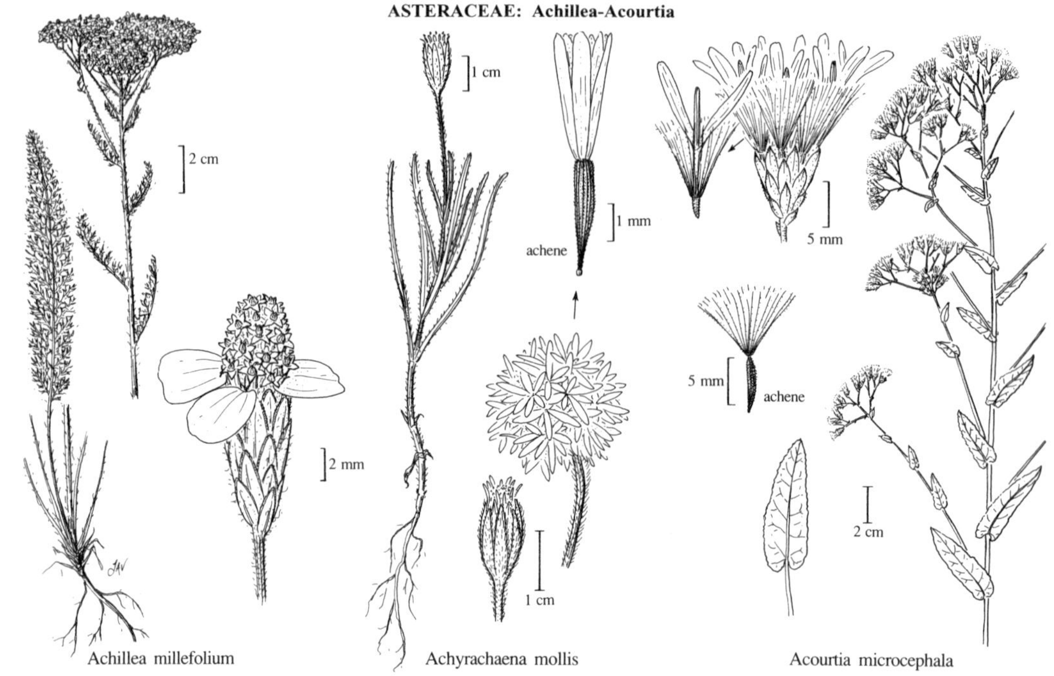


disk flowers 10-ribbed, with pappus of conspicuously white, shiny scales in 2 series of 5 each. Mar-May. Occasional; mostly on grassy, n-facing slopes; coastal scrub, grasslands; 200-1360 ft (60-415 m). North side between Prisoners' Harbor and Campo del Norte; Central Valley; s side in Cañada de los Sauces, near mouth of Willows Canyon, and along South Ridge. Santa Rosa, Santa Catalina, and San Clemente islands; s OR to n Baja CA. Achenes were eaten by Native Americans.

## ACOURTIA D.Don

Perennial herbs. Leaves alternate, entire, often clasping. Heads ± discoid but appearing radiate, solitary or in panicles; involucres cylindrical to campanulate; phyllaries in 3-5 series, unequal; receptacles naked. Corollas bilabiate, upper lip deeply 2-lobed, lower lip 3-toothed; anthers with tail-like, basal appendages. Achenes cylindrical to fusiform; pappus composed of many barbellate bristles. Ca. 40 species; N. America.

**A. microcephala** DC. [*Perezia m.* (DC.) A.Gray] SACAPELLOTE Stems to 1 m tall, striate, glandular. Leaves lanceolate, 3-20 cm long, gradually reduced upwards, margins denticulate, base cordate. Involucres cylindrical to turbinate, 8-12 mm long; phyllaries oblanceolate, short-acuminate to mucronate. Corollas lavender to white. Achenes 1.5-2 mm long, densely glandular. Apr-Aug. Occasional; grassy slopes and disturbed hillsides; chaparral, coastal scrub, grasslands, oak woodland; 100-950 ft (30-290 m). Widely scattered locations in w and central portions of island. Santa Rosa and Santa Catalina islands; Monterey Co. to n Baja CA. This species was used medicinally by Native Americans.

## AGOSERIS Raf. MOUNTAIN DANDELION

Acaulescent annuals or perennial herbs, with milky sap. Leaves in a basal rosette, entire to pinnatifid, linear to oblanceolate, sessile. Heads ligulate, solitary on naked or bracteate scapes; involucres cylindrical to hemispherical; phyllaries in 2-4 series, subequal, imbricated; receptacles naked. Flowers ligulate, bisexual, corollas yellow to reddish. Achenes cylindrical to fusiform, with long, slender beak between fruit and pappus; pappus composed of many fine, capillary bristles. Ca. 10 species; N. and S. America.

**A. grandiflora** (Nutt.) Greene Perennial herbs, 2-5 dm tall. Leaves variable, linear, lanceolate, or oblanceolate in outline, 7-25 cm long, toothed to irregularly lobed, glabrous to pubescent, apices acute to acuminate. Involucre 1.5-2.5 cm long, outer phyllaries broadly ovate, with white-ciliate margins, inner ones narrowly lanceolate, with scarious margins. Corolla ligules not exserted or slightly so. Achene (excluding beak) 3-5 mm long, glabrous,

ASTERACEAE: Agoseris-Anthemis

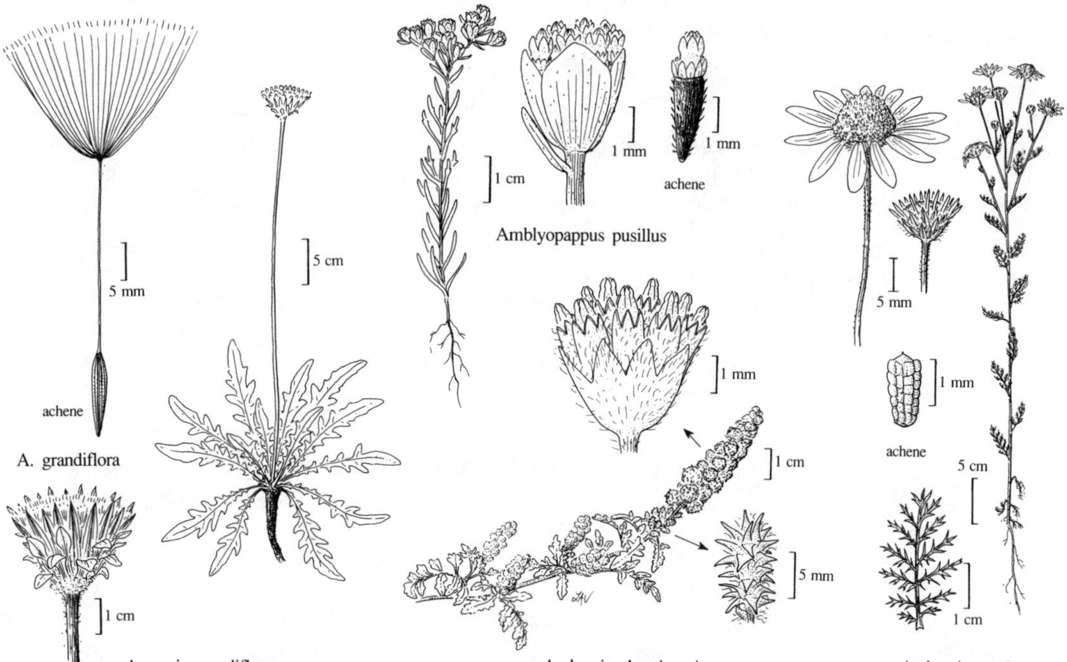


beak 2-4 times longer than achene. May-Jun. Scarce; grassy n-facing slopes, roadcuts, exposed ridgetops; grasslands, coastal scrub; 100-1350 ft (30-410 m). Cañada Christy and Cañada de la Portezuela in Central Valley; s side in Cañada de los Sauces. San Miguel and Santa Rosa islands; British Columbia to n Baja CA, e to ID and NV.

## **AMBLYOPAPPUS** Hook. & Arn.

With only 1 species.

**A. pusillus** Hook. & Arn. PINEAPPLE WEED Annuals. Leaves alternate, simple, linear to linear-oblanceolate, ± entire or pinnately lobed. Heads discoid or minutely radiate, in open, ± flat-topped to rounded panicles; involucres cylindrical to campanulate; phyllaries 3-8, in ± 1 series, ± equal, free; receptacles naked. Outer flowers ray-like, pistillate, ligules minute, 3-8, yellow to reddish brown; inner disk flowers bisexual, many, yellow. Achenes striate, puberulent, black; pappus composed of 7-12, obovate, chartaceous scales. Mar-Jul. Scarce; coastal flats, bluffs, dry streambanks; coastal scrub; 10-250 ft (3-75 m). Widely scattered locations around perimeter of island. All CA Channel islands; all Baja CA islands except San Geronimo; San Francisco Bay region to Baja CA and S. America.

## **AMBROSIA** L. RAGWEED or BUR-SAGE

Monoecious perennial herbs. Leaves pinnately lobed to divided, lower ones often opposite, upper ones alternate. Staminate heads many, in terminal spikes or racemes; pistillate heads 1-few, in axillary clusters; involucres of staminate heads cup-shaped, phyllaries fused; receptacles flat; involucres of pistillate heads bur-like, phyllaries fused, spiny, often enclosing 1 flower. Staminate flowers with tubular corolla, stamens free; pistillate flowers with corolla absent. Achene(s) enclosed within bur-like involucre; pappus absent. Ca. 24 species; mostly N. America. Pollen of some species causes severe allergies.

**A. chamissonis** (Less.) Greene SILVER BEACHBUR Plants often sprawling or procumbent, 1.5-4 dm tall, 1-3 m across, sparsely pubescent to silvery canescent. Leaves ovate to oblanceolate, often deeply lobed or dissected, blades 1.5-5 cm long, canescent; petioles to 3.5 cm long. Staminate involucres with 5-12 lobes; pistillate involucres 4-7 mm long, glandular-puberulent, spines ca. 2 mm long. Staminate corollas ca. 1.5 mm long. Achenes ovoid, rugose, glabrous. Mar-Oct. Occasional; sandy beaches, sand dunes; coastal strand, coastal dunes; 3-25 ft (1-8 m). Widely scattered locations around perimeter of island. All CA Channel islands except Santa Barbara; San Martin and Cedros islands; British Columbia to Baja CA and S. America.

## **ANTHEMIS** L. CHAMOMILE

Aromatic annuals. Leaves alternate, pinnately lobed to divided. Heads radiate, solitary; involucres hemispherical; phyllaries in 2-5 series, subequal, imbricated; receptacles ± conic, chaffy. Ray flowers pistillate or sterile, white to yellow; disk flowers bisexual. Achenes cylindrical, conspicuously ribbed; pappus a minute crown or absent. Ca. 200 species; temperate, Europe, w Asia, n Africa.

**A. cotula** L. MAYWEED Plants to 1-4 dm tall, often with many branches from base. Leaves 1,5-6 cm long, finely divided into linear segments, hirsute, glandular. Peduncles 2-6 cm long; involucres 4-6 mm long; phyllaries lanceolate, center green, edges straw-colored, hirsute, margins hyaline, apices acute. Ray corollas white; disk corollas yellow. Achenes ca. 1.5 mm long, with ca. 10 ribs, minutely tuberculate. Jun-Oct. Introduced. Rare; not seen recently; disturbed sites near marsh; 5-10 ft (2-3 m). North side at Pelican Bay and Prisoners' Harbor. Santa Rosa and Santa Catalina islands; Cedros Island; throughout N. America; native of Europe. First collected at Prisoners' Harbor in June 1930.

## **ARTEMISIA** L. SAGEBRUSH, WORMWOOD or MUGWORT

Aromatic annuals, perennial herbs, or shrubs. Leaves alternate, entire to pinnately dissected. Heads discoid, in axillary to terminal racemes or panicles; involucres hemispherical to subglobose; phyllaries in several series, imbricated, inner ones with scarious margins; receptacles flat to convex, naked. Outer flowers pistillate, often glandular; inner flowers bisexual, sometimes staminate. Achenes obovoid to fusiform, 3-5 ribbed; pappus a minute crown or absent. Ca. 300 species; mostly temperate, n hemisphere, S. America. Many species used for medicinal or herbal purposes.

1\. Leaves pinnately divided, segments linear; subshrub or shrub ................................................ **A. californica**
1' Leaves entire or apices 3-5-lobed; herbaceous perennial ...................................................... **A. douglasiana**

**A. californica** Less. COASTAL SAGEBRUSH Rounded, much branched shrubs, to 1.8 m tall, to 3 m wide, young stems pubescent, grayish to white. Leaves 1-2 times pinnately dissected, segments linear, upper leaves sometimes entire and in clusters. Involucres hemispheric or subglobose, 2-3 mm long. Pistillate flowers 6-10; bisexual flowers 15-30. May-Oct. Abundant; mostly on rocky, s-facing slopes; coastal scrub, grasslands; 20-1250 ft (6-380 m). Throughout island. All CA Channel islands; Los Coronados, Todos Santos, Guadalupe, and Cedros islands; San Francisco Bay region to Baja CA. This species was an important medicinal plant for Native Americans.

**A. douglasiana** Besser in Hook. MUGWORT Perennial herbs; stems 0.4-1.5 m tall, usually tomentose. Leaves entire or apices with 3-5 lobes, 3-13 cm long. Involucres campanulate, 3-5 mm long. Pistillate flowers 6-10; bisexual flowers 10-25. Aug-Nov. Occasional; moist canyons and floodplains; coastal scrub, riparian communities; 20-1400 ft (6-425 m). North side in canyons above Cueva Valdez and in Cañada del Puerto; Central Valley between lower Cañada Christy and Stanton Ranch. Santa Rosa and Santa Catalina islands; WA to Baja CA, e to Rocky Mountains. Achenes and peeled shoots were eaten by Native Americans; this species also was used medicinally. Leaves have reportedly been used to prevent dermatitis attributed to poison oak.

## **ASTER** L. ASTER

Annuals or perennial herbs, often with rhizomes. Leaves alternate, simple, entire to toothed. Heads radiate, solitary or in axillary to terminal, cymose clusters; involucres hemispherical to turbinate; phyllaries in several series, imbricated, outer ones usually ± oblanceolate, each with tan to whitish, chartaceous base and green apex, sometimes green throughout, margins scarious; receptacles flat to convex, naked. Ray flowers pistillate, corollas white to purple; disk flowers bisexual, yellow. Achenes obovoid to cylindrical, 4-5 veined; pappus composed of many, white to brownish bristles. Ca. 250 species; mostly temperate, n hemisphere.

1\. Annuals; phyllaries narrowly lanceolate, with a green midrib and ± scarious margins ..................................
........................................................................................................................ **A. subulatus** var. **ligulatus**
1' Perennial herbs; phyllaries ± oblanceolate, with tan to whitish, chartaceous bases and green apices
    2. Leaves broadly lanceolate to oblanceolate, sessile, margins entire to crenate ...................... **A. chilensis**
    2' Leaves broadly elliptic to ovate, lower ones petiolate, margins serrate ............................ **A. radulinus**

ASTERACEAE: Artemisia, Aster

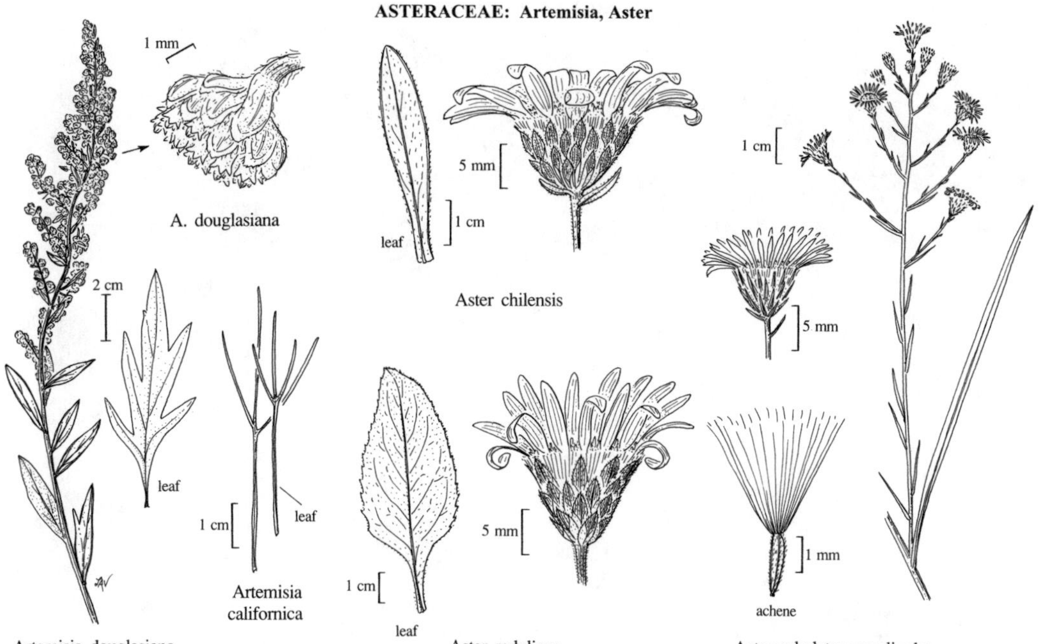


**A. chilensis** Nees COMMON CALIFORNIA ASTER Perennial herbs, 5-14 dm tall; stems ascending, villous. Leaves broadly lanceolate to oblanceolate, to 12 cm long, reduced upward, villous, sessile, margins entire to crenate. Heads many, 8-10 mm long; involucres broadly turbinate, ca. 6 mm long; phyllaries in 3-5 series, outer ones oblanceolate, green with brown midribs, inner ones lanceolate. Ray corollas to 12 mm long, pink or purplish. Jul-Sep. Rare; moist banks and seasonally wet depressions; coastal scrub, riparian communities; 20-100 ft (6-30 m). North side at mouth of Cañada del Agua and at water hole near China Harbor; Central Valley in lower Cañada Christy. Santa Rosa Island; British Columbia to s CA.

**A. radulinus** A.Gray ROUGH-LEAVED ASTER Perennial herbs, 3-5 dm tall; stems erect, glabrate. Leaves broadly elliptic to ovate, to 13 cm long, sparsely puberulent, lower ones petiolate, margins sharply serrate. Heads few, to 15 mm long; involucre broadly turbinate, 7-10 mm long; phyllaries in 4-5 series, outer ones ovate-elliptic, greenish with translucent midribs, inner ones oblong. Ray corollas to 13 mm long, white to pale lavender. Jul-Sep. Scarce; mostly on shaded n-facing slopes under *Quercus pacifica*; chaparral, oak woodland; 300-1350 ft (120-410 m). Central Valley between lower Cañada Christy and Rancho del Sur; s side in Cañada de los Sauces drainage. Santa Rosa Island; British Columbia to Santa Barbara Co. This species is rarely seen in flower on Santa Cruz Island; most reproduction is vegetative.

**A. subulatus** Michx. var. **ligulatus** Shinners [*A. exilis* Elliott] SLIM ASTER Annuals, 3-10 dm tall; stems erect glabrous. Leaves linear to oblanceolate, to 11 cm long, glabrous, ± sessile, margins entire. Heads many, ca. 8 mm long; involucre cylindrical to turbinate, ca. 6 mm long; phyllaries in 3-4 series, narrowly lanceolate, with green midribs and scarious margins. Ray corollas to 4 mm long, light pink or purple. Aug-Nov. Scarce; moist streamsides, floodplains; riparian communities; 20-400 ft (6-120 m). North side in Cañada del Agua and at water hole near China Harbor; Central Valley in lower Cañada Christy; s side in canyons between Fraser Point and Christy Ranch and in Cañada de Malva Real. Santa Rosa and Santa Catalina islands; n CA to Mexico, e to central N. America.

## BACCHARIS L.

Suffrutescent perennial herbs, shrubs, or subshrubs, dioecious. Leaves alternate, entire to toothed. Heads discoid, in axillary to terminal racemes or panicles, pistillate ones often larger than staminate ones; involucres hemispherical to turbinate; phyllaries in 3-6 series, imbricated, usually lanceolate, green, margins usually scarious and ciliate to fimbriate; receptacles naked or chaffy. Corollas white to pinkish; pistillate corollas filiform, staminate corollas ± salverform. Achenes cylindrical, 5-10 ribbed; pappus composed of many barbellate, capillary bristles. Ca. 350 species; N. and S. America. Leaves were used medicinally by Native Americans.

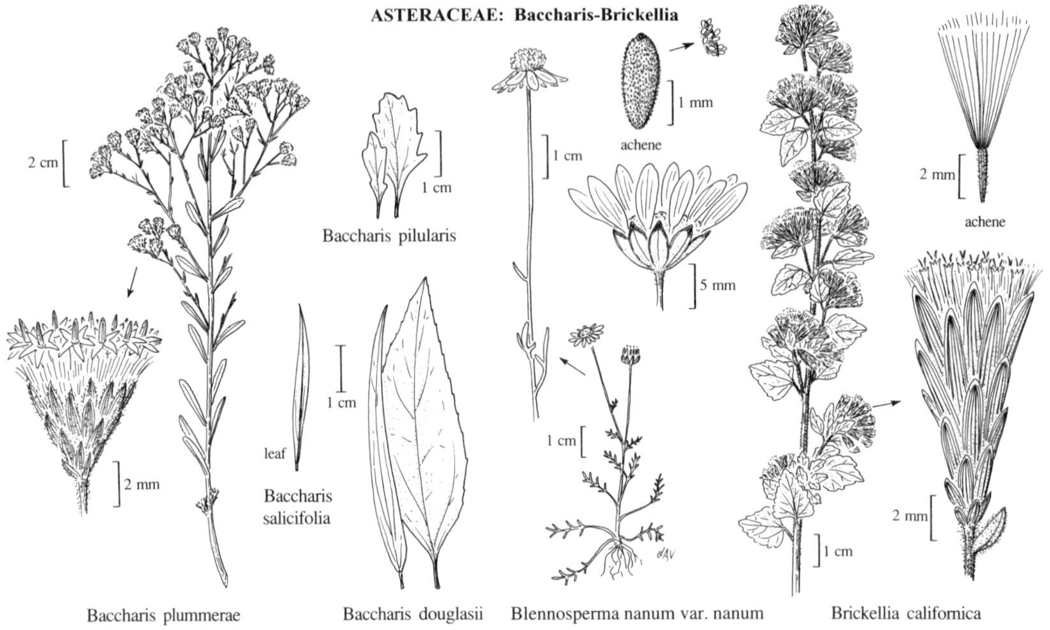


1. Shrubs, 1-4 m tall; achenes ± glabrous
   2. Leaves oblanceolate to obovate, 1-3 cm long; achenes with 8-10 ribs .................................. **B. pilularis**
   2' Leaves lanceolate, 2-14 cm long; achenes with 4-5 ribs ...................................................... **B. salicifolia**

1' Suffrutescent perennials or subshrubs, often less than 1.5 m tall; achenes puberulent or ± glandular
   3. Stems usually solitary, erect; stems and leaves glabrate, often densely glandular; achenes ± glandular ........................................................................................................................ **B. douglasii**
   3' Stems usually much branched; stems and leaves pubescent to tomentose; achenes puberulent ............. ........................................................................................................................ **B. plummerae**

**B. douglasii** DC. STICKY BACCHARIS Suffrutescent perennial herbs or subshrubs, 0.3-0.6 m tall; stems glandular, resinous. Leaves lanceolate, to 10 cm long, with short petioles, margins subentire. Heads in terminal, compound clusters; staminate and pistillate involucres 3-6 mm long. Achenes 0.5-1.5 mm long, ± glandular, with 4-5 ribs. May-Nov. Occasional; moist streamsides, seeps, marshes; riparian communities; 5-1250 ft (2-380 m). Widely scattered locations throughout island. San Miguel, Santa Rosa, and Santa Catalina islands; nw CA to s CA, e to Sierra Nevada foothills.

**B. pilularis** DC. [*B. p.* var. *consanguinea* (DC.) C.B.Wolf] COYOTE BRUSH Shrubs, 1-3.5 m tall; stems glandular, resinous. Leaves oblanceolate to obovate, 1-3 cm long, subsessile, margins toothed or upper half with short lobes. Heads in axillary or terminal panicles; staminate involucres 3-5 mm long, pistillate ones 4-6 mm long. Achenes 1-2 mm long, glabrous, with 8-10 ribs. Jul-Nov. Abundant; mostly on gentle slopes and floodplains; coastal scrub, grasslands; 20-1450 ft (6-440 m). Throughout island. All CA Channel islands; OR to n Baja CA. Stems were used by Native Americans for dwelling construction.

**B. plummerae** A.Gray subsp. **plummerae** PLUMMER'S BACCHARIS Suffrutescent perennial herbs, 0.5-1.5 m tall; stems pubescent. Leaves lanceolate to oblanceolate, to 5 cm long, sessile, distal margins often serrate. Heads in flat-topped panicles; staminate involucres 4-6 mm long, pistillate involucres 5-7 mm long. Achenes 3-4 mm long, ± glandular, puberulent, with 4-5 ribs. May-Oct. A species of special environmental concern. Occasional; mostly on n-facing slopes and bluffs; chaparral, coastal scrub, oak woodland; 15-1300 ft (5-395 m). Widely scattered locations throughout central portion of island, mostly on n side and in Central Valley. Anacapa Island; San Francisco Bay region to s CA mountains. Subspecies *glabrata* Hoover occurs in nw San Luis Obispo Co.

**B. salicifolia** (Ruiz & Pav.) Pers. [*B. glutinosa* Pers.] MULE FAT Shrubs, 2-4 m tall; stems glabrate, resinous. Leaves lanceolate, 2-14 cm long, subsessile, margins entire to shallowly toothed. Heads compound, in corymbose panicles, axillary or terminal; staminate and pistillate involucres 3.5-6 mm long. Achenes ca. 1 mm long, glabrous, with 4-5 ribs. Feb-Nov. Abundant; floodplains; riparian communities; 10-500 ft (3-150 m). Throughout island. All CA Channel islands except Santa Barbara; CA to S. America, e to UT and TX.

## **BLENNOSPERMA** Less.

Annuals, often branched from base. Leaves alternate, entire to pinnatifid. Heads radiate, solitary at ends of branches; involucres ± cylindrical; phyllaries in 1 series, ± equal; receptacles flat, naked. Corollas yellow, ligules sometimes purple below; ray flowers pistillate or absent; disk flowers staminate. Achenes obovoid, papillate; pappus absent. Three species; CA and Chile.

**B. nanum** (Hook.) S.F.Blake var. **nanum** Plants 6-20 cm tall, simple to diffusely branched; stems glabrous. Lower leaves pinnately divided into 5-10 linear to narrowly oblanceolate segments, 0.5-5 cm long, leaves reduced upwards. Involucres 4.5-6 mm long; phyllaries oblong to oblanceolate, apices reddish brown to purplish. Ray corollas oblong, lower surfaces purplish brown, upper surfaces yellow; disk corollas yellow. Achenes 2-3.5 mm long. Feb-Apr. Rare; open exposed slopes; coastal scrub, grasslands; 500-700 ft (150-215 m). Ridge ca. 0.8 miles (1.3 km) se of West Point, near Peak 724. Coastal n CA to s CA. Another variety occurs on the Point Reyes Peninsula, Marin Co. Achenes were eaten by Native Americans.

## **BRICKELLIA** Elliot BRICKELL BUSH

Shrubs or subshrubs. Leaves alternate or opposite, simple. Heads discoid, in axillary or terminal racemes or panicles; involucres campanulate to cylindrical; phyllaries in several series, linear to oblong, subequal, ± imbricated, striate; receptacles flat, naked. Flowers bisexual, usually white. Achenes cylindrical, 10-ribbed, pubescent; pappus composed of many barbellate, capillary bristles. Ca. 100 species; warm temperate N. & S. America.

**B. californica** (Torr. & A.Gray) A.Gray CALIFORNIA BONESET Shrubs, 6-13 dm tall; stems with many branches, puberulent or glabrate. Leaves deltoid to broadly ovate, 1-4 cm long, base obtuse to cordate, margins crenate to ± serrate, apices obtuse or acute. Involucres 9-11 mm long, cylindrical to narrowly turbinate; phyllaries in 3-5 series, striate, obtuse. Flowers 8-18. Achenes ca. 3 mm long, puberulent; pappus ca. 6 mm long. (Mar) Jul-Oct. Common; mostly on dry, rocky, s-facing slopes and gravel floodplains; chaparral, coastal scrub; 15-2000 ft (5-610 m). Throughout island on igneous substrates. Anacapa and Santa Catalina islands; WA to n Mexico, e to ID and TX. The flowers are very aromatic at night and are pollinated by moths. Leaves were used medicinally by Native Americans.

## **CENTAUREA** L. STAR THISTLE

Annuals or perennial herbs. Leaves alternate, lower ones pinnately lobed, upper ones entire to lobed, often sessile or decurrent. Heads discoid, solitary or in racemes or panicles; involucres ovoid to subglobose; phyllaries many, imbricated, spine-tipped; receptacles flat, with many long bristles. Outer flowers often sterile, corollas sometimes ray-like; inner flowers bisexual, corollas white, yellow, blue, or purple, anthers with tailed basal appendages. Achenes obovoid to broadly cylindrical, attached obliquely at base; pappus composed of stiff bristles or scales. Ca. 450 species; temperate, mostly Europe and Asia, 2 species native of N. America.

1. Apical spines of largest phyllaries 5-10 mm long .......... **C. melitensis**
1' Apical spines of largest phyllaries 10-25 mm long .......... **C. solstitialis**

**C. melitensis** L. TOCALOTE Stems erect, to 6 dm tall. Cauline leaves lanceolate to oblanceolate, decurrent, sessile. Involucres 10-15 mm long; phyllaries spinose-appendaged, each with 5-9 segments, apical spines 5-10 mm long, lateral ones shorter. Corollas yellow. Achenes ca. 2.5 mm long, brownish or mottled. Apr-Jun. Introduced. Abundant; disturbed areas, pastures, floodplains; coastal scrub, grasslands, riparian communities; 5-1900 ft (2-580 m). Throughout island. All CA Channel islands; Todos Santos, Guadalupe, and Cedros islands; throughout most of N. America; native of Europe. Greene (1887) reported that this species was "not at all prevalent" in 1886.

**C. solstitialis** L. YELLOW STAR THISTLE Stems erect, to 10 dm tall. Cauline leaves lanceolate to oblanceolate, very much reduced and scale-like. Involucres 10-15 mm long; phyllaries spinose-appendaged, each with ca. 7 segments, apical spines 10-25 mm long, lateral ones shorter. Corollas yellow. Achenes ca. 2.5 mm long, brownish or mottled. Jul-Dec. Introduced. Common; disturbed areas, pastures, floodplains; coastal scrub, grasslands, riparian communities; 10-800 ft (3-245 m). North side in Cañada del Puerto; Central Valley; s side in Cañada de los Sauces and Laguna Canyon drainages. Santa Catalina Island; throughout most of w N. America; native of Europe. First collected by Hoffmann from a population of "a few plants" at Stanton Ranch in July 1930. This species is toxic to horses.

## **CHAMOMILLA** Gray

Annuals, often aromatic. Leaves alternate, 1-2 times pinnatifid, petiolate or sessile. Heads discoid, solitary or 2-3 in terminal clusters; involucres broadly hemispheric; phyllaries in 2-3 series, subequal, imbricated, membranous, margins scarious; receptacles hemispheric to conic, naked. Flowers many, corolla often 4-lobed. Achene cylindrical, with 3-5 ribs on ± 1 side of achene; pappus absent or a minute, crown-like ring of fused scales. Ca. 5 species; Europe, Asia, and N. America.

**C. suaveolens** (Pursh) Rydb. [*Matricaria matricarioides* (Less.) Porter] PINEAPPLE WEED Plants 7-30 cm tall; stems often branched from base, pineapple-scented. Leaves 1-5 cm long, grayish green; segments linear, mucronate. Heads conic in outline, 7-10 mm long; phyllaries 3-5 mm long, spreading with age. Corollas 3-5 mm long. Achenes ca. 1 mm long, often with 2 reddish marginal ribs. May-Jun. Introduced. Occasional; roadsides, other disturbed areas; coastal scrub, grasslands, pine forest; 10-750 ft (3-230 m). North side at Prisoners' Harbor; Central Valley, especially near buildings and airfields; s side in Cañada de los Sauces. Santa Rosa, Anacapa, Santa Catalina islands; Guadalupe Island; much of CA e to Sierra Nevada foothills; native of nw North America and Asia. This species is substituted locally for *C. recutita* as a medicinal tea. First reported for Santa Cruz Island by Brandegee (1888).

## **CHRYSANTHEMUM** L.

Annuals. Leaves alternate, toothed to pinnately lobed. Heads radiate, solitary at ends of branches; involucres hemispherical; phyllaries many, in 2-3 series, margins hyaline to scarious; receptacles ± flat, naked. Ray flowers pistillate, corollas white or yellow; disk flowers bisexual, yellow. Achenes of ray flowers 3-angled, those of disk flowers cylindrical; pappus a minute crown or absent. Ca. 5 species; Mediterranean Europe and n Africa.

**C. coronarium** L. GARLAND or CROWN DAISY Plants 0.6-1 m tall; stems several, often branched from base. Leaves oblong to oblanceolate in outline, 2-6 cm long, segments linear to oblanceolate. Involucres 7-10 mm long. Ray corollas 11-15 mm long, broadly elliptic, yellow; disk corollas yellow. Ray achenes 2-3 mm long, with most angles slightly winged; disk achenes with only 1 rib prominently winged. Aug. Introduced. Rare; not seen recently; persisting where planted; 30-100 ft (10-30 m). Pelican Bay. San Nicolas and Santa Catalina islands; Todos Santos Island; c and s CA, nw Baja CA; native of Europe. First collected by M. Dunkle at Pelican Bay in August 1939; he considered his collection (8622) to have been planted and not yet naturalized. Leaves and tender stems are eaten as a vegetable in Asia.

## **CICHORIUM** L. CHICORY

Annual or perennial herbs, with milky sap. Leaves alternate, deeply toothed to pinnately lobed, sessile, often clasping, petiole often winged. Heads ligulate, often solitary, axillary or terminal; involucres cylindrical; phyllaries in 2 series, outer ones much shorter than inner ones, spreading, inner ones ± erect; receptacles flat to slightly convex, naked. Flowers ligulate, bisexual, corollas white or blue to purple. Achenes oblong, angled, truncate; pappus composed of many, minute scales in 2-3 series. Ca. 8 species; Europe and n Africa.

**C. intybus** L. Stems erect, to 1 m tall, with deep, woody taproot; stems branched, glabrous to hispid. Leaves oblanceolate to spatulate, basal and lower cauline to 20 cm long. Involucres 10-15 mm long; phyllaries with stalked glands, outer ones broadly lanceolate, acute to acuminate, inner ones narrowly lanceolate to lanceolate. Corollas blue, sometimes white, much longer than involucre. May-Jul. Introduced. Scarce; disturbed areas, roadsides, pastures; coastal scrub, grasslands; 10-250 ft (3-75 m). Central Valley between Christy Beach and Christy Ranch airfield. Throughout N. America; native of Europe. First collected near Christy Ranch in June 1930. The dried roots are used medicinally and as a substitute for coffee after being roasted. This plant is also grown as a vegetable and as a fodder plant for livestock.

## **CIRSIUM** Mill. THISTLE

Annuals or monocarpic perennial herbs. Leaves alternate, lobed to pinnatifid, spiny, sessile, clasping to decurrent. Heads discoid, solitary at ends of branches or axillary, sometimes sessile; involucres cylindrical to hemispherical; phyllaries many, imbricated, outer ones usually spine-tipped; receptacles flat to ± conic, densely bristly. Flowers bisexual, corollas radial to slightly bilateral. Achenes ovoid to obovoid; pappus composed of many plumose bristles, persistent or deciduous. Ca. 200 species; temperate N. America and Europe.

**ASTERACEAE: Centaurea-Cirsium**

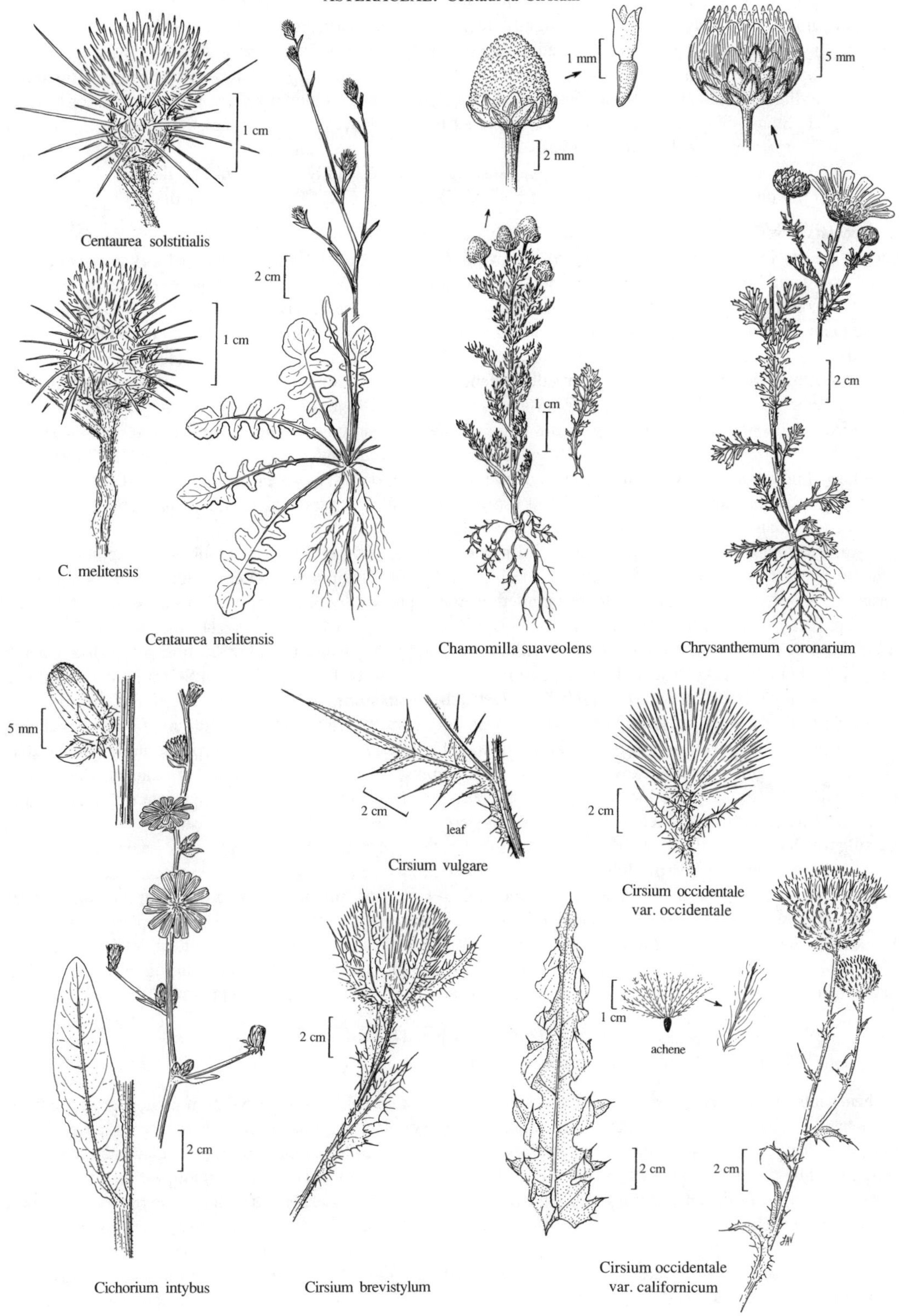

1. Stems appearing winged from long, decurrent, spiny leaf bases; upper surfaces of leaves hispid **C. vulgare**
1' Stems not winged, leaves clasping, to slightly decurrent; upper surfaces of leaves glabrate, pilose, or woolly
  2. Phyllaries S-shaped, bases appressed, ± recurved at middle, tips curved or hooked upwardly or inwardly ........ **C. occidentale** var. **californicum**
  2' Phyllaries not S-shaped, bases erect, spreading rigidly from middle or reflexed, tips straight or reflexed
    3. Stems thinly villous to glabrate; upper surfaces of leaves with scattered, crinkled trichomes; leaf margins densely spinose-ciliate ........ **C. brevistylum**
    3' Stems tomentose to woolly; upper surfaces of leaves thinly woolly, rarely glabrate; leaf margins with evenly spaced, short spines between lobes ........ **C. occidentale** var. **occidentale**

**C. brevistylum** Cronquist INDIAN THISTLE Stems to 2.5 m tall, thinly villous to glabrate. Leaves to 25 cm long, lobed to incised, clasping, upper surfaces with scattered, crinkled hairs, lower surfaces gray-canescent, spines to 20 mm long. Heads hemispheric, 3-5 cm long; involucres 2.5-4 cm long; phyllaries narrowly lanceolate or linear, erect, most with short, spinose tips. Corollas purplish or rose, rarely white. Apr-Jun. Scarce; sheltered slopes, canyon bottoms; chaparral, coastal scrub, pine forest; 50-1500 ft (15-460 m). Central Valley in upper Cañada Christy; s side in Cañada de los Sauces and on Sierra Blanca ridge. British Columbia to coastal s CA.

**C. occidentale** (Nutt.) Jeps. Stems 1-2 m tall, tomentose to woolly, sometimes glabrate with age. Leaves to 30 cm long, lobed to incised, clasping, upper surfaces sericeous, woolly to thinly woolly, lower surfaces grayish or white-canescent, spines 1-10 mm long. Heads campanulate to broadly hemispheric, 3-5 cm long; involucres 2-5 cm long; phyllaries linear to narrowly lanceolate, erect to curved upward, apices with spines to 7 mm long. Corollas white to reddish. Mar-Jul. Scarce to occasional; grassy slopes, pastures, canyon bottoms, stream banks; chaparral, coastal scrub, grasslands; 100-2100 ft (30-650 m). Widely scattered locations throughout island. All CA Channel islands except Anacapa and Santa Barbara; c and s CA.

var. **californicum** (A.Gray) D.J.Keil & Turner CALIFORNIA THISTLE Leaf lobes with spines to 3 mm long; upper surfaces of leaves thinly woolly to glabrate. Heads hemispheric, 3-4 cm long; phyllaries S-shaped, bases appressed, ± recurved at middle, tips curved or hooked upwardly or inwardly. Corollas white to lavender. Mar-Apr. Scarce; grassy slopes, pastures, stream banks; chaparral, coastal scrub, grasslands; 100-400 ft (30-125 m). North side near Scorpion Ranch; Central Valley e of Christy Ranch and at Valley Anchorage; s side at Coches Prietos Anchorage. Santa Rosa and Santa Catalina islands; Monterey Co. to s CA, e to Sierra Nevada foothills.

var. **occidentale** COBWEBBY THISTLE Leaf lobes with spines 8-10 mm long; upper surfaces of leaves sericeous to thinly woolly. Heads campanulate to broadly hemispheric; phyllaries with erect bases, spreading rigidly from middle or reflexed, tips straight or reflexed. Corollas usually reddish to purplish, rarely white. Mar-Jul. Occasional; grassy slopes, canyon bottoms, stream banks; chaparral, coastal scrub, grasslands; 100-2100 ft (30-650 m). Widely scattered locations in w and c parts of island. All CA Channel islands except Anacapa and Santa Barbara; c and s CA. Peeled roots and young stems were eaten by Native Americans.

**C. vulgare** (Savi) Ten. BULL THISTLE Stems 0.5-2 m tall, sparsely pubescent to villous. Leaves to 15-35 cm long, coarsely toothed to sharply lobed, decurrent, upper surface hispid, lower surface gray-canescent, spines to 15 mm long. Heads hemispheric to broadly cylindrical, 3-4 cm long; involucres 2-3.5 cm long; phyllaries linear to narrowly lanceolate, outer ones reflexed, middle ones squarrose, inner ones erect. Corollas purple. Aug-Sep. Introduced. Scarce; canyon bottoms, rocky stream banks; coastal scrub, riparian communities; 50-500 ft (15-150 m). Canyons in nw part of island. San Miguel, Santa Rosa, and Santa Catalina islands; throughout N. America, native of Europe. First collected in the first major drainage w of Profile Point in September 1989.

## **CNICUS** L. BLESSED THISTLE

With only 1 species.

**C. benedictus** L. Annuals, to 6 dm tall. Leaves alternate, oblong to oblanceolate, 5-30 cm long, coarsely toothed to pinnately lobed, bases decurrent or with short, winged petioles, margins spiny. Heads discoid, sessile, subtended by leafy, spiny bracts; involucres ± globose, 2-4 cm long; phyllaries imbricated, outer ones ovate, appressed, with spreading, spinose tips, inner ones lanceolate, with pinnately divided, spinulose tips; receptacles densely bristly. Corollas 1.5-2 cm long, yellow; outermost flowers sterile, inner ones bisexual, style with a pubescent ring below stigmatic lobes. Achenes ± cylindrical, ribbed, attached obliquely near base, apices toothed; pappus composed of awn-like bristles in 2 series, outer ones 9-10 mm long, inner ones 2-5 mm long, ciliate. May. Introduced. Rare; sandy creek banks; ca. 400 ft (120 m). Known from a single collection made in May 1962 in the Central Valley ca. 2 miles (3 km) w of Stanton Ranch. Throughout U.S; native of Europe. A medicinal tea can be made from this species and the seeds have been used as a source of oil.

## CONYZA Less.

Annuals or perennial herbs. Leaves alternate, linear to oblanceolate, entire or toothed. Heads discoid or minutely radiate, in terminal racemes or panicles; involucres hemispherical; phyllaries in 2-3 ± equal series, linear to lanceolate, subequal; receptacles naked. Corollas white to pinkish; ray flowers, when present, minute, pistillate; disk flowers either pistillate or bisexual, pistillate ones many, bisexual flowers few. Achenes ellipsoid, compressed, puberulent; pappus composed of capillary bristles. Ca. 50 species; temperate and subtropical, worldwide.

1. Outermost flowers ± tubular; involucres 4-5 mm long, 5-7 mm wide; phyllaries hirsute ....... **C. bonariensis**

1' Outermost flowers with minute ligules; involucres 3-4 mm long, 2.5-4 mm wide; phyllaries ± glabrous ...... ................................................................................................................................. **C. canadensis**

**C. bonariensis** (L.) Cronquist FLAX-LEAVED FLEABANE Annuals; stems erect, 3-17 dm tall, simple or with erect, leafy branches, densely strigose to hirsute, grayish green. Leaves narrowly lanceolate to oblanceolate, to 13 cm long, to 2.5 cm wide, lower ones toothed or shallowly lobed in upper half, upper ones subentire, densely strigose. Involucres 4-5 mm long, 5-7 mm wide; phyllaries hirsute, outer ones leaf-like, inner ones with scarious margins. Flowers 100-250, pistillate ones 100-180, staminate ones ca. 30. Achenes 1.2-1.7 mm long. Apr-Nov. Introduced. Occasional; disturbed areas, grassy slopes, canyon bottoms, and gravel flood plains; many plant communities; 5-1300 ft (2- 396 m). Widely scattered locations throughout island. All CA Channel islands except San Miguel; throughout N. America; native of S. America. First collected at Pelican Bay in June 1930.

**C. canadensis** (L.) Cronquist HORSEWEED Annuals; stems erect, 2-15 dm tall, simple or with lax to erect branches, sparsely strigose to glabrate, green. Leaves narrowly lanceolate to oblanceolate, to 9 cm long, to 1 cm wide, entire to shallowly lobed in upper half, sparsely strigose to glabrate. Involucres 3-4 mm long, 2.5-4 mm wide; phyllaries glabrous, with scarious margins. Flowers 40-60, pistillate ones 25-40, staminate ones 6-15. Achenes 1.2-1.5 mm long. Jul-Nov. Occasional; disturbed areas, grassy slopes, and gravel flood plains; many plant communities; 5-1550 ft (2-475 m). Widely scattered locations in w and central parts of island. All CA Channel islands; ± worldwide. Leaves were eaten by Native Americans; a medicinal tea also was made from this species.

## COREOPSIS L. TICKSEED

Annuals, perennial herbs, or shrubs. Leaves often basal and alternate, pinnately dissected. Heads radiate, solitary to many in terminal flat-topped racemes or panicles; involucres campanulate to hemispherical; phyllaries in ± 2 unequal series, outer ones herbaceous, spreading, inner ones erect, ± membranous; receptacles flat, paleaceous, paleae scarious. Ray flowers bisexual to sterile, corollas yellow; disk flowers bisexual, corollas yellow to brownish. Achenes ± ovoid, compressed with flattened sides parallel to plane of phyllaries, those of ray and disk flowers sometimes different in shape, size, and kind of pappus; pappus absent or of 2 awns or scales. Ca. 115 species; N. and S. America, Africa.

**C. gigantea** (Kellogg) H.M.Hall GIANT COREOPSIS Shrubs, to 1.5 m tall; stems succulent, glabrous. Leaves clustered at ends of branches, succulent, deciduous, ultimate leaf segments linear. Involucres 1.5-2.5 cm wide; phyllaries striate, outer ones 5-8, 1.2-2 cm long, inner ones 8-15, 1-1.5 cm long. Ray corollas to 3 cm long, yellow; disk corollas ca. 6 mm long, yellow or brownish. Achenes ca. 4 mm long, brown, glabrous. Feb-Apr (Sep). Common; rocky bluffs and canyon walls; chaparral, coastal scrub, riparian woodland; 5-1500 ft (2-460 m). Perimeter of island, especially on n side; widely scattered locations in interior part of island, mostly on igneous substrates. All CA Channel islands; Guadalupe Island; San Luis Obispo Co. to Los Angeles Co.

## COTULA L.

Annuals or perennial herbs, ± aromatic. Leaves alternate, entire to pinnately dissected. Heads discoid, solitary, axillary or at ends of branches; involucres broadly hemispheric; phyllaries in 2-5 series, margins scarious; receptacles flat to convex, naked. Outer flowers pistillate, corollas absent, short-pedicellate in fruit; inner flowers bisexual, subsessile, corollas 4-lobed, yellow. Achenes compressed, those of outermost flowers winged, those of inner flowers wingless or with narrow wing; pappus absent. Ca. 80 species; temperate and tropical, ± worldwide.

1. Annuals; leaves 1-2 times pinnately dissected ........................................................................... **C. australis**

1' Perennial herbs; leaves entire to pinnately lobed ............................................................ **C. coronopifolia**

**ASTERACEAE: Cnicus-Cotula**

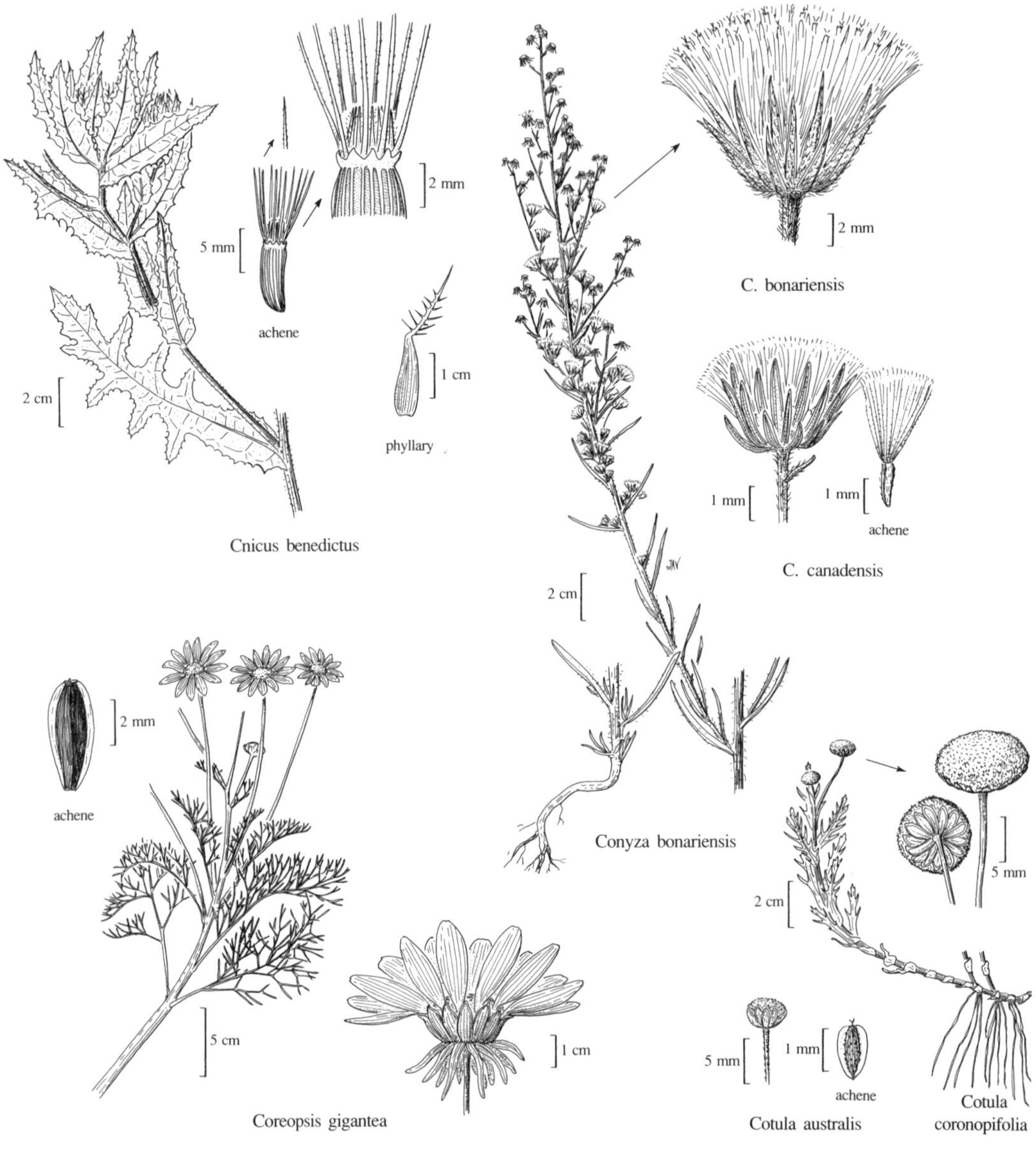


**C. australis** (Sieber) Hook.f. AUSTRALIAN BRASS BUTTONS Annuals; stems filiform, to 15 cm tall. Leaves 1-2 times pinnately dissected, to 3 cm long, segments linear or lanceolate. Heads 3-5 mm wide; phyllaries in 2 series, oblong to ovate, tips brown. Outer flowers in 2-3 series. Outer achenes ca. 1 mm long, slightly shorter than pedicels, winged, inner ones subsessile, with scarious margins. Feb-Jun (Oct). Introduced. Common; disturbed areas, grassy slopes, gravel flood plains, and moist canyon bottoms; coastal scrub, grasslands, oak woodland, pine forest; 20-1300 ft (6-395 m). Throughout island. Santa Rosa, Anacapa, Santa Barbara, and Santa Catalina islands; native of Australia. First collected at an unspecified location in March 1925.

**C. coronopifolia** L. BRASS BUTTONS Perennial herbs; stems fleshy, to 40 cm tall, rooting at nodes. Leaves entire to pinnately lobed, to 7 cm long. Heads 5-10 mm wide; phyllaries in 3-4 series, elliptic to broadly lanceolate, tips straw-colored. Outer flowers in 1 series. Outer achenes 1.5-2 mm long, equal to pedicels, broadly winged, inner ones subsessile, with scarious margins. Feb-Jun (Oct). Introduced. Occasional; moist flats, seasonal ponds, canyon bottoms, and marshes; coastal strand, riparian communities; 5-900 ft (2-275 m). Widely scattered localities throughout island. San Miguel, Santa Rosa, Anacapa, San Nicolas, and Santa Catalina islands; coastal n CA to s CA; native of s Africa. First collected at Prisoners' Harbor in June 1930.

## ENCELIA Adans.

Shrubs. Leaves alternate, entire to toothed. Heads radiate or discoid, solitary or in panicles; involucres hemispherical; phyllaries in 2-3 series, subequal, imbricated, outer ones foliaceous; receptacles convex, paleaceous, paleae chaffy, enfolding disk achenes. Ray flowers sterile, corollas yellow; disk flowers bisexual, corollas yellow to brownish. Achenes strongly compressed with edges facing phyllaries, flattened sides perpendicular to plane of phyllaries; pappus absent or of 2 scales. Thirteen species; w N. America and S. America.

**E. californica** Nutt. BUSH or CHAPARRAL SUNFLOWER Plants to 1 m tall; stems striate, pubescent. Leaves lanceolate to narrowly ovate, 2-6 cm long, 0.7-2.5 cm wide, puberulent; petioles 5-15 mm long. Involucres 15-25 mm wide; phyllaries 10-12 mm long, tomentose. Ray corollas 12-20 mm long; disk corollas 4-6 mm long, glandular, purplish. Achenes 4.5-6 mm long. Mar-Jun (Sep). Occasional; mostly on rocky s-facing slopes; coastal scrub, chaparral; 40-500 ft (12-150 m). North side from Scorpion Canyon area to San Pedro Point; Central Valley near Stanton Ranch airfield; s side from Coches Prietos area to San Pedro Point. Anacapa, Santa Barbara, Santa Catalina, and San Clemente islands; Los Coronados, Todos Santos, and San Martin islands; Monterey Co. to n Baja CA. The foliage of this species causes contact dermatitis in some people.

## ERECHTITES Raf.

Annual or perennial herbs. Leaves alternate, entire to pinnately dissected. Heads discoid, many, often in terminal, rounded to flat-topped panicles; involucres cylindrical; phyllaries in ± 1 equal series, linear to lanceolate; receptacles flat, naked. Corollas yellow; disk flowers either pistillate or bisexual, sometimes staminate, outer ones pistillate, inner ones bisexual. Achenes cylindrical, ribbed; pappus composed of many fine, capillary bristles. Ca. 50 species; temperate and subtropical, worldwide.

**E. glomerata** (Poir.) DC. [*E. arguta* (A.Rich.) DC.] AUSTRALASIAN FIREWEED Annuals; stems erect, to 2 m tall, tomentose to glabrate. Leaves lanceolate in outline, 6-15 cm long, toothed to pinnatifid, upper surfaces glabrate, lower surfaces woolly, lower leaves petiolate, upper ones sessile. Heads in flat-topped panicles; involucres 5-6 mm long; phyllaries ± tomentose. Corollas pale yellow. Achenes ca. 1.5 mm long, sparsely puberulent. Apr-Oct. Introduced. Scarce at present but spreading; disturbed areas and moist slopes; coastal scrub, pine forest, riparian communities; 70-1200 ft (20-365 m). Central Valley in upper reaches of Cañada Christy drainage; s side in Cañada de los Sauces, upper portion of Pozo Canyon, and on Sierra Blanca ridge. San Miguel Island; OR to Santa Barbara Co. First collected on San Miguel Island in May 1984 and on Santa Cruz Island on the e side of Sierra Blanca ridge in April 1989.

## ERIGERON L. FLEABANE

Annuals or perennial herbs. Leaves alternate, linear to oblanceolate, entire to subentire, sessile or tapered to petioles. Heads radiate, subscapose or in open panicles; involucres hemispheric; phyllaries in 1-4 series, subequal to evidently imbricate, narrowly lanceolate, herbaceous throughout or margins scarious; receptacles flat to slightly convex, naked. Ray flowers pistillate, purple or bluish; disk flowers bisexual, yellow. Achenes flattened, 2-nerved, villous; pappus composed of white to tan, barbellate, capillary bristles, in 1-2 series, outer series, if present, inconspicuous. Ca. 200 species; worldwide, but mostly N. America.

1. Leaves, especially basal ones, ± thick, fleshy, obovate; plants of coastal bluffs .......... **E. glaucus**
1' Leaves herbaceous, not fleshy, lower ones linear, linear-lanceolate, or oblanceolate; plants of the interior
  2. Phyllaries evidently graduated; heads often in open corymbs; leaves linear to linear-lanceolate .......... **E. foliosus**
  2' Phyllaries ± equal; heads 1-3 per stem; lower leaves oblanceolate .......... **E. sanctarum**

**E. foliosus** Nutt. var. **stenophyllus** (Nutt.) A.Gray Perennial herbs; stems erect, to 7(10) dm tall, 1-several with herbaceous base, simple, striate, glabrous to sparsely strigose. Leaves linear to narrowly oblanceolate, to 6 cm long, 1-5 mm wide, strigose. Involucres 4-7 mm long, to 15 mm wide; phyllaries graduated, hirsute, outer ones herbaceous, inner ones scarious-margined. Ray flowers 20-60, purple or bluish. Pappus usually in 2 series, outer series composed of few inconspicuous setae. Mar-Sep. Occasional; rocky slopes, open and wooded flats; chaparral, coastal scrub, grassland, oak woodland; 30-2000 ft (10-610 m). Throughout much of the central and e portion of island. San Miguel, Santa Rosa, Anacapa, and Santa Catalina islands; sw OR to n Baja CA. Plants with narrowly oblanceolate leaves 3-5 mm wide, known from Pelican Bay, Christy Canyon, and Cañada de los Sauces, belong to var. *foliosus*. Although the 2 varieties are apparently distinct on the island, they intergrade extensively on the mainland. Roots were used medicinally by Native Americans.

**E. glaucus** Ker Gawl. SEASIDE DAISY Perennial herbs; stems erect or sprawling, with a fleshy caudex, often forming mats, 1-3.5 dm tall, villous, glabrate. Leaves widely oblanceolate, to 9 cm long, to 3 cm wide, entire, hirsute, lower surfaces glandular, basal ones in a rosette, cauline ones reduced abruptly above rosette. Involucres ca. 1 cm long, 1.5-2.5 cm wide; phyllaries herbaceous, subequal, canescent. Ray flowers 60-100, pale lavender to bluish. Pappus in 1 series, sometimes with few outer setae. Mar-Aug. Occasional; rocky coastal bluffs and headlands; coastal scrub; 10-200 ft (3-60 m). North side between West Point and Scorpion Canyon, mostly on igneous substrates. San Miguel, Santa Rosa, and Anacapa islands; sw OR to San Luis Obispo Co.

**E. sanctarum** S.Watson SAINT'S DAISY Perennial herbs; stems erect, to 1 dm tall, slender, puberulent. Leaves lanceolate to oblanceolate, to 2.5 cm long, to 5 mm wide, strigose. Involucres ca. 7 mm long; phyllaries subequal, villous, outer ones herbaceous with brown midrib, inner ones tan, brown midrib, and scarious margins. Ray flowers ca. 60, purple to lavender. Pappus of 1 series or with inconspicuous outer series of short setae. Jun. A species of special environmental concern. Rare; shaded ravine; ca. 1150 ft (350 m). Known from a single collection made in June 1965 on the n side of Ragged Mountain. Santa Rosa Island; San Luis Obispo Co. to coastal foothills of Santa Ynez Mtns.

## ERIOPHYLLUM Lag.

Annuals or shrubs; stems tomentose. Leaves alternate, simple to ± compound. Heads usually radiate, axillary to terminal, solitary or in flat-topped clusters; involucres campanulate to hemispherical; phyllaries usually in 1 series, free or ± fused; receptacles flat to narrowly conic, often naked. Ray flowers, when present, pistillate, yellow; disk flowers bisexual, yellow. Outer achenes ± 4-angled, inner ones ± compressed; pappus composed of 4-15 paleaceous scales or absent. Ca. 15 species; w N. America.

1. Phyllaries 4-7 per involucre; ray flowers 4-7 ........ **E. confertiflorum**
1' Phyllaries 8-11 per involucre; ray flowers 6-9 ........ **E. staechadifolium**

**E. confertiflorum** (DC.) A.Gray var. **confertiflorum** GOLDEN-YARROW Shrubs, to 1 m tall. Leaves oblanceolate in outline, 1-2 times pinnately dissected, 2-7.5 cm long, segments linear, 1-3 mm wide; upper surfaces glabrate, lower surfaces tomentose. Heads 4-6 mm long, 3-5 mm wide, in flat-topped clusters; involucres 4-5 mm long; phyllaries 4-7, broadly elliptic to ovate, convex, tomentose, edges strongly overlapping. Ray flowers 4-7, corollas 3-6 mm long; disk flowers 2-3.5 mm long. Achenes 2-3 mm long. Apr-Jun. Common; rocky coastal bluffs, slopes, and ridgetops; chaparral, coastal scrub, grassland; 50-1200 ft (15-365 m). Throughout island. All CA Channel islands except Santa Barbara and San Nicolas; Los Coronados and Cedros islands; coastal n CA to n Baja CA. Another variety occurs in the Sierra Nevada foothills. Stems were boiled and used medicinally by Native Americans.

**E. staechadifolium** Lagasca SEASIDE WOOLLY SUNFLOWER Shrubs, to 1 m tall. Leaves obovate in outline, unequally pinnately lobed, 2-6 cm long, segments linear, 1-3 mm wide; upper surfaces glabrate, lower surfaces white-tomentose. Heads 7-10 mm long, 4-7 mm wide, in flat-topped clusters; involucres 5-7 mm long; phyllaries 8-11, broadly narrowly oblanceolate, keeled, glandular-puberulent, distinct to base or upper edges overlapping. Ray flowers 6-9, corollas 3-5 mm long; disk flowers 3.5-4 mm long. Achenes 3-4 mm long. Mar-Nov. Occasional; n-facing coastal bluffs and slopes; coastal scrub; 15-250 ft (5-75 m). North side between West Point and Potato Harbor. San Miguel, Santa Rosa, and Anacapa islands; s OR to Santa Barbara Co.

## FILAGO L.

Annuals; stems woolly or tomentose. Leaves alternate, entire, ± sessile. Heads discoid, in small capitate clusters; involucres cylindrical to ± hemispherical; outermost phyllaries leaf-like, inner ones chaff-like to ± herbaceous, partly enfolding outer flowers, innermost ones subtending flowers; receptacles convex to obconic, chaffy. Outer 3-8 series of flowers pistillate, inner series of flowers bisexual, not enfolded by receptacular chaff scales. Achenes obovoid, ± compressed, outer ones with no pappus, smooth, shiny, enfolded by persistent chaff scales, inner ones with deciduous pappus composed of bristles, dull, papillate.

1. Leaves linear to oblanceolate, ± spreading, flexible ........ **F. californica**
1' Leaves linear to subulate, appressed, stiff ........ **F. gallica**

## ASTERACEAE: Encelia-Filago

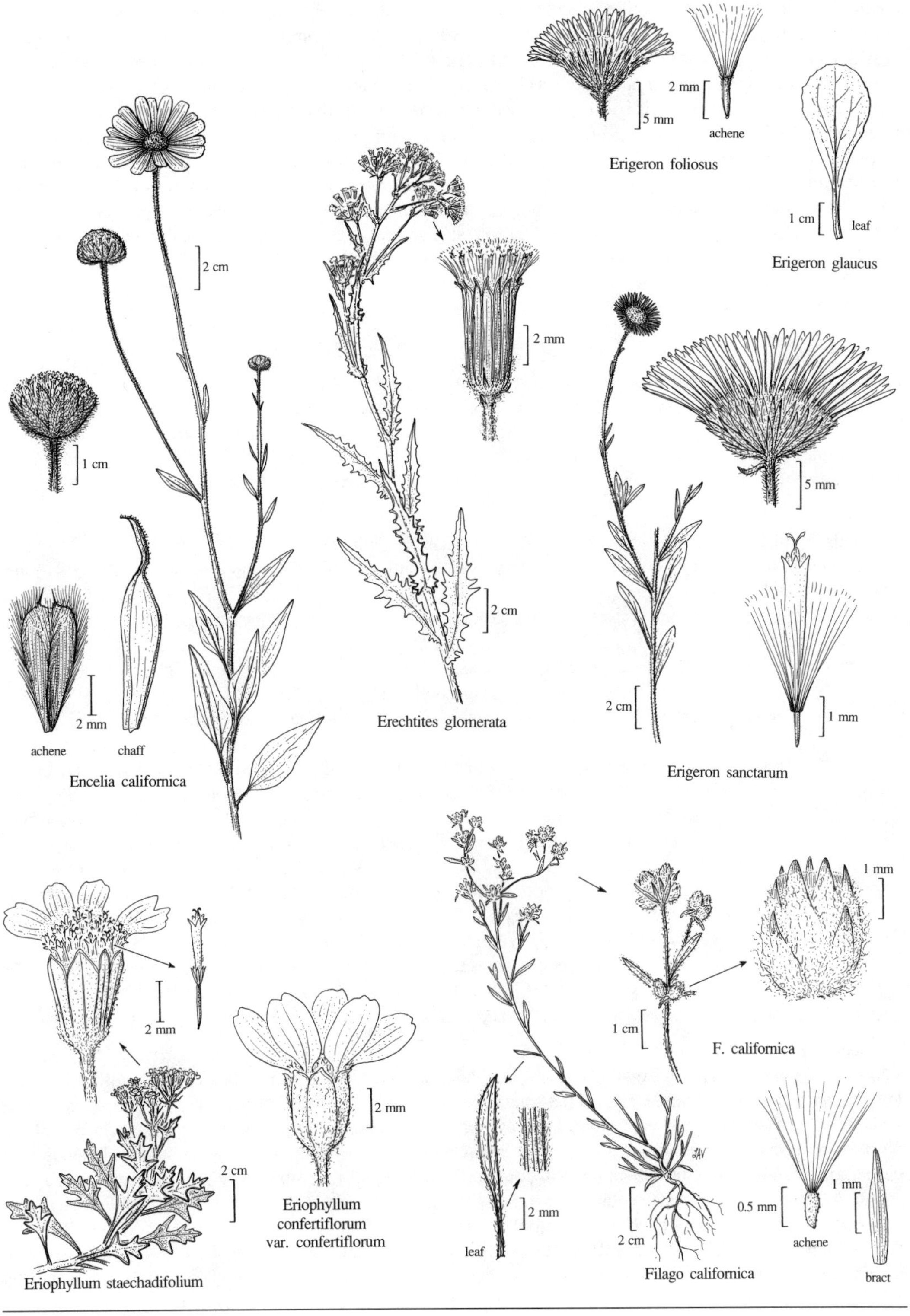

**F. californica** Nutt. CALIFORNIA FILAGO Stems to 3 dm tall, simple or branched from base, grayish-woolly. Leaves linear to oblanceolate, to 2 cm long, ± spreading, flexible, grayish villous. Heads ca. 4 mm long, in compact clusters, terminal or axillary, with leafy bracts that do not exceed heads; outer flowers enfolded by keeled, tomentose bracts. Outer pistillate flowers ca. 10. Mar-May. Common; rocky slopes, ridgetops, grassy hillsides; chaparral, coastal scrub, grassland, pine forest; 40-1700 ft (12-520 m). Scattered throughout island. All CA Channel islands except San Nicolas; Los Coronados, Todos Santos, Guadalupe, and Cedros islands; s OR to n Mexico, e to UT and TX. This species was used medicinally by Native Americans.

**F. gallica** L. NARROW-LEAVED FILAGO Stems to 2 dm tall, simple to branched, with soft, appressed hairs. Leaves linear to subulate, to 1.5 cm long, appressed, stiff, upper surfaces villous, lower surfaces tomentose, margins revolute. Heads ca. 4 mm long, in compact, terminal clusters, with leafy bracts 2-3 times as long as heads; outer flowers enfolded ± densely woolly bracts. Outer pistillate flowers ca. 5. Apr-Jun. Introduced. Common; rocky slopes, ridgetops, and grassy hillsides; chaparral, coastal scrub, grassland, pine forest; 80-2100 ft (25-640 m). Widely scattered locations in central portion of island. Santa Catalina and San Clemente islands; throughout CA, n Baja CA, Hawaii, and Chile; native of Europe. First collected on the island on El Tigre ridge in May 1969.

## GNAPHALIUM L. CUDWEED or EVERLASTING

Annuals or perennial herbs, often aromatic; stems woolly. Leaves alternate, entire, sessile, often decurrent or clasping. Heads discoid, ± sessile, often in dense clusters, these in racemes, spikes, or panicles; involucres cylindrical to subglobose; phyllaries in several series, imbricated, margins hyaline to scarious, often throughout, reflexed in fruit; receptacles flat, naked. Corollas white to purplish, pistillate ones many, corollas filiform, bisexual ones few, tubular. Achenes oblong, angled; pappus in 1 series of many fine bristles. Ca. 120 species; worldwide.

1. Heads in spikate or capitate clusters, inflorescence with leaf-like bracts
    2. Heads in cymose or subglobose, axillary clusters; involucres 2.5-4 mm long; pappus bristles falling separately ........ **G. palustre**
    2' Heads in dense spikate clusters; involucres 4-6 mm long; pappus bristles fused at base, falling as 1 unit ........ **G. purpureum**

1' Heads in rounded to flat-topped panicles, inflorescence bracts not conspicuously leaf-like
    3. Leaves white to grayish, persistently woolly on both surfaces
        4. Perennial herbs with basal leaves present at time of flowering; heads with 30-50 flowers ........ **G. canescens**
        4' Annuals with basal leaves absent or withered at time of flowering; heads with 40-100 flowers
            5. Heads 3-4 mm long; pappus bristles falling in clusters; pistillate corollas 1.5-2 mm long, lobes red ........ **G. luteo-album**
            5' Heads 4-6 mm long; pappus bristles falling separately; pistillate corollas 2-2.5 mm long, lobes yellow ........ **G. stramineum**
    3' Leaves green and glandular above or grayish green and thinly tomentose, becoming glabrous with age
        6. Heads turbinate to ovoid; phyllaries usually pink or pink-tinged ........ **G. ramosissimum**
        6' Heads campanulate or globose; phyllaries usually white, yellow, or tan
            7. Stems and lower surfaces of leaves densely woolly; leaves not strongly scented .... **G. bicolor**
            7' Stems and lower surfaces of leaves thinly villous or glabrate; leaves strongly scented ........ **G. californicum**

**G. bicolor** Bioletti BICOLORED EVERLASTING Suffrutescent, short-lived perennial herbs, to 9 dm tall; stems white-woolly. Leaves linear, oblanceolate, or oblong, 1.5-8 cm long, upper surfaces thinly woolly to glabrate, lower surfaces densely woolly. Heads in flat-topped panicles; involucres campanulate, ca. 6 mm long; phyllaries white to straw-colored, outer ones ovate, scarious, shiny, inner ones narrowly oblong. Pistillate flowers 50-65, bisexual ones 8-12. Pappus bristles free. Feb-Sep. Common; rocky slopes, dry washes, gravel floodplains; chaparral, coastal scrub, pine forest, riparian communities; 25-1500 ft (8-460 m). Widely scattered throughout island. All CA Channel islands; Todos Santos, San Martin, Guadalupe, and Cedros islands; San Francisco Bay region to n Baja CA, e to Sierra Nevada foothills. This species was used medicinally by Native Americans.

**G. californicum** DC. GREEN EVERLASTING Annuals or biennials, to 9 dm tall; stems sparsely pubescent to glabrate. Leaves linear to oblanceolate, 3-15 cm long, thinly villous or glabrate, glandular, strongly scented. Heads in flat-topped panicles; involucres subglobose, ca. 6 mm long; phyllaries white to straw-colored, outer ones ovate, scarious, shiny, inner ones linear. Pistillate flowers 40-100, bisexual ones 7-12. Pappus bristles free. Feb-Jun (Sep). Common; rocky slopes, ridgetops, dry washes, gravel floodplains; chaparral, coastal scrub, pine forest, riparian communities; 20-2250 ft (6-690 m). Widely scattered locations throughout much of the island. All CA Channel islands; OR to Baja CA. Leaves and heads were used medicinally by Native Americans.

**G. canescens** DC. Short-lived perennial herbs, to 15 dm tall; stems sparsely pubescent to glabrate. Leaves linear to oblanceolate or oblong, 1.5-8 cm long, white to grayish, tomentose, glandular, strongly scented or not. Heads in flat-topped panicles; involucres campanulate to obconic, 5-6 mm long; phyllaries white to straw-colored, outer ones ovate, ± tomentose, inner ones narrowly oblong. Pistillate flowers 25-35, bisexual ones 5-15. Pappus bristles free. Feb-Jun (Sep). Common; rocky slopes, ridgetops, canyon walls, dry washes, gravel floodplains; chaparral, coastal scrub, riparian communities; 50-1300 ft (15-400 m). Widely scattered throughout central and e portions of island. Santa Rosa, Anacapa, Santa Barbara, Santa Catalina, and San Clemente islands; Los Coronados and Todos Santos islands; San Francisco Bay region to n Baja CA.

1. Leaves linear to narrowly lanceolate, cauline ones ascending to erect, upper ones decurrent ........................ subsp. **beneolens**

1' Leaves narrowly lanceolate, oblanceolate, or oblong, cauline ones ± spreading, upper ones sessile but not decurrent ........................ subsp. **microcephalum**

subsp. **beneolens** (Davidson) Stebbins & D.J.Keil [*G. beneolens* Davidson] FRAGRANT EVERLASTING Plants 3.5-12 dm tall, often strongly scented. Leaves linear to narrowly lanceolate, 1.5-5 cm long, cauline ones ± ascending to erect, upper ones decurrent. Rare; rocky slopes; chaparral, coastal scrub; ca. 800 ft (245 m). Upper portion of Cañada de los Sauces. Santa Rosa, Santa Catalina, and San Clemente islands; Todos Santos Island; coastal CA w of Great Valley.

subsp. **microcephalum** (Nutt.) Stebbins & D.J.Keil [*G. microcephalum* Nutt.] WHITE EVERLASTING Plants 3.5-15 dm tall, odorless. Leaves narrowly lanceolate, oblanceolate, or oblong, 1.5-8 cm long, cauline ones ± spreading, upper ones sessile but not decurrent. Common; rocky slopes, ridgetops, canyon walls, dry washes, gravel floodplains; chaparral, coastal scrub, riparian communities; 50-2400 ft (15-730 m). Widely scattered locations throughout island. Santa Rosa, Anacapa, Santa Barbara, and Santa Catalina islands; Los Coronados Island; San Francisco Bay region to n Baja CA. This species was used medicinally by Native Americans.

**G. luteo-album** L. WEEDY CUDWEED Annuals; stems erect to decumbent, to 5 dm tall, white or grayish-woolly. Leaves linear-oblanceolate to narrowly oblong, 1-4 cm long, gray to grayish green, thinly woolly to canescent, glandular. Heads in dense, flat-topped clusters; involucres campanulate to subglobose, 3-4 mm long; phyllaries straw-colored, outer ones ovate, scarious, shiny, inner ones narrowly lanceolate. Pistillate flowers 60-140, bisexual ones 5-10. Pappus bristles free. Jan-Dec. Introduced. Occasional; disturbed sites, moist creek banks, floodplains, seasonal ponds, pastures; chaparral, riparian communities; 10-1500 ft (3-460 m). Widely scattered locations throughout w and central portions of island. All CA Channel islands; WA to n Baja CA; native of Europe and Asia. First collected at Prisoners' Harbor in October 1958.

**G. palustre** Nutt. LOWLAND CUDWEED Annuals; stems erect to spreading, to 2.5 dm tall, simple or branched from near base, white or grayish-woolly. Leaves oblanceolate to oblong, 1-3.5 cm long, thinly to densely woolly or canescent. Heads subtended by leaf-like bracts, in cymose or subglobose, often terminal clusters; involucres campanulate, 2.5-4 mm long; phyllaries linear to narrowly oblong, green in lower half, margins white to straw-colored. Pistillate flowers 90-110, bisexual ones 5-9. Pappus bristles free. May-Oct. Scarce; moist creek banks, fresh water seeps, seasonal ponds, marshes; riparian communities; 5-1400 ft (2-430 m). North side at Prisoners' Harbor and in pools near Mount Pleasant; Central Valley in Hooker and Two-Fork canyons. Santa Catalina and San Clemente islands; throughout much of w N. America. This species was used medicinally by Native Americans.

**G. purpureum** L. PURPLE CUDWEED Annuals or short-lived perennial herbs; stems erect to decumbent, to 4.5 dm tall, usually branched from base, sometimes suffrutescent, white woolly, usually odorless. Leaves obovate, oblanceolate or spatulate, 1.5-8 cm long, upper surfaces thinly woolly to glabrate, lower surfaces white-canescent. Heads subtended by leaf-like bracts, in dense spikate clusters; involucres campanulate, 4-6 mm long; phyllaries green at base, upper part brown, purplish, or straw-colored, outer ones ovate, shiny, inner ones narrowly oblong to linear. Pistillate flowers 80-120, bisexual ones 5-8. Pappus bristles fused at base. Mar-May. Occasional; mostly on rocky, n-facing slopes; chaparral, coastal scrub, grasslands, pine forest; 5-1300 ft (2-400 m). Widely scattered locations throughout island. San Miguel, Santa Rosa, Anacapa, and Santa Catalina islands; throughout much of N. America.

**G. ramosissimum** Nutt. PINK EVERLASTING Short-lived perennial herbs, to 15 dm tall; stems dark green, pubescent, glandular. Leaves linear to lanceolate or oblong, 1.5-7 cm long, villous to glabrate, glandular. Heads in subcylindrical or pyramidal panicles; involucres turbinate to ovoid, 4-5 mm long; phyllaries pink or pink-tinged, outer ones ovate, scarious, shiny, inner ones linear to narrowly oblong. Pistillate flowers 45-60, bisexual ones 5-8. Pappus bristles free. Jul-Sep. Scarce; rocky slopes, canyon bottoms; coastal scrub, oak woodland, pine forest; 50-1500 ft (15-460 m). Widely scattered on n side between Profile Point and Prisoners' Harbor; Central Valley in upper reaches of Cañada Christy drainage; s side in Willows Canyon. Los Coronados Island; coastal n CA to s CA.

**G. stramineum** Kunth [*G. chilense* Spreng.] COTTON-BATTING Annuals or short-lived perennial herbs; stems erect to decumbent, to 6 dm tall, grayish-woolly. Leaves linear-lanceolate to oblanceolate, 1.5-7 cm long, grayish to yellowish green, ± woolly. Heads in dense, flat-topped clusters; involucres campanulate to subglobose, 4-6 mm long; phyllaries white to straw-colored, outer ones ovate, scarious, shiny, inner ones linear. Pistillate flowers 160-200, bisexual ones 20-28. Pappus bristles free. Feb-Sep. Scarce; coastal bluffs, moist creek banks, floodplains, sandy beaches; chaparral, coastal scrub, coastal strand; 20-2400 ft (6-730 m). North side at West Point, Cueva Valdez, Picacho Diablo, canyon w of Potato Harbor; Central Valley at Christy Beach; scattered locations on s side. All CA Channel islands except Santa Barbara and San Clemente; Guadalupe and Cedros islands; British Columbia to Mexico, e to Great Plains. Leaves and heads were used medicinally by Native Americans.

## GRINDELIA Willd. GUMPLANT

Perennial herbs or subshrubs. Leaves alternate, linear to oblanceolate, entire to pinnately lobed, sessile, often clasping, glandular, often viscid. Heads usually radiate, solitary or in small terminal clusters; involucres hemispheric to subglobose; phyllaries in 4-many series, evidently imbricated, ± appressed to strongly reflexed, viscid or resinous; receptacles flat to convex, naked. Ray flowers usually present, pistillate, yellow; disk flowers bisexual, yellow. Achenes cylindrical to slightly compressed, glabrous, often shiny; pappus composed of 1-6, white to tan, awn-like bristles. Ca. 80 species; N. and S. America.

**G. camporum** Greene var. **bracteosum** (J.T.Howell) M.A.Lane [*G. latifolia* Kellogg] Stems erect, to 1-1.5 dm tall, succulent, often suffrutescent. Basal leaves spatulate, to 10 cm long; cauline leaves lanceolate, ovate, or oblong, 2-8 cm long, glandular-punctate, margins subentire to sharply toothed, apices acute to obtuse, bases subcordate to clasping. Heads 1-4; involucres 9-12 mm long, outer phyllaries usually reflexed, leaf-like. Corollas yellow. Achenes 3.5-4 mm long, brown, with 4-5 ribs. Aug-Oct. Occasional; grassy flats and slopes, ridgetops, disturbed sites; chaparral, grasslands; 15-1200 ft (5-365 m). North side at Prisoners' Harbor and in Cañada del Puerto; Central Valley between Islay Canyon and Valley Peak area; s side in upper portion of Laguna Canyon. All CA Channel islands except Santa Barbara and San Clemente; coastal CA. Dried leaves and heads were used medicinally by Native Americans; leaves also were eaten. Plants from Santa Cruz Island appear to be hybrids with *G. hirsutula* Hook. & Arn., according to annotations by M.A. Lane (SBBG).

## HAZARDIA Greene

Perennial herbs or shrubs. Leaves alternate, cuneate to oblanceolate, entire to toothed, sessile or subsessile. Heads discoid or radiate, solitary, in small terminal clusters, or in spikes; involucres hemispheric, turbinate, or campanulate; phyllaries in 4-many series, imbricated, subequal; receptacles flat to slightly convex, naked. Ray flowers, when present, pistillate, yellow; disk flowers bisexual, yellow, corolla tube tapering gradually to throat. Achenes cylindrical, 4-5-angled or ribbed; pappus composed of many white to brown, barbellate bristles. 13 species; w N. America.

1. Heads usually radiate; stems and leaves densely tomentose; cauline leaves 4-14 cm long ............ **H. detonsa**
1' Heads usually discoid; stems and leaves thinly tomentose to glabrous; cauline leaves 1.5-5 cm long ............ ............................................................ **H. squarrosa** var. **grindelioides**

**H. detonsa** (Greene) Greene [*Haplopappus d.* (Greene) Raven] NORTHERN ISLAND HAZARDIA Shrubs; stems erect, 0.5-3 dm tall, densely tomentose. Leaves ovate to obovate, 4-14 cm long, to 5 cm wide, coriaceous, margins subentire to sharply serrate, upper surfaces tomentose, glandular, lower surfaces canescent. Heads usually radiate, turbinate, 12-20 mm long; involucres 10-15 mm long; outer phyllaries canescent, inner ones tomentose near ± erect apices. Disk flowers 30-40 per head. Achenes 3-5 mm long, puberulent, with 4-5 ribs. (Apr) Jul-Nov. Island endemic. A species of special environmental concern. Common; coastal bluffs, rocky cliffs,

ASTERACEAE: Gnaphalium, Grindelia

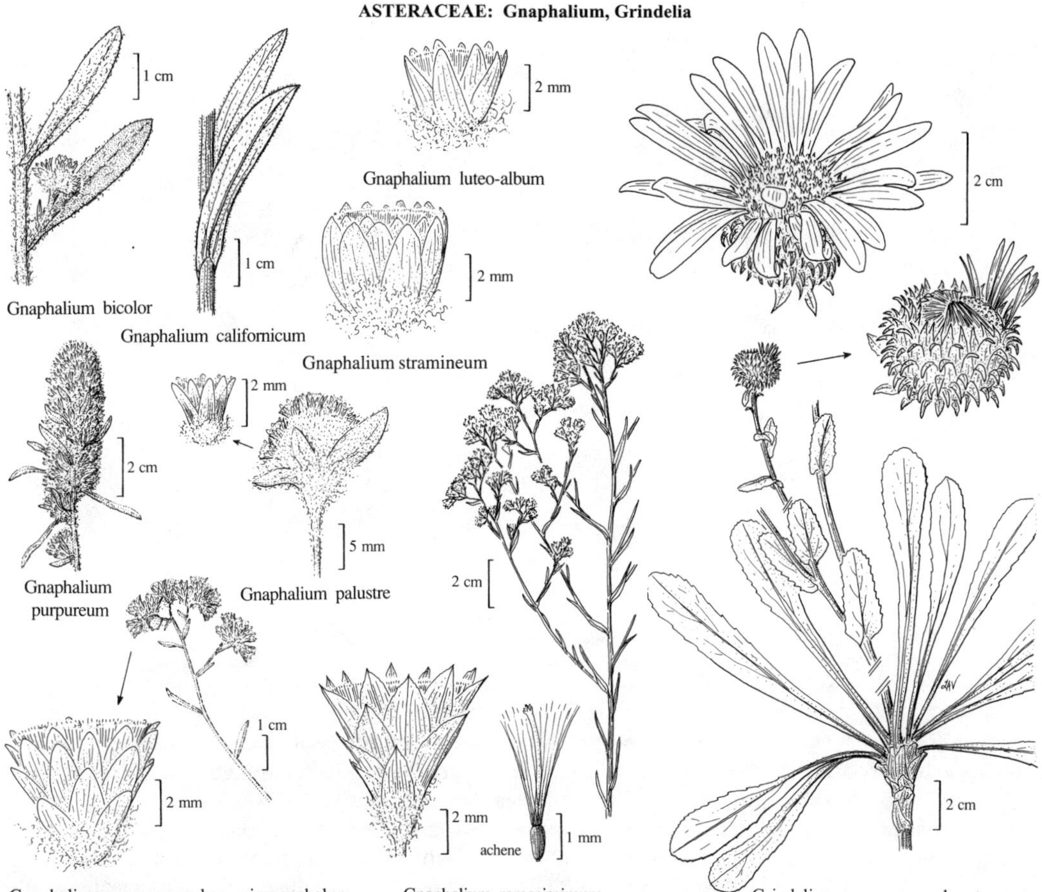


canyon walls, ridgetops; chaparral, coastal scrub, grasslands, pine forest; 20-1900 ft (6-580 m). Scattered locations throughout, but most abundant on bluffs around perimeter of island. Santa Rosa and Anacapa islands. Many populations show evidence of hybridization with *H. squarrosa* var. *grindelioides*.

**H. squarrosa** (Hook. & Arn.) Greene var. **grindelioides** (DC.) W.D.Clark [*Haplopappus s.* Hook. & Arn. subsp. *g.* (DC.) D.D.Keck] SAW-TOOTHED GOLDENBUSH Shrubs; stems erect, 0.3-10 dm tall, pubescent and glabrate with age. Leaves oblanceolate to ovate, 1-5 cm long, to 1.5 cm wide, coriaceous, margins sharply serrate, upper surfaces ± glabrous, glandular, lower surfaces tomentulose or glabrous. Heads usually discoid, turbinate, 12-18 mm long; involucres 10-14 mm long; phyllaries straw-colored, puberulent and glandular near recurved apices. Disk flowers 10-20 per head. Achenes 3-5 mm long, glabrous to sparsely puberulent, with 5-10 inconspicuous ribs. Jun-Dec. Occasional; coastal bluffs, rocky slopes, ridgetops; chaparral, coastal scrub, grasslands, pine forest; 100-1400 ft (30-430 m). Widely scattered locations throughout island. San Miguel, Santa Rosa, Anacapa, and Santa Catalina islands; Monterey Co. to n Baja Ca. Some plants have been confused with the mainland var. *obtusa*, which is characterized by strictly erect, subtruncate to obtuse phyllaries and 18-30 flowers per head.

## HELIANTHUS L. SUNFLOWER

Annuals or perennial herbs. Leaves alternate or opposite, often petiolate, often scabrous. Heads radiate, axillary or terminal, solitary or in open, cymose clusters; involucres hemispheric to campanulate; phyllaries in 2-4 series, ± equal, free; receptacles flat to ± conic, with chaffy bracts subtending disk flowers. Ray flowers much longer than disk flowers, sterile, yellow; disk flowers bisexual, yellow to red. Achenes ± obovoid, ± compressed, 4-5-angled, black; pappus composed of 2 deciduous scales. 67 species; N. and S. America.

ASTERACEAE: Hazardia, Helianthus

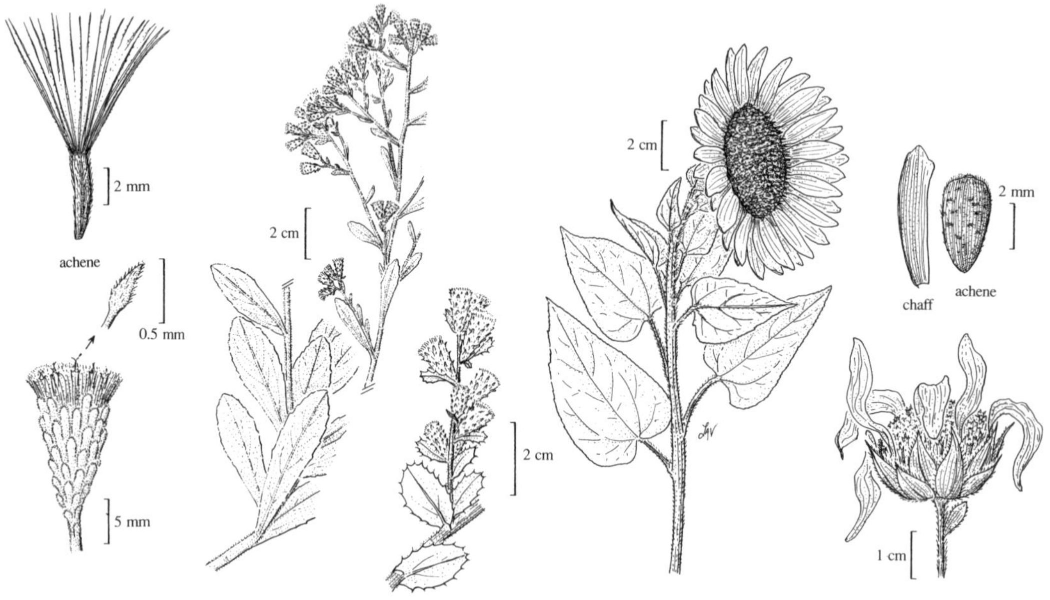


Hazardia detonsa Hazardia squarrosa var. grindelioides Helianthus annuus

**H. annuus** L. COMMON SUNFLOWER Annuals; stems stout, to 2 m tall, hispid to densely pubescent. Leaves lanceolate to broadly ovate, 7-15 cm long, ± scabrous, margins serrate. Involucres to 5 cm wide; phyllaries leaf-like, broadly lanceolate, pubescent. Ray flowers 1.5-2.5 cm long, corollas yellow with red veins; disk flowers puberulent near base. Achenes ca. 6 mm long, puberulent, brown. Jul-Aug. Introduced. Rare; seasonal ponds, canyon bottoms; chaparral, grasslands; 200-700 ft (60-215 m). North side at seasonal pond above China Harbor; Central Valley at Mission Spring near the Stanton Ranch. Anacapa, Santa Catalina, and San Clemente islands; throughout N. America. This species is the source of sunflower seeds and oil; yellow, purple, and black dyes have been made from the flowers.

## HEMIZONIA DC. TARWEED

Annuals, perennial herbs, or shrubs, often strongly aromatic. Leaves usually alternate, sometimes also basal, linear to oblanceolate, entire to pinnately lobed. Heads radiate, solitary or in terminal, flat-topped clusters, subsessile to pedunculate; involucres hemispheric to campanulate; phyllaries often in 1 series, partly enfolding ray achenes; receptacles flat to slightly convex, with chaffy bracts in 1 series between ray and disk flowers. Ray flowers pistillate or sterile, white to yellow; disk flowers usually bisexual, sometimes staminate, white to yellow; anthers reddish to black. Achenes of ray flowers ± 3-angled, those of disk flowers cylindrical to obconic; pappus absent or composed of ± 10 thin scales. Ca. 25 species; sw N. America.

1. Shrubs .......... **H. clementina**
1' Annuals
    2. Leaf lobes and apices with spinulose tips; phyllaries and leaves sericeous to thinly lanate .... **H. fitchii**
    2' Leaf lobes and apices not spinulose; phyllaries and leaves glandular-puberulent to hirsutulous or ± glabrous
        3. Ray flowers ± 5; disk flowers 6; phyllaries ± glabrous, often with sessile glands ... **H. fasciculata**
        3' Ray flowers 8-11; disk flowers 14-20; phyllaries glandular or glandular-puberulent **H. increscens**

**H. clementina** Brandegee ISLAND TARWEED Shrubs, 2-5 dm tall; stems erect to decumbent, puberulent to sparsely hirsute. Leaves 1.5-3.5 cm long, entire to remotely toothed, hirsutulous, ± glandular. Heads hemispheric; involucres 4-6 mm long; phyllaries ± glabrous to hirsutulous near bases. Ray flowers 11-13; disk flowers 20-30. Ray achenes 2-3.5 mm long. Apr-May. Island endemic. A species of special environmental concern. Rare; eroded s-facing coastal flats and bluffs; coastal scrub; ca. 300 ft (91 m). South side of island, ca. 1 mi (1.5 km) w of Sandstone Point. All CA Channel islands except San Miguel and Santa Rosa.

ASTERACEAE: Hemizonia, Heterotheca

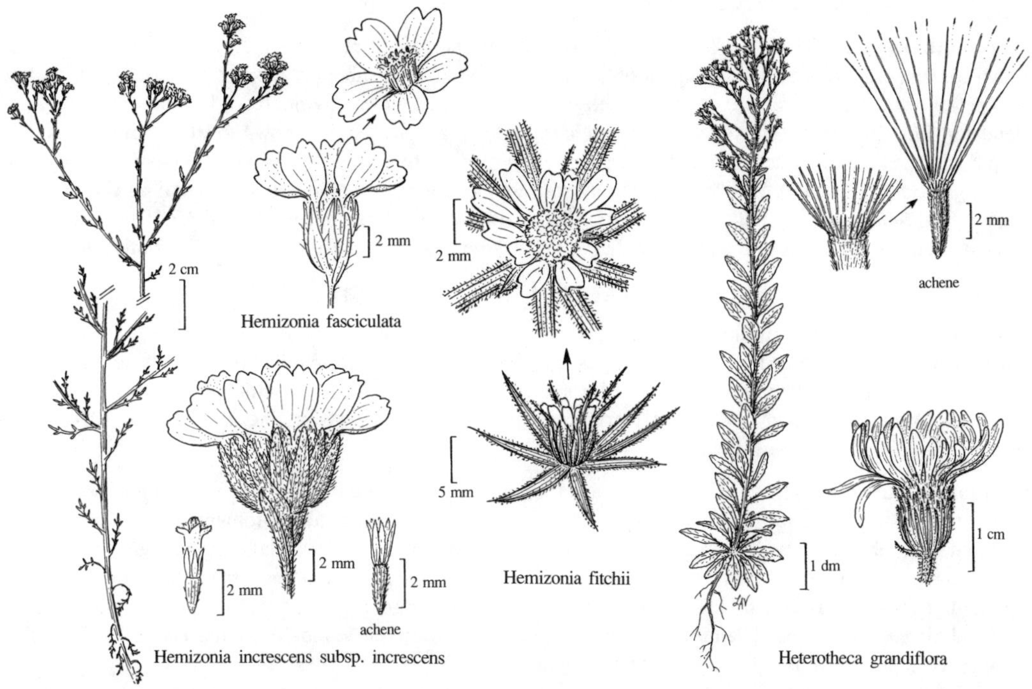


**H. fasciculata** (DC.) Torr. & A.Gray COMMON TARWEED Annuals, 1-5 dm tall; stems erect or decumbent, ± glabrous to hirsute. Leaves ± glandular to glandular-puberulent, lower ones subentire to pinnately lobed, cauline ones usually subentire, 0.3-2 cm long. Heads campanulate; involucres 4-7 mm long; phyllaries ± glabrous, usually with sessile glands. Ray flowers ± 5; disk flowers 6. Ray achenes 2-3 mm long. Apr-Oct. Occasional; coastal flats, grassy slopes, pastures; coastal scrub, grasslands; 20-850 ft (6-260 m). Widely scattered locations throughout island, but most abundant and conspicuous near Fraser Point and near Smugglers' Cove. All CA Channel islands except San Miguel and San Nicolas; Todos Santos and Cedros islands; Monterey Co. to Baja CA. This species was boiled and eaten as an emergency food by Cahuilla Tribe of s CA.

**H. fitchii** A.Gray Annuals, 1-6 dm tall; stems erect or decumbent, usually with many branches, densely lanate to tomentose. Leaves sericeous to thinly lanate, with stalked glands, lower ones subentire to pinnately lobed, lobes with spinulose tips, cauline ones entire, 1-3 cm long. Heads hemispheric; involucres 5-10 mm long; phyllaries glandular-puberulent to thinly lanate. Ray flowers 10-13; disk flowers many. Ray achenes 2-3 mm long. Jun-Nov. Introduced. Occasional; grassy slopes, pastures, disturbed sites, ridgetops; chaparral, coastal scrub, grasslands; 230-950 ft (70-290 m). Central Valley near University of California field station and along Islay Canyon Road; s side along S Ridge Road between Islay Canyon and Coches Prietos drainages. Southern OR to s CA. First collected about 0.75 mile w of Stanton Ranch in Islay Pasture in October 1958. Achenes and roots were eaten by Native Americans.

**H. increscens** (H.M.Hall ex Keck) Tanowitz subsp. **increscens** [*H. paniculata* H.M.Hall ex Keck] Annuals, 3-9 dm tall; stems erect or spreading, hirsutulous to glabrate. Leaves hirsutulous to glabrate, deep green, lower ones entire to lobed, cauline ones entire, 0.5-2.5 cm long. Heads hemispheric; involucres 5-9 mm long; phyllaries hirsutulous and glandular-puberulent. Ray flowers 8-11; disk flowers 15-20. Ray achenes 2-3 mm long. May-Jul. Scarce; grassy fields and pastures; coastal scrub, grasslands; 50-150 ft (15-45 m). Central Valley in lower portion of Cañada Christy. Santa Rosa Island; coastal CA. Two other subspecies occur on the mainland in CA.

## HETEROTHECA Cass.

Annuals or perennial herbs. Leaves alternate, entire to toothed, reduced upward. Heads radiate, often in flat-topped panicles; involucres hemispheric to campanulate; phyllaries in 3-6 series, subequal, ± imbricated; receptacles flat to slightly convex, naked. Ray flowers pistillate, yellow; disk flowers bisexual, yellow. Achenes ± obconic, those of ray flowers 3-angled, those of disk flowers compressed; pappus of ray flowers absent or composed of capillary bristles, pappus of disk flowers in ± 2 series, outer one composed of minute, linear scales, inner one composed of capillary bristles. Ca. 30 species; N. and S. America.

**H. grandiflora** Nutt. TELEGRAPH WEED Annuals or short-lived perennials, to 1 m tall; stems simple to branched, lanate. Leaves ovate, oblong, or oblanceolate, to 6 cm long, subentire to serrate, densely villous, lower ones petiolate, upper ones sessile. Heads on densely glandular peduncles; involucres 7-11 mm long; phyllaries ± glandular. Ray flowers 25-35. Achenes 3-4 mm long, villous; pappus reddish brown to straw-colored. Feb-Oct. Common; beaches, dry washes, floodplains, eroded slopes, disturbed sites; chaparral, coastal scrub, coastal strand, grasslands; 20-2450 ft (5-745 m). Widely scattered locations throughout w and central portions of island. San Nicolas, Santa Catalina, and San Clemente islands; nw CA to n Mexico, e to UT and AZ. The Luiseño Tribe of s CA used the tall stems to make arrow shafts; this species also was used medicinally by Native Americans..

## HIERACIUM L. HAWKWEED

Perennial herbs, often with milky sap. Leaves alternate, entire to toothed, pilose to tomentose. Heads ligulate, few to many, in rounded or flat-topped panicles; involucres ± cylindrical; phyllaries in 2-4 series, subequal, ± imbricated; receptacles slightly convex, naked, sometimes puberulent. Flowers ligulate, bisexual, white, yellow, or orange. Achenes narrowly cylindrical, striate; pappus composed of many fine, fragile or brittle, capillary bristles. Ca. 250 species; worldwide.

**H. argutum** Nutt. Plants to 1.5 m tall; stems erect, branched, tomentose below. Leaves oblong to oblanceolate, lower ones densely clustered, 8-20 cm long, petiolate, upper ones much reduced, tomentose, sessile; involucres 7-10 mm long; phyllaries lanceolate to linear-lanceolate, herbaceous, with dark, stalked glands. Achenes 2-3 mm long, brown, glabrous. Jun-Oct. Occasional; shaded n-facing slopes, canyon bottoms; chaparral, pine forest; 200-1300 ft (60-395 m). North side between Fry's Harbor and China Harbor; Central Valley; s side in Coches Prietos drainage. Santa Rosa Island; Monterey Co. to s CA, e to Sierra Nevada foothills. The milky latex found in this species has been used to make a chewing gum.

## HYPOCHAERIS L. CAT'S EAR

Annual or perennial herbs, often with milky sap. Leaves usually in basal rosettes, entire to lobed. Heads ligulate, solitary to few in ± flat-topped clusters; involucres campanulate to narrowly cylindrical; phyllaries in 1-4 series, imbricated; receptacles flat, with thin, chaffy scales partly enfolding achenes. Flowers ligulate, bisexual, corollas usually yellow. Achenes ± cylindrical, outer ones truncate, without a beak, inner ones with beak between achene and pappus; pappus in 2 series, outer one often composed of short, simple bristles, inner one composed of longer, plumose bristles. Ca. 50 species; Europe, Asia, and S. America.

**H. glabra** L. SMOOTH CAT'S EAR Annuals; stems 1-few, to 5 dm tall, erect or decumbent. Leaves oblong to oblanceolate, 2-15 cm long, shallowly lobed, margins sparsely ciliate. Involucres 10-15 mm long; phyllaries ovate to lanceolate, glabrous. Outer achenes ca. 4 mm long, inner ones ca. 7 mm long. Mar-Jun. Introduced. Common; grassy slopes, disturbed flats, pastures; chaparral, coastal scrub, grasslands, pine forest; 5-2200 ft (2-670 m). Scattered locations throughout island. Santa Rosa, Anacapa, Santa Catalina, and San Clemente islands; Todos Santos, Guadalupe, and Cedros islands; throughout N. America; native of Europe. Introduced to Santa Cruz Island prior to 1886 (Greene 1887). Leaves of this species are edible and can be used for greens.

## ISOCOMA Nutt.

Shrubs. Leaves alternate, sometimes clustered in axils, entire to serrate, sessile or subsessile. Heads discoid, in paniculate or flat-topped clusters; involucres subcylindrical to turbinate; phyllaries in many series, subequal, imbricated; receptacles flat to slightly convex, naked. Flowers bisexual, corolla tube abruptly expanded to wide, cylindrical throat, corollas yellow. Achenes narrowly obconic, compressed or angled; pappus of many white to tan bristles. Ca. 10 species; sw N. America and Mexico.

**I. menziesii** (Hook. & Arn.) G.L.Nesom [*Haplopappus venetus* (Kunth) S.F.Blake] Stems erect to prostrate, thinly tomentose to glabrate. Leaves oblong, ovate, or obovate, 1-5 cm long, 0.5-2 cm wide, subentire to sharply serrate, sparsely puberulent or glabrous, clasping to subsessile. Heads ± hemispheric; involucres 6-8 mm long; phyllaries lanceolate, puberulent to glabrate. Achenes 2-3 mm long, densely villous to glabrate. Feb-Oct. Common; coastal bluffs, beaches, alluvial fans, rocky slopes and flats, canyon walls; chaparral, coastal scrub, coastal strand, grasslands; 6-1400 ft (2-430 m). Scattered locations throughout island. All CA Channel islands except Santa Barbara; coastal n CA to Baja CA.

ASTERACEAE: Hieracium-Jaumea

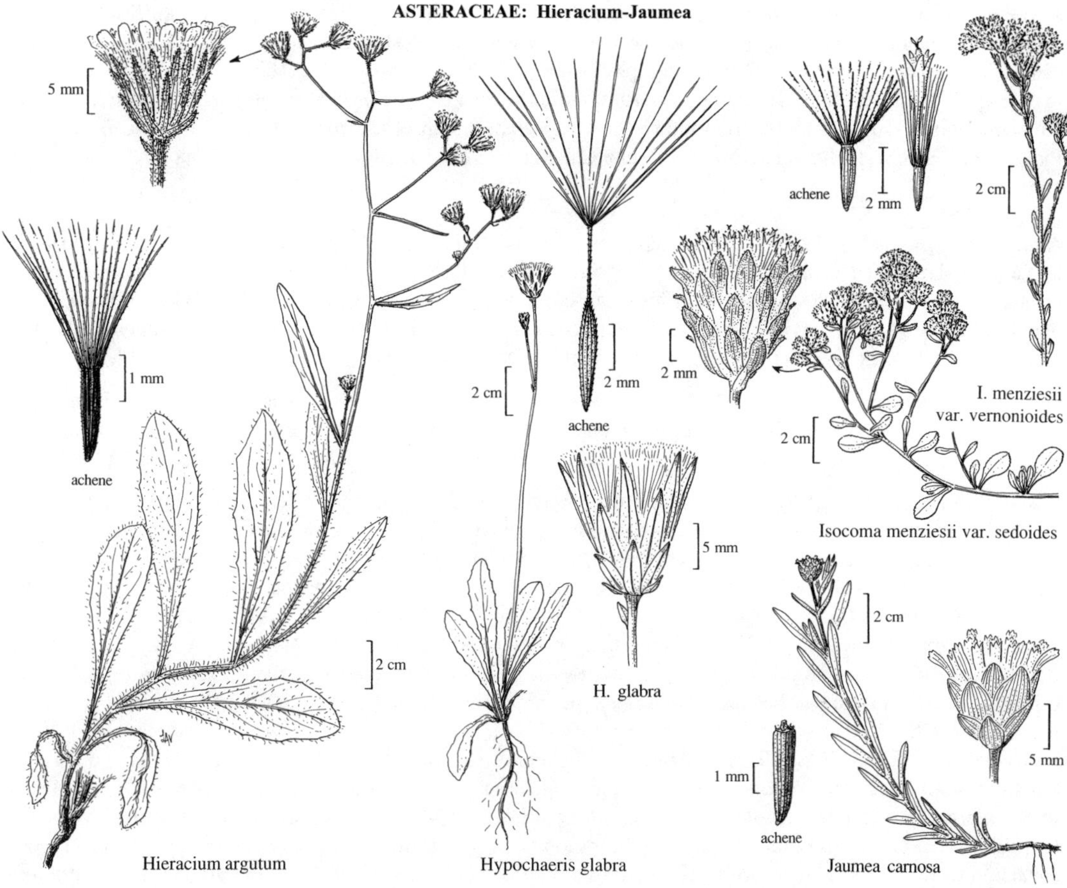


1. Stems decumbent to prostrate; leaves broadly obovate, fleshy, usually not densely clustered .... var. **sedoides**
1' Stems erect; leaves oblong, ovate, or obovate, not very fleshy, usually in dense, axillary clusters ................ ........................ var. **vernonioides**

var. **sedoides** (Greene) G.L.Nesom [*H. v.* var. *sedoides* (Greene) Munz] PROSTRATE COASTAL GOLDENBUSH Plants less than 0.5 dm tall; stems prostrate to decumbent. Leaves broadly obovate, fleshy, usually not densely clustered. Feb-Oct. Occasional; coastal bluffs, beaches, alluvial fans; coastal scrub, coastal strand, grasslands; 6-300 ft (2-90 m). North side near West Point; w and sw coast from Fraser Point to mouth of Laguna Canyon. San Miguel, Santa Rosa, and Anacapa islands; San Francisco Bay region to n Baja CA.

var. **vernonioides** (Nutt.) G.L.Nesom [*H. v.* subsp. *vernonioides* (Nutt.) H.M.Hall] Plants to 1.5 m tall; stems erect. Leaves oblong, ovate, or obovate, not very fleshy, usually in dense, axillary clusters. Feb-Oct. Common; coastal bluffs, rocky slopes and flats, canyon walls; chaparral, coastal scrub, grasslands; 6-1400 ft (2-430 m). Scattered locations throughout island. All CA Channel islands except Santa Barbara; coastal n CA to Baja CA.

## JAUMEA Pers.

Perennial herbs, with stolons or rhizomes; aerial stems prostrate to ascending. Leaves opposite, entire. Heads radiate or discoid, usually solitary, terminal; involucres cylindrical to ovoid; phyllaries in 3-5 series, imbricated, subequal, erect to appressed; receptacles conic, naked. Ray corollas, when present, pistillate, yellow; disk corollas bisexual, yellow. Achenes ± fusiform, angled or ribbed; pappus absent or composed of short bristles. Ca. 2 species; Pacific coasts of N. and S. America.

**J. carnosa** (Less.) A.Gray Plants with creeping rhizomes 1-5 dm long; stems simple or with few branches, decumbent, glabrous. Leaves linear to oblanceolate, 1-4 cm long, fleshy, glabrous. Heads solitary; involucres campanulate to hemispheric, 9-12 mm long; phyllaries ovate. Ray flowers 6-10. Achenes 2-5 mm long, with 4 main angles, glabrous. May-Sep. Scarce; coastal bluffs, freshwater seeps, canyon bottoms; coastal scrub, riparian communities; 10-30 ft (3-10 m). Southwestern portion at mouth of Cañada de los Sauces, Pozo Beach, and Playa Larga. San Miguel, Santa Rosa, and Santa Catalina islands; British Columbia to n Baja CA.

## LACTUCA L.

Annuals or perennial herbs, with milky sap. Leaves basal and alternate, cauline ones often clasping. Heads ligulate, in flat-topped panicles; involucres ± cylindrical; phyllaries in 2-5 series, unequal, imbricated, erect; receptacles flat, naked. Flowers ligulate, bisexual, corollas pale to deep yellow or blue. Achenes fusiform to ovoid, compressed, ribbed, with short to long, slender beak between fruit and pappus; pappus composed of many soft, capillary bristles, each falling separately. Ca. 100 species; temperate, worldwide.

1. Cauline leaves linear to lanceolate, margins entire to ± toothed; flowers 8-12 per head ................ **L. saligna**
1' Cauline leaves oblong to elliptic, margins dentate and often ± deeply lobed; flowers 15-20 per head ........... ........................................................ **L. serriola**

**L. saligna** L. WILLOW LETTUCE Annuals to 1.3 m tall; stems several, erect, often decumbent at base. Leaves mostly linear to lanceolate, cauline ones 3-20 cm long, midveins sparsely hispid, margins entire to ± toothed. Heads in narrow, spike-like panicles; involucres 10-15 mm long. Flowers 8-12 per head. Corollas pale yellow. Achenes (excluding beaks) 3-3.5 mm long; beaks 3-4 mm long. Jul-Oct. Introduced. Occasional; coastal bluffs, disturbed grasslands, floodplains, eroded slopes; coastal scrub, grasslands; 15-1250 ft (5-380 m). North side near West Point, Valdez Anchorage, between Cavern Point and San Pedro Point; Central Valley near Christy Ranch; s side in Cañada de los Sauces, upper portion of Pozo Canyon, on Sierra Blanca Ridge, and in Laguna Canyon. San Miguel and San Nicolas islands; N. America; native of Europe. First collected at Valdez Anchorage in August 1988.

**L. serriola** L. PRICKLY LETTUCE Annuals or biennials, to 2 m tall; stems several, erect. Leaves oblong to elliptic, cauline ones 2-15 cm long, densely hispid to glabrate, margins dentate and often ± deeply lobed. Heads in open, branched panicles; involucres 8-12 mm long. Flowers 15-20 per head. Corollas pale yellow. Achenes (excluding beaks) 3-3.5 mm long; beaks 3.5-4 mm long. Jun-Aug. Introduced. Occasional; grassy slopes, seasonal ponds, flood plains, disturbed sites; coastal scrub, grasslands; 100-1200 ft (30-365 m). North side at Platts Harbor, Cañada del Puerto, and above China Harbor; Central Valley near Christy and Stanton ranches. All CA Channel islands except Santa Barbara; ± worldwide; native of Europe. First collected at Stanton Ranch in May 1962. Leaves of this species are edible and can be cooked as greens; seeds are the source of Egyptian lettuce seed oil.

## LASTHENIA Cass. GOLDFIELDS

Annuals or perennial herbs. Leaves opposite, entire to pinnatifid. Heads radiate, solitary to few, axillary or terminal; involucres ± hemispherical; phyllaries in 1-2 series, subequal, erect; receptacles conic to hemispherical, naked. Ray flowers pistillate, yellow; disk flowers bisexual, yellow. Achenes cylindrical to obovoid, smooth to angled; pappus absent or composed of scales or awns. 17 species; w N. America and Chile.

**L. californica** Lindl. [*L. chrysostoma (*Fisch. & C.A.Mey.) Greene] Annuals, to 30 cm tall; stems simple or branched, puberulent. Leaves linear to narrowly oblanceolate, 0.5-5 cm long, margins hispidulous. Involucres 3-8 mm long; phyllaries 8-13, ovate to broadly lanceolate, puberulent. Ray flowers 8-12, 4-9 mm long. Achenes 2.5-3 mm long, puberulent. Mar-Jul. Common; coastal bluffs and flats, grassy slopes, ridgetops; coastal scrub, grasslands; 30-1650 ft (9-505 m). Widely scattered locations throughout island, but most abundant and conspicuous near Fraser Point and near the Scorpion and Smugglers' ranches. All CA Channel islands; Los Coronados, Todos Santos, Guadalupe, and Cedros islands; OR to n Mexico, e to AZ.

## LAYIA Hook. & Arn. ex DC.

Annuals. Leaves sessile, lower ones sometimes opposite, upper ones usually alternate, entire to pinnately lobed, often strongly glandular. Heads usually radiate, solitary to several in open, terminal clusters; involucres broadly campanulate to hemispheric; phyllaries in ± 1-2 series, subequal, inner ones enfolding ray achenes; receptacles flat to slightly convex, with chaffy bracts subtending disk flowers. Ray flowers pistillate, corollas white to yellow; disk corollas bisexual, corollas yellow. Achenes of ray flowers compressed, with flattened sides parallel

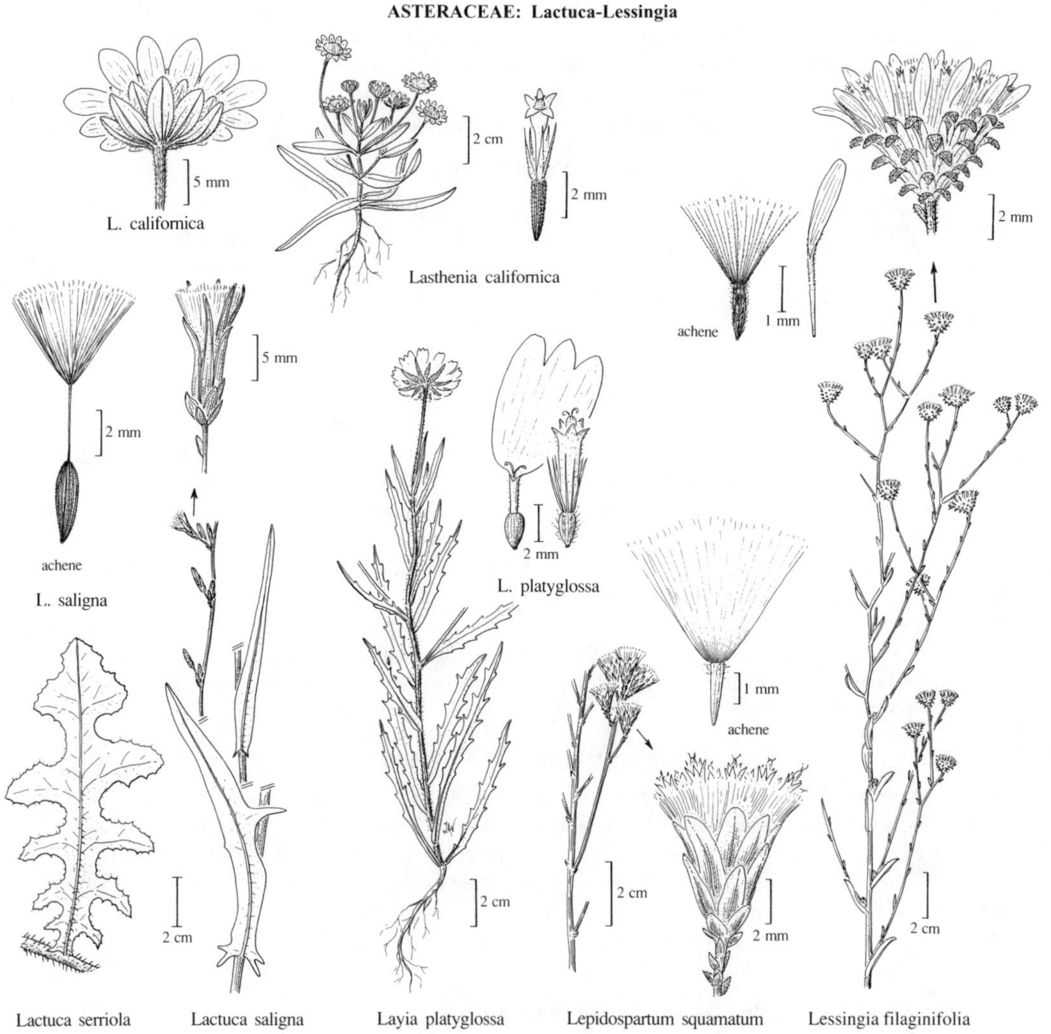


to plane of phyllaries, those of disk flowers obconic; pappus absent from ray flowers, that of disk flowers composed of awns or bristles. Ca. 15 species; w N. America.

**L. platyglossa** (Fisch. & Mey.) A.Gray TIDY TIPS Plants to 50 cm tall; stems simple or with many branches. Leaves ± lanceolate, 3-10 cm long, entire to toothed, hispid, sessile. Outer phyllaries lanceolate, 8-12 mm long, hispid, apices rounded; inner phyllaries 3.5-6 mm long, apices acute. Ray flowers 8-14, yellow, often with white tips, apices with 3 teeth; disk flowers with black anthers. Ray achenes ca. 3 mm long, glabrous; disk achenes 2.5-3 mm long; disk pappus 3-4 mm long. Mar-May. Common; grassy slopes, ridgetops; coastal scrub, grasslands, pine forest; 20-2100 ft (6-655 m). Widely scattered locations throughout island. San Miguel, Santa Rosa, Santa Catalina, and San Clemente islands; Guadalupe Island; much of CA w of Sierra Nevada. Achenes were eaten by Native Americans.

## LEPIDOSPARTUM A.Gray

Shrubs. Leaves alternate, linear to scale-like, entire. Heads discoid, in open to dense racemes or panicles; involucres cylindrical to narrowly campanulate; phyllaries in several series, subequal, imbricated; receptacles ± flat, naked. Flowers bisexual, corollas yellow. Achenes ± cylindrical, smooth to angled, striate; pappus in 2-3 series, composed of many, barbellate, capillary bristles. 3 species; w N. America.

**L. squamatum** (A.Gray) A.Gray SCALE BROOM Shrubs rounded, to 1 m tall; young stems white-tomentose, becoming glabrous. Lower leaves oblanceolate, 5-20 mm long, early deciduous, upper ones scale-like, 1-4 mm long. Heads 10-13 mm long; phyllaries 5-7 mm long, subtended by leaf-like bracts. Achenes 3-4 mm long, with ca. 10 ribs, brown. Aug-Nov. Occasional; floodplains; riparian communities; 25-250 ft (7-75 m). North side in Cañada del Puerto; Central Valley near Islay Canyon. Santa Clara Co. to n Baja CA. Reportedly toxic to humans.

## LESSINGIA Cham.

Annuals, perennial herbs, or subshrubs. Leaves alternate, entire to pinnately lobed, reduced upward, glabrous to glandular. Heads discoid or radiate, axillary to terminal, solitary or in clusters; involucres hemispherical to turbinate; phyllaries in several series, imbricated, erect to spreading or reflexed; receptacles ± flat, naked. Ray flowers, when present, fertile to sterile, white, purple, or yellow; disk flowers bisexual, yellow. Achenes obovoid, pubescent; pappus of ray achenes absent, pappus of disk achenes composed of many white to brownish bristles. Ca. 15 species; w N. America.

**L. filaginifolia** (Hook. & Arn.) M.A.Lane var. **filaginifolia** [*Corethrogyne f.* (Hook. & Arn.) Nutt.] CUDWEED ASTER Perennial herbs to 6 dm tall, much branched from base; stems canescent, glandular, glabrate. Leaves linear-lanceolate to spatulate, 1.5-6 cm long, canescent to tomentose. Involucres 6-9 mm long; phyllaries ± linear. Ray flowers 1-1.5 cm long, corollas white to purple; disk flowers 8-10 mm long. Achenes ca. 3 mm long. Apr-Nov. Common; dry sandy washes, floodplains, rocky slopes, coastal bluffs; coastal scrub, grasslands; 20-1200 ft (6-365 m). North side between West Point and Profile Point; Central Valley in Cañada Christy drainage and near Stanton Ranch; s side between Cañada de los Sauces and Sierra Blanca ridge. All CA Channel islands except Santa Barbara and San Clemente; OR to Baja CA. Most of the varieties described under *Corethrogyne filaginifolia* appear to intergrade and need further study; however, plants referrable to *C. f.* var. *viscidula* (Greene) D.D.Keck appear to be distinctive on Santa Cruz Island. This race, with leaves green and glandular throughout and only obscurely tomentose, occurs near the Christy Pines and in the vicinity of the Stanton Ranch at elevations between 250-900 ft (75-275 m).

## MADIA Molina TARWEED

Annuals or perennial herbs, often aromatic or strongly scented. Lower leaves opposite, often in basal rosette, upper ones usually alternate, linear to lanceolate, entire to toothed. Heads radiate, solitary, terminal or axillary, in flat-topped clusters or dense racemes; involucres hemispherical to subglobose; phyllaries in ± 1 series, subequal, enfolding ray achenes; receptacles flat to slightly convex, naked except for 1 series of chaffy bracts between ray and disk flowers. Ray flowers pistillate, corollas yellow; disk flowers bisexual or staminate, yellow or reddish. Achenes compressed to cylindrical and angled; pappus absent or composed of awns or bristles. Ca 20 species; w N. and S. America.

1. Involucres 2-4.5 mm long; disk flowers 1-2 per head, glabrous ........ **M. exigua**
1' Involucres 8-12 mm long; disk flowers 2-13, puberulent
    2. Stems with stalked glands in upper half ........ **M. gracilis**
    2' Stems with stalked glands throughout ........ **M. sativa**

**M. exigua** (Smith) A.Gray THREADSTEM TARWEED Annuals, 5-35 cm tall; stems simple or branched, slender, puberulent, with stalked glands in upper half. Leaves 0.5-3 cm long. Heads on stiff, filiform peduncles, solitary or in open panicles; involucres 2-4.5 mm long. Ray flowers 5-8, ligules ca. 1 mm long. Disk flowers 1-2, glabrous. Ray achenes 2-3 mm long; disk achenes 1.5-3 mm long. Apr-Sep. Occasional; dry rocky slopes, ridgetops, washes; chaparral, coastal scrub, pine forest; 40-2000 ft (12-610 m). Widely scattered locations throughout island. Santa Catalina Island; British Columbia to Baja CA, e to MT.

**M. gracilis** (Smith) D.D.Keck SLENDER TARWEED Plants to 100 cm tall; stems simple or branched, slender, pubescent, with stalked glands in upper half. Leaves 3.5-8 cm long. Heads on flexuous peduncles, solitary or in open panicles; involucres 8-10 mm long. Ray flowers 8-12, ligules 1.5-4 mm long. Disk flowers 2-10, puberulent. Ray and disk achenes all alike, 3-5 mm long. Mar-Jun. Scarce; disturbed pastures and open sites; grasslands; 5-250 ft (2-75 m). North side between Platts and Prisoners' harbors; Central Valley near University of California field station. Santa Catalina Island; British Columbia to w S. America, e to MT and UT.

**M. sativa** Molina COAST TARWEED Annuals, 30-160 cm tall; stems simple or branched, pubescent, with stalked glands throughout. Leaves 3-14 cm long. Heads on flexuous peduncles, in congested to dense panicles; involucres 8-12 mm long. Ray flowers 7-10, ligules 1.5-3 mm long. Disk flowers 10-13, puberulent. Ray and disk achenes all alike, 3-5 mm long. Apr-Aug. Occasional; disturbed sites, rocky bluffs, rocky slopes, stream banks; oak woodland, pine forest; 5-750 ft (2-230 m). Widely scattered locations throughout western and central portions of island. Santa Catalina and and San Clemente islands; Alaska to w S. America. This species has a long history as an oil crop and is cultivated in Europe. Oil from the achenes is used as a substitute for olive oil, in the production of soap, and for the lubrication of machinery.

## **MALACOTHRIX** DC.

Annuals or perennial herbs, scapose to branched throughout and leafy. Leaves alternate, subentire to pinnatifid, sessile. Heads ligulate, terminal or axillary, solitary or in panicles; involucres campanulate to hemispheric; phyllaries in several series, sometimes imbricated, outer ones shorter than inner ones or graduated, inner ones subequal; receptacles flat, naked or with short bristles. Flowers ligulate, bisexual, corollas yellow or white, often with purplish stripe. Achenes ± fusiform, with 15 veins, of which 5 are often prominent; outer pappus either absent or of 1-18 scales or smooth, persistent bristles; inner pappus of soft bristles, barbellate in lower third, fused and falling in 1 unit. 22 species; w N. America, Argentina, and Chile.

1. Perennial herbs, either suffrutescent or with woody, underground base
    2. Corollas yellow; leaves white-tomentose, especially when young; inner phyllaries obtuse .... **M. incana**
    2' Corollas white; leaves ± glabrous; inner phyllaries attenuate ...................... **M. saxatilis** var. **implicata**

1' Annuals, stems herbaceous throughout
    3. Heads subglobose; outer phyllaries orbicular, margins widely scarious ............................... **M. coulteri**
    3' Heads campanulate to hemispheric; outer phyllaries lanceolate to ovate, margins narrowly scarious
        4. Outer corollas 10-20 mm long, exserted 5 or more mm beyond phyllaries ................. **M. squalida**
        4' Outer corollas 4-8 mm long, exserted only 1-4 mm
            5. Outer pappus absent ........................................................................................ **M. indecora**
            5' Outer pappus of ± 18 white, scarious teeth and 1 persistent bristle ....................... **M. similis**

**M. coulteri** A.Gray SNAKES-HEAD Annuals, to 1.5 (5) dm tall. Leaves linear to lanceolate, dentate to pinnatifid, glabrous. Heads subglobose, 12-15 mm long; outer phyllaries strongly imbricated, orbicular to widely ovate, margins conspicuously scarious; receptacles sparsely pubescent to glabrate. Corollas 8-12 mm long, pale yellow, rarely white, partly exserted. Achenes 1.5-3 mm long; outer pappus of 2-6, persistent, brownish to tan bristles. Mar. Rare; not seen recently; coastal slopes and bluffs; coastal scrub; ca. 15 ft (5 m). Known only from 3 historic collections: in an unspecified location on Santa Cruz Island in 1888, on dry s-facing slope at Smugglers' Cove in March 1932, and at "base of cliff above beach" at Valley Anchorage in March 1929. Santa Rosa Island; Monterey Co. and San Joaquin Valley to s CA, e to UT & AZ.

**M. incana** (Nutt.) Torr. & A.Gray DUNE MALACOTHRIX or DUNEDELION Perennial herbs; stems several, to 2.5 dm tall, glabrous to densely tomentose. Leaves oblong to narrowly obovate, entire to lobed, somewhat fleshy, ± tomentose. Heads hemispheric, 15-20 mm long, slightly exceeding leaves; outer phyllaries lanceolate to ovate, inner ones obtuse; receptacles glabrous. Corollas 10-20 mm long, yellow, exserted. Achenes 1.5-2 mm long; outer pappus absent. Jan-Oct. A species of special environmental concern. Rare; not seen recently; sand dunes, coastal slopes, and canyons; coastal sand dunes, coastal scrub, coastal strand; ca. 5-100 ft (2-30 m). Previously known from "the western extremity of Santa Cruz, where it was growing in small quantity on a shaded sandstone terrace a little above the beach" (Greene 1886) and from n side in Cañada del Puerto (Jepson 1927). San Miguel, Santa Rosa, and San Nicolas islands; San Luis Obispo Co. to Ventura Co. (previously to San Diego Co.).

**M. indecora** Greene SANTA CRUZ ISLAND MALACOTHRIX Annuals, 2-4 dm tall. Leaves oblong to oblanceolate, fleshy, ± pinnatifid, glabrous to sparsely tomentose below, lobes rounded. Heads hemispheric, 8-10 mm long; outer phyllaries ± ovate, inner ones linear-lanceolate; receptacles with inconspicuous paleae. Corollas 4-8 mm long, greenish yellow, slightly exserted. Achenes 1-1.5 mm long; outer pappus absent. May-Jul. Island endemic. A species of special environmental concern. Rare; coastal bluffs, coastal flats; coastal scrub; ca. 25 ft (8 m). Currently known only from s side at Black Point; previously collected on "two or three execrable islets, nesting places of innumerable cormorants and gulls, close by the northern shore of Santa Cruz Island" (Greene 1886) and at Twin Harbor in July 1939. San Miguel Island.

**M. saxatilis** (Nutt.) Torr. & A.Gray var. **implicata** (Eastw.) H.M.Hall CLIFF MALACOTHRIX Suffrutescent perennial herbs; stems several, to 15 dm tall, glabrous to slightly pubescent. Leaves linear to oblanceolate, glabrous, 1-2 times pinnately dissected, lobes linear to lanceolate, occasionally toothed. Heads hemispheric, 15-20 mm long; phyllaries in several series, linear-lanceolate, acute; receptacles glabrous. Corollas 12-20 mm

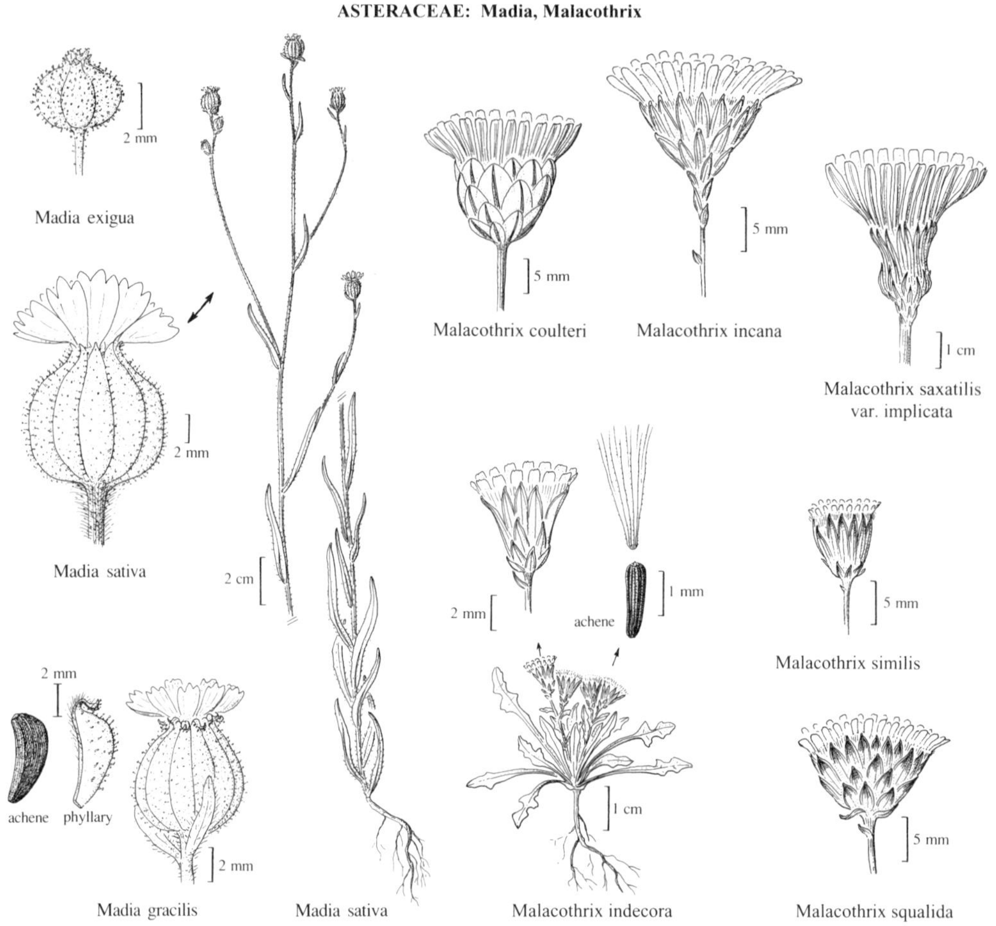


long, white with purple or pink stripes, exserted. Achenes 2-2.5 mm long; outer pappus of inconspicuous, unequal teeth. (Feb) Mar-Nov. Island endemic. Common; coastal bluffs, rocky slopes, canyon walls; coastal scrub, pine forest; 10-1300 ft (3-395 m). Widely scattered locations throughout island. San Miguel, Santa Rosa, Anacapa, and San Nicolas islands. Other varieties occur on the mainland from Monterey Co. to Santa Barbara Co.

**M. similis** W.S.Davis & Raven Annuals, 8-30 cm tall. Leaves linear-lanceolate, entire to lobed or pinnatifid, glabrous. Heads narrowly campanulate, 6-10 mm long; phyllaries lanceolate to ovate, acute to obtuse; receptacles ± naked. Corollas 4-8 mm long, pale yellow, slightly exserted. Achenes 1.5-2 mm long, slightly curved; outer pappus a crown of ca. 18 white-scarious, irregular, lance-deltoid teeth and 1 persistent bristle. Mar-May. Rare; not seen recently; the only known collections from Santa Cruz Island were made in 1888 and did not include specific locality data. San Miguel Island; Los Coronados, Todos Santos, and Cedros islands; Port Hueneme, Ventura Co. and n Baja CA.

**M. squalida** Greene ISLAND MALACOTHRIX Annuals, to 9 cm tall. Leaves linear to widely lanceolate, glabrous to ± tomentose below, irregularly toothed or lobed, lobes lanceolate, acute. Heads hemispheric, 12-15 mm long; outer phyllaries ovate, inner ones lanceolate; receptacles inconspicuously pubescent. Corollas 10-20 mm long, light yellow, exserted. Achenes 1.5-2 mm long; outer pappus of irregular teeth, persistent bristles absent. May. Island endemic. A species of special environmental concern. Rare; rocky coastal bluffs and flats; coastal scrub; ca. 50 ft (15 m). Collected just w of Potato Harbor in May 1968; previously known from "two or three execrable islets, nesting places of innumerable cormorants and gulls, close by the northern shore of Santa Cruz Island" and from a "rocky promontory above Prisoners' Harbor" (Greene 1886). Anacapa Island.

ASTERACEAE: Micropus-Osteospermum

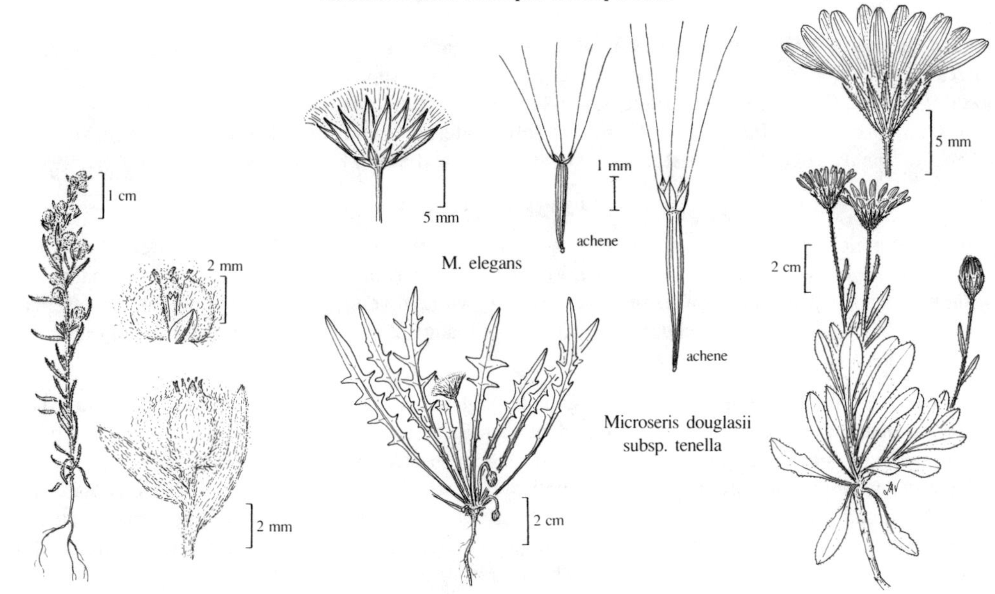


Micropus californicus var. californicus

Microseris elegans

Osteospermum fruticosum

## MICROPUS L.

Annuals, usually densely woolly or tomentose. Leaves alternate, entire, sessile. Heads discoid, solitary or 2-5 in dense, subglobose, bracteate clusters; involucres broadly hemispheric; phyllaries 4-6, equal ± rounded, persistent, midveins green, margins scarious; receptacles ± cylindrical, with chaffy bracts enclosing pistillate flowers. Flowers staminate or pistillate, outer ones pistillate, 4-10, enclosed by swollen, chaffy bracts that fall with fruit, styles attached laterally; inner flowers staminate, 2-5, usually not subtended by a bract. Achenes obovoid, compressed, brown, shiny, pappus absent, ovary of staminate flowers not developing but sometimes with a pappus composed of capillary bristles. 5 species; sw N. America and Europe.

**M. californicus** Fisch. & C.A.Mey. var. **californicus** SLENDER COTTONWEED Plants 5-20 cm tall; stems simple or branched at base. Leaves linear to narrowly lanceolate, 0.5-1.5 cm long. Heads 2.5-3.5 mm long; outer chaffy bracts 3-4 mm long; phyllaries densely woolly. Outer pistillate flowers 5-6. Achenes 1.5-2 mm long, obconic, gibbous. Jan-May. Common; rocky slopes and ridgetops; chaparral, coastal scrub, grasslands; 40-2000 ft (12-610 m). Throughout island. Santa Rosa and Santa Catalina islands; Guadalupe Island; OR to n Baja CA. Var. *subvestitus*, with outer chaffy bracts 2-3 mm long, occurs on the mainland in c CA.

## MICROSERIS D.Don

Annuals. Leaves ± basal, upper ones alternate, entire to pinnatifid, sessile. Heads ligulate, solitary on naked or bracteate scapes, ± nodding in bud; involucres cylindrical to hemispheric; phyllaries green, inner faces dark-pubescent, margins scarious, outer ones shorter than inner ones, ± imbricate, inner ones subequal, lanceolate to lance-elliptic; receptacles ± flat, naked. Flowers ligulate, bisexual, corollas yellow-orange, yellow, or white, slightly exserted. Achenes cylindrical or fusiform, ca. 10-ribbed; pappus composed of ca. 5 awn-tipped scales, awns minutely barbellate. Ca. 15 species; w N. America, Chile, Australia, and New Zealand.

1. Achenes ± fusiform, 3-6.5 mm long; pappus awns 3-7 mm long ........................ **M. douglasii** subsp. **tenella**
1' Achenes ± cylindrical, 1.5-3.5 mm long; pappus awns 1-4 mm long ............................................ **M. elegans**

**M. douglasii** (DC.) Sch.Bip. subsp. **tenella** (A.Gray) K.L.Chambers Plants to 1.5 dm tall. Leaves linear to lanceolate, to 10 cm long, scurfy to glabrate. Involucres 7-15 mm long. Achenes 3-6.5 mm long, ± fusiform; pappus scales less than 1 mm long, awns 3-7 mm long. Mar-Apr. Scarce; grassy mesas and slopes; grasslands; ca. 200 ft (60 m). South side near Christy Ranch. San Miguel and Santa Rosa islands; central California coast ranges and delta region of Great Valley, to s CA. The typical subspecies occurs widely on the mainland from OR to Baja CA.

**M. elegans** Greene ex A.Gray Plants to 2 dm tall. Leaves linear to lanceolate, to 10 cm long, scurfy to glabrate. Involucres 6-10 mm long. Achenes 1.5-3 mm long, ± cylindrical; pappus scales less than 2 mm long, awns 1-4 mm long. Mar-Apr. Occasional; open ridgetops and slopes, grassy mesas, rock faces in canyons; coastal scrub, grasslands, pine forest; 100-1600 ft (30-490 m). Widely scattered locations throughout island. San Miguel and San Clemente islands; central CA coast ranges, Great Valley, and Sierra Nevada foothills to s CA.

## OSTEOSPERMUM L.

Annuals or perennial herbs. Leaves alternate, entire to pinnately lobed, sessile or petiolate. Heads radiate, usually terminal, solitary; involucres broadly hemispheric; phyllaries in 1-4 series, subequal, imbricated; receptacles flat to convex, naked. Ray flowers pistillate, corollas white to yellow or reddish; disk flowers bisexual or staminate, corollas blue to purple. Achenes ± ovoid, smooth, angled, or tuberculate; pappus usually absent. Ca. 70 species; Africa and w Asia.

**O. fruticosum** L. Annuals; stems decumbent or sprawling, glandular. Lower leaves obovate to spatulate, 3-7 cm long, glabrous, margins entire or crenate, ciliate, upper ones linear, sessile. Involucres 10-12 mm long; phyllaries lanceolate, attenuate, midrib villous. Ray flowers 12-15, upper surfaces of ligules purple, lower surfaces white to pink; disk corollas purple. Achenes to 7 mm long, with 4-5 angles, grayish brown. Mar. Introduced. Rare; sandy beach; coastal strand; ca. 5 ft (1.5 m). South end of Christy Beach. Coastal s CA; native of s Africa. Collections were made from a single clump in March 1989; the population was then removed by Peter Schuyler of The Nature Conservancy to prevent further invasion. This species is toxic to livestock.

## PERITYLE Benth.

Annuals, perennial herbs, or shrubs. Lower leaves opposite, upper ones alternate, entire to pinnatifid, sessile to petiolate. Heads discoid to radiate, often terminal, solitary or 2-5 in clusters; involucres cylindrical to hemispheric; phyllaries in 1-3 series, ± equal; receptacles flat to ± conic, naked. Ray flowers pistillate, white or yellow; disk flowers bisexual, usually 4-lobed, yellow. Achenes linear to oblanceolate, strongly compressed, margins thickened, often pubescent; pappus absent or composed of minute scales or bristles. Ca. 75 species; sw N. America.

**P. emoryi** Torr. EMORY'S ROCK DAISY Annuals or perennial herbs, to 4 dm tall; stems simple to branched, sometimes suffrutescent, glandular, often glabrate below. Leaves ovate to cordate, 0.3-1.5 cm long, 0.5-1.5 cm wide, sharply and irregularly lobed and toothed. Heads usually radiate; involucres campanulate to hemispheric; phyllaries lanceolate, glandular, villous near apices. Ray flowers 10-12, corollas white. Achenes 2-3 mm long. Mar-Sep. Occasional; coastal bluffs; coastal scrub; 6-500 ft (2-150 m). Scattered locations around perimeter of island. All CA Channel islands except San Miguel and San Nicolas; all Baja CA islands except San Geronimo; coastal s CA, where uncommon, but widespread in deserts of CA, NV, AZ, and n Mexico.

## PICRIS L. OX-TONGUE

Annuals or perennial herbs; stems with milky sap. Leaves basal and cauline, alternate, entire to pinnatifid, hispid and often with pustulate trichomes. Heads ligulate, in racemes or panicles or solitary and terminal; involucres cylindrical to hemispheric; phyllaries in 1-3 series, imbricated; receptacles ± flat, naked. Flowers usually yellow. Achenes ± cylindrical, with 5-10 ribs, with beak between achene and pappus; pappus composed of plumose or simple bristles. Ca. 45 species; Europe, Asia, and Africa.

**P. echioides** L. BRISTLY OX-TONGUE Plants 3-10 dm tall; stems usually branched. Leaves oblong to oblanceolate, 4-18 cm long, dentate to irregularly lobed, upper ones sessile or clasping. Heads many, terminal and axillary, often in clusters, 1.5-3.5 cm long; involucres ± hemispheric; phyllaries ± erect, outer ones ovate, leaf-like, inner ones lanceolate. Achenes 2.5-3.5 mm long, beak 2.5-3.5 mm long; pappus plumose, white. May-Jun. Rare; eroded creek bank; coastal scrub; ca. 30 ft (9 m). Known from a single collection made in May 1992 in lower Alegria Canyon. San Nicolas and Santa Catalina islands; throughout N. America, native of Europe. Cooked leaves are edible.

## PLUCHEA Cass.

Annuals, perennial herbs, or shrubs. Leaves alternate, entire to toothed, sessile or petiolate. Heads discoid, usually many in terminal panicles; involucres ± hemispheric to campanulate; phyllaries 3-5 series, unequal,

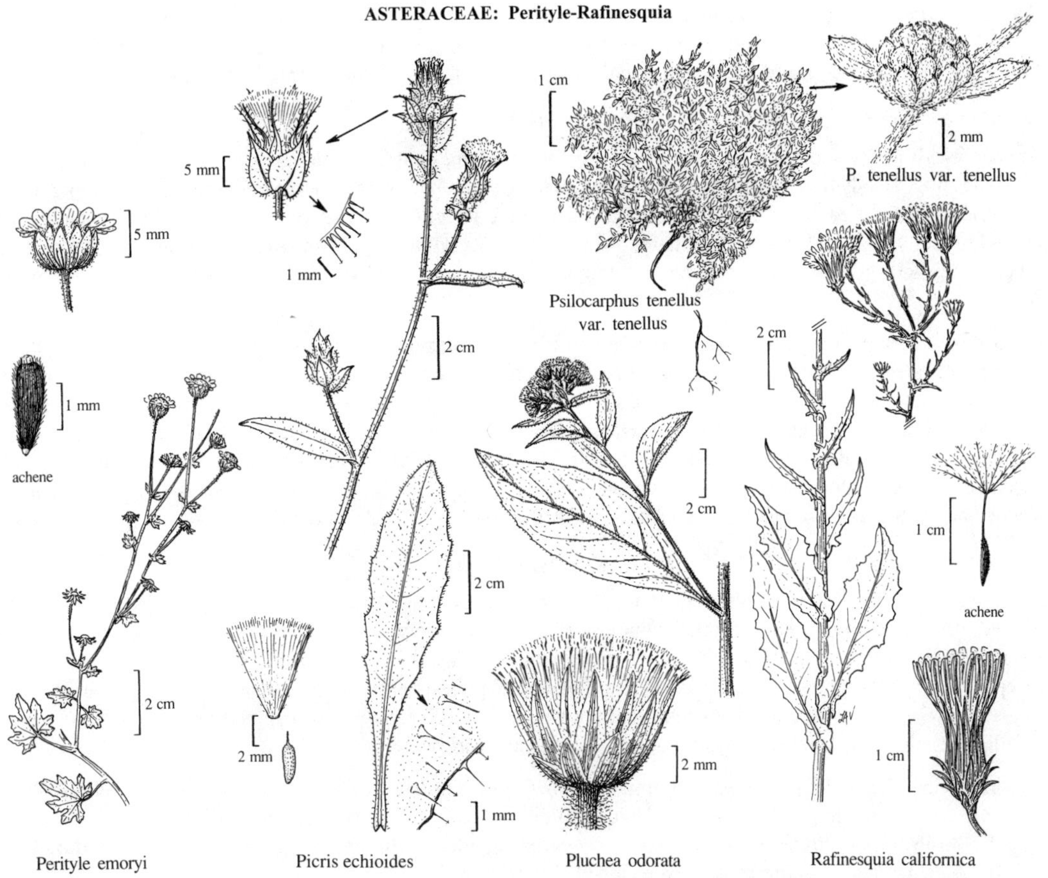


imbricated, outer ones thicker than inner ones; receptacles naked. Outer flowers pistillate, many, inner ones bisexual or staminate, few; corollas usually pink to purple. Achenes cylindrical to ± fusiform, with 4-5 grooves; pappus in 1 series, composed of capillary bristles. Ca. 40 species; warm temperate, worldwide.

**P. odorata** (L.) Cass. SALT MARSH FLEABANE Annuals or perennial herbs, to 12 dm tall; stems branched above base, glandular-pubescent, strongly scented. Leaves lanceolate to ovate, 5-15 cm long, glandular-puberulent, margins ± serrate, apices acute to acuminate. Heads many; involucres campanulate to hemispheric, 5-6 mm long; outer phyllaries lanceolate to ovate, inner ones linear to linear-lanceolate. Corollas purplish. Jul-Nov. Scarce; margins of creeks and lagoons; riparian communities; 10-200 ft (3-60 m). North side at Prisoners' Harbor and near China Harbor; s side in Cañada de Malva Real, Laguna Canyon, and at Valley Anchorage. Santa Catalina Island; San Francisco Bay region to n S. America, e to Atlantic Coast.

## **PSILOCARPHUS** Nutt. WOOLLY-HEADS

Annuals, often woolly or tomentose. Leaves mostly opposite, entire, sessile. Heads discoid, often in dense, subglobose, bracteate clusters; involucres broadly hemispheric; true phyllaries absent, outer flowers subtended by a whorl of phyllary-like bracts; receptacles subglobose to obovoid, chaffy. Outer flowers pistillate, each subtended by a woolly, often minutely beaked bract, inner ones staminate, 2-10, not subtended by bracts. Achenes cylindrical to obovoid, slightly compressed, smooth, shiny; pappus absent. Ca. 5 species; w N. America and S. America.

**P. tenellus** Nutt. var. **tenellus** SLENDER WOOLLY-HEADS Plants forming mats to 30 cm wide; stems with many branches, thinly woolly to villous. Leaves spatulate to oblong, 3-10 mm long, woolly. Heads solitary, terminal or in forks of branches. Pistillate flowers 20-35, staminate ones 5-7, corollas 5-lobed. Achenes obovoid, ca. 1 mm long, brown, shiny. Mar-Jun. Common; coastal bluffs, rocky ridgetops and slopes, disturbed flats, floodplains; coastal scrub, grasslands; 3-2000 ft (1-610 m). Scattered locations throughout island. Santa Rosa, Santa Catalina, and San Clemente islands; British Columbia to n Baja CA and Chile. Variety *globiferus* (DC.) Morefield, with 4-lobed staminate corollas and ovate leaves, occurs on the mainland of n CA.

## **RAFINESQUIA** Nutt.

Annuals, with milky sap. Leaves in basal rosettes or alternate, dentate to pinnately lobed, sessile, upper ones often clasping. Heads ligulate, solitary or in open, flat-topped clusters; involucres cylindrical to ± campanulate; phyllaries in 2-5 series, outer 2-3 series unequal, spreading to reflexed, inner ones equal, erect; receptacles flat to slightly convex, naked. Flowers ligulate, bisexual, corollas white to pinkish. Achenes ± fusiform, smooth to tuberculate, with beak between achene and pappus; pappus composed of plumose bristles. 2 species; sw N. America.

**R. californica** Nutt. CALIFORNIA CHICORY Plants 2-12 dm tall; stems erect, branched, glabrous, hollow. Leaves lanceolate to elliptic or oblong, 4-20 cm long, green, often purple tinged, glabrous. Outer phyllaries 3-6 mm long, inner ones 12-17 mm long. Corollas white, 5-8 mm long. Achenes 3-4 mm long, outer ones pubescent, dark mottled, inner ones glabrous, tan. Mar-May. Occasional; shaded slopes and floodplains; chaparral, island woodland, oak woodland, pine forest; 50-1100 ft. (15-335 m). Widely scattered locations throughout island, especially in the Central Valley. All CA Channel islands except San Miguel and San Nicolas; Los Coronados, Todos Santos, San Martin, and Cedros islands; n CA to Baja CA, e to UT and AZ.

## **SENECIO** L. GROUNDSEL or RAGWORT

Annuals or perennials. Leaves alternate, entire to dissected, petiolate to sessile and clasping. Heads radiate or discoid, often in terminal, ± flat-topped panicles; involucres cylindrical to campanulate or hemispherical; phyllaries in 2 series, linear to lanceolate, outer series much shorter than inner one, inner phyllaries equal, erect; receptacles flat to slightly convex, naked. Ray flowers, if present, pistillate, corollas yellow; disk flowers bisexual, corollas yellow to orange. Achenes cylindrical, with 5-10 veins or ribs; pappus composed of many fine, capillary bristles. Ca. 1500 species; temperate to tropical, worldwide.

1. Plants vine-like or climbing; leaves palmately lobed, base cordate ........................................ **S. mikanioides**
1' Plants annual or perennial but not vine-like or climbing; leaves pinnately veined to lobed, base not cordate
    2. Shrubs or subshrubs; leaves pinnately dissected, segments filiform ............. **S. flaccidus** var. **douglasii**
    2' Annuals; leaves entire to toothed or lobed
        3. Outer, shorter phyllaries with green tips; inner phyllaries ± 8 or ± 13 ...................... **S. aphanactis**
        3' Outer, shorter phyllaries with black tips; inner phyllaries ± 21 ..................................... **S. vulgaris**

**S. aphanactis** Greene RAYLESS RAGWORT Slender annuals, 4-25 cm tall; stems simple or branched, glabrous. Leaves coarsely toothed to lobed, 1-3.5 cm long, ± glabrous. Involucres 5-7 mm long; outer phyllaries with green tips, inner ones erect, usually 8, sometimes 13. Ray flowers inconspicuous, corollas less than 2 mm long; disk flowers equal to or slightly exceeding involucre. Feb-Apr. A species of special environmental concern. Rare; coastal bluffs, rocky ridgetops, open slopes; chaparral, coastal scrub; 50-1320 ft (15-400 m). Currently known only from ridge on n side of Cañada de los Sauces and Sierra Blanca Ridge; previously collected near Fry's Harbor in March 1934 and on ridge s of Rancho Portezuelo, near head of Willows Canyon, in March 1932. Santa Rosa and Santa Catalina islands; central CA to Baja CA.

**S. flaccidus** Less. var. **douglasii** (DC.) B.L.Turner & T.M.Barkley [*S. douglasii* DC.) BUSH GROUNDSEL Shrubs or subshrubs, 4.5-15 dm tall; stems with many branches, tomentose to glabrate. Leaves pinnately dissected, 1.5-8 cm long, segments filiform, upper surfaces glabrate, lower surfaces grayish tomentose. Involucres 7-10 mm long; outer phyllaries short, inner ones erect, ± 21. Ray flowers showy, corollas 1-1.5 cm long; disk flowers slightly exceeding involucre. Jun-Nov. Common; rocky slopes, ridgetops, floodplains; coastal scrub, chaparral; 25-2000 ft (8-610 m). Scattered locations throughout island. Santa Catalina and San Clemente islands; throughout CA e of Sierra Nevada foothills to Baja CA. Other varieties occur throughout w N. America. This species was used medicinally by Native Americans; reportedly toxic to livestock.

**S. mikanioides** Walp. GERMAN IVY Plants vine-like or climbing; stems to 6 m long, glabrous. Leaves palmately lobed, 5-17 cm long, glabrous, base cordate. Involucres 3-4 mm long; outer phyllaries short, inner ones erect, ± 21. Ray flowers absent; disk flowers exceeding involucre. Dec-Mar. Introduced. Rare, not seen recently; disturbed site; ca. 1300 ft (395 m). Known from a single collection made in January 1977 on S Ridge Road 0.3 mi e of Christy Pines. Santa Rosa Island; coastal CA; native of s Africa. Reportedly toxic to livestock.

**S. vulgaris** L. COMMON GROUNDSEL Annuals, 5-30 cm tall; stems simple or branched, glabrous or puberulent. Leaves coarsely toothed to lobed, 0.5-5 cm long, ± glabrous. Involucres 6-8 mm long; outer phyllaries with black tips, inner ones erect, usually 21, sometimes 13. Ray flowers absent; disk flowers equal to or slightly exceeding involucre. Feb-Apr. Introduced. Common; disturbed sites, rocky slopes; chaparral, grasslands, oak woodland, pine forest; 15-2000 ft (5-610 m). Throughout island. All CA Channel islands except Santa Barbara; throughout N. America; native of Europe. Reported for Santa Cruz Island by Brandegee (1888). This species has been used in folk medicine; toxic to livestock.

## SILYBUM Adans. MILK THISTLE

Annuals or biennials. Leaves alternate, sessile, coarsely toothed to lobed, dark green with white mottling, margins spiny. Heads discoid, solitary at ends of branches; involucres broadly ovoid to globose; phyllaries many, imbricated, margins and tips spiny, outer ones spreading to reflexed; receptacles flat to ± conic, densely bristly. Flowers bisexual, corollas radial, narrowly cylindrical, style tip with a swollen node. Achenes ± ovoid, slightly compressed; pappus composed of many ± flattened, barbellate bristles, deciduous. 2 species; Mediterranean Europe.

**S. marianum** (L.) Gaertn. Plants to 2 m tall, branched. Leaves oblanceolate to elliptic, basal clustered, to 50 cm long and 40 cm wide, petiolate, upper ones reduced in size, spinose, clasping. Heads 1.5-5 cm long; phyllaries to 6 cm long, coriaceous, brownish, apical spines green. Achenes 5-8 mm long, brown, with black spots. Apr-Jul. Introduced. Common; canyon bottoms, disturbed pastures and slopes; coastal scrub, grasslands; 20-1300 ft (6-395 m). Throughout island. Santa Rosa, Santa Barbara, San Nicolas, and Santa Catalina islands; throughout w N. America; native of Europe. Introduced to Santa Cruz Island prior to 1886 and already "abundant in the sandy beds of the broader canyons, both north and south" and "forming thickets impenetrable at the growing season of the year" (Greene 1887). Very young leaves and peeled upper stems are edible uncooked, roasted achenes have been used as a coffee substitute, and an infusion of the plant has been used in folk medicine.

## SOLIDAGO L. GOLDENROD

Perennial herbs, often with rhizomes or erect, underground stems. Leaves alternate, entire to toothed, subsessile. Heads radiate, often in flat-topped or one-sided panicles; involucres cylindrical to campanulate; phyllaries in 3-5 series, subequal, imbricated; receptacles flat to slightly convex, naked. Ray flowers pistillate, corollas short, yellow; disk flowers bisexual, corollas yellow. Achenes obconic, slightly compressed; pappus in 1 series, composed of many, barbellate bristles. Ca. 150 species; N. and S. America, Europe, and Asia.

**S. californica** Nutt. CALIFORNIA GOLDENROD Stems erect, 0.5-1.5 m tall, simple, canescent. Cauline leaves lanceolate, oblanceolate, or oblong, 4-12 cm long, upper ones elliptic, reduced, canescent. Heads in one-sided panicles; involucres 3-4 mm long; phyllaries glabrous to puberulent, acute to obtuse, margins scarious. Ray flowers 8-13; disk flowers 4-12. Achenes 1-1.5 mm long, puberulent. Aug-Nov. Occasional; rocky slopes, often in moist n-facing sites; chaparral, coastal scrub, pine forest; 20-1200 ft (6-365 m). Widely scattered locations throughout island. Santa Rosa and Santa Catalina islands; OR to Baja CA. Young leaves were eaten by Native Americans; leaves and roots were used medicinally.

## SONCHUS L. SOW THISTLE

Annuals, perennial herbs, or shrubs, with milky sap. Leaves alternate, ± entire to pinnatifid, sessile, upper ones clasping. Heads ligulate, often few in terminal, open, flat-topped or ± umbellate clusters; involucres campanulate; phyllaries in ± 3 series, phyllaries of outer series short, triangular, those of inner 2 series linear; receptacles flat to slightly convex, naked. Flowers ligulate, bisexual, corollas yellow. Achenes compressed ribbed to striate; pappus composed of many soft, capillary bristles. Ca. 55 species; Europe, Asia, and Africa.

1. Lobes at leaf base rounded; leaf margins dentate and spinulose; achenes with 3 ribs on each side, ribs smooth ........ **S. asper**

1' Lobes at leaf base acute; leaf margins irregularly toothed; achenes with 2-4 ribs on each side, ribs transversely wrinkled ........ **S. oleraceus**

**S. asper** (L.) Hill PRICKLY SOW THISTLE Annuals, to 2 m tall; stems angled, leafy. Leaves lanceolate to obovate in outline, to 20 cm long, margins dentate and spinulose, lobes at leaf base with rounded apices. Involucres 12-15 mm long; inner phyllaries narrowly ovate. Corolla ligule shorter than tube. Achenes 2.5-3 mm long, with 3 smooth ribs on each side. Mar-May (Sep). Introduced. Common; disturbed pastures, roadsides, floodplains, rocky slopes and ridgetops; coastal scrub, grasslands, riparian communities; 10-2450 ft (3-745 m). Scattered locations throughout island. All CA Channel islands except Santa Barbara; Cedros Island; throughout N. America; native of Europe. Introduced to Santa Cruz Island prior to 1886 (Greene 1887).

**S. oleraceus** L. COMMON SOW THISTLE Annuals, to 1.5 m tall; stems terete, striate, sparingly leafy. Leaves lanceolate to ovate in outline, to 25 cm long, margins, lobes at leaf base with acute apices. Involucres 10-14 mm long; inner phyllaries lanceolate. Corolla ligule length equal to tube. Achenes 2-2.5 mm long, with 2-4, transversely wrinkled ribs on each side. Mar-May (Sep). Introduced. Common; disturbed pastures, roadsides, floodplains, rocky slopes and ridgetops; coastal scrub, grasslands, riparian communities; 10-2450 ft (3-745 m). Scattered locations throughout island. All CA Channel islands; Los Coronados, Todos Santos, San Martin, Guadalupe, and Cedros islands; throughout N. America; native of Europe. Introduced to Santa Cruz Island prior to 1886 (Greene 1887). This species is edible and can be cooked for greens.

## STEBBINSOSERIS K.L.Chambers

Annuals. Leaves mostly in basal rosettes, entire to pinnatifid, linear to oblanceolate, sessile. Heads ligulate, solitary on naked or bracteate scapes, ± nodding in bud; involucres ± hemispheric; phyllaries in ± 2 series, green to tan, often striped or dotted with red, inner faces glabrous or puberulent, margins scarious, glabrous or puberulent, outer ones shorter than inner ones, inner ones subequal, ± lanceolate; receptacles ± flat, naked. Flowers ligulate, bisexual, corollas yellow or white, lower surfaces often reddish. Achenes fusiform, slender, with ca. 10 ribs; pappus of ca. 5, awn-tipped scales, awns minutely barbellate. 2 species; w N. America.

**S. heterocarpa** (Nutt.) K.L.Chambers [*Microseris heterocarpa* (Nutt.) K.L.Chambers] Plants to 2.5 dm tall. Leaves to 10 cm long, densely scurfy to glabrate. Involucres 10-20 mm long. Achenes 6-8 mm long; pappus scales 5-8 mm long, scarious, tan, apices bifid; awns 5-8 mm long. Apr-May. Scarce; mostly on grassy slopes; coastal scrub, grasslands; 250-900 ft (75-275 m). Central Valley in lower portion of Cañada Christy; s side along road to Fraser Point, on ridge s of Christy airstrip, and in upper portion of Pozo Canyon drainage. Santa Rosa, Anacapa, San Nicolas, Santa Catalina, and San Clemente islands; Guadalupe Island; n CA coast Ranges, Great Valley, and Sierra Nevada foothills to n Mexico, e to AZ.

## STEPHANOMERIA Nutt.

Annuals, perennial herbs, or shrubs, sap milky. Leaves in basal rosettes or alternate, often much reduced upwards, entire to pinnatifid, upper ones subsessile. Heads ligulate, terminal or axillary, solitary or in panicles; involucres cylindrical; phyllaries in 2-several series, ± imbricated, outer ones shorter than inner ones, inner ones subequal; receptacles flat to slightly convex, ± naked. Flowers ligulate, bisexual, corollas white to pink. Achenes cylindrical to ± fusiform, with ± 5 ribs, apices truncate; pappus composed of basally flattened, plumose bristles. 24 species; w N. America.

1. Phyllaries in several series, imbricated; involucres 10-13 mm long; plants perennial ............. **S. cichoriacea**

1' Phyllaries in ± 2 series, outer ones shorter than inner ones; involucres 6-9 mm long; plants annual

  2. Achenes smooth or ± rugose, but not grooved lengthwise ................................ **S. virgata** subsp. **virgata**

  2' Achenes grooved lengthwise

    3. Outer series of phyllaries reflexed; achenes 1.5-2.5 mm long; pappus bristles free ..... **S. diegensis**

    3' Outer series of phyllaries erect; achenes 2.5-3.5 mm long; pappus bristles often fused at base into groups of 4-5 ........................................................................................ **S. exigua** subsp. **coronaria**

**S. cichoriacea** A.Gray TEJON MILK-ASTER Suffrutescent perennial herbs; stems erect, 4-12 dm tall, branched, woolly, glabrate. Leaves woolly, glabrate, lower ones spatulate in outline, pinnatifid, cauline ones oblong to oblanceolate, entire to toothed. Involucres 10-13 mm long; phyllaries imbricated. Flowers 10-12, corollas pink. Achenes 3-6 mm long, striate, smooth; pappus plumose throughout, brownish. Aug-Oct. Occasional; coastal bluffs, rocky slopes and ridgetops, floodplains; chaparral, coastal scrub; 10-2450 ft (3-745 m). North side between Cueva Valdez area and Eagle Canyon; Central Valley between upper portion of Cañada Christy and Stanton Ranch, especially on slopes near Picacho del Diablo; s side at mouth of Horqueta Canyon and near Valley Anchorage. Santa Rosa Island; Monterey Co. to San Diego Co.

**S. diegensis** Gottlieb SAN DIEGO MILK-ASTER Annuals; stems erect, to 6 dm tall, branched, ± glabrous. Leaves ± glabrous or glabrate, lower ones narrowly oblong, coarsely toothed to pinnatifid, cauline ones much reduced. Involucres 6-8 mm long; phyllaries in ± 2 series. Flowers 10-13, corollas pink to white. Achenes

## ASTERACEAE: Senecio-Stylocline

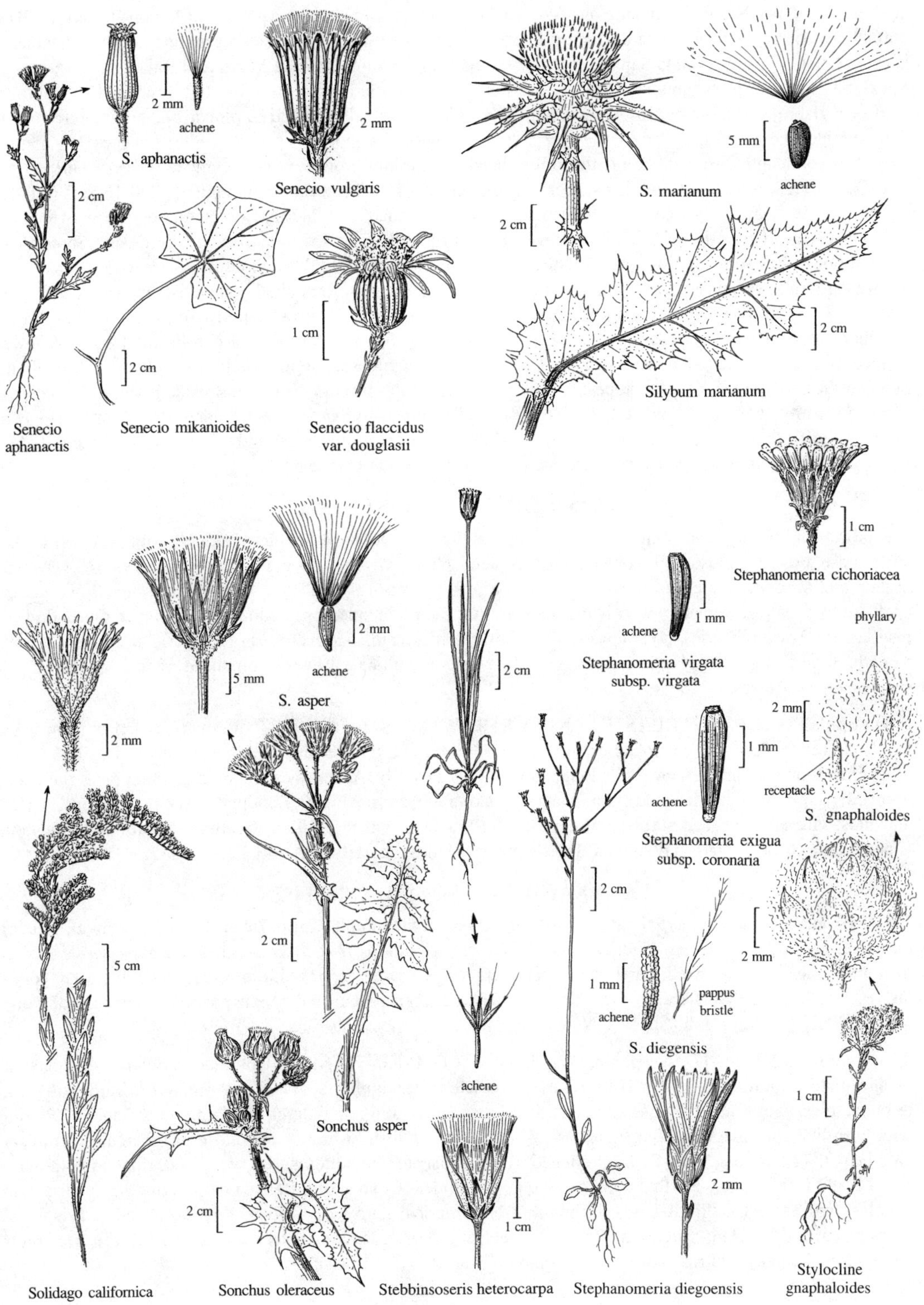

1.5-2.5 mm long, grooved lengthwise; pappus bristles free, plumose in upper half, white. Jul-Sep. Scarce; coastal bluffs, floodplains, rocky canyon walls and slopes; coastal scrub, oak woodland, riparian communities; 20-300 ft (6-90 m). North side in canyon above Lady's Harbor; s side in lower portion of Laguna Canyon and on bluffs between Laguna Canyon and Coches Prietos Anchorage. Santa Rosa, Santa Catalina, and San Clemente islands; Los Coronados, Todos Santos, and Guadalupe islands; Santa Monica Mtns. to n Baja CA.

**S. exigua** Nutt. subsp. **coronaria** (Greene) Gottlieb Annuals; stems erect, to 6 dm tall, branched, ± glabrous. Leaves ± glabrous or glabrate, lower ones narrowly oblong, coarsely toothed to pinnatifid, cauline ones bract-like. Involucres 6-8 mm long; phyllaries in ± 2 series. Flowers 5-6, corollas pink to white. Achenes 2.5-3.5 mm long, grooved lengthwise; pappus bristles often fused at base into groups of 4-5, plumose in upper half, white. Aug-Oct. Scarce; roadcuts, floodplains, rocky slopes; coastal scrub, riparian communities; 200-1600 ft (60-490 m). North side along Navy Road near Eagle Canyon and in Cañada del Puerto; Central Valley just w of Bosque Mano and near Stanton Ranch; s side in lower portion of Cañada de los Sauces. San Miguel and Santa Rosa islands; OR to Los Angeles Co., e to ID. Other subspecies occur widely throughout w N. America.

**S. virgata** Benth. subsp. **virgata** Annuals; stems erect, to 20 dm tall, branched, tomentose or glabrous. Leaves tomentose or glabrous, lower ones oblong to spatulate, sinuate to shallowly pinnatifid, cauline ones entire. Involucres 6-9 mm long; phyllaries in ± 2 series. Flowers 6-10, upper surfaces of corolla ligules white, lower surfaces pink to lavender. Achenes 2-3.5 mm long, smooth or ± rugose; pappus bristles free, plumose throughout. Sep-Oct. Scarce; roadcuts, rocky slopes; coastal scrub; 200-1300 ft (60-395 m). North side between Eagle Canyon and vicinity of Mount Pleasant. Santa Rosa, Santa Catalina, and San Clemente islands; San Luis Obispo Co. to San Diego Co. Subsp. *pleurocarpa* (Greene) Gottlieb occurs on the mainland ± away from the coast. The milky latex found in this species was used medicinally by Native Americans.

## STYLOCLINE Nutt. NEST STRAW

Annuals, often loosely woolly or tomentose. Leaves alternate, entire, ± sessile. Heads discoid, often in dense, subglobose, bracteate clusters; involucres broadly hemispheric to subglobose; phyllaries absent or 1-5, unequal, minute and scarious, scale-like or minute and scarious; receptacles slender, ± cylindrical, outer flowers each enclosed by a chaffy, winged, ± woolly bract, inner ones each subtended by a reduced chaffy bract. Outer flowers pistillate, in 3-several series, style apical; inner flowers staminate, 2-5. Achenes obovoid, slightly compressed, smooth, shiny; pappus of pistillate flowers absent, that of staminate flowers composed of few bristles. Ca. 6 species; sw N. America.

**S. gnaphaloides** Nutt. EVERLASTING NEST STRAW Stems 3-10 cm tall, with many branches from base. Leaves linear to narrowly lanceolate or spatulate, 2-9 mm long, those subtending heads longer. Heads 4-6 mm long; bracts subtending fruits ca. 3 mm long, outer surfaces woolly. Achenes ca. 1 mm long, inner faces flattened, grooved; pappus bristles thickened at tip. Mar-Aug. Common; coastal bluffs, floodplains, rocky ridgetops, slopes, and flats; chaparral, coastal scrub, grasslands; 10-1450 ft (3-440 m). Widely scattered locations throughout island. Santa Rosa, Santa Catalina, and San Clemente islands; coastal n CA to s CA, e to AZ.

## TARAXACUM F.H.Wigg. DANDELION

Perennial herbs, with milky sap. Leaves in basal rosettes, toothed to pinnatifid. Heads ligulate, terminal, solitary on naked scapes; involucres cylindrical to campanulate; phyllaries in 2-5 series, unequal, outer series shorter than inner ones, often reflexed, inner ones erect; receptacles ± convex, naked. Flowers ligulate, bisexual, corollas yellow. Achenes cylindrical to fusiform, with beak between achene and pappus; pappus composed of many soft, capillary bristles. Ca. 60 species; N. and S. America, Europe, and Asia.

**T. officinale** F.H.Wigg. [*T. laevigatum* (Willd.) DC.] COMMON DANDELION Leaves oblong, spatulate, or oblanceolate in outline, 5-25 cm long, lobes lanceolate to triangular, spreading or pointed toward leaf base, terminal lobes with acute to obtuse apices. Peduncles to 30 cm long; involucres 16-22 mm long; inner phyllaries with horn-like appendages at tips. Achenes 2-4 mm long, reddish to brown, spinulose or tuberculate in upper half, beak 6-11 mm long. Feb-Sep. Introduced. Rare; moist canyon bottoms; oak woodland, riparian communities; 20-500 ft (6-150 m). North side in canyon above Trident Cove and in Cañada del Puerto. Santa Rosa, San Nicolas, and Santa Catalina islands; throughout N. America; native of Europe. First collected at Prisoners' Harbor in May 1962. This species is often cultivated for its edible leaves and as a source of wine; the dried roots are used medicinally and roasted roots are sometimes used as a coffee substitute.

ASTERACEAE: Taraxacum-Xanthium

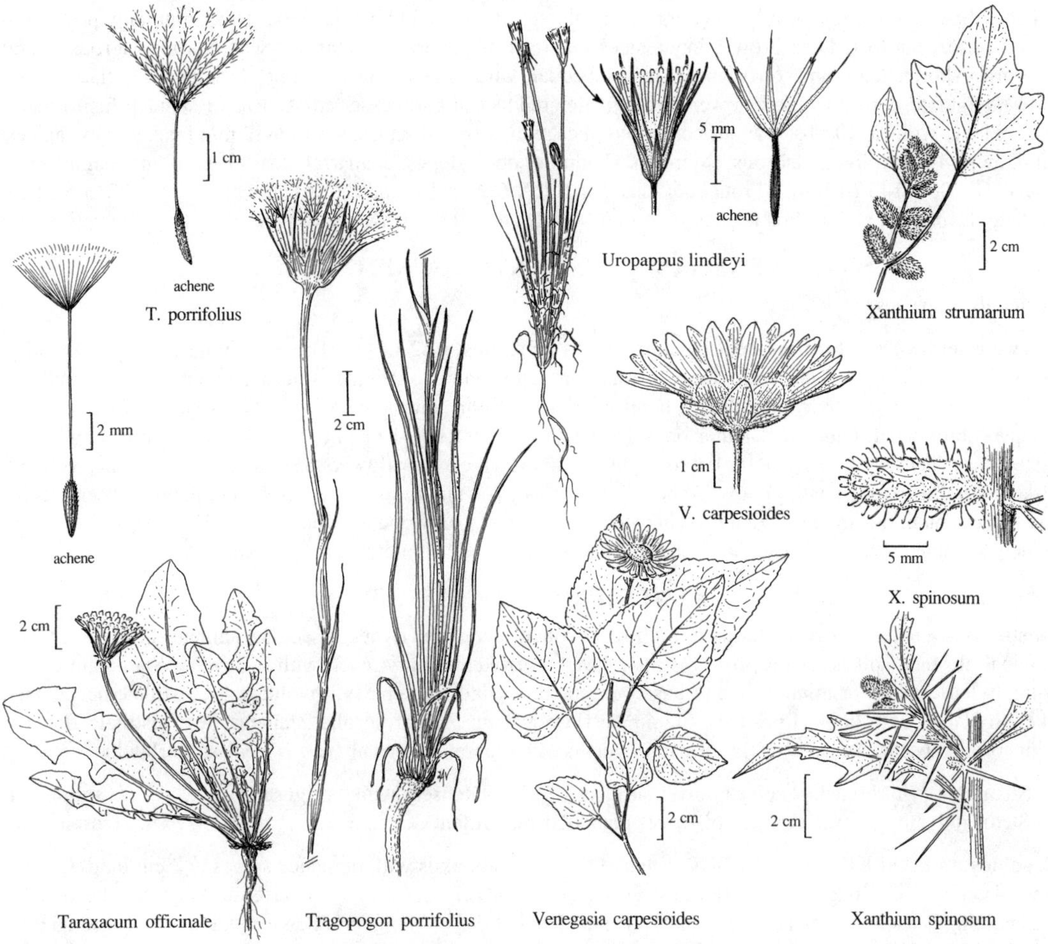


## **TRAGOPOGON** L. GOAT'S BEARD

Perennial herbs, often short-lived, with milky sap. Leaves basal and alternate, entire, linear, with parallel veins. Heads ligulate, terminal, solitary; involucres cylindrical to narrowly conic; phyllaries in ± 1 series, subequal; receptacles flat to slightly convex, naked. Flowers ligulate, bisexual, corollas yellow to purple. Achenes fusiform, with beak between achene and pappus; pappus composed of plumose bristles, bristles flattened near base. Ca. 45 species; Europe and Asia.

**T. porrifolius** L. SALSIFY or OYSTER PLANT Plants 5-9 dm tall. Leaves linear-lanceolate, 10-40 cm long, tapered to bases. Involucres 3.5-7 cm long; phyllaries glabrous. Corollas purple. Achenes (excluding beak) 9-14 mm long, beaks 15-40 mm long; pappus brownish. May-Jun. Introduced. Scarce; disturbed slopes; coastal scrub, grasslands; 20-500 ft (6-150 m). North side at Prisoners' Harbor; Central Valley between Bosque Mano area and Stanton Ranch airfield. San Nicolas and San Clemente islands; throughout N. America; native of Europe. First collected at the Stanton Ranch in June 1930. This species is widely cultivated for its edible roots; the latex from the roots has been used by Native Americans in British Columbia for chewing gum.

## **UROPAPPUS** Nutt.

With only 1 species.

**U. lindleyi** (DC.) Nutt. [*Microseris lindleyi* (DC.) A. Gray; *M. linearifolia* (DC.) Sch.Bip.] SILVER PUFFS Annuals, 1-4 dm tall; stems simple or branched. Leaves mostly in basal rosettes, linear to narrowly lanceolate, to

30 cm long, entire to pinnatifid, glabrous, puberulent, or sparsely scurfy, sessile. Heads ligulate, solitary on naked or bracteate scapes, usually erect in bud; involucres cylindrical to hemispheric, 10-25 mm long; phyllaries in ± 2 series, tan to reddish brown, inner faces glabrous or puberulent, margins scabrous to glabrous, outer phyllaries shorter than inner ones, inner ones subequal, lanceolate to lance-elliptic; receptacles ± flat, naked. Flowers ligulate, corollas yellow, lower surfaces often reddish, slightly exserted. Achenes ± slender, fusiform, 8-10 mm long, with ca. 10 ribs; pappus composed of ca. 5 awn-tipped scales, scales 6-12 mm long, silvery, apices bifid; awns 4-5 mm long, glabrous. Mar-Jun. Common; open slopes; chaparral, coastal scrub, grasslands, pine forest; 10-2000 ft (3-610 m). Throughout island. All CA Channel islands; Los Coronados, Todos Santos, San Martin, Guadalupe, and Cedros islands; WA to n Mexico, e to ID, UT, & TX.

### VENEGASIA DC. CANYON SUNFLOWER

With only 1 species.

**V. carpesioides** DC. Suffrutescent perennial herbs or shrubs, to 2 m tall. Leaves alternate, simple, broadly ovate, 2-13 cm long, 1-9 cm wide, bases cordate, margins entire to dentate, apices acuminate. Heads radiate, axillary or terminal, solitary; involucres hemispheric, 8-10 mm long; phyllaries in ± 2 series, 5-15 mm long, unequal, imbricated, puberulent, outer ones shorter than inner ones, leaf-like, broadly ovate, spreading to reflexed, inner ones erect; receptacles flat to slightly convex, naked. Ray flowers pistillate, 12-20, corollas yellow; disk flowers bisexual, many, yellow. Achenes cylindrical, 2-3 mm long, ribbed or striate; pappus absent. Jan-Aug. Occasional; mostly on shaded n-facing slopes; chaparral, oak woodland, pine forest; 10-1300 ft (3-400 m). Widely scattered locations throughout island. Santa Rosa Island; Monterey Co. to n Baja CA.

### XANTHIUM L. COCKLEBUR

Monoecious annuals. Leaves alternate, toothed to lobed, petiolate. Heads discoid; staminate heads many, in terminal clusters, spikes, or racemes; pistillate heads bur-like, axillary, each with 2 flowers. Staminate heads with phyllaries short or minute, free to base; receptacles cylindrical, chaffy; involucres of pistillate heads bur-like, phyllaries with hooked prickles. Staminate flowers with tubular corolla, stamens free; pistillate flowers with corolla absent. Achenes enclosed within bur-like involucre; pappus absent. 2 species; worldwide.

1. Stem axils with branched spines; lower surfaces of leaves densely white-strigose ..................... **X. spinosum**
1' Stems not spiny; lower surfaces of leaves glandular-puberulent ............................................ **X. strumarium**

**X. spinosum** L. SPINY COCKLEBUR Stems 1.5-10 dm tall, axils with branched spines to 2 cm long. Leaves lanceolate, 4-12 cm long, upper surfaces strigose, lower surfaces densely white-strigose. Staminate heads 0.9-1.2 cm long, densely glandular, puberulent; pistillate heads 1-1.5 cm long, prickles ± equal in length throughout. Apr-Dec. Introduced. Occasional; sandy beaches, disturbed pastures, dry washes, and floodplains; coastal scrub, coastal strand; 5-1500 ft (2-460 m). North side at Prisoners' Harbor and mouth of Scorpion Canyon; Central Valley near Christy Beach and Ranch and at foot of Portezuela Grade; s side in Pozo Canyon, along Playa Larga, and near mouth of Laguna Canyon. San Miguel, Santa Rosa, Santa Barbara, and Santa Catalina islands; worldwide. A weedy species first described from Europe, but region of origin uncertain. First collected at Prisoners' Harbor in October 1958; eradication efforts began at that time but were unsuccessful.

**X. strumarium** L. COCKLEBUR Stems 2-15 dm tall, axillary spines absent. Leaves lanceolate to orbicular, 4-12 cm long, bases triangular or cordate, upper surfaces densely glandular, lower surfaces ± glandular-puberulent. Staminate heads 1-1.5 cm long, puberulent; pistillate heads 1.2-2.3 cm long, prickles near apices longer and thicker than those below. Sep-Dec. Probably introduced. Rare; dry washes and disturbed pastures; coastal scrub, grasslands; 5-450 ft (2-140 m). Central Valley at mouth of Cañada Christy and e of Stanton Ranch; s side "at a Chinese fishing camp near the southern shore" (Greene 1887). San Nicolas and Santa Catalina islands; worldwide. A weedy species first described from Europe, but region of origin uncertain. This species can cause contact dermatitis in some humans; can be toxic to livestock.

## BERBERIDACEAE Juss.

### Barberry Family

Perennial herbs with rhizomes or shrubs. Leaves alternate, simple to compound, deciduous or evergreen; stipules absent. Flowers bisexual, radial, in racemes, spikes, or panicles. Perianth parts and stamens free, usually in multiples of 3; sepals 6-18 or absent; petals often 6; stamens 6-12; ovary superior, locule 1, style and stigma 1.

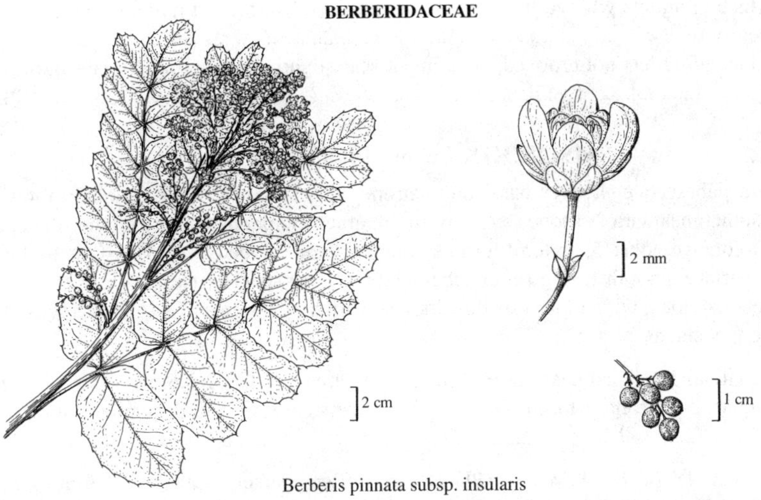

Berberis pinnata subsp. insularis

Fruit a berry, capsule, or achene. Ca. 15 genera and 700 species; temperate to tropical, worldwide. Some genera are cultivated as ornamentals.

### BERBERIS L. BARBERRY

Shrubs; inner bark and wood often yellow. Leaves simple to pinnately compound with 5-11 leaflets. Sepals in 3 whorls of 3; petals in 1-2 whorls of 3; stamens 6-12, anthers opening by flap-like valves; ovary with several ovules per locule; stigma more or less capitate. Fruit a berry, more or less globose. 450-600 species; temperate, worldwide. Some species widely cultivated as ornamentals.

**Berberis pinnata** Lagasca subsp. **insularis** Munz [*Mahonia p.* (Lagasca) Fedde subsp. *i.* (Munz) Roof] ISLAND BARBERRY Shrubs 2-4 m tall; stems erect to sprawling or reclining on other plants. Leaves 8-25 cm; leaflets ovate to lanceolate, subentire to serrate. Perianth yellow. Fruit bluish black. Feb-Apr. Island endemic. A subspecies of special environmental concern. Rare; n-facing sites in canyons and on rocky slopes; chaparral, oak woodland, pine forest; 450-1100 ft (140-335 m). North side in upper parts of Hazard and Trident canyons; Central Valley in upper reaches of Cañada Christy. Santa Rosa (known from a single collection in 1930) and Anacapa (known from a single population on West Anacapa) islands. Subspecies *pinnata* occurs on the mainland. The leaves and bark were used medicinally by Native Americans; a brilliant yellow dye can be made from the roots and inner bark of the stems.

## BORAGINACEAE Juss.
### BORAGE FAMILY

Annuals, perennial herbs, or shrubs, often coarsely pubescent. Leaves in basal rosettes or cauline, alternate or opposite, simple, often entire; stipules absent. Flowers bisexual, radial in one-sided spikes or spike-like panicles, spikes usually coiled in bud, uncoiling in fruit. Sepals 5, ± free to fused; petals 5, fused; stamens 5, adnate to corolla, alternate lobes; pistil 1, ovary superior, usually deeply 4-lobed, locules 2, ovules 2-4, styles 1-2, inserted between lobes. Fruit ± a schizocarp, separating into 1-4 nutlets. Ca. 150 genera and ca. 2500 species; worldwide.

1. Perennials; ovary not strongly lobed; style absent, stigma at ovary apex, disc-like ................. **Heliotropium**
1' Annuals; ovary strongly lobed; style inserted between ovary lobes
   2. Nutlets widely spreading at maturity, margins and dorsal surface with hooked bristles; flowers axillary ................................ **Pectocarya**
   2' Nutlets erect at maturity, margins entire, dorsal surface smooth to tuberculate, but without hooked bristles; flowers in terminal spikes
      3. Corollas yellow to orange ................................ **Amsinckia**
      3' Corollas white

4. Ventral surface of nutlets with vertical groove or slit; dorsal surface of nutlet smooth to tuberculate but not prominently rugose .......... **Cryptantha**
4' Ventral surface of nutlets not grooved, attachment scar slightly to prominently raised; dorsal surface of nutlet ± rugose .......... **Plagiobothrys**

## **AMSINCKIA** Lehm. FIDDLENECK

Annuals, with hispid pubescence. Leaves basal and cauline, alternate, sessile or petiolate. Flowers in coiled spikes. Sepals ± 5, sometimes with 2-4 lobes strongly fused, erect and persistent in fruit; corolla campanulate to salverform, yellow to orange, lobes 5; style filiform, stigma capitate, 2-lobed. Nutlets 1-4, deltoid to ovoid, with ventral scar, dorsal surface smooth to rugose or tuberculate. Ca. 10 species; temperate w N. and S. America. Some species are heterostylous, with 2 kinds of flowers, one with long styles and ± short stamens, the other with short styles and ± long stamens.

1. Calyx lobes ± equal, all 5 ± fused near base; stems usually erect; nutlets 2-3.5 mm long ......... **A. menziesii**
1' Calyx lobes unequal, 2-3 strongly fused, others ± free; stems spreading to decumbent; nutlets 1.5-2.5 mm long .......... **A. spectabilis**

**A. menziesii** (Lehm.) A.Nelson & J.F.Macbr. Plants erect, often sparsely branched, 1-6 dm tall, hirsute and often strigose. Leaves 0.7-16 cm long, 0.2-2 cm wide, hirsute, lower leaves lanceolate, upper ones linear. Inflorescence bracteate below or bractless, bracts oblanceolate, 4-7 mm long. Calyx lobes 4-8 mm long, hispid; corolla 4-10 mm long, yellow, tube equal to calyx or exserted, limb 2-6 mm wide. Nutlets tan to brown, 2-3.5 mm long, tuberculate. Mar-May. Common; hillsides, ridgetops, and sandy flats; coastal scrub, grasslands; 25-1600 ft (8-490 m). Widely scattered locations throughout island. All CA Channel islands; British Columbia to Baja CA, e to ID and AZ.

1. Corolla strongly exserted from calyx, tube 7-10 mm long .......... var. **intermedia**
1' Corolla equal to calyx or slightly exserted, tube 4-7 mm long .......... var. **menziesii**

var. **intermedia** (Fisch. & C.A.Mey.) Ganders Leaves 2-16 cm long. Inflorescence ± bractless. Calyx lobes 4-8 mm long; corolla strongly exserted from calyx, bright yellow, tube 7-10 mm long, limb 2.5-6 mm wide. Nutlets ± brown, ± 3 mm long. Mar-May. Common; grassy hillsides and sandy flats; coastal scrub, grasslands; 25-1500 ft (8-460 m). Widely scattered locations throughout island. All CA Channel islands; Todos Santos and Cedros islands; British Columbia to CA, e to ID and AZ; British Columbia to Baja CA, e to ID. Seeds were pounded into flour, made into cakes, and eaten raw by Native Americans.

var. **menziesii** Leaves 0.7-9 cm long. Inflorescence bracteate near base, bracts oblanceolate. Calyx lobes 5-8 mm long; corolla equal to calyx or slightly exserted, pale yellow, tube 4-7 mm long, limb 2-3 mm wide. Nutlets ± tan, 2-3 mm long. Mar-May. Scarce; rocky hillsides and ridgetops; grasslands; 200-1600 ft (60-490 m). Central Valley near Lagunitas Secas and in Cañada del Medio near La Cascada, University of California Field Station, and Stanton Ranch. Anacapa and Santa Catalina islands; Guadalupe and Cedros islands; British Columbia to Baja CA, e to ID and AZ.

**A. spectabilis** Fisch. & C.A.Mey. var. **spectabilis** Plants with ± spreading to decumbent stems, stems 1-6 dm long, sparsely hispid. Leaves lanceolate, 1-15 cm long, 0.2-1.5 cm wide, lower surface ± pustulate, upper surface sparsely hispid. Inflorescence bracteate, bracts lanceolate, 5-10 mm long. Calyx 3-6 mm long, hispid, 2-3 lobes strongly fused; corolla 6-15 mm long, golden-yellow to orange, tube exserted, limb 2-5 mm wide. Nutlets dark brown, 1.5-2.5 mm long, tuberculate and weakly papillate. Mar-Apr. Scarce; sandy beaches and dunes; coastal strand and coastal dunes; 10-40 ft (3-12 m). South side at Forney's Cove, and at the mouths of Pozo and Willows canyons. All CA Channel islands except Santa Catalina; British Columbia to Baja CA. Variety *microcarpa* (Greene) Jepson & Hoover, with erect stems and free calyx lobes, occurs in coastal Santa Barbara and San Luis Obispo cos.

## **CRYPTANTHA** Lehm. ex G.Don

Annuals or perennial herbs, with hispid or pustulate pubescence. Leaves basal and cauline, sessile or petiolate, lower ones opposite, upper ones alternate. Flowers in coiled spikes. Sepals 5, erect and persistent or deciduous in fruit; corolla rotate to salverform, usually white, lobes 5, mouth of throat with white or yellow appendages; style filiform, stigma ± capitate, 2-lobed. Nutlets 1-4, deltoid to ovoid, with ventral, often grooved scar, dorsal surface smooth and shiny to rugose or tuberculate. Ca. 160 species; temperate w N. and S. America.

1. Calyx lobes 1-2 mm long in fruit; nutlets of 2 kinds, 1 usually smooth, shiny, other 2-3 papillate ................................ **C. micromeres**
1' Calyx lobes 2-6 mm in fruit; nutlets all alike, either all smooth or all tuberculate
   2. Calyx lobes linear to narrowly lanceolate in fruit; nutlets smooth, shiny, white to grayish, often brown-mottled ................ **C. clevelandii**
   2' Calyx lobes lanceolate in fruit; nutlets tuberculate, brown ................ **C. muricata**

**C. clevelandii** Greene Annuals, 1-5 dm tall, stems erect to decumbent. Leaves linear, 1-5 cm long, ± appressed-pubescent. Calyx lobes linear in fruit, 2-6 mm long, strigose and hispid; corolla limb 1-3 mm wide. Nutlets 1-4, ± equal in shape and surface, 1-2 mm long, smooth, shiny, white to grayish, often brown-mottled. Feb-May. Common; sandy sites, rocky slopes, ridgetops, and floodplains; chaparral, coastal scrub, grasslands, and oak woodland; 15-1650 ft (5-503 m). Widely scattered locations throughout island. All CA Channel islands except San Nicolas; Todos Santos Island; n CA coast ranges to Baja CA.

**C. micromeres** (A.Gray) Greene Annuals, 1-5 dm tall, stems erect. Leaves linear to oblong, 1-4 cm long, ± hispidulous. Calyx lobes linear to narrowly lanceolate in fruit, 1-2 mm long, hispidulous; corolla limb 0.5-1 mm wide. Nutlets 3-4, 0.5-1 mm long, of 2 kinds, 1 usually smooth, shiny, brown, often longer than other 2-3, other 2-3 papillate, grayish brown. Mar-May. Scarce; rocky slopes, disturbed sites, and floodplains; chaparral and oak woodland; 150-1300 ft (45-395 m). North side near Pelican Bay; s side on South Ridge between Willows and Islay Canyon roads and near Coches Prietos Anchorage. Santa Rosa and Santa Catalina islands; San Francisco Bay region to Baja CA, e to Sierra Nevada foothills. This species is usually more common after fires.

**C. muricata** (Hook. & Arn.) A.Nelson & J.F. MacBr. [*C. m.* var. *jonesii* (A.Gray) I.M.Johnst.] Annuals, 2-10 dm, stems erect. Leaves linear to narrowly lanceolate, 0.5-4 cm long, strigose to hispid, some trichomes pustulate. Calyx lobes lanceolate in fruit, 2-4 mm long, strigose and hispid; corolla limb 1.5-5 mm wide. Nutlets 4, ± equal in shape and surface, 1-2 mm long, tuberculate, brown. Apr. Rare; rocky slopes; chaparral; elevation unknown. Known from a single collection on a rocky ridge near Pelican Bay in 1932. Much of CA to Baja CA.

## **HELIOTROPIUM** L. HELIOTROPE

Annuals, perennial herbs, or shrubs, glabrous to pubescent. Leaves alternate, petiolate to sessile. Flowers in coiled spikes or solitary. Sepals 5, erect and persistent in fruit; corolla rotate to campanulate, white to purple, lobes 5, sinuses between lobes plicate; ovary globose, not lobed, style ± absent, stigma linear to disc-like. Nutlets 2 or 4, ovoid to subglobose, with ventral, lateral scar, dorsal surface smooth to tuberculate, glabrous or pubescent. Ca. 250 species; temperate and tropical, worldwide.

**H. curassavicum** L. [*H. c.* var. *oculatum* (A.Heller) I.M.Johnst.] Perennial herbs, 1.5-3 dm tall, stems erect to prostrate, branching diffusely. Leaves cauline, oblanceolate to spatulate, 1-6 cm long, 0.5-2 cm wide, glabrous, succulent. Flowers in solitary or paired spikes 1-4 cm long, bractless. Calyx 2-4 mm long; corolla 2-6 mm wide, white to bluish, with yellow spots in throat, spots aging purplish. Nutlets subglobose, 2-2.5 mm long, brown. Mar-Sep. Occasional; sandy beaches, floodplains, and moist grassy flats; coastal dunes, coastal strand, grasslands; 10-200 ft (3-60 m). North side at Prisoners' Harbor; Central Valley at Christy Beach and in Cañada Cebada; s side in Cañada de los Sauces, on beaches from Pozo Canyon to Punta Arena, at mouth of Alamos Canyon, in Horqueta Canyon, and at Smugglers' Cove. All CA Channel islands except Santa Barbara; San Martin and Cedros islands; CA, e to UT and AZ, s to subtropical America. Seeds were eaten by Native Americans, who also used the leaves and roots medicinally.

## **PECTOCARYA** DC. ex Meisn.

Annuals with strigose to canescent pubescence; stems often spreading. Leaves alternate, ± linear, sessile to subsessile. Flowers usually solitary, axillary; pedicels short. Sepals 5, spreading to reflexed and persistent in fruit; corolla rotate to salverform, white, lobes 5, throat with white appendages; style short, stigma ± capitate. Nutlets 4, in pairs, spreading, ± compressed, margins entire to toothed, with hooked bristles or prickles. Ca. 15 species; temperate w N. and S. America.

1. Nutlet margins dentate, teeth often confluent, white ................ **P. linearis** subsp. **ferocula**
1' Nutlet margins ± entire to weakly toothed ................ **P. penicillata**

**BORAGINACEAE: Amsinckia-Heliotropium**

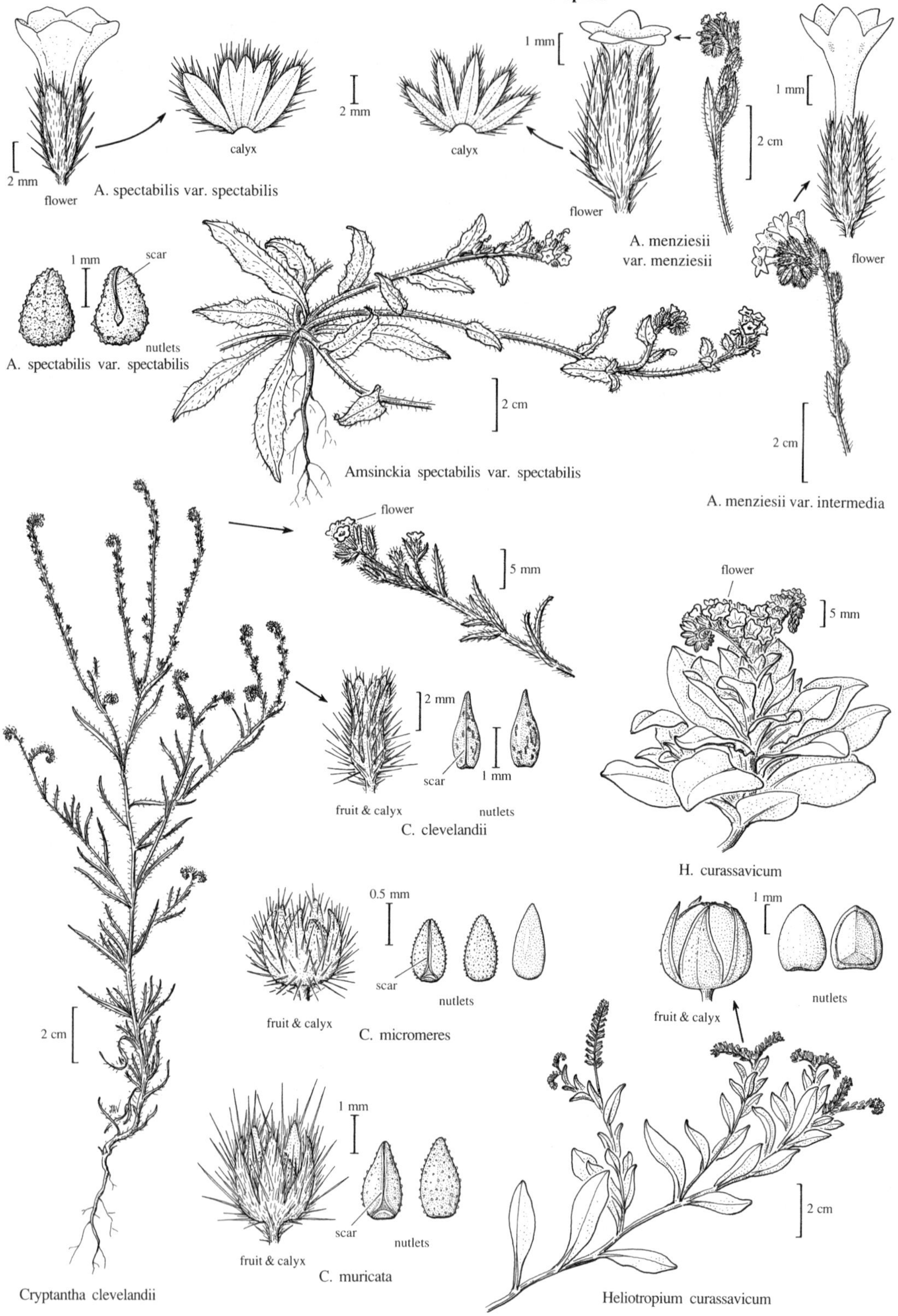

BORAGINACEAE: Pectocarya, Plagiobothrys

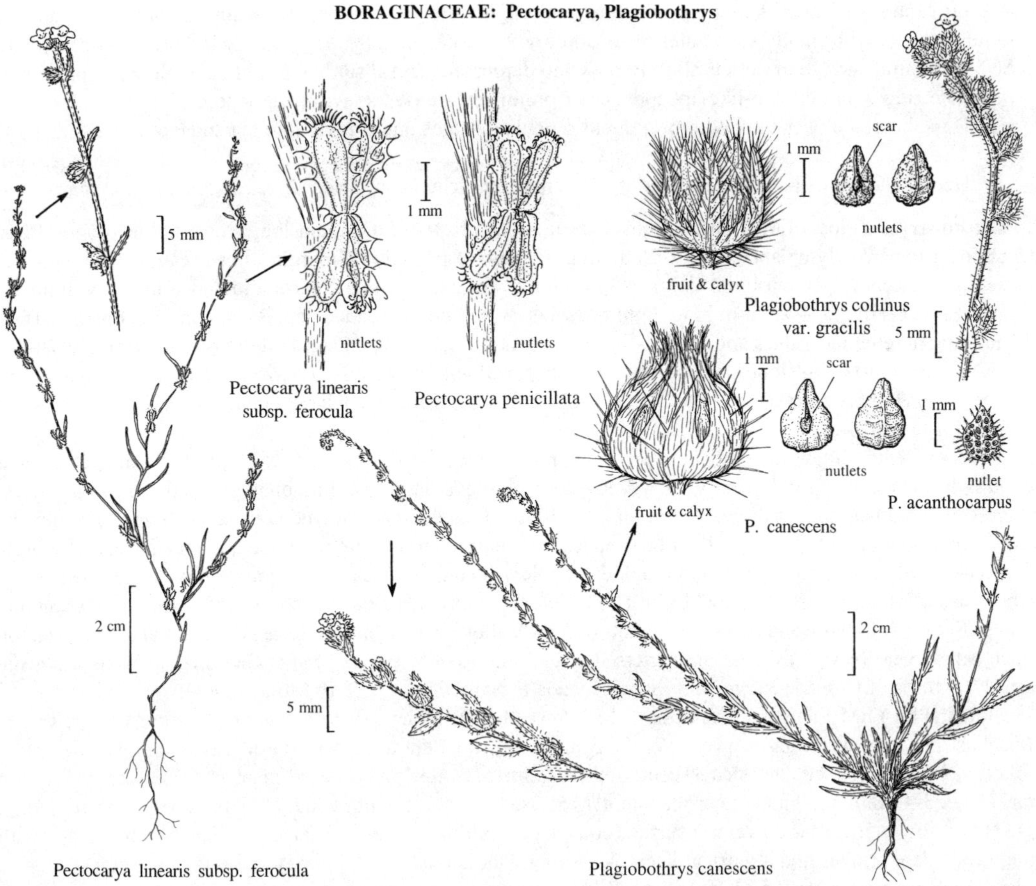


**P. linearis** (Ruiz and Pav.) DC. subsp. **ferocula** (I.M.Johnst.) Thorne Stems decumbent to spreading. Leaves filiform to narrowly oblanceolate, 0.5-2.5 cm long. Calyx lobes shorter than nutlets; corolla ± 2 mm long. Nutlets linear to ± oblong, 2-3.5 mm long, 0.5-1 mm wide, margins incurved, dentate, teeth often confluent, white, with hooked bristles. Feb-May. Common; sandy flats, rocky slopes, ridgetops, and floodplains; chaparral and coastal scrub; 40-2100 ft (12-640 m). Scattered locations throughout island. All CA Channel islands except San Miguel and Santa Barbara; Guadalupe and Cedros islands; San Francisco Bay region to Baja CA, e to Great Central Valley and Mojave Desert. Subsp. *linearis* occurs in S. America.

**P. penicillata** (Hook. & Arn.) A.DC. Stems prostrate to widely spreading. Leaves linear to narrowly spatulate, 1-2 cm long. Calyx lobes slightly shorter than nutlets; corolla ± 2 mm long. Nutlets oblong, 1.5-2.5 mm long, 0.5-0.8 mm wide, margins incurved, ± entire to weakly toothed, with hooked bristles. Mar-Apr. Rare; coastal sand dunes; ca. 20-50 ft (6-15 m). Known from a single collection near Fraser Point in 1960. British Columbia to n Baja CA, e to WY and AZ.

## PLAGIOBOTHRYS Fisch. & C.A.Mey. POPCORN FLOWER

Annuals or perennial herbs, glabrous or with strigose pubescence. Leaves usually basal and cauline, sessile or petiolate, lower ones in rosettes or opposite, upper ones alternate. Flowers in coiled spikes. Sepals 5, sometimes partly fused, ± erect or appressed and persistent in fruit; corolla ± salverform, usually white, lobes 5, throat with white or yellow appendages; style filiform, stigma ± capitate, 2-lobed. Nutlets usually 4, deltoid to ovoid, with basal or ventral scar, dorsal surface usually rugose, sometimes also tuberculate or hispidulous. Ca. 65 species; temperate w N. and S. America.

1. Scar on ventral surface of nutlet conspicuously raised, minutely stipe-like; dorsal surface of nutlets papillate to tuberculate, with faintly reticulate ridges and a vertical keel ............................... **P. collinus** var. **gracilis**
1' Scar on ventral surface of nutlets slightly raised to depressed; dorsal surface of nutlets with either prominent reticulate ridges and prickle-like appendages or prominent transverse wrinkles or folds
 2. Dorsal surface of nutlets with prominent reticulate ridges and prickle-like appendages ........................ ................................................................................................................ **P. acanthocarpus**
 2' Dorsal surface of nutlets with prominent transverse wrinkles or folds ............................... **P. canescens**

**P. acanthocarpus** (Piper) I.M.Johnst. Stems ascending to prostrate, 0.8-2 dm long; roots tan to brown. Basal leaves linear to narrowly oblanceolate, 1-2 cm long, strigose, especially on margins, upper cauline ones reduced in size. Inflorescence 2-10 cm long. Calyx 3-5 mm long, strigose to hirsute; corolla included in calyx, limb 1-3 mm wide. Nutlets ovoid, 1-2.5 mm long, light brown, scar on ventral surface depressed, concave, dorsal surface with prominent reticulate ridges and prickle-like appendages, longest appendages minutely and retrorsely barbed. Mar. Rare; rocky n-facing slopes and grassy flats; chaparral and grasslands; 100-750 ft (30-230 m). North side between Twin Harbors and Pelican Bay. Santa Catalina Island; Guadalupe Island; Great Valley and coastal Santa Barbara Co. to San Diego Co.

**P. canescens** Benth. Stems decumbent to prostrate, rarely erect, 1-5 dm long; roots purplish. Basal leaves linear to narrowly oblanceolate, 0.7-4 cm long, villous to hispidulous, upper cauline ones gradually reduced in size. Inflorescence 5-30 cm long. Calyx 3-5 mm long, ± villous-hirsute; corolla included in calyx, limb 2-4 mm wide. Nutlets round-ovoid, 1-2 mm long, light brown, scar on ventral surface slightly raised, dorsal surface obscurely tuberculate, with prominent, transverse wrinkles or folds. Feb-Jun. Occasional; grassy slopes and flats, flood-plains; coastal scrub, grasslands; 200-1650 ft (60-503 m). North side near Picacho Diablo, Mount Pleasant and between Scorpion Canyon and San Pedro Point; Central Valley near Lagunitas Secas, Cañada Larga, and Stanton Ranch; s side near Navy Base and Smugglers' Cove. Santa Rosa, Anacapa, Santa Catalina, and San Clemente islands; foothills of Cascade Range and Sierra Nevada to San Diego Co., e to Mojave Desert.

**P. collinus** (Phil.) I.M.Johnst. var. **gracilis** (I.M.Johnst.) L.C.Higgins [*P. californicus* (A.Gray) Greene var. *g.* I.M.Johnst.] Stems spreading or prostrate, 0.5-3 dm long; roots tan to brown. Basal leaves spatulate to oblanceolate, 1-4 cm long, ± hirsute to canescent, upper cauline ones few and reduced in size. Inflorescence 5-18 cm long. Calyx 3-4 mm long, hirsute; corolla slightly exserted, limb 2-2.5 mm wide. Nutlets ovoid, 1-1.5 mm long, light brown to ± white, scar on ventral surface conspicuously raised, dorsal surface papillate to tuberculate, with faintly reticulated ridges and a vertical keel. Mar-May. Occasional; rocky slopes and ridgetops, grassy slopes; coastal scrub, grassland; 20-1950 ft (6-595 m). Widely scattered locations throughout island. San Miguel, Anacapa, Santa Catalina, and San Clemente islands; Cedros Island; San Diego Co. Several other varieties occur elsewhere in CA and S.America.

## **BRASSICACEAE** Burnett [**CRUCIFERAE** Juss.]

### MUSTARD FAMILY

Annuals, perennial herbs, shrubs. Leaves basal and alternate, simple to pinnatifid; stipules absent. Flowers bisexual, radial, in racemes, spikes, or panicles. Perianth parts free; sepals 4; petals 4; stamens 6, tetradynamous; pistil 1, ovary superior, locules 2, usually separated by a thin septum, placentation parietal, ovules in 1-2 rows per locule, style 1, lobes 2. Fruit a silique (4 or more times longer than wide) or silicle (1-4 times longer than wide). Ca. 400 genera and 3000 species; temperate to tropical, worldwide. Some genera cultivated as ornamentals or for food (e.g., broccoli, brussel sprouts, cabbage, cauliflower, mustard, radish, water cress).

1. Fruit a silicle, 1-6 mm long, length usually less than 3 times width
 2. Silicles subterete to subglobose, sometimes inflated
  3. Leaves thick, succulent; silicles jointed, with 2 segments; plants of beaches and coastal dunes ..... ................................................................................................................ **Cakile**
  3' Leaves ± thin, not succulent; silicles not jointed into 2 segments, plants usually not on beaches and coastal dunes
   4. Petals white; leaf margins entire to dentate; cauline leaves usually clasping stem; silicles broadly ovate to cordate in outline ............................................................... **Cardaria**
   4' Petals yellow; leaves, especially lower ones, 1-3 times dissected; cauline leaves short-petiolate to sessile, not clasping stem; silicles oblong or narrowly elliptic in outline ........ **Descurainia**

2' Silicles flat, often strongly so, or cup-shaped, with winged margins in *Thysanocarpus*
5. Silicles oblong to broadly elliptic in outline
6. Silicle margins winged, wings entire, perforated, or lobed; seed 1 per silicle .................................................................. **Thysanocarpus**
6' Silicle margins not winged; seeds 2-many per silicle
7. Silicle apex notched; seeds 2 per silicle; stems and leaves with simple trichomes .................................................................. **Lepidium**
7' Silicle apex obtuse to truncate, entire; seeds 5-many per silicle; stems and leaves with branched trichomes
8. Stems usually solitary, erect; leaves mostly basal, entire; silicles usually stellate-puberulent, smooth or with inconspicuous veins .................................. **Draba**
8' Stems many, spreading to weakly erect; leaves basal and cauline, entire to deeply lobed; silicles usually glabrous, usually with conspicuously raised veins .................................................................. **Hutchinsia**
5' Silicles round to obcordate in outline
9. Silicles spreading to ascending or erect
10. Silicles triangular to obcordate in outline, apex truncate to broadly notched; seeds 4 or more per silicle .................................................................. **Capsella**
10' Silicles ± orbicular to broadly obovate in outline, apex narrowly notched; seeds 2 per silicle .................................................................. **Lepidium**
9' Silicles reflexed to pendulous
11. Silicles puberulent (trichomes hooked), margins without wings; leaves puberulent, trichomes branched .................................................................. **Athysanus**
11' Silicles glabrous, margins winged; leaves glabrous to sparsely puberulent, trichomes simple .................................................................. **Thysanocarpus**
1. Fruit a silique, 8-90 mm long, 3 to many times longer than width
12. Siliques transversely jointed, indehiscent lengthwise (breaking transversely into 2 segments in *Cakile*)
13. Leaves succulent; silique with 2 segments, each with 1 seed, apical beak, if present, flat; plants of beaches and coastal dunes .................................................................. **Cakile**
13' Leaves not succulent; silique usually with 3 or more segments, beak terete, 1-2 cm long; plants of interior habitats or coastal bluffs .................................................................. **Raphanus**
12' Siliques dehiscent lengthwise, not transversely jointed
14. Siliques with apical beaks (elongated, seedless segment between ovary and style) 3-20 mm long
15. Siliques hirsutulous; cauline leaves with conspicuous basal lobes .................... **Sisymbrium**
15' Siliques glabrous to sparsely puberulent; cauline leaves without basal lobes
16. Silique valves with 1 prominent midvein .................................................. **Brassica**
16' Silique valves with 3-7 prominent veins
17. Sepals erect; siliques appressed, 8-10 mm long, beaks 3-5 mm long. **Hirschfeldia**
17' Sepals ± spreading to reflexed; siliques ascending to spreading, 20-30 mm long; beaks 8-12 mm long .................................................................. **Sinapis**
14' Silique apices without beaks, styles sometimes persistent, beak-like, but less than 3 mm long
18. Basal and lower cauline leaves simple, margins entire to dentate, sometimes deeply toothed in *Arabis* and *Guillenia*
19. Siliques strongly compressed, ± flat
20. Siliques linear to narrowly oblong in outline, 20-90 mm long; leaves basal and cauline .................................................................. **Arabis**
20' Siliques elliptic in outline, 4-6 mm long; leaves mostly basal ..................... **Draba**
19' Siliques slightly compressed to terete
21. Leaf margins entire to denticulate; leaves with appressed, forked trichomes; siliques spreading to ascending .................................................................. **Erysimum**
21' Leaves dentate to lobed with narrowly triangular teeth; leaves with spreading, simple trichomes; fully mature siliques pendulous ............................................. **Guillenia**

18' Basal and lower cauline leaves very deeply and pinnately lobed or compound
22. Stems hispidulous to hirsute; petals yellow
23. Lower cauline leaves densely pubescent, 1-3 times dissected, terminal and lateral segments ± equal in size; pedicels spreading in fruit ........ **Descurainia**
23' Lower cauline leaves glabrous to sparsely pubescent, pinnately lobed, terminal lobe much longer than lateral ones; pedicels erect to appressed ........ **Sisymbrium**
22' Stems glabrous to sparsely pubescent; petals white to lavender or purple, sometimes yellowish in *Guillenia*
24. Stems prostrate to decumbent, rooting at nodes; plants of aquatic habitats ........ **Rorippa**
24' Stems usually erect, not rooting at nodes; plants of dry habitats
25. Lateral leaf segments linear, margins entire ........ **Sibara**
25' Lateral leaf segments lanceolate to ovate or orbicular, margins of larger segments usually toothed
26. Lateral leaf segments broadly ovate to orbicular; siliques spreading to ascending, compressed, flat ........ **Cardamine**
26' Lateral leaf segments narrowly lanceolate to triangular; siliques pendulous, terete ........ **Guillenia**

## **ARABIS** L.

Short- to long-lived perennial herbs, glabrous to stellate-pubescent. Leaves basal and cauline, margins usually entire to dentate, basal leaves petiolate, cauline ones usually sessile. Pedicels in fruit erect to reflexed. Petals white to purple. Fruit a silique, linear, flattened parallel to septum between locules, smooth to inconspicuously veined; seeds many, in 1 or 2 rows per locule. Ca. 80 species; temperate N. America.

1. Stems slender; cauline leaves lanceolate to ovate, glabrous; petals 3-6 mm long; siliques ca. 1 mm wide ........ **A. glabra**
1' Stems thick, often suffrutescent; cauline leaves narrowly oblong, puberulent; petals 6-10 mm long; sliques 2-4 mm wide ........ **A. hoffmannii**

**A. glabra** (L.) Bernh. TOWER MUSTARD Short-lived perennial herbs, 4-7 dm tall; stems slender throughout. Basal leaves oblong to broadly lanceolate, 4-12 cm long, margins lobed to dentate, cauline ones lanceolate to ovate, 1.5-6 cm long, glabrous, bases clasping, auriculate, margins entire. Pedicels in fruit erect. Petals 3-6 mm long, yellowish white. Siliques linear, 2-8 cm long, ca. 1 mm wide, flattened to subterete. Mar-Apr. Rare, not seen recently; open rocky slopes, grassy hillsides; chaparral, grasslands, oak woodland; 200-450 ft (60-140 m). Central Valley between Stanton Ranch and El Sobrante. Much of N. America. Last seen on Santa Cruz Island in March 1932.

**A. hoffmannii** (Munz) Rollins HOFFMANN'S ROCK CRESS Perennial herbs, 1.5-7 dm tall; stems thick, often suffrutescent. Leaves gradually reduced upward, basal ones oblong to narrowly oblanceolate, 4-10 cm long, margins entire to dentate, upper ones narrowly oblong, 1.5-5 cm long, puberulent, sessile to ± clasping, margins entire. Pedicels in fruit erect. Petals 6-10 mm long, white to lavender. Siliques narrowly oblong, 6-9 cm long, 2-4 mm wide, ± flat. Feb-Apr. Island endemic. A species of special environmental concern. Rare; steep, n-facing, volcanic rock faces and rocky hillsides, coastal bluffs; chaparral, coastal scrub, grasslands; 200-1250 ft (60-380 m). North side on seacliffs at Platts Harbor and in canyon nearby; Central Valley n of Centinela and near El Sobrante. Santa Rosa Island, where last seen in December 1930. Restricted to igneous and metamorphic substrates.

## **ATHYSANUS** Greene

With only 1 species.

**A. pusillus** (Hook.) Greene Annuals, 5-30 cm tall; stems stellate-pubescent. Leaves mostly basal, oblanceolate to obovate, 3-20 mm long, sparsely stellate-pubescent, short-petiolate to sessile, margins entire to remotely toothed. Pedicels in fruit 1-3 mm long, recurved. Sepals erect, purplish; petals 1-2 mm long, white, sometimes absent. Fruit a silicle, round in outline, flat, 1-2.5 mm wide, reflexed, puberulent, trichomes hooked; locule 1; seed 1 per silicle. Feb-Apr (May). Occasional; rocky n-facing slopes, grassy hillsides; grasslands, oak woodland; 15-2400 ft (4-730 m). North side between Lady's and Prisoners' harbors and near El Montañon; Central Valley in Cañada Christy and Cañada de la Portezuela; s side in Coches Prietos drainage. Santa Catalina and San Clemente islands; Cedros Island; British Columbia to Mexico, e to MT.

## **BRASSICA** L. MUSTARD

Annuals; stems glabrous to stellate-pubescent. Leaves basal and cauline, simple to pinnately lobed or divided, basal ones petiolate, cauline ones short-petiolate to sessile. Pedicels in fruit erect to spreading. Petals usually yellow. Fruit a silique, linear to narrowly oblong, erect to reflexed, terete or slightly compressed parallel to septum between locules, smooth to angled; seeds many, usually in 1 row per locule. Ca. 35 species; temperate, Europe and Asia.

1. Cauline leaves short-petiolate to sessile; fruits stiffly ascending to appressed, 8-20 mm long .......... **B. nigra**
1' Cauline leaves clasping, auriculate; fruits ascending to somewhat spreading, 25-60 mm long .......... **B. rapa**

**B. nigra** (L.) Koch Plants 4.5-10 dm tall; stems glabrous to pubescent. Leaves reduced upward, basal ones 9-22 cm long, pinnately lobed to deeply dentate, lateral lobes spreading, upper cauline ones ovate to elliptic, short-petiolate to sessile. Pedicels in fruit 2-5 mm long. Petals 7-12 mm long, yellow. Siliques stiffly ascending to appressed, 8-20 mm long, glabrous, apical beak 3-7 mm long. Mar-Apr (Jun). Introduced. Occasional; disturbed pastures, grassy hillsides, talus slopes; coastal scrub, grasslands; 30-1000 ft (9-305 m). North side near Scorpion Canyon; Central Valley between Cañada de la Portezuela and Valley Anchorage; s side near Smugglers' Cove. All CA Channel islands; Guadalupe Island; N. America; native of Europe and Asia. Can be extremely abundant in years with heavy rainfall. First collected at Smugglers' Cove in June 1930. Seeds of this species are used medicinally and to produce commercial mustard; leaves are edible raw or cooked.

**B. rapa** L. [*B. campestris* L.] Plants 3-8 dm tall; stems glabrous to sparsely pubescent. Leaves reduced upward, basal ones 10-25 cm long, pinnately lobed, lateral lobes spreading, upper cauline ones narrowly ovate to oblong, clasping, auriculate. Pedicels in fruit 8-20 mm long. Petals 5-10 mm long, yellow. Siliques ascending to somewhat spreading, 25-60 mm long, glabrous, apical beak 5-15 mm long. Mar-Apr. Introduced. Scarce; disturbed pastures, grassy hillsides; coastal scrub, grasslands; 20-400 ft (6-120 m). North side near Prisoners' Harbor; Central Valley between Bosque Mano and Stanton Ranch. San Miguel, Santa Rosa, Santa Catalina, and San Clemente islands; Guadalupe Island; N. America; native of Europe. First collected near Stanton Ranch in June 1930. This species has been cultivated for its edible root; leaves and flowers also are edible; rape-seed oil is obtained from seeds.

## **CAKILE** J.Mill.

Annuals or perennial herbs; stems usually branched from base, glabrous. Leaves cauline, lobed to deeply divided, thick, usually succulent, usually glabrous, usually petiolate. Pedicels in fruit short, thick, widely spreading to reflexed. Petals white to purple. Fruit a silique, cylindrical to lanceolate in outline, thick, ± succulent, corky when dry, jointed, with 2 indehiscent segments; seeds 2 per silicle, 1 per segment. Ca. 7 species; N. America, Europe, and Asia.

**C. maritima** Scop. SEA ROCKET Annuals or short-lived perennial herbs, 0.5-6 dm tall; stems erect to prostrate. Leaves 2-13 cm long, deeply pinnately lobed. Petals obovate, 3-5 mm wide, pale purple to lavender, rarely white. Siliques 1.5-2.5 cm long. lower segment with 2 lateral lobes or horns. Jan-Dec. Introduced. Occasional; beaches, sandy areas; beach and dune vegetation; 0-50 ft (0-15 m). Sandy beaches around perimeter of island, especially on s side. All CA Channel islands except Santa Barbara; Todos Santos, San Martin, San Benito, Cedros, and Natividad islands; British Columbia to Baja CA; native of Europe. First collected on beach at the mouth of Cañada de los Sauces in October 1958. Entire plant is edible after cooking; young plants are edible raw. *C. edentula* (Bigelow) Hook. was reported from the island by Smith (1976).

## **CAPSELLA** Medik.

Annuals or short-lived perennial herbs, glabrous to sparsely pubescent, trichomes simple to stellate. Leaves mostly in basal rosettes, simple to deeply lobed, lower leaves usually petiolate, upper leaves entire, sessile, clasping the stem. Pedicels in fruit widely spreading. Petals white to pink. Fruit a silicle, usually obcordate in outline, apex truncate to notched, flat, compressed perpendicular to septum between locules; seeds 4 or more per locule. Ca. 4 species; temperate, Europe.

**BRASSICACEAE: Arabis-Cakile**

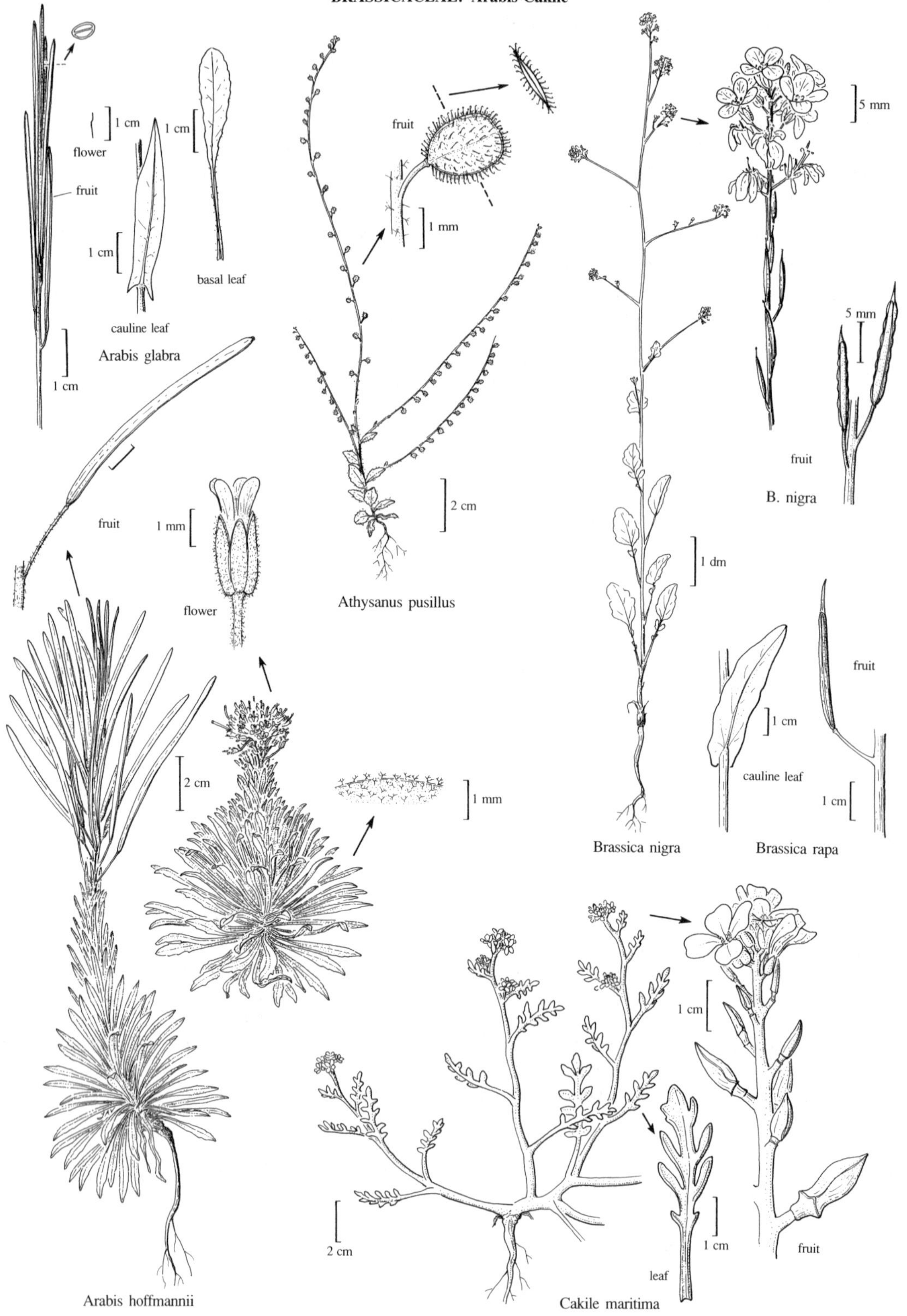

**C. bursa-pastoris** (L.) Medik. SHEPHERD'S PURSE Annuals, 7-40 cm tall; stems glabrous to sparsely pubescent. Leaves mostly basal, oblanceolate, 1.5-8 cm long, petiolate, margins subentire to pinnately lobed, cauline ones absent or reduced, entire, sessile to clasping. Pedicels in fruit 8-12 mm long, spreading. Petals 1-3 mm long, white. Silicles triangular to obcordate in outline, 4-6 mm long, apex truncate to broadly notched. Feb-May (Aug). Introduced. Common; disturbed sites, grassy hillsides, streambanks, and vernal ponds; coastal scrub, grasslands, riparian communities; 40-1850 ft (12-565 m). Widely scattered locations throughout island. San Miguel, Santa Rosa, Santa Catalina, and San Clemente islands; Guadalupe Island; N. America; native of Europe. First reported on island by Greene (1887). Leaves and seeds were eaten by the Cahuilla Tribe; root can be dried and used as a substitute for ginger; entire plant is an important source of vitamin K.

## CARDAMINE L.

Annuals or perennial herbs; stems glabrous to pubescent (trichomes simple), with thick taproots or thick rhizomes. Leaves basal and cauline, simple to pinnately or palmately compound, basal ones usually long-petiolate, cauline ones petiolate to subsessile. Pedicels in fruit ascending to widely spreading. Petals white to red. Fruit usually a silique, linear to oblong in outline, usually flat, compressed perpendicular to septum between locules; seeds many, in 1 row per locule. Ca. 170 species; temperate, worldwide.

1. Basal leaves trifoliolate, petioles 9-15 cm long; plants with deep-seated rhizomes; pedicels in fruit spreading ........ **C. californica**

1' Basal leaves pinnately compound, petioles 1-2 cm long; plants with taproots; pedicels in fruit ascending ........ **C. oligosperma**

**C. californica** (Torr. & A.Gray) Greene var. **californica** [*Dentaria californica* Nutt. ex Torr. & A.Gray] MILK MAIDS Perennial herbs, 25-45 cm tall, with rhizomes. Leaves of 2 kinds, either 1-few from rhizomes or basal and cauline and borne on erect stems with flowers and fruits. Leaves from rhizomes trifoliolate, leaflets broadly ovate to ± orbicular, 6-10 cm long, petioles 9-15 cm long; basal leaves of fertile stems similar to rhizome leaves, cauline ones trifoliolate, sometimes with 5 leaflets. Pedicels in fruit 1-2 cm long, spreading. Petals 8-12 mm long, white to pale lavender. Siliques 2-5 cm long. Feb-Mar (May). Occasional; shaded, n-facing slopes and canyon bottoms; oak woodland, pine forest; 150-1150 ft (45-350 m). North side between Fry's and Prisoners' harbors; Central Valley between Cañada Christy and near Stanton Ranch; s side in Cañada Larga and on Sierra Blanca ridge. San Miguel, Santa Rosa, Anacapa, and Santa Catalina islands; CA coast ranges to n Baja CA. Four other varieties occur from sw OR to Baja CA. Roots were eaten by the Maidu Tribe.

**C. oligosperma** Torr. & A.Gray BITTER CRESS Annuals, 8-15 cm tall. Leaves mostly basal, pinnately compound, leaflets broadly obovate to ± orbicular, 0.5-1 cm long, petioles 1-2 cm long; upper cauline leaves few, reduced in size. Siliques in fruit 3-10 mm long, ascending. Petals 1.5-4 mm long, white. Fruits 1-2 cm long. Mar. Rare, not seen recently; habitat, community, and elevation unknown. Based on a single collection by Hoffmann at an unspecified location in March 1925. Santa Catalina Island (introduced in nursery); British Columbia to Baja CA, e to MT and CO. This species was used medicinally and eaten raw or cooked by the Maidu Tribe.

## CARDARIA Desv.

Perennial herbs, often with rhizomes; stems glabrous to sparsely pubescent, trichomes simple. Leaves cauline, simple to deeply lobed, short-petiolate to sessile or clasping. Pedicels in fruit ascending to erect. Petals white. Fruit a silicle, subglobose to ovoid, indehiscent to weakly dehiscent; seeds usually 1 per locule. Ca. 5 species; temperate, Europe and w Asia.

**C. draba** (L.) Desv. HOARY CRESS Plants 2.5-6 dm tall, with well developed rhizomes; stems glabrous to sparsely pubescent. Leaves oblanceolate to obovate, 1.5-12 cm long, short-petiolate to sessile and clasping, margins dentate. Pedicels in fruit 5-10 mm long, spreading. Petals 2-4 mm long, white. Silicles broadly ovate to cordate in outline, 2-4 mm long, inflated, indehiscent. Mar-May. Introduced. Scarce; disturbed pastures, floodplains, grassy flats; grasslands; 20-750 ft (6-230 m). North side at Prisoners' Harbor and in Cañada del Puerto; scattered in Central Valley between upper portion of Cañada Christy and near Stanton Ranch. Santa Catalina Island; N. America; native of Europe and Asia. First collected "in grass fields, south of main ranch" in April 1931. Leaves and stems are edible.

## **DESCURAINIA** Webb & Berthel.

Annuals or short-lived perennial herbs; stems stellate-pubescent. Leaves cauline, reduced upward, 1-3 times divided or compound, subglabrous to stellate-pubescent, basal leaves withering in fruit, lower cauline ones usually petiolate, upper ones usually sessile. Pedicels in fruit ascending to erect. Petals usually yellow. Fruit a silique or silicle, linear or club-shaped, terete to subterete; seeds many, in 1 or 2 rows per locule. Ca. 40 species; temperate, worldwide.

**D. pinnata** (Walter) Britton subsp. **menziesii** (DC.) Detling TANSY MUSTARD Annuals, 0.6-7 dm tall; stems hispidulous, sometimes glabrous. Leaves pinnately dissected, 2-7 cm long, densely stellate-pubescent. Pedicels in fruit 4-10 mm long, spreading. Petals 1-3 mm long, yellow. Fruits oblong or narrowly elliptic in outline, 5-12 mm long; seeds usually in 2 rows per locule, often crowded. Mar-Apr. Scarce; disturbed sites, dry, rocky slopes and blufftops; coastal scrub; 40-100 ft (12-30 m). North side at Pelican Bay and at mouth of Scorpion Canyon; s side at Valley Anchorage and Smugglers' Cove. Santa Rosa, Anacapa, Santa Catalina, and San Clemente islands; Los Coronados, Todos Santos, and Guadalupe islands; San Francisco Bay region to s CA, e to Mojave Desert. Seven other subspecies occur in w N. America. Seeds and young leaves were eaten by Native Americans; ground seeds were used medicinally by the Cahuilla Tribe.

## **DRABA** L.

Annuals or perennial herbs; stems usually pubescent, trichomes forked to stellate. Leaves mostly in basal rosettes, petiolate to subsessile, simple, margins entire to dentate, cauline ones few, reduced upward. Pedicels in fruit widely spreading to erect. Petals white to yellow. Fruit a silique or silicle, linear to round in outline, usually compressed parallel to septum between locules; seeds 2-many, in 2 rows per locule. Ca. 350 species; temperate, worldwide.

**D. cuneifolia** Torr. & A.Gray var. **integrifolia** S.Watson Annuals, 2-10 cm tall; stems stellate-puberulent. Leaves mostly basal, ovate to broadly oblanceolate, 5-15 mm long, densely stellate-pubescent, margins entire to toothed. Pedicels in fruit 2-5 mm long, spreading to ascending. Petals 1-3 mm long, white. Fruit usually a silicle, oblong to narrowly elliptic in outline, 4-6 mm long; sparsely stellate-pubescent. Mar-Apr. Rare; rocky slopes and ridgetops; chaparral, coastal scrub; 50-1950 ft (15-595 m). North side on e face of Picacho Diablo; Central Valley in canyon just e of Christy Beach. San Nicolas, Santa Catalina, and San Clemente islands; Cedros Island; w N. America. Two other varieties occur in w North America.

## **ERYSIMUM** L.

Annuals, short-lived perennial herbs, or shrubs; stems stellate-pubescent. Leaves cauline, reduced upward, subglabrous to pubescent, trichomes forked to stellate, often appressed, margins entire to dentate or lobed. Pedicels in fruit widely spreading to erect. Petals usually yellow. Fruit a silique, linear, terete or 4-angled; seeds many, usually in 1 row per locule. Ca. 180 species; temperate, N. America, Europe, Asia.

**E. insulare** Greene ISLAND WALLFLOWER Perennial herbs, 1.5-3 dm tall; stems thick, often suffrutescent. Leaves reduced upward, crowded below middle, linear to linear-oblanceolate, 4-8 cm long, margins entire to denticulate. Pedicels in fruit stiffly spreading to ascending. Petals 10-20 mm long, yellow. Siliques 3-9 cm long, compressed to 4-angled. Mar-May. Island endemic. A species of special environmental concern. Rare, not seen recently; rocky slopes; coastal scrub; 20-60 ft (6-18 m). Known only from fragmentary collections made at an unspecified location in 1888 and on the outer rock at Willows Anchorage in 1966. San Miguel, Santa Rosa, and Anacapa islands. The closely related *E. suffrutescens* (Abrams) Rossbach, with mostly 3-branched trichomes, occurs on the coastal mainland of San Luis Obispo and Santa Barbara cos.

## **GUILLENIA** Greene

Annuals; stems glabrous to pubescent, trichomes simple. Leaves basal and cauline, simple to lobed, lower ones in rosettes, withering in fruit, cauline ones reduced upward, subglabrous to pubescent, trichomes forked to stellate. Pedicels in fruit reflexed or erect. Petals white to yellow. Fruit a silique, linear, terete; seeds many, usually in 1 row per locule. Ca. 2 species; temperate, w N. America.

**G. lasiophylla** (Hook. & Arn.) Greene [*Caulanthus lasiophyllus* (Hook. & Arn.) Payson] CALIFORNIA MUSTARD Annuals, 2-6 dm tall; stems glabrous to sparsely hirsute. Leaves cauline, oblanceolate to oblong in outline, 5-15 cm long, lower ones spreading, dentate to deeply pinnately lobed, lobes narrowly lanceolate to

**BRASSICACEAE: Capsella-Erysimum**

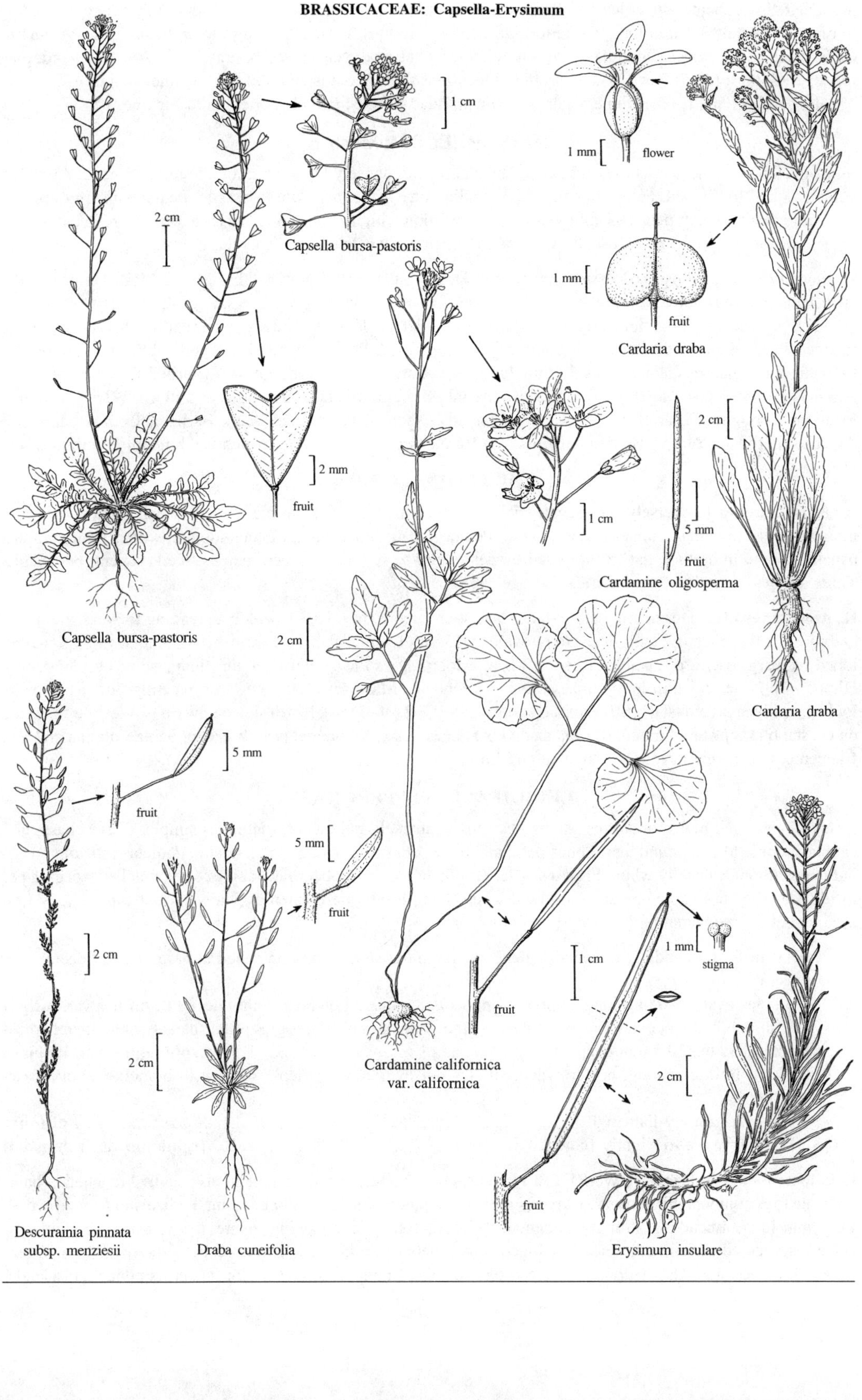

triangular, apices acute; upper leaves often ascending, entire to dentate. Pedicels in fruit 1-3 mm long, spreading to recurved. Petals 3-6 mm long, white to light yellow. Siliques linear, 25-45 mm long, pendulous. Mar-May. Occasional; rocky slopes, floodplains, grassy hillsides; chaparral, coastal scrub, grasslands, oak woodland, pine forest; 70-2300 ft (20-700 m). Widely scattered locations throughout island. All CA Channel islands except San Nicolas; Todos Santos, Guadalupe, Cedros, and Natividad islands; British Columbia to Mexico, e to UT.

## HIRSCHFELDIA Moench

Short-lived perennials, sometimes annual; stems glabrous to densely hispid, trichomes simple. Leaves basal and cauline, lobed to divided, basal ones petiolate, cauline ones short-petiolate to sessile. Pedicels in fruit erect to spreading. Sepals erect; petals usually yellow. Fruit a silique, linear to narrowly oblong, erect to reflexed, terete, smooth to angled; seeds many, usually in 1 row per locule. Two species; temperate, Europe.

**H. incana** (L.) Lagr.-Foss. [*Brassica geniculata* (Desf.) Ball] SHORT-PODDED MUSTARD Plants 3-9 dm tall; stems glabrous above middle, hispid near base. Leaves reduced upward, basal ones pinnately lobed, 9-15 cm long, petiolate, terminal lobe larger than lateral ones, lateral lobes spreading, upper cauline leaves lanceolate to lanceolate, sessile. Pedicels in fruit 2-4 mm long, appressed. Petals 6-8 mm long, yellow. Siliques appressed, 8-10 mm long, glabrous, apical beak 3-5 mm long. Apr-Oct. Introduced. Common; disturbed sites, grassy slopes and flats, rocky slopes, ridgetops, floodplains; coastal scrub, grasslands; 10-1900 ft (3-580 m). Widely scattered locations throughout island. All CA Channel islands except Anacapa and Santa Barbara; Cedros Island; N. America; native of Europe. Sometimes mistaken for *Sisymbrium officinale*, which has hirsutulous siliques.

## HUTCHINSIA R.Br.

Annuals, glabrous to sparsely pubescent, trichomes branched. Leaves simple, mostly basal, reduced upward, margins entire to dentate, upper leaves sessile. Pedicels in fruit ascending. Petals minute, white. Fruit a silicle, usually elliptic in outline, flat, compressed perpendicular to septum between locules; seeds several per locule. Three species; temperate, N. America, Europe, Asia.

**H. procumbens** (L.) Desv. Annuals, 3-8 cm tall; stems many from based, widely spreading to weakly erect, ± glabrous. Leaves mostly basal, linear to obovate, 5-15 mm long, ± glabrous, short-petiolate to sessile, pinnately lobed or margins entire. Pedicels in fruit 3-8 mm long, spreading. Petals ± 1 mm long, white. Fruit broadly elliptic in outline, 2-3 mm long, apices blunt to rounded. Mar-May. Rare, not seen recently but easily overlooked; sandy areas, coastal bluffs; coastal scrub; 100-250 ft (30-75 m). North side on nw tip of West Point; s side on coastal bluffs just n of mouth of Cañada de los Sauces. All CA Channel islands except Santa Catalina and San Clemente; Guadalupe Island; N. America and Europe.

## LEPIDIUM L. PEPPERGRASS

Annuals, perennial herbs, or shrubs; stems glabrous to sparsely pubescent, trichomes simple. Leaves basal and cauline, simple to compound, basal ones petiolate, upper ones short-petiolate to sessile. Pedicels in fruit spreading to erect. Petals usually white. Fruit a silicle, elliptic to round or obcordate, apices often notched; seeds 1 per locule. Ca. 175 species; temperate, worldwide. Seeds and leaves were eaten by Native Americans. The Maidu used leaves for medicinal purposes and washing hair.

1. Pedicels in fruit ascending, crowded; silicles overlapping; stems often branched at base, usually decumbent to prostrate
   2. Silicles ovate, 4-6 mm long, appressed-puberulent, apical notch ca. 1 mm deep . **L. latipes** var. **latipes**
   2' Silicles broadly obovate to ± orbicular, 2-3 mm long, glabrous, margins often sparsely puberulent, apical notch less than 0.5 mm deep .......... **L. oblongum** var. **insulare**

1' Pedicels in fruit spreading, not crowded, silicles not overlapping; stems solitary to branched above base, usually erect
   3. Pedicels strongly flattened .......... **L. nitidum**
   3' Pedicels terete to slightly flattened .......... **L. virginicum** var. **robinsonii**

**L. latipes** Hook. var. **latipes** DWARF PEPPERGRASS Annuals, 1.5-8 cm tall; stems usually branched at base, prostrate to decumbent, hirsutulous. Leaves simple to pinnately dissected, 2-7 cm long. Pedicels in fruit crowded, 1-2.5 mm long, ascending, strongly flattened. Silicles strongly overlapping, ovate, 4-6 mm long, reticulately veined, appressed-puberulent, apically winged, notch between wings ca. 1 mm deep. Feb-Apr (May). Occasional; disturbed sites, flats with heavy or compacted soils, floodplains, grassy slopes, vernal ponds; grasslands,

**BRASSICACEAE: Guillenia-Raphanus**

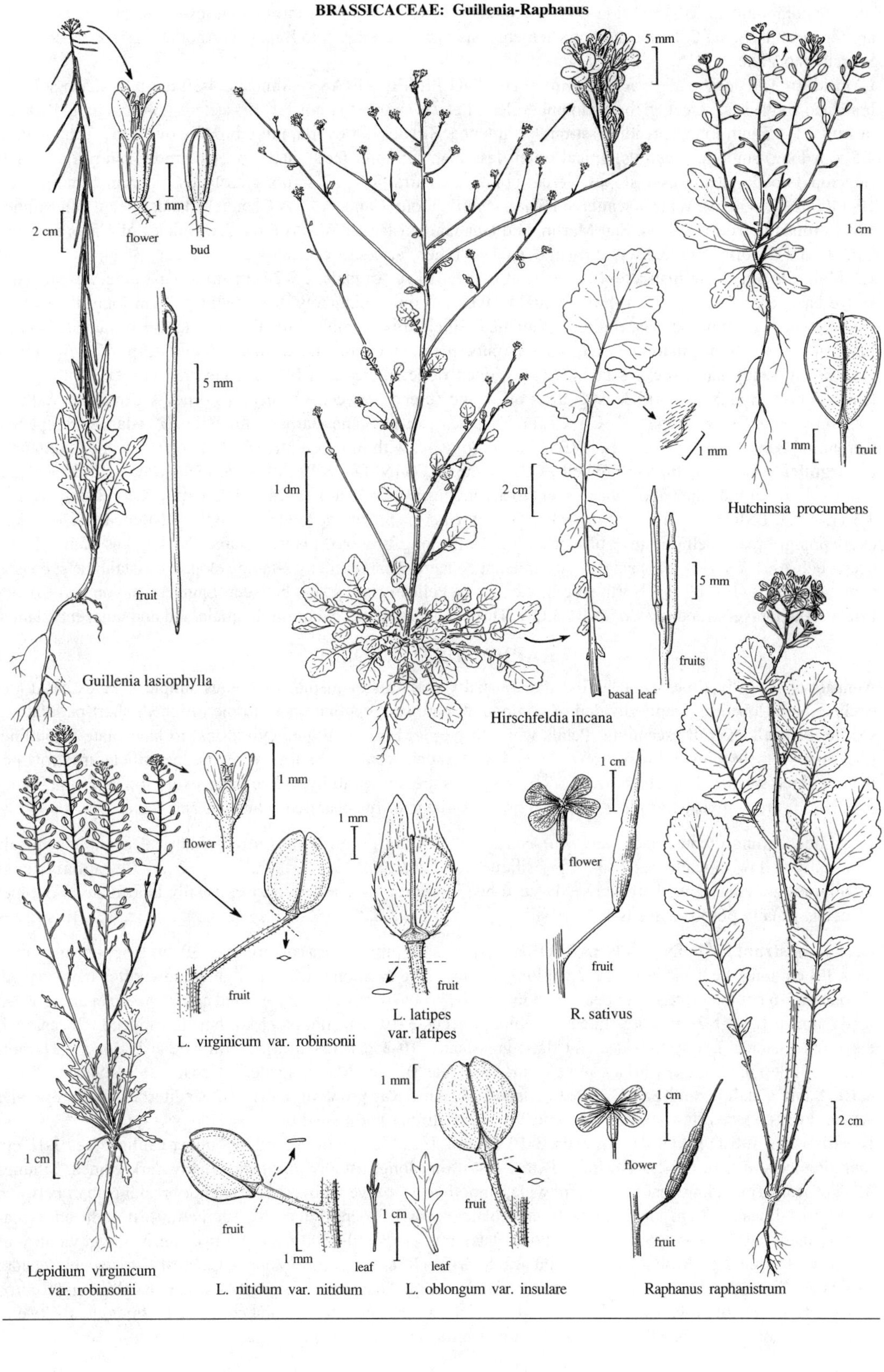

riparian communities; 10-1300 ft (3-395 m). Widely scattered locations throughout lower elevations, especially at w and e ends. Santa Catalina and San Clemente islands; northern CA to Baja CA. Another variety occurs in the Sacramento Valley of CA.

**L. nitidum** Torr. & A.Gray var. **nitidum** SHINING PEPPERGRASS Annuals, 4-30 cm tall; stems solitary, branched above base, erect, glabrous to puberulent. Leaves simple to pinnately dissected, 1.5-8 cm long. Pedicels in fruit 2.5-4.5 mm long, spreading, strongly flattened. Silicles not overlapping, broadly ovate to ± orbicular, 3-4.5 mm long, smooth, glabrous, apical notch less than 0.5 mm deep. Jan-Apr. Common; sandy flats, open ridgetops, floodplains, grassy slopes, vernal ponds; chaparral, coastal scrub, grasslands, riparian communities; 50-1400 ft (15-430 m). Widely scattered locations throughout island. All CA Channel islands except San Miguel; Los Coronados, Todos Santos, San Martin, and Guadalupe islands; WA to Baja CA. Silicles of plants on Santa Cruz Island are often mottled with brown dots. Two other varieties occur in w N. America.

**L. oblongum** Small var. **insulare** C.L.Hitchc. LENTEJILLA Annuals, 2.5-20 cm tall; stems solitary to branched above base, erect to decumbent, hirsutulous. Leaves 1-2 times pinnately dissected, 0.5-3 cm long. Pedicels in fruit crowded, 1-3 mm long, ascending, ± flattened. Silicles overlapping, broadly obovate to ± orbicular, 2-3 mm long, smooth, glabrous, margins often sparsely puberulent, apical notch less than 0.5 mm deep. Feb-Apr (Jun). Occasional; sandy sites, rocky s-facing slopes, open ridgetops, coastal bluffs; chaparral, coastal scrub, grasslands; 5-1100 ft (1.5-335 m). Coastal areas on w and se end between vicinity of Forney's Cove and Willows Canyon; Sierra Blanca ridge. All CA Channel islands except Santa Barbara; all Baja CA islands except San Geronimo; Santa Barbara Co. to Baja CA. Var. *oblongum*, with more elliptic silicles, occurs on the mainland.

**L. virginicum** L. var. **robinsonii** (Thell.) C.L.Hitchc. ROBINSON'S PEPPERGRASS Annuals, 5-20 cm tall; stems solitary or branched above base, erect to ascending, puberulent. Leaves simple to pinnately dissected, 1.5-4.5 cm long. Pedicels in fruit not crowded, 3-5 mm long, spreading, terete to slightly flattened. Silicles not overlapping, broadly elliptic to ± orbicular, 2.5-4 mm long, smooth, glabrous, apical notch less than 0.5 mm deep. Feb-Apr. A variety of special environmental concern. Scarce; rocky s-facing slopes, coastal bluffs; coastal scrub; 40-320 ft (12-100 m). North side in vicinity of Pelican Bay; s side between Laguna Canyon and Coches Prietos Anchorage. Monterey Co. to Baja CA. Three other varieties occur on the mainland and adjacent islands.

## RAPHANUS L. RADISH

Annuals or short-lived perennial herbs; stems usually scabrous or hispid, trichomes simple. Leaves basal and cauline, basal lobed to deeply divided or compound, petiolate, cauline ones simple to lobed, short-petiolate to sessile. Pedicels in fruit ascending. Petals white to purple. Fruit a silique, cylindrical to lanceolate in outline, thick, transversely jointed, often grooved lengthwise, indehiscent; seeds few to several per silicle, in 1 row per locule. Three species; Europe, n Africa. The 2 species are known to hybridize extensively in California; many populations are composed of plants that can not be satisfactorily separated (Panetsos and Baker 1967).

1. Siliques strongly constricted between seeds, seed-bearing portion uniformly cylindrical in outline; petals usually yellow to white; seeds 3-10 per silique ........ **R. raphanistrum**

1' Siliques not constricted between seeds, seed-bearing portion widest near base, usually lanceolate in outline; petals usually purplish; seeds 1-3 per silique ........ **R. sativus**

**R. raphanistrum** L. JOINTED CHARLOCK Annuals, 3-8 dm tall. Basal leaves 10-30 cm long, upper cauline ones 3-8 cm long. Pedicels in fruit 1-2 cm long. Petals 12-18 mm long, white to light yellow, often dark-veined. Siliques 2.5-6 cm long (including beak), 4-5 mm wide, grooved, strongly constricted and breaking apart between seeds; apical beak 1-2 cm long, linear to oblong; seeds 3-10 per silique. Mar-Jun. Introduced. Occasional; disturbed pastures, grassy hillsides and flats; grasslands; 10-250 ft (3-75 m). North side at Prisoners' Harbor; Central Valley in Cañadas Christy and Cebada, and near Bosque Mano; s side in Cañada de los Sauces. Santa Rosa, Santa Catalina, and San Clemente islands; N. America; probably native of Mediterranean Europe and Africa. First collected in Cañada Christy in July 1979. Entire plant is edible.

**R. sativus** L. WILD RADISH Annuals, 3-10 dm tall. Basal leaves 6-15 cm long, upper cauline ones 3-10 cm long. Pedicels in fruit 0.6-1.8 cm long. Petals 15-20 mm long, usually purplish, usually dark-veined. Siliques 3.5-8 cm long (including beak), 5-10 mm wide, smooth to ± grooved, not constricted nor breaking apart between seeds; apical beak 1-2 cm long, narrowly lanceolate; seeds 1-3 per silique. Mar-Jun (Aug). Introduced. Occasional; disturbed pastures, grassy hillsides and flats; grasslands; 10-1300 ft (3-395 m). North side in vicinity of Prisoners' Harbor, near Mount Pleasant, and near Scorpion Ranch; Central Valley in Cañadas Christy and Cebada; s side near Smugglers' Cove. All CA Channel islands except Anacapa and Santa Barbara; Guadalupe and Cedros islands; N. America; probably native of Mediterranean Europe and Africa. First collected at Smugglers' Cove in June 1930. Entire plant is edible but roots can be fibrous and tough; young siliques are very tasty.

## **RORIPPA** Scop.

Annuals or perennial herbs; stems prostrate to erect, glabrous to pubescent, trichomes simple. Leaves basal and cauline, simple to compound, reduced upward, subglabrous to pubescent, petiolate to sessile. Pedicels in fruit spreading to erect. Petals white to yellow. Fruit a silique or silicle, linear in outline to ellipsoid or globose; seeds many, in 1 or 2 rows per locule. Ca. 75 species; temperate, worldwide.

**R. nasturtium-aquaticum** (L.) Hayek [*Nasturtium officinale* R.Br.] WATER CRESS Aquatic, perennial herbs; stems usually emergent, rooting at nodes, segments above water-line 10-30 cm tall. Leaves pinnately compound, oblong to oblanceolate in outline, 3-15 cm long, leaflets ovate to broadly oblong. Pedicels in fruit 5-10 mm long, spreading to ascending. Petals 2.5-4.5 mm long, white. Siliques 8-20 mm long; seeds in 2 rows per locule. Apr-Jun (Sep). Probably introduced. Rare, not seen recently; moist streambeds; riparian communities; 5-25 ft (1.5-7.5 m). Known from only two sterile collections made at Prisoners' Harbor in July 1930 and July 1939. San Nicolas and Santa Catalina islands; N. America, Europe. Widely cultivated for edible stems and leaves; seeds are used in mustard.

## **SIBARA** Greene

Annuals or short-lived perennial herbs; stems glabrous to sparsely pubescent, trichomes simple or branched. Leaves basal and cauline, petiolate, lower ones lobed to compound, upper ones entire to deeply lobed. Pedicels in fruit ascending to widely spreading. Petals white to purple. Fruit usually a silique, linear in outline, terete or flat and compressed parallel to septum between locules; seeds many, in 1 row per locule. Ten species; temperate, N. America.

**S. filifolia** (Greene) Greene Annuals, 5-20 cm tall; stems glabrous. Leaves mostly cauline, reduced upward, pinnately lobed, 2-6 cm long, lobes linear. Pedicels in fruit 8-12 mm long, ascending. Petals 2.5-5 mm long, lavender to purple. Siliques 2-5 cm long, less than 1 mm wide, flat, glabrous. Mar. Island endemic. A species of special environmental concern. Rare, not seen recently; open ridgetops; coastal scrub; elevation unknown. Collected on "stony ridge above Pelican Bay" in June 1930 and on "stony ridge above Laguna Canyon" in March 1932. Santa Catalina and San Clemente islands.

## **SINAPIS** L.

Annuals; stems glabrous to sparsely pubescent, trichomes simple. Leaves cauline, lower ones lobed to deeply divided, basal ones petiolate, upper ones simple to lobed, short-petiolate to sessile. Pedicels in fruit erect to spreading. Petals usually yellow. Fruit a silique, linear to narrowly oblong, usually terete, smooth to angled; seeds many, usually in 1 row per locule. Seven species; temperate, Europe, n Africa, w Asia.

**S. arvensis** L. [*Brassica kaber* (DC.) Wheeler] CHARLOCK Plants 3-7 dm tall; stems glabrous to sparsely hispidulous. Leaves reduced upward, basal ones 5-15 cm long, pinnately lobed, terminal lobe ovate to oblanceolate, much larger than lateral ones, lateral lobes spreading; upper cauline leaves 1.5-4 cm long, sessile. Pedicels in fruit 2-5 mm long, ascending. Petals 4-6 mm long, yellow. Siliques 20-30 mm long, slightly constricted between seeds, apical beak 8-12 mm long. Mar-Apr. Introduced. Scarce; disturbed pastures, grassy slopes; grasslands; 100-250 ft (30-75 m). Central Valley near Stanton Ranch. Santa Catalina Island; N. America; native of Mediterranean Europe. First collected at Valley Anchorage in March 1929. Seeds are edible.

## **SISYMBRIUM** L.

Annuals or short-lived perennial herbs; stems glabrous to hirsute. Leaves basal and cauline, reduced upward, usually deeply, pinnately lobed, subglabrous to pubescent, lower ones petiolate, upper ones short-petiolate to sessile. Pedicels in fruit ascending to erect. Petals usually white to yellow. Fruit a silique, usually linear, terete to subterete; seeds many, in 1 row per locule. Ca. 90 species; temperate, worldwide.

**S. officinale** L. HEDGE MUSTARD Plants 3-8 dm tall; stems hirsutulous, trichomes spreading to retrorse. Leaves short-petiolate, pinnately lobed, reduced upward, lower ones oblanceolate in outline, 6-20 cm long, upper ones simple to pinnately lobed, 1-4 cm long, terminal lobe oblong to narrowly lanceolate, lateral lobes reflexed. Pedicels in fruit 1.5-3 mm long, erect. Petals 3-4 mm long, yellow. Siliques 8-12 mm long, appressed, hirsutulous, style sometimes beak-like, 3-6 mm long. Apr-Jun. Introduced. Common; disturbed sites, floodplains, grassy hillsides, shaded n-facing slopes; chaparral, coastal scrub, grasslands, oak woodland, riparian communities; 15-2300 ft (5-700 m). Widely scattered locations throughout island. Santa Catalina Island; N. America; native of Europe. Reported "in cultivated lands only" on Santa Cruz Island by Greene (1887). This species is edible raw or cooked and also has been used medicinally.

**BRASSICACEAE: Rorripa-Thysanocarpus**

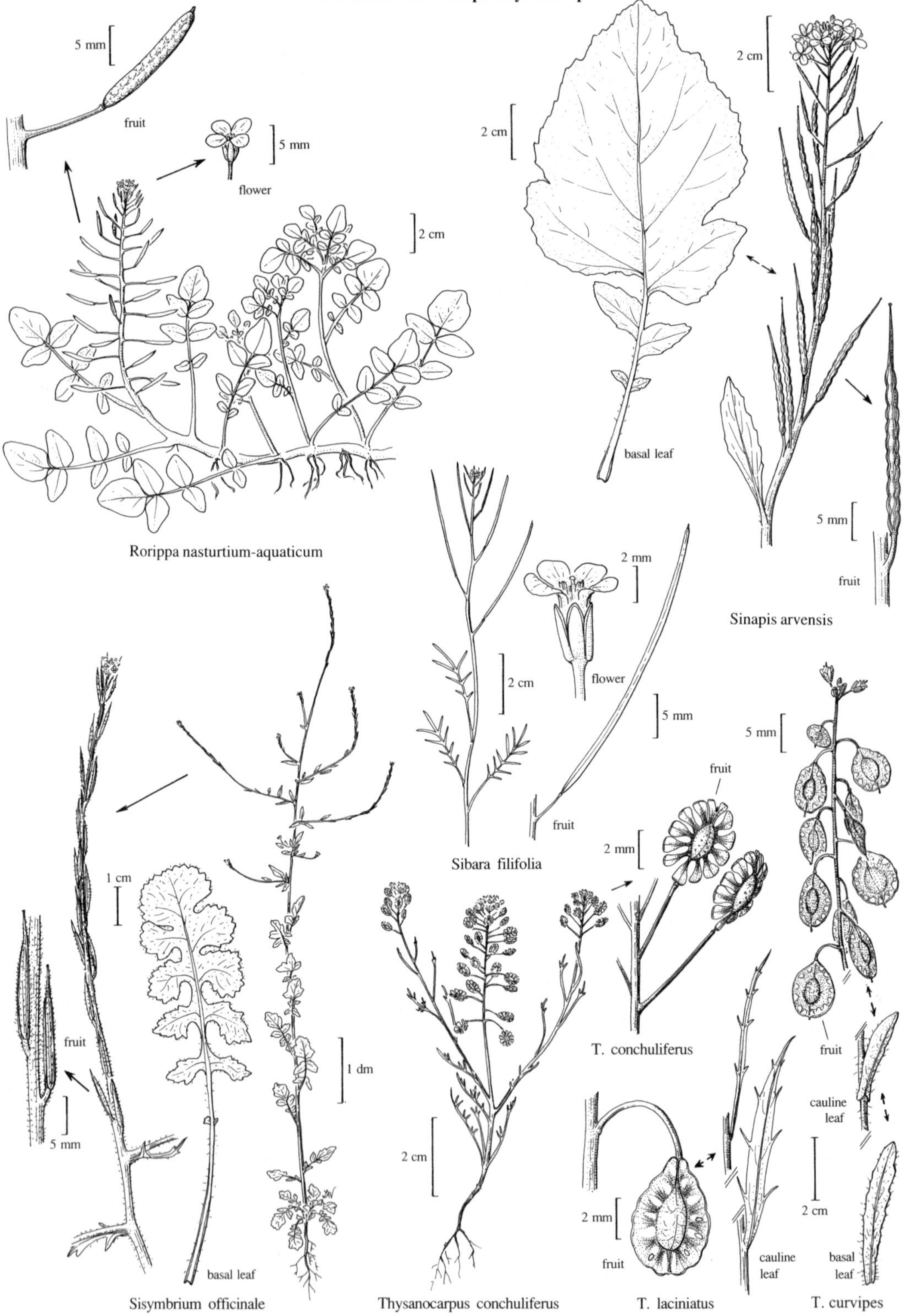


Rorippa nasturtium-aquaticum

Sinapis arvensis

Sibara filifolia

T. conchuliferus

Sisymbrium officinale

Thysanocarpus conchuliferus

T. laciniatus

T. curvipes

## THYSANOCARPUS Hook. LACEPOD or FRINGEPOD

Annuals; stems glabrous to pubescent, trichomes simple. Leaves cauline, reduced upward, entire, dentate, or pinnately lobed, upper ones short-petiolate to clasping. Pedicels in fruit reflexed to pendulous. Petals white to purple. Fruit a silicle, round to oval, strongly flattened, sometimes cup-like, margins winged, entire to perforated or lobed; locule 1; seed 1 per silicle. Five species; w N. America.

1. Silicles cup-shaped, one side concave, other side convex, wing margin perforated to lobed; pedicels spreading .......... **T. conchuliferus**

1' Silicles flat (or nearly so), wing margin entire to sometimes perforated; pedicels arched or recurved
  2. Upper cauline leaves clasping, bases auriculate; lower leaves hirsute, often in rosettes ...... **T. curvipes**
  2' Upper leaves sessile, not clasping; lower leaves glabrous, not in rosettes .......... **T. laciniatus**

**T. conchuliferus** Greene SANTA CRUZ ISLAND LACEPOD Plants 4-20 cm tall; stems glabrous, ± glaucous. Leaves linear to lanceolate, lower ones 4-9 cm long, deeply dentate to pinnately lobed, lobes linear to oblong, 1-6 mm long, upper leaves 1-5 cm long, dentate. Pedicels in fruit 4-8 mm long, spreading to ascending. Petals 2-3 mm long, pink to purplish. Fruits elliptic in outline, cup-shaped, one side concave, other side convex, 3-4 mm long, wing margin perforated to lobed. Feb-Apr (May). Endemic to Santa Cruz Island. A species of special environmental concern. Feb-Apr (May). Scarce; mostly on shaded n-facing rock faces, slopes, and canyon bottoms; chaparral, oak woodland; 800-1800 ft (245-550 m). North side between Lady's and Prisoners' harbors; Central Valley near Lagunitas Secas and in Cañada de la Portezuela; s side on Sierra Blanca ridge. This species has not been seen on the s side since 1932.

**T. curvipes** Hook. HAIRY LACEPOD Plants 4-30 cm tall; stems glabrous above middle, hirsute near base. Leaves linear, lower ones in rosettes, 2-8 cm long, dentate to shallowly lobed, hirsute, upper leaves 1-5 cm long, entire to dentate, bases auriculate, clasping. Pedicels in fruit 5-8 mm long, usually arched or recurved. Petals 1-2.5 mm long, white to pinkish. Fruits elliptic to orbicular in outline, flat, 4-6 mm long, wing entire to sometimes perforated. Feb-Apr. Rare; grassy hillsides, shaded n-facing slopes; grasslands, oak woodland; 200-2200 ft (60-670 m). North side on Picacho Diablo, near upper portion of Twin Harbor Canyon, near Pelican Bay, and w of the China Pines. Santa Catalina Island; British Columbia to Baja CA, e to Rocky Mountains. Seeds were eaten by Native Americans; a medicinal tea also was made from the leaves.

**T. laciniatus** Nutt. ex Torr. & A.Gray [*T. l.* var. *ramosus* (Greene) Munz] NARROW-LEAVED LACEPOD Plants 10-50 cm tall; stems glabrous, sometimes sparsely pubescent. Leaves linear, lower ones not in rosettes, 2-8 cm long, usually lobed, lobes linear, 1-5(10) mm long, upper leaves 1-5 cm long, entire to dentate, sessile, not auriculate. Pedicels in fruit 4-10 mm long, usually arched or recurved. Petals 1-2.5 mm long, white. Fruits elliptic to orbicular in outline, flat, 4-6 mm long, wing entire to perforated. Jan-Apr (Jul). Common; grassy hillsides, shaded n-facing slopes; chaparral, grasslands, oak woodland, pine forest; 10-2400 ft (3-730 m). Santa Rosa, Santa Catalina, and San Clemente islands; Cedros Island; San Francisco Bay region to Baja CA.

# CACTACEAE Juss.
## CACTUS FAMILY

Shrubs or trees; stems branched or unbranched, succulent, flat, cylindrical, or globose, smooth to tuberculate or ribbed; new branches and flowers arising from highly modified, condensed nodes (areoles). Leaves usually modified into spines and glochids (minute spines) or much reduced and caducous, clustered on areoles; stipules absent. Flowers usually solitary, ± sessile. Hypanthium present; perianth parts many, sepals and petals often intergrading, outer ones often scale-like, inner ones petaloid, fused to hypanthium; stamens many, adnate to lower part of hypanthium; pistil 1, ovary inferior, locule 1, ovules many, styles or stigmatic lobes usually many. Fruit a berry, often leathery, often with areoles. Ca. 130 genera and ca. 1600 species; N. and S. America.

## OPUNTIA J.Mill. PRICKLY PEAR, CHOLLA

Stems branched, appearing jointed, segments between joints flat to cylindrical. Areoles usually circular; spines and glochids (minute spines) present; leaves present on young, growing stems, minute, terete, succulent, caducous. Flowers produced on stem segments of previous year. Hypanthium tube cup-shaped; stamens sensitive to touch, collapsing around style and stigma; ovary with areoles. Fruit fleshy to dry; seeds bony, flattened. Ca. 300 species; N. & S. America. Stem segments were cooked and eaten by Native Americans; fruits were eaten raw; seeds also were eaten by the Luiseño Tribe.

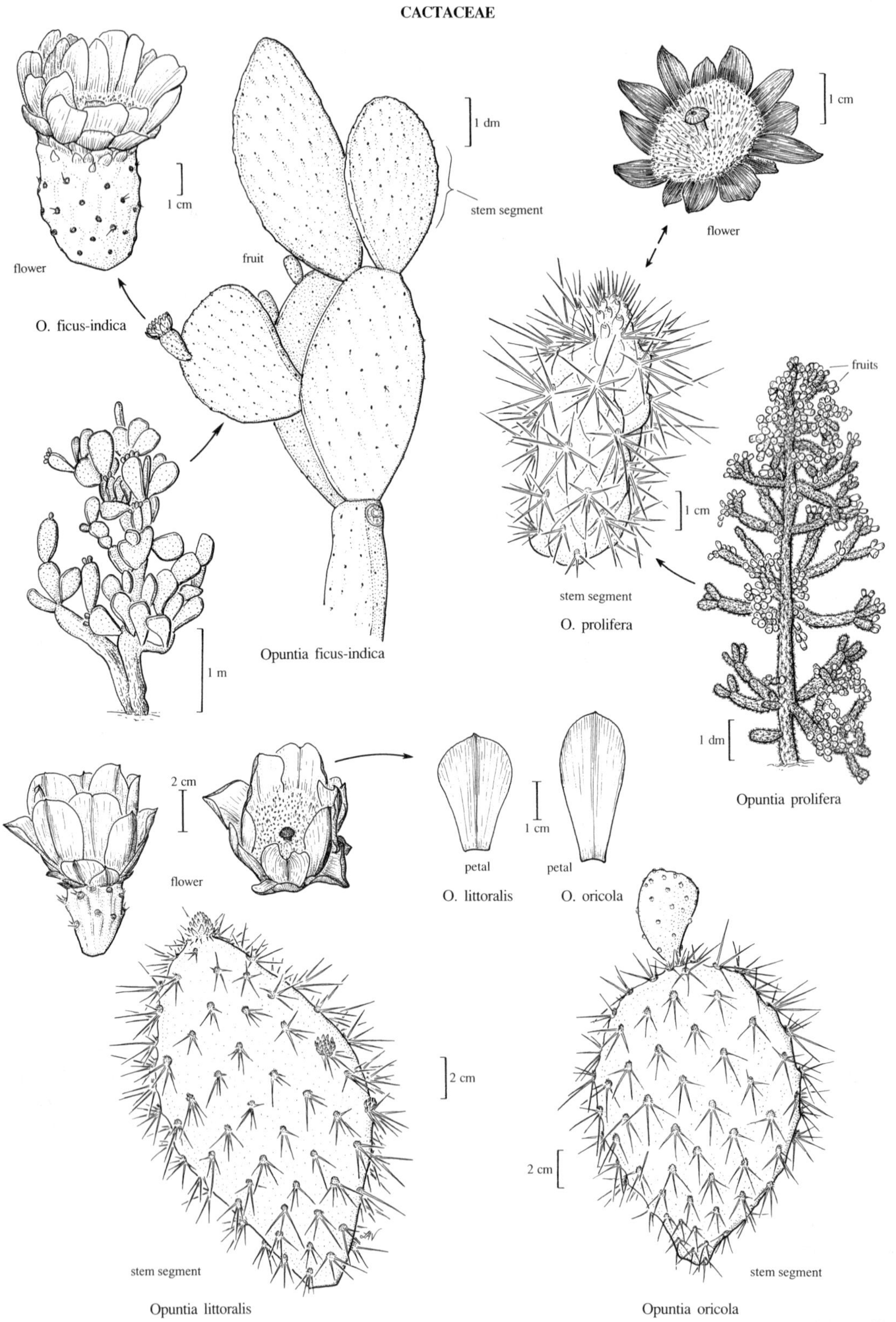
CACTACEAE
flower
1 cm
O. ficus-indica
fruit
1 dm
stem segment
1 m
Opuntia ficus-indica
1 cm
flower
1 cm
stem segment
O. prolifera
fruits
1 dm
Opuntia prolifera
2 cm
flower
petal
1 cm
petal
O. littoralis
O. oricola
2 cm
2 cm
stem segment
Opuntia littoralis
stem segment
Opuntia oricola

1. Stem segments cylindrical, tuberculate; perianth parts purple to reddish purple ........................ **O. prolifera**
1' Stem segments flat, not tuberculate; perianth parts yellow to orange
   2. Stems erect, segments mostly 25-55 cm long; areoles usually 3-4.5 cm apart .............. **O. ficus-indica**
   2' Stems erect to sprawling, segments mostly 15-25 cm long; areoles usually 1.5-2.5 cm apart
      3. Stem segments elliptic to obovate, their length 1.7-2.3 times the width; unweathered spines bone white, sometimes darkened at base; innermost perianth parts 1.2-1.7 longer than width; style pink ........................................................................................................................ **O. littoralis**
      3' Stem segments orbicular to broadly elliptic, their length 1-1.5 times the width; mature but unweathered spines yellow and mostly translucent; innermost perianth parts 1.8-2.6 longer than width; style red ........................................................................................................ **O. oricola**

**O. ficus-indica** (L.) J.Mill. MISSION CACTUS Plants 1-2 m tall; stems erect, segments elliptic to obovate, 25-55 cm long, 18-40 cm wide, green; areoles usually 2-4.5 cm apart, spines 3-6 per areole, sometimes absent, ± flat in cross-section, longest ones 2-3 cm long, straight, spreading to ± reflexed. Petaloid perianth parts yellow to orange. Fruit ellipsoid, 5-8.5 cm long. May-Jun. Introduced. Rare; rocky ridgetops and slopes; coastal scrub; ca. 50 ft (15 m). North side at Pelican Bay. Anacapa, San Nicolas, Santa Catalina, and San Clemente islands; Todos Santos Island; cultivated and naturalized ± worldwide; probably native of Mexico or C. America.

**O. littoralis** (Engelm.) Cockerell var. **littoralis** COASTAL PRICKLY PEAR. Plants 0.5-1.5 m tall; stems clumped, sprawling, segments narrowly elliptic to obovate, 15-20 cm long, 8-10 cm wide, yellowish green; areoles usually 1.5-3 cm apart, spines 4-10 per areole, ± flat to narrowly elliptic in cross-section, longest ones 2.5-4 cm long, straight, spreading to ± reflexed. Petaloid perianth parts yellow. Fruit ellipsoid to obovoid, 3-5 cm long. Jun-Jul (Sep). Abundant; rocky s-facing slopes, open ridgetops, grassy hillsides, coastal bluffs; chaparral, coastal scrub, grasslands; 10-2000 ft (3-610 m). Widely scattered locations throughout island. All CA Channel islands; Los Coronados and Todos Santos islands; Santa Barbara Co. to n Baja CA. Three other varieties occur on the mainland from Ventura Co. to Baja CA, e to AZ and UT. Hybridizes freely with *O. oricola*; hybrids have stem segments intermediate in size and shape between parents and white styles (Philbrick 1963).

**O. oricola** Philbrick TALL PRICKLY PEAR Plants 1-2 m tall; stems erect, clumped, sometimes sprawling, segments orbicular to broadly elliptic to obovate, 15-25 cm long, 14-22 cm wide, dark green; areoles usually 1-2 cm apart, spines 6-12 per areole, round to elliptic in cross-section, longest ones 2.5-4 cm long, straight to curved, spreading to reflexed. Petaloid perianth parts yellow. Fruit subglobose to broadly obovoid, 3-5 cm long, red. Jun-Jul (Oct). Common; rocky s-facing slopes, open ridgetops, grassy hillsides, coastal bluffs; chaparral, coastal scrub, grasslands; 50-1500 ft (15-460 m). Widely scattered locations throughout island. All CA Channel islands; Los Coronados, Todos Santos, and Cedros islands; Santa Barbara Co. to Baja CA.

**O. prolifera** Engelm. COASTAL CHOLLA Plants 0.5-2 m tall; stems erect, clumped, segments cylindrical, 7-15 cm long, 4-5 cm wide, dark green; areoles usually 0.5-1 cm apart, spines 5-12 per areole, elliptic in cross-section, longest ones 1-2.5 cm long, straight, ascending to reflexed. Petaloid perianth parts purple to reddish purple. Fruits persistent, often in pendulous chains of 2-4, subglobose to broadly obovoid, 2-3 cm long, green. May-Jun. Occasional; rocky s-facing slopes, open ridgetops, grassy hillsides, coastal bluffs, sandy flats; coastal scrub, grasslands; 10-550 ft (3-170 m). South side between Cañada de Malva Real and Horqueta Canyon. All CA Channel islands except San Miguel; Los Coronados, San Martin, and Guadalupe islands; Ventura Co. to Baja CA.

## CAMPANULACEAE L.
### Bellflower Family

Annuals or perennials. Leaves usually alternate, simple; stipules absent. Flowers bisexual, radial, solitary or in racemes, spikes, or panicles. Hypanthium often present; calyx adnate to ovary, with 5 lobes; corolla usually with 5 lobes; stamens 5, free; pistil 1, ovary inferior to half-inferior, locules 1-3, style 1, with 2-5 lobes. Fruit a capsule. Ca. 90 genera and 2000 species; temperate to tropical, worldwide. Some genera (Campanula) are cultivated as ornamentals.

1. Flowers terminal, corolla salverform; most leaves oblanceolate ................................................... **Githopsis**
1' Flowers axillary and terminal, corolla rotate; most leaves ovate to orbicular ................................ **Triodanis**

CAMPANULACEAE

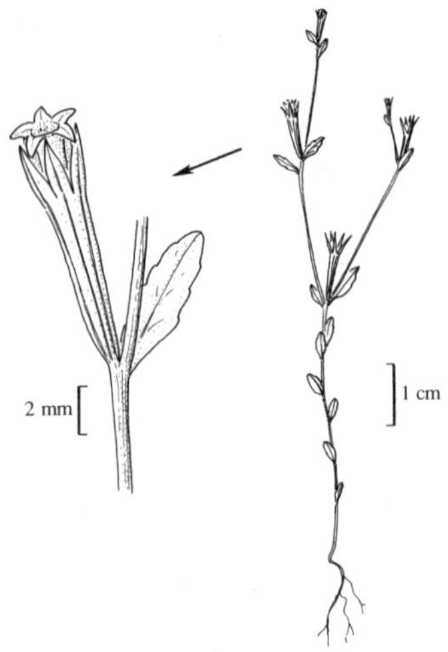


Githopsis diffusa subsp. diffusa

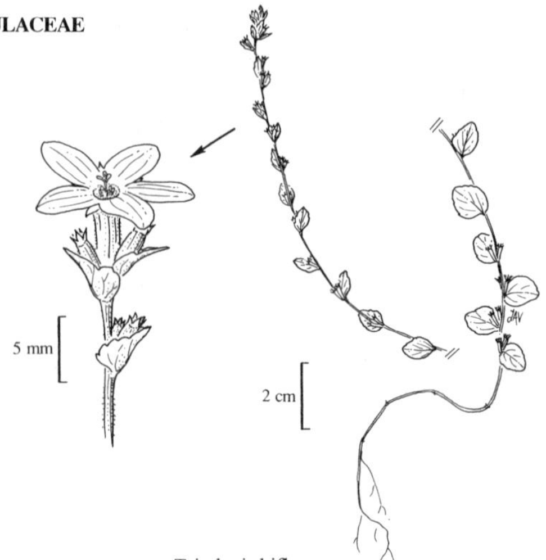


Triodanis biflora

## GITHOPSIS Nutt. BLUECUP

Annuals. Leaves oblong, oblanceolate, or ovate, toothed. Flowers usually solitary, terminal, pedicellate or sessile. Sepals linear to lanceolate; corolla campanulate to salverform, tube and throat usually white, lobes white to purple; ovary inferior, with 2-3 locules. Fruit dehiscing by an apical pore formed from deciduous style base. 4 species; w N. America.

**G. diffusa** A. Gray subsp. **diffusa** SOUTHERN BLUECUP Plants with ascending to decumbent stems 3-30 cm long. Leaves oblanceolate, 5-10 mm long, dentate to serrate. Calyx lobes 1-3 mm long; corolla 3-5 mm long, lobes blue. Fruit 5-9 mm long, with 10 vertical ribs. Apr-May. Scarce; moist n-facing slopes; chaparral, pine forest; 200-2400 ft (60-730 m). North side between Picacho Diablo and Prisoners' Harbor; Central Valley between Cañada de la Portezuela and Stanton Ranch. Santa Lucia Mountains to n Baja CA, e to Sierra Nevada foothills. Three other subspecies occur on the mainland of CA and n Baja CA; another is restricted to Guadalupe Island.

## TRIODANIS Raf. ex Greene VENUS LOOKING-GLASS

Annuals. Leaves broadly lanceolate to cordate, toothed. Flowers usually solitary, axillary and terminal, sessile, axillary ones often cleistogamous and much smaller than terminal, chasmogamous ones. Sepals narrowly triangular; corolla rotate, deep blue to purplish; ovary inferior, with 1-3 locules. Fruit dehiscing by lateral pores. Ca. 8 species; w N. America and Europe.

**T. biflora** (Ruiz & Pav.) Greene Plants with erect, stems 10-50 cm long. Leaves ovate to orbicular, 6-40 mm long, entire to dentate. Calyx lobes 1.5-4 mm long; corolla 5-10 mm long, blue. Fruit 6-10 mm long, lateral pores near apex. Apr-Aug. Occasional; n-facing rocky ledges and slopes, disturbed sites, and grassy flats; chaparral, coastal scrub, grasslands; 30-1400 ft (9-427 m). Widely scattered locations on n side and in Central Valley; s side in Laguna and Smugglers' canyons. Santa Rosa and Santa Catalina islands; Guadalupe Island; w and c N. America to S. America.

# CAPRIFOLIACEAE Juss.

## HONEYSUCKLE FAMILY

Shrubs, trees, or vines. Leaves opposite, usually simple; stipules usually absent. Flowers bisexual, radial or bilateral, in axillary or terminal clusters. Calyx tubular, ± adnate to ovary, with 5 lobes; corolla rotate to salverform or strongly bilateral, usually with 5 lobes; stamens 5, adnate to corolla tube; pistil 1, ovary inferior, locules 1-5, style 1, stigma often capitate. Fruit a berry, drupe, or capsule. Ca. 15 genera and 400 species; temperate to tropical, worldwide.

CAPRIFOLIACEAE

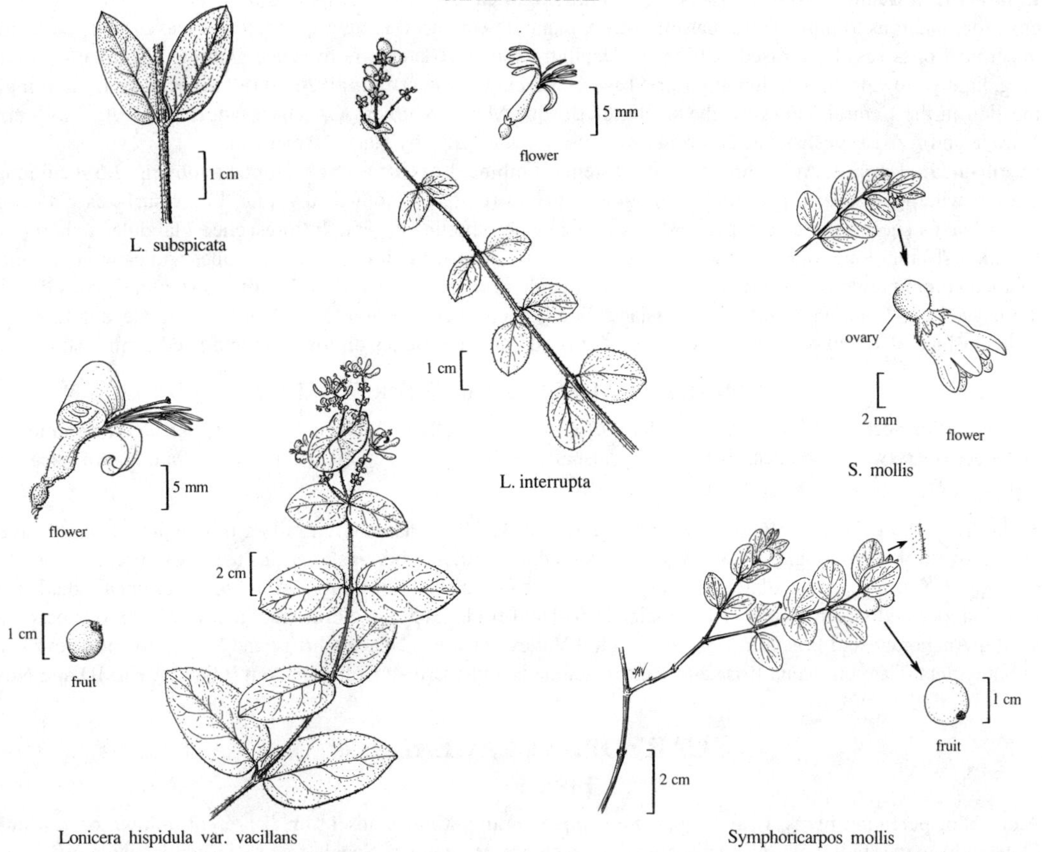


1. Corolla bilateral, swollen or spurred at base; fruit red to yellow .......... **Lonicera**
1' Corolla radial; fruit white .......... **Symphoricarpos**

## **LONICERA** L. HONEYSUCKLE

Shrubs or vines. Flowers often in pairs at upper nodes or in terminal spikes. Corolla ± bilateral to bilabiate, upper lip ± 4-lobed, tube with basal swelling or spur, stamens slightly exserted; ovary with 2-3 locules. Fruit a berry, usually with 2 or more seeds. Ca. 200 species; temperate to subtropical N. America, Europe, Asia, and N. Africa.

1. Corolla pink to purplish, 10-15 mm long; upper leaves sometimes with stipules .......... **L. hispidula**
1' Corolla light yellow or cream, 8-10 mm long; upper leaves without stipules
    2. Leaves broadly elliptic to orbicular, margins plane, lower surfaces light green .......... **L. interrupta**
    2' Leaves narrowly elliptic to oblong, margins revolute .......... **L. subspicata**

**L. hispidula** Douglas ex Torr. & A. Gray var. **vacillans** A.Gray CALIFORNIA HONEYSUCKLE Shrubs; stems often sprawling or climbing. Leaves broadly elliptic to ovate, 4-8 cm long, 2.5-5 cm wide, puberulent, margins entire, plane, lower leaves short-petiolate, uppermost ones sessile, ± fused at base, ± clasping, lower surfaces glaucous; upper leaves sometimes with stipules. Inflorescence glandular-puberulent. Corolla 10-15 mm long, pink to purplish. Fruit red. Mar-Aug. Scarce; moist n-facing slopes, shaded canyon walls; oak woodland, riparian communities; 50-500 ft (15-150 m). North side in several canyons between Valdez Anchorage and Prisoners' Harbor; s side at Las Torres ridge. Santa Catalina and San Clemente islands; s OR to s CA. Fruit of this variety was eaten by Native Americans. Two other varieties occur on the mainland.

**L. interrupta** Benth. Shrubs or vines; stems climbing. Leaves broadly elliptic to orbicular, 2-3 cm long, 1-2.5 cm wide, glabrous to puberulent, margins entire, plane, lower surfaces light green, lower leaves short-petiolate, uppermost ones sessile, ± fused at base, ± clasping; stipules absent. Inflorescence glabrous. Corolla 8-10 mm long, light yellow. Fruit red. Jun-Jul. Rare; base of n-facing slope; chaparral; 260 ft (80 m). Known from a single location in the Central Valley, at the w end of Bosque Mano. Northern CA coast ranges to s CA, e to Sierra Nevada and AZ. Leaves, stems, and roots were used medicinally by Native Americans.

**L. subspicata** Hook. & Arn. Shrubs or vines; stems climbing. Leaves narrowly elliptic to oblong, 1.5-4 cm long, 1-2 cm wide, glabrous to puberulent, margins entire to remotely toothed, revolute, lower surfaces glaucous, lower leaves short-petiolate, uppermost ones subsessile; stipules absent. Inflorescence glandular-puberulent. Corolla 8-10 mm long, light yellow. Fruit red to yellow. Mar-May. Rare; n-facing slopes and canyon bottoms; riparian communities; 50-500 ft (15-152 m). North side in canyon above Baby's Cove; s side in Coches Prietos drainage. Santa Catalina Island; Cedros Island; San Francisco Bay region to s CA, e to Sierra Nevada foothills. Relationships of island populations to those of several named varieties on the mainland need further study.

### SYMPHORICARPOS Duhamel SNOWBERRY

Shrubs. Inflorescence of few flowers, axillary or terminal. Corolla radial, campanulate to salverform, stamens ± included; ovary with 4 locules. Fruit a berry, usually with 2 seeds. Ca. 10 species; temperate N. America and Asia.

**S. mollis** Nutt. in Torr. & A.Gray CREEPING SNOWBERRY Plants with trailing or creeping stems. Leaves oval to ovate, 1-4 cm long, 0.5-2 cm wide, subsessile, puberulent on veins, margins usually entire. Flowers 2-8 per cluster. Corolla campanulate, 3-5 mm long, pink to reddish. Fruit white. Apr-Jun. Occasional; shaded n-facing slopes and streambanks; oak woodland; 50-1000 ft (15-305 m). North side in several canyons between Valdez Anchorage and Prisoners' Harbor; Central Valley between Cañada Christy and Valley Anchorage; s side in Smugglers' Canyon. Santa Rosa and Santa Catalina islands; British Columbia to n Baja CA, e to ID and NM.

## CARYOPHYLLACEAE Juss.
## PINK FAMILY

Annuals or perennial herbs. Leaves opposite, simple, entire; stipules absent or, if present, minute or scarious. Flowers bisexual, radial, axillary or terminal, in racemes or panicles. Sepals free or fused, 5; petals usually 5, sometimes small or absent; stamens usually 10, free; pistil 1, ovary superior, locule 1, styles 2-5 or 1 with 2-3 lobes. Fruit a capsule or utricle, capsule often opening by valves or apical teeth. Ca. 85 genera and 2400 species; temperate to tropical, worldwide. Some genera cultivated as ornamentals, including carnations and babys-breath.

1. Ovary with 1 ovule; fruit a utricle, with 1 seed; petals absent .......... **Cardionema**
1' Ovary with several to many ovules; fruit a capsule, usually with many seeds; petals usually present
  2. Sepals fused into a cylindrical or cup-like calyx; petals clawed; ovary separated from receptacle by a short stalk .......... **Silene**
  2' Sepals free; petals, when present, usually not clearly clawed; ovary sessile
    3. Stipules present, scarious
      4. Leaves oblanceolate to broadly obovate; style 1, with 3 lobes .......... **Polycarpon**
      4' Leaves filiform to linear; styles 3 and distinct to base .......... **Spergularia**
    3' Stipules absent
      5. Capsule with number of valves (6-10) or apical teeth twice number of styles; styles opposite sepals
        6. Styles 5; capsule cylindrical, with 10 teeth at apex .......... **Cerastium**
        6' Styles 3; capsule ovoid to ellipsoid, usually with 6 teeth at apex .......... **Stellaria**
      5' Capsule with number of valves (3-5) or apical teeth equal to number of styles; styles alternate sepals
        7. Sepals 2.5-4 mm long; petals 3-6 mm long .......... **Minuartia**
        7' Sepals 1.5-2.5 mm long; petals absent, minute, or equal in length to sepals .......... **Sagina**

## CARDIONEMA DC.

Prostrate or tufted perennial herbs; stems many branched, densely woolly. Leaves filiform, spine-tipped; stipules lanceolate to ovate, scarious or papery. Flowers 1-5, axillary. Sepals free, ± unequal, densely woolly, tips spreading and ± spine-tipped; petals minute, scale-like; stamens 3-5, attached at base of sepals; styles 2. Fruit a utricle, with 1 seed, enclosed by persistent calyx. 6 species; w N. America and Chile.

**C. ramosissimum** (Weinm.) A.Nelson & J.F. MacBr. SANDMAT Stems 3-20 cm long, prostrate. Leaves 5-12 mm long, glabrous, spinulose; stipules 4-6 mm long, narrowly acuminate. Sepals 4-6 mm long, yellowish green. Fruit 4-6 mm long; seed ovoid, ca. 1.5 mm long. Apr-Oct. Occasional; sandy sites, rocky ridgetops, slopes, and floodplains; chaparral, coastal scrub, pine forest; 50-2400 ft (15-730 m). Widely scattered locations throughout w and central portions of island. San Miguel and Santa Rosa islands; WA to Mexico and Chile.

## CERASTIUM L. CHICKWEED

Annuals or perennial herbs; stems usually erect, simple. Leaves linear to ovate; stipules absent. Flowers few-many in axillary or terminal cymes. Sepals free, 4-5, equal, pubescent; petals 5 and usually bifid or absent; stamens 5 or 10; styles 5. Fruit a capsule, cylindrical, with 10 apical teeth. 60 species; worldwide.

**C. glomeratum** Thuill. MOUSE-EAR CHICKWEED Annuals; stems erect, 3-25 cm tall, viscid. Leaves lanceolate, oblanceolate or ovate, 5-20 mm long, 4-10 mm wide, pubescent, apex rounded to obtuse. Sepals ovate to lanceolate, 4-5 mm long, glandular puberulent, margins scarious; petals ± equal to sepals in length; stamens 10. Fruit cylindrical, 6-8 mm long; seeds ca. 0.5 mm long, minutely muricate. Mar-Jun. Introduced. Common; shaded n-facing slopes and streambanks; chaparral, oak woodland; 20-1200 ft (6-365 m). Widely scattered locations throughout island. All CA Channel islands except San Nicolas and Santa Barbara; Guadalupe Island; N. America; native of Europe. First collected from a location "between Prisoners' Harbor and Christy Ranch" in March 1925.

## MINUARTIA L.

Annuals or perennial herbs; stems erect to prostrate. Leaves filiform to oblong; stipules absent. Flowers few-many in axillary or terminal cymes or solitary. Sepals ± free, equal, glabrous to puberulent; petals 5, entire to bifid, or absent; stamens 10; styles 3. Fruit a capsule, ovoid to ellipsoid, opening by 3 valves or with 3 apical teeth. 120 species; N. America, Europe, and Asia.

**M. douglasii** (Torr. & A.Gray) Mattf. [*Arenaria douglasii* Torr. & A.Gray] DOUGLAS' SANDWORT Annuals, 4-10 cm tall; stems simple to branched, glabrous to glandular-puberulent. Leaves filiform, 5-30 mm long, glabrous. Flowers in open, terminal cymes. Sepals ovate, 2.5-4 mm long, margins scarious; petals 3-6 mm long, white. Fruit subglobose, 4.5-5.5 mm long; seeds compressed and appearing almost circular, 1-1.5 mm long. Apr-May. Rare; rocky n-facing slopes; chaparral; 1000-1650 ft (305-505 m). North side between upper w fork of Orizaba Canyon and slopes n of Red Peak. Santa Rosa, Santa Catalina, and San Clemente islands; s OR to n Baja CA, e to AZ.

## POLYCARPON L.

Prostrate or matted annuals; stems many branched, glabrous. Leaves opposite to ± whorled, oblanceolate to obovate; stipules lanceolate to triangular, scarious. Flowers few-many, in axillary cymes. Sepals free, glabrous, tips often apiculate; petals shorter than sepals; stamens 3-5, ± fused at base; style 1, with 3 lobes. Fruit a capsule, ovoid to globose, opening by 3 valves. 6 species; worldwide.

1. Stipules 0.5-1 mm long; sepals 1-1.5 mm long; leaves apparently opposite ............................. **P. depressum**
1' Stipules 2-2.5 mm long; sepals ca. 2 mm long; leaves opposite to apparently whorled ........ **P. tetraphyllum**

**P. depressum** Nutt. Plants prostrate, less than 5 cm tall. Leaves opposite, oblanceolate to broadly obovate, 3-5 mm long, 1-2.5 mm wide; stipules ovate to deltate, 0.5-1 mm long. Sepals elliptic to ovate, 1-2 mm long; petals linear. Fruit globose, ca. 1.5 mm long; seeds light brown, ca. 0.5 mm long. Jun. Rare; floodplains; riparian communities; ca. 40 ft (12 m). Known from a single collection made from a "stony wash" in Willows Canyon in June 1930. Santa Catalina Island; Todos Santos and Cedros islands; coast ranges of Monterey Co. to n Baja CA.

**P. tetraphyllum** (L.) L. FOUR-LEAVED ALL-SEED Plants prostrate, less than 5 cm tall. Leaves opposite to whorled, oblanceolate to obovate, 3-5 mm long, 1-2.5 mm wide; stipules lanceolate to deltate, 2-2.5 mm long. Sepals lanceolate to ovate, ca. 2 mm long; petals linear to narrowly elliptic. Fruit globose, ca. 1.5 mm long; seeds

CARYOPHYLLACEAE: Cardionema-Polycarpon

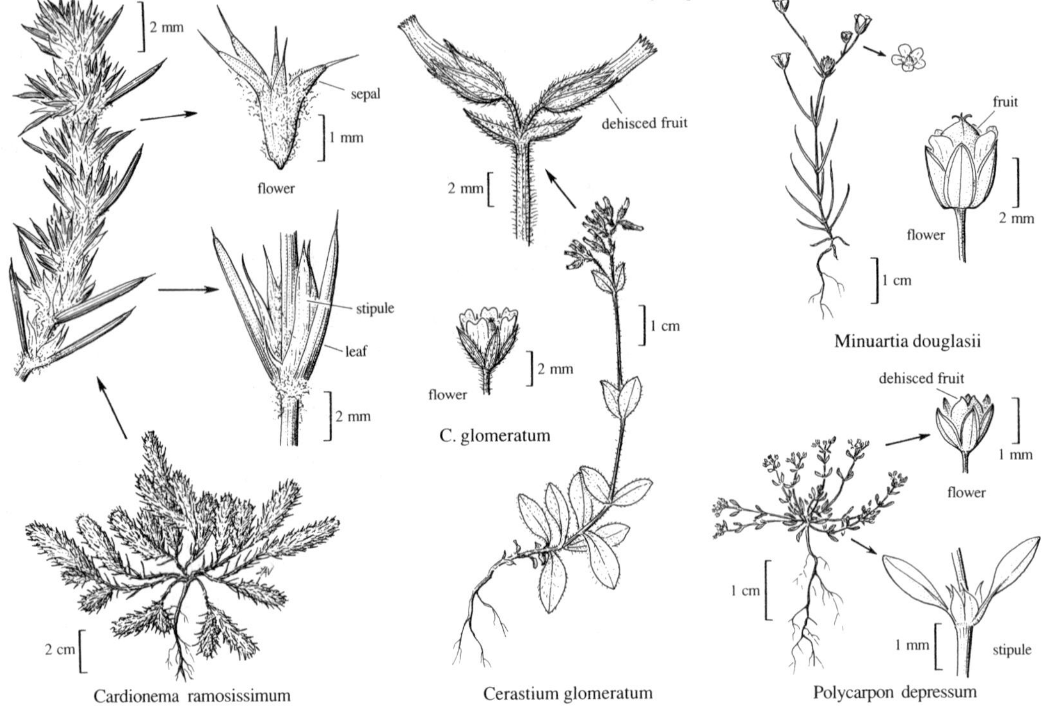


light brown, ca. 0.5 mm long. Jun. Introduced. Rare; disturbed sites; ca. 200 ft (60 m). Central Valley at Stanton Ranch. Anacapa and Santa Catalina islands; n CA coast ranges and Great Valley to s CA; native of Europe. First collected at Stanton Ranch in June 1994.

## SAGINA L. PEARLWORT

Annuals or perennial herbs, often matted or tufted. Leaves filiform to subulate, apex aristate to apiculate; stipules absent. Flowers solitary, terminal or axillary. Sepals 4-5, free, glabrous to glandular-puberulent; petals 4-5 or absent; stamens 4-10; styles 4-5, Fruit a capsule, ovoid, opening by 4-5 valves. Ca. 25 species; temperate, n hemisphere and tropical mountains.

1. Petals usually absent; stamens 4; leaf margins ciliolate near base .................................................. **S. apetala**
1' Petals 5; stamens 5 or 10; leaf margins glabrous ...................................... **S. decumbens** subsp. **occidentalis**

**S. apetala** Ard. DWARF PEARLWORT Plants 1-3 cm tall, with ascending to decumbent stems. Leaves linear, 4-8 mm long, margins ciliolate. Sepals usually 4, 1.5-2 mm long, glandular-puberulent; petals usually absent, rarely 4, but then minute; stamens 4. Fruit ± 1.5-2 mm long, opening by ± 4 valves; seeds ± compressed, ± rugose to ± papillate, ca. 0.2 mm long. Apr. Rare; shaded n-facing rock faces, disturbed flats; grasslands, riparian communities; 100-1600 ft (30-490 m). North side in upper w fork of Orizaba Canyon and at Pelican Bay; Central Valley in Cañada de la Siesta. San Miguel and San Nicolas islands; WA to s CA; e to Atlantic Coast.

**S. decumbens** (Elliott) Torr. & A.Gray subsp. **occidentalis** (S.Watson) G.Crow WESTERN PEARLWORT Plants 3-13 cm tall, with ascending to decumbent stems. Leaves filiform, 3-20 mm long, margins glabrous. Sepals 5, 1.5-2.5 mm long, glabrous to glandular-puberulent; petals 5, ± equal to sepals in length, elliptic; stamens 5 or 10. Fruit 1.5-3 mm long, opening by ± 5 valves; seeds ± compressed, smooth to faintly rugose, 0.3-0.4 mm long. Mar-May (Nov). Scarce; moist n-facing slopes, seeps, and floodplains; chaparral, riparian communities; 10-1500 ft (3-460 m). North side in several canyons between Valdez Anchorage and Prisoners' Harbor; Central Valley near La Cascada and in Islay Canyon; s side in Coches Prietos drainage. San Miguel, Santa Rosa, and Santa Catalina islands; British Columbia to n Baja CA. Subsp. *decumbens* occurs in e N. America.

## SILENE L. CATCHFLY or CAMPION

Annuals or perennial herbs; stems usually erect, often glandular. Leaves linear to oblanceolate; stipules absent. Flowers solitary or in terminal cymes. Sepals fused, tube ± cylindrical, usually with 10 or more prominent veins, lobes 5; petals 5, blade often with minute appendages, with long claw; stamens 10, fused with petals to stalk below ovary; styles 3-5. Fruit a capsule, cylindrical to ovoid, with 3-6 apical teeth. 500 species; N. America, Europe, and Asia.

1. Calyx with 15-30 prominent veins ........................................ **S. multinervia**
1' Calyx with 10 prominent veins
  2. Petals bright red, each with 4 lobes; calyx tube 15-25 mm long .................... **S. laciniata** subsp. **major**
  2' Petals white, pink, or lavender, each entire or bifid; calyx tube 4-10 mm long
    3. Stems and leaves glabrous to puberulent, upper internodes viscid; calyx glabrous; petals bifid ..... ........................................ **S. antirrhina**
    3' Stems and leaves densely pubescent; calyx glandular-puberulent; petals entire or toothed ............ ........................................ **S. gallica**

**S. antirrhina** L. SLEEPY CATCHFLY Annuals; stems erect, 1-6 dm tall, glabrous, upper internodes glandular-puberulent and viscid. Leaves 1-3.5 cm long, 2-6 mm wide. Calyx tube 4-9 mm long, with 10 veins, usually glabrous; petals white to pink, bifid, appendages minute or absent. Capsule ovoid, 4-5.5 mm long; seeds with 3-4 rows of dorsal papillae, ca. 0.5 mm long, black. Mar-May. Scarce; open slopes, disturbed sites, and creek banks; chaparral, coastal scrub; 30-1100 ft (9-335 m). North side near Pelican Bay; Central Valley near Christy Ranch, Stanton Ranch, and Rancho del Sur; s side near Rancho Nuevo and Black Point peninsula, and in Laguna, Willows, and Coches Prietos drainages. San Miguel, Santa Rosa, Santa Catalina, and San Clemente islands; Todos Santos and Guadalupe islands; British Columbia to Mexico, e to Atlantic Coast. This species is usually seen only after fires or in years with heavy rainfall.

**S. gallica** L. WINDMILL PINK Annuals; stems erect to decumbent, 1-4 dm tall, densely pubescent, sometimes glandular. Leaves 0.5-4 cm long, 1-8 mm wide. Calyx tube 6-10 mm long, with 10 veins, glandular-puberulent; petals white to pink, entire or toothed, appendages ca. 1 mm long. Capsule ovoid, 5-7 mm long; seeds with 7-9 rows of dorsal papillae, ca. 1 mm long, grayish. Mar-Jun (Sep). Introduced. Abundant; open slopes, disturbed sites, and floodplains; chaparral, coastal scrub, grasslands, oak woodland, pine forest; 5-2200 ft (1-670 m). Throughout island. All CA Channel islands; Los Coronados, Todos Santos, Guadalupe, and Cedros islands; British Columbia s to n Baja CA, e to Atlantic Coast; native of Europe. Greene (1887) reported that this species was present on Santa Cruz Island in 1886 and was "quite as common as on the mainland".

**S. laciniata** Cav. subsp. **major** C.L.Hitchc. & Maguire INDIAN PINK Perennial herbs; stems erect to decumbent, 1.5-4 dm tall, pubescent to glandular-puberulent. Leaves 2-8 cm long, 0.2-1.5 cm wide. Calyx tube 15-25 mm long, with 10 veins, pubescent to villous; petals red to pink, deeply cleft into 4 linear to lanceolate lobes, appendages 1-1.5 mm long. Capsule oblong to ovoid, 15-18 mm long; seeds with 3-4 rows of dorsal papillae, ca. 1-1.5 mm long, reddish brown. Apr-Aug. Common; coastal bluffs, rocky n-facing slopes; chaparral, coastal scrub, grasslands, oak woodland, pine forest; 10-1400 ft (3-425 m). Widely scattered locations throughout island. San Miguel, Santa Rosa, Anacapa, and San Clemente islands; Los Coronados Island; coast ranges of Monterey Co. to n Baja CA. Subsp. *laciniata* occurs in the mountains of n Mexico and adjacent U.S.

**S. multinervia** S.Watson Annuals; stems erect, 2-6 dm tall, villous and glandular-pubescent. Leaves 0.5-5 cm long, 2-8 mm wide. Calyx tube 8-12 mm long, with 15-30 veins, glandular-pubescent; petals white to pink, bifid, appendages absent. Capsule ovoid, 6-9 mm long; seeds with 3-4 rows of dorsal papillae, 0.5-1 mm long, black. Apr. Rare; burned sites; chaparral; elevation unknown. Known from a single collection on a burned site above Pelican Bay in April 1932. Santa Catalina Island; n CA coast ranges to Baja CA.

## SPERGULARIA (Pers.) J.Presl & C.Presl SAND-SPURREY

Annuals or perennial herbs; stems erect to spreading. Leaves filiform to linear, often succulent, ± terete; stipules lanceolate to triangular, scarious. Flowers few-many in terminal cymes. Sepals 5, free or fused at base, lanceolate to ovate, glabrous to glandular; petals 5, entire; stamens 2-10; styles 3. Fruit a capsule, ovoid, opening by 3 valves. Ca. 40 species; worldwide.

1. Perennial herbs, with a thick, fleshy root; cauline leaves usually with axillary leaf clusters; stipules 4-8 mm long ........................................ **S. macrotheca** var. **macrotheca**
1' Annuals, with slender roots; cauline leaves usually without axillary leaf clusters; stipules 1.5-4 mm long

CARYOPHYLLACEAE: Sagina-Stellaria

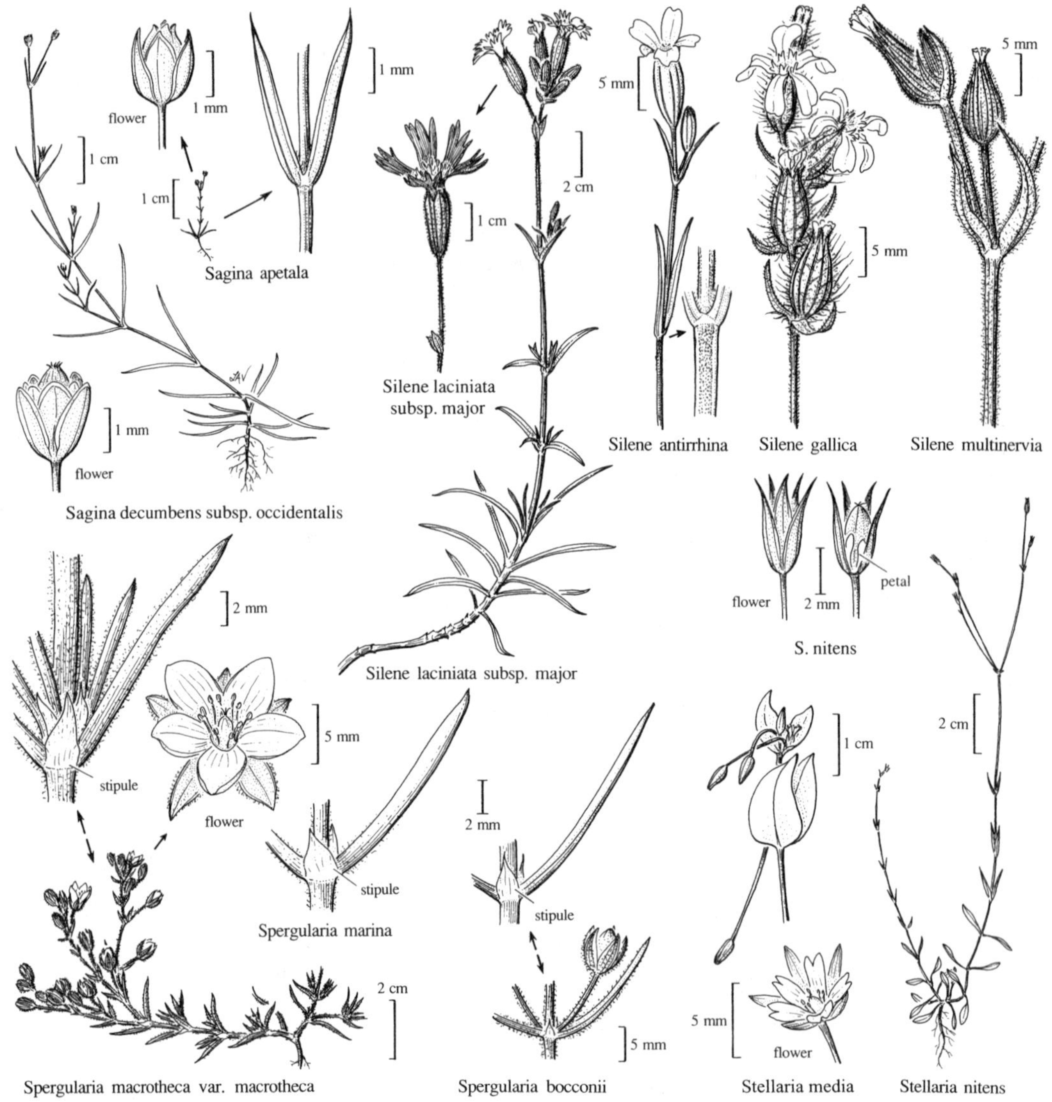


2. Stems with glabrous lower internodes; stamens 6-10; stipules short-acuminate .................. **S. bocconii**

2' Stems with glandular-puberulent lower internodes; stamens 2-5; stipules acute .................... **S. marina**

**S. bocconii** (Scheele) Merino Annuals, 5-20 cm tall; stems erect to decumbent, lower internodes glabrous, upper ones glandular-puberulent. Leaves 10-20 mm long, without axillary leaf clusters; stipules deltate, 2-4 mm long, short-acuminate. Sepals 2-4 mm long; petals white to pink, shorter than sepals; stamens 6-10. Fruit 2.5-5 mm long; seeds ca. 0.5 mm long, ± rugose to papillate, without a wing, light brown. Mar-Jun (Oct). Introduced. Occasional; disturbed sites, roadsides, vernal ponds, grassy pastures; grasslands; 10-1250 ft (3-380 m). Widely scattered locations throughout island, especially in e portion. Anacapa, San Nicolas, Santa Barbara, Santa Catalina, and San Clemente islands; OR to s CA; native of Europe.

**S. macrotheca** (Hornem.) Heyn var. **macrotheca** Perennial herbs, 8-30 cm tall, with thick, fleshy roots; stems many branched, ascending, glabrous to glandular-puberulent. Leaves 10-30 mm long, often with axillary leaf clusters; stipules narrowly deltate, 4-8 mm long, long-acuminate. Sepals 5-8 mm long; petals pink, shorter than sepals; stamens 9-10. Fruit 5-8 mm long; seeds 0.5-1 mm long, smooth to ± rugose but not papillate, narrowly winged. Mar-Oct. Common; coastal beaches, bluffs, dunes, flats, grassy slopes, and rocky ridgetops; coastal strand, coastal scrub, grasslands; 10-700 ft (3-215 m). Scattered locations around perimeter of island. All CA Channel islands; Los Coronados and Guadalupe islands; British Columbia s to Baja CA. Other varieties occur on the mainland of CA.

**S. marina** (L.) Griseb. Annuals, 5-15 cm tall; stems erect, glandular-puberulent. Leaves 10-25 mm long, usually without axillary leaf clusters; stipules broadly deltate, 1.5-3.5 mm long, acute. Sepals 2.5-5 mm long; petals white to pink, shorter than sepals; stamens 2-5. Fruit 3-5 mm long; seeds 0.6-0.8 mm long, smooth to ± rugose or papillate, with or without a wing, brownish. Apr-Oct. Scarce; sandy flats, creek mouths, floodplains; coastal scrub, grasslands, riparian communities; 10-50 ft (3-15 m). North side at mouth of Dick's Canyon, Prisoners' Harbor, and Scorpion Anchorage. All CA Channel islands except San Miguel; Guadalupe Island; N. & S. America, Europe, and Asia.

### STELLARIA L. CHICKWEED or STARWORT

Annuals or perennial herbs; stems erect to prostrate. Leaves linear to ovate; stipules absent. Flowers solitary and axillary or few-many in axillary or terminal cymes. Sepals free, lanceolate to ovate, 4-5, glabrous to glandular-puberulent; petals absent or 5 and deeply bifid; stamens usually 10; styles 3. Fruit a capsule, ovoid to cylindrical, with 3 apical teeth. Ca. 120 species; worldwide.

1. Cauline leaves well-spaced, blades mostly ovate; floral bracts leaf-like .......... **S. media**
1' Cauline leaves crowded near base, blades oblanceolate to linear; floral bracts scarious .......... **S. nitens**

**S. media** (L.) Vill. COMMON CHICKWEED Annuals, often growing over one winter, 5-40 cm tall; stems erect to prostrate, glabrous or pubescent in lines. Leaves ovate, blades 5-40 mm long, 2-20 mm wide, lower ones petiolate; petioles often ciliate. Floral bracts leaf-like. Sepals 5, lanceolate to ovate, 3-5 mm long, acute, usually pubescent; petals shorter than sepals. Capsule longer than calyx; seeds ca. 1 mm long, papillate to rugose, brownish. Feb-May (Aug). Introduced. Common; shaded, n-facing slopes, seeps, disturbed sites, moist floodplains, and creekbanks; grasslands, oak woodland, riparian communities; 20-1300 ft (6-395 m). Throughout island. San Miguel, Santa Rosa, Anacapa, Santa Catalina, and San Clemente islands; n CA coast ranges to Baja CA, e to Sierra Nevada; native of Europe. Greene (1887) reported that this species was present on Santa Cruz Island in 1886. The entire plant is edible and can be used for greens.

**S. nitens** Nutt. SHINING CHICKWEED Annuals, 5-15 cm tall; stems erect, glabrous or lower internodes slightly pubescent. Leaf blades 5-12 mm long, 1-5 mm wide, lower ones oblanceolate, petiolate, upper ones linear to narrowly lanceolate, ± sessile. Floral bracts scarious. Sepals 5, lanceolate, 3-4 mm long, acuminate, usually glabrous; petals absent or ca. half sepal length. Capsule ± equal to the calyx; seeds ca. 0.5 mm long, papillate to rugose, brown. Feb-May. Scarce; shaded, n-facing slopes, seeps, moist ravines; chaparral, grasslands, oak woodland; 15-1700 ft (4-520 m). North side between Hazard's Canyon and near Navy Base; Central Valley between Cañada de la Portezuela and Stanton Ranch. Santa Rosa and Santa Catalina islands; Guadalupe Island; British Columbia to Mexico, e to MT and UT.

## CHENOPODIACEAE Vent.
## GOOSEFOOT FAMILY

Annuals or perennials, sometimes monoecious or dioecious; stems often fleshy or succulent. Leaves usually alternate, simple, entire to toothed or lobed; stipules absent; pubescence often composed of globose trichomes that collapse when dry, appearing mealy or scurfy. Flowers bisexual or unisexual, radial, axillary or terminal, in racemes, spikes, or panicles. Sepals free or fused, 1-5 or absent, often persistent in fruit; petals absent; stamens 1-5 when present, usually equal to number of sepals, free; pistil 1, ovary superior, locule 1, ovule 1, styles 1-3. Fruit a utricle, globose to compressed and often lenticular, fruit wall thick to thin and translucent; seeds either vertical (with compressed side parallel to floral axis) or horizontal (with compressed upper side perpendicular to floral axis). Ca. 100 genera and 1300 species; temperate to subtropical, often dry climates and often in alkaline or saline habitats, worldwide. Some genera are cultivated for food (e.g., beets, chard, and spinach).

1. Leaves scale-like, opposite; stems and inflorescence often fleshy .......... **Salicornia**
1' Leaves not scale-like, usually alternate (except in *Atriplex watsonii*); stems and inflorescence not fleshy
    2. Leaves often fleshy, either terete and more or less linear or thin, rigid, and spine-tipped; ovary pear-shaped
        3. Leaves tipped with a spine or bristle; calyx keeled and winged in fruit, but not fleshy .......... **Salsola**
        3' Leaves not tipped with a spine or bristle; calyx fleshy, plane, and wingless in fruit .......... **Suaeda**

2' Leaves with wide, flat blades, neither terete nor rigid, usually not fleshy; ovary usually ovoid to globose, sometimes ± compressed, but not pear-shaped
4. Flowers unisexual, pistillate ones enclosed by 2 more or less accrescent bracts ................ **Atriplex**
4' Flowers bisexual, not enclosed by bracts
5. Ovary appearing partly inferior, sunken in receptacle; sepals thickened and swollen in fruit; fruits adherent in globose clusters ........................................................................................ **Beta**
5' Ovary superior; sepals more or less thin, not swollen in fruit; fruits not adherent to each other
6. Sepals 3-5; stamens 5; flowers in dense, terminal panicles ...................... **Chenopodium**
6' Sepal 1-3; stamens 1-3; flowers 1-few in axillary clusters
7. Leaves ovate to obovate; style branches 3 ............................................. **Aphanisma**
7' Leaves lanceolate to spatulate; styles 2 .................................................. **Monolepis**

## **APHANISMA** Nutt. ex Moq.

The only species in the genus.

**A. blitoides** Nutt. ex Moq. Annuals; stems often branched at base, often sprawling, succulent, 1-5 dm long. Leaves alternate, 1-4 cm long, 0.5-1 cm wide, glabrous, lower ones obovate, petiolate, upper ones ovate, sessile. Flowers bisexual, 1-3 per axillary cluster, ± sessile. Calyx usually with 3 lobes, ca. 1 mm long; stamen 1; style 1, with 3 stigmatic branches. Fruit depressed-globose, 1-2 mm in diameter; seed compressed, lenticular, horizontal. Mar-Jun. A species of special environmental concern. Rare; not seen recently; rocky slopes, sea bluffs, stream banks; coastal scrub; 10-200 ft (3-60 m). Known from only four collections made between 1930 and 1939; in Central Valley near mouth of Cañada Christy; s side in Cañada de los Sauces. All CA Channel islands except San Miguel; all Baja California islands except San Geronimo; Santa Barbara Co. to n Baja CA. This species is usually associated with *Opuntia* clumps.

## **ATRIPLEX** L. SALTBUSH

Annuals, perennial herbs, or shrubs, monoecious or dioecious, glabrous or with farinose or scurfy pubescence. Leaves alternate or opposite, sessile or petiolate, entire to dentate. Flowers unisexual, solitary, clustered in leaf axils, or in terminal spikes or panicles. Staminate flowers with 3-5 calyx lobes and 3-5 stamens, bracts absent; pistillate flowers with perianth absent, subtended by 2, distinct to fused bracts, their margins entire to toothed, ovary ovoid to globose, with 2 style branches. Fruit often lenticular, enclosed by persistent bracts; seed usually vertical. Ca. 250 species; temperate to subtropical, worldwide. Difficult to identify without mature fruits and bracts.

1. Shrubs, with woody stems and branches, usually 1-3 m tall; plants dioecious ......................... **A. lentiformis**
1' Annuals or perennial herbs, sometimes woody at base, usually less than 1 m tall; plants monoecious (except *A. watsonii*)
2. Leaves sparsely scurfy when young, becoming green and glabrous with age
3. Fruiting bracts fused to middle or above; leaves lanceolate to oblong or oval; staminate flowers in terminal spikes or panicles ........................................................................................ **A. serenana**
3' Fruiting bracts fused only near base; leaves broadly triangular to hastate; staminate flowers either in very short spikes or mixed with pistillate flowers ............................................... **A. triangularis**
2' Leaves scurfy, often densely so, grayish to white
4. Fruiting bracts broadest below middle
5. Fruiting bracts thick, fleshy, red, with 3-5 teeth above middle ....................... **A. semibaccata**
5' Fruiting bracts ± compressed, neither fleshy nor red, margins mostly entire
6. Leaves mostly alternate, lanceolate to oblanceolate; fruiting bracts mostly free, ovate, 2-4 mm long ........................................................................................ **A. californica**
6' Leaves mostly opposite, broadly elliptic to ovate; fruiting bracts fused to above middle, 3-7 mm long ........................................................................................ **A. watsonii**
4' Fruiting bracts broadest at or above middle
7. Plants erect; leaves narrowly ovate to hastate; staminate and pistillate flowers usually mixed in same clusters ........................................................................................ **A. argentea**
7' Plants prostrate; leaves lanceolate to ovate, elliptic, or oblong; staminate and pistillate flowers usually in different clusters

**CHENOPODIACEAE: Aphanisma, Atriplex**

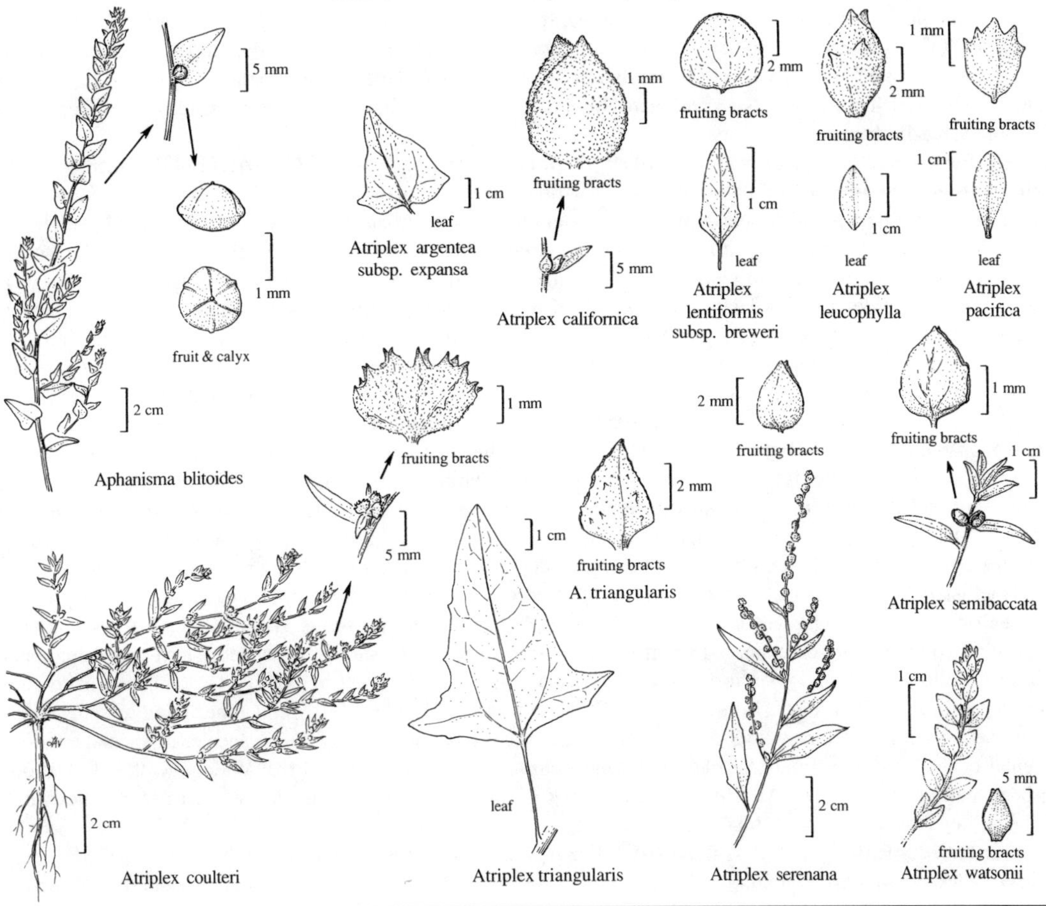


8. Fruiting bracts globose to 4-angled, 4-8 mm long; seeds 2.5-3 mm long. **A. leucophylla**

8' Fruiting bracts ± compressed, 1-3 mm long; seeds 1-2 mm long

   9. Perennial herbs; leaves elliptic, lanceolate, or ovate; fruiting bracts 2-3 mm long, free margins deeply dentate ........ **A. coulteri**

   9' Annuals; leaves lanceolate to elliptic or spatulate; fruiting bracts 1-1.5 mm long, margins with 3-5 teeth at apices ........ **A. pacifica**

**A. argentea** Nutt. subsp. **expansa** (S.Watson) H.M.Hall & Clem. SILVERSCALE Erect annuals, 1.5-6 dm tall. Leaves alternate, narrowly ovate to hastate, 2-8 cm long, 2-5 cm wide, grayish, farinose or scurfy, petiolate, margins entire to undulate. Staminate and pistillate flowers mixed, axillary or in terminal spikes. Fruiting bracts fused to middle or above, obovate, 4-7 mm long, compressed, free margins laciniate. Seeds 1.5-2 mm long. Jun-Aug. Rare; dry ravines, sea bluffs; coastal scrub; 10-50 ft (3-5 m). South side in vicinity of Smugglers' Cove. Santa Rosa, Santa Catalina, and San Clemente islands; San Francisco Bay region to n Mexico, e to TX.

**A. californica** Moq. in DC. CALIFORNIA SALTBUSH Perennial herbs; stems usually prostrate, 2-5 dm long. Leaves alternate, crowded, lanceolate to oblanceolate, 5-15 mm long, 2-4 mm wide, gray, scurfy, sessile, margins ± entire. Staminate flowers in terminal spikes or mixed with pistillate flowers in axillary clusters. Fruiting bracts mostly free, ovate, 2-4 mm long, margins entire. Seeds 1-2 mm long. Apr-Nov. Common; rocky sea cliffs, slopes, ridgetops, and alluvial flats; coastal scrub, grasslands; 3-450 ft (1-140 m). Around perimeter of island, especially at western and eastern ends. All CA Channel islands; Los Coronados, Todos Santos, Guadalupe, and Cedros islands; coastal n CA to Baja CA.

**A. coulteri** (Moq.) D.Dietr. COULTER'S SALTBUSH Perennial herbs; stems usually prostrate, 1-4 dm long. Leaves usually alternate, elliptic, lanceolate, or ovate, 7-14 mm long, 2-4 mm wide, grayish green, scurfy, sessile, margins ± entire. Staminate flowers in short terminal spikes, pistillate ones in axillary clusters. Fruiting bracts fused to middle, obovate, 2-3 mm long, compressed, free margins sharply dentate. Seeds 1-2 mm long. Mar-Oct. A species of special environmental concern. Common; sandy flats, rocky sea cliffs, slopes, ridgetops, floodplains, and alkaline flats; coastal scrub, grasslands; 10-1500 ft (3-460 m). Widely scattered locations throughout island. All CA Channel islands except San Nicolas and Santa Barbara; San Benito, Cedros, and Natividad islands; Santa Barbara Co. to n Baja CA.

**A. lentiformis** Nutt. subsp. **breweri** (S.Watson) H.M.Hall & Clem. BREWER'S SALTBUSH Shrubs, 1-2 m tall; stems erect to ascending. Leaves alternate, narrowly lanceolate to ± ovate, 2-4.5 cm long, 0.5-2.5 cm wide, grayish, scurfy, short-petiolate, margins ± entire. Staminate and pistillate flowers mixed in terminal panicles or sometimes on separate plants. Fruiting bracts fused to middle, ovate, 4-7 mm long, slightly compressed, margins entire. Seeds 1-2 mm long. Jul-Oct. Scarce; sea cliffs, alkaline flats, and floodplains; coastal scrub; 10-30 ft (3-10 m). North side at China Harbor and mouth of Scorpion Canyon; s side near mouth of Willow Canyon, Valley Anchorage, and mouth of Aguaje Canyon near Yellowbanks Anchorage. Anacapa, Santa Catalina, and San Clemente islands (planted on San Nicolas Island); coastal Monterey Co. to n Baja CA. Subsp. *lentiformis*, from the interior of CA, UT, and AZ, differs only by its smaller leaves and fruiting bracts ranging from 2.5-5 mm long. Flowers, stems, and leaves of subsp. *lentiformis* were used medicinally by Native Americans.

**A. leucophylla** (Moq.) D.Dietr. SEA SCALE Perennial herbs; stems usually prostrate, 3-5 dm long. Leaves alternate, ovate to oblong, 0.8-3.5 cm long, 0.6-2 cm wide, white-scurfy, sessile, ± entire. Staminate flowers in dense terminal panicles, pistillate ones in axillary clusters. Fruiting bracts fused to apex, ovate, 4-angled to globose, 4-8 mm long, tuberculate, margins entire to dentate. Seeds 2.5-3 mm long. Mar-Oct. Scarce; sandy beaches and dunes; coastal strand; 3-10 ft (1-3 m). Widely scattered locations around perimeter of island. All CA Channel islands except Santa Barbara; San Martin, Cedros, and Natividad islands; coastal n CA to n Baja CA.

**A. pacifica** A.Nelson SOUTH COAST SALTSCALE Annuals; stems prostrate, 3-10 dm long. Leaves alternate, lanceolate to elliptic or spatulate, 5-18 mm long, 2-8 mm wide, subsessile or lower ones short-petiolate, upper surfaces yellowish green, lower surfaces ± scurfy, margins entire. Flowers in axillary clusters, staminate ones usually above pistillate ones. Fruiting bracts fused to middle, obovate, 1-1.5 mm long, compressed, with a prominent midline, margins with 3-5 teeth at apex. Seeds ca. 1 mm long. May-Jul. A species of special environmental concern. Scarce; sandy beaches and dunes; coastal strand; 3-10 ft (1-3 m). Widely scattered locations around perimeter of island. All CA Channel islands except San Miguel; San Martin, Cedros, and Natividad islands; Los Angeles Co. to n Baja CA.

**A. semibaccata** R.Br. AUSTRALIAN SALTBUSH Perennial herbs; stems usually prostrate, 2-12 dm long. Leaves alternate, elliptic to oblong, 1-2.5 cm long, 0.3-1 cm wide, grayish green, scurfy to glabrate, subsessile, margins entire to dentate. Staminate flowers in terminal spikes, pistillate ones in axillary clusters. Fruiting bracts fused to middle or above, rhombic, 4-7 mm long, fleshy, often bright red, ± compressed, free margins laciniate. Seeds 1.5-2 mm long. Apr-Dec. Introduced. Common; rocky sea cliffs, slopes, floodplains, and alkaline flats; coastal scrub, grasslands; 10-1500 ft (3-460 m). Widely scattered locations throughout island. All CA Channel islands; Los Coronados, Todos Santos, San Martin, and Cedros islands; throughout much of CA to n Mexico, e to UT and TX; native of Australia.

**A. serenana** A.Nelson var. **davidsonii** (Standl.) Munz DAVIDSON'S BRACTSCALE Annuals, 3-10 dm tall; stems erect to decumbent. Leaves alternate, lanceolate to oblong or oval, 2-4 cm long, 0.2-2 cm wide, ± grayish green, sparsely scurfy, green and glabrate with age, subsessile, margins usually dentate. Staminate flowers in globose clusters arranged in terminal panicles, pistillate ones in axillary clusters. Fruiting bracts fused to ± middle, cuneate to orbicular, 3-5 mm long, slightly compressed, free margins deeply dentate. Seeds 1-1.5 mm long. Jun. Rare, not seen recently; disturbed sites; grasslands; ca. 50 ft (15 m). Known from a single collection made in June 1930 on "waste ground" at Pelican Bay. Santa Rosa and Santa Catalina islands; Los Coronados islands; Great Valley to Baja CA, e to NV. Var. *serenana*, with spikate inflorescences, occurs on the adjacent mainland.

**A. triangularis** Willd. [*A. patula* L. subsp. *hastata* (L.) H.M.Hall & Clem. misapplied to this species] SPEAR-LEAF SALTBUSH or SPEARSCALE Erect annuals, 3-10 dm tall. Leaves alternate, broadly triangular or hastate, 2-8 cm long, 1-7 cm wide, grayish and scurfy when young, green and glabrate with age, long-petiolate, margins entire to shallowly toothed. Staminate and pistillate flowers mixed in terminal spikes. Fruiting bracts fused only at base, triangular, 2-7 mm long, slightly compressed, surfaces tuberculate, margins ± entire. Seeds 1-2 mm long. Aug-Oct. Scarce; coastal flats, floodplains, sandy creekbeds; coastal scrub, coastal strand, riparian communities; 5-50 ft (1-15 m). North side at Prisoners' Harbor and at mouth of Scorpion Canyon; Central Valley at Christy

Beach; s side just e of Forney's Cove, at mouth of Cañada de los Sauces, and at mouth of Willows Canyon. San Miguel, San Nicolas, and Santa Catalina islands; throughout much of N. America, Europe and Asia. Leaves are edible and can be used as a substitute for spinach.

**A. watsonii** A.Nelson MATSCALE Perennial herbs; stems prostrate, forming mats 1-3 m in diameter. Leaves mostly opposite, broadly elliptic to ovate, 5-20 mm long, 3-7 mm wide, white-scurfy, sessile, margins entire to crenate near apices. Staminate flowers in terminal spikes, pistillate ones in axillary clusters. Fruiting bracts fused to above middle, ovate, 3-7 mm long, thick, margins entire to crenate near apices. May-Jul. Scarce; coastal bluffs, alkaline flats, and floodplains; coastal scrub, coastal strand; 20-70 ft (6-21 m). West end between West and Near points. All CA Channel islands except Santa Barbara; San Martin Island; Santa Barbara Co. to n Baja CA.

## **BETA** L. BEET

Annuals or biennials. Leaves basal and alternate, petiolate, entire. Inflorescence a spike or spikate panicle. Flowers bisexual; sepals 5, swollen and persistent in fruit; stamens 5; ovary sunken in receptacle, with 2-3 style branches. Fruit adherent to receptacle and swollen calyx, often adherent or fused at base to adjacent fruits; seed horizontal. Ca. 5 species of Mediterranean Europe. Cultivated for food and fodder (beets, Swiss chard).

**B. vulgaris** L. BEET Plants 2-10 dm tall. Basal leaves with petioles 7-20 cm long and blades 12-15 cm long, lower blades ovate to more or less sagittate, margin undulate; cauline leaves shorter than lower ones. Sepals 2-3 mm long, incurved and keeled in fruit, margins scarious. Utricle 2-3 mm long, usually in globose clusters. Mar-May. Introduced. Scarce; sandy beaches, coastal flats, floodplains; coastal scrub, coastal strand, grasslands; 20-320 ft (6-100 m). North side near Cavern Point and at mouth of Scorpion Canyon. All CA Channel islands except San Nicolas and Santa Barbara; Cedros Island; coastal CA, e to Atlantic States; native of Europe. Leaves and roots are edible; leaves are sometimes used as a substitute for tobacco.

## **CHENOPODIUM** L. GOOSEFOOT or PIGWEED

Annuals or perennial herbs, usually farinose. Leaves alternate, usually petiolate, entire to lobed. Flowers bisexual, clustered in leaf axils or in terminal spikes or panicles. Sepals 4-5, free or partly fused into a tube, plane to keeled, usually persistent; stamens usually 5; ovary ± globose; styles or style branches 2-5. Fruit lenticular, enclosed by persistent sepals, fruit wall thin, often adherent to seed; seed horizontal or vertical. Ca. 150 species; temperate, worldwide. Some species are used for food (leaves, seeds), including the protein-rich Quinoa (*C. quinoa*).

1. Plants with glandular-puberulent leaves, but not farinose .................. **C. ambrosioides**

1' Plants with farinose leaves, never glandular

  2. Perennials; calyx tube longer than lobes; most leaves broadly deltate .................. **C. californicum**

  2' Annuals; calyx tube much shorter than lobes or absent; most leaves rhombic-ovate, upper ones often lanceolate

    3. Upper surfaces of leaves usually shiny, bright green .................. **C. murale**

    3' Upper surfaces of leaves usually farinose, dull green .................. **C. berlandieri**

**C. ambrosioides** L. EPAZOTE or MEXICAN TEA Annuals, 4-10 dm tall. Leaf blades lanceolate to narrowly oblong, 2-10 cm long, 0.2-2 cm wide, tapered to petiolate bases, glandular-puberulent, margins subentire to repand-dentate or pinnately lobed. Sepals plane, glandular, enclosing fruit. Fruit wall thin, deciduous; seed horizontal or vertical, ca. 0.7 mm wide, surface smooth, black, shiny. May-Nov. Introduced. Common; sandy flats, floodplains, moist creekbeds, and around seasonal ponds; 5-700 ft (1-213 m). Numerous drainages and canyons throughout island. Santa Rosa, San Nicolas, and Santa Catalina islands; throughout much of N. America; native of tropical America. Present in small numbers but not yet "thoroughly established" on Santa Cruz Island in 1886 (Greene 1887). This species has been used medicinally and is the source of essential oils; leaves are also used as a condiment in soups and bean dishes.

**C. berlandieri** Moq. [*C. b.* Moq. var. *sinuatum* (Murray) Murray] PITSEED GOOSEFOOT Annuals, 2-20 dm tall. Leaf blades lanceolate to rhombic-ovate, 1-5 cm long, 0.5-2 cm wide, farinose, margins dentate to serrate; petioles 1-2 cm long. Sepals prominently keeled, farinose, enclosing fruit. Fruit wall adherent to seed; seed horizontal, 1.2-1.5 mm wide, surface foveolate, black, dull if fruit wall removed. Mar-Sep. Scarce; rocky slopes, disturbed flats, sandy sites; coastal scrub, riparian communities; 10-300 ft (3-90 m). North side at Pelican Bay, in Cañada del Puerto, and at mouth of Scorpion Canyon; s side at Coches Prietos and Valley anchorages. Santa Rosa, Anacapa, and Santa Catalina islands; throughout much of N. America. Often confused with

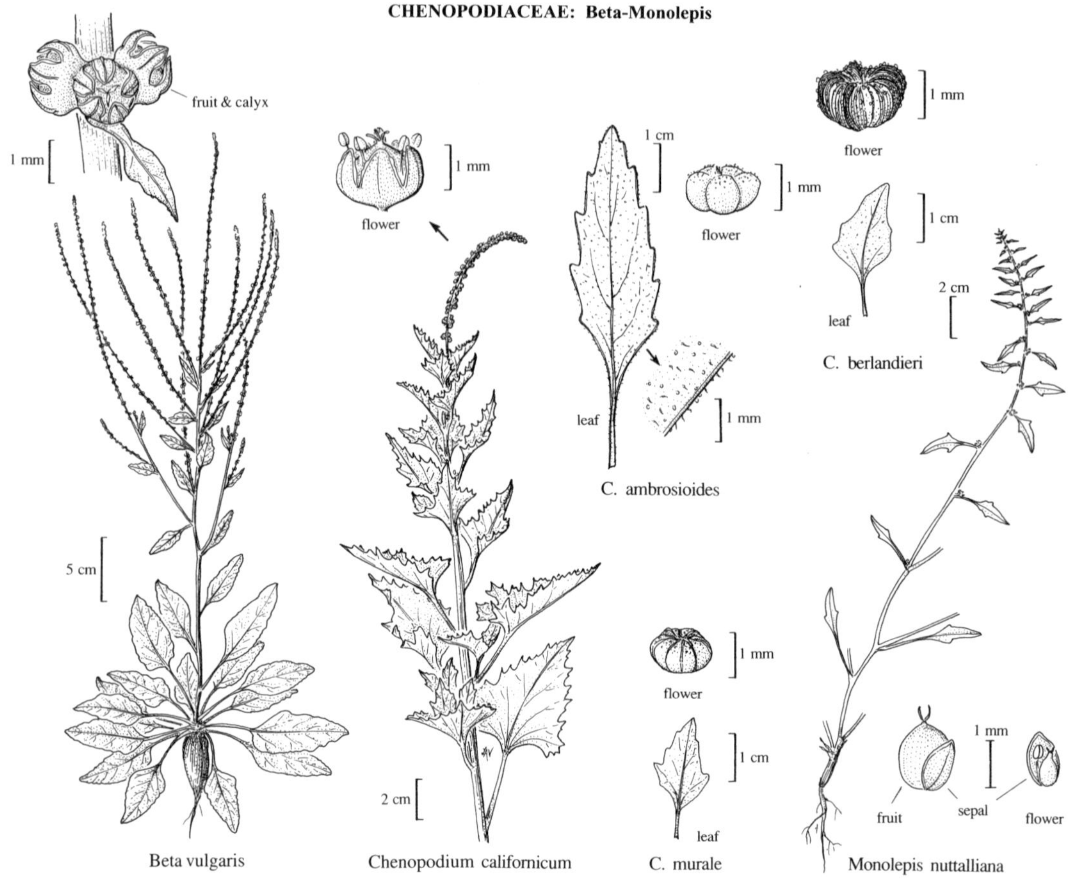


*C. album* L., a native of Europe, which has seeds with a smooth to rugulose (but not foveolate) surface. Leaves and seeds are edible.

**C. californicum** (S.Watson) S.Watson SOAP ROOT Perennial herbs, with stout, fleshy roots; stems ascending to decumbent, 3-8 dm long. Leaf blades deltate, 3-10 cm long, 2-8 cm wide, green and ± glabrous to glaucous, bases truncate to cordate, margins sharply and irregularly sinuate-dentate; petioles 3-13 cm long. Sepals fused into tube that encloses fruit, ± glabrous. Fruit wall adherent to seed, minutely ridged; seed vertical, ca. 2 mm long, surface ± smooth, dark brown. Feb-Aug. Common; moist slopes, grassy hillsides, streambanks; coastal scrub, grasslands, oak woodland; 20-1550 ft (6-475 m). Widely scattered locations throughout island. All CA Channel islands; Los Coronados and Todos Santos islands; throughout much of CA and n Baja CA. Roots can be dried and grated to make soap; Native Americans also used the leaves and root medicinally.

**C. murale** L. NETTLE-LEAF GOOSEFOOT Annuals, 2-5 dm tall. Leaf blades ovate to rhombic-ovate, 2-6 cm long, 0.4-1.2 cm wide, upper surfaces bright green and shiny, lower surfaces ± farinose, margins irregularly sinuate-dentate; petioles 0.5-2 cm long. Sepals slightly keeled, farinose, partly enclosing fruit. Fruit wall adherent to seed, with sharp edge; seed horizontal, ca. 1.5 mm wide, surface ± smooth, brown to black. Mar-Jul. Introduced. Occasional; disturbed sites, sandy beaches, rocky sea cliffs, and ridgetops; coastal scrub, coastal strand; 5-1150 ft (1-350 m). Widely scattered locations throughout island, especially on s side. All CA Channel islands; all Baja CA islands; throughout much of N. America; native of Europe. Present in small numbers but not yet "thoroughly established" on Santa Cruz Island in 1886 (Greene 1887). Leaves are edible.

## MONOLEPIS Schrad.

Annuals; stems often branched at base, usually fleshy or succulent. Leaves alternate, elliptic to oblanceolate, usually entire, sometimes with hastate bases, sessile or petiolate. Flowers usually bisexual, in axillary or terminal clusters. Sepals 1-3, often persistent; stamens absent (flowers then pistillate) or only 1 per flower; styles or style branches 2. Fruit ovoid to globose, wall thin and adherent to seed, smooth to tuberculate; seed erect. Ca. 3-4 species; temperate N. and S. America, Europe, and Asia.

**M. nuttalliana** (Schult.) Greene Stems ascending to decumbent, 5-15 cm long, succulent, slightly farinose. Leaf blades lanceolate to spatulate, 1-4 cm long, 0.3-1 cm wide, hastate or with 2 teeth at middle and tapered to petiolate bases. Flowers in dense axillary clusters, often reddish. Sepal spatulate to obovate, ca. 1 mm long. Fruit wall minutely pitted; seed ca. 1 mm wide, with an acute margin, dark brown. Mar-Apr. Rare; disturbed sites, rocky slopes, and grassy hillsides; coastal scrub, grasslands; 20-350 ft (6-105 m). West end between vicinity of Black Point peninsula and Christy Beach. San Miguel, Santa Rosa, Santa Barbara, and San Clemente islands; throughout much of w N. America. Washed roots, cooked with salt and fat, were eaten by Native Americans in Arizona; seeds are also edible.

## SALICORNIA L. GLASSWORT or PICKLEWEED

Annuals, perennial herbs, or shrubs; stems jointed, succulent. Leaves opposite, scale-like, sessile, slightly farinose. Flowers bisexual, subtended by minute scale-like bracts, usually in clusters of 3, clusters in fleshy, cylindrical, terminal spikes. Calyx with truncate or toothed margin, fleshy, often deciduous in fruit; stamens 1-2; ovary ovoid, styles or style branches 2-3. Fruit ovoid to oblong in outline, wall usually free from seeds; seed vertical. Ca. 20 species; temperate to tropical, ± worldwide.

1. Terminal portion of spike without flowers, flowering portion of spike 2-3 mm wide; seed glabrous ............. **S. subterminalis**
1' Spike with flowers to apex, 3-4 mm wide; seed puberulent ............. **S. virginica**

**S. subterminalis** Parish [*Arthrocnemum s.* (Parish) Standl.] PARISH'S GLASSWORT Perennial herbs; stems spreading to erect, 1.5-3 dm tall. Flowering spikes 1.5-3.5 cm long, 2-3 mm wide, distal portion without flowers. Seeds glabrous. Mar-Sep. Rare; coastal bluffs and alkaline flats; coastal scrub; 20-50 ft (6-15 m). North side near West Point and Potato Harbor; s side at Forney's Cove. San Miguel, Santa Rosa, Santa Catalina, and San Clemente islands; San Martin Island; San Francisco Bay region and Great Valley to n Mexico. Seeds were ground into a meal and eaten by Native Americans; the fleshy stems can be eaten raw or pickled and have also been used in the manufacture of glass and soap because of their high soda content.

**S. virginica** L. PICKLEWEED Perennial herbs, sometimes shrub-like; stems spreading to decumbent, 2-6 dm long, often rooting at nodes. Flowering spikes 1.5-5 cm long, 3-4 mm wide, with flowers to apex. Seeds puberulent. May-Sep. Rare; moist creekbanks and brackish lagoon margins; riparian communities; 10-50 ft (3-15 m). South side at mouth of Cañada de los Sauces. All CA Channel islands except Santa Barbara; San Martin Island; throughout much of N. America. Seeds and stems have been used in the same ways as those of *S. subterminalis*.

## SALSOLA L.

Annuals, perennial herbs, or shrubs. Leaves usually alternate, filiform to oblong, entire, usually tipped with a bristle or spine, sessile or clasping. Flowers bisexual, solitary or clustered in leaf axils, often subtended by 1-2 bracts. Calyx with 4-5 lobes, thickened and often keeled or winged in fruit; stamens usually 5; styles branches 2-3. Fruit globose, apex ± depressed or orbicular and compressed, enclosed by persistent calyx; seed usually horizontal. Ca. 100 species; temperate to tropical, worldwide, often of saline or alkaline habitats.

**S. tragus** L. [*S. australis* R.Br., *S. pestifer* A.Nelson, *S. kali* L.] RUSSIAN THISTLE or TUMBLEWEED Annuals, 3-10 dm tall, often densely branched and rounded, central stem breaking naturally at its base in fruit, upper stems often striate. Leaves linear, 1-5 cm long, glabrous. Fruiting bracts subcylindrical, spiny. Calyx 2.5-3 mm long in fruit, 2-5 mm wide (including wings); wings often toothed or crenate. Seeds 1.5-2 mm wide, black, shiny. Jul-Oct. Probably introduced. Occasional; sandy beaches, eroded ravines, sea cliffs, ridgetops, and slopes; coastal scrub, coastal strand; 10-1250 ft (3-380 m). North side at Prisoners' Harbor and near Cañada del Agua; Central Valley at Christy Beach and near Valley Anchorage; scattered locations on s side between Playa Larga and Smugglers' Cove. Santa Rosa, San Nicolas, Santa Catalina, and San Clemente islands; Los Coronados, Todos Santos, and Cedros islands; The correct name for this species is in much doubt. If this species is restricted

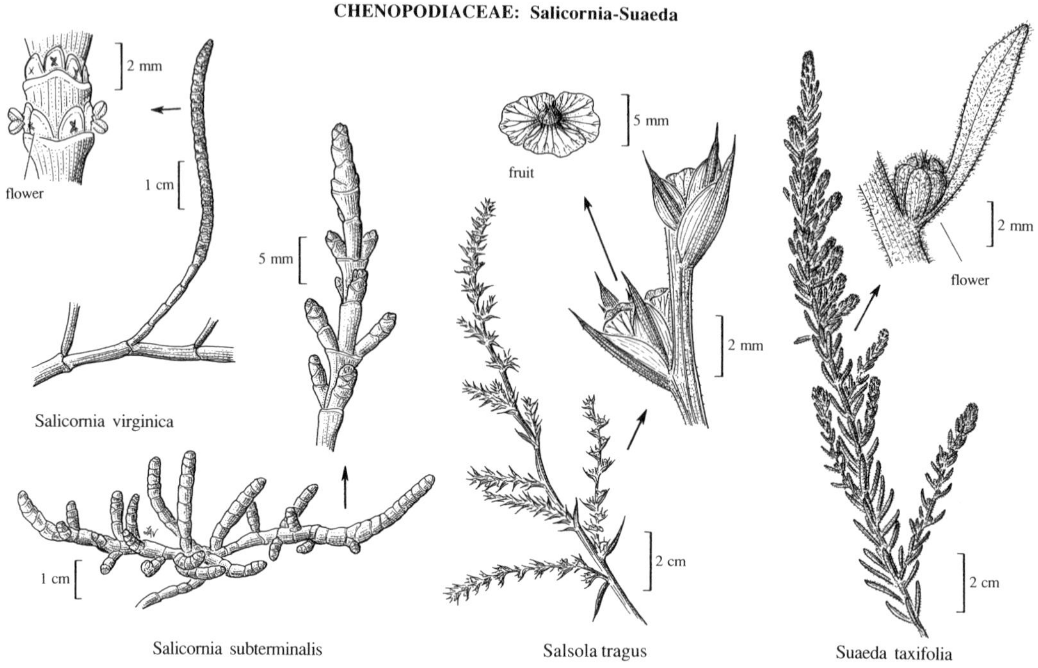


to N. America, the correct name is *S. pestifer* A.Nelson. First collected at Prisoners' Harbor in August 1939; rarely seen again until the early 1990s but now spreading rapidly. Tender branch tips can be cooked and eaten.

### SUAEDA Forssk. ex Scop.

Annuals, perennial herbs, or shrubs. Leaves alternate, cylindrical to oblong, entire, fleshy. Flowers bisexual, solitary or clustered in leaf axils, subtended by 1-3 bracts. Calyx with 5 fleshy lobes, lobes equal to unequal in size, sometimes keeled or with horn-like appendages; stamens 5; ovary subglobose, stigmatic lobes 2-4. Fruit lenticular, enclosed by calyx; seed horizontal or vertical. Ca. 110 species; worldwide, usually on saline or alkaline soils.

**S. taxifolia** (Standl.) Standl. SEA BLITE or SEEP WEED Suffrutescent perennial herbs; stems ascending to decumbent, 2-5 dm long, often densely pubescent. Leaves terete, 0.5-2 cm long, densely glandular-puberulent. Calyx lobes 1-3 mm long, equal, not appendaged; ovary pear-shaped. Seeds 1-2 mm long, black to brown, shiny. Mar-Oct. A species of special environmental concern. Scarce; sandy beaches, coastal bluffs, and alkaline flats; coastal scrub, coastal strand; 10-160 ft (3-50 m). North side in vicinity of Scorpion and Little Scorpion canyons; Central Valley at Christy Beach; s side at Valley Anchorage and Smugglers' Cove. All CA Channel islands; Los Coronados, San Martin, Guadalupe, and Cedros islands; Santa Barbara Co. to n Baja CA. Seeds and boiled leaves were eaten by Native Americans; stems and leaves were steeped in water to produce a durable black dye.

## CISTACEAE Juss.
## ROCKROSE FAMILY

Perennial herbs or shrubs. Leaves alternate, opposite, or whorled, simple, usually entire; stipules present or absent. Flowers bisexual, radial, axillary or terminal, often in racemes or panicles. Sepals free, often persistent in fruit, 3 or 5, if 5 then outer 2 sepals bract-like; petals usually 5; stamens many, free; pistil 1, ovary superior, locule 1, style usually 1, stigma entire to lobed. Fruit a capsule, with 3-10 valves; seeds many. Ca. 7 genera and 175 species; temperate, mostly n hemisphere. Some genera are cultivated for ornament, including *Cistus* and *Helianthemum*.

CISTACEAE

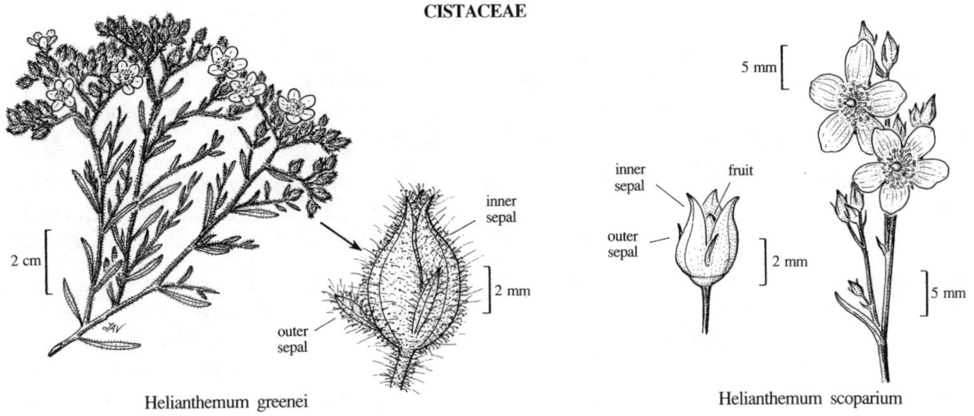


Helianthemum greenei

Helianthemum scoparium

## HELIANTHEMUM Miller RUSH-ROSE

Perennial herbs or shrubs; stems often many, erect, evergreen. Leaves usually alternate, evergreen; trichomes in sparse to dense, stellate clusters. Sepals 5, outer 2 sepals reduced, narrow, ± bract-like; petals yellow, wrinkled in bud; ovary with 3 placentas. Fruit with 3 valves. Ca. 110 species; temperate n hemisphere.

1. Inflorescence strongly glandular-pubescent, reddish brown; outer sepals ± lanceolate ................ **H. greenei**
1' Inflorescence stellate-puberulent to glabrate, not glandular; outer sepals linear ...................... **H. scoparium**

**H. greenei** B.L.Rob. ISLAND RUSH-ROSE Stems 1-3 dm long, stellate-pubescent. Leaves linear to oblanceolate, 1-3 cm long, 1-8 mm wide. Inflorescence with many flowers, strongly glandular-pubescent, often tinged reddish brown. Outer sepals ± lanceolate, 2.5-4 mm long, inner ones 5-7 mm long. Fruit 4.5-6 mm long. Mar-May (Jun). Island endemic. A species of special environmental concern. Occasional; burned sites, open ridgetops and slopes, grassy hillsides; chaparral, coastal scrub, grasslands, oak woodland, pine forest; 100-1600 ft (30-490 m). Widely scattered locations throughout island. San Miguel, Santa Rosa, and Santa Catalina islands. Hybrids with *H. scoparium* are common where the 2 species occur together.

**H. scoparium** Nutt. COMMON RUSH-ROSE Stems 1-3 dm long, stellate-pubescent to glabrate. Leaves linear, 1-2.5 cm long, 1-3 mm wide. Inflorescence with few flowers, stellate-puberulent but not glandular, rarely white-canescent. Outer sepals linear, 0.5-4 mm long, inner ones 3-7 mm long. Fruit 2.5-4 mm long. Mar-May. Occasional; burned sites, open ridgetops, and rocky slopes; chaparral, coastal scrub, pine forest; 500-1300 ft (150-395 m). North side between Fry's Harbor and Cañada del Puerto; Central Valley near Christy Pines, in Islay Canyon and near Stanton Ranch; s side between Cañada de los Sauces and Valley Anchorage. Santa Rosa and Santa Catalina islands; n CA coast ranges to San Diego Co., e to Sierra Nevada foothills. Flowers were boiled by Native Americans to make a medicinal tea.

# CONVOLVULACEAE Juss.
## MORNING-GLORY FAMILY

Annuals, perennial herbs, or shrubs; stems often twining or trailing. Leaves alternate, simple, entire to lobed; stipules absent. Flowers bisexual, radial, axillary or terminal, solitary or few in clusters. Sepals 5, free to fused at base, often overlapping, often persistent in fruit; petals 5, fused, corolla campanulate to funnelform; stamens 5, adnate to corolla; pistil 1, ovary superior, locules 2, styles 1-3, stigma entire to lobed. Fruit a capsule, often globose; seeds usually 2-4. Ca. 60 genera and 1650 species; worldwide.

1. Style 1, entire or with 2 minute lobes; corolla lobes usually much shorter than tube
   2. Calyx enclosed by 2 large bracts; stigmas oblong to elliptic, slightly or not at all flattened .. **Calystegia**
   2' Calyx not enclosed by 2 large bracts; stigmas linear, ± flattened ...................................... **Convolvulus**
1. Styles 2, distinct or fused near base; corolla lobes much longer than tube
   3. Ovary entire; stems erect ........................................................................................................ **Cressa**
   3' Ovary with 2 lobes; stems often creeping, usually underground ........................................... **Dichondra**

## CALYSTEGIA R.Br. MORNING-GLORY

Perennial herbs; stems twining or prostrate. Leaves entire to lobed, glabrous, bases sagittate or hastate, petiolate. Flowers solitary, axillary, subtended by 2 large, leaf-like bracts. Sepals enclosed by bracts; corollas campanulate to funnelform, with 5 short lobes, white to purple; style 1, with 2 oblong to elliptic stigmatic lobes, apices blunt. Fruit with 1 locule; seeds usually 4. Ca. 150 species; temperate, worldwide.

1. Leaves triangular-ovate to triangular-lanceolate; corolla white to cream; plants of dry places ........................................ **C. macrostegia**
1' Leaves reniform; corolla pink, rose, or purple; plants of sandy soils near beach ........................ **C. soldanella**

**C. macrostegia** (Greene) Brummitt subsp. **macrostegia** ISLAND MORNING-GLORY Stems twining or trailing, to 3 m long. Leaves triangular-ovate to triangular-lanceolate, 5-10 cm long, base with ± spreading lobes, apices acute to acuminate. Floral bracts ovate, 1-2.5 cm long. Corolla 4-6.6 cm long, white to cream-colored. Apr-Jun (Dec). Island endemic. Common; sandy flats and dunes, grassy hillsides, coastal bluffs, rocky canyon walls, creekbanks; chaparral, coastal scrub, grasslands, riparian communities; 10-1500 ft (3-460 m). Scattered locations throughout island. San Miguel, Santa Rosa, Anacapa, and Santa Catalina islands; San Martin and Guadalupe islands. Some plants on s side are similar to subsp. *cyclostegia.*

**C. soldanella** (L.) R.Br. BEACH MORNING-GLORY Stems prostrate, 1-5 m long. Leaves reniform, 1.5-4 cm long, often ± fleshy, base sometimes lobed, apices rounded. Floral bracts ovate, 1-1.5 cm long, often notched at apices. Corolla 3.5-5 cm long, pink, rose, or purple. Mar-Jun. Rare; sandy beaches and dunes; coastal strand; 10-30 ft (3-10 m). North side at mouth of Cañada del Agua; Central Valley at Christy Beach; s side at mouth of Cañada de los Sauces. All CA Channel islands except Anacapa and Santa Barbara; oceanic beaches, ± worldwide.

## CONVOLVULUS L. MORNING-GLORY

Annuals or perennial herbs; stems twining or prostrate. Leaves entire, glabrous to pubescent, petiolate with sagittate base or tapered. Flowers solitary or in small clusters, axillary, subtended by 2 slender bracts or without bracts. Corollas campanulate to funnelform, with 5 short lobes, white to pink; style 1, with 2 linear, flattened stigmatic lobes, apices blunt or acute. Fruit with 2 locules; seeds 2-4. Ca. 250 species; temperate, worldwide.

1. Leaves oblong-elliptic to triangular-ovate; corollas 15-23 mm long; perennials ........................... **C. arvensis**
1' Leaves linear to oblanceolate; corollas 5-6 mm long; annuals ..................................................... **C. simulans**

**C. arvensis** L. BINDWEED Perennial herbs; stems to 1 m long, glabrate. Leaves 1.5-3 cm long, blades oblong-elliptic to triangular-ovate, bases sagittate or rounded. Floral bracts linear. Sepals oblong, 3-5 mm long; corollas funnelform, 15-23 mm long, white or pink-tinged. Apr-Jun (Oct). Introduced. Occasional; disturbed sites and grassy flats; chaparral, coastal scrub, grasslands; 40-1000 ft (12-300 m). Throughout Central Valley; s side in Cañada de los Sauces, and near Coches Prietos and Smugglers' Cove. Santa Rosa, San Nicolas, and Santa Catalina islands; throughout much of N. America; native of Europe. Greene (1887) saw "only a few plants in a field near the principal settlement" on Santa Cruz Island and considered the species to be a recent introduction.

**C. simulans** L.M.Perry Annuals; stems to 0.3 m long, puberulent to hispidulous. Leaves 2.5-5 cm long, blades linear to oblanceolate, narrowed to petiolate bases. Floral bracts linear to narrowly oblanceolate. Sepals oblong to narrowly ovate, 3-4 mm long; corollas campanulate, 5-6 mm long, pink-tinged. Apr-Jul. A species of special environmental concern. Rare; dry slopes with heavy clay soil; coastal scrub, grasslands; 350-400 ft (105-120 m). Known only from the vicinity of Black Point Canyon. Santa Catalina and San Clemente islands; San Francisco Bay region and Great Valley to Baja CA.

## CRESSA L. ALKALI WEED

Perennial herbs; stems erect to decumbent. Leaves entire, usually appressed-pubescent, sessile. Flowers solitary, axillary. Corollas ± funnelform, deeply 5-lobed, white; styles 2, stigmas capitate. Fruit with 2 locules; seeds 1-4. Ca. 250 species; temperate and tropical, worldwide.

**C. truxillensis** Kunth Stems to 2 dm tall, densely grayish-pubescent. Leaves narrowly ovate to elliptic, 5-10 mm long, grayish-woolly. Sepals 4-5 mm long, densely puberulent; corolla ca. 6 mm long, lobes spreading, ca. 2 mm long, white or purple-tinged. Fruit 2-3 mm long, puberulent. May-Sep. Scarce; coastal bluffs and alkaline flats; coastal scrub, coastal strand; 10-100 ft (3-30 m). North side at mouth of Scorpion Canyon; s side at Playa Larga and just e of Punta Arena. All CA Channel islands except San Nicolas and Santa Barbara; OR to Mexico, e to Great Plains.

**CONVOLVULACEAE**

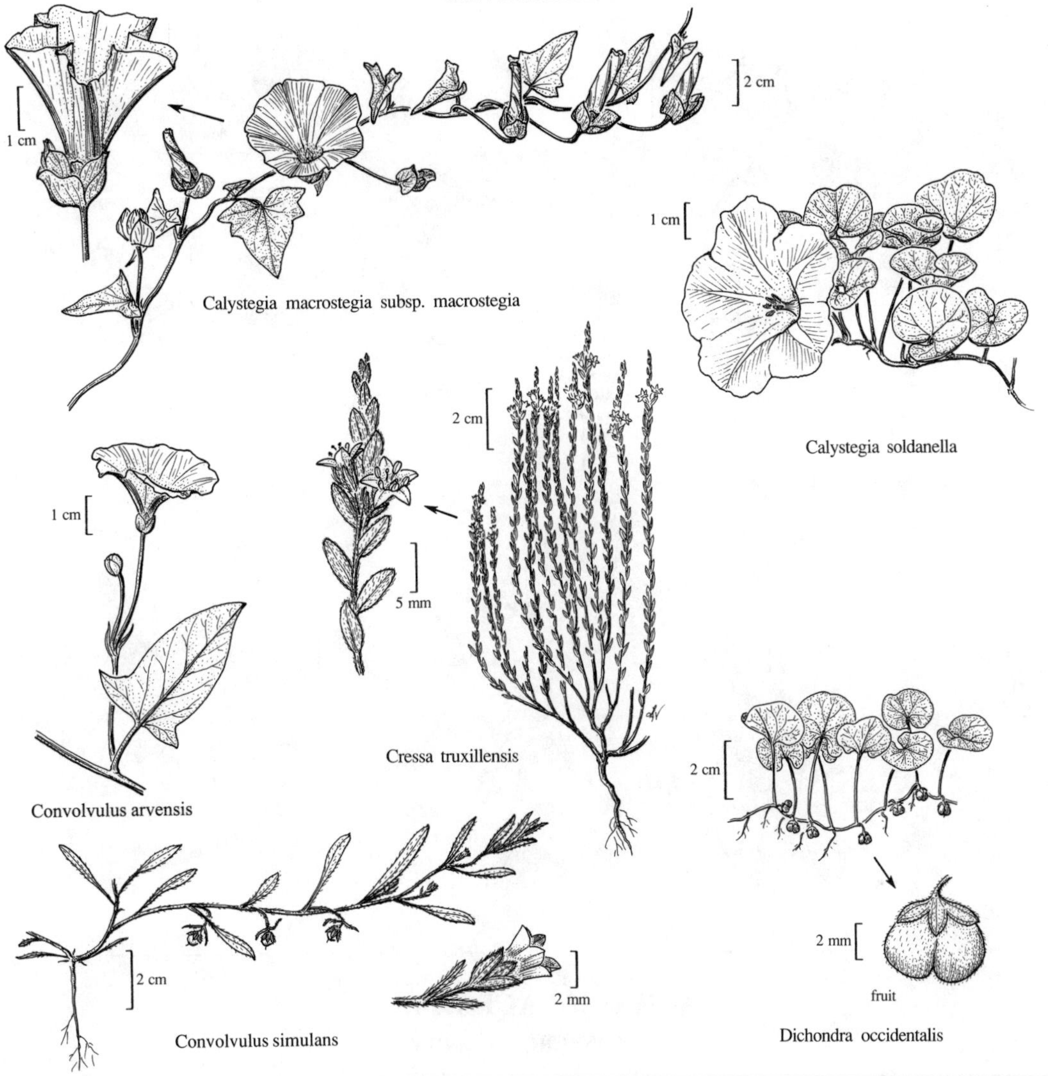


## **DICHONDRA** J.R.Forst. & G.Forst.

Perennial herbs, often matted, with stolons or prostrate stems rooting at nodes. Leaves entire, orbicular to reniform, glabrous to pubescent, petiolate. Flowers solitary, axillary. Corollas ± rotate, deeply 5-lobed, white; styles 2, stigmas capitate. Fruit deeply lobed, with 2 locules; seeds 2-4. Ca. 10 species; temperate, worldwide.

**D. occidentalis** House WESTERN DICHONDRA Stems creeping, usually partially buried, pubescent. Leaves 1-2.5 cm wide, sparsely pubescent; petioles 1-6 cm long. Calyx lobes obovate, 2-3 mm long; corolla 3-3.5 mm long, lobes ± erect, white. Fruit 3-3.5 mm long, puberulent. Mar-Jul. A species of special environmental concern. Scarce; coastal bluffs, rocky slopes and ridgetops; chaparral, coastal scrub, grasslands; 250-1100 ft (75-335 m). North side on ridge just s of West Point; s side on slopes n of Forney's Cove, near Black Point peninsula, near Kinton Point, in upper portions of Pozo and Johnson canyons, on Sierra Blanca ridge, and in Laguna Canyon. San Miguel, Santa Rosa, and Santa Catalina islands; Los Coronados and Todos Santos islands; Santa Barbara Co. to n Baja CA.

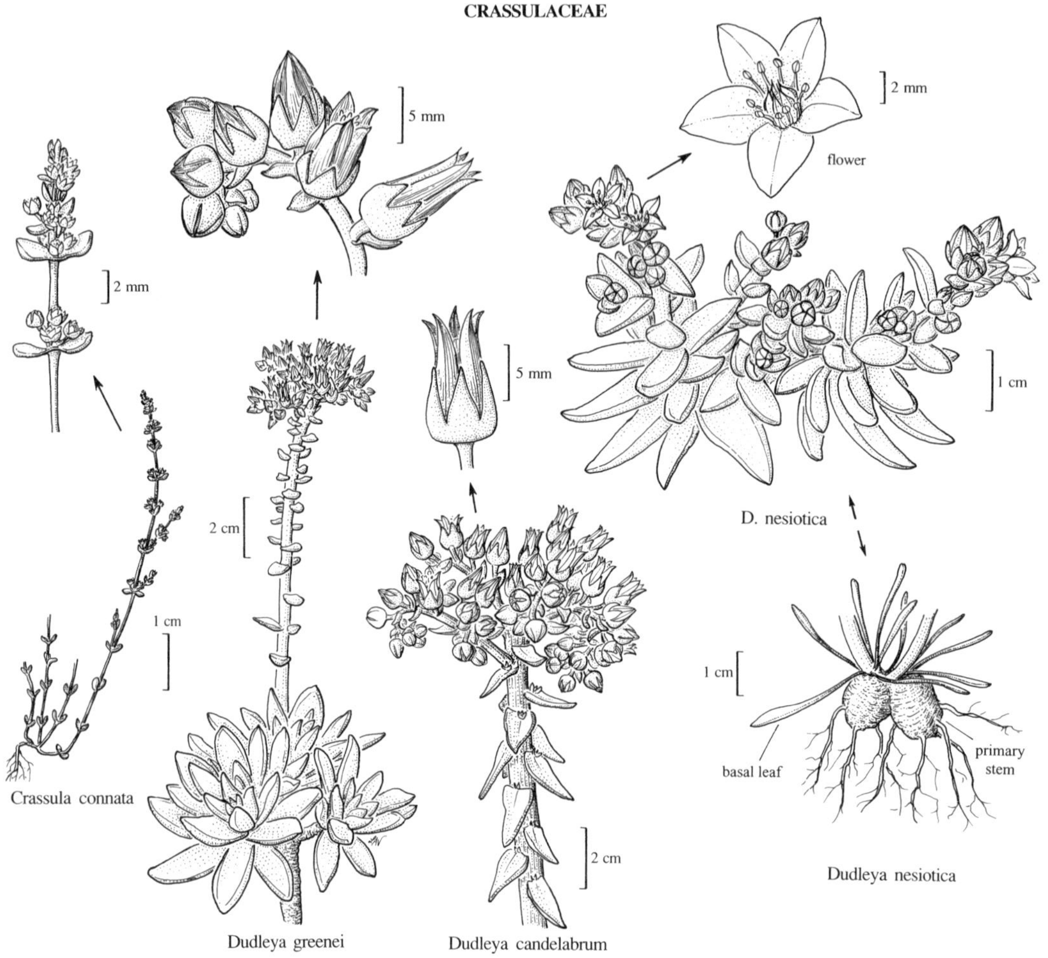


# CRASSULACEAE DC.

## Stonecrop Family

Annuals or perennial herbs, glabrous, succulent. Leaves mostly basal, sometimes cauline, alternate or opposite, simple; stipules absent. Flowers in terminal cymes or flowers solitary and axillary. Flower bisexual, radial; sepals 3-5, free or fused; petals 3-5, free or fused; stamens 3-10; pistils usually 3-5, usually free, sometimes fused at base, ovary with 1 locule and 1 style. Fruit a follicle; seeds 1-many. Ca. 30 genera and 1500 species; temperate to subtropical.

1. Annuals, less than 7 cm tall; leaves cauline, opposite, less than 3 mm long .................................... **Crassula**
1' Perennial herbs, usually taller than 10 cm; leaves mostly basal, in rosettes, 20 or more mm long ... **Dudleya**

## CRASSULA L.

Annuals or perennial herbs. Leaves cauline, opposite, usually connate at base. Flowers 1-few per node, axillary and terminal. Stamens 3-5. Fruits 3-5. Ca. 300 species; ± worldwide.

**C. connata** (Ruiz & Pav.) A.Berger [*Tillaea erecta* Hook. & Arn.] PYGMY WEED Annuals; stems erect, 2-7 cm tall. Leaves ovate to oblong, 1.5-3 mm long, fused at base. Flowers solitary or in axillary clusters. Sepals usually 4, ca. 1 mm long; petals lanceolate, usually 4, shorter than sepals. Feb-May. Common; sandy flats, rocky ridgetops and slopes, coastal bluffs; coastal scrub, grasslands; 50-1500 ft (15-460 m). Widely scattered locations throughout island. All CA Channel islands; all Baja CA islands except San Geronimo and Natividad; OR to S. America, e to TX.

### **DUDLEYA** Britton & Rose

Succulent perennial herbs, often suffrutescent. Leaves thick, succulent, basal ones in rosettes, cauline ones alternate, much reduced, often bract-like. Flowers in terminal cymes or panicles. Sepals 5, fused at base; petals 5, fused at base; stamens 10, adnate to petals; pistils 5. Ca. 50 species; sw N. America. Leaves of many species were eaten by Native Americans.

1. Primary stem under ground, corm-like; basal rosette leaves vernal, drying in summer .............. **D. nesiotica**
1' Primary stem above ground, often branched, not corm-like; basal rosette leaves evergreen
    2. Primary stem 2-8 cm wide, usually simple, usually swollen at base; sepals 5-8 mm long .................................... **D. candelabrum**
    2' Primary stem 2-5 cm wide, often branched dichotomously, base of primary stem or branches not swollen; sepals 1.5-5 mm long .............................. **D. greenei**

**D. candelabrum** Rose CANDLEHOLDER DUDLEYA Primary stem above ground, 2-8 cm wide, usually simple, usually swollen at base. Basal leaves persistent, ± evergreen, oblong to linear-oblong, 6-17 cm long, 2-7 cm wide. Inflorescence 15-35 cm tall. Sepals 5-8 mm long; petals 8-12 mm long, pale yellow. Apr-Jul. Island endemic. A species of special environmental concern. Common; coastal bluffs, rocky ridgetops and slopes; chaparral, coastal scrub, pine forest; 20-2000 ft (6-610 m). North side between West Point and Pelican Bay; Central Valley between Christy Beach and Cañada del Puerto; s side between Cañada de los Sauces and Smugglers' Cove. San Miguel and Santa Rosa islands.

**D. greenei** Rose GREENE'S DUDLEYA Primary stem above ground, 2-5 cm wide, often branched dichotomously, base of primary stem or branches not swollen. Basal leaves persistent, ± evergreen, linear, lanceolate, or oblanceolate, 3-15 cm long, 1.5-3 cm wide. Sepals 1.5-5 mm long; petals 8-12 mm long, pale yellow to white. May-Jul. Island endemic. A species of special environmental concern. Common; coastal bluffs and rocky slopes; coastal scrub; 5-800 ft (1-245 m). Around perimeter of island, primarily on north side on igneous substrates. San Miguel, Santa Rosa, and Santa Catalina islands.

**D. nesiotica** (Moran) Moran SANTA CRUZ ISLAND LIVE-FOREVER Primary stem under ground, corm-like, 0.5-2.5 cm wide. Basal leaves vernal, drying in summer but persistent, oblanceolate to spatulate, 2.5-5 cm long, 0.5-2.5 cm wide. Sepals 3-4 mm long; petals 7-14 mm long, white. Apr-Jun. Endemic to Santa Cruz Island. A species of special environmental concern. Rare; coastal flats; coastal scrub, grasslands; 30-100 ft (10-30 m). Known only from the vicinity of Fraser Point and Forney's Cove.

## **CUCURBITACEAE** Juss.
### GOURD FAMILY

Annual or perennial herbs, usually monoecious; stems vine-like, twining, or trailing, usually with tendrils. Leaves alternate, simple, often palmately veined and lobed. Flowers radial, unisexual, solitary or in racemes or panicles. Calyx with 5 lobes; corolla campanulate to funnelform, with 5 lobes; stamens 3-5; ovary inferior, locule 1. Fruit a hard-walled berry (pepo). Ca. 120 genera and 760 species; mostly tropical and warm temperate, worldwide. Some species cultivated for fruit (gourds, melons, squashes).

1. Leaves deltate to ovate, ± entire; corollas 8-12 cm long .................................... **Cucurbita**
1' Leaves orbicular-cordate, palmately lobed; corollas 0.3-2.5 cm long .............................. **Marah**

### **CUCURBITA** L. GOURD

Perennial herbs, with large, fleshy roots; stems prostrate and trailing. Leaves deltate to ovate, bases ± truncate, margins entire to irregularly toothed. Flowers usually solitary, staminate ones slightly smaller than pistillate ones. Calyx free from corolla; corolla ± campanulate to funnelform; style branches 3, each with 2 lobes. Fruit globose, smooth, green, mottled or striped; seeds many. Ca. 30 species; tropical and temperate America. Cultivated species and their cultivars include squash, gourds, and pumpkins.

**C. foetidissima** Kunth CALABAZILLA Leaf blades 18-24 cm long; petioles 8-11 cm long. Corollas 8-12 cm long, broadly campanulate, yellow. Fruit 6-9 cm in diameter, faint orange-brown with age. May-Sep. Rare;

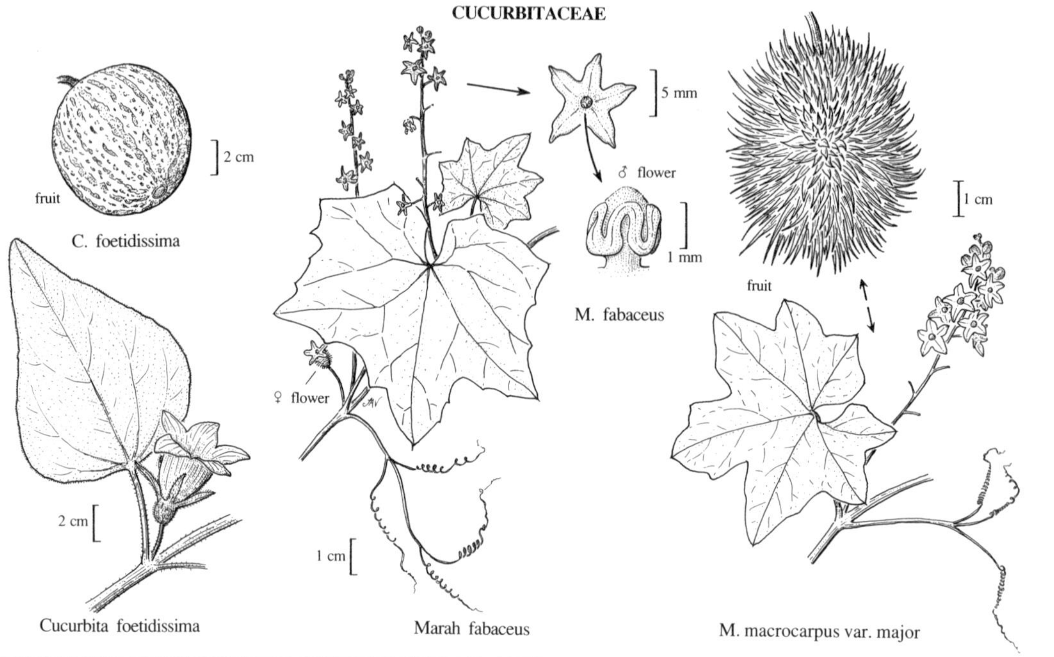


sandy sites; coastal scrub, coastal strand; 10-40 ft (3-12 m). North side at Fry's Harbor (collected there only once in 1930); Central Valley at Christy Beach. Great Valley to Baja CA, e to NE, TX, and n Mexico.

### MARAH Kellogg MAN-ROOT or WILD CUCUMBER

Perennial herbs, with large, fleshy roots; stems often vine-like or twining. Leaves ± orbicular-cordate, palmately lobed. Staminate flowers in slender, erect racemes or panicles, slightly smaller than solitary pistillate flowers. Calyx adnate to corolla, sepals obscure; corolla ± rotate, white to yellow-green; style 1, stigma capitate. Fruit globose to ellipsoid, prickly, dark green, often striped; seeds 2-many. Ca. 7 species; w North America.

1. Corollas 3-8 mm wide, yellowish to greenish white; ovules 2-4; fruit globose .......................... **M. fabaceus**
1' Corollas 12-30 mm wide, white; ovules 8-many; fruit oblong to cylindrical in outline ....... **M. macrocarpus**

**M. fabaceus** (Naudin) Greene CALIFORNIA MAN-ROOT Leaf blades 3.5-15 cm wide; petioles 2-8 cm long. Corollas 3-8 mm wide, rotate. Fruit 3-5 cm wide; seeds 2-4. Mar-May. Occasional; rocky slopes and canyon bottoms; chaparral, coastal scrub, oak woodland; 20-800 ft (2-245 m). Central Valley between Cañada de la Portezuela and Cañada del Puerto. Northern CA to San Diego Co., e to Sierra Nevada foothills.

**M. macrocarpus** (Greene) Greene var. **major** (Dunn) Stocking WILD CUCUMBER Leaf blades 7-23 cm wide; petioles 2.5-6 cm long. Corollas 12-30 mm wide, shallowly rotate to cup-like. Fruit 4.5-10 cm wide; seeds 8-many. Feb-May (Oct). May be Island endemic; needs further study. Common; coastal bluffs, rocky slopes, and canyon bottoms; chaparral, coastal scrub, grasslands, oak woodland; 40-600 ft (12-185 m). Scattered locations throughout island. All CA Channel islands; Los Coronados, Todos Santos, San Martin, and Cedros islands. Var. *macrocarpus*, with smaller flowers and fruits, occurs on the adjacent mainland from Santa Barbara Co. to Baja CA. Seeds and roots were widely used for medicinal purposes by Native Americans; oil from seeds was mixed with pigments to make paints; roots were crushed and put in streams to stupefy fish.

## CUSCUTACEAE Dumort.
## DODDER FAMILY

With only 1 genus.

### CUSCUTA L.

Parasitic plants; stems yellowish green to orange, filiform, glabrous, usually twining. Leaves absent or scale-like; stipules absent. Flowers bisexual, radial, in axillary, capitate to spike-like clusters. Calyx with 4-5 lobes;

CUSCUTACEAE

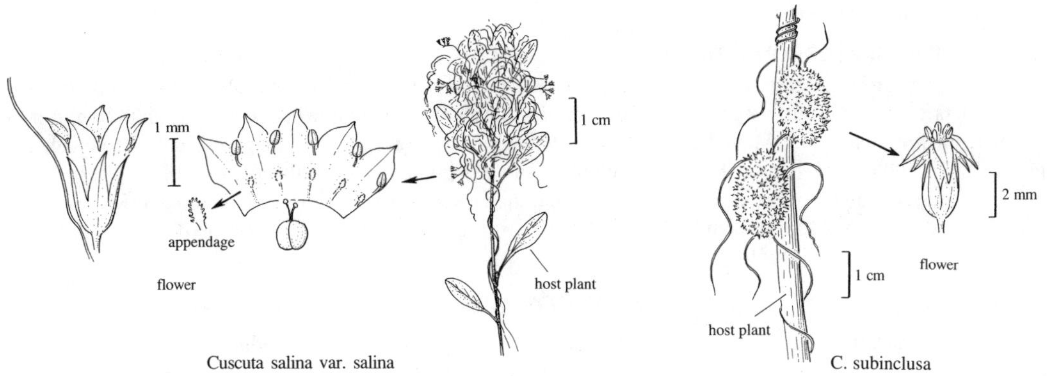


corolla cylindrical to rotate, with 5 lobes and 5 scales alternating with lobes; stamens 4-5, adnate to corolla; pistil 1, ovary superior, locules 2, each with 2 ovules; styles usually 2. Fruit a capsule or indehiscent; seeds 1-4. Ca. 150 species; temperate to tropical, mostly N. and S. America.

1. Anthers with short filaments; corolla 4-5 mm long .......................................................... **C. salina** var. **major**
1' Anthers sessile; corolla 2-3 mm long ............................................................................................ **C. subinclusa**

**C. salina** Engelm. var. **major** Yunck. Flowers with short-pedicels. Calyx tube 0.8-1 mm long, lobes 1-1.2 mm long, not overlapping; corolla tube ca. 2 mm long, lobes ca. 1 mm long; anthers with short filaments; styles 0.5-1 mm long. Mar-Apr. Rare; coastal flats, sandy sites; coastal scrub, coastal strand, grasslands; 10-30 ft (3-10 m). Known only from vicinity of Fraser Point and Forney's Cove. Santa Rosa and Anacapa islands; British Columbia to Baja CA, e to UT and AZ. Parasitic on *Ambrosia, Atriplex, Erodium, Frankenia, Lasthenia,* and *Mesembryanthemum*. Two other varieties occur on the mainland.

**C. subinclusa** Durand & Hilg. [*C. ceanothi* Behr] Flowers sessile. Calyx tube 1-3 mm long, lobes 2-2.5 mm long, overlapping; corolla tube 4-5 mm long, lobes 1-1.5 mm long; anthers sessile; styles 1-1.5 mm long. Sep. Rare; disturbed sites, rocky slopes; coastal scrub; ca. 200 ft (60 m). Central Valley near Stanton Ranch. Oregon to n Baja CA, e to Sierra Nevada. Parasitic on *Artemisia*, *Foeniculum*, and *Salvia.*

## ERICACEAE Juss.
### Heath Family

Perennials. Leaves alternate or opposite, simple, evergreen or deciduous; stipules absent. Flowers usually bisexual, radial, in racemes, panicles, or flowers solitary, often subtended by bracts. Sepals 4-5, usually free; petals 4-5, free or fused; stamens 8-10, usually free, anthers opening by pores or slits, often with tail-like appendages; pistil 1, ovary superior, locules 1-5, style 1, stigma capitate or lobed. Fruit a capsule or berry; seeds few to many. Ca. 100 genera and 3000 species; temperate to subtropical, ± worldwide.

1. Ovary inferior; fruit blue to black, pulp fleshy ........................................................................ **Vaccinium**
1' Ovary superior; fruit red to brown, pulp usually mealy
   2. Leaves with both surfaces ± similar in color and surface features; ovary and fruit smooth ........................................................................................................................................ **Arctostaphylos**
   2' Leaves with upper surfaces different from lower surfaces in color and surface features
      3. Leaves ovate to oblong, 7-14 cm long, lower surfaces ± glabrous; inflorescence a panicle; fruit reddish orange ........................................................................................................ **Arbutus**
      3' Leaves elliptic to obovate, 2-6 cm long, lower surfaces puberulent; inflorescence a raceme; fruit red ........................................................................................................................ **Comarostaphylis**

### ARBUTUS L. MADRONE

Shrubs or trees; bark smooth, reddish, shredding with age. Leaves alternate, evergreen, coriaceous, color and pubescence of upper surface different from lower surface. Flowers in panicles. Sepals 5, fused at base; corolla urn-shaped, with 5 lobes; stamens 10, anthers opening by broad slits, appendages slender; ovary superior, papillate, locules 5. Fruit ± berry-like, globose, papillate; seeds few, with a hard coat, often adherent. Ca. 15 species; temperate; w N. America and Europe.

**A. menziesii** Pursh MADRONE Trees to 10 m tall. Leaf blades ovate to oblong, 7-14 cm long, 3.5-7 cm wide, glabrous, upper surfaces green, lower surfaces glaucous, apices usually rounded, petioles 2-4 cm long. Inflorescence bracts ± ovate, 3-5 mm long. Pedicels straight. Corollas 4.5-7 mm long, white to pink. Fruit 6-10 mm wide, bright reddish orange. Mar-May. Rare; n-facing slopes and canyon bottoms; chaparral, oak woodland, pine forest; 1000-1500 ft (300-460 m). Known from only 3 small populations on north side, in Little Valdez Canyon, in a small side canyon sw of Fry's Harbor, and on slopes ne of Red Mountain. British Columbia to San Diego Co. Ripe fruits were eaten raw or cooked by Native Americans; roots, bark, and leaves were used medicinally; wood was used for tools and lodgepoles.

## ARCTOSTAPHYLOS Adans. MANZANITA

Shrubs, sometimes tree-like; bark smooth to rough, reddish to grayish, often shredding with age. Leaves alternate, evergreen, coriaceous, both surfaces similar. Flowers in racemes or panicles. Sepals usually 5, fused at base; corolla urn-shaped, with 4-5 lobes; stamens usually 10, anthers opening by pores, appendages slender, recurved; ovary superior, smooth, locules 2-10. Fruit drupe- or berry-like, smooth; seeds 2-10, with hard coats, free or fused together. Ca. 60 species; temperate; w N. America (mostly CA) and Europe. Fruits of many species were eaten by Native Americans.

1. Inflorescence bracts scale-like, 1.5-3 mm long; pedicels glandular-puberulent; ovary glabrous to sparsely tomentulose .......... **A. insularis**

1' Inflorescence bracts leaf-like, lanceolate, 6-15 mm long; pedicels tomentulose; ovary densely tomentulose
    2. Twigs with setose trichomes; leaves not crowded, not strongly overlapping, lower surfaces tomentulose to glandular-puberulent; inflorescence usually a panicle with 2-8 branches; plants with a basal burl .......... **A. tomentosa**
    2' Twigs lacking setose trichomes; leaves crowded, overlapping, lower surfaces puberulent to glabrous; inflorescence usually a raceme, sometimes with 2 branches; plants lacking a burl .......... **A. viridissima**

**A. insularis** Greene SANTA CRUZ ISLAND MANZANITA Shrubs, 1-5 m tall; mature stem bark reddish, twigs glabrous to sparsely tomentulose, burl absent. Leaves not crowded, blades elliptic to oblong, 2-4.5 cm long, 1-3 cm wide, ± glabrous, upper surfaces darker green than lower surfaces, bases truncate to rounded, margin ± entire, apices obtuse to rounded, petioles 4-9 mm long. Inflorescence open, branches 6-10, bracts scale-like, 1.5-3 mm long. Pedicels glandular-puberulent. Corollas 4-6 mm long, white; ovary glabrous to sparsely tomentulose. Fruit 7-15 mm wide, brown to orange-brown, glabrate. Nov-May. Endemic to Santa Cruz Island. Abundant; coastal bluffs, rocky slopes, canyon walls; chaparral, oak woodland, pine forest; 30-1700 ft (9-520 m). North side between Painted Cave area and Eagle Canyon; Central Valley between Cañada Christy and vicinity of Rancho del Sur; s side between Cañada de los Sauces and vicinity of Coches Prietos Anchorage. Mostly restricted to igneous substrates.

**A. tomentosa** (Pursh) Lindl. Shrubs, 1-3 m tall; mature stem bark reddish, twigs tomentulose, sometimes also glandular-pubescent, burl present. Leaves not crowded, blades ovate, oblong, or lanceolate, 2-5 cm long, 1-2.5 cm wide, upper surfaces green, ± shiny, glabrous, lower surfaces ± tomentose to glandular-puberulent, bases truncate to rounded, margins ± entire, apices obtuse to ± acute, petioles 2-5 mm long. Inflorescence branches 2-8, crowded, bracts leaf-like, lanceolate, 8-15 mm long. Pedicels tomentulose, sometimes also glandular-pubescent. Corollas 3-5 mm long, white; ovary densely tomentulose. Fruit 5-10 mm wide, brown, ± puberulent. Jan-Jun. Abundant; rocky slopes and ridgetops; chaparral, pine forest; 300-1700 ft (90-520 m). Scattered at several locations on island. Santa Rosa Island; San Francisco Bay region to s CA. The two subspecies intergrade at several localities throughout the island. At least 7 other subspecies are recognized and occur on the mainland.

1. Twigs densely puberulent to tomentose, sometimes sparsely glandular; lower surfaces of leaves smooth, ± tomentose .......... subsp. **insulicola**

1' Twigs tomentulose and glandular-pubescent; lower surfaces of leaves papillate, tomentulose to sparsely glandular-puberulent .......... subsp. **subcordata**

subsp. **insulicola** P.V.Wells ISLAND MANZANITA Twigs densely puberulent to tomentose, sometimes sparsely glandular. Lower surfaces of leaves smooth, ± tomentose. Pedicels tomentulose. Feb-Jun. Island endemic. A subspecies of special environmental concern. Common; rocky slopes and ridgetops; chaparral, pine forest; 300-1700 ft (90-520 m). North side between Hazard's Canyon and Pelican Bay, on slopes above Cañada del

# ERICACEAE

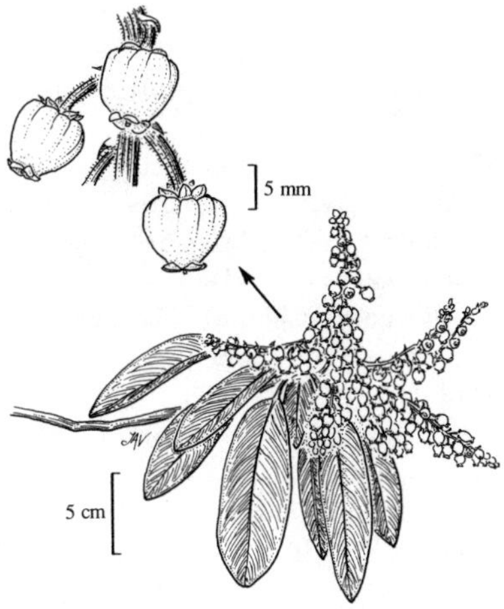


Arbutus menziesii

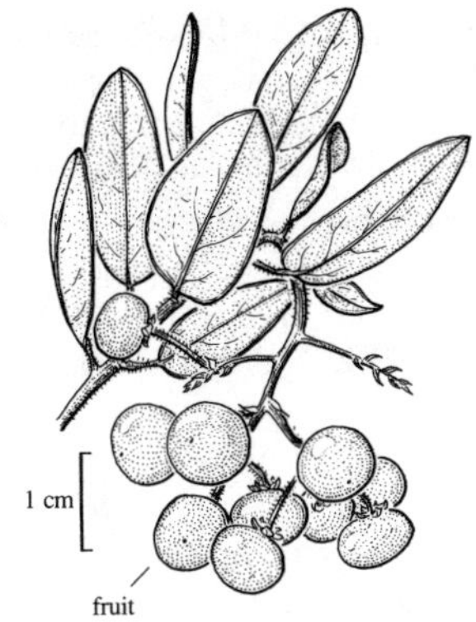


Arctostaphylos insularis

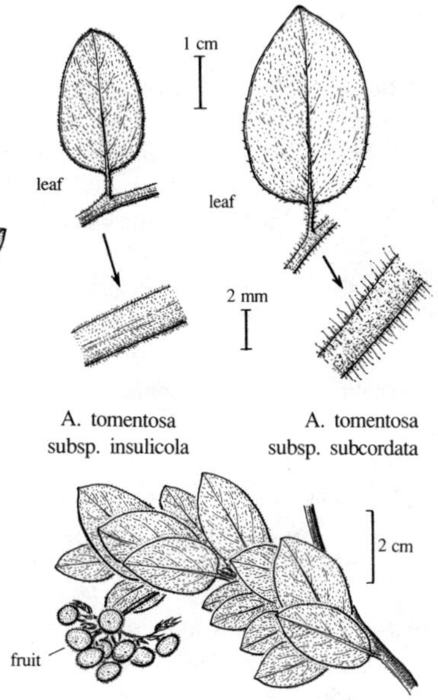


Arctostaphylos tomentosa subsp. insulicola

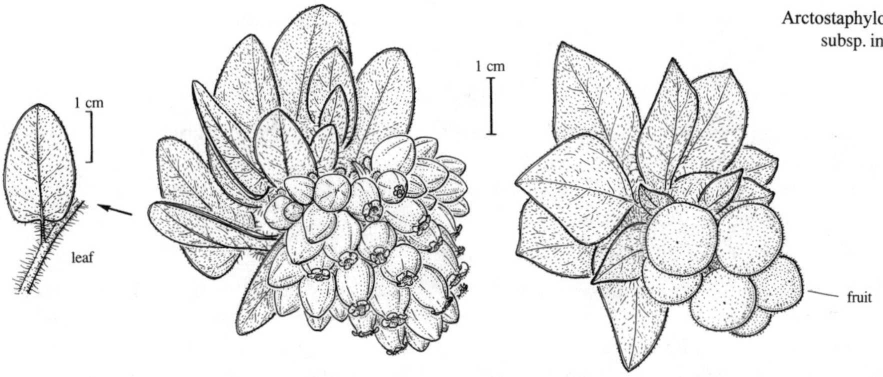


Arctostaphylos viridissima

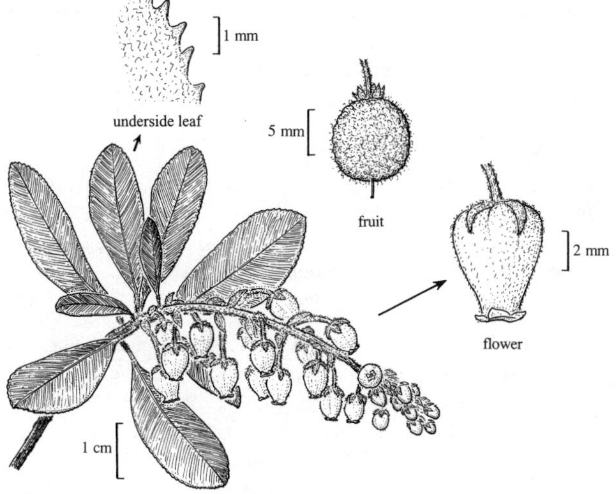


Comarostaphylis diversifolia subsp. planifolia

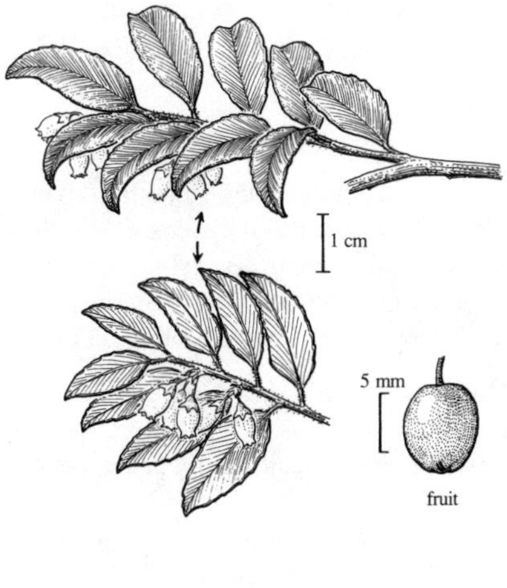


Vaccinium ovatum

Puerto, and above China Harbor; in Central Valley mostly on n-facing slopes between upper part of Cañada Christy and vicinity of Rancho del Sur; s side on South Ridge between upper part of Cañada de los Sauces and vicinity of Albert's Anchorage, with isolated populations in Laguna, Willows, and Coches Prietos drainages. Santa Rosa Island.

subsp. **subcordata** (Eastw.) P.V.Wells SUBCORDATE MANZANITA Twigs tomentulose and glandular-pubescent. Lower surfaces of leaves papillate, tomentulose to sparsely glandular-puberulent. Pedicels glandular-puberulent. Jan-Apr. A subspecies of special environmental concern. Abundant; rocky slopes and ridgetops; chaparral, pine forest; 500-2400 ft (150-730 m). North side between Hazard's Canyon and Prisoners' Harbor, on slopes above Cañada del Puerto, in Eagle Canyon and Cañada del Muro, and above China Harbor; in Central Valley mostly on n-facing slopes between upper part of Cañada Christy and vicinity of Rancho del Sur; s side on South Ridge between upper part of Cañada de los Sauces and vicinity of Albert's Anchorage, with isolated populations on Sierra Blanca and El Tigre ridges. Santa Rosa Island; coastal San Luis Obispo and Santa Barbara cos.

**A. viridissima** (Eastw.) McMinn MCMINN'S MANZANITA Shrubs, 1-4 m tall; mature stem bark reddish, twigs tomentulose and white-setose, burl absent. Leaves crowded, overlapping, blades ovate to oblong, 2-3.5 cm long, 1-2.5 cm wide, upper surfaces green, ± shiny, puberulent to glabrous, bases truncate to lobed, margins entire or toothed near base, apices obtuse to acute, petioles 1-3.5 mm long. Inflorescence a raceme, sometimes with 2 branches, dense, bracts leaf-like, lanceolate, 6-10 mm long. Pedicels tomentulose. Corollas 3-5 mm long, white; ovary densely tomentulose. Fruit 8-15 mm wide, brown, sparsely puberulent. Dec-May. Endemic to Santa Cruz Island. A species of special environmental concern. Abundant; rocky slopes and ridgetops; chaparral, pine forest; 350-1300 ft (100-400 m). North side between vicinity of Mount Pleasant and upper part of Scorpion Canyon; Central Valley on small knoll s of Stanton Ranch airfield; s side between Ceanothus Canyon area and Aguaje Canyon near Yellowbanks Anchorage. Mostly restricted to Monterey Shale.

## **COMAROSTAPHYLIS** Zucc.

Shrubs, sometimes tree-like; bark smooth to rough, reddish to grayish, often shredding with age. Leaves alternate, evergreen, coriaceous, color and pubescence of upper surfaces different from lower surfaces. Flowers in racemes or panicles. Sepals usually 5, fused; corolla urn-shaped, with 4-5 lobes; stamens usually 10, anthers opening by short slits, appendages slender; ovary superior, papillate, locules 4-6. Fruit berry-like, papillate; seeds 4-6, with hard coat, fused together. Ca. 10 species; mostly tropical, N. and S. America.

**C. diversifolia** (Parry) Greene subsp. **planifolia** (Jeps.) G.D.Wallace SUMMER HOLLY Plants to 6 m tall. Leaf blades broadly elliptic to obovate, 2-6 cm long, 1-3.5 cm wide, upper surfaces green, glossy, lower surfaces puberulent, glaucous, margins entire to serrate, apices usually obtuse to acuminate, petioles 3-8 cm long. Inflorescence bracts ovate to lanceolate, 3-7 mm long. Pedicels recurved. Corollas 4-7 mm long, white. Fruit 5-7 mm wide, bright red, papillate. Apr-Jun (Jul). Common; n-facing slopes and shaded canyon walls; chaparral, pine forest; 100-1750 ft (30-535 m). North side between vicinity of Profile Point and Eagle Canyon, and between China Harbor and Scorpion Canyon; Central Valley between Cañada Christy and La Cascada; s side near Sierra Blanca ridge and on n side of Las Torres ridge. Santa Rosa, Anacapa, and Santa Catalina islands; Santa Barbara and Los Angeles cos. Subsp. *diversifolia*, with revolute leaf margins and linear bracts, occurs from San Diego Co. to Baja CA.

## **VACCINIUM** L. BLUEBERRY, CRANBERRY, or HUCKLEBERRY

Shrubs or subshrubs; bark smooth to rough, brown to grayish, rough. Leaves alternate, deciduous or evergreen, color and pubescence of upper surfaces different from lower surfaces. Flowers solitary or in umbel-like clusters. Sepals 4-5, fused; corolla cup- to urn-shaped, with 4-5 lobes; stamens 8-10, anthers opening by pores, appendages present or absent; ovary inferior, smooth, locules 4-5. Fruit a berry, smooth; seeds many. Ca. 10 species; mostly tropical, N. and S. America. Fruits of many species were eaten by native Americans.

**V. ovatum** Pursh CALIFORNIA HUCKLEBERRY Shrubs, to 2.5 m tall. Leaves evergreen, blades elliptic to lanceolate, 1.5-3.5 cm long, 0.4-1.2 cm wide, upper surfaces green, glossy, lower surfaces pale green, margins usually serrate, apices usually acute, petioles 1-4 mm long. Inflorescence bracts scale-like, 1-2 mm long. Corollas urn-shaped, 5-8 mm long, white. Fruit 4-7 mm wide, blue to black, smooth. Mar-Jun. Scarce; n-facing slopes and shaded canyon walls; chaparral, grasslands, pine forest; 100-1500 ft (30-460 m). North side near Pelican Bay and China Harbor; Central Valley and s side in vicinity of Christy Pines. Santa Rosa Island; British Columbia to San Diego Co. Fruits are edible and can be substituted in recipes for blueberries, currants, or gooseberries.

# EUPHORBIACEAE Juss.

## Spurge Family

Monoecious or dioecious annuals or perennials, with milky or acrid sap. Leaves alternate or opposite, simple, entire; stipules present or absent. Flowers usually unisexual; inflorescence a raceme, spike, panicle, or a flower-like cyathium (1 naked, stalked, pistillate flower surrounded by few to many naked, staminate flowers, together subtended by an involucre of appendaged bracts). Calyx, if present, composed of 3-5 free or fused sepals; corolla absent; stamens 1-many; pistil 1, locules 1-3, ovules 1-2 per locule. Fruit a capsule; seeds 3-6. Ca. 325 genera and 8000 species; worldwide. Some species are used for rubber, source of starch (cassava, tapioca), tung oil, and castor oil.

1. Inflorescence a cyathium
   2. Leaf blade bases oblique; stems prostrate to spreading ........ **Chamaesyce**
   2' Leaf blade bases symmetrical; stems erect ........ **Euphorbia**

1' Inflorescence not a cyathium; flowers in small clusters, elongated spikes, or solitary, not subtended by an involucre
   3. Leaves ovate, with stiff, simple and stellate hairs; flowers solitary or in small clusters... **Eremocarpus**
   3' Leaves linear, glabrate; inflorescence a spike ........ **Stillingia**

### CHAMAESYCE A.Gray

Annuals or perennial herbs, with opaque sap. Leaves usually opposite, blade bases oblique; stipules present. Inflorescence a cyathium; involucral bracts with or without glandular, petaloid appendages; calyx and corolla absent. Staminate flowers 3-many per cyathium; pistillate flower 1 per cyathium, stalked, often recurved; ovary with 3 locules and 3 styles. Ca. 250 species; warm temperate to subtropical, worldwide.

**C. maculata** (L.) Small SPOTTED SPURGE Annuals; stems prostrate, 3-20 cm long. Leaves opposite, ovate to oblong, 0.4-1.2 cm long. Cyathia axillary, 1-4 per node; bract appendages rounded to crenate, white to pink. Capsule 1-1.5 mm wide. Jun-Jul. Introduced. Rare; disturbed sites; 200 ft (60 m). Known only from Stanton Ranch. Widespread in w N. America; native of e N. America. First collected in June 1994. This species can cause dermatitis in humans and can be toxic to livestock.

### EREMOCARPUS Benth.

With only 1 species.

**E. setigerus** (Hook.) Benth. DOVE WEED or TURKEY MULLEIN Strongly scented annuals with both stellate and stiff, simple trichomes; stems 5-25 cm tall. Leaves alternate, ovate to rhombic-ovate, 1.5-6 cm long, entire, grayish; petioles slender; stipules absent. Staminate flowers in terminal clusters, 2-5 mm long, sepals 5-6, stamens 6-10, exserted; pistillate flowers 3-6 mm long, 1-3 per cluster, often axillary, sepals absent; ovary with 1 locule, 1 ovule, and 1 filiform style. Capsule 3-5 mm long. Aug-Oct. Common; disturbed sites, open ridgetops and slopes, grassy hillsides, floodplains; coastal scrub, grasslands; 10-1500 ft (3-460 m). Widely scattered locations throughout island. Santa Rosa, Santa Catalina, and San Clemente islands; WA to n Baja CA, e to ID, UT, and AZ. This species was used medicinally by Native Americans; plants were also crushed and thrown into pools of water to stupefy fish, making them easier to catch. Toxic to livestock.

### EUPHORBIA L. SPURGE

Annuals or perennial herbs, with opaque sap. Leaves usually alternate, blade bases symmetrical; stipules present. Inflorescence a cyathium; involucral bracts with glandular, petaloid appendages; calyx and corolla absent. Staminate flowers 10-many per cyathium; pistillate flower 1 per cyathium, stalked, often recurved; ovary with 3 locules and 3 styles. Ca. 1600 species; mostly warm temperate, worldwide. Many species were used medicinally by Native Americans. Sap of all species is caustic and can irritate skin and eyes.

1. Shrubs; cyathium appendages rounded, white ........ **E. misera**

1' Annuals; cyathium appendages attenuate, yellowish ........ **E. peplus**

**E. misera** Benth. CLIFF SPURGE Shrub to 1.2 m tall. Leaves alternate, elliptic to obovate, 1-1.5 cm long. Cyathia terminal, 2-5 per cluster; bract appendages rounded, red at base, tips white. Capsule 4-5 mm wide. May-Jun. Rare; rocky s-facing slopes, coastal bluffs; coastal scrub; 100-150 ft (30-45 m). South side at small

EUPHORBIACEAE

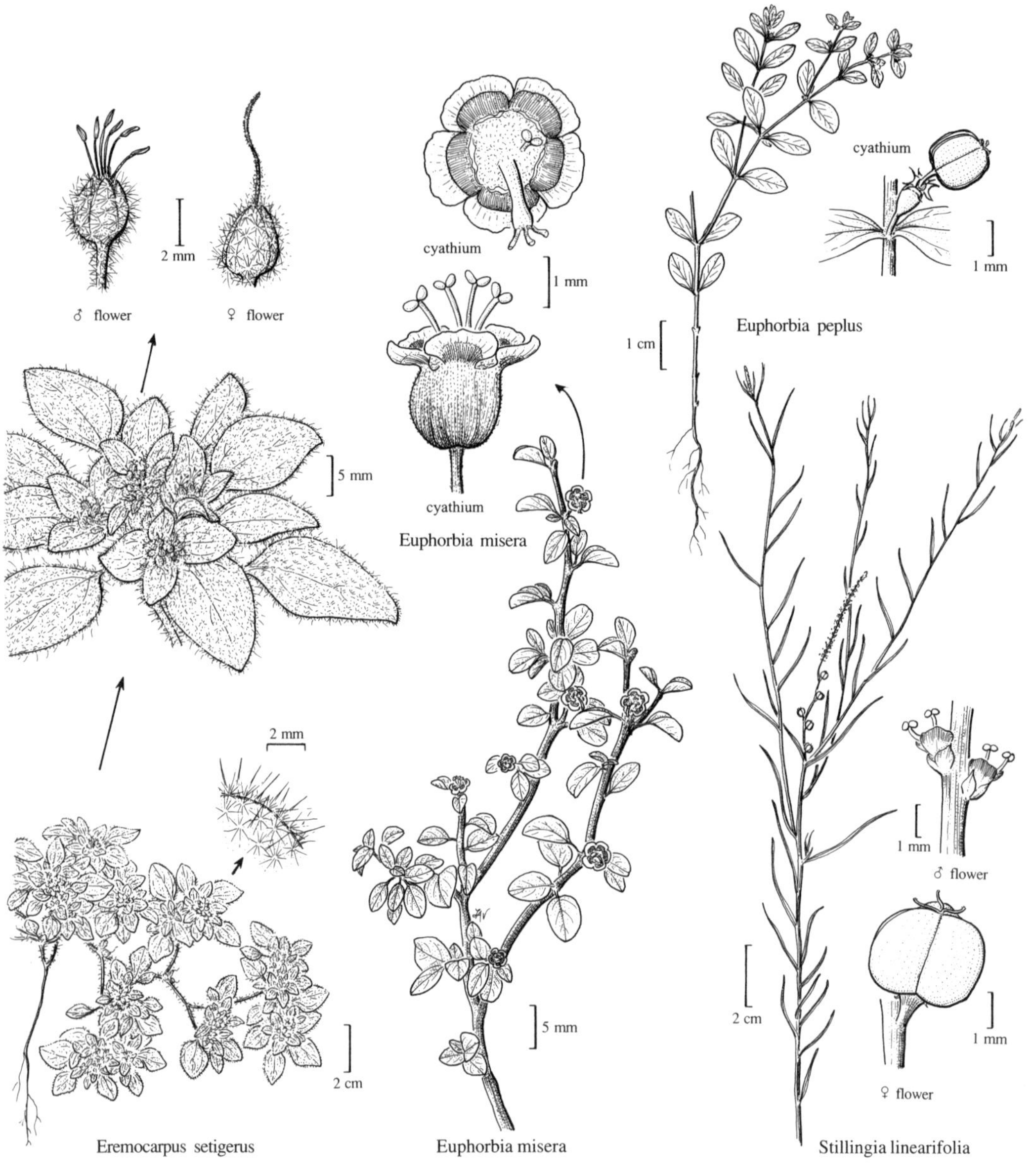


cove just w of Corral Point. Santa Catalina and San Clemente islands; all Baja CA islands except San Geronimo; Orange Co., CA to n Baja CA, disjunct near Whitewater, Riverside Co., CA.

**E. peplus** L. PETTY SPURGE Annuals, 3-40 cm tall. Leaves opposite, orbicular to widely ovate, 0.5-3 cm long. Cyathia terminal, usually solitary; bract appendages attenuate, yellowish green. Capsule 1-2 mm wide. Feb-May. Introduced. Scarce; disturbed sites, grassy flats; grasslands; 20-200 ft (6-60 m). North side at Fry's Harbor, Pelican Bay, and Scorpion Ranch; Central Valley at Stanton Ranch. Santa Rosa, Santa Catalina, and San Clemente islands; Canada to Mexico, e to the Atlantic Coast; native of Europe. Toxic if eaten.

## STILLINGIA Garden ex L.

Annuals or perennial herbs with opaque to clear sap. Leaves alternate, simple; stipules minute or absent. Flowers in axillary or terminal spikes; staminate flowers usually many, each with a 2-3 stamens; pistillate flowers

1-6, usually near base of spike, ovary with 3 locules. Ca. 30 species; tropical and warm temperate America, e Malaysia, Madagascar.

**S. linearifolia** S.Watson Suffrutescent perennials, 5-7 dm tall, with several to many branches from base. Leaves linear, 2-3 cm long, ± 2 mm wide, glabrate. Inflorescence 2.5-4 cm, terminal. Staminate flowers many, reddish, each with a 2-lobed calyx and 2 stamens; pistillate flowers few, each subtended by stalked glands. May-Jun. Rare; rocky s-facing slopes, coastal bluffs; coastal scrub; 200-400 ft (60-120 m). South side between mouth of Laguna Canyon and Corral Point. Ventura Co to n Baja CA, e to AZ and n Mexico. Sap is caustic and can irritate skin and eyes.

## FABACEAE Lindl. [Leguminosae Juss.]
### Legume Family

Annuals or perennials. Leaves alternate, pinnately or palmately compound; stipules present. Flowers bisexual, radial or bilateral, in racemes, panicles, umbels, or solitary. Sepals 5, usually fused; petals usually 5, similar in shape or not, free or fused, if unequal then papilionaceous and composed of an upper banner, 2 lateral wings, and 2 lower petals either free or fused into a keel; stamens usually 10, free or partly fused; pistil 1, ovary superior, with 1 locule, 1 placenta, 1 style, and 1 stigma. Fruit a legume (splitting along 2 sutures when dry); seeds 1-many. Ca. 650 genera and 16,500 species; worldwide. Many species are important foods and sources of protein (e.g., peanuts, soybeans, peas, beans, and alfalfa). Many other species are cultivated as ornamentals.

1. Mature leaves simple, with 2-5 parallel veins; flowers radial; corollas inconspicuous; stamens many, conspicuous ........ **Acacia melanoxylon**
1' Mature leaves pinnately or palmately compound; flowers bilateral; corolla conspicuous, papilionaceous; stamens 10, usually included in keel
  2. Most leaves either palmately compound or with 3 leaflets (trifoliolate)
    3. Shrubs
      4. Leaves palmately compound, leaflets 5 or more; all stamens fused ........ **Lupinus**
      4' Leaves usually trifoliolate (sometimes pinnately compound in *Lotus*); 9 stamens fused, 1 free or all free in *Pickeringia*
        5. Plants not intricately branched, woody near base only; corolla yellow to white; 9 stamens fused, 1 free ........ **Lotus**
        5' Plants intricately branched, woody throughout; corolla deep pink or purple; stamens all free ........ **Pickeringia**
    3' Annuals or herbaceous perennials
      6. All 10 stamens fused; leaves palmately compound, leaflets 5 or more ........ **Lupinus**
      6' 9 stamens fused, 1 free; leaves trifoliolate (sometimes pinnately compound in *Lotus*)
        7. Flowers in racemes
          8. Fruit clearly coiled 2-5 times, sometimes sickle-shaped; corolla 4-10 mm long ........ **Medicago**
          8' Fruit ovoid, straight; corolla 2.5-5 mm long ........ **Melilotus**
        7' Flowers in umbels or capitate clusters
          9. Fruit clearly coiled, prickly ........ **Medicago polymorpha**
          9' Fruit straight, smooth
            10. Leaflets usually with entire margins; fruits usually dehiscent, much exceeding shriveled perianth parts ........ **Lotus**
            10' Leaflets usually with wavy or toothed margins; fruits usually indehiscent, ± equal to and sometimes included in persistent corolla ........ **Trifolium**
  2' Most leaves pinnately compound, leaflets 4 or more
    11. Trees
      12. Corollas 4-5 mm long, radial, yellow-green; stipules minute, deciduous ........ **Albizia**
      12' Corollas 15-20 mm long, papilionaceous, white; stipules spine-like ........ **Robinia**
    11' Annuals or herbaceous perennials, stems sometimes suffrutescent; stipules herbaceous or membranous
      13. Leaves odd-pinnately compound, terminated by a leaflet; stems erect to prostrate, not conspicuously twining, vine-like, or climbing
        14. Inflorescence a raceme ........ **Astragalus**
        14' Inflorescence an umbel ........ **Lotus**

13' Leaves apparently even-pinnately compound, terminated by a bristle or tendril; stems often vine-like, twining or climbing
15. Style terete, puberulent in a ring or tuft near tip ........ **Vicia**
15' Style flattened, puberulent along 1 side
16. Leaflets 8-12, narrowly to broadly elliptic; flowers (3)6-18 per raceme ........ **Lathyrus**
16' Leaflets 4-6, broadly elliptic to round-ovate; flowers 1-3 per raceme ........ **Pisum**

## ACACIA Miller

Shrubs or trees. Leaves pinnately compound or apparently simple, each composed of flattened petiole and midrib; stipules present or absent. Flowers radial, in capitate or cylindrical spikes, spikes often in racemose or paniculate clusters. Calyx and corolla small, inconspicuous; stamens many, free, filaments colorful. Fruit linear to oblong in outline, terete to flat, usually dehiscent. Ca. 1200 species; subtropical to tropical, worldwide.

**A. melanoxylon** R.Br. BLACKWOOD ACACIA Trees 2-8 m tall. Mature leaves apparently simple, lanceolate, 5-15 cm long, young leaves of seedlings (sometimes young branches) twice pinnately compound, glabrous. Flowers in capitate spikes, spikes 2-8, in racemose clusters. Calyx rotate, ca. 1 mm long; corolla lobes ca. 1 mm long, white to yellowish; stamens yellow. Fruit 5-10 cm long, less than 1 cm wide, curved, flat. Apr. Introduced. Scarce; canyon bottoms, rocky slopes, exposed ridgetops; chaparral, riparian communities; 150-1200 ft (45-365 m). North side in Cañada del Puerto and at Campo del Norte; Central Valley in Islay Canyon and near Stanton Ranch; s side on South Ridge near Willows Canyon. Santa Catalina Island; San Francisco Bay region to Baja CA; native of Australia. Wood is used for furniture.

## ALBIZIA Durazz.

Shrubs or trees. Leaves even-pinnately compound; stipules small, deciduous. Flowers radial, in capitate or cylindrical spikes, spikes often in racemose or paniculate clusters. Calyx and corolla small, inconspicuous, petals fused, tube longer than lobes; stamens many, filaments fused into tube below middle, free above, colorful. Fruit oblong in outline, flat, usually dehiscent. Ca. 150 species; subtropical to tropical, worldwide.

**A. lophantha** (Willd.) Benth. [*A. distachya* J.F.Macbr.] PLUME ACACIA Trees 3-7 m tall. Leaves twice pinnately compound, 8-20 cm long; ultimate leaflets many, oblong, 5-10 mm long, glabrous, upper surface dark green, lower surface light green. Racemes 6-8 cm long. Calyx rotate, 1.5-2 mm long, puberulent; corolla 4-5 mm long, yellow-green; stamens yellow. Fruit 6-9 cm long, 1.5-2 cm wide, straight, flat. Sep-Nov. Introduced. Scarce; disturbed sites, floodplains; chaparral, riparian communities; 10-350 ft (3-105 m). North side at Prisoners' Harbor; Central Valley at Rancho del Sur. Monterey Co. to s CA; native of Australia.

## ASTRAGALUS L. LOCOWEED, MILK VETCH, or RATTLEPOD

Annuals or perennial herbs; stems and leaves with simple to forked trichomes. Leaves odd-pinnately compound; stipules free, sometimes fused near stem base. Flowers in axillary or terminal racemes. Calyx lobes equal to subequal; corolla papilionaceous, keel with a blunt tip; stamens 10, 9 fused and 1 free. Fruit sessile or stipitate, sometimes with 2 locules formed by an internal septum derived from enfolded valves. Ca. 2000 species; mostly temperate, N. America, Europe, and Asia. Several species are known to be toxic, especially to horses. The commercially important Gum of Tragacanth is derived from several Asian species.

1. Annuals; leaflets 7-17 per leaf; calyx 2-4 mm long; corolla, including keel, 3-6 mm long; fruits not inflated
2. Fruits shorter than or ± equal to calyx; flowers and fruits strongly ascending, inflorescence remaining dense in fruit; flowers subsessile ........ **A. didymocarpus** var. **didymocarpus**
2' Fruits exserted from calyx; flowers ascending, fruits spreading to deflexed, inflorescences dense in flower, ± open in fruit; flowers pedicellate ........ **A. gambelianus**
1' Herbaceous perennials; leaflets 17-39 per leaf; calyx 4-6 mm long; corolla, including keel, 9-15 mm long; fruits inflated, bladder-like
3. Stems and leaves densely white-tomentulose; fruits sessile ........ **A. miguelensis**
3' Stems and leaves strigulose, greenish; fruits on stipes 6-12 mm long ....... **A. trichopodus** var. **lonchus**

**A. didymocarpus** Hook. & Arn. var. **didymocarpus** [*A. catalinensis* Nutt.] COMMON DWARF LOCOWEED Annuals; stems erect to prostrate, to 3 dm long. Leaves 3-5 cm long; leaflets 7-17, cuneate-oblanceolate, 3-8 mm long, thinly strigulose, upper surfaces often glabrous; stipules lanceolate to deltate, 1.5-3 mm long. Flowers 7-30, subsessile, crowded in globose or ovoid heads. Calyx 3-4 mm long, villous with black hairs and a band of

FABACEAE: Acacia, Astragalus

white hairs on upper side; corolla 5-6 mm long, banner violet, wings reddish purple with white tips, keel reddish purple. Fruit ovoid to subglobose, 1.5-2 mm long, erect, sessile, divided dorsally by a narrow groove into 2 wrinkled, sack-like, 1-seeded lobes, outer surface thinly strigulose. Mar-May. Occasional; sandy flats, coastal bluffs, grassy hillsides; coastal scrub, grasslands; 10-1500 ft (3-460 m). Widely scattered locations throughout island. All CA Channel islands except Santa Barbara; San Francisco Bay region, Great Valley, and Sierra Nevada foothills to Mojave Desert, e to NV. Often associated with Chumash shell middens. Three other varieties occur in sw U. S.

**A. gambelianus** E.Sheld. [*A. nigrescens* Nutt.; *A. elmeri* Greene] GAMBEL'S DWARF LOCOWEED Annuals; stems erect to ascending, to 2.5 dm long. Leaves 1.5-4 cm long; leaflets 9-15, narrowly cuneate-oblanceolate, 3-8 mm long, thinly strigulose, upper surfaces often glabrous; stipules lanceolate-ovate to deltate, 1.5-2.5 mm long. Flowers 7-13, pedicellate, at first ascending and crowded in dense heads, inflorescence becoming more open in fruit, flowers spreading to deflexed in fruit. Calyx 2-2.5 mm long, villous with black hairs; corolla 3-4 mm long, light purple. Fruit ovoid, 3-4 mm long, deflexed, sessile, with a dorsal groove, outer surface obscurely wrinkled, white-villous. Mar-May. Occasional; grassy and rocky slopes, exposed ridgetops; chaparral, coastal scrub, grasslands; 100-2000 ft (30-610 m). Widely scattered locations on slopes in Central Valley between Christy and Stanton Ranches; s side at Forney's Cove and Coches Prietos. Santa Catalina Island; OR to n Baja CA.

**A. miguelensis** Greene SAN MIGUEL ISLAND LOCOWEED Perennial herbs; stems decumbent to ascending, to 8 dm long, densely white-tomentulose. Leaves 5-12 cm long; leaflets 17-27, oblong to obovate, 5-18 mm long, white-tomentulose; stipules broadly deltate, 3.5-5 mm long, fused at base. Flowers 10-30, pedicellate, at first erect to spreading and crowded in ± dense, axillary racemes, becoming somewhat deflexed in fruit. Calyx 4-6 mm long, tomentulose; corolla 9-14 mm long, greenish to creamy white, drying yellowish. Fruit ovoid, 25-35 mm long, spreading, sessile, inflated and bladder-like, walls papery, outer surface finely tomentulose. Mar-Jun (Nov). Island endemic. A species of special environmental concern. Scarce; sandy flats, sand dunes, coastal bluffs; coastal dunes, coastal scrub; 10-300 ft (3-90 m). North side between Potato Harbor and Cavern Point; s side near Fraser Point. San Miguel, Santa Rosa, Anacapa, and San Clemente islands.

**A. trichopodus** (Nutt.) A.Gray var. **lonchus** (M.E.Jones) Barneby [*A. t.* subsp. *leucopsis* (Torr. & A.Gray) Thorne] SOUTHERN CALIFORNIA LOCOWEED Perennial herbs; stems erect to ascending, to 6 dm long, strigulose. Leaves 6-12 cm long; leaflets 21-39, oblong to broadly oblanceolate, 6-18 mm long, thinly strigulose to villous, upper surfaces often glabrous; stipules lanceolate to deltate, 3-5 mm long, free or fused at base. Flowers 15-30, pedicellate, at first in dense racemes, spreading to somewhat deflexed in fruit. Calyx 5-6 mm long, strigulose; corolla 13-15 mm long, greenish to creamy or yellowish white. Fruit ellipsoid, 25-40 mm long, spreading, on a stipe 6-12 mm long, inflated and bladder-like, wall papery, outer surface thinly strigulose to glabrous. Mar-Jul (Sep). Occasional; dry ravines, loose s-facing talus slopes, floodplains, grassy flats; chaparral, coastal scrub, riparian communities; 10-770 ft (3-235 m). Widely scattered locations around perimeter of island. Santa Rosa, Anacapa, and Santa Catalina islands; Los Coronados and Todos Santos islands; Santa Barbara Co. to n Baja CA. Two other varieties occur on the mainland from San Luis Obispo Co. to n Baja CA.

## LATHYRUS L. WILD PEA

Annuals or perennial herbs; stems erect or vine-like, twining or climbing, often angular or winged. Leaves apparently even-pinnately compound, terminal leaflet modified into a tendril or bristle; stipules free, persistent. Flowers in axillary racemes. Calyx lobes unequal, upper ones shorter than lower ones; corolla papilionaceous; stamens 10, 9 fused and 1 free; style flattened, puberulent along 1 side. Fruit oblong in outline, flat, usually dehiscent. Ca. 150 species; temperate N. America, Europe, Asia, and Africa. Several species cultivated as ornamentals. Seeds of many species are toxic if eaten in large quantities.

**L. vestitus** Nutt. var. **vestitus** Perennial herbs, stems to 2 m long, angled, glabrous to puberulent. Leaves 4-18 cm long (including tendril); leaflets 8-12, narrowly to broadly elliptic, 15-45 mm long, tendril branched, coiled. Flowers (3)6-18 per raceme; calyx tube 4-6 mm long, upper lobes 2-3 mm long, lower ones 5-7 mm long; corolla 15-22 mm long, banner lavender to purple, often purple-streaked, wings and keel pale lavender to whitish, all petals aging light golden brown. Fruits 3-5 cm long. (Jan) Mar-May (Nov). Occasional; n-facing slopes, canyon bottoms, shaded hillsides; chaparral, oak woodland, pine forest; 15-1300 ft (5-395 m). Widely scattered locations on n side; throughout Central Valley; s side in Coches Prietos drainage and in Smugglers' Canyon. Santa Rosa, Anacapa, Santa Catalina, and San Clemente islands; n CA coast ranges to Ventura Co. Four other varieties occur on the mainland from WA to Baja CA. Roasted seeds have been used as a coffee substitute.

## LOTUS L. TREFOIL

Annual, perennial herbs, or shrubs. Leaves usually trifoliolate or odd-pinnately compound; leaflets usually 3-many, sometimes solitary, lowest pair sometimes stipule-like; stipules usually gland-like, minute. Flowers few, usually in umbel-like clusters, axillary, sessile or on bracteate peduncles. Calyx lobes equal to subequal; corolla papilionaceous; stamens 10, 9 fused and 1 free; style forming a beak in fruit, usually persistent, straight to coiled. Fruit sessile, terete to flattened, apex often beaked. Ca. 100 species; mostly temperate, N. America, Europe, and Asia.

1. Shrubs or suffrutescent perennial herbs
    2. Stems and leaves densely canescent to silvery-tomentose; fruit ellipsoid, 4-6 mm long (excluding beak), beak 4-6 mm long; seed usually 1 per fruit ........ **L. argophyllus** var. **niveus**
    
    2' Stems and leaves glabrous to silvery-strigulose; fruit linear to narrowly oblong, 10-55 mm long (excluding beak), beak 0.5-5 mm long; seeds 2-many per fruit
        3. Peduncles less than 0.5 cm long, umbels usually subsessile; corollas 9-11 mm long, yellow, aging reddish; leaflets 3-5 ........ **L. dendroideus** var. **dendroideus**
        
        3' Peduncles 2.5-10 cm long; corollas 15-21 mm long, creamy white, sometimes rose or pinkish; leaflets 7-9 ........ **L. grandiflorus** var. **grandiflorus**

1' Annuals or slender perennial herbs with rhizomes
4. Leaflets 5, lower pair at petiole base, stipule-like; slender perennial herbs with rhizomes; petals yellow, aging greenish ........ **L. corniculatus**
4' Leaflets 3-9, if 5 then lower pair neither at petiole base nor stipule-like; annuals; petals white to pink or yellow, if yellow then aging reddish
5. Calyx lobes much longer than tube, lobes villous; banners pink, wings white to pinkish, keel cream with yellow tips ........ **L. purshianus**
5' Calyx lobe length ± equal to tube length, lobes subglabrous to strigulose; banners, wings, and keel ± yellow throughout (cream to pinkish in *L. micranthus*), usually aging reddish
6. Fruit ± indehiscent, curved at base, tapered to a conspicuously coiled beak ........ **L. hamatus**
6' Fruit dehiscent, usually straight, beak straight to hooked
7. Most leaves with 3-5 leaflets
8. Calyx 3-5 mm long, villosulous to strigulose; corolla yellow, aging reddish, 4-7 mm long ........ **L. wrangelianus**
8' Calyx 1.5-2.5 mm long, subglabrous; corolla cream to pinkish ....... **L. micranthus**
7' Most leaves with 5-9 leaflets
9. Wing length shorter than or equal to keel length; most leaflets broadly elliptic to orbicular ........ **L. salsuginosus**
9' Wing length longer than keel length; most leaflets linear-oblong to obovate ........ **L. strigosus**

**L. argophyllus** (A.Gray) Greene var. **niveus** (Greene) Ottley SANTA CRUZ ISLAND SILVER LOTUS. Suffrutescent perennial herbs; stems erect to spreading, to 5 dm long, branches arching to nearly prostrate, densely canescent to silvery-tomentose. Leaves 1-4 cm long; leaflets 3-6, broadly oblong to obovate, 6-25 mm long, densely canescent to silvery-tomentose. Flowers 6-18 per umbel, umbel subsessile. Calyx 6-8 mm long, densely canescent, lobes 2-3.5 mm long; corolla 8-12 mm long, yellow, aging reddish brown. Fruit ellipsoid, 4-6 mm long (excluding beak), straight to slightly curved, ± indehiscent, beak 4-6 mm long, stout, tip curved or hooked; seed 1. Mar-Jun (Sep). Endemic to Santa Cruz Island. A variety of special environmental concern. Common; rocky slopes, exposed ridgetops, sandy flats and washes, gravel floodplains; chaparral, coastal scrub, riparian communities; 50-2250 ft (15-680 m). Widely scattered on igneous and metamorphic substrates throughout island, especially on n side. Four other varieties occur on other Channel Islands or mainland CA.

**L. corniculatus** L. BIRD'S-FOOT TREFOIL Perennial herbs; stems decumbent to prostrate, 1-5 dm long. Leaves 10-20 mm long; leaflets usually 5, oblong to broadly oblanceolate, 5-18 mm long, glabrous to sparsely strigose, lower pair at petiole base, stipule-like. Flowers 3-8 per umbel, peduncle 2.5-6 cm long. Calyx 2-5 mm long, glabrous, lobes 2-3 mm long; corolla 10-13 mm long, yellow, aging greenish. Fruit linear, 20-28 mm long (excluding beak), straight, dehiscent, beak 5 mm long, fragile, deciduous; seeds 8-18. May-Sep. Introduced. Rare; moist flats, stream banks; grasslands, riparian communities; 10-50 ft (3-15 m). North side at Prisoners' Harbor; Central Valley near mouth of Cañada Christy. Santa Catalina Island; throughout N. America; native of Europe. This species is grown for livestock food.

**L. dendroideus** (Greene) Greene var. **dendroideus** [*L. scoparius* (Nutt.) Ottley var. *dendroideus* (Greene) Ottley] ISLAND DEERWEED Shrubs; stems erect to nearly prostrate, to 13 dm tall, silvery-strigulose, glabrate. Leaves 5-18 mm long; leaflets 3-5, oblong to narrowly obovate, 8-13 mm long, silvery-strigulose, glabrate. Flowers 3-12 per umbel, umbel subsessile or peduncles 2-5 mm long. Calyx 4.5-5.5 mm long, strigulose, lobes 0.5-1 mm long; corolla 9-11 mm long, yellow, aging reddish. Fruit linear to narrowly oblong, 10-12 mm long (excluding beak), curved, ± indehiscent, beak 4-5 mm long, curved; seeds 2-4. Jan-Jul (Oct). Island endemic. A variety of special environmental concern. Common; coastal bluffs, sandy washes, floodplains, rocky slopes, exposed ridgetops; chaparral, coastal scrub, pine forest; 10-1500 ft (3-460 m). Widely scattered locations throughout island. Santa Rosa, Anacapa, and Santa Catalina islands. Some plants with more densely strigulose leaves, on bluffs above West Point, are similar to var. *veatchii*, known only from San Miguel Island and nw Baja CA.

**L. grandiflorus** (Benth.) Greene var. **grandiflorus** CHAPARRAL LOTUS Suffrutescent perennial herbs, to 7 dm tall; stems sprawling, strigose. Leaves 30-75 mm long, glabrous to strigose; leaflets 7-9, elliptical to obovate, 17-28 mm long. Flowers 2-9 per umbel, peduncles 2.5-10 cm long. Calyx 6-9 mm long, strigose to villosulous, lobes 3-4 mm long; corolla 15-21 mm long, creamy white, sometimes rose- or pink-tinged. Fruit linear to narrowly oblong, 30-55 mm long (excluding beak), straight, dehiscent, beak 0.5-3 mm long; seeds many. Jan-Jun (Oct). Occasional; floodplains, open flats, rocky slopes, exposed ridgetops; chaparral, coastal scrub, pine forest, riparian communities; 30-1500 ft (10-460 m). Widely scattered locations, mostly in western and central portions of island. Santa Rosa and Santa Catalina islands; Guadalupe Island; throughout much of CA w of Sierra Nevada to Baja CA. Variety *macranthus* (Greene) Isely, with glabrous stems and leaves, occurs in the n Sierra Nevada and Cascade Ranges of CA.

**L. hamatus** Greene SMALL-FLOWERED LOTUS Annuals; stems prostrate, to 5 dm long, villosulous to glabrous. Leaves 5-22 mm long, villosulous; leaflets 5-7, elliptical to obovate, 2-9 mm long. Flowers 2-8 per umbel, umbels subsessile or peduncles 1-8 mm long. Calyx 1.5-2.5 mm long, villosulous, lobes ca. 0.5 mm long; corolla 3-3.5 mm long, yellow, aging reddish. Fruit linear 10-14 mm long (including coiled beak), curved near base, indehiscent; seeds 2-4. Feb-Jun. Occasional; grassy hillsides, rocky slopes; coastal scrub, grasslands; 20-1400 ft (6-430 m). Central Valley near Cañada Larga and Islay Canyon; s side between Pozo Canyon and Coches Prietos Anchorage. Santa Rosa, Santa Catalina, and San Clemente islands; coastal s Monterey Co. to Baja CA. Can be locally abundant in years with heavy rainfall.

**L. micranthus** Benth. SAN DIEGO LOTUS Annuals; stems erect to prostrate, to 3 dm long, sparsely strigulose to glabrous. Leaves 6-15 mm long, strigulose to glabrous; leaflets 3-5, elliptical to obovate, 3-8 mm long. Flowers solitary, subsessile. Calyx 1.5-2.5 mm long, subglabrous, lobes 0.5-1 mm long; corolla 9-11 mm long, cream, aging reddish. Fruit linear, 12-15 mm long (excluding beak), straight, dehiscent, beak 1-2 mm long; seeds 3-6. Feb-Apr. Scarce; grassy flats, moist slopes, canyon bottoms; coastal scrub, grasslands; 25-1350 ft (8-410 m). North side between Pelican Bay and Prisoners' Harbor; Central Valley between La Cascada and Stanton Ranch. British Columbia to s CA.

**L. purshianus** (Benth.) Clem. & E.G.Clem. SPANISH CLOVER Annuals; stems erect to nearly prostrate, to 6 dm long, villous, glabrate. Leaves 10-35 mm long, villous to subglabrous; leaflets 1 or 3, broadly elliptical to ovate, 6-32 mm long. Flowers solitary, subsessile or peduncles 1-10 mm long. Calyx 3-5.5 mm long, villous, lobes 3-5 mm long; corolla 5-6 mm long, banner pink, wings pinkish white, keel whitish with yellow tips. Fruit linear, 18-25 mm long (excluding beak), straight, dehiscent, beak 0.5 mm long; seeds 4-6. Jun-Sep. Occasional; floodplains; coastal scrub, riparian communities; 200-750 ft (60-230 m). North side in Cañada del Puerto; Central Valley between La Cascada and Stanton Ranch. Santa Catalina Island; throughout w N. America.

**L. salsuginosus** Greene var. **salsuginosus** COASTAL LOTUS Annuals; stems prostrate, to 5 dm long, sparsely strigulose to glabrate. Leaves often succulent. Leaves 15-25 mm long, succulent; leaflets 5-6, elliptic to broadly obovate or orbicular, 4-12 mm long. Flowers solitary or 2-4 per umbel, umbels usually subsessile. Calyx 2.5-4.5 mm long, lobes 1.5-3 mm long; corolla 7-9 mm long, yellow, aging reddish orange, banner marked with red. Fruit linear, 15-28 mm long (excluding beak), straight, dehiscent, beak 0.5 mm long, curved or hooked; seeds 6-8. Mar-Jun (Oct). Occasional; coastal bluffs, alluvial flats, loose talus slopes, creekbeds, floodplains; coastal scrub, riparian communities; 10-850 ft (3-260 m). Widely scattered locations throughout island. San Miguel, Santa Rosa, Anacapa, and Santa Catalina Islands; s San Francisco Bay region to n Baja CA, e to Sierra Nevada. Var. *brevivexillus* Ottley, with corollas 3-5 mm long and wings shorter than the keel, occurs on San Benito and Cedros islands and in the interior deserts of sw N. America.

**L. strigosus** (Nutt. in Torr. & A.Gray) Greene BISHOP'S LOTUS Annuals; stems ascending to decumbent, 1-3 dm long, sparsely strigose. Leaves 5-30 mm long, sparsely strigose to glabrous; leaflets 5-9, linear-oblong to obovate, 1-10 mm long. Flowers solitary or 2 per umbel, peduncles 7-28 mm long. Calyx 3-5.5 mm long, sparsely strigose to glabrous, lobes 1.5-2.5 mm long; corolla 10-12 mm long, yellow, aging reddish, back of banner usually reddish orange. Fruit narrowly oblong, 15-32 mm long (excluding beak), straight to slightly curved, dehiscent, beak 0.5 mm long; seeds 6-10. Feb-May (Oct). Common; coastal bluffs, coastal flats, floodplains, rocky s-facing slopes; chaparral, coastal scrub, grasslands, riparian communities; 20-1800 ft (6-395 m). Widely scattered locations throughout island, especially on s side. All CA Channel islands except Santa Barbara and San Nicolas; Todos Santos, San Martin, and Cedros islands; San Francisco Bay region and Great Valley to Mexico, e to AZ.

**L. wrangelianus** Fisch. & C.A.Mey. [*L. subpinnatus* Lagasca has been misapplied to this species in CA] Annuals; stems ascending to prostrate, 8-30 cm long, usually sparsely strigulose. Leaves 12-20 mm long, villosulous to strigulose, leaflets 3-5, elliptical to obovate, 3-15 mm long. Flowers solitary, subsessile. Calyx 3-5 mm long, sparsely strigulose, lobes 1.5-3 mm long; corolla 4.5-9 mm long, yellow, aging reddish. Fruit narrowly oblong, straight, 9-18 mm long (excluding beak), dehiscent, beak 0.5-1 mm long, hooked; seeds 3-5. Mar-May (Sep). Occasional; coastal bluffs, coastal flats, floodplains, rocky slopes; coastal scrub, riparian communities; 60-1200 ft (18-365 m). Scattered locations throughout Central Valley; s side between Fraser Point and Valley Anchorage. Santa Rosa, Anacapa, and Santa Catalina islands; throughout CA w of Sierra Nevada to s CA.

**FABACEAE: Lathyrus, Lotus**

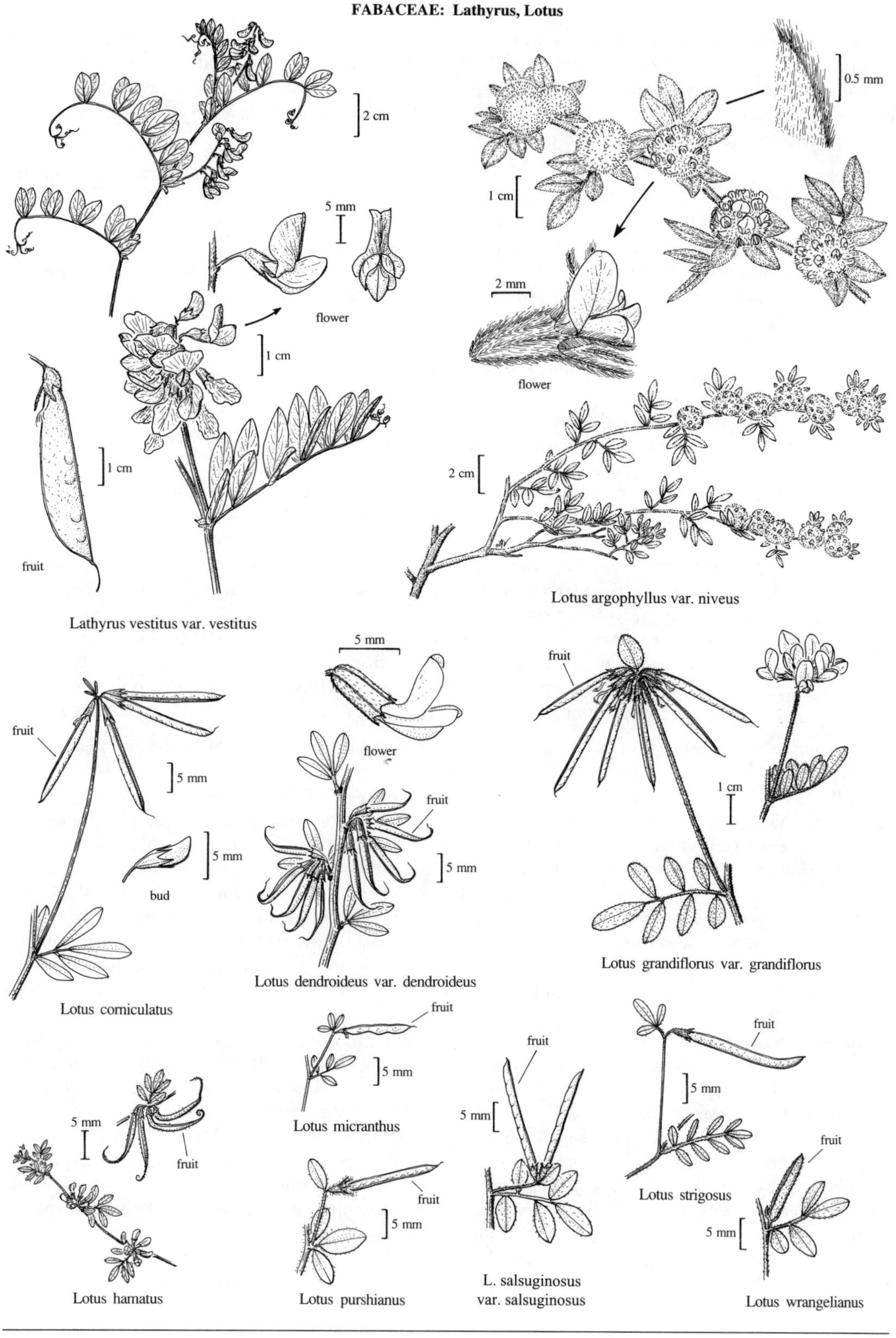

## LUPINUS L. LUPINE

Annuals, perennial herbs, or shrubs; stems decumbent to erect. Leaves palmately compound; leaflets 5-many, usually oblanceolate, margins entire; stipules usually fused to petiole. Flowers distinctly to loosely whorled, in cylindrical racemes. Calyx apparently bilabiate, lobes strongly fused, obscure; corolla papilionaceous; stamens 10, fused. Fruit usually oblong in outline, dehiscent. Ca. 200 species; temperate N. and S. America, Europe, and Africa. Several species are cultivated as ornamentals; many are toxic to livestock.

1. Shrubs ........ **L. albifrons** var. **douglasii**
1' Annuals
  2. Flowers, at least lower ones, in distinct whorls
    3. Floral bracts persistent, reflexed; pedicels 0.5-2 mm long; banner ± uniformly colored, white to yellowish; seeds 1-2 per fruit ........ **L. microcarpus**
    3' Floral bracts deciduous; lower pedicels 2-6 mm long; banner usually with white vertical band that ages purplish; seeds 4-8 per fruit
      4. Keel 4-6 mm long, upper margin ciliate near apex; blades villosulous ........ **L. bicolor**
      4' Keel 10-14 mm long, upper margin glabrous near apex, ciliate near claw; blades glabrous to sparsely strigose ........ **L. succulentus**
  2' Flowers not in distinct whorls, lower ones evenly distributed along rachis
    5. Stem and inflorescence rachis stiffly hirsute; leaflets broadly obovate, 10-20 mm wide ........ **L. hirsutissimus**
    5. Stem and inflorescence rachis subglabrous to villosulous or hirsutulous, trichomes not stiff, not stinging; leaflets narrowly oblanceolate to oblong, 1-4 mm wide
      6. Upper keel margins ciliate near base, glabrous near apex; leaflets oblong to narrowly oblanceolate, larger ones with truncate apices ........ **L. truncatus**
      6' Upper keel margins glabrous near base, glabrous to ciliate near apex; leaflets oblanceolate to narrowly oblanceolate, rounded to acuminate
        7. Upper keel margins usually ciliate near apex; inflorescence bracts deciduous; lower calyx lip longer than upper one; seeds 4-8 per fruit ........ **L. bicolor**
        7' Upper keel margins glabrous; inflorescence bracts persistent in fruit; calyx lips subequal, seeds 2-5 per fruit ........ **L. concinnus**

**L. albifrons** Benth. var. **douglasii** (J.Agardh) C.P.Sm. SILVER LUPINE Shrubs, 4-10 dm tall. Leaflets 6-8, narrowly oblanceolate, 10-40 mm long, 3-8 mm wide, villosulous to silvery-villous. Inflorescence 5-20 cm long, upper flowers in whorls, lower ones in loose whorls or scattered, pedicels 4-8 mm long, floral bracts 4-15 mm long, deciduous. Calyx 5-10 mm long, lower lip longer than upper one; banner and wings lavender to blue, banner with white vertical band, aging reddish purple; keel 8-11 mm long, whitish, apex blackish or purple, upper margins ciliate, lower one glabrous. Fruit 2-4 cm long, seeds 4-8. Mar-Oct. Common; coastal bluffs, creek beds, floodplains, rocky slopes; chaparral, coastal scrub, oak woodland, riparian communities; 10-2200 ft (3-670 m). Widely scattered locations throughout western and central portions of island. All CA Channel islands except Santa Barbara and San Clemente; San Francisco Bay region to Santa Barbara Co. Two to three other varieties occur from s OR to s CA. Some plants, with leaflets longer than 30 mm and bracts equal to or shorter than the flower buds, are similar to var. *albifrons* on the adjacent mainland. A medicinal tea was made from the species by the Karok Tribe.

**L. bicolor** Lindl. [subspp. *microphyllus* (S.Watson) D.B.Dunn, *tridentatus* (C.P.Sm.) D.B.Dunn, *umbellatus* (Greene) D.B.Dunn; var. *trifidus* (S.Watson) C.P.Sm.] DOVE LUPINE Annuals, 0.6-3 dm tall; stems spreading to ascending. Leaflets 5-8, narrowly oblanceolate, 6-20 mm long, 1-4 mm wide, villosulous, grayish green, upper surface sometimes subglabrous. Inflorescence 2-7 cm long, upper flowers in whorls, lower ones in loose whorls or scattered, pedicels 1-4 mm long, floral bracts 3-5 mm long, deciduous. Calyx 2-5 mm long, lower lip longer than upper one; banner and wings blue, banner with white vertical band, black-dotted, aging purple, keel 4-6 mm long, white, apex dark purple, upper margin ciliate near apex. Fruit 1-2 cm long, seeds 4-8. Feb-Jun. Common; canyon bottoms, floodplains, grassy flats and hillsides, rocky slopes, exposed ridgetops; chaparral, coastal scrub, grasslands, pine forest; 10-2250 ft (3-685 m). Widely scattered locations throughout island. All CA Channel islands except Santa Barbara; Guadalupe Island; British Columbia to Baja CA.

**L. concinnus** J.Agardh [*L. agardhianus* A.Heller] Annuals, 0.4-1.5 dm tall; stems ascending to erect. Leaflets 5-7, oblanceolate, 5-16 mm long, 2-4 mm wide, sparsely villous to hirsutulous, greenish. Inflorescence 1-3 cm long, flowers crowded, not in whorls, lower ones sometimes axillary, pedicels 1-2 mm long, floral bracts 2-4 mm

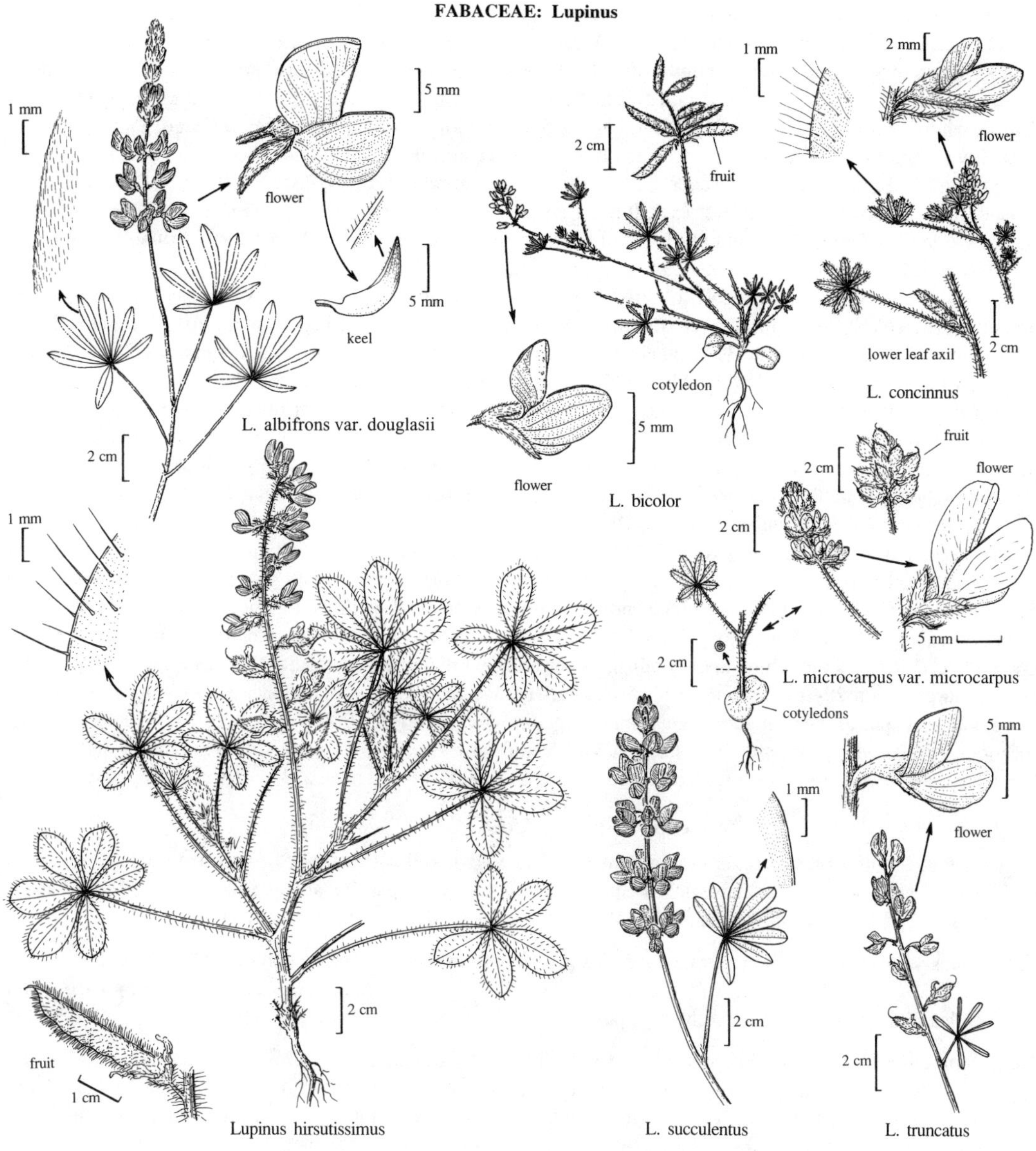


long, deciduous. Calyx 2.5-5 mm long, lips subequal; banner rose to purple, with white vertical band, black-dotted, aging purple, wings blue, keel 2-4 mm long, white, apex rose to purple, upper margins glabrous. Fruit 1-1.5 cm long, seeds 2-5. Mar-May. Occasional; rocky slopes and flats, dry ravines, floodplains; coastal scrub, grasslands, pine forest; 10-1950 ft (3-595 m). Widely scattered locations throughout island, especially around perimeter. Santa Rosa, Santa Catalina, and San Clemente islands; Cedros Island; San Francisco Bay region to Mexico, e to UT and TX.

**L. hirsutissimus** Benth. STINGING LUPINE Annuals, 1.5-6 dm tall; stems erect, stiffly hirsute. Leaflets 5-7, broadly oblanceolate, 20-40 mm long, 10-20 mm wide, stiffly hirsute, greenish. Inflorescence 6-10 cm long, flowers not whorled, pedicels 4-6 mm long, floral bracts 4-7 mm long, ± persistent, spreading. Calyx 5-10 mm long, lips subequal; banner and wings reddish purple, aging blue, banner spot usually yellowish, keel 9-12 mm long, white, apex reddish purple, upper margins glabrous. Fruit 2-3 cm long, seeds 3-6. Feb-May. Occasional; rocky s-facing coastal bluffs and slopes, floodplains; chaparral, coastal scrub, riparian communities; 100-650 ft (30-200 m). North side in Cañada del Puerto; scattered in Central Valley between La Cascada and Valley Anchorage; s side in Laguna, Willows, and Smugglers' canyons. Santa Rosa, Santa Catalina, and San Clemente islands; San Francisco Bay region to Baja CA. Mostly seen only in years with heavy rainfall or after fires.

**L. microcarpus** Sims var. **microcarpus** Annuals, 0.5-4 dm tall; stems spreading to erect. Leaflets 5-10, elliptic to oblanceolate, 8-25 mm long, 4-8 mm wide, villous, greenish. Inflorescence 3-12 cm long, flowers in 2-4 distinct whorls, pedicels 0.5-2 mm long, floral bracts 5-12 mm long, villous, persistent, reflexed. Calyx 3-10 mm long, lower lips longer than upper ones; petals white to cream, aging yellowish, keel 8-12 mm long, upper margins ciliate. Fruit 1-2 cm long, seeds 1-2. Mar-Jun. Scarce; sand dunes, coastal bluffs, clay flats; chaparral, coastal dune, coastal scrub; 20-400 ft (6-120 m). Central Valley at Christy Beach and near west end of Bosque Mano; s side near Fraser Point and at mouth of Cañada de los Sauces. Santa Rosa Island; British Columbia to Baja CA and S. America. Several other varieties occur in w N. America.

**L. succulentus** Koch SUCCULENT LUPINE Annuals, 0.7-6 dm tall; stems erect. Leaflets 5-8, narrowly to broadly oblanceolate, 20-40 mm long, 4-12 mm wide, glabrous to sparsely strigose, green. Inflorescence 4-20 cm long, flowers in 2-9 distinct whorls, pedicels 2-6 mm long, floral bracts 2-5 mm long, deciduous. Calyx 4-8 mm long, lips subequal; banner and wings blue to bluish purple, banner with vertical white band, black-dotted, aging reddish purple, keel 10-14 mm long, white, apex blackish to purple, upper margin glabrous near apex, ciliate near claw. Fruit 1.5-3 cm long, seeds 4-8. Feb-Jun. Occasional; grassy hillsides, rocky slopes, creekbeds, floodplains; chaparral, coastal scrub, grasslands, riparian communities; 10-1400 ft (3-425 m). Widely scattered locations throughout island. All CA Channel islands except Santa Barbara (introduced on San Nicolas); Los Coronados Islands; Northern CA and Central Valley to Baja CA.

**L. truncatus** Hook. & Arn. TRUNCATE LUPINE Annuals, 1.5-6 dm tall; stems erect. Leaflets 5-7, narrowly oblanceolate to oblong, 10-40 mm long, 2-3.5 mm wide, subglabrous to sparsely strigose, green. Inflorescence 5-15 cm long, flowers evenly distributed, not in whorls, pedicels 2-4 mm long, floral bracts 2-5 mm long, deciduous. Calyx 3-5 mm long, lips subequal; banner and wings blue to reddish purple, banner spot white, black-dotted; keel 7-9 mm long, white, apex reddish, upper margin ciliate near base, glabrous near apex; petals aging reddish purple. Fruit 20-30 mm long, seeds 5-8. Mar-Jun. Occasional; grassy hillsides, rocky slopes, creekbeds, floodplains; chaparral, coastal scrub, grasslands, riparian communities; 10-450 ft (3-135 m). Widely scattered locations throughout island, mostly in Central Valley and on s side. Santa Rosa, Anacapa, Santa Catalina, and San Clemente islands; Los Coronados and Todos Santos islands; San Francisco Bay region to Baja CA

## MEDICAGO L.

Annuals or perennial herbs; stems prostrate to erect. Leaves trifoliolate; stipules fused to petiole. Flowers in umbellate clusters or racemes. Calyx lobes equal to unequal; corolla papilionaceous; stamens 10, 9 fused and 1 free. Fruit coiled to sickle-shaped, indehiscent. Ca. 55 species; Mediterranean and Asia.

1. Corollas yellow, 3-6 mm long; annuals, stems prostrate to ascending; fruits with or without prickles .......... **M. polymorpha**

1' Corollas violet to purple, 8-10 mm long; perennial herbs, stems erect; fruit smooth .......... **M. sativa**

**M. polymorpha** L. [*M. denticulata* Willd.] BUR-CLOVER Annuals, stems prostrate to ascending, 1.5-3 dm long. Leaflets 5-12 mm long, 3-10 mm wide, broadly obovate to obcordate, margins entire to denticulate. Flowers 2-5, in umbellate clusters. Calyx 2-3.5 mm long; corollas 3-6 mm long, yellow. Fruit coiled several times, ± globose, prickles present or absent. Mar-Jun (Aug). Introduced. Common; disturbed sites, rocky slopes, moist canyon bottoms, grassy hillsides, floodplains, alluvial flats; most plant communities; 10-1000 ft (3-305 m). Throughout island. All CA Channel islands; Guadalupe and Cedros islands; throughout s N. America, ± worldwide; native of Mediterranean Europe. First reported by Greene (1887). Can cause photosensitivity in livestock.

**M. sativa** L. ALFALFA Perennial herbs, stems ascending to erect, 5-7 dm tall. Leaflets 6-15 mm long, 2-6 mm wide, lanceolate to obovate, margins entire to denticulate near apex. Flowers 5-20, in umbellate or racemose clusters. Calyx 4-5 mm long; corollas 8-10 mm long, violet to purple. Fruit coiled several times, round in outline or sickle-shaped, smooth. Apr-Sep (Nov). Introduced. Scarce; disturbed sites, roadsides; coastal scrub, grasslands; 10-400 ft (3-120 m). North side at Prisoners' Harbor and in Cañada del Muro; Central Valley in upper portion of Cañada Christy and between La Cascada and Stanton Ranch airfield. All CA Channel islands except Santa Barbara; Cedros Island; native of Europe and Asia. Reported as "occasionally persistent in fields" by Hoffmann (1932b). Can be toxic to livestock.

## MELILOTUS Miller SWEET CLOVER

Annuals or short-lived perennial herbs; stems usually erect. Leaves trifoliolate; stipules fused to petiole. Flowers in racemes. Calyx lobes subequal; corolla papilionaceous; stamens 10, 9 fused and 1 free. Fruit ovoid, straight, indehiscent. Ca. 20 species; warm temperate to subtropical, Europe and Asia.

1. Corollas white, 3.5-6 mm long; inflorescences (including peduncles) 6-12 cm long ........................ **M. albus**
1' Corollas yellow, 2-3 mm long; inflorescences (including peduncles) 2-5 cm long ........................ **M. indicus**

**M. albus** Medik. WHITE SWEET CLOVER Annuals or biennials, stems ascending, 6-10 dm tall. Leaflets 8-30 mm long, 3-15 mm wide, elliptic to obovate, margins serrulate. Inflorescences (including peduncles) 6-12 cm long. Calyx 1-2.5 mm long; corollas 3.5-6 mm long, white. Fruit ovoid, 2-5 mm long. May-Jun. Introduced. Rare; disturbed sites, roadsides; coastal scrub, riparian communities; 10-50 ft (3-15 m). North side at Prisoners' Harbor and in Cañada del Puerto. Anacapa, San Nicolas, Santa Catalina, and San Clemente islands; throughout N. America; native of Europe and Asia. First collected at Prisoners' Harbor in May 1962. This species is the original source of the anticoagulant drug dicoumarol; improperly dried plants in hay can be toxic to livestock.
**M. indicus** (L.) All. [*M. parviflorus* Desf.] YELLOW SWEET CLOVER Annuals, stems ascending to erect, 1.5-8 dm tall. Leaflets 10-25 mm long, 5-10 mm wide, cuneate-obovate, margins denticulate to serrulate. Inflorescences (including peduncles) 2-5 cm long. Calyx 1-2 mm long; corollas 2-3 mm long, yellow. Fruit ovoid, 1.5-4 mm long. Mar-May (Aug). Introduced. Common; disturbed sites, moist creekbeds, canyon bottoms, open slopes, floodplains; coastal scrub, riparian communities; 10-1000 ft (3-305 m). Scattered locations throughout island. All CA Channel islands; Guadalupe and Cedros islands; throughout N. America; native of Mediterranean Europe. First reported by Greene (1887). Can be toxic to livestock.

## PICKERINGIA Nutt. ex Torr. & A.Gray

With only 1 species.

**P. montana** Nutt. ex Torr. & A.Gray var. **montana** CHAPARRAL PEA Shrubs, 1-3 m tall; stems intricately branched, spreading to erect, terminal ones spinose. Leaves evergreen, usually trifoliolate, sessile; leaflets 1-3, elliptic to oblanceolate, 10-25 mm long, glabrous to sparsely strigose, upper surface dark green, lower surface glaucous. Flowers in racemes. Calyx campanulate, 4-6 mm long, lobes subequal, minute; corolla papilionaceous, 12-18 mm long, pink to purple, keel petals free; stamens 10, free. Fruit oblong in outline, 3-5 cm long, often constricted between seeds. Apr-May. Rare; rocky n-facing slopes; chaparral, pine forest; 200-750 ft (60-230 m). North side near Pelican Bay. Northern CA coast ranges to Baja CA. Variety *tomentosa* (Abrams) I.M.Johnst., with strigose to tomentose leaves, occurs on the mainland of s CA and Baja CA.

## PISUM L. PEA

Annuals; stems erect to vine-like, twining or climbing, terete. Leaves apparently even-pinnately compound, terminal leaflet modified into simple or branched tendril; stipules leaf-like, fused or free. Flowers few, in axillary racemes. Calyx lobes subequal; corolla papilionaceous; stamens 10, 9 fused and 1 free; style folded lengthwise, puberulent along upper side. Fruit oblong in outline, flat, dehiscent. Five species; Mediterranean and w Asia. Several species are cultivated for edible fruits and seeds.

**P. sativum** L. Annuals, stems decumbent to vine-like, 2-6 dm tall. smooth, glabrous. Leaves 6-15 cm long (including tendril); leaflets 4-6, broadly elliptic to round-ovate, 15-25 mm long, tendril branched, coiled; stipules 30-40 mm long, bases oblique. Flowers 1-3 per raceme; calyx 8-10 mm long; corolla 15-20 mm long, purple. Fruits 4-6 cm long. May. Introduced. Rare; canyon bottom; chaparral; ca. 275 ft (85 m). Known from a single collection made in Islay Canyon in May 1982. Cultivated for edible fruits and naturalized worldwide; native of Mediterranean Europe.

## ROBINIA L. LOCUST

Shrubs or trees. Leaves odd-pinnately compound, deciduous; stipules spine-like, persistent. Flowers many, in axillary racemes. Calyx campanulate, lobes equal; corolla papilionaceous; stamens 10, 9 fused and 1 free. Fruit oblong in outline, flat to subterete, dehiscent. Ca. 5 species; N. America.

**R. pseudoacacia** L. BLACK LOCUST Trees 3-10 m tall. Leaves pinnately compound, 15-25 cm long; leaflets narrowly ovate to elliptic, 2.5-4.5 cm long. Racemes 10-15 cm long. Calyx 5-7 mm long, puberulent, upper lobes shorter than lower ones; corolla 15-20 mm long, white. Fruit oblong, 5-10 cm long, flat. Apr. Introduced. Scarce;

## FABACEAE: Medicago-Robinia

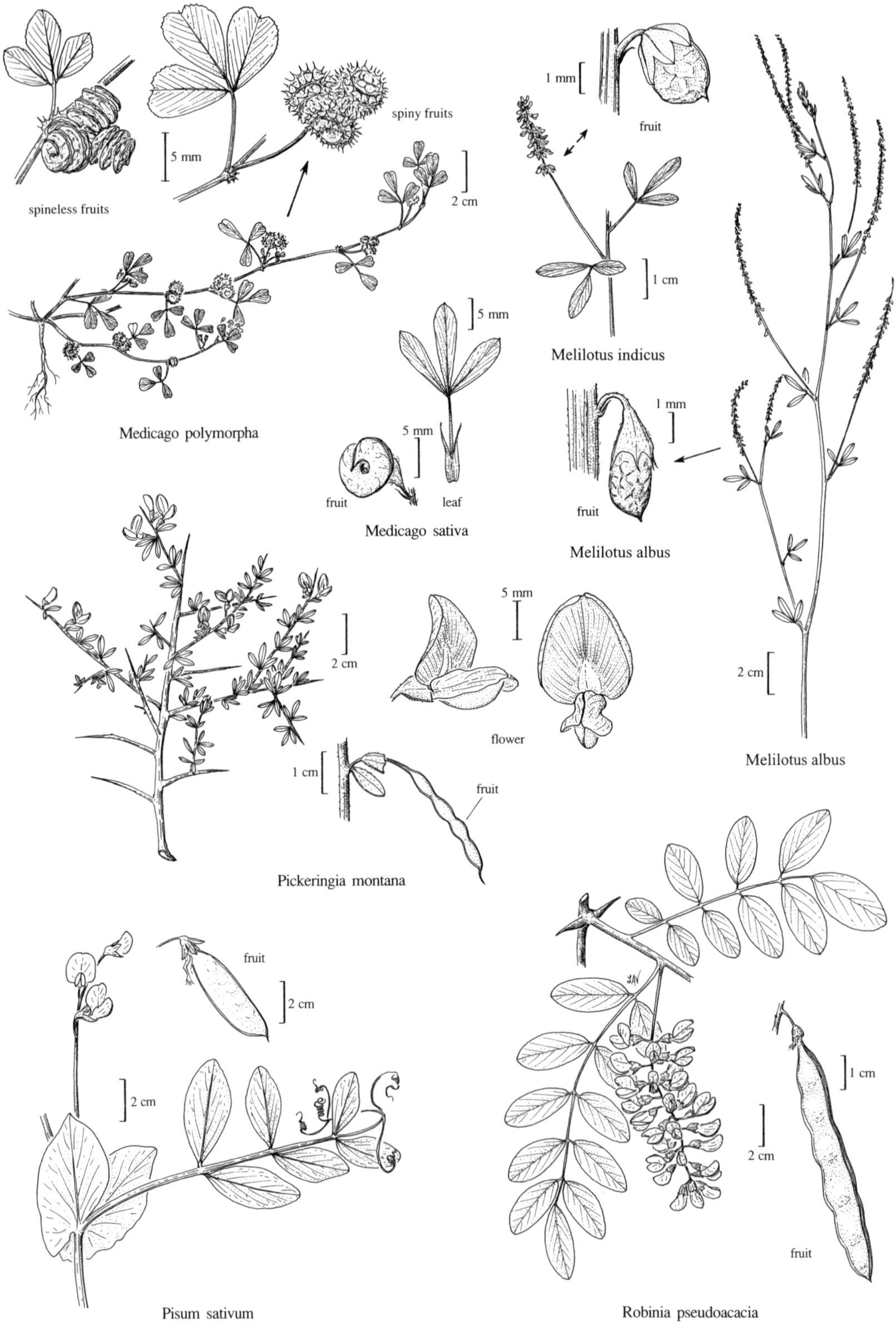

disturbed areas, canyon bottoms, floodplains; coastal scrub, riparian communities; 10-350 ft (3-105 m). North side at Prisoners' Harbor, in Cañada del Puerto, and at Scorpion Ranch; Central Valley between Stanton Ranch and Rancho del Sur. San Nicolas Island; widely cultivated and naturalized throughout w N. America; native of e N. America.

## TRIFOLIUM L. CLOVER

Annuals or perennial herbs; stems decumbent to erect. Leaves usually trifoliolate; stipules fused to petiole, often conspicuous. Flowers in axillary or terminal, cylindrical to globose spikes, spikes often subtended by involucre of free to fused bracts. Calyx lobes equal to unequal; corolla papilionaceous; stamens 10, 9 fused and 1 free. Fruit ovoid to oblong in outline, terete, often stipitate, usually equal to persistent corolla. Ca. 240 species; temperate to subtropical, ± worldwide.

1. Spike not subtended by involucre of bracts
    2. Flowers pedicellate, often reflexed with age; calyx lobes glabrous to ciliate
        3. Calyx lobes ciliate, with conspicuous flat appendages; corollas 8-9 mm long ............ **T. ciliolatum**
        3' Calyx lobes glabrous; corollas 5-6 mm long ............ **T. gracilentum**
    2' Flowers sessile, not reflexed; calyx lobes plumose-villous
        4. Spikes pedunculate, peduncles 1-10 cm long, not closely subtended by upper leaves or stipules ............ **T. albopurpureum**
        4' Spikes sessile, usually in pairs, closely subtended by upper leaves and stipules ............ **T. macraei**
1' Spike subtended by involucre of bracts (bracts sometimes inconspicuous in *T. depauperatum*)
    5. Banner conspicuously inflated in fruit; involucral bracts ± free
        6. Involucral lobes 1.5-2 mm long; corollas 5-8 mm long, petals rose to deep-purple ............ **T. depauperatum** var. **truncatum**
        6' Involucral lobes 8-9 mm long; corollas 12-15 mm long, petals cream to pinkish ............ **T. fucatum**
    5' Banner not inflated in fruit; involucral bracts fused
        7. Involucre campanulate ............ **T. microcephalum**
        7' Involucre rotate to bowl-shaped
            8. Leaflets oblong to obovate; involucre 4-8 lobed, lobes 3-7-toothed; corolla 5-7 mm long ............ **T. variegatum**
            8' Leaflets linear to lanceolate-oblong; involucre unevenly laciniate; corolla 11-14 mm long ............ **T. willdenovii**

**T. albopurpureum** Torr. & A.Gray RANCHERIA CLOVER. Stems erect to decumbent, to 25 cm long, villous. Leaflets oblanceolate to cuneate-oblong, 4-16 mm long, margins denticulate; stipules lanceolate to ovate, margins entire, apices acute. Peduncles 1-10 cm long; involucre absent; flowers sessile. Calyx 6-8 mm long, lobes subulate, densely plumose-villous; corolla 6-7 mm long, banner whitish to light purple, wings and keel reddish purple, apices white. Seed 1 per fruit. Mar-May (Jun). Common; grassy hillsides, rocky slopes, streambeds, floodplains; coastal scrub, grasslands, riparian communities; 10-1900 ft (3-580 m). Widely scattered locations throughout western and central portions of island. Santa Rosa, San Nicolas, and Santa Catalina islands; British Columbia to Baja CA.

**T. ciliolatum** Benth. TREE CLOVER Stems erect, to 40 cm long, mostly glabrous. Leaflets oblanceolate, elliptic, or obovate, 8-35 mm long, margins serrulate; stipules, lanceolate, margins sparsely ciliate, apices acuminate. Peduncles 2.5-8 cm long; involucre absent; flowers pedicellate, pedicels ca. 2 mm long. Calyx 6-7 mm long, lobes narrowly acuminate, ciliate with short, flat appendages; corolla 8-9 mm long, banner pink, wings and keel dark pink to purple, apices whitish. Seed 1 per fruit. Mar-May (Jun). Occasional; grassy hillsides, rocky slopes, floodplains; chaparral, coastal scrub, grasslands, riparian communities; 10-1000 ft (3-305 m). Scattered locations on n side between Hazard's Anchorage and Prisoners' Harbor; Central Valley between Cañada de la Portezuela and Valley Anchorage; s side between Laguna Canyon and Valley Anchorage. Santa Rosa and Santa Catalina islands; WA to Baja CA.

**T. depauperatum** Desv. var. **truncatum** (Greene) Isely [*T. amplectens* Torr. & A.Gray var. *truncatum* (Greene) Jepson] DWARF BLADDER CLOVER Stems erect to spreading, to 18 cm long, mostly glabrous. Leaflets oblanceolate, elliptic, or linear, 2-13 mm long, margins serrate; stipules lanceolate to narrowly ovate, margins entire, apices subulate. Peduncles 2-3.5 cm long; involucres present, deeply lobed, lobes 1.5-2 mm long, entire or with 2 apical teeth; flowers sessile. Calyx 2.5-3 mm long, lobes subulate, unequal; corolla 5-8 mm long, banner rose to purple, conspicuously inflated in fruit, wings and keel rose to deep purple, apices whitish. Seeds

1-2 per fruit. Mar-May (Jun). Common; grassy hillsides, rocky slopes and ridgetops, seasonal ponds, floodplains; coastal scrub, grasslands, riparian communities; 100-1950 ft (30-595 m). Widely scattered locations throughout island. All CA Channel islands except Santa Barbara; CA w of Sierra Nevada and Peninsular Ranges. Three other varieties occur in w N. America.

**T. fucatum** Lindl. BULL CLOVER Stems decumbent to prostrate, to 25 cm long, glabrous. Leaflets rhombic-ovate to obovate, 5-15 mm long, margins denticulate; stipules lanceolate to narrowly ovate, margins entire, apices subulate. Peduncles 2.5-8 cm long; involucres present, deeply 5-9-lobed, lobes 8-9 mm long, scarious, entire, apices subulate; flowers pedicellate, pedicels ca. 1 mm long. Calyx 7-9 mm long, scarious, lobes subulate, apices 2-3-cleft; corolla 12-15 mm long, semi-transparent, petals cream to pinkish, keel apex with purplish-black dots, banner conspicuously inflated in fruit. Seeds 2-4 per fruit. Mar-May. Scarce; sandy sites, dry ravines, rocky slopes; coastal scrub; 30-150 ft (9-45 m). Central Valley at Christy Beach; widely scattered locations on s side between Fraser Point and mouth of Cañada de los Sauces. All CA Channel islands except Santa Barbara and San Nicolas; OR to s CA.

**T. gracilentum** Torr. & A.Gray var. **gracilentum** PINPOINT CLOVER Stems erect to decumbent, to 35 cm long, subglabrous. Leaflets obovate to obcordate, 6-15 mm long, margins serrulate; stipules lanceolate-ovate, margins entire, apices acute. Peduncles 1.5-5 cm long; involucres absent; flowers pedicellate, pedicels 2-3 mm long. Calyx 4-6 mm long, lobes lanceolate, apices acuminate; corolla 5-6 mm long, pink to purplish, wing and keel apices whitish. Seeds 1-2 per fruit. Mar-May. Occasional; coastal bluffs, coastal flats, grassy hillsides, exposed ridgetops; coastal scrub, grasslands; 30-1500 ft (9-460 m). Widely scattered locations around perimeter of island. All CA Channel islands; Todos Santos and Guadalupe islands; WA to Baja CA, e to AZ. Var. *palmeri* (S.Watson) L.F.McDermott, with lanceolate to elliptic leaflets, is an island endemic restricted to the Southern Channel Islands and Guadalupe Island.

**T. macraei** Hook. & Arn. CHILEAN CLOVER Stems erect to decumbent, to 15 cm long, villous. Leaflets obovate, oblanceolate, or elliptic, 3-13 mm long, margins serrulate; stipules ovate, margins entire, apices acuminate. Spikes usually in pairs, sessile; involucres absent; flowers sessile. Calyx 6-8 mm long, lobes subulate, densely plumose-villous; corolla 7-10 mm long, banner white, pink, or lavender, wings and keel reddish-purple, apices white. Seed 1 per fruit. Mar-Jun. Common; coastal bluffs, coastal flats, rocky slopes, exposed ridgetops, floodplains; coastal scrub, grasslands; 40-1650 ft (12-500 m). Widely scattered locations, mostly around perimeter of island. Santa Rosa and Santa Catalina islands; northern CA coast to Santa Barbara Co.; disjunct in S. America.

**T. microcephalum** Pursh SMALL-HEADED CLOVER Stems erect to decumbent, to 30 cm long, villous. Leaflets oblanceolate to obovate, 3-12 mm long, margins serrulate; stipules ovate, margins entire, apices acuminate. Peduncles 1.5-5 cm long; involucres campanulate, deeply 7-10-lobed, lobes lanceolate-ovate, 3-4 mm long, entire or apex 2-lobed; flowers sessile. Calyx ca. 4 mm long, lobes subulate; corolla 5-6 mm long, whitish to pink. Seed 1 per fruit. Mar-May (Aug). Common; grassy hillsides, rocky slopes, exposed ridgetops, moist swales, floodplains; coastal scrub, grasslands, pine forest; 80-1600 ft (25-490 m). Widely scattered locations throughout island. All CA Channel islands except Santa Barbara; Guadalupe Island; British Columbia to s CA, e to ID and AZ.

**T. variegatum** Nutt. Stems decumbent to prostrate, to 50 cm long, mostly glabrous. Leaflets oblong to obovate, 4-22 mm long, margins serrulate; stipules ovate to ± round, margins laciniate. Peduncles 1.2-8 cm long; involucres rotate, deeply 4-8-lobed, lobes 2-2.5 mm long, each with 3-7 irregular teeth; flowers short-pedicellate to subsessile. Calyx ca. 4 mm long, lobes subulate to setaceous; corolla 5-7 mm long, banner and wings pink to reddish purple, apices whitish, keel reddish purple. Seeds 2 per fruit. May-Aug (Oct). Scarce; moist creekbeds and seeps; riparian communities; 200-1400 ft (60-430 m). North side in Cañada del Puerto; canyons on n side of Central Valley between Picacho Diablo and Red Mountain. British Columbia to Baja CA, e to Rocky Mtns.

**T. willdenovii** Spreng. [*T. tridentatum* Lindl.] TOMCAT CLOVER Stems erect to decumbent, to 40 cm long, mostly glabrous. Leaflets linear to lanceolate-oblong, 15-40 mm long, margins serrulate (rarely entire); lower stipules lanceolate, margins entire or few-toothed near apex, upper ones ovate to ± round, margins laciniate. Peduncles 4-8 cm long; involucres rotate to bowl-shaped, unevenly laciniate, 10-25 mm wide; flowers sessile to subsessile. Calyx ca. 8 mm long, lobes entire to few-toothed near apex, subulate to setaceous; corolla 11-14 mm long, banner rose-pink, wings and keel rose-purple, apices white. Seeds 1-2 per fruit. Mar-May. Common; grassy hillsides, rocky slopes; chaparral, coastal scrub, grasslands, oak woodland, pine forest; 60-1500 ft (20-460 m). Widely scattered locations throughout island. All CA Channel islands; Los Coronados and Todos Santos islands; British Columbia to Mexico, e to Rocky Mtns.

## VICIA L. VETCH

Annuals or perennial herbs; stems sprawling to vine-like, twining or climbing, often angular or slightly winged. Leaves apparently even-pinnately compound, terminal leaflet modified into branched, coiled tendril; stipules free, often lobed or toothed. Flowers in axillary, often 1-sided racemes. Calyx lobes unequal to equal; corolla papilionaceous; stamens 10, 9 fused and 1 free; style usually terete, apex with tuft of trichomes. Fruit oblong in outline, flat, dehiscent. Ca. 140 species; N. America, Europe, and Asia.

1. Flowers and fruits subsessile; calyx tube 5-7 mm long ........ **V. sativa**
1' Flowers and fruits pedunculate; calyx tube 1.5-5 mm long
  2. Flowers 3-14 per inflorescence; corollas 10-20 mm long; peduncles in fruit 25-70 mm long
    3. Flowers evenly distributed around rachis; calyx lobes broadly lanceolate, glabrous to sparsely pubescent ........ **V. americana**
    3' Flowers on one side of rachis; calyx lobes conspicuously ciliate ........ **V. benghalensis**
  2' Flowers 1-2 per inflorescence; corollas 4-6.5 mm long; peduncles in fruit 12-30 mm long
    4. Stylar trichomes densely tufted, concentrated on lower side of style; fruits sparsely pubescent ........ **V. hassei**
    4' Stylar trichomes evenly distributed around style; fruits glabrous ........ **V. ludoviciana** subsp. **ludoviciana**

**V. americana** Willd. Perennial herbs; stems 3-10 dm long. Leaflets 8-16, narrowly oblong to elliptic, 12-50 mm long, 3-7 mm wide, apices acute to slightly notched. Flowers 3-8, peduncles in fruit 25-70 mm long. Calyx tube 3-5 mm long, lobes 1-2 mm long; corolla 10-15 mm long, lavender to reddish purple, stylar trichomes densely tufted, evenly distributed around style. Fruit 2.5-3 cm long, stipe 2-4 mm long. Feb-May. Occasional; grassy hillsides, rocky slopes, canyon bottoms; chaparral, coastal scrub, grasslands, oak woodland; 80-1400 ft (25-425 m). North side between Pelican Bay and China Harbor; Central Valley between upper part of Cañada Christy and Stanton Ranch airfield; s side between Laguna Canyon and Coches Prietos drainage. San Miguel and Santa Rosa islands; throughout N. America. Stems, leaves, and seeds were eaten by Native Americans; roots were used for binding materials together.

**V. benghalensis** L. Annuals; stems 4-16 dm long. Leaflets 8-16, narrowly oblong to elliptic, 16-25 mm long, 5-9 mm wide, apices rounded to abruptly acuminate. Flowers 6-14, peduncles in fruit 40-60 mm long. Calyx tube 4-5 mm long, lobes 5-7 mm long; corolla 15-20 mm long, purple to red-purple, stylar trichomes densely tufted, evenly distributed around style. Fruit 2.5-3 cm long, stipe 1-2 mm long. May. Rare; disturbed sites; coastal scrub; ca. 320 ft (97 m). Known from a single collection made in the Central Valley just w of Bosque Mano in May 1992. Anacapa, San Nicolas, and Santa Catalina islands; throughout N. America; native of Europe and Asia.

**V. hassei** S.Watson Annuals; stems 3-6 dm long. Leaflets 4-10, narrowly oblong to narrowly elliptic, 10-20 mm long, 2-4 mm wide, apices truncate to slightly notched. Flowers 1-2, peduncles (pedicel-like if flower solitary) in fruit 15-30 mm long. Calyx tube 1.5-2 mm long, lobes 0.5-1 mm long; corolla 4-6.5 mm long, banner pale lavender, wings and keel whitish; stylar trichomes densely tufted, concentrated on lower side of style. Fruit 2-3 cm long, sparsely pubescent, stipe 1-2 mm long. Mar-May. Scarce; grassy hillsides, moist swales, rocky slopes, canyon bottoms; coastal scrub, grasslands, riparian communities; 150-600 ft (45-185 m). North side near Pelican Bay; s side near Black Point and in Coches Prietos drainage. All CA Channel islands except Santa Barbara; Todos Santos and Guadalupe islands; s OR to Baja CA.

**V. ludoviciana** Nutt. subsp. **ludoviciana** [*V. exigua* Nutt. in Torr. & A.Gray] Annuals; stems 2-8 dm long. Leaflets 4-10, linear to narrowly oblong or narrowly elliptic, 10-30 mm long, 1-3 mm wide, apices usually acute. Flowers 1-2, peduncles (pedicel-like if flower solitary) in fruit 12-25 mm long. Calyx tube 1.5-2 mm long, lobes 1-1.5 mm long; corolla 4-6 mm long, banner pale lavender, wings and keel whitish, petals aging blue; stylar trichomes evenly distributed around style. Fruit 2-2.5 cm long, glabrous, stipe 1-2 mm long. Apr-May. Scarce; grassy hillsides, moist swales, rocky slopes, canyon bottoms; coastal scrub, grasslands, riparian communities; 60-500 ft (18-150 m). Central Valley in Cañada de la Portezuela; s side in Johnson Canyon, Willows Canyon, and Coches Prietos drainage. Santa Rosa, Anacapa, Santa Catalina, and San Clemente islands; Guadalupe Island; San Luis Obispo Co. to Baja CA. Often associated with *Opuntia* patches.

**V. sativa** L. Annuals; stems 2-16 dm long. Leaflets 8-12, oblong to ovate, 10-25 mm long, 4-12 mm wide, apices usually truncate to notched. Flowers 1-3, axillary, subsessile in fruit. Calyx tube 5-7 mm long, lobes 5-6 mm long; corolla 15-25 mm long, banner lavender, wings reddish purple, keel whitish, apex reddish purple; stylar

**FABACEAE: Trifolium, Vicia**

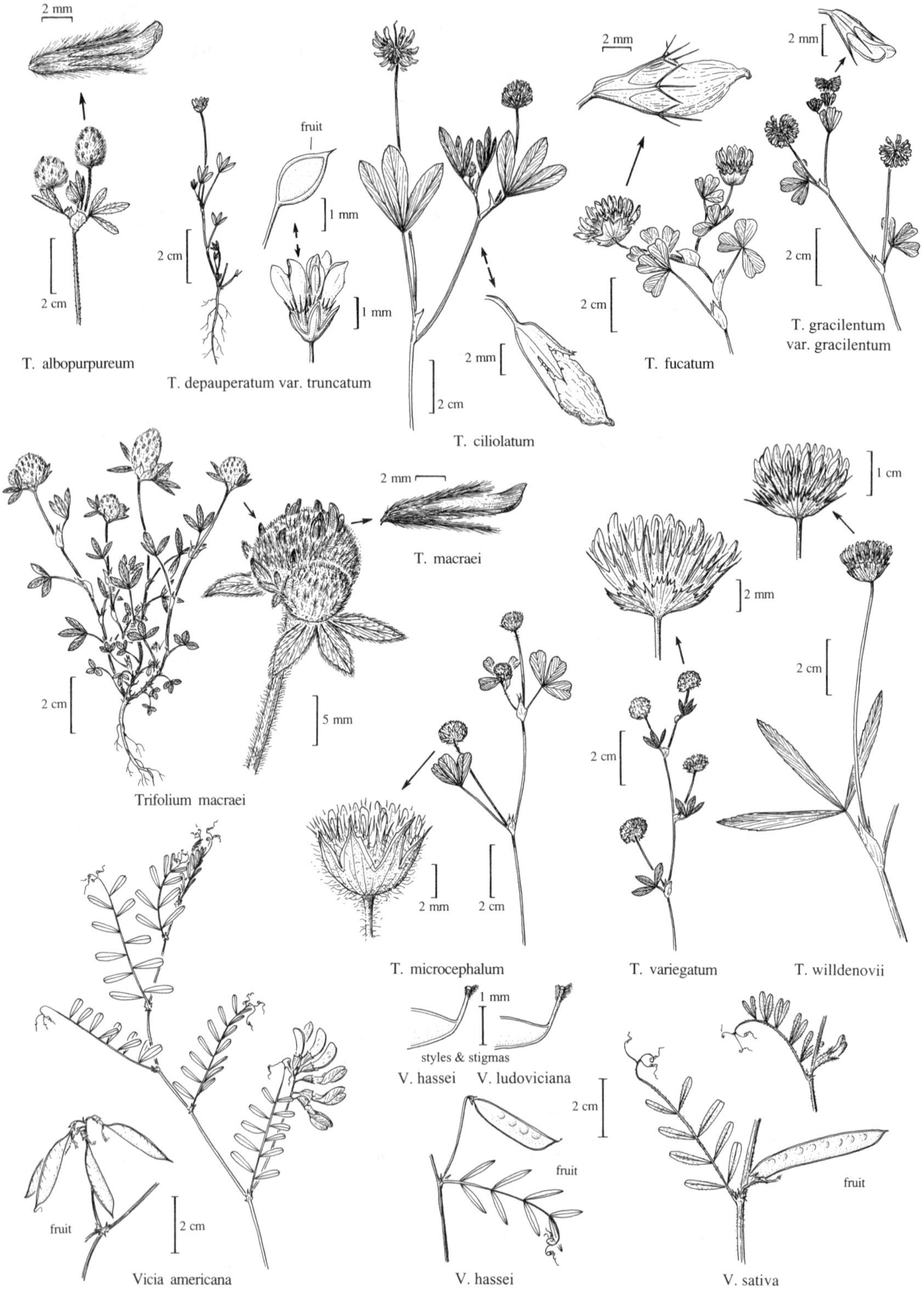

trichomes densely tufted, concentrated on lower side of style. Fruit 3-5 cm long, stipe absent. Apr. Introduced. Rare; disturbed sites; coastal scrub, grasslands; 200-250 ft (60-75 m). Central Valley between La Cascada and La Selva. Anacapa Island; throughout N. America; native of Europe. First collected near La Selva in April 1992. Young stems, leaves, and fruits can be cooked and eaten; ground seeds can be used for making flour.

## **FAGACEAE** Dumort.

### OAK FAMILY

Monoecious shrubs or trees. Leaves usually alternate, simple, entire to lobed, evergreen or deciduous; stipules present, persistent or caducous. Flowers unisexual, staminate and pistillate ones usually in separate inflorescences or staminate flowers below pistillate ones; staminate flowers usually many in catkins or stiff inflorescences; pistillate flowers solitary or 2-3 per cluster, each flower or cluster subtended by bracteate or spinose cupule. Calyx minute, with 3-6 lobes; petals absent. Staminate flowers with 4-many stamens; pistillate flowers with inferior ovary, 3-6 locules, 3-6 styles. Fruit a nut, subtended by a deeply lobed or undivided cupule, together often referred to as an acorn. Nine genera and ca. 1000 species, mostly temperate, some tropical and subtropical, n hemisphere and se Asia. One of the most important sources of wood in the n hemisphere and also used for edible nuts, dyes, and cork. Treatment contributed by Kevin C. Nixon

### **QUERCUS** L. OAK

Shrubs or trees. Leaves entire to toothed or pinnately lobed, glabrous, tomentose, or with branched trichomes; stipules scarious, caducous. Staminate inflorescences in catkins. Pistillate flowers solitary, each subtended by a cup (acorn cup), its outer surface covered with scales. Calyx irregularly lobed. Nut round in cross-section. Ca. 600 species; N. America to n S. America, Europe, and Asia. Many populations on Santa Cruz Island are composed of hybrids involving different species. Some of these hybrids are common, including *Q. ×macdonaldii.*

1. Leaves lobed, with at least some sinuses between lobes extending more than 1/4 the distance from margin to midvein
  2. Deepest sinuses of leaf extending more than half the distance from margin to midvein
    3. Lower surfaces of leaves glabrous, shiny at maturity, marginal teeth with aristate bristles 2-3 mm long ........ **Q. kelloggii**
    3' Lower surfaces of leaves puberulent or stellate, dull to whitish, marginal teeth mucronate, mucros ca. 0.1-0.2 mm long ........ **Q. lobata**
  2' Deepest sinuses of leaf extending less than 1/2 the distance from margin to midvein
    4 Lobes tapering to soft, aristate teeth, these 2-3 mm long; mature bark smooth and gray or black and furrowed; acorns maturing in 2 years ........ **Q. parvula × Q. kelloggii**
    4' Lobes without teeth or lobes mucronate; mature bark scaly to flaky, whitish to light gray; acorns maturing in 1 year
      5. Leaves bluish green; lower surfaces of leaves with 4-6 rayed trichomes, rays erect; moderate to large trees ........ **Q. douglasii**
      5' Leaves green; lower surfaces of leaves with 8-10 rayed trichomes, rays spreading; shrubs or small trees ........ **Q. ×macdonaldii**

1' Leaves entire to toothed, without clearly defined lobes
  6. Lower surfaces of leaves densely glandular and tomentose, often glabrous or glaucous with age; acorn cups with dense tomentum obscuring scales, lateral wall often thicker than 5 mm; trees
    7. Leaves planar, rarely cupped, but not revolute, usually entire but margin sometimes with spinulose teeth; lower leaf surfaces strongly glaucous, with some glandular trichomes ........ **Q. chrysolepis**
    7' Leaves with conspicuously revolute margins or, if flat, then with regularly spaced mucronate teeth; lower leaf surfaces persistently yellow-tomentose ........ **Q. tomentella**
  6' Lower surfaces of leaves mostly glabrous or lower surfaces either with minutely stellate trichomes or with tufts of trichomes in vein axils; acorn cups with thin lateral walls, scales puberulent to glabrate; trees or shrubs
    8. Leaves glabrous and shiny on both surfaces at maturity; inner surface of fruit wall tomentose

9. Leaves cupped or revolute, orbicular, ovate, or obovate; lower surfaces of leaves glabrate except for minute tufts of trichomes in vein axils; trees; nuts oblong-cylindrical, sharply acute ........................................................................................................ **Q. agrifolia** var. **agrifolia**
9' Leaves planar or moderately convex, ovate to lanceolate; lower surfaces of leaves completely glabrous; shrubs; nuts globose or oblong-cylindrical, obtuse .................................. **Q. parvula** var. **parvula**

8' Leaves not glabrous on both surfaces, variously pubescent, at least lower surfaces glaucous or waxy; inner surface of fruit wall glabrate

10. Leaves light green to bluish green, thin, bases rounded to rounded-attenuate; petioles often more than 5 mm long; trees or large shrubs ........................................................................ **Q. douglasii**
10' Leaves green or yellowish green, leathery, bases cuneate, tapered, or strongly cordate; petioles less than 5 mm long

11. Leaf bases cordate to truncate; lower surfaces of leaves velvet-like, covered with erect, curly trichomes .................................................................................................... **Q. dumosa**
11' Leaf bases cuneate or attenuate; lower surfaces of leaves glabrate or with minute, appressed, stellate hairs .................................................................................................... **Q. pacifica**

**Q. agrifolia** Nee var. **agrifolia** COAST LIVE OAK Trees, to 15 m tall; bark dark gray, checkered to furrowed; twigs brown to grayish brown, puberulent. Leaves evergreen; blades orbicular, ovate, or obovate, 3-6 cm long, 2-4 cm wide, cupped or revolute, bases obtuse to rounded, margins toothed, apices rounded, upper surfaces of mature leaves green, glabrous, lower surfaces ± glabrous except for minute tufts of trichomes in vein axils. Acorn maturing in 1 year; cup hemispheric, 10-15 mm wide, 6-10 mm deep, scales thin, flat, appressed; nut cylindrical, 25-35 mm long, 6-12 mm wide, tapered to tip, inner surface of nut wall tomentose. Mar-Apr. Abundant; mostly in canyons and on n-facing slopes; grasslands, oak woodland, riparian woodland; 10-2400 ft (3-730 m). Throughout island, but most abundant on n side between Cueva Valdez and Eagle Canyon, and in the Central Valley. Santa Rosa Island; n CA Coast Ranges to n Baja CA. Hybrids with *Q. parvula* are known in Los Pinos de la Calera above China Harbor. Variety *oxyadenia* (Torr.) J.T.Howell, with lower surfaces of leaves densely tomentulose, occurs in the peninsular ranges of s CA and n Baja California.

**Q. chrysolepis** Liebm. in Benth. CANYON LIVE OAK Trees, sometimes shrubs, to 25 m tall; bark light gray-brown, with closely appressed scales; twigs golden brown, densely puberulent. Leaves evergreen; blades oblong, 2-7 cm long, 1-3.5 cm wide, usually planar, bases obtuse to rounded, margins slightly inrolled, entire, spinulose-dentate, or irregularly toothed, apices mucronate to ± spinose, upper surfaces of mature leaves yellowish green, with 8-12-rayed trichomes, lower surfaces bluish white, glabrate. Acorn maturing in 2 years; cup saucer-shaped, 20-40 mm wide, 4-8 mm deep, scales tuberculate, appressed, deeply embedded and fused or swollen; nut ovoid, 15-30 mm long, 10-20 mm wide, inner surface of nut wall lightly tomentose. Apr-May. Occasional; mostly on exposed n-facing slopes; chaparral, oak woodland; 1500-2350 ft (460-715 m). North side between Peak 1848 West and Red Peak, and on El Montañon ridge; s side on El Tigre ridge. Anacapa, Santa Catalina, and San Clemente Islands; OR to Baja CA, e to AZ. Hybridizes with *Q. tomentella.*

**Q. douglasii** Hook. & Arn. BLUE OAK Trees, to 20 m tall; bark gray, scaly; twigs reddish or yellowish, densely puberulent to glabrate. Leaves deciduous; blades obovate to elliptic in outline, 2.5-6.5 cm long, 2-3 cm wide, bases tapered to rounded, margins shallowly lobed or irregularly toothed, sometimes entire, lobes mucronate or rounded, apices rounded, upper surfaces of mature leaves bluish green, lower surfaces light green, with 4-6-rayed trichomes. Acorn maturing in 1 year; cup hemispheric, 10-15 mm wide, 5-10 mm deep, scales flat to strongly and irregularly tuberculate, especially near base of cup; nut fusiform or subcylindrical, 20-30 mm long, 10-16 mm wide, inner surface of nut wall glabrous. Apr-May. Occasional; mostly in canyons and on exposed n-facing slopes; chaparral, grasslands, oak woodland; 100-2000 ft (30-610 m). North side near Peak Alta 2, Peak 1848 West, and in Cañada del Puerto; scattered locations in Central Valley from Christy Pines to near Stanton Ranch; s side near Alameda de los Coches Prietos and near Los Pinos del Sur. Santa Catalina Island; much of CA to n Ventura and Los Angeles cos. Hybridizes with *Q. lobata* and *Q. pacifica.*

**Q. dumosa** Nutt. NUTTALL'S SCRUB OAK Shrubs, 1-2 m tall, with many widely branching stems; bark grayish, smooth, becoming scaly; twigs reddish or grayish, glabrous to sparsely puberulent. Leaves ± evergreen; blades oblong, elliptic, or orbicular, 1-2 cm long, 0.6-2 cm wide, undulate to ± cupped, bases cordate, margins shallowly lobed to irregularly toothed, sometimes entire, apices rounded to spinose-acute, upper surfaces of mature leaves shiny green, lower surfaces light green, with 4-6-rayed trichomes. Acorn maturing in 1 year; cup hemispheric, 8-15 mm wide, 5-8 mm deep, scales moderately to barely tuberculate, ± flat near base of cup; nut fusiform or subcylindrical, 10-20 mm long, 5-10 mm wide, inner surface of nut wall glabrous. Aug. Rare, not

## FAGACEAE

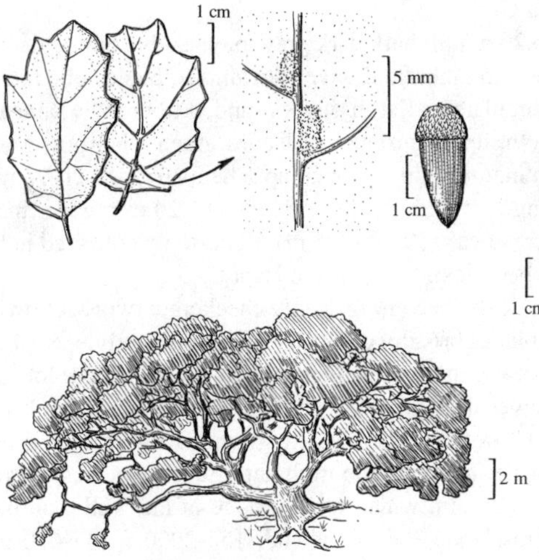

Quercus agrifolia

Quercus lobata

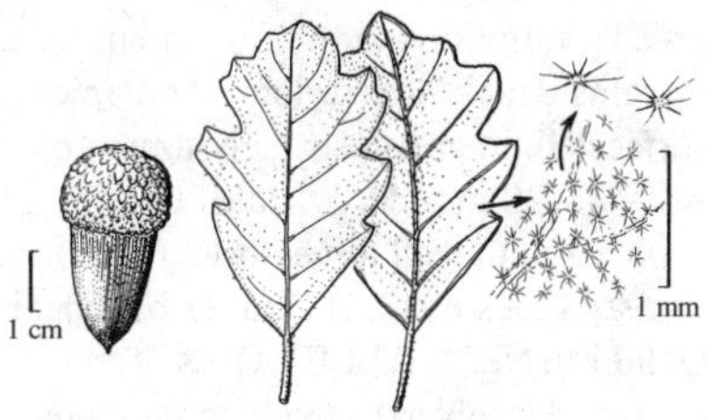

Quercus ×macdonaldii

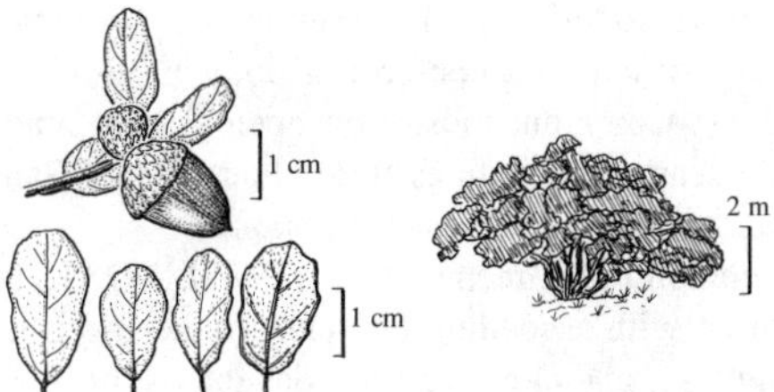

Quercus pacifica

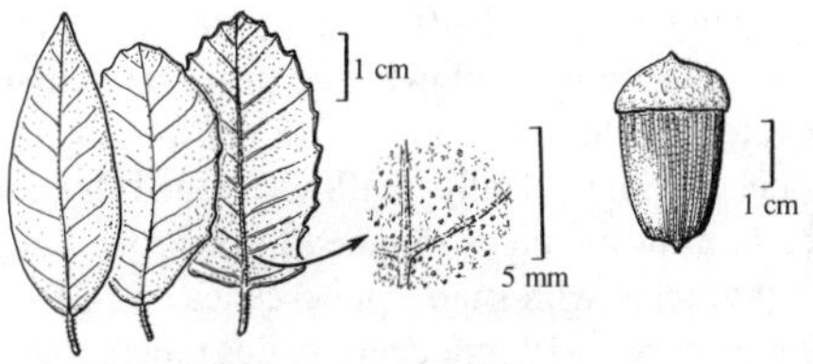

Q. chrysolepis

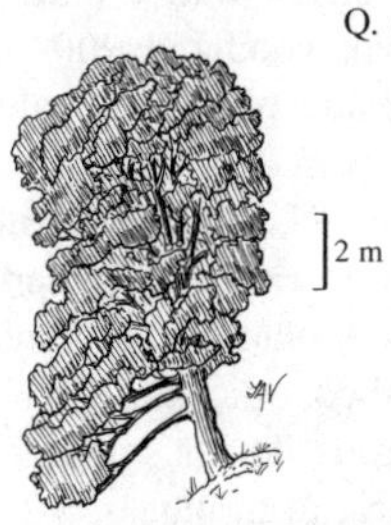

Quercus chrysolepis

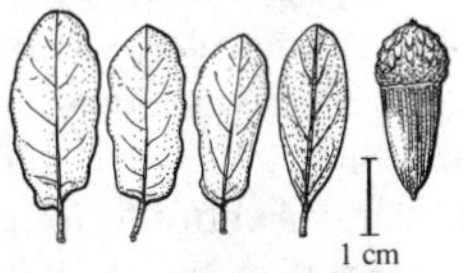

Quercus dumosa

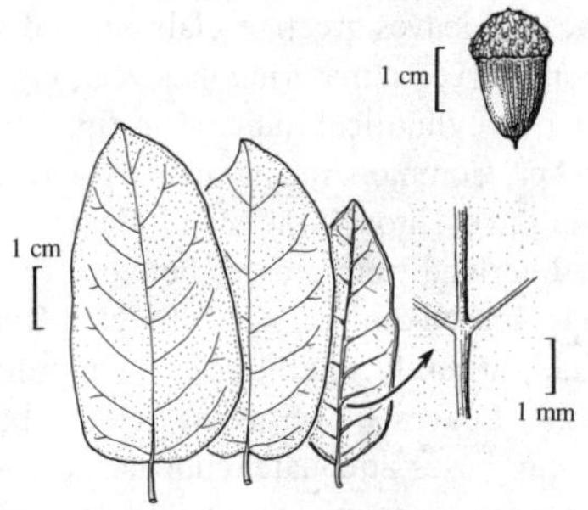

Quercus parvula

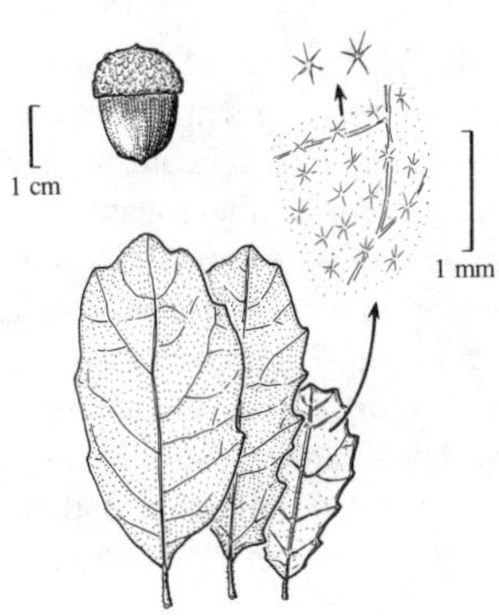

Quercus douglasii

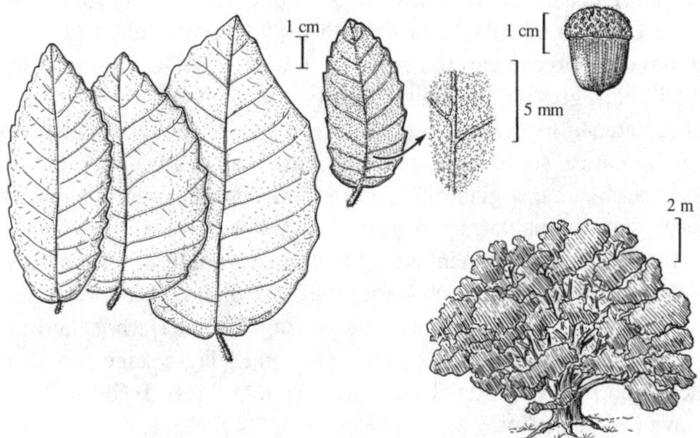

Quercus tomentella

seen recently; habitat, plant community, and elevation unknown. Known from a single collection without locality by L. G. Yates in 1893. Coastal Santa Barbara Co. to n Baja CA.

**Q. kelloggii** Newb. CALIFORNIA BLACK OAK Trees, to 20 m tall; bark dark gray to black, becoming deeply checkered or furrowed; twigs reddish or brownish, puberulent to glabrate. Leaves deciduous; blades obovate to broadly elliptic in outline, 7-15 cm long, 5-12 cm wide, ± undulate or flat, bases ± rounded to broadly cuneate, margins deeply lobed, lobes and apices with 2-4 aristate teeth, upper surfaces of leaves green, glabrous, lower surfaces light green, sparsely tomentulose to glabrate. Acorn maturing in 2 years; cup ± hemispheric, 20-25 mm wide, 15-20 mm deep, scales thin, flat, appressed; nut oblong to ovoid, 20-30 mm long, 14-20 mm wide, inner surface of nut wall tomentose. Rare; n-facing slopes; pine forest; ca. 1000 ft (305 m). Reportedly collected in the Christy Pines by C. H. Muller but specimen apparently has been lost. Oregon to n Baja CA.

**Q. lobata** Nee VALLEY OAK Trees, to 20 m tall; bark gray, scaly, becoming deeply checkered; twigs yellowish or gray, densely to sparsely tomentulose. Leaves deciduous; blades broadly obovate to elliptic in outline, 5-10 cm long, 3-6 cm wide, ± undulate or flat, bases rounded to cuneate, margins moderately to deeply lobed, lobes ± oblong, obtuse, apices broadly rounded, upper surfaces of leaves dark green or grayish, shiny, ± glabrous, lower surfaces light green, with sparse to dense pubescence of 8-10-rayed trichomes. Acorn maturing in 1 year; cup hemispheric to turbinate, 14-30 mm wide, 10-30 mm deep, scales strongly and irregularly tuberculate, especially near base of cup; nut fusiform or oblong, 30-60 mm long, 15-25 mm wide, inner surface of nut wall glabrous. Mar-May. Common; mostly on open slopes; pine forest, grasslands, oak woodland; 150-2000 ft (45-610 m). Widely scattered localities throughout island. Santa Catalina Island; n CA to Los Angeles Co. Hybrids with *Q. pacifica* are called *Q. ×macdonaldii.*

**Q. ×macdonaldii** Greene MACDONALD OAK [*Q. ×morehus* Kellogg misapplied] Shrubs or trees, 2-6 (10) m tall, often with ascending branches; bark grayish, scaly to checkered; twigs yellowish or gray, puberulent to tomentulose, glabrate. Leaves ± deciduous; blades broadly obovate to oblong in outline, 3-8 cm long, 1.5-4 cm wide, ± undulate or flat, bases subacute, rounded, or subcordate, margins usually shallowly lobed or toothed above middle, sinuses between lobes less than half the distance to midrib, apices rounded to mucronate, upper surfaces of leaves green, ± glabrous and shiny, lower surfaces light green, with sparse pubescence of 8-10-rayed trichomes. Acorn maturing in 1 year; cup ± hemispheric, 10-20 mm wide, 7-10 mm deep, scales tuberculate; nut oblong or cylindrical, tapered to tip, 20-30 mm long, 10-15 mm wide, inner surface of nut wall ± tomentose. Mar-Apr. Common; mostly in canyons and on sheltered slopes; chaparral, grasslands, oak woodland; 200-1400 ft (60-430 m). Throughout island. Santa Rosa and Santa Catalina Islands. Generally considered to be a stabilized hybrid derived from crosses between *Q. lobata* and *Q. pacifica*, but other island species may be involved.

**Q. pacifica** Nixon & C.H.Mull. [*Q. dumosa* Nutt. var. *polycarpa* Greene] ISLAND SCRUB OAK Shrubs, to 2 m tall, or rarely small trees, to 5 m tall; bark grayish, becoming scaly; twigs reddish or brownish, puberulent, glabrate. Leaves ± evergreen; blades obovate or oblong, 1.5-4 cm long, 0.7-2 cm wide, planar to ± convex or undulate, bases attenuate, cuneate, or rounded, margins entire to irregularly few-toothed, apices usually blunt and rounded, sometimes mucronate, upper surfaces of mature leaves green, ± glabrous, lower surfaces with minute, appressed, stellate hairs or glabrate. Acorn maturing in 1 year; cup hemispheric to turbinate, 8-15 mm wide, 5-8 mm deep, scales moderately to strongly tuberculate; nut fusiform or cylindrical, 20-30 mm long, 9-15 mm wide, inner surface of nut wall glabrate. Mar-Apr. Island endemic. Abundant; canyons, slopes, and ridges; chaparral, coastal scrub, oak woodland, pine forest; 150-1500 ft (45-460 m). Throughout island. Santa Rosa and Santa Catalina Islands. Hybrids with *Q. lobata* are called *Q. ×macdonaldii.*

**Q. parvula** Greene var. **parvula** [*Q. wislizeni* A.DC. misapplied] SANTA CRUZ ISLAND OAK Shrubs, to 2 m tall; bark grayish, becoming ± scaly; twigs greenish or brownish, glabrous. Leaves evergreen; blades ovate to lanceolate, 4-10 cm long, 2-4 cm wide, planar or moderately convex above, bases rounded to subcordate, margins usually entire, sometimes spinose-toothed, apices usually acute to acuminate, sometimes mucronate, surfaces green and shiny, ± glabrous. Acorn maturing in 2 years; cup broadly hemispheric, 15-20 mm wide, 6-10 mm deep, scales thin, flat, appressed; nut globose, oblong, or cylindrical, 15-40 mm long, 10-15 mm wide, inner surface of nut wall tomentose. Apr-Jun. A variety of special environmental concern. Common; mostly on exposed slopes and ridges; chaparral, oak woodland, pine forest; 300-1800 ft (90-550 m). North side above Cueva Valdez, above Fry's Harbor, Pelican Bay, China Harbor, and near Scorpion Anchorage; Central Valley near Christy Pines and Stanton Ranch; s side on El Tigre ridge and near Los Pinos del Sur. Coastal Santa Barbara Co. (near Lompoc, Hollister Ranch, and Gaviota Pass). Hybrids with *Q. agrifolia* occur in Los Pinos de la Calera above China Harbor.

**Q. parvula** Greene × **Q. kelloggii** Newb. [*Q. ×morehus* Kellogg misapplied] Shrubs or small trees, 2-5 m tall; bark gray or black, smooth, becoming furrowed. Leaves ± deciduous; blades obovate to broadly elliptic in outline,

FRANKENIACEAE

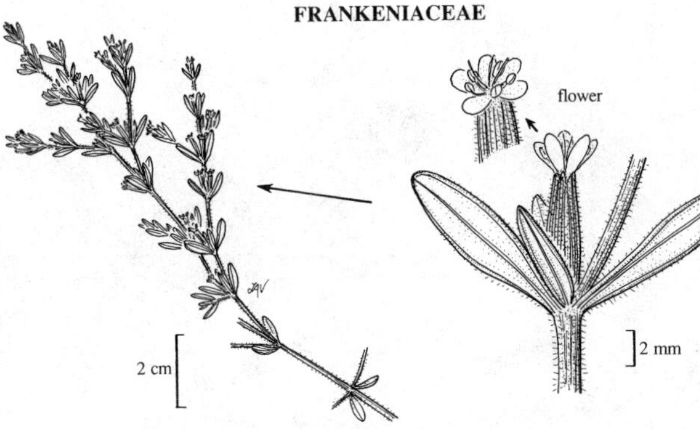


Frankenia salina

7-13 cm long, 4-10 cm wide, bases tapered to truncate, margins serrate or shallowly toothed to lobed, lobes with soft aristate tips, apices ± aristate, surfaces dark green, shiny. Acorn maturing in 2 years, cup broadly hemispheric, scales thin, scaly, moderately pubescent, nut globose, oblong, or cylindrical. Apr. Rare, not seen recently; isolated individuals in valley bottom; grasslands, oak woodland; 750-1250 ft (230-380 m). Central Valley in Cañada de la Portezuela. Coastal Santa Barbara Co. (Hollister Ranch).

**Q. tomentella** Engelm. ISLAND OAK Trees, to 20 m tall; bark light gray, flaky; twigs tomentose. Leaves evergreen; blades oblong-lanceolate to elliptic, 7-10 cm long, 2.5-4 cm wide, undulate or sometimes concave beneath, bases obtuse to cordate, margins entire or crenate to dentate with 10-12 mucronate teeth on each side, often revolute, apices acute to obtuse, surfaces dark green, shiny, upper surfaces sparsely glandular-pubescent, lower surfaces densely yellow-tomentose. Acorn maturing in 2 years; cup hemispheric, 20-30 mm wide, 4-8 mm deep, scales tuberculate, appressed, deeply embedded and fused or swollen; nut ovoid, 20-30 mm long, 15-20 mm wide, inner surface of nut wall lightly tomentose. May-Jul. Island endemic. A species of special environmental concern. Common; mostly in canyons and on n-facing slopes; chaparral, oak woodland, pine forest, riparian woodland; 100-2400 ft (30-730 m). North side between Peak Alta 2 and Prisoners' Harbor, and above China Harbor; Central Valley near Christy Pines; s side on Sierra Blanca ridge, near El Tigre Ridge, and Los Pinos del Sur. Santa Rosa, Anacapa, Santa Catalina, and San Clemente Islands; Guadalupe Island. Hybridizes with *Q. chrysolepis*.

## FRANKENIACEAE L.

### FRANKENIA FAMILY

Perennial herbs or shrubs; stems erect to prostrate. Leaves opposite, simple, often coriaceous or fleshy; stipules absent. Flowers bisexual, radial, solitary or in axillary clusters. Sepals 4-7, fused, persistent; petals 4-6, free, overlapping, with conspicuous claws and scale-like appendages; stamens 3-12, in 2 whorls; pistil 1, ovary superior, locules 1-4, style 1, stigmatic branches 1-4. Fruit a capsule; seeds 1-many. Ca. 3 genera and 30 species; mostly saline habitats, temperate and subtropical, worldwide.

### FRANKENIA L.

Leaves in axillary clusters, ± in ranks of 4. Calyx tubular, lobes shorter than tube; petal blades ± spreading; style with 2-3 stigmatic branches. Capsule oblong to linear, angled. Ca. 25 species; worldwide. Some species are used medicinally or as a source of salt.

**F. salina** (Molina) I.M.Johnst. [*F. grandifolia* Cham. & Schltdl.] ALKALI HEATH Stems herbaceous to suffrutescent, 1.5-3 dm tall. Leaves fused at base, blades elliptic to obovate, 5-15 mm long, puberulent, subsessile, margins revolute, lower surfaces glaucous. Calyx 6-7 mm long; petals 6-12 mm long, white to pinkish, slightly exserted; stamens 5-6. Fruit 4-5 mm long; seeds ca. 1 mm long, brown. Jun-Oct. Occasional; sandy beaches, coastal flats, creekbanks, and floodplains; coastal scrub, coastal strand, grasslands, riparian communities; 3-300 ft (1-90 m). North side near Scorpion Canyon; s side between Fraser Point and Punta Arena, and near Smugglers' Cove. All CA Channel islands except Santa Barbara; Todos Santos and Guadalupe islands; San Francisco Bay

GARRYACEAE

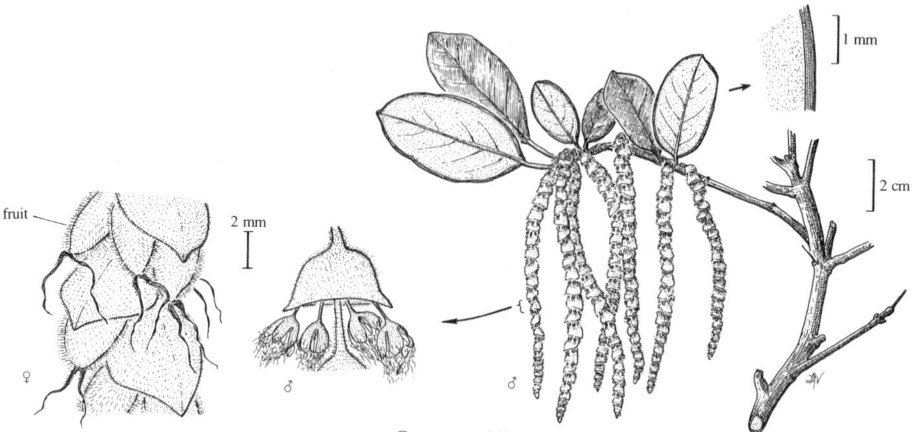


Garrya veatchii

region and Great Valley to Mexico and S. America, e to NV. This species was used medicinally by Native Americans.

## GARRYACEAE Lindl.
## SILK TASSEL FAMILY

With only 1 genus.

### GARRYA Lindl.

Dioecious shrubs or small trees. Leaves opposite, simple, entire, evergreen; stipules absent. Flowers unisexual, many, in pendulous, bracteate spikes. Staminate flowers usually in clusters of 3-5 per bract; sepals 4; petals absent; stamens 4. Pistillate flowers usually in clusters of 3 per bract, sometimes solitary, sepals and petals absent or sepals 2 and vestigial; ovary inferior, locule 1, stigmatic branches usually 2. Fruit a berry or dry and indehiscent, ovoid to globose, fleshy, greenish to gray or black; seeds 2. Ca. 15 species; temperate to subtropical, w N. America and C. America. Leaves and bark of several species were used medicinally by Native Americans.

**G. veatchii** Kellogg Shrubs or small trees, to 6 m tall. Leaf blades ovate to broadly elliptical, 3.5-12 cm long, 1.5-6.5 cm wide, bases rounded, margins entire to slightly undulate, sometimes slightly revolute, apices rounded to acute, upper surfaces green, shiny, lower surfaces densely gray-tomentulose; petioles to 1 cm long. Spikes 3-5 per cluster, 5-7 cm long, terminating lateral twigs. Pistillate flowers usually 1 per bract. Fruit subglobose, 5-8 mm long, gray-tomentulose. Jan-Mar. Scarce; rocky slopes; chaparral, island woodland; 500-1900 ft (150-580 m). North side between Cueva Valdez and Pelican Bay, in widely scattered populations on upper slopes; s side in Willows Canyon near Peak 678. Cedros Island; s San Luis Obispo Co. to n Baja CA.

## GENTIANACEAE Juss.
## GENTIAN FAMILY

Annuals or perennial herbs. Leaves usually opposite or whorled, simple, entire, often sessile. Flowers bisexual, radial, usually in racemes or panicles, sometimes solitary. Sepals 4-5, fused, persistent; petals 4-5, fused; stamens 4-5, adnate to corolla; pistil 1, ovary superior, locule 1, style with 2 stigmatic branches. Fruit a capsule; seeds many. Ca. 75 genera and 1200 species; temperate to tropical worldwide.

### CENTAURIUM Hill CENTAURY

Annuals. Leaves opposite. Flowers solitary or in panicles. Corolla salverform to funnelform, bright pink; stamens exserted, filaments filiform, anthers twisted after dehiscence; stigma oblong to fan-shaped. Ca. 30 species; temperate N. America and Europe, Chile, and Australia. Stems and leaves of several species were used medicinally by Native Americans.

GENTIANACEAE

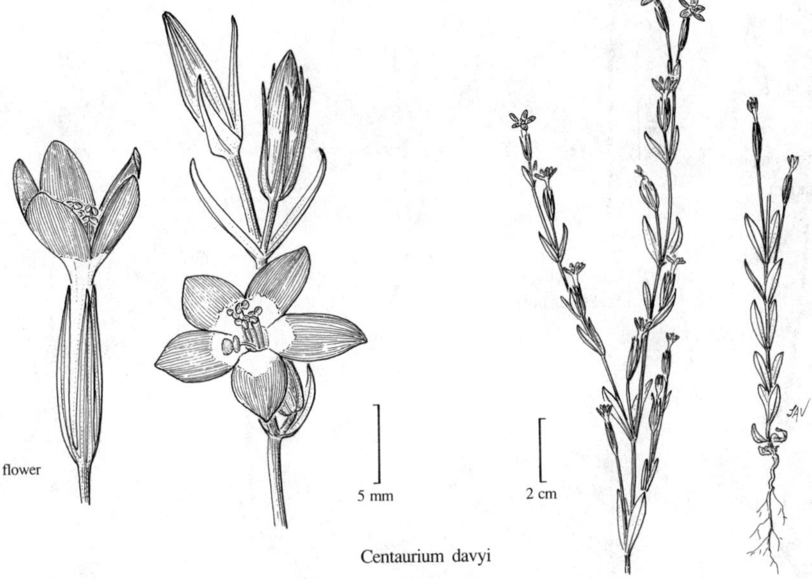


Centaurium davyi

**C. davyi** (Jeps.) Abrams Plants 3-30 cm tall. Leaves ovate to ± elliptic, 8-20 mm long. Sepals linear to narrowly oblong, 4-8 mm long; corolla tube 7-11 mm long, lobes 3-6 mm long; anthers 1-1.5 mm long before dehiscence; stigma fan-shaped. Apr-Jul. Occasional; grassy hillsides and rocky slopes; coastal scrub, grasslands, pine forest; 50-1300 ft (15-400 m). North side between Twin Harbors and Eagle Canyon; Central Valley near Christy Ranch, Christy Pines, and Valley Anchorage; s side near Christy Pines and Coches Prietos drainage. Santa Rosa and San Clemente islands; Humboldt Co. to coastal Santa Barbara Co.

# GERANIACEAE Juss.

## GERANIUM FAMILY

Annuals, perennial herbs, or shrubs. Leaves basal and cauline, opposite or alternate, simple to compound; stipules present. Flowers bisexual, radial to ± bilateral, usually in terminal cymes or umbels. Sepals 5, free; petals 5, free, sometimes with 5 basal nectar glands; stamens 5 or 10; pistil 1, usually 5-lobed, ovary superior, locules 5, styles 5 and fused to central axis or column, persistent to locules in fruit, locules and style segments falling as separate units (mericarps). Fruit a schizocarp composed of 5 mericarps; seeds 1-2 per mericarp. Ca. 14 genera and 730 species; mostly temperate, worldwide. Some species are cultivated for oils or as ornamentals.

1. Flowers ± bilateral; petals unequal in length, without basal nectar glands; upper sepal with a nectar spur adnate to pedicel ........................................ **Pelargonium**

1' Flowers radial, petals all alike, alternating with nectar glands; sepals without nectar spur

    2. Leaves pinnately veined to pinnatifid; fertile stamens 5, alternating with 5 staminodes; style segments spirally coiled in fruit ........................................ **Erodium**

    2' Leaves palmately lobed; fertile stamens 10, staminodes absent; style segments recurved in fruit ........................................ **Geranium**

### ERODIUM L'Hér. FILAREE or STORKSBILL

Annuals or biennials. Leaves simple to pinnately compound, basal ones in dense rosettes, cauline ones alternate or opposite. Flowers in umbels or umbel-like clusters. Corolla usually radial, petals alternating with scale-like nectar glands; stamens 5, alternating with 5 scale-like staminodes; style column elongate, segments spirally coiled in fruit, puberulent to hispidulous on side (inner) facing column. Ca. 75 species; temperate, ± worldwide.

1. Leaves orbicular to reniform, bases cordate, margins crenate to shallowly lobed ............. **E. macrophyllum**

1' Leaves oblong in outline, bases not cordate, lobed to pinnately compound

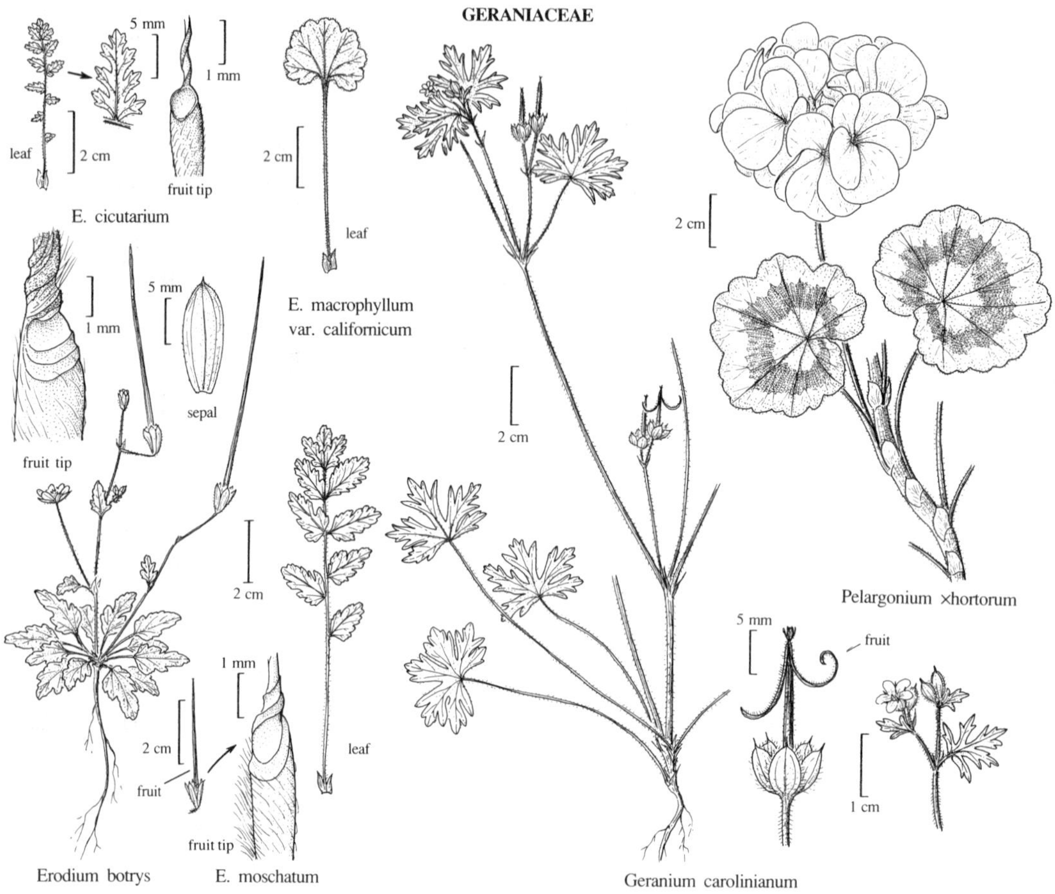


2. Lower leaves lobed to ± pinnatifid; style column 8-12.5 cm long ........ **E. botrys**
2' Lower leaves pinnately compound; style column 2.5-4 cm long
  3. Leaflets serrate to coarsely toothed; sepal tips with 1-2 white bristles; petal claws ciliate ........ **E. cicutarium**
  3' Leaflets pinnately lobed or divided; sepal tips without bristles; petal claws glabrous ........ **E. moschatum**

**E. botrys** (Cav.) Bertol. BROAD-LEAF FILAREE Annuals; stems 10-70 cm long. Leaves simple, shallowly to deeply lobed or pinnatifid, blades ovate to oblong-ovate in outline, lower ones 3-12 cm long, glabrous to sparsely puberulent; stipules ovate. Pedicels 15-25 mm long. Sepals 5-10 mm long, 13-15 mm long in fruit; petals 10-15 mm long, lavender; style column 8-12.5 cm long. Mericarps 8-10 mm long, with apical concave depression subtended by 2 transverse folds. Mar-May. Introduced. Common; disturbed sites, floodplains, grassy hillsides, rocky slopes, and open ridgetops; chaparral, coastal scrub, grasslands; 10-2400 ft (3-730 m). Widely scattered locations throughout island. Santa Rosa, San Nicolas, Santa Catalina, and San Clemente islands; throughout CA; native of Europe. First collected west of Stanton Ranch in April 1960.

**E. cicutarium** (L.) L'Hér. REDSTEM FILAREE Annuals; stems 10-50 long. Leaves pinnately compound, blades ovate to oblanceolate in outline, sparsely pubescent, lower ones 3-10 cm long; stipules lanceolate. Pedicels 8-12 mm long. Sepals 3-6 mm long, 5-8 mm long in fruit, densely hispid near base, apices with 1-2 white bristles; petals 5-7 mm long, red-lavender; style column 2.5-4 cm long. Mericarps 4-5 mm long, with apical concave depression, depression sometimes subtended by 1 transverse fold. Feb-Jun. Introduced. Abundant; disturbed and open sites; many plant communities; 10-1500 ft (3-460 m). Throughout island. All CA Channel islands; Los Coronados, Todos Santos, San Martin, Guadalupe, and Cedros islands; throughout N. America;

native of Europe and Asia. Reported by Rothrock (Wheeler 1876) and Greene (1887). Leaves were eaten by Native Americans.

**E. macrophyllum** Hook. & Arn. Annuals or biennials, usually scapose, 4-15 cm tall. Leaves simple, blades orbicular to reniform, 1-3 cm long, puberulent, margins crenate to shallowly lobed; stipules ovate. Pedicels 3-10 mm long. Sepals 4-6 mm long, 10-12 mm long in fruit, puberulent; petals 6-8 mm long, white, tinged red to lavender; style column 3-4 cm long. Mericarps 8-10 mm long, with apical, concave, puberulent depression. Apr. Rare, not seen recently; habitat, plant community, and elevation unknown. Known from a single collection without locality by T. S. Brandegee in 1888. Great Valley of CA to n Mexico, e to UT.

**E. moschatum** (L.) L'Hér. WHITESTEM FILAREE Annuals or biennials; stems 10-60 cm long. Leaves pinnately compound, blades oblong to obovate in outline, 3-12 cm long, puberulent to strigose, leaflets serrate to lobed; stipules ovate. Pedicels 6-20 mm long. Sepals 3-6 mm long, 6-8 mm long in fruit; petals 7-12 mm long, red to violet, claws glabrous; style column 2-4 cm long. Mericarps 4-5 mm long, with apical, concave, depression subtended by 1 transverse fold. Mar-May. Introduced. Occasional; disturbed and open sites; many plant communities; 10-1500 ft (3-460 m). Widely scattered locations throughout island. All CA Channel islands; Los Coronados, Todos Santos, San Martin, and Guadalupe islands; throughout CA; native of Europe. First reported by Greene (1887).

### **GERANIUM** L. GERANIUM or CRANESBILL

Annuals or perennial herbs. Leaves simple to deeply divided, with palmate veins or divisions, basal ones in rosettes, cauline ones alternate or opposite. Flowers in few-flowered umbel-like cymes. Corolla radial, petals alternating with scale-like nectar glands, upper 2 usually longer than lower 3; stamens 10; style column elongate, segments spirally erect to recurved in fruit, glabrous to puberulent on side facing column. Ca. 250 species; temperate, ± worldwide.

**G. carolinianum** L. CAROLINA CRANESBILL Annuals; stems erect, 2-6 dm tall. Leaf blades 1-3.5 cm wide, with 5-7 oblong to wedge-shaped segments. Pedicels 2-7 mm long. Sepals 5-7 mm long, apices mucronate, with short bristle; petals 4.5-7 mm long, white to pink, apices sometimes notched; style column 10-14 mm long. Mericarps 2-3 mm long. Mar-May. Scarce; disturbed sites, grassy hillsides, and canyon bottoms; grasslands, oak woodland; 40-550 ft (12-170 m). North side between Pelican Bay and Prisoners' Harbor; Central Valley in Cañada de la Mina; s side in Coches Prietos drainage. Santa Rosa and Santa Catalina islands; throughout N. America. This species was used medicinally by Native Americans.

### **PELARGONIUM** L'Hér. GARDEN GERANIUM

Annuals, perennial herbs, or shrubs, often strong-smelling. Leaves simple to deeply divided, with palmate veins or divisions, lower ones alternate, upper ones alternate to opposite. Flowers in umbels. Upper sepal with a nectar spur adnate to pedicel; corollas ± radial to bilateral, petals without basal nectar glands; stamens 10, some sterile; style column elongate, segments spirally erect to recurved in fruit, puberulent to hispidulous on side facing column. Ca. 250 species; temperate, s Africa and Australia. Many species are cultivated as ornamentals or for aromatic oils. Leaves of several species are used for food.

**P. ×hortorum** L.H.Bailey Subshrubs; stems 3-5 dm tall, with soft wood. Leaf blades orbicular to reniform, 12-20 cm wide, often with different colors of circular bands, margins crenate. Flowers 13-25 per inflorescence. Sepals 5-8 mm long; petals 15-20 mm long, red; style column 25-30 mm long. Mericarps fusiform, 4-6 mm long. May-Nov. Introduced. Scarce, persisting near plantings; rocky slopes; coastal scrub; 40-250 ft (12-75 m). North side at Fry's Harbor, Pelican Bay, Prisoners' Harbor, and Scorpion Ranch; Central Valley near Christy Ranch. Santa Rosa, San Nicolas, Santa Catalina, and San Clemente islands; Todos Santos Island; Santa Barbara Co. to San Diego Co. A fertile hybrid derived from *P. inquinans* (L.) L'Hér. and *P. zonale* (L.) L'Hér., native of s Africa.

## **GROSSULARIACEAE** DC.
## CURRANT OR GOOSEBERRY FAMILY

With only 1 genus.

### **RIBES** L. CURRANT or GOOSEBERRY

Shrubs; stems spiny or not. Leaves alternate, often clustered on short, lateral twigs, simple, usually orbicular to cordate, palmately veined and lobed; stipules present. Flowers bisexual, radial, in racemes. Hypanthium tube

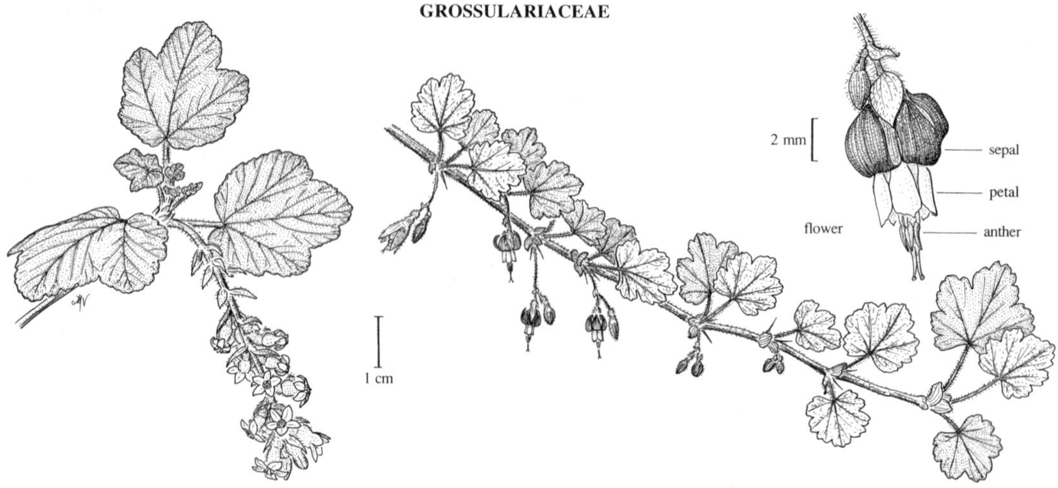


Ribes malvaceum var. malvaceum

Ribes thacherianum

present; sepals 5, often petaloid; petals 5; stamens 5, alternate petals; ovary inferior, locule 1, style 1, with 2 stigmatic lobes. Fruit a glabrous to spiny berry. Ca. 150 species; temperate N. and S. America, Eurasia. Some species are cultivated as ornamentals or for fruit.

1. Spines absent from stems; inflorescence a raceme of 8-21 flowers; fruit puberulent ........................................ **R. malvaceum** var. **malvaceum**

1' Spines present at stem nodes; inflorescence of 1-2 flowers; fruit bristly to spiny ............... **R. thacherianum**

**R. malvaceum** Sm. var. **malvaceum** CHAPARRAL CURRANT Plants 1-2.5 m; stems without spines. Leaves 3-8 cm long, upper surfaces dark olive-green, rugose, glandular-puberulent, lower surfaces gray-tomentulose. Raceme with 8-21 flowers. Hypanthium bright rose; sepals 3-6 mm long. Fruit 5-7 mm wide, purple-black, glaucous, puberulent. Jan-Apr. Scarce; coastal bluffs, shaded slopes, and canyon bottoms; coastal scrub, pine forest; 30-1300 ft (10-400 m). North side at mouth of Eagle Canyon; Central Valley in upper part of Cañada Christy; s side near Christy Pines and on Sierra Blanca ridge. Santa Rosa, Anacapa, and San Clemente islands; n CA coast ranges to Los Angeles Co. Fruits were eaten by Native Americans; roots were used medicinally. Var. *viridifolium* Abrams, with upper surfaces of leaves bright green, occurs in the mountains of mainland s CA.

**R. thacherianum** (Jeps.) Munz [*R. menziesii* Pursh var. *t.* Jeps.] SANTA CRUZ ISLAND GOOSEBERRY Plants 0.5-2.5 m tall; stem nodes with 1-2 spines. Leaves 2.5-7 cm long, upper surfaces green, smooth, ± glabrous, lower surfaces pilose to strigose. Raceme with 1-3 flowers. Hypanthium pink; sepals 7-10 mm long. Fruit 7-10 mm wide, purple, bristly to spiny. (Nov) Feb-Jun (Sep). Endemic to Santa Cruz Island. A species of special environmental concern. Occasional; shaded slopes, canyon walls, and floodplains; oak woodland, pine forest, riparian woodland; 30-1400 ft (10-425 m). North side between Valdez and Diablo canyons; Central Valley in upper part of Cañada Christy; s side in upper part of Cañada de los Sauces, on Sierra Blanca ridge, and in upper part of Laguna Canyon. Fruits are edible.

## HYDROPHYLLACEAE R.Br.
### Waterleaf Family

Annuals, perennial herbs, or shrubs. Leaves cauline, alternate or opposite, simple to pinnatifid; stipules absent. Flowers bisexual, radial, solitary or in racemes or spikes, spikes often coiled in bud and one-sided, uncoiling in fruit. Sepals 5, ± fused, sometimes alternating with sepaloid appendages at base of sinuses; petals 5, fused; stamens 5, alternate and adnate to corolla lobes, included or exserted; pistil 1, ovary superior, locule 1, sometimes appearing as 2 by intrusion of placentas, styles 1-2. Fruit a capsule; seeds 1-many. Ca. 20 genera and 300 species; worldwide, but mostly N. America.

1. Ovary apparently with 2 locules
   2. Flowers pendulous; corollas pale yellow, persistent in fruit ........................................... **Emmenanthe**
   2' Flowers erect; corollas white, blue, or purplish, ± deciduous in fruit ..................................... **Phacelia**

1' Ovary with 1 locule
   3. Stems and leaves viscid, ± strongly scented; calyx without appendages between lobes; ovules borne on both sides of placenta .......................................................................................................... **Eucrypta**
   3' Stems and leaves not viscid, not scented; calyx with sepaloid appendages between lobes; ovules borne on one side of placenta
      4. Stems and fruits puberulent or glabrous, but trichomes neither bristly nor prickly; seeds with colorless appendage at one end ........................................................................................... **Nemophila**
      4' Stems and fruits with bristle-like or prickly trichomes; seeds without appendages ...... **Pholistoma**

## **EMMENANTHE** Benth. WHISPERING BELLS

With only 1 species.

**E. penduliflora** Benth. var. **penduliflora** Annuals, to 6 dm tall; stems erect, glandular and villous. Leaves alternate, oblong in outline, 3-10 cm long, toothed to pinnately lobed, lower ones sessile to subsessile. Inflorescence terminal; flowers pendulous; pedicels filiform. Calyx lobes lanceolate to ovate, 4.5-7 mm long, glandular and villous, sepaloid appendages absent; corollas campanulate, 7-12 mm long, pale yellow, persistent in fruit; stamens included; ovary with ± 2 locules. Fruits oblong in outline, 4-7 mm long; seed surface reticulate, reddish brown. Mar-May. Scarce; burned or disturbed sites, coastal bluffs, rocky slopes, and floodplains; chaparral, coastal scrub; 50-250 ft (15-75 m). North side at Pelican Bay (after 1931 fire); Central Valley near University of California Field Station and Rancho del Sur; s side between mouth of Laguna Canyon and Valley Anchorage. Santa Catalina and San Clemente islands; Guadalupe Island; throughout CA, e to UT and AZ. Var. *rosea* Brand, with white to reddish corollas, occurs in the mainland mountains from San Francisco to Mount Pinos, Ventura Co. Usually seen only in years with heavy rainfall or after fires.

## **EUCRYPTA** Nutt.

Annuals, viscid, ± strongly scented; stems erect, simple or branched. Leaf blades toothed to pinnatifid, lower leaves opposite, petiolate, upper ones alternate, sessile to clasping. Inflorescence axillary or terminal; flowers ± erect; pedicels filiform. Calyx lobes oblong to spatulate; corollas campanulate, white, yellow, or bluish, deciduous in fruit; stamens included; ovary with 1 locule. Fruits globose to ovoid; seeds smooth to wrinkled, brown to black. Two species; sw N. America.

**E. chrysanthemifolia** (Benth.) Greene var. **chrysanthemifolia** Plants 1-5 dm tall. Leaves oblong to ovate in outline, 3-10 cm long, pinnatifid, primary segments 5-9, each deeply lobed. Calyx lobes ovate, 1.5-3 mm long, hirsute; corollas 3-5 mm long, yellowish white. Fruits 2-3.5 mm wide, enclosed by calyx; seeds of 2 kinds, either orbicular and smooth or ovoid and wrinkled. Mar-Jun. Occasional; shaded n-facing slopes, coastal bluffs, canyon bottoms, floodplains; chaparral, coastal scrub, oak woodland, pine forest; 10-1300 ft (3-400 m). Widely scattered locations throughout island. All CA Channel islands except San Nicolas; Los Coronados, Todos Santos, and Cedros islands; San Francisco Bay region and Great Valley to n Baja CA. Var. *bipinnatifida* (Torr.) Constance, with shorter corolla, occurs on Guadalupe Island and on mainland e to NV and AZ.

## **NEMOPHILA** Nutt.

Annuals; stems simple to diffusely branched, glabrous to sparsely pubescent. Leaves alternate or opposite, petiolate, blades toothed to lobed. Flowers solitary, axillary; pedicels recurved. Calyx lobes alternating with spreading to reflexed sepaloid appendages; corollas rotate to campanulate, white, blue, or purple, deciduous in fruit; stamens included; ovary with 1 locule. Fruits globose to ovoid; seeds smooth, wrinkled, or pitted, with colorless appendage at one end. Ca. 11 species; temperate w and se N. America.

**N. pedunculata** Douglas. ex Benth. Stems to 3 dm tall, sharply angled. Leaves opposite, oblong to ovate in outline, 0.5-3.5 cm long, deeply, pinnately lobed, appressed-hispid. Calyx lobes lanceolate to ovate, 2-4 mm long, hirsute, sepaloid appendages 1.5-3 mm long; corollas 3.5-6 mm long, white, usually with blue lines or spots. Fruit 3-5 mm long; seeds smooth to pitted. Mar-May. Occasional; shaded n-facing slopes and canyon bottoms, grassy hillsides; chaparral, grasslands, oak woodland; 15-2200 ft (5-670 m). North side between Hazard's Anchorage and Prisoners' Harbor, especially near Picacho Diablo; Central Valley near Lagunitas Secas and in

**HYDROPHYLLACEAE: Emmenanthe-Nemophila**

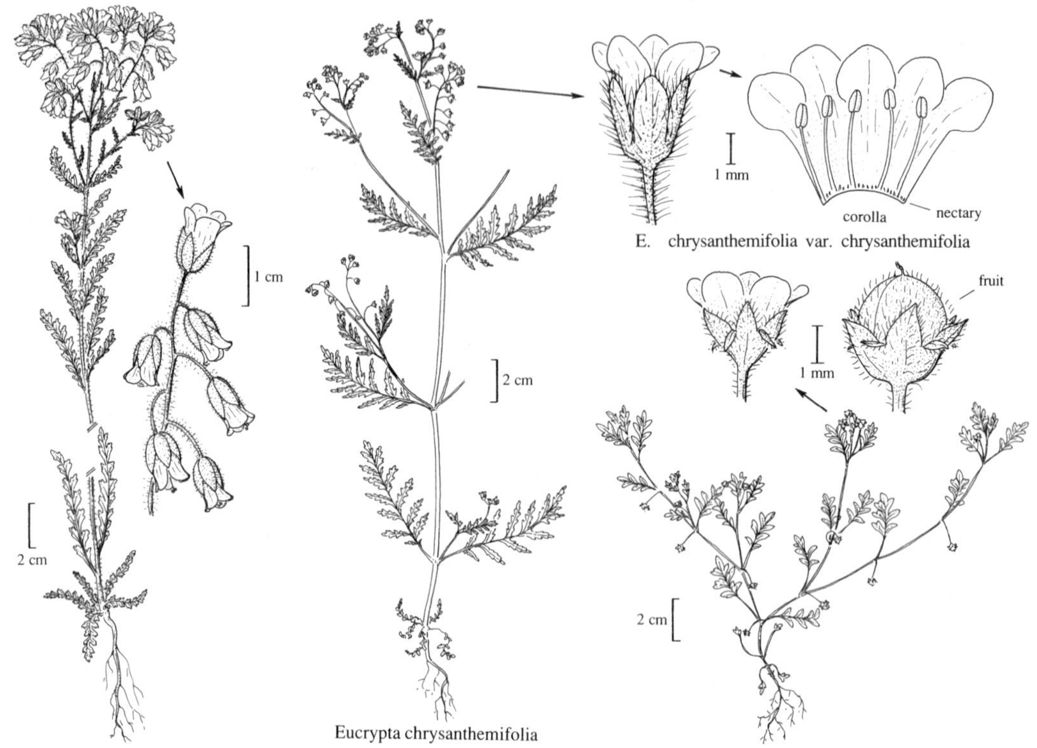


Emmenanthe penduliflora

Eucrypta chrysanthemifolia var. chrysanthemifolia

Nemophila pedunculata

Cañada de la Portezuela. San Miguel, Santa Rosa, and San Nicolas islands; British Columbia to Baja CA, e to ID and NV.

## PHACELIA Juss.

Annual or perennial herbs; stems erect to prostrate, simple or branched. Leaves alternate, entire to pinnatifid, petiolate to sessile, pubescent to hispid. Inflorescence axillary or terminal, usually one-sided, coiled in flower; flowers ± erect, sessile to pedicellate. Calyx lobes linear to spatulate; corollas rotate to campanulate, white, yellow, or blue to purple, deciduous or persistent in fruit; stamens included or exserted; ovary with ± 2 locules. Fruits globose to ovoid; seeds smooth to wrinkled, gray or brown. Ca. 175 species; N. and S. America.

1. Perennial herbs, often suffrutescent; stems often decumbent to sprawling, with stiff, pustulate trichomes ........ **P. ramosissima** var. **austrolitoralis**

1' Annuals; stems usually erect, glandular to hispid, but trichomes not pustulate

 2. Leaves irregularly toothed, entire to shallowly lobed; corollas 10-15 mm long; pedicels 3-10 mm long in fruit ........ **P. viscida**

 2' Leaves deeply pinnately lobed to pinnatifid; corollas 6-13 mm long; flowers subsessile, pedicels to 3 mm long in fruit

  3. Stems densely long-hispid; calyx lobes linear to narrowly oblanceolate in fruit; fruit minutely pustulate ........ **P. cicutaria**

  3' Stems hispidulous to glabrate; calyx lobes lanceolate to oblanceolate; fruit sericeous ... **P. distans**

**P. cicutaria** Greene CATERPILLAR PHACELIA. Annuals, 2-8 dm tall; stems densely long-hispid. Leaves lanceolate to ovate in outline, 2-15 cm long, deeply pinnately lobed to pinnatifid. Flowers in dense coiled spikes; pedicels 0.5-3 mm long. Calyx 6-10 mm long in flower, 9-12 mm long in fruit, lobes linear to narrowly oblanceolate; corollas broadly campanulate, 7-13 mm long, yellowish to lavender or bluish. Fruits ± globose, 2.5-4 mm long, hispidulous to ± pustulose. Mar-Jun. Common; coastal bluffs, rocky slopes, shaded canyon bottoms;

chaparral, coastal scrub, oak woodland; 10-850 ft (3-260 m). Widely scattered locations throughout island. Santa Rosa, Anacapa, Santa Catalina, and San Clemente islands; Todos Santos Island; Monterey Co. to n Baja CA, e to Sierra Nevada foothills and Mojave Desert. Var. *cicutaria* occurs in the Sierra Nevada foothills.

1. Inflorescence rachis with stiff trichomes; fruit apex ± pustulose .......... var. **hispida**
1' Inflorescence rachis with ± wavy trichomes; fruit apex hispidulous .......... var. **hubbyi**

var. **hispida** (A.Gray) J.T.Howell Inflorescence ± open at base in fruit, rachis with stiff trichomes; fruit apex ± pustulose. Mar-Jun. Common; coastal bluffs, rocky slopes, shaded canyon bottoms; chaparral, coastal scrub, oak woodland; 10-850 ft (3-260 m). Widely scattered locations throughout island. Santa Rosa, Anacapa, Santa Catalina, and San Clemente islands; Todos Santos Island; Monterey Co. to n Baja CA, e to Sierra Nevada foothills and Mojave Desert.

var. **hubbyi** (J.F.Macbr.) J.T.Howell Inflorescence dense in fruit, rachis with stiff trichomes; fruit apex hispidulous. May. Scarce; coastal bluffs, loose talus slopes; coastal scrub; 10-100 ft (3-30 m). North side at China Harbor; s side at Coches Prietos Anchorage and Smugglers' Cove. Santa Barbara to Los Angeles Co.

**P. distans** Benth. Annuals, 2-8 dm tall; stems hispidulous to glabrate. Leaves lanceolate to ovate in outline, 2-12 cm long, deeply pinnately lobed to pinnatifid. Flowers in dense coiled spikes, subsessile. Calyx 5-7 mm long, 6-9 mm long in fruit, lobes lanceolate to oblanceolate; corollas campanulate to funnelform, 6-8 mm long, blue to pale blue. Fruits ± globose, 2-3 mm long, sericeous. Mar-Jun. Occasional; coastal bluffs, loose talus slopes, rocky slopes, grassy hillsides; coastal scrub, grasslands; 50-650 ft (15-200 m). Widely scattered locations around perimeter of island. All CA Channel islands except San Nicolas; Los Coronados and Todos Santos islands; Great Valley to n Mexico, e to NV.

**P. ramosissima** Lehm. var. **austrolitoralis** Munz Suffrutescent perennials; stems prostrate to ascending, 3-10 dm tall, with pustulate trichomes. Leaves oblong to broadly ovate in outline, 4-15 cm long, deeply lobed, lobes irregularly toothed to shallowly lobed. Flowers in dense coiled spikes, subsessile. Calyx 5-7 mm long in flower, 6-8 mm long in fruit, lobes lanceolate to oblanceolate; corollas campanulate, 5-8 mm long, white to purplish. Fruits ovoid, 2.5-4 mm long, puberulent. Mar-Jun. Scarce; coastal bluffs, sandy flats; coastal dunes, coastal scrub; 25-200 ft (6-60 m). Widely scattered locations around perimeter of island. Santa Rosa Island; San Francisco Bay region to s CA. Five other varieties occur on the mainland and other Channel Islands.

**P. viscida** (Benth.) Torr. Annuals, 1.5-7 dm tall. Leaves ± ovate in outline, 2-9 cm long, irregularly toothed or pinnately lobed. Flowers in racemes, becoming open in fruit; pedicels 3-10 mm long. Calyx 6-8 mm long in flower, 7-11 mm long in fruit, lobes linear to oblanceolate; corollas rotate to broadly campanulate, 10-15 mm long, lavender or blue with white center. Fruits ovoid, 8-13 mm long, glandular-puberulent. Mar-Jun. Occasional; coastal bluffs, dry rocky slopes, floodplains; chaparral, coastal scrub; 20-1000 ft (6-300 m). Widely scattered locations throughout island. San Miguel, Santa Rosa, Anacapa, and Santa Catalina islands; Monterey Co. to n Baja CA. Can be locally abundant in years with heavy rainfall or after fires. Foliage can cause contact dermatitis in humans.

## PHOLISTOMA Lilja ex Lindblom

Annuals; stems simple to diffusely branched, decumbent to prostrate, with bristle-like or prickly hairs. Upper leaves opposite, lower ones alternate, petioles ± winged, clasping, blades pinnately lobed. Flowers solitary or in open clusters, axillary; pedicels straight to recurved. Calyx lobes alternating with spreading to reflexed sepaloid appendages; corollas rotate, white or blue to purple, deciduous in fruit; stamens included; ovary with 1 locule. Fruits globose to ovoid, prickly or bristly; seeds smooth, wrinkled, or pitted, brown to black. Ca. 3 species; sw N. America.

1. Petioles broadly winged and auriculate-clasping; corollas 10-30 mm wide, blue to purple; styles 5-8 mm long .......... **P. auritum**
1' Petioles narrowly winged, scarcely if at all auriculate-clasping; corollas 6-10 mm wide, white; styles 1-3 mm long .......... **P. racemosum**

**P. auritum** (Lindl.) Lilja var. **auritum** FIESTA FLOWER Stems to 10 dm long. Leaves to 15 cm long, blades oblong to lanceolate in outline, petioles broadly winged and auriculate clasping. Calyx 6-10 mm long, lobes lanceolate to lanceolate-ovate; corollas 7-20 mm long, 10-30 mm wide, blue to purple; styles 5-8 mm long. Fruits 6-9 mm long. Mar-May. Scarce; canyon bottoms, coastal bluffs, shaded n-facing slopes, grassy hillsides; chaparral, coastal scrub, grasslands, oak woodland; 6-1300 ft (2-400 m). North side near Prisoners' Harbor, in Cañada del Puerto, and near China Harbor. Santa Barbara, Santa Catalina, and San Clemente islands; Los Coronados

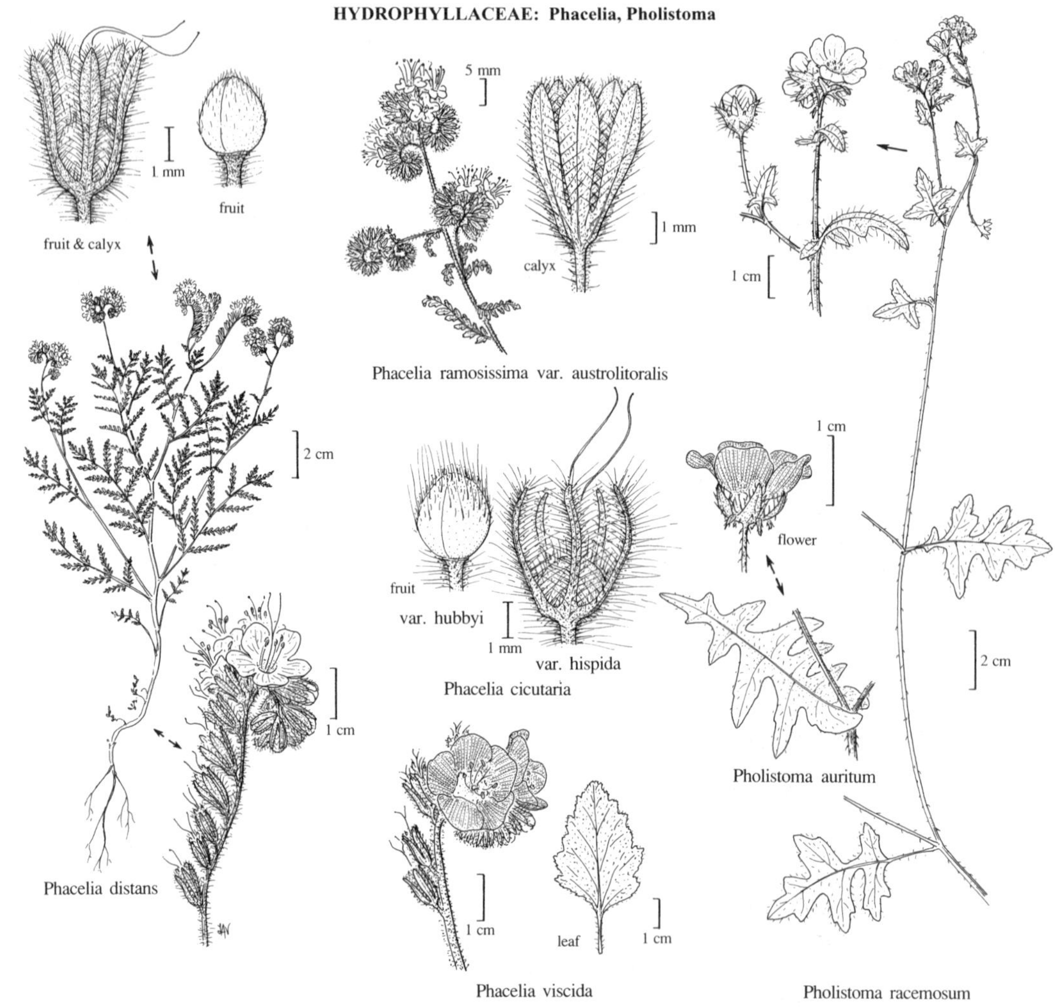


and Todos Santos islands; Great Valley to s CA, e to AZ. Variety *arizonicum* (M.E.Jones) Constance, with corollas 3-7 mm wide, occurs in se CA, AZ, and on Cedros Island.

**P. racemosum** (Nutt.) Constance Stems to 6 dm long. Leaves to 12 cm long, lanceolate-ovate to ovate in outline, petioles narrowly winged, scarcely if at all auriculate-clasping. Calyx 5-7 mm long, lobes lanceolate-ovate; corollas 7-10 mm long, 6-10 mm wide, corollas white, aging brownish-orange; styles 1-3 mm long. Fruits 5-7 mm long. Mar-May. Occasional; shaded n-facing hillsides and canyon bottoms; chaparral, grasslands, oak woodland; 15-1200 ft (5-365 m). North side between Lady's Cove and Prisoners' Harbor and in Scorpion Canyon; Central Valley in Cañada de la Portezuela. Santa Barbara, Santa Catalina, and San Clemente islands; all Baja CA islands except San Geronimo and San Benito; Santa Barbara Co. to n Baja CA.

## **LAMIACEAE** Lindl. [**Labiatae** Juss.]
### MINT FAMILY

Annuals, perennial herbs, or shrubs, strongly scented; stems usually square in cross-section. Leaves opposite, simple to dissected. Flowers bisexual, ± bilateral, sometimes solitary, but usually in dense, capitate clusters surrounding stem, clusters axillary and terminal, disposed in raceme- or spike-like inflorescences. Calyx radial, often with 5 lobes, or bilateral, persistent in fruit; petals fused, corollas usually bilabiate; stamens 2 or 4 and didynamous; pistil 1, ovary superior, usually deeply 4-lobed, locules 2, ovules 4, style 1, inserted between lobes.

Fruit ± a schizocarp, separating into 1-4 nutlets. Ca. 220 genera and ca. 5600 species; worldwide. Many Mediterranean species are cultivated for volatile oils, which are used in medicine and as flavorings for foods or teas.

1. Calyx ± bilateral, irregularly 3-5-lobed or with 2 lips
   2. Calyx with 2 lips, lips entire, without spinescent lobes, tube with dorsal crest; fertile stamens 4 .......... **Scutellaria**
   2' Calyx irregularly 3-5-lobed, some or all with spinescent tips, tube cylindrical, without dorsal crest; fertile stamens 2 .......... **Salvia**

1' Calyx ± radial, lobes ± equal in length and shape
   3. Calyx lobes 10, tips spinose, spreading or recurved to hooked .......... **Marrubium**
   3' Calyx lobes 5, acute to spinescent, tips ± straight
      4. Flowers in short racemes or solitary, in axils of leaves
         5. Shrubs; corollas 30-35 mm long; leaves broadly lanceolate to ovate .......... **Lepechinia**
         5' Perennial herbs with trailing stems; corollas 8-10 mm long; leaves broadly ovate to ± orbicular .......... **Satureja**
      4' Flowers in capitate clusters, either subtended by axillary leaves or in terminal spike-like inflorescences
         6. Flower clusters subtended by sessile leaves or leaf-like bracts; nutlets angular, with 3 sides, apices truncate .......... **Lamium**
         6' Flower clusters in terminal spike-like inflorescences, subtended by reduced bracts; nutlets ± ovoid, apices rounded
            7. Corollas 4-5 mm long; calyx lobes without spinose tips .......... **Mentha**
            7' Corollas 15-20 mm long; calyx lobes with spinose tips .......... **Stachys**

## **LAMIUM** L. HENBIT or DEAD-NETTLE

Annuals or perennial herbs, not strongly scented. Flowers in axillary or terminal capitate clusters, subtended by leaves or leaf-like bracts. Calyx ± radial, tube with 5 veins and 5 sharp-pointed lobes; corollas bilabiate, upper lip galeate, entire, lower lip with 3 lobes, lateral lobes broad and shorter than central lobes; stamens 4, ascending under upper lip, lower pair longer than upper one; anthers conspicuously pubescent. Nutlets angular, with 3 sides, smooth or tuberculate, apices truncate. Ca. 40 species; temperate Europe, Asia and n Africa. Some species used medicinally.

**L. amplexicaule** L. HENBIT Annuals, 10-25 cm tall; stems branched from base, decumbent, sparsely pubescent. Leaves broadly ovate to ± orbicular, 0.8-2.5 cm long, 0.7-2.5 cm wide, bases truncate to cordate, margins coarsely crenate, upper ones sessile, clasping, lower ones with petioles 8-15 mm long. Calyx 5-6 mm long, hirsute, lobes erect, equal to tube; corollas 12-15 mm long, puberulent, reddish purple. Nutlets 2-2.5 mm long, usually mottled, brownish. Feb-Apr. Introduced. Rare, not seen recently; shaded floodplains, roadsides, disturbed sites; chaparral, oak woodland; 500-1000 ft (150-305 m). Central Valley in Cañada de la Portezuela and in Cañada de la Mina. Throughout N. America; native of Europe. First collected at Rancho Portezuelo in March 1932.

## **LEPECHINIA** Willd. PITCHER SAGE

Perennial herbs or shrubs, strongly aromatic. Flowers in short racemes, lower bracts leaf-like, gradually reduced in size upward. Calyx ± radial, campanulate, tube with 10 veins and 5 unequal lobes, usually inflated or enlarged in fruit; corollas bilabiate, upper lip with 4 broad, flat, erect to slightly reflexed lobes, lower lip entire, longer than upper one, usually parallel to long axis of tube; stamens 4, in 2 pairs, included. Nutlets ovoid, smooth, apices rounded. Ca. 30 species; w N. America, S. America, and Hawaii.

**L. fragrans** (Greene) Epling FRAGRANT PITCHER SAGE Shrubs, to 2 m tall. Lower leaf blades broadly lanceolate to deltate, 2-10 cm long, 1-5 cm wide, bases truncate to subsagittate, margins crenate, apices acute to obtuse, surfaces conspicuously veined, rugose, villous to tomentose, with petioles 5-15 mm long; upper leaves sessile, reduced in size. Pedicels 5-15 mm long. Calyx 12-18 mm long in flower, villous, lobes lanceolate, usually longer than tube, 20-25 mm long in fruit, membranous and reticulate with purple veins; corollas 30-35 mm long, yellowish in bud, pale lavender when open. Nutlets ca. 4 mm long, black. May-Oct. A species of special environmental concern. Rare; dry ravines, rocky slopes, exposed ridgetops; chaparral; 200-1250 ft (60-380 m). North side just w of Pelican Bay; s side in upper Willows Canyon and on El Tigre ridge. Santa Rosa and Santa Catalina islands; Santa Monica and San Gabriel Mountains.

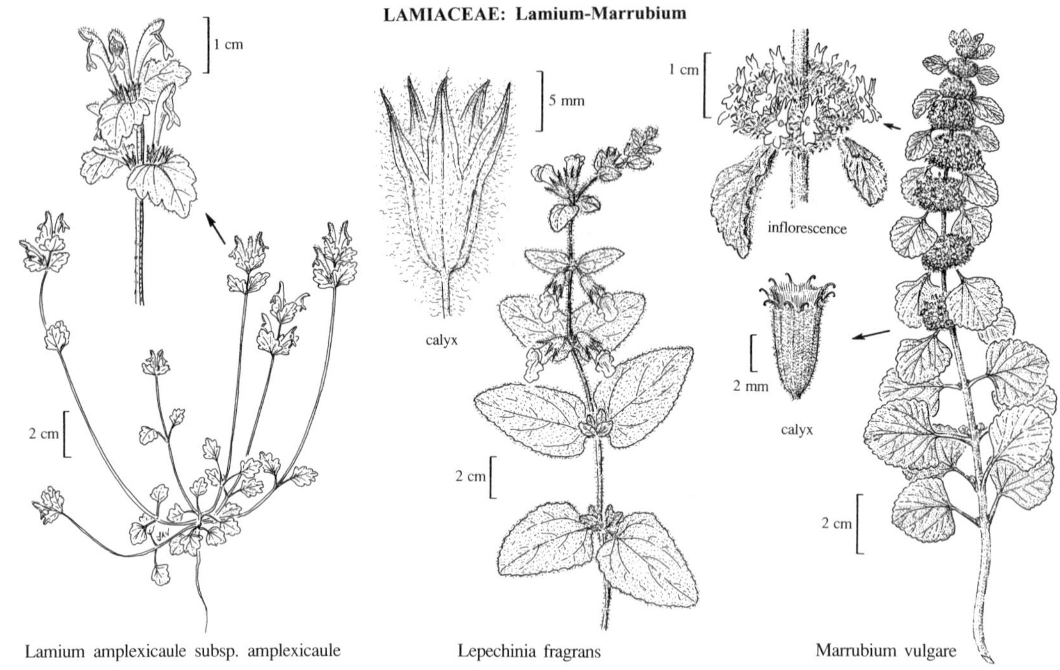


## MARRUBIUM L. HOREHOUND

Perennial herbs, not strongly scented. Flowers in capitate clusters, axillary or terminal, subtended by leaves or leaf-like bracts. Calyx ± radial, tube with 5-10 veins and 5-10 sharp-pointed, spreading to recurved lobes; corollas bilabiate, upper lip erect, flat, with 2 lobes, lower lip with 3 lobes, spreading, lateral lobes smaller than central lobes; stamens 4, included in tube, lower pair longer than upper one. Nutlets smooth, apices truncate. Ca. 30 species; temperate Europe.

**M. vulgare** L. Plants 0.6-1 m tall; stems several, decumbent to ± erect, conspicuously white-woolly. Leaf blades broadly elliptic to broadly ovate, 2-4 cm long, rugose, canescent-tomentose, often glabrate; petioles 1-2 cm long. Calyx tube 4.5-5.5 mm long, with 10 veins and lobes, lobes ca. 2 mm long, usually recurved or hooked at tip; corollas 5-7 mm long, white. Nutlets ca. 2 mm long, brownish. Apr-Oct. Introduced. Common; disturbed flats, roadsides, canyon bottoms, ridgetops; chaparral, coastal scrub, grasslands, oak woodland; 10-1500 ft (3-460 m). Scattered locations throughout island. All CA Channel islands except Anacapa and Santa Barbara; Todos Santos Island; throughout N. America; native of Europe. First collected near Pelican Bay and in "canyon west of Gull Rock" in June 1930. This species is used medicinally and to make horehound candy.

## MENTHA L. MINT

Perennial herbs, often with rhizomes, strongly scented. Flowers in capitate clusters, axillary or terminal, often in spike-like inflorescences, subtended by reduced bracts. Calyx ± radial, tube with 10 veins and 5, sharp-pointed, ciliate lobes; corollas ± bilabiate, lips usually equal, upper one slightly lobed, lower one with 3 lobes; stamens 4, included in tube or exserted, pairs usually equal. Nutlets ovoid to ellipsoid, smooth. Ca. 25 species; temperate, N. America and Europe.

**M. spicata** L. SPEARMINT Plants 0.3-1 m tall, glabrous to sparsely pubescent with retrorse trichomes. Leaf blades elliptic to lanceolate-ovate, 3-4 cm long, bases rounded to truncate, margins serrate, apices acute; petioles less than 4 mm long. Pedicels 0.5-1 mm long. Calyx 2.5-3 mm long, glandular, lobes ca. 1 mm long; corollas 4-5 mm long, glabrous, white to pink. Jul-Oct. Introduced. Rare; disturbed sites, moist flats; grasslands; 10 ft (3 m). North side at Prisoners' Harbor. Santa Catalina Island; throughout N. America; native of Europe. First collected in August 1976. This species is the commercial source of spearmint oil.

**LAMIACEAE: Mentha-Stachys**

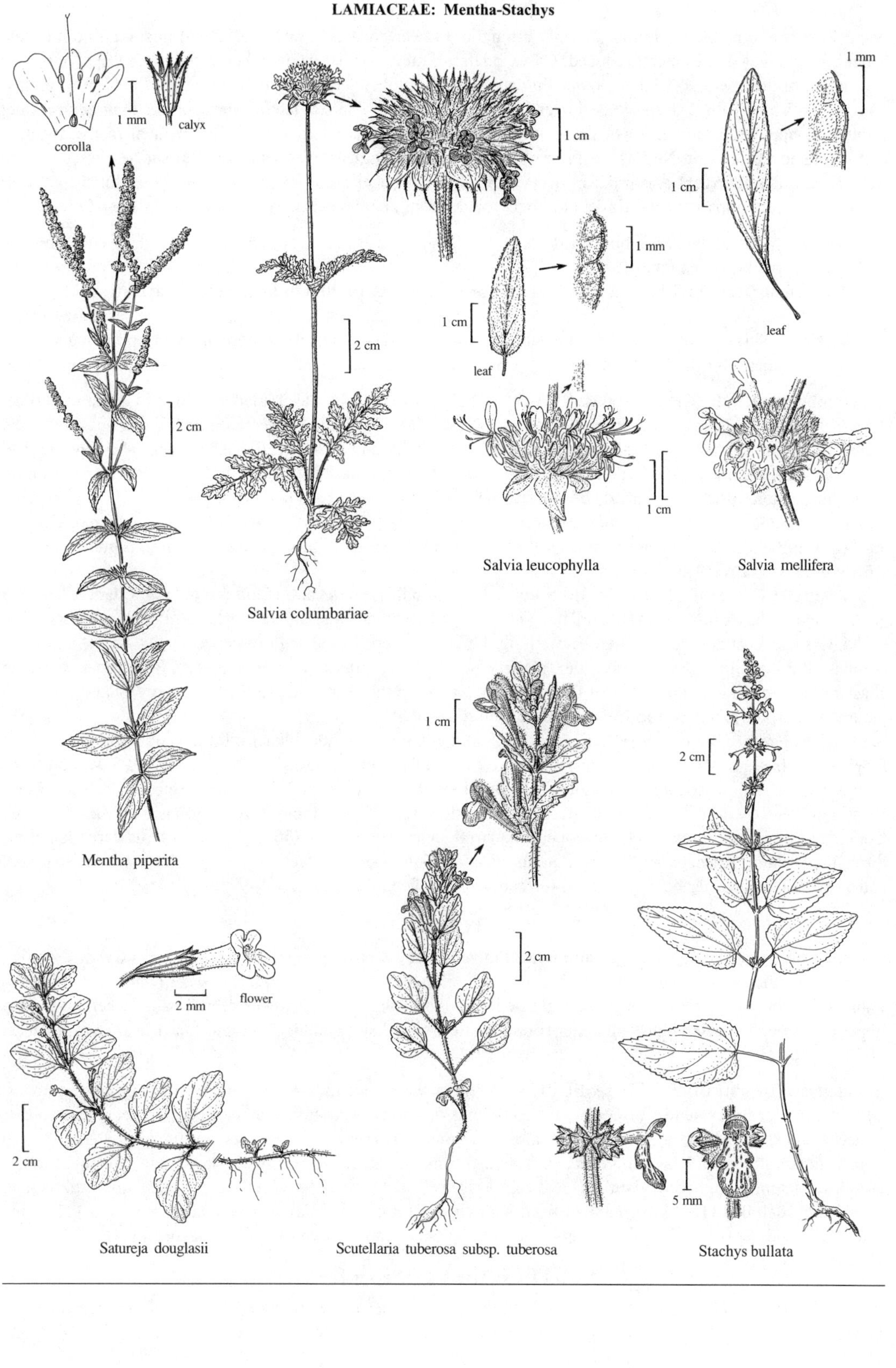

## SALVIA L. SAGE

Annual, perennial herbs, or shrubs, strongly aromatic. Flower clusters capitate, arranged in spike-like inflorescences; bracts leaf-like to much reduced. Calyx bilateral, tube with 10 veins, lobes with spinose tips, upper lip concave or arched, with 1-3 lobes, lower lip usually with 2 lobes; corollas bilabiate, upper lip erect or arched forward, entire or with 2 deep lobes, lower lip spreading, with 3 lobes, lateral ones shorter than central one; stamens 4, upper pair sterile, lower pair with anther sacs well separated, attached to filament-like connective that is jointed to filament. Nutlets ovoid to oblong in outline, angled to smooth. Ca. 900 species; temperate to tropical, worldwide except Australia. Flowers of several CA species are favored by honeybees for their nectar and consequently impart special flavors to honey made from them; seeds of all island species are edible.

1. Annuals; most leaves basal, pinnatifid ........ **S. columbariae**
1' Shrubs; most leaves cauline, entire
    2. Leaf surfaces grayish, with branched trichomes; corollas 10-20 mm long, reddish lavender ........ **S. leucophylla**
    2' Upper surfaces of leaves green, glabrous, lower surfaces densely pubescent with appressed, simple trichomes; corollas 9-12 mm long, bluish ........ **S. mellifera**

**S. columbariae** Benth. CHIA Annuals, 0.1-0.4 m tall. Leaves mostly basal, blades oblong to ovate in outline, 4-8 cm long, deeply pinnatifid, sparsely villous; petioles 2-9 cm long. Flower clusters 1-3 per inflorescence. Calyx 8-10 mm long; corollas 12-16 mm long, blue. Nutlets 1.5-2 mm long, yellowish brown. Mar-Apr. Scarce; floodplains, talus slopes, rocky s-facing slopes and canyon walls; coastal scrub; 100-1250 ft (30-380 m). Central Valley in Cañada Christy and Cañada de la Portezuela; s side between Rancho Nuevo and Black Point peninsula. Santa Rosa, Santa Catalina, and San Clemente islands; Cedros Island; CA Coast Ranges to n Mexico, e to UT and AZ. Can be locally abundant in years with heavy rainfall. Seeds of this species are rich in protein and were widely used by Native Americans.

**S. leucophylla** Greene PURPLE SAGE Shrubs, 1-1.5 m tall. Leaves cauline, blades oblong to lanceolate, 2-5 cm long, entire, bases truncate to subcordate, margins entire, toothed, or crenulate, surfaces grayish, with branched trichomes; sessile or petioles less than 5 mm long. Flower clusters 4-6 per inflorescence. Calyx 7.5-12 mm long; corollas 10-20 mm long, reddish lavender. Nutlets 2-3 mm long, gray to brown. Apr-Jun. Rare; steep, n-facing coastal bluffs; coastal scrub; 130-180 ft (40-55 m). Known only from seacliffs (on Monterey Shale) just w of mouth of Scorpion Canyon. San Luis Obispo Co. to n Baja CA.

**S. mellifera** Greene BLACK SAGE Shrubs, 1-2.5 m tall. Leaves cauline, blades elliptic to oblanceolate, 3-5 cm long, entire, bases usually tapered, margins entire to crenulate, upper surfaces green, glabrous, lower surfaces with appressed, simple trichomes; petioles 5-15 mm long. Flower clusters 3-7 per inflorescence. Calyx 6-8 mm long; corollas 9-12 mm long, pale to deep blue. Nutlets 1.5-2.5 mm long, light yellowish brown. Feb-Jun. Common; floodplains, rocky s-facing slopes; chaparral, coastal scrub; 40-750 ft (12-230 m). Scattered locations throughout island. Anacapa and Santa Catalina islands; San Francisco Bay region to n Baja CA. Seeds were eaten by the Chumash Tribe; a medicinal tea can be made from the leaves.

## SATUREJA L.

Perennial herbs or shrubs, strongly aromatic. Flowers 1-few, axillary, subtended by leaves. Calyx ± radial, campanulate, tube with 15 veins and 5 lobes; corollas bilabiate, upper lip erect, ± flat, slightly notched, lower lip longer than upper one, with 3 lobes, central one larger than lateral ones; stamens 4, in 2 pairs, ascending under upper lip, lower pair often longer than upper one. Nutlets with 3 sides, angled, smooth. Ca. 150 species; temperate, ± worldwide.

**S. douglasii** (Benth.) Briq. YERBA BUENA Perennial herbs; stems 2-6 dm long, trailing, rooting at nodes. Leaf blades broadly ovate to ± orbicular, 1-2.5 cm long, bases obtuse to truncate, margins crenate, apices obtuse, punctate and sparsely pubescent, especially along margins and veins stems; petioles 4-7 mm long. Pedicels 7-10 mm long, usually with 2 lanceolate bracts. Calyx 5-6 mm long, pubescent, with short, acute, unequal lobes; corollas 8-10 mm long, white. Nutlets ca. 1 mm long, yellow-brown. Apr. Rare; shaded canyon bottoms; oak woodland; 50-100 ft (15-30 m). Known from a single collection near Valley Anchorage in April 1982. Santa Catalina Island; British Columbia to Los Angeles Co. Leaves can be used to make a flavorful tea.

## SCUTELLARIA L. SKULLCAP

Perennial herbs, usually not strongly scented, often with rhizomes or tubers. Flowers 1-6, axillary, subtended by leaves. Calyx bilateral, ± bilabiate, lips entire, tube with dorsal, transverse crest or swelling; corollas bilabiate, upper lip ± galeate, lower lip with 3 lobes, central one broader than lateral ones, with reflexed margins; stamens 4, in 2 pairs, ascending under upper lip, lower pair longer than upper one, each with only 1 anther sac. Nutlets subglobose, papillate, brown or black. Ca. 300 species; temperate, ± worldwide.

**S. tuberosa** Benth. [*S. t.* Benth. subsp. *australis* Epling] Plants 5-25 cm tall, usually branched at base, with tuber-like rhizomes; stems sparsely villous. Leaf blades ovate, 1.5-2 cm long, sparsely villous, margins subentire to crenate, lower ones with petioles 5-15 mm long. Pedicels 1-2 mm long. Calyx 4-5 mm long in flower, pilose, purple-tinged; corollas 18-20 mm long, deep blue, throat sometimes white. Nutlets ca. 1.5 mm long, black. Feb-Apr. Occasional; grassy hillsides, rocky slopes; chaparral, oak woodland; 40-550 ft (12-170 m). North side between Hazard's Anchorage and Prisoners' Harbor; Central Valley between La Cascada and Stanton Ranch; s side between Willows Canyon and Coches Prietos drainage. Oregon to Baja CA. This species was used medicinally by Native Americans.

### STACHYS L. HEDGE NETTLE or WOOD MINT

Annual or perennial herbs, often aromatic. Flowers in axillary or terminal clusters arranged in spike-like inflorescences. Calyx ± radial to bilateral, tube with 5-10 veins and 5 lobes; corollas bilabiate, upper lip ± galeate, entire to slightly notched, lower lip longer than upper one, spreading, with 3 lobes, central one larger than lateral ones; stamens 4, in 2 pairs, ascending under upper lip, lower pair often longer than upper one. Nutlets ovoid to oblong in outline, smooth. Ca. 300 species; temperate, ± worldwide, except Australia. Several species have edible tubers and others have been used in folk medicine.

**S. bullata** Benth. Perennial herbs, with rhizomes, usually forming colonies; stems 0.4-1.2 m tall, decumbent, rooting at nodes, sparsely retrorse-hispid on angles, glandular. Leaf blades lanceolate to narrowly ovate, 3-18 cm long, bases truncate to subcordate, margins crenate-serrate, apices acute, petioles to 7 cm long. Flowers usually 6 per cluster; pedicels ca 1 mm long. Calyx 5-6 mm long, glandular, pilose, lobes triangular, ca. 2 mm long, with spinose tips; corollas 15-20 mm long, light rose with purple markings on lower lip. Nutlets ca. 2 mm long, dark reddish brown. Mar-Jun. Common; coastal bluffs, rocky slopes, canyon bottoms, floodplains; chaparral, coastal scrub, oak woodland, riparian communities; 6-1100 ft (2-335 m). Scattered locations throughout island. Santa Rosa and Anacapa islands; San Francisco Bay region to Orange Co. Island plants have non-aromatic foliage, have larger leaves and flowers than mainland plants, and have been called *S. acuminata* Greene. Leaves were used medicinally by Native Americans; a flavorful tea can be made from the leaves.

## LOASACEAE Dumort.
## STICK-LEAF FAMILY

Annual, perennial herbs, or shrubs; pubescence composed of barbed or stinging trichomes. Leaves alternate to opposite, entire to pinnately lobed; stipules absent. Flowers bisexual, radial, solitary or in cymes or racemes. Sepals 4-5, usually persistent; petals 4-5, free or fused; stamens 5-many, outer ones sometimes sterile and petaloid; pistil 1, ovary inferior, with 1 locule and 1-5 parietal placentas, style 1. Fruit a capsule; seeds many. Ca. 15 genera and 200 species; temperate and subtropical, N. and S. America, Africa.

### MENTZELIA L. STICK-LEAF

Plants with barbed, non-stinging trichomes. Leaves alternate, entire to lobed, sessile or short-petiolate. Flowers solitary, axillary or in terminal, bracteate cymes. Sepals 5; petals 5, white, yellow, or orange; stamens many, outer ones sometimes sterile and petaloid; ovary with 3-5 placentas. Seeds irregularly angled, surfaces smooth, striate, or tuberculate. Ca. 50 species; temperate and subtropical, w N. America and S. America. Seeds of several species were ground and eaten by Native Americans.

1. Floral bracts ± lanceolate, shorter than ovary; outer filaments filiform ........................................ **M. affinis**
1' Floral bracts broadly ovate, concealing flowers; outer filaments wider and forked at apex..... **M. micrantha**

**M. affinis** Greene HYDRA STICK-LEAF Annuals, 10-30 cm tall. Leaves oblanceolate to lanceolate in outline, 2-6 cm long, 0.5-2 cm wide, pinnately lobed. Flowers in terminal, bracteate clusters; bracts ± lanceolate, shorter than ovary. Petals 3-8 mm long; filaments of outer stamens filiform, ± like inner ones. Fruit narrowly cylindrical, 15-25 mm long. Mar-May. Rare; rocky s-facing slopes, loose talus slopes; coastal scrub; 15-450 ft (5-140 m). South side at Smugglers' Cove (last seen in March 1932) and just w of Los Pinos del Sur. San Nicolas, Santa Catalina, and San Clemente islands; San Francisco Bay region to Baja CA, e to Sierra Nevada and AZ.

LOASACEAE

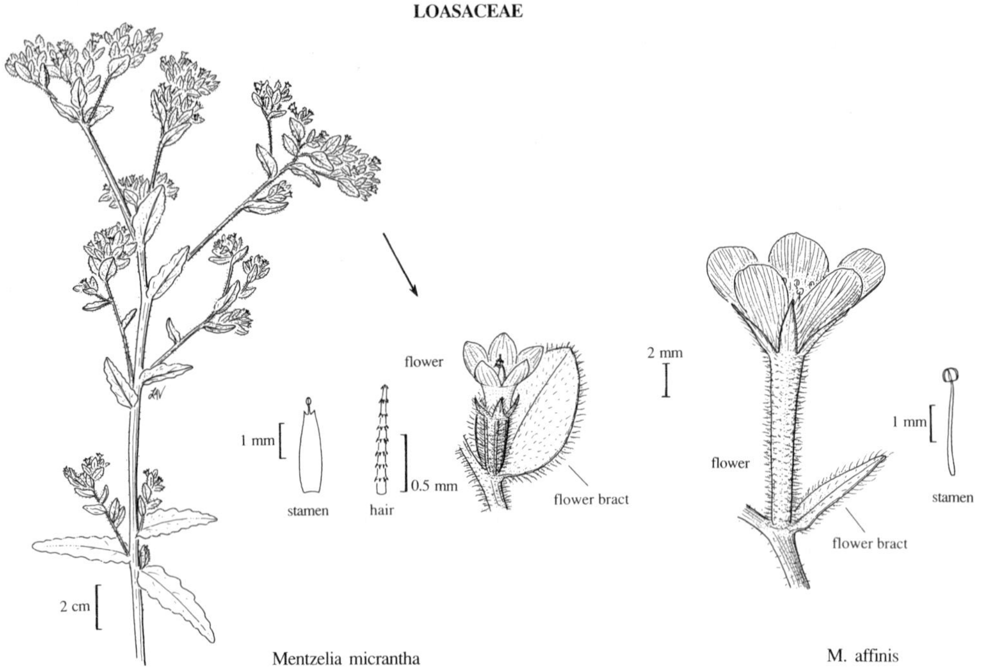


Mentzelia micrantha M. affinis

**M. micrantha** (Hook. & Arn.) Torr. & A.Gray SMALL-FLOWERED STICK-LEAF Annuals, 15-70 cm tall. Leaves ovate, 2-4.5 cm long, 1-3 cm wide, entire or irregularly toothed. Flowers in terminal, bracteate clusters; bracts broadly ovate, concealing flowers. Petals 2-5 mm long; filaments of outer 5 stamens wider and forked at apex. Fruit narrowly cylindrical, 6-12 mm long. May-Jun. Scarce; rocky s-facing slopes, loose talus slopes; coastal scrub; 50-450 ft (15-140 m). North side in Cañada del Puerto; s side between Los Pinos del Sur and Sandstone Point. Santa Catalina and San Clemente islands; Guadalupe Island; n CA coast ranges to Baja CA.

## LYTHRACEAE J.St.-Hil.
### Loosestrife Family

Annuals or perennials. Leaves alternate, opposite, or whorled, simple; stipules absent. Flowers bisexual, radial, axillary or terminal, solitary or in racemes, spikes, or panicles. Hypanthium present, campanulate to cylindrical, persistent in fruit; sepals 4-6, minute, often alternate with minute appendages in sinuses; petals 4-6 or absent, inserted on hypanthium; stamens 4-6 or twice number of petals; pistil 1, ovary superior, locules 2-6, style filiform, stigma capitate. Fruit a capsule, sometimes opening irregularly; seeds many, minute, flat or angled. Ca. 25 genera and 580 species; temperate to tropical, worldwide.

### LYTHRUM L. LOOSESTRIFE

Annuals or perennial herbs; stems erect to prostrate. Leaves sessile or short-petiolate, lower ones usually opposite, upper ones usually alternate. Flowers subtended by 2 bracts, usually solitary in axils of upper leaves. Hypanthium cylindrical, with 8-12 ribs; stamens usually 4-6. Fruit usually cylindrical, dehiscent by 2 valves. Ca. 35 species; temperate, worldwide.

1. Petals 4-6 mm long; sepals ± equal to appendages in length; pedicels ca. 2 mm long .......... **L. californicum**

1' Petals 1-2 mm long; sepals longer than appendages in length; flowers sessile or pedicels 1-1.5 mm long ..... .......... **L. hyssopifolium**

**L. californicum** Torr. & A.Gray CALIFORNIA LOOSESTRIFE Perennial herbs; stems erect, 6-12 dm tall, suffrutescent at base. Leaves linear to linear-oblong, 1-4.5 cm long. Pedicels ca. 2 mm long. Hypanthium tube 5-6 mm long, sepals broadly triangular, ca. 0.5 mm long, appendages triangular, 0.5-0.7 mm long; petals 4-6 mm long, pink to purple; style 4-5 mm long, exserted. Jun-Oct. Rare; seeps and moist creek beds; chaparral, riparian

LYTHRACEAE

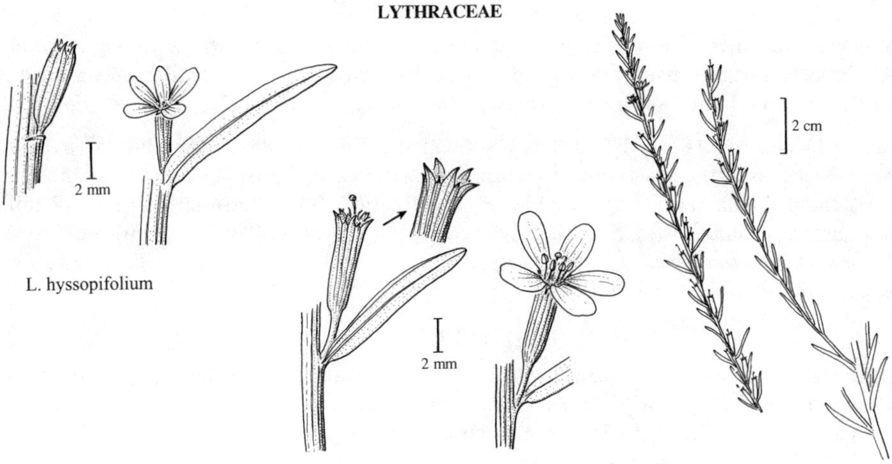


L. hyssopifolium

Lythrum californicum

communities; 10-400 ft (3-160 m). North side in Cañada del Puerto; Central Valley in Cottonwood Canyon. Throughout CA and w N. America. This species is heterostylous, with 2 floral forms, one of them with long, exserted styles, the other with short, included styles. Only the long-styled form is known on Santa Cruz Island.

**L. hyssopifolium** L. Annuals or short-lived perennials; stems decumbent to prostrate, 1-8 dm long. Leaves linear to oblong, 1-1.5 cm long. Flowers sessile or pedicels 1-1.5 mm long. Hypanthium tube 3-4 mm long, sepals broadly triangular, ca. 0.5 mm long, appendages lanceolate, 1-1.2 mm long; petals 1-2 mm long, pink to pale blue; style ca. 1.5 mm long, included. Apr-Oct. Introduced. Occasional; moist creek beds; riparian communities; 5-400 ft (2-120 m). North side at Prisoners' Harbor and in Cañada del Puerto; Central Valley at mouth of Cañada Christy and between La Cascada and Valley Anchorage; s side in canyons between Rancho Nuevo and Black Point peninsula. Santa Rosa and San Nicolas islands; Throughout N. America; native of Europe. First collected at Valley Anchorage in July 1967.

## MALVACEAE Juss.

## MALLOW FAMILY

Annuals or perennials, often with stellate pubescence. Leaves alternate, simple, blades usually palmately lobed to divided; stipules present. Flowers bisexual, radial, in racemes or solitary, axillary or terminal. Calyx often subtended by involucre of free or fused bracts, sepals 5, fused; petals 5, free from each other but adnate to base of filaments; stamens many, filaments fused into columnar tube surrounding style; pistil 1, ovary superior, locules 5-many, styles 5 or more, stigmas linear or capitate. Fruit a schizocarp, composed of 5-many usually indehiscent mericarps. Ca. 100 genera and 2000 species; temperate to tropical, worldwide. Some species are cultivated for fiber (e.g., *Gossypium*, cotton) or are used as ornamentals (e.g., *Hibiscus*).

1. Styles or style branches terminated by capitate stigmas
    2. Annuals; stems prostrate to decumbent; petals 0.4-0.6 cm long .......... **Eremalche**
    2' Shrubs; stems erect; petals 0.8-1.2 cm long .......... **Malacothamnus**

1' Styles or style branches and stigmas confluent and linear
    3. Leaves of 2 forms, lower ones crenate to shallowly lobed, upper ones with dissected to deeply lobed blades; stamen column divided near apex into 2 series of filaments .......... **Sidalcea**
    3' Lower and upper leaves usually alike, but not with dissected or deeply lobed blades; stamen column simple, not divided near apex into 2 distinct series
        4. Flowers in a terminal, spike-like inflorescence; involucre subtending calyx composed of 6-9 fused bracts; fruit with 20-40 mericarps .......... **Alcea**
        4' Flowers axillary; involucre subtending calyx composed of ± 3 free to fused bracts; fruit with 6-15 mericarps; flowers axillary
            5. Shrubs; petals 25-40 mm long .......... **Lavatera**
            5' Annuals or biennial herbs; petals 4-15 mm long .......... **Malva**

## **ALCEA** L. HOLLYHOCK

Biennial or perennial herbs. Flowers in terminal spikes or racemes. Involucre subtending calyx composed of 6-9, fused, broadly triangular bracts. Petals large, showy; stamen column simple, all filaments arising from near apex; stigmas filiform. Fruit disk-shaped; mericarps 15 or more. Ca. 50 species; temperate Europe and Asia.

**A. rosea** L. [*Althaea r.* (L.) Cav.] HOLLYHOCK Stems 1-2.5 m tall. Leaf blades orbicular in outline, 7-15 cm long, with 5-7 shallow lobes, bases cordate, margins crenate; petioles 2-5 cm long. Lobes of involucre 8-12 mm long. Petals 35-50 mm long, white to red; styles 20-40. Mericarps 20-40, horseshoe-shaped, 5-7 mm long. Jun. Introduced. Scarce; disturbed sites, floodplains; riparian communities; 10-250 ft (3-75 m). North side in Cañada del Puerto and at Scorpion Ranch; Central Valley at Stanton Ranch. Santa Catalina Island; widely cultivated in N. America; native of Europe or w Asia.

## **EREMALCHE** Greene

Annuals. Flowers solitary or in terminal clusters. Involucre subtending calyx composed of 3-4 free, linear bracts. Petals small to large, usually showy; stamen column simple, all filaments arising from near apex; stigmas capitate. Fruit subglobose; mericarps 10-35. Ca. 4 species; sw N. America.

**E. exilis** (A.Gray) Greene [*Malvastrum e.* A.Gray] Stems prostrate to decumbent, usually less than 30 cm long. Leaf blades 1-2 cm long, 1-2.5 cm wide, petioles 1-4 cm long. Involucral bracts linear, 2.5-4 mm long. Petals 4-6 mm long, white to pale purple. Mericarps 10-12, 1-1.5 mm long, outer surface transversely ridged. Rare, not seen recently; exact habitat, plant community, and elevation unknown. Known from a single collection by T. S. Brandegee in April 1888. Santa Catalina and San Clemente islands; Todos Santos, San Benito, and Natividad islands; deserts of s CA to Baja CA, e to AZ.

## **LAVATERA** L. TREE MALLOW

Annuals, perennial herbs, or shrubs. Flowers usually solitary or in pairs. Involucre subtending calyx composed of 3, basally fused, ovate to lanceolate bracts. Petals large, showy; stamen column simple, all filaments arising from near apex; stigmas filiform. Fruit disk-shaped; mericarps 6-10. Ca. 25 species; temperate to subtropical, ± worldwide.

**L. assurgentiflora** Kellogg MALVA ROSA or ISLAND TREE MALLOW Shrubs, 1-3 m tall, with soft wood. Leaf blades orbicular in outline, with 5-7 lobes, 5-15 cm long, usually glabrous, bases cordate, margins irregularly toothed; petioles 5-13 cm long. Involucral bracts 6-10 mm long. Petals 2.5-4 cm long, purple to white with purple veins; styles 6-8. Mericarps 6-8, ca. 5 mm long. Feb-Oct. Introduced. A subspecies of special environmental concern. Rare; roadsides, disturbed sites, shaded hillsides, rocky slopes, floodplains; chaparral, oak woodland, riparian communities; 200-600 ft (60-185 m). Central Valley. Native of San Miguel, Anacapa, Santa Catalina and San Clemente islands. Subsp. *assurgentiflora* has been widely planted on the Channel islands and on the mainland.

1. Upper leaf surface stellate-puberulent; stamen tube puberulent ................................ subsp. **assurgentiflora**
1' Upper leaf surface glabrous to subglabrous; stamen tube glabrous to subglabrous .................. subsp. **glabra**

Subsp. **assurgentiflora** MALVA ROSA or NORTHERN ISLAND TREE MALLOW. Upper leaf surface stellate-puberulent. Petal apex subentire; stamen tube puberulent. Feb-Oct. Introduced. Rare; roadsides, disturbed sites, rocky slopes; chaparral; 200-600 ft (60-185 m). Central Valley along Islay Canyon Road, University of California field station, Stanton Ranch, and along road between Stanton Ranch airfield and South Ridge Road. Native to San Miguel and Anacapa islands; apparently planted on Santa Rosa, San Nicolas, Todos Santos, and Cedros islands. First collected in Islay Canyon in December 1981.

Subsp. **glabra** Philbrick MALVA ROSA or SOUTHERN ISLAND TREE MALLOW. Upper leaf-surface glabrous to subglabrous. Petal apex usually erose; stamen tube glabrous to subglabrous. Apr-Jun. Rare; roadsides, shaded hillsides, floodplains; oak woodland, grasslands, riparian communities; 200-400 ft (60-120 m). Central Valley between La Cascada and University of California field station, and Stanton Ranch. Native to Santa Catalina and San Clemente islands. First collected near La Cascada in April 1990.

## **MALACOTHAMNUS** Greene

Shrubs. Flowers usually in axillary clusters, these arranged in panicles. Involucre subtending calyx composed of 3-4 free, linear to elliptic bracts. Petals large, showy; stamen column simple, all filaments arising from near apex; stigmas capitate. Fruit subglobose; mericarps 7-14. Ca. 20 species; sw N. America and Chile.

# MALVACEAE

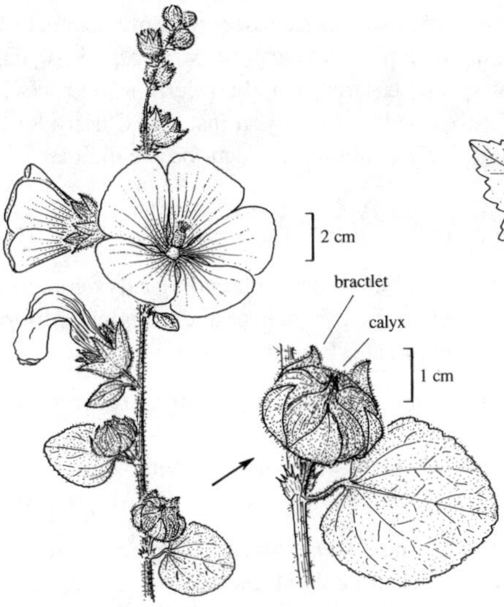


Alcea rosea

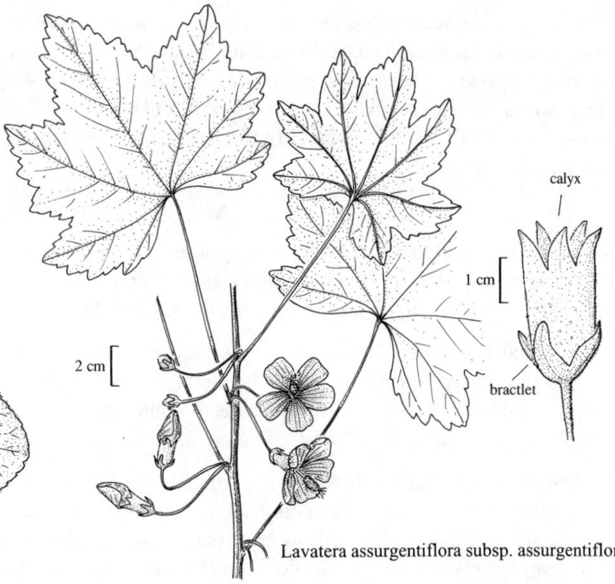


Lavatera assurgentiflora subsp. assurgentiflora

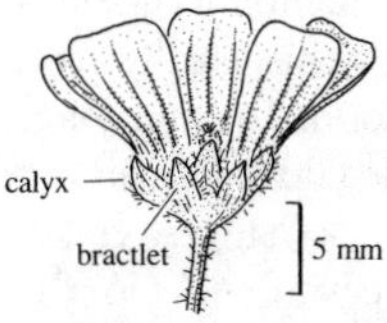


Malva nicaeensis

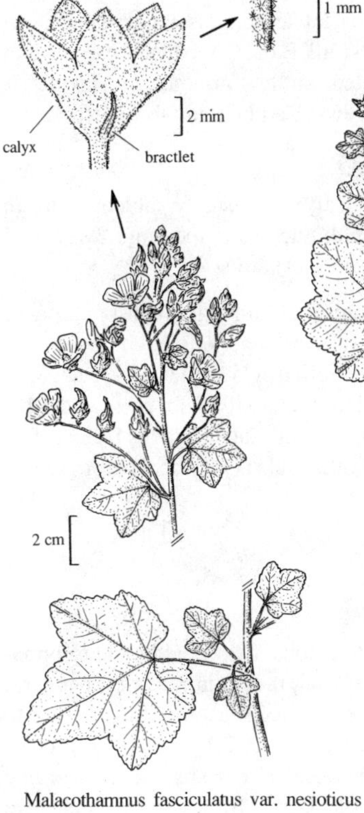


Malacothamnus fasciculatus var. nesioticus

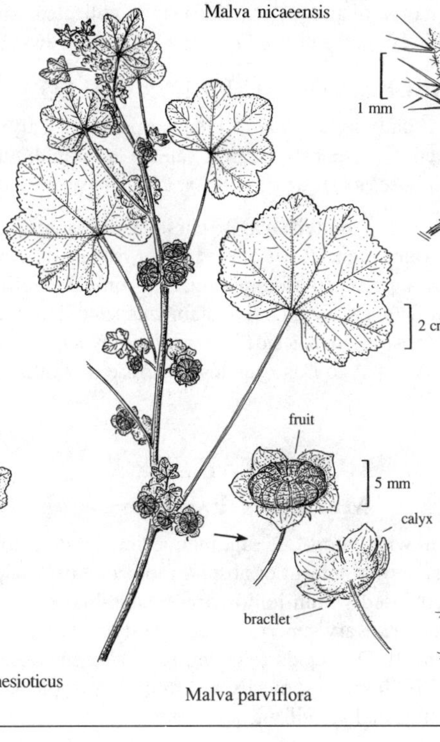


Malva parviflora

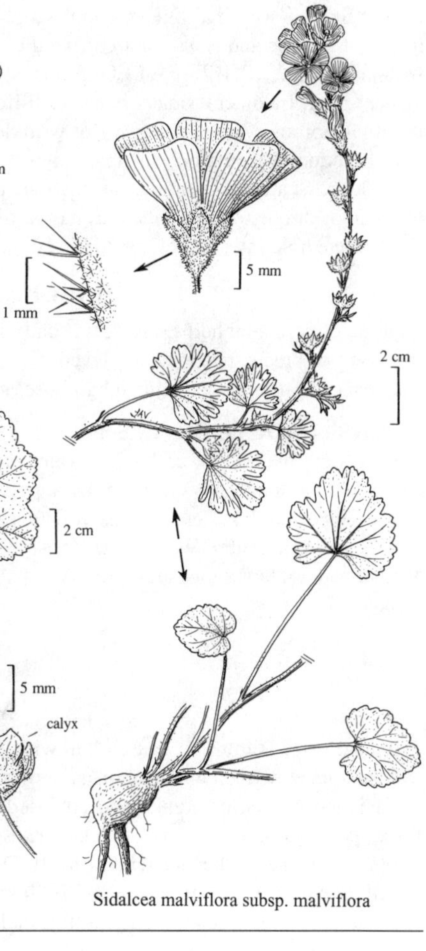


Sidalcea malviflora subsp. malviflora

**M. fasciculatus** (Torr. & A.Gray) Greene var. **nesioticus** (B.L.Rob.) Kearney SANTA CRUZ ISLAND BUSH MALLOW Plants 1-2 m tall. Leaf blades ovate, 3-4 cm long, with 3-5 lobes, bases truncate to cuneate, margins crenate, upper surfaces pale green, sparsely puberulent to glabrate, lower surfaces densely tomentulose; petioles 1-3 cm long. Involucral bracts 2-4 mm long. Petals 0.8-1.2 cm long, rose-pink; styles 8-10. Mericarps 8-10, ca. 5 mm long. Apr-Jun. Endemic to Santa Cruz Island. A variety of special environmental concern. Rare; rocky, s-facing slopes; chaparral, coastal scrub; 60-450 ft (18-140 m). Central Valley in canyon just s of Christy Ranch and on slopes n of University of California field station. At least 4-5 other subspecies occur on the mainland and other Channel islands.

## MALVA L. MALLOW

Annuals or perennial herbs. Flowers usually solitary or in axillary clusters. Involucre subtending calyx composed of 2-3, linear to ovate, free bracts. Stamen column simple, all filaments arising from near apex; stigmas filiform. Fruit disk-shaped to subglobose; mericarps 8-15. Ca. 100 species; temperate, Europe, Asia, and Africa.

1. Involucral bracts broadly lanceolate to ovate; petals 8-15 mm long, claws usually ciliate; mericarps with sharp (not winged) angles between lateral and dorsal walls ........ **M. nicaeensis**

1' Involucral bracts linear to filiform; petals 4-5 mm long, claws usually glabrous; mericarps with winged angles between lateral and dorsal walls ........ **M. parviflora**

**M. nicaeensis** All. BULL MALLOW Annuals or short-lived perennials; stems spreading to prostrate. Leaf blades 1-4 cm long, 2-9 cm wide, petioles 1.5-7 cm long. Involucral bracts broadly lanceolate to ovate, 2.5-5 mm long. Petals 8-15 mm long, pink to lavender, claws usually ciliate. Mericarps smooth or slightly to densely hirsute, angle between lateral and dorsal walls sharp but not winged. May-Jun. Introduced. Scarce; disturbed flats, roadsides; coastal scrub, grasslands; 200-1200 ft (60-365 m). North side at intersection of Navy Road and road to Stanton Ranch airfield; Central Valley between La Cascada and El Sobrante. Throughout w N. America; native of Europe and Asia. First collected at Stanton Ranch in June 1930.

**M. parviflora** L. CHEESEWEED Annuals; stems usually erect. Leaf blades 1-7 cm long, 2-9 cm wide, petioles 1-2 cm long. Involucral bracts linear to filiform, 1-4 mm long. Petals 4-5 mm long, white to lavender, claws usually glabrous. Mericarps rugose or wrinkled dorsally, angle between lateral and dorsal walls ± winged. Mar-Sep. Introduced. Common; disturbed sites, rocky slopes, grassy hillsides; coastal scrub, grasslands; 10-1500 ft (3-460 m). Scattered locations throughout island. All CA Channel islands; all Baja CA islands except San Geronimo; throughout N. America; native of Europe and Asia. First collected on "low ground" in May 1927. Fruits are edible raw; leaves are edible cooked and can be used to thicken stews. Can be toxic to livestock.

## SIDALCEA A.Gray CHECKER MALLOW

Annuals or perennial herbs. Flowers usually in terminal racemes or panicles. Calyx not usually subtended by an involucre of bracts; petals white, lavender, or purple, showy; stamen column divided near apex into 2 series of filaments; stigmas linear. Fruit subglobose; mericarps 5-10. Ca. 25 species; w N. America.

**S. malviflora** (DC.) A.Gray ex Benth. subsp. **malviflora** Perennial herbs; stems 2-5 dm long, often spreading or trailing. Leaf blades 2-5 cm long, margins crenate; petioles of lower leaves 3-7 cm long, petioles of upper ones less than 1 cm long. Calyx 6-9 mm long; petals 10-25 mm long, rose-pink, with ciliate claws. Mericarps ca. 8, 3-4 mm long. Mar-Jun. Scarce; coastal flats, grassy hillsides, stabilized sand dunes; coastal scrub, grasslands; 10-250 ft (3-75 m). Central Valley near Christy Ranch; s side between Forney's Cove and Cañada de los Sauces. San Miguel and Santa Rosa islands; Coastal n CA to Los Angeles Co. At least 7 other subspecies occur from OR to n Baja CA.

# MORACEAE Link
## MULBERRY FAMILY

Perennial herbs, shrubs, or trees, often with an opaque sap, monoecious or dioecious. Leaves alternate to opposite, simple, evergreen to deciduous; stipules present or absent. Flowers unisexual, radial, in racemes, panicles, in capitate clusters, or minute and enclosed within hollow, receptacle-like structure; sepals usually 4, free to fused; petals absent; stamens 2-6; pistil 1, ovary superior, locule 1, style simple or with 2 branches. Fruit composed of achenes either subtended by fleshy sepals (e.g. mulberry) or enclosed in a fleshy, receptacle-like structure (e.g. fig). Ca. 40 genera and 1000 species; mostly subtropical or tropical, worldwide. Some species are important sources of food, including figs and breadfruit.

MORACEAE

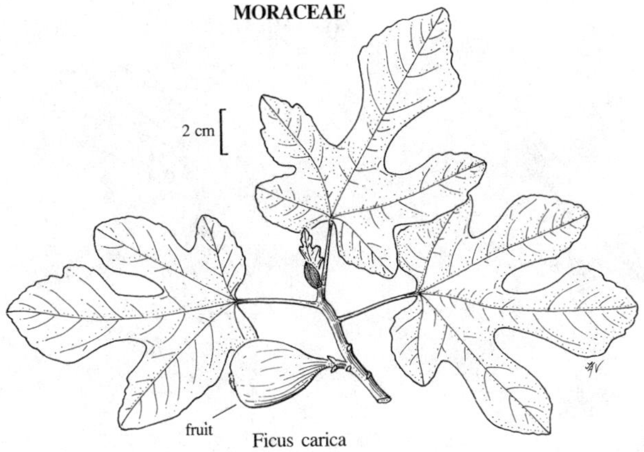


Ficus carica

## FICUS L. FIG

Monoecious shrubs or trees. Leaves alternate, entire to lobed, often with palmate venation; stipules present, often fused. Flowers many, minute, enclosed within hollow receptacle-like structure that enlarges, becomes fleshy in fruit, and opens at apex. Staminate perianth with 2-6 lobes; pistillate perianth reduced or absent; style simple. Ca. 500 species; subtropical to tropical, worldwide.

**F. carica** L. COMMON FIG Trees, 3-5 m tall, often with multiple trunks. Leaves orbicular to broadly ovate in outline, blades 5-25 cm long, 5-20 cm wide, with 3-5 undulate lobes, upper surfaces glabrous, lower surfaces puberulent, petioles 3-10 cm long. Flowering receptacle greenish. Fruiting receptacle obovoid, 2.5-6 cm long, 2-3 cm wide, black to dark brown. Apr-Jul. Introduced. Scarce; disturbed sites, coastal flats; coastal scrub; 10-100 ft (3-30 m). Planted on n side at Pelican Bay (where persisting without cultivation) and at Scorpion Ranch; Central Valley at Christy Ranch; s side at Coches Prietos Anchorage and at Smugglers' Ranch. San Miguel, Santa Catalina, and San Clemente islands; Cedros Island; cultivated throughout much of s CA; native of Europe and n Africa. Planted at Smugglers' Ranch as early as 1889. Ripe fruits are edible; sap can cause dermatitis in humans.

# MYRTACEAE Juss.
## MYRTLE FAMILY

Evergreen shrubs or trees, often aromatic. Leaves alternate or opposite, simple, entire; stipules absent. Flowers bisexual, radial, in racemes, umbels, or corymbs. Hypanthium present; perianth parts free or fused into a deciduous cap; sepals 4-5; petals 4-5; stamens many, often in bundles opposite petals; ovary inferior, locules 1-many, style 1. Fruit a berry or capsule. Ca. 120 genera and 3850 species, tropical and temperate Africa, Australia, and S. America. Many species are cultivated as ornamentals.

## EUCALYPTUS L.'HÄr. GUM TREE

Trees; bark smooth to deeply furrowed, often peeling as strips or flakes. Leaves of 2 forms, juvenile ones often opposite, mature ones alternate. Inflorescence usually umbellate. Flowers with sepals and petals fused to form a lid or cap, deciduous at anthesis; stamens variously colored; ovary with 3-6 locules; seeds many. Fruit a woody capsule, dehiscent by 3-6 apical valves. Ca. 450 species native of Australia. Several species cultivated as ornamental trees. Several species contain toxins.

1. Flowers solitary in leaf axils; fruit 4-angled, warty, glaucous, 15-20 mm wide .......... **E. globulus**
1' Flowers in umbels; fruit cylindrical, smooth, not especially glaucous, 5-10 mm wide
   2. Bark persistent, thick, furrowed at least to lower branches; stamens usually pink; fruit ovoid to pear-shaped, with 5-6 non-protruding valves .......... **E. sideroxylon**
   2' Bark shed in patches, persistent only at base of trunk, smooth; stamens white; fruit hemispheric to ovoid, with 4-5 protruding valves

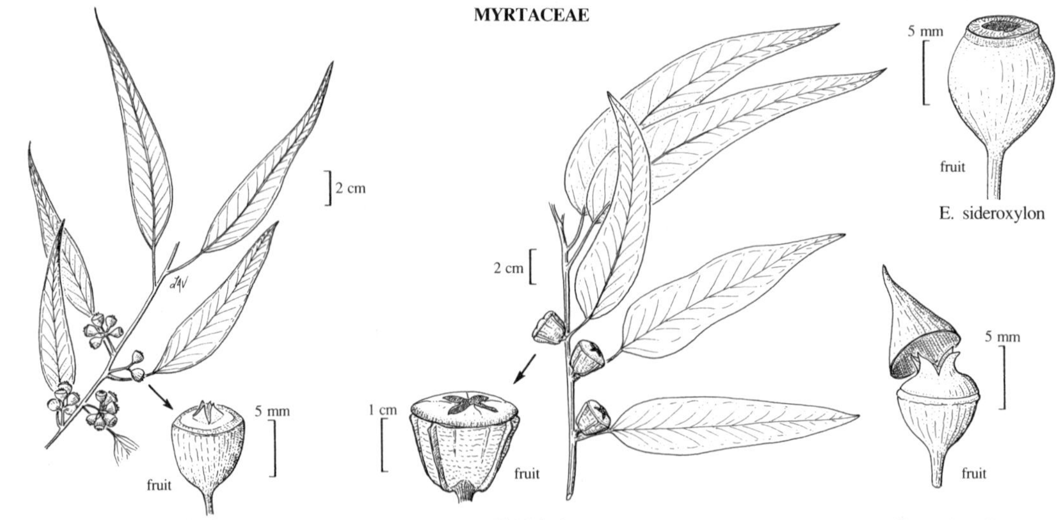


Eucalyptus camaldulensis E. globulus E. tereticornis

3. Seeds yellow, mixed with brownish black chaff; leaves dull green, often grayish; flower buds yellow-green, cap hemispherical to blunt-conic and beaked .................................................... **E. camaldulensis**

3' Seeds and chaff brownish black; leaves glossy green; flower buds light green to pale pink, cap horn-shaped or long-conic ........................................................................................................ **E. tereticornis**

**E. camaldulensis** Dehnh. RIVER RED GUM Bark dull white, with red or creamy white blotches, smooth and shed in irregular flakes. Mature leaves lanceolate, 8-25 cm long, 0.7-2 cm wide, dull green to grayish. Flowers 15 cm wide, buds yellow-green, smooth, cap hemispherical or conical; stamens white. Fruit hemispherical to ovoid, 5-8 mm wide, with 4 protruding valves. Apr-Sep. Introduced. Scarce; disturbed sites, roadsides; 200-1500 ft (60-460 m). Central Valley between Stanton Ranch and La Selva; s side at Navy Base. Santa Rosa Island; San Francisco Bay region to San Diego; native of Australia. Leaves and sap have been used medicinally. Plants in the Central Valley, with a beaked, hemispheric bud cap and fruiting valves 1 mm long belong to var. *camaldulensis*. Plants near the Navy installation, with blunt, conic or bullet-shaped bud caps and fruiting valves 2-3 mm long, belong to var. *obtusa* Blakely.

**E. globulus** Labill. TASMANIAN BLUE GUM Bark gray-brown to black, rough and persistent at base of trunk, peeling in long strips above base to expose smooth, gray or cream surface. Mature leaves lanceolate to sickle-shaped, pendulous, 12-30 cm long, 2-5 cm wide, glossy, dark green. Flowers 3-5 cm wide, buds bluish green, glaucous, warty, cap flattened, pointed in bud; stamens creamy white to yellow. Fruit conical or cup-shaped, 15-20 mm wide, with 3-5 enclosed or protruding valves. Feb-Oct. Introduced. Occasional; disturbed sites, roadsides, canyon bottoms, floodplains; riparian communities; 10-1500 ft (3-460 m). Scattered locations throughout island, especially on n side and in Central Valley. All CA Channel islands except San Miguel and Santa Barbara (apparently planted on Santa Barbara Island but did not persist); n CA coast ranges to Baja CA; native of Australia. Leaves have been used medicinally and as an insecticide. Refined oils can be toxic to humans if ingested.

**E. sideroxylon** A.Cunn. ex Woolls RED IRON BARK Bark dark gray to brown or black, thick, rough, deeply furrowed, persistent to lower branches. Mature leaves lanceolate, 6-14 cm long, 1-3 cm wide, dull dark green or grayish blue. Flowers 15-20 mm wide, buds yellowish green, cap conical, pointed in bud; stamens usually pink. Fruit ovoid to pear-shaped, 8-10 mm wide, with 5-6 enclosed valves. Apr-Jun. Introduced. Rare; disturbed sites, roadsides; 200-300 ft (60-90 m). Central Valley between Stanton Ranch and La Selva. Santa Barbara Co. to San Diego; native of Australia.

**E. tereticornis** Smith FOREST RED GUM Bark gray-mottled with irregular, gray or bluish patches, usually rough and persistent at base of trunk, shed above in large flakes, exposed surface smooth. Flowers 15 mm wide, buds light green to pale pink, smooth, cap long conical to horn-shaped; stamens white, rarely pink. Fruit ovoid, 7-9 mm wide, with 4-5 protruding valves. Apr-Jun. Introduced. Scarce; disturbed sites, canyon bottoms, flood-plains; pine forest, riparian communities; 20-300 ft (6-90 m). North side at Pelican Bay (and escaping into

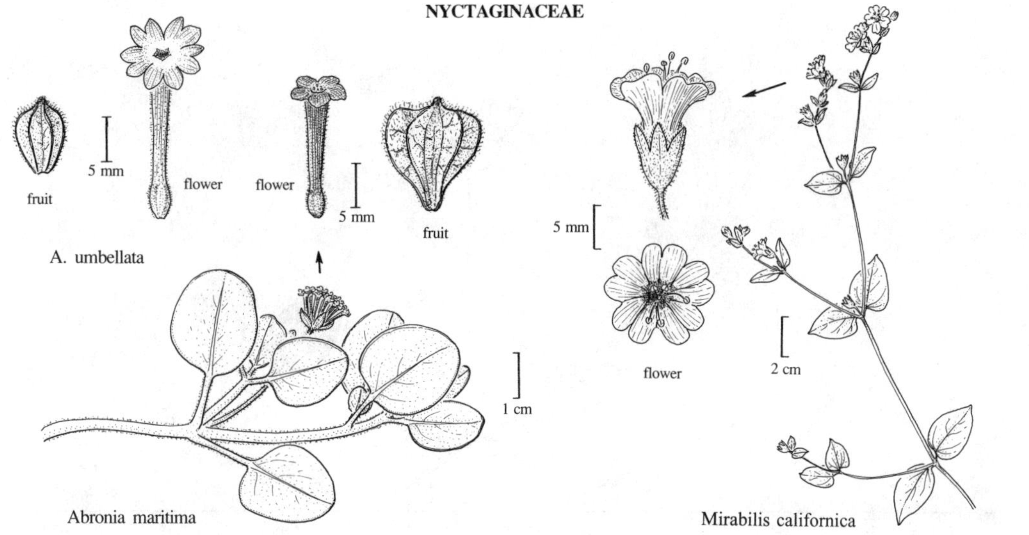


nearby pine forest) and in Cañada del Puerto; Central Valley at La Selva. Santa Rosa Island; San Joaquin Valley to San Diego; native of Australia.

## NYCTAGINACEAE Juss.
## FOUR O'CLOCK FAMILY

Annuals or perennials. Leaves opposite, simple; stipules absent. Flowers bisexual, radial, usually in umbels or spikes, axillary or terminal, flower clusters often subtended by an involucre. Perianth parts fused into 1 whorl, campanulate to salverform, lobes 4-5, base persistent and hardened in fruit; stamens 1-many; ovary superior but appearing inferior, because of adherent perianth tube, locule 1, ovule 1, style 1. Fruit an achene. Ca. 35 genera and 350 species; temperate and subtropical, ± worldwide. Many species are cultivated as ornamentals, including *Bougainvillea* and *Mirabilis* (FOUR O'CLOCK).

1. Involucre composed of 4-8, ± free bracts; stigma linear, included in tube; fruit with wing-like lobes .......... **Abronia**

1' Involucre cup-like, composed of 5 fused bracts; stigma capitate, usually exserted; fruit smooth to wrinkled .......... **Mirabilis**

### ABRONIA Juss. SAND-VERBENA

Annuals or perennial herbs, often glandular. Leaves lanceolate to round, often fleshy. Flowers in capitate clusters or umbels, clusters or umbels; involucre composed of 4-8 scarious, lanceolate to ovate bracts. Perianth salverform, with slender tube constricted above ovary and enlarged at throat, limb with 4-5 lobes; stamens 4-5, included; stigma linear, included. Fruit, (including persistent perianth) turbinate, deeply lobed or winged. Ca. 35 species; temperate w N. America. Roots of some species were eaten by Native Americans.

1. Perianth dark crimson to reddish purple; fruit with coarsely and reticulately veined, wing-like lobes .......... **A. maritima**

1' Perianth light rose to white; fruit with faintly net-veined, wing-like lobes .......... **A. umbellata**

**A. maritima** S.Watson STICKY SAND-VERBENA Perennial herbs; stems prostrate, to 2.5 m long, densely glandular. Leaf blades elliptic to oblong, 4.5-7 cm long, thick. Perianth dark crimson to reddish purple, 12-14 mm long, limb 5-9 mm wide. Fruit 12-20 mm wide, with 5, coarsely and reticulately veined, wing-like lobes. Feb-Oct. A species of special environmental concern. Occasional; sandy beaches and dunes; beach and dune communities; 0-50 ft (0-15 m). North side at Prisoners' Harbor; Central Valley at Christy Beach; s side between Forney's Cove and beach at mouth of Coches Prietos drainage. All CA Channel islands except Santa Barbara; San Martin, Cedros, and Natividad islands; San Luis Obispo Co. to n Baja CA.

OLEACEAE

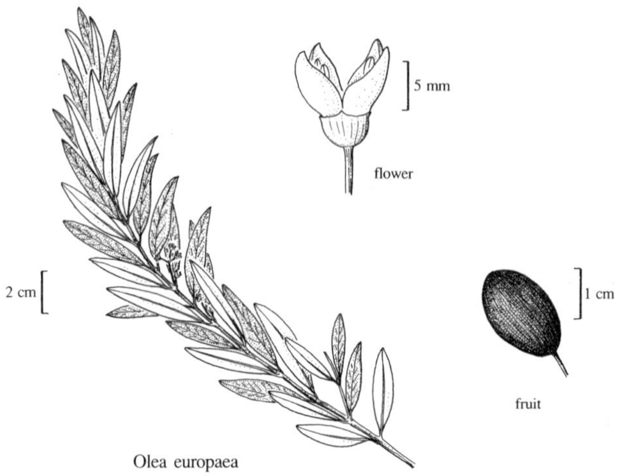


Olea europaea

**A. umbellata** Lam. BEACH SAND-VERBENA Perennial herbs; stems prostrate, to 1.5 m long, sparsely to moderately glandular. Leaf blades ovate to rhombic, 2-6.5 cm long. Perianth light rose to white, 12-15 mm long, limb 8-12 mm wide. Fruit 5-8 mm wide, with 2-5, usually 4, faintly net-veined, wing-like lobes. Mar-Oct. Occasional; beaches and dunes, sandy slopes; beach and dune communities; 0-200 ft (0-60 m). Central Valley at Christy Beach; s side between Fraser Point and mouth of Laguna Canyon. All CA Channel islands except Anacapa and Santa Barbara; San Francisco Bay region to Los Angeles Co.

### MIRABILIS L. FOUR O'CLOCK

Perennial herbs or subshrubs. Leaves linear to round. Flowers in terminal clusters; involucre composed of 5 fused bracts. Perianth campanulate to funnelform, limb with 5 lobes; stamens 3-5, usually exserted; stigma usually capitate, usually exserted. Fruit (including persistent perianth) subglobose to turbinate, smooth to ridged. Ca. 60 species; temperate, N. and S. America. Some species cultivated as ornamentals; others used for dyes or food.

**M. californica** A.Gray WISHBONE BUSH Subshrubs; stems 1.5-8 dm long, ascending to prostrate, ± glabrous to glandular-puberulent. Leaf blades ovate, 1-3.5 cm long, 1-3 cm wide. Perianth 1-1.5 cm long, rose-purple to deep pinkish red. Fruit broadly ovoid, 4-6 mm long, smooth or wrinkled. Feb-Jul. Common; rocky s-facing slopes; coastal scrub; 10-1500 ft (3-460 m). Widely scattered locations throughout island, especially on s side. All CA Channel islands except San Miguel and San Nicolas; all Baja CA islands except San Geronimo; San Francisco Bay region to n Baja CA.

## OLEACEAE Hoffmanns. & Link
## OLIVE FAMILY

Shrubs or trees. Leaves alternate or opposite, simple to pinnately compound. Flowers bisexual or unisexual, usually radial, axillary or terminal, in short racemes or panicles, sometimes solitary. Calyx with 4-15 lobes; petals 4-6, sometimes absent, fused; stamens usually 2, adnate to corolla; pistil 1, ovary superior, locules 2. Fruit a drupe, achene, or samara; seeds usually 1 per locule. Ca. 25 genera and 900 species; ± worldwide. Some species are cultivated as ornamentals, including jasmine, privet, and lilac.

### OLEA L.

Shrubs or trees. Leaves opposite, simple. Flowers bisexual, radial, in racemes or panicles or solitary. Calyx with 4 teeth or lobes; corolla with 4 lobes. Fruit a drupe. Ca. 20 species; tropical to warm temperate Europe, Asia, and Africa.

**O. europaea** L. OLIVE Trees, to 5 m tall. Leaves oblong to oblanceolate, 3.5-6 cm long, 0.8-1.2 cm wide, entire, upper surfaces dark green, ± glabrous, lower surfaces pale green, densely pubescent with scale-like trichomes. Corollas rotate, 2.5-4 mm long, white to yellowish. Fruit a drupe, 10-20 mm long, green to black, oily. Apr-Jun. Introduced. Scarce; disturbed sites, rocky slopes, ridgetops; chaparral, coastal scrub, grasslands, pine forest; 50-725 ft (15-220 m). North side near Pelican Bay; s side near Smugglers' Cove. Anacapa Island; naturalized at scattered localities in CA and Baja CA; native of Europe. Planted at Smugglers' Ranch as early as 1889. This species has been cultivated for many centuries for its fruit and oils.

## ONAGRACEAE Juss.
### Evening Primrose Family

Annuals or perennial herbs. Leaves alternate or opposite, simple; stipules absent. Flowers bisexual, radial, axillary or terminal, in clusters or racemes, sometimes solitary. calyx and corolla each with 4 lobes, adnate to ovary, usually with a hypanthium; stamens 4, 8, or 10, adnate to hypanthium; ovary inferior, locules 4, style 1, with 4 stigmatic branches or 1 capitate stigma. Fruit a capsule; seeds 1-many. Ca. 24 genera and 650 species; temperate and subtropical, worldwide. Many species are cultivated as ornamentals, including *Fuchsia.*

1. Sepals 5, persistent; petals 5; stamens 10 .......... **Ludwigia**
1' Sepals 4, deciduous; petals 4; stamens 4 or 8
  2. Seeds with tufts of long, soft trichomes; sepals erect to spreading; petals ± notched .......... **Epilobium**
  2' Seeds without tufts of trichomes; 1 or more sepals reflexed; petals not usually notched
    3. Style simple, stigma capitate .......... **Camissonia**
    3' Style with 4 branches, stigmas ± linear or terete, not capitate
      4. Filament attached to anther base, anthers erect .......... **Clarkia**
      4' Filament attached to middle of anthers, anthers pivoting freely .......... **Oenothera**

### CAMISSONIA Link

Annuals or perennial herbs. Leaves basal and cauline, alternate. Flowers in spikes or racemes, usually opening at dawn. Sepals reflexed; stamens 4 or 8, anthers attached at base; style simple, with capitate stigma. Fruit straight to coiled or sinuous, sessile. 62 species; temperate w N. America, 1 species in S. America. *C. hirtella, C. ignota, C. intermedia, C. micrantha,* and *C. robusta* comprise a group of strongly selfing diploids and polyploids, which differ from each other primarily with respect to minor or overlapping differences in habit, fruit shape and size, and leaf shape. Consequently, individual specimens may be difficult to identify with certainty.

1. Perennial herbs; stems prostrate to decumbent; leaves densely silver-strigose .......... **C. cheiranthifolia**
1' Annuals; stems erect, ascending or decumbent; leaves glabrous to villous, but not silver-strigose
  2. Basal leaves pinnately lobed; cauline leaves few, much reduced; fruit straight, reflexed, 45-90 mm long .......... **C. californica**
  2' Basal leaves entire, serrulate, or denticulate; cauline leaves gradually reduced above basal rosette; fruit straight to coiled (contorted when dry), spreading to erect, 10-30 mm long
    3. Fruit 1.5-2 mm wide near base; most pollen grains with 4 pores, rarely with 5 pores; hexaploid (n=21) .......... **C. robusta**
    3' Fruit 0.7-1.3 mm wide near base; most pollen grains with 3 or 5 pores; diploid or tetraploid (n=7 or n=14)
      4. Cauline leaves petiolate, petioles 5-20 mm long, leaf bases attenuate; fruit strongly coiled (up to 4-5 times) .......... **C. ignota**
      4' Cauline leaves subsessile, bases cordate, truncate, or rounded; fruit straight to coiled only 1-2 times
        5. Most upper cauline leaves ovate to narrowly ovate; fruit 0.7-0.9 mm wide near base .......... **C. hirtella**
        5' Most upper cauline leaves narrowly elliptic, narrowly lanceolate, or narrowly ovate; fruit 1.1-1.3 mm wide near base
          6. Stems erect; upper cauline leaves narrowly lanceolate to narrowly ovate; inflorescence glandular puberulent .......... **C. intermedia**
          6' Stems decumbent, rarely erect; upper cauline leaves elliptic to elliptic-lanceolate; inflorescence not glandular .......... **C. micrantha**

**C. californica** (Torr. & A.Gray) Raven Annuals; stems ± simple, erect, to 1.5 m tall. Leaves narrowly elliptic, lower ones petiolate, pinnately lobed, upper ones few, 1-5 cm long, entire, subsessile. Sepals 4-8 mm long; petals 6-14 mm long, yellow, with maroon flecks near bases. Fruit 50-90 mm long, 1-1.5 mm wide, straight, cylindrical to 4-angled when dry. Apr-Jul. Occasional; coastal bluffs, rocky s-facing slopes, loose talus slopes, open ridgetops, floodplains; coastal scrub, riparian communities; 15-600 ft (5-305 m). North side in Cañada del Puerto; Central Valley northeast of Christy Ranch and between top of Portezuela Grade and Bosque Mano; s side between Pozo Beach and Coches Prietos Anchorage. Santa Rosa and Santa Catalina islands; San Martin Island; San Francisco Bay region to Baja CA, e to AZ and n Mex.

**C. cheiranthifolia** (Hornem. ex Spreng.) Raim. subsp. **cheiranthifolia** BEACH EVENING PRIMROSE Perennial herbs, to 10 cm tall. Stems prostrate to decumbent, to 0.5 m long. Leaves ovate to obovate, 1-5 cm long, entire to remotely serrulate, densely silver-strigose. Sepals 5-11 mm long; petals 6-10 mm long, yellow; styles 5-10 mm long. Fruit 10-30 mm long, 2.5-5 mm wide, 4-angled, usually coiled. Mar-Jun. Occasional; sandy beaches and dunes; beach and dune communities; 10-50 ft (3-15 m). Central Valley at Christy Beach; s side between Fraser Point and mouth of Willows Canyon. All CA Channel islands except Anacapa and Santa Catalina; sw OR to Jalama Beach, Santa Barbara Co. Subsp. *suffruticosa* (S.Watson) Raven, a subshrub with petals 10-20 mm long and styles 13-23 mm long, occurs on the mainland of s CA and n Baja CA and on San Nicolas and San Martin islands.

**C. hirtella** (Greene) Raven [*Oenothera hirtella* Greene] Annuals, to 50 cm tall; stems erect to ascending, villous. Leaves subsessile, ovate, 10-25 mm long, denticulate, bases cordate to truncate. Sepals 2-5 mm long; petals 2-10 mm long, yellow, reddish with age. Fruit 10-20 mm long, 0.7-0.9 mm wide, coiled 1-2 times. Apr-Jun. Rare; rocky slopes, floodplains; chaparral, coastal scrub, riparian communities; 50-400 ft (15-120 m). North side near Pelican Bay; Central Valley in Islay Canyon. Northern CA coast ranges to Baja CA, e to the Sierra Nevada foothills. Diploid (n=7).

**C. ignota** (Jeps.) Raven Annuals; stems erect to ascending, to 50 cm tall, sparsely villous to strigulose. Leaves with petioles 5-20 mm long, blades narrowly elliptic to narrowly ovate, 15-50 mm long, serrulate, bases attenuate. Sepals 3-6 mm long; petals 4-8 mm long, yellow, reddish with age. Fruit 20-30 mm long, 0.8-1 mm wide, cylindrical, 4-angled when dry, strongly coiled. Mar-Apr. Rare; coastal bluffs, rocky s-facing slopes; coastal scrub; 20-200 ft (6-60 m). South side in Coches Prietos drainage and near Valley Anchorage. Santa Rosa Island; n CA coast ranges and s Sacramento Valley to n Baja CA, e to Madera Co. Diploid (n=7).

**C. intermedia** Raven Annuals, 4-30 cm tall; stems erect to ascending, villous. Leaves subsessile, narrowly ovate to lanceolate, 10-50 mm long, denticulate, sparsely villous, bases truncate to rounded. Sepals 1-3 mm long; petals 1.5-4 mm long, yellow, reddish with age. Fruit 10-20 mm long, 1.1-1.3 mm wide, 4-angled when dry, straight to coiled once. Apr-Jun. Occasional; rocky slopes, sandy floodplains; chaparral, coastal scrub, grasslands, oak woodland, riparian communities; 70-1400 ft (20-430 m). Widely scattered locations throughout island. San Miguel and Santa Catalina islands; n CA coast ranges to Baja CA. Tetraploid (n=14).

**C. micrantha** (Hornem. ex Spreng.) Raven Annuals; stems 4-40 cm long, ascending to decumbent, villous. Leaves subsessile, elliptic to elliptic-lanceolate, 15-70 mm long, denticulate, bases cuneate to rounded. Sepals 1-2 mm long; petals 1.5-3 mm long, yellow, usually with 1-2 red spots at bases. Fruit 12-20 mm long, 1.1-1.2 mm wide, 4-angled when dry, straight to coiled once. Mar-May. Occasional; disturbed sites, coastal bluffs, rocky slopes, loose talus slopes, floodplains; coastal scrub, oak woodland, riparian communities; 15-200 ft (5-60 m). Widely scattered locations throughout island, especially on s side. All CA Channel islands except Santa Barbara and San Nicolas; n CA coast ranges to San Diego Co. Diploid (n=7).

**C. robusta** Raven Annuals, 6-40 cm tall; stems erect, villous. Leaves subsessile, narrowly elliptic to elliptic-lanceolate, 10-90 mm long, denticulate, bases truncate to cordate. Sepals 2.5-4 mm long; petals 3-6 mm long, yellow, reddish with age. Fruit 10-20 mm long, 1.5-2 mm wide, 4-angled, coiled once. Apr-May. Rare; rocky slopes; chaparral, coastal scrub; 20-450 ft (6-140 m). North side between Fry's Harbor and Pelican Bay. San Miguel, Santa Catalina, and San Clemente islands; Guadalupe Island; San Diego Co. to Baja CA. Hexaploid (n=21).

## CLARKIA Pursh CLARKIA, GODETIA, or FAREWELL-TO-SPRING

Annuals. Leaves cauline, alternate or opposite. Flowers in racemes, opening at dawn. Sepals reflexed; stamens 8, anthers attached at base, erect; style with 4 stigmatic branches. Fruit sessile to pedicellate, straight or coiled, cylindrical to quadrangular. Seeds glabrous. Ca. 40 species, temperate western N. and S. America. Some species and their hybrids are commonly cultivated as ornamentals.

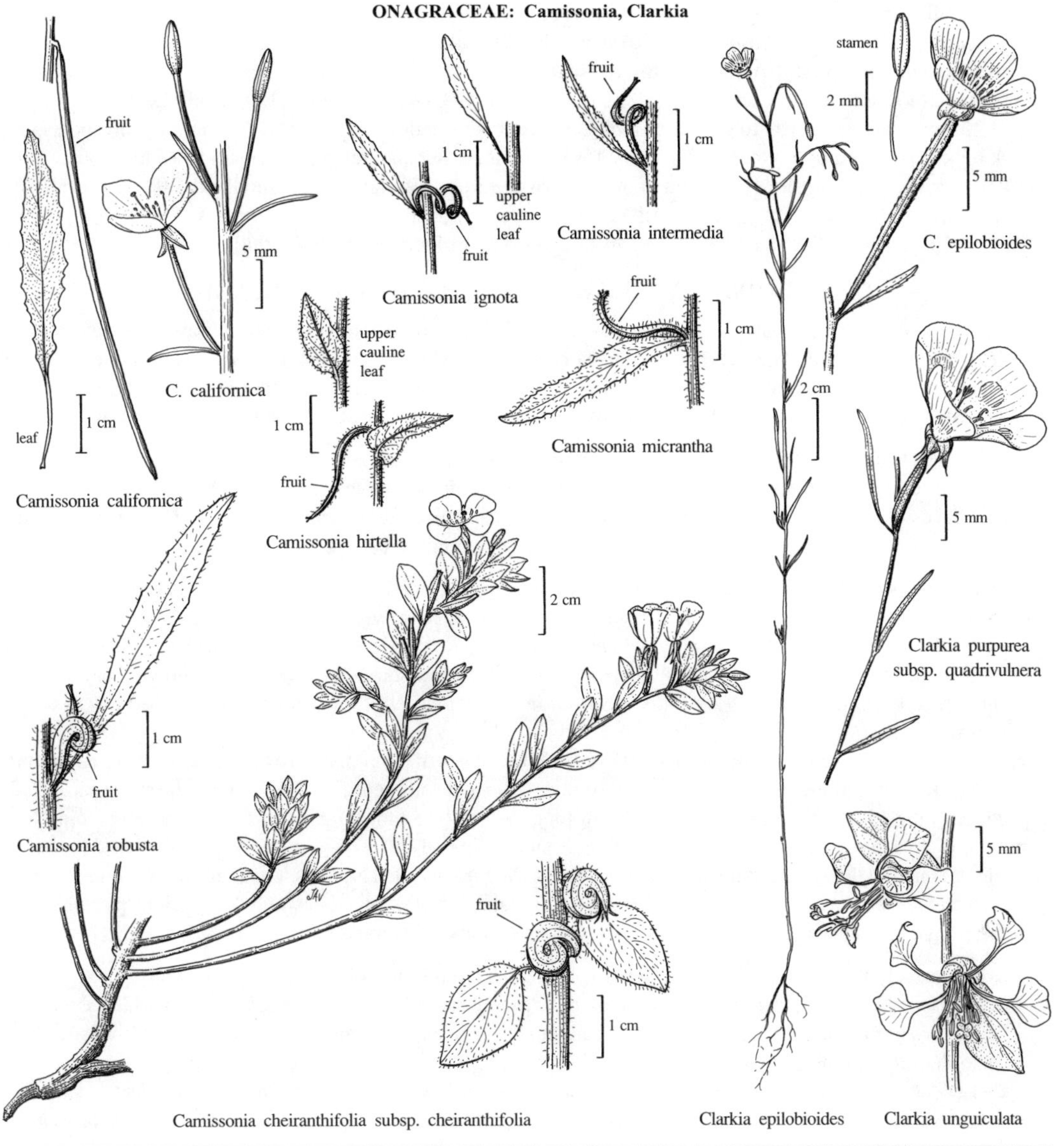


1. Petals with a distinct claw ........ **C. unquiculata**
1' Petals not clawed, tapered to base
  2. Flower buds pendulous; petals white, aging pink; fruit pedicellate ........ **C. epilobioides**
  2' Flower buds erect; petals purple; fruit subsessile ........ **C. purpurea**

**C. epilobioides** (Nutt.) A.Nelson & J.F.Macbr. Plants erect, to 70 cm tall. Leaves linear to narrowly lanceolate or oblanceolate, 15-25 mm long, 2-4 mm wide. Flower buds pendulous; petals 5-10 mm long, white, aging pink. Fruit cylindrical, 10-25(30) mm long, pedicel to 11 mm long. Mar-May. Common; shaded n-facing slopes, grassy hillsides; chaparral, grasslands, oak woodland, pine forest, riparian communities; 20-2350 ft (6-715 m). Scattered locations throughout island. Santa Rosa, Anacapa, Santa Catalina, and San Clemente islands; San Francisco Bay region to n Baja CA.

**C. purpurea** (Curtis) A.Nelson & J.F.Macbr. subsp. **quadrivulnera** (Douglas) F.H.Lewis and M.R.Lewis Plants erect, to 7.5(10) cm tall. Leaves linear to narrowly lanceolate or oblanceolate, 20-70 mm long, 2-5(12) mm wide. Flower buds erect; petals 5-25 mm long, deep purple. Fruit quadrangular, 10-20(30) mm long, sessile to subsessile.

May. Rare; disturbed flats, grassy areas; coastal scrub, grasslands; ca. 250 ft (75 m). Central Valley just west of University of California field station. Santa Rosa and Santa Catalina islands; WA to Baja CA, e to AZ. Two other subspecies occur throughout CA w of the Sierra Nevada.

**C. unquiculata** Lindl. [*C. elegans* Douglas] ELEGANT CLARKIA Plants erect, to 1 m tall. Leaves 10-60 mm long, 5-20 mm wide, lanceolate to elliptic or ovate. Flower buds pendulous; petals 10-20 mm long, lavender-pink to dark red-purple, with or without a spot. Fruit 15-30 mm long, subquadrangular, pedicel to 10 mm long. Mar-Jun. Rare; shaded n-facing slopes, canyon bottoms; coastal scrub, grasslands, oak woodland; 200-650 ft (60-200 m). North side in Scorpion Canyon; Central Valley in upper part of Cañada de la Siesta and near El Sobrante. Santa Catalina Island; n CA coast ranges to San Diego Co., e to Sierra Nevada foothills.

## **EPILOBIUM** L. WILLOW HERB or CALIFORNIA FUCHSIA

Annuals, perennial herbs, or subshrubs. Leaves basal and cauline, usually opposite below, alternate above. Flowers in racemes. Sepals 4, erect; petals 4, usually notched, white to purplish or scarlet-red; stamens 8, in 2 series, anthers attached at middle; style 1, stigma capitate or 4-lobed. Fruit cylindrical to clavate, straight or coiled; seeds with a deciduous tuft of long, soft trichomes. Ca. 170 species, temperate to subtropical, worldwide.

1. Petals scarlet-red; plants subshrubs or at least suffrutescent; hypanthium tube 20-35 mm long ..... **E. canum**
1' Petals white to purplish, but not scarlet-red; plants herbaceous; hypanthium tube 0.5-8 mm long
  2. Leaves linear, oblong, or narrowly elliptic, entire, early deciduous; annuals ............. **E. brachycarpum**
  2' Leaves lanceolate to ovate, serrulate, persistent; perennial herbs ........................................ **E. ciliatum**

**E. brachycarpum** C.Presl [*E. paniculatum* Torr. & A.Gray] Annuals, 30-70 cm tall; stems erect, bases with exfoliating epidermis. Leaves mostly cauline, linear, oblong, or narrowly elliptic, 1-3.5 cm long, entire, early deciduous. Hypanthium tube 1-7.5 mm long; sepals 2-6 mm long; petals 3-10 mm long, white to purplish. Fruit 15-30 mm long. Sep. Rare; rocky slopes; chaparral, oak woodland; ca. 1000 ft (300 m). Known from a single collection in east fork of canyon above Lady's Harbor in September 1989. British Columbia to San Diego Co., e to c N. America.

**E. canum** (Greene) Raven CALIFORNIA FUCHSIA Suffrutescent perennial herbs or subshrubs; stems erect to decumbent. Leaves cauline, linear to narrowly elliptic, 0.5-5 cm long, entire to denticulate. Flower scarlet-red; hypanthium tube 15-25 mm long; sepals 5-12 mm long; petals 7-15 mm long. Fruit 15-30 mm long. Jun-Nov. Common; rocky slopes, canyon walls; chaparral, coastal scrub; 20-1350 ft (6-410 m). Widely scattered locations throughout island. All CA Channel islands except Santa Barbara and San Nicolas; OR to n Mexico, e to WY and NM. Three races recognized as subspecies occur on the island. The following key and descriptions pertain to the extreme forms, among which there appears to be extensive local intergradation.

1. Leaves sparsely to moderately villous; lower leaves 20-50 mm long, 3-8 mm wide ......... subsp. **mexicanum**
1' Leaves moderately to densely villous or canescent; lower leaves 10-30 mm long, 0.5-4 mm wide
  2. Leaves narrowly elliptic to narrowly lanceolate, lower ones 10-30 mm long, 1.5-4 mm wide, moderately to densely villous or canescent ........................................................................ subsp. **angustifolium**
  2' Leaves ± filiform, lower ones 10-15 mm long, 0.5-2 mm wide, densely villous or canescent ................ ........................................................................................................................ subsp. **canum**

subsp. **angustifolium** (D.D.Keck) Raven [*Zauschneria californica* C.Presl; *Z. c.* subsp. *angustifolia* D.D. Keck] Leaves narrowly elliptic to narrowly lanceolate, moderately to densely villous or canescent, lower ones 10-30 mm long, 1.5-4 mm wide. Jun-Nov. Common; coastal bluffs, rocky slopes; chaparral, coastal scrub; 20-1300 ft (6-400 m). Widely scattered locations throughout island. San Miguel, Santa Rosa, Anacapa, and Santa Catalina islands; Monterey Co. to San Diego Co.

subsp. **canum** [*Zauschneria cana* Greene] Leaves ± filiform, often appearing to be in dense fascicles, densely villous to canescent, lower ones 10-15 mm long, 0.5-2 mm wide. Aug-Nov. Common; rocky s-facing slopes; chaparral, coastal scrub; 20-1350 ft (6-410 m). Widely scattered locations throughout island, especially on s side. Anacapa and Santa Catalina islands; Monterey Co. to Los Angeles Co.

subsp. **mexicanum** (C.Presl) Raven [*Zauschneria mexicana* C.Presl; *Z. californica* subsp. *mexicana* (C.Presl) Raven] Leaves narrowly elliptic to lanceolate or oblanceolate, villous to sparsely villous, lower ones 20-50 mm long, 3-8 mm wide. Jun-Oct. Occasional; rocky slopes, canyon walls; chaparral, coastal scrub; 20-450 ft (6-140 m). North side between Cueva Valdez and Pelican Bay; Central Valley. Santa Catalina and San Clemente islands; Monterey Co. to Baja CA.

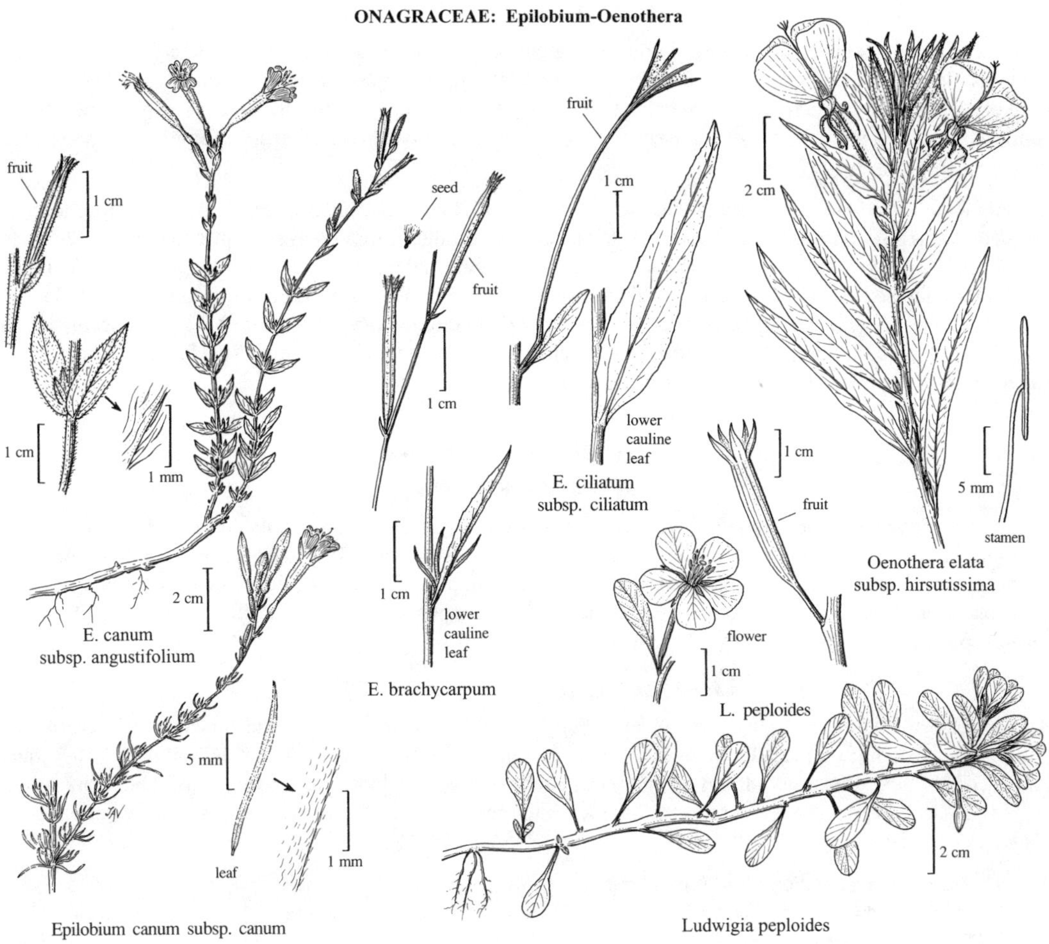


**E. ciliatum** Raf. subsp. **ciliatum** WILLOW HERB Perennial herbs, 12-120 cm tall; stems erect, without exfoliating epidermis. Leaves basal and cauline, lanceolate to ovate, 1-5 cm long, serrulate, persistent. Hypanthium tube 0.5-3 mm long, sepals 2-6 mm long; petals 2-15 mm long; white to purplish. Fruit 20-70 mm long. Feb-Nov. Common; moist streambeds, seeps; riparian communities; 15-1250 ft (6-380 m). Scattered locations throughout western and central portions of island. San Miguel and Santa Catalina islands; N. America, s S. America, e Asia, introduced elsewhere. Two other subspecies occur in w N. America. Foliage was used medicinally by the Maidu Tribe.

## LUDWIGIA L.

Perennial herbs, often in moist or aquatic habitats. Leaves cauline, alternate to opposite. Flowers in spikes or solitary. Hypanthium absent; sepals 4-6, persistent, erect to spreading; petals 4-5, white to yellow; stamens 4 or 10 (in 2 series of 5 each), anthers attached at middle; style 1, stigma capitate. Fruit straight, dehiscing irregularly or by a terminal pore. 82 species, temperate to subtropical, worldwide. Some species are edible; one was used medicinally by the Maidu Tribe.

**L. peploides** (Kunth) Raven subsp. **peploides** MARSH PURSLANE Plants usually matted; stems prostrate to decumbent, glabrous. Leaves alternate, 10-42 mm long, blades oblong to obovate, bases tapered to petiole. Sepals usually 5, 3-9 mm long, petals 10-15 mm long, yellow. Capsule cylindrical, 15-23 mm long, pedicellate, spreading to reflexed. Jul-Oct. Rare; moist streambeds; riparian communities; 5-15 ft (1.5-4.5 m). North side at mouth of Cañada del Puerto; Central Valley at mouth of Cañada Christy. Oregon to Baja CA, e to se N. America, Eurasia, Australia. Subsp. *montevidensis* (Spreng.) Raven, with pubescent stems, occurs on the mainland.

### **OENOTHERA** L. EVENING PRIMROSE

Annuals or perennial herbs. Leaves basal and cauline, alternate. Flowers solitary or in racemes, usually opening at sunset. Sepals 4, reflexed; stamens 8, anthers attached at middle; style 1, with 4 ± linear stigmatic branches. Fruit straight to curved, sessile or pedicellate. 119 species, temperate N. and S. America. Some species are cultivated as ornamentals. Seeds of some species were eaten by Native Americans; stems and flowers were used medicinally by the Diegueño Tribe.

**O. elata** Kunth subsp. **hirsutissima** (A.Gray ex S.Watson) W.Dietr. [*O. hookeri* Torr. & A.Gray subsp. *venusta* (Bartlett) Munz] Biennials, 1-2 m tall, upper stems densely strigulose. Basal leaves elliptic to oblong, 12-18 cm long, cauline ones gradually reduced upwards, lanceolate, entire to denticulate, 3-12 cm long. Hypanthium tube 30-43 mm long; sepals 25-35 mm long, reddish; petals 25-45 mm long, yellow. Fruit straight, erect, 30-35 mm long. May-Dec. Scarce; floodplains, stream banks; riparian communities; 10-900 ft (2-275 m). North side in Cañada del Puerto; Central Valley between Cañada de la Portezuela and Stanton Ranch. Throughout w N. America, to C. America. Two other subspecies occur in w N. America.

## **OROBANCHACEAE** Vent.
## BROOM-RAPE FAMILY

Plants parasitic, annuals or perennial herbs; stems usually fleshy, yellow, brown, or purplish. Leaves reduced, scale-like. Flowers bisexual, bilateral, in spikes or racemes. Calyx with 4-5 equal to unequal lobes; corolla ± bilabiate, with 5 lobes; stamens 4, in 2 pairs, adnate to corolla; pistil 1, ovary superior, locule 1. Fruit a capsule; seeds many, minute. Ca. 15 genera and 200 species; mostly temperate, N. America, Europe, and Asia. Roots of several species were eaten by Native Americans.

### **OROBANCHE** L. BROOM-RAPE

Annuals or perennial herbs, usually glandular-puberulent; underground stem and attachment to host-root thick, bul- or tuber-like. Flowers usually in dense racemes or spikes; bracts lanceolate to deltate. Calyx with 4-5 equal lobes; corolla tube often curved, upper lip erect to reflexed, with 2 lobes, lower lip with 3 spreading lobes; stigmas 2. Ca. 150 species; temperate, N. America, Europe, and Asia. Roots of several species were eaten by Native Americans.

1. Inflorescence open; pedicels 15-90 mm long
    2. Flowers 3-14 per stem; calyx lobes triangular to deltate, subequal to tube ...................... **O. fasciculata**
    2' Flowers 1-3 per stem; calyx lobes ± linear to narrowly triangular, longer than tube ............ **O. uniflora**

1' Inflorescence dense, spike-like; pedicels 2-6 mm long
    3. Stems dark purple; calyx tube with a deep, ventral sinus; corolla papillose ......................... **O. bulbosa**
    3' Stems yellowish to light brown; calyx tube not deeply divided ventrally; corolla glandular-puberulent ............................................................................................................ **O. parishii** subsp. **brachyloba**

**O. bulbosa** Beck Plants 7-20 cm tall; underground stem scales imbricated; aerial stems dark purple. Inflorescence dense, spike-like, flowers many; pedicels 2-6 mm long. Calyx tube 4-6 mm long, with a deep, ventral sinus, lobes narrowly triangular, 5-8 mm long; corolla 16-19 mm long, brownish to purplish, papillose; anthers glabrous to sparsely puberulent. Apr-Jun (Sep). Scarce; rocky n-facing slopes, exposed ridgetops; chaparral, pine forest; 500-1700 ft (150-520 m). Widely scattered locations on n side between Peak 1848 West and Cañada del Puerto; s side near Christy Pines. Santa Rosa and Santa Catalina islands; n CA coast Ranges to Baja CA. On Santa Cruz Island this species is apparently parasitic on *Adenostoma*.

**O. fasciculata** Nutt. Plants 10-18 cm tall; stems brown to purplish near base. Inflorescence open, flowers 3-14 per stem; pedicels 15-60 mm long. Calyx tube 3-6 mm long, lobes triangular to deltate, 3-7 mm long; corolla 12-20 mm long, yellow to purplish, glandular-puberulent; anthers puberulent. May-Jun. Scarce; loose s-facing talus slopes, exposed ridgetops; chaparral, coastal scrub; 600-950 ft (185-290 m). Central Valley on Portezuela Grade; s side on ridges near Alberts Anchorage and Los Pinos del Sur. All CA Channel islands except Santa Barbara and San Clemente; AK to Mexico, e to Great Plains. On Santa Cruz Island this species is apparently parasitic on *Eriogonum*. Entire plants were eaten by the Maidu Tribe.

**O. parishii** (Jeps.) Heckard subsp. **brachyloba** Heckard Plants 8-15 cm tall; stems yellowish to light brown. Inflorescence dense, spike-like, flowers many; pedicels 2-4.5 mm long. Calyx tube 3-5 mm long, lobes narrowly triangular, 6-10 mm long; corolla 15-22 mm long, cream to purplish, with yellow markings, glandular-puberulent;

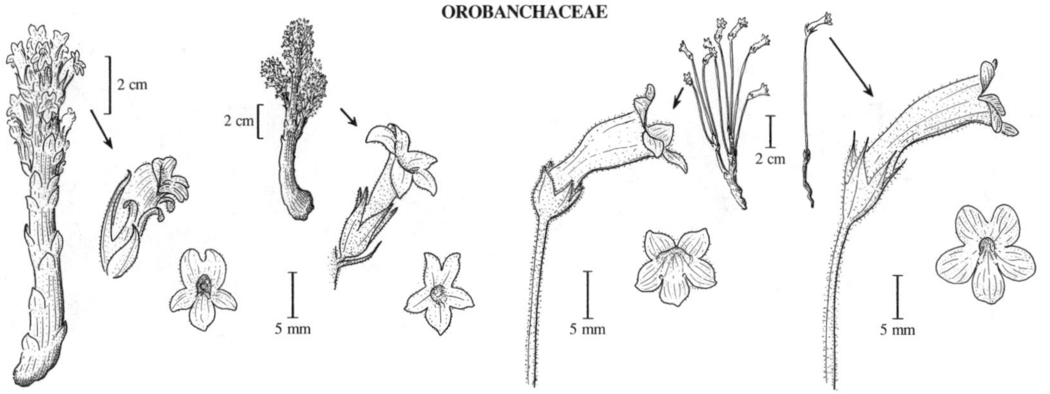


Orobanche parishii subsp. brachyloba    O. bulbosa    O. fasciculata    O. uniflora subsp. occidentalis

anthers ± glabrous. Jun-Oct. A subspecies of special environmental concern. Scarce; coastal flats, loose s-facing talus slopes, exposed ridgetops, floodplains; 10-560 ft (3-170 m). North side near east end of China Harbor; scattered in coastal areas on s side between West Point and w end of Pozo Beach. San Miguel, Santa Rosa, San Nicolas, and Santa Catalina islands; San Luis Obispo Co. to n Baja CA. On Santa Cruz Island this subspecies is apparently parasitic on *Isocoma*. Subsp. *parishii*, with calyx lobes 10-16 mm long, occurs from the s Sierra Nevada to Baja CA.

**O. uniflora** L. Plants 6-14 cm tall; stems yellowish to brown. Inflorescence open, flowers 1-3 per stem; pedicels 30-90 mm long. Calyx tube 2-3.5 mm long, lobes ± linear to narrowly triangular, 3-7 mm long; corolla 12-23 mm long, blue with yellow markings to purple, glandular-puberulent; anthers puberulent. Mar-Apr. Occasional; shaded canyon bottoms, rocky n-facing slopes; chaparral; 10-1200 ft (3-365 m). North side between Lady's and Prisoners' harbors; Central Valley between La Cascada and Stanton Ranch. Throughout N. America. On Santa Cruz Island this species is apparently parasitic on *Saxifraga*. This species was used medicinally by the Maidu Tribe.

## OXALIDACEAE R.Br.
### OXALIS OR WOOD SORREL FAMILY

Annuals or perennial herbs. Leaves basal or alternate, compound; stipules present. Flowers bisexual, radial, in racemes or umbels. Sepals 5, free or fused; petals 5, free or fused; stamens 10 or 15, fused near bases; pistil 1, ovary superior, locules 3-5, styles 3-5. Fruit a capsule; seeds many. Ca. 8 genera and 575 species; temperate to tropical, ± worldwide.

### OXALIS L. WOOD SORREL

Annuals or perennial herbs, often with rhizomes, bulbs, or tubers. Leaflets 3, often obcordate. Inflorescence axillary or terminal, solitary or in umbel-like clusters. Sepals and petals free; stamens 10, in 2 unequal series, filaments fused near bases; locules 5; styles 5. Fruit explosively dehiscent. Ca. 500 species; worldwide. Some species are cultivated as ornamentals or for edible tubers; other species are important weeds.

1. Leaflets 10-15 mm long; petioles 4.5-14 cm long; petals 18-25 mm long, bright yellow ........ **O. pes-caprae**
1' Leaflets 5-9 mm long; petioles 1.6-6 cm long; petals 4-12 mm long, light yellow
  2. Petals often 8-12 mm long; stems decumbent, not rooting at nodes; taproot often woody ........................ **O. albicans** subsp. **californica**
  2' Petals often 4-8 mm long; stems rooting at nodes; taproot slender or fleshy, not especially woody ........ **O. corniculata**

**O. albicans** Kunth subsp. **californica** (Abrams) Eiten CALIFORNIA WOOD SORREL Perennial herbs, with ± woody taproots; stems decumbent, 1-4 dm long, ± glabrous. Leaflets 5-9 mm long; petioles 2-5 cm long. Flowers 1-5 per peduncle. Petals often 8-12 mm long, light yellow. Fruit 6-14 mm long. Feb-May. Rare; rocky slopes, canyon bottoms, grassy hillsides; coastal scrub, grasslands; 10-185 ft (3-60 m). North side in Cañada del

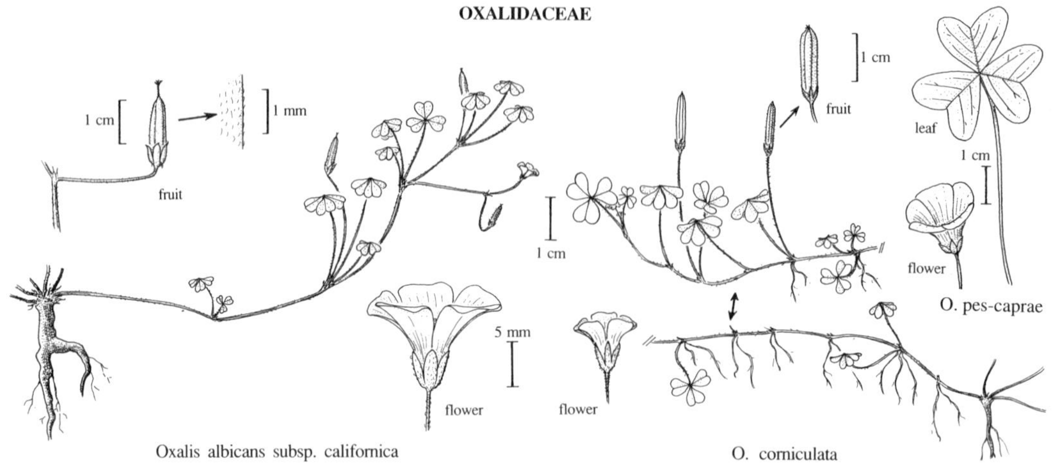


Puerto; s side in lower part of Smugglers' Canyon. Santa Catalina Island; Santa Barbara Co. to Baja CA. Subsp. *pilosa* (Nutt.) G.Eiten, with slightly curved trichomes on the stems, occurs on the mainland from Mendocino Co. to n Baja CA, e to TX.

**O. corniculata** L. CREEPING WOOD SORREL Perennial herbs, with a ± fleshy taproot; stems prostrate or creeping, often rooting at nodes, 0.5-3 dm long. Flowers 1-5 per peduncle. Leaflets 5-8 mm long; petioles 1.5-6 cm long. Petals often 4-8 mm long, light yellow. Fruit 6-18 mm long. Jun-Nov. Introduced. Rare; disturbed sites; coastal scrub, grasslands; 20-200 ft (6-60 m). North side in lower part of Valdez Canyon and at Prisoners' Harbor; Central Valley at Christy and Stanton ranches. San Nicolas, Santa Catalina, and San Clemente islands; throughout N. America; native of Europe. First collected at Stanton Ranch in June 1930.

**O. pes-caprae** L. BERMUDA BUTTERCUP Acaulescent, ± scapose perennial herbs, with short, fleshy rhizomes and bulblets. Flowers 5-14 per peduncle. Leaflets 10-15 mm long; petioles 4.5-14 cm long. Petals 18-25 mm long, bright yellow. Fruits and seeds not produced. Apr. Introduced. Rare; sandy flats; coastal beach and dune; ca. 10 ft (3 m). Known from a single collection near west end of Pozo Beach in April 1993. San Nicolas, Santa Catalina, and San Clemente islands; Mendocino Co. to n Baja CA; native of s Africa. This species apparently reproduces and spreads only by bulblets. Toxic to sheep.

## PAPAVERACEAE Juss.

### Poppy Family

Annuals, perennial herbs, or shrubs; stems with clear or opaque sap. Leaves basal or cauline, usually alternate, simple to deeply dissected; stipules absent. Flowers bisexual, radial or bilateral, usually solitary. Sepals 2 or 4, often caducous; petals usually 4 or 6, free; stamens 4-many; pistil 1, ovary superior, locule 1, placentation parietal; style 1 or stigmas sessile. Fruit usually a capsule, dehiscent by valves or apical pores; seeds many. Ca. 23 genera and 210 species; mostly temperate, N. America, Europe, and Asia. Many species are cultivated as ornamentals; Opium Poppy is the natural source of morphine.

1. Shrubs with evergreen leaves .......... **Dendromecon**

1' Annuals or perennial herbs
  2. Sepals usually 3; petals usually 6; leaves entire to minutely toothed, cauline ones, if present, opposite
    3. Stamens 4-6; fruit a capsule, linear; seeds many .......... **Meconella**
    3' Stamens many; fruit breaking apart into 6-20 distinct segments, each segment with several seeds separated by constrictions .......... **Platystemon**
  2' Sepals usually 2; petals usually 4; leaves deeply lobed to dissected, cauline ones alternate
    4. Sepals fused into a cap that separates and falls from flower as petals expand; receptacle ± cup-shaped; fruit linear, dehiscent by 2 valves .......... **Eschscholzia**
    4' Sepals free from each other, caducous; receptacle not cup-shaped; fruit obovoid, dehiscent by terminal pores

5. Style absent, stigma sessile, disc-like ........ **Papaver**
5' Style short, slender, stigma capitate ........ **Stylomecon**

## **DENDROMECON** Benth. TREE POPPY

Shrubs or small trees, with clear sap. Leaves alternate, simple, entire, coriaceous, evergreen, glabrous. Flowers terminal, on short branches. Sepals 2, free; petals 4, yellow; stamens many; style absent, stigmatic lobes 2. Fruit a capsule, dehiscent from base by 2 valves. Ca. 2-3 species; much of CA w of the Sierra Nevada, Baja CA.

**D. rigida** Benth. subsp. **harfordii** (Kellogg) Raven [*D. harfordii* Kellogg] NORTHERN ISLAND BUSH POPPY Plants 1-3 m tall; stems green, glaucous. Leaves elliptic, rarely oblong, 3-8 cm long, 1.5-4.5 cm wide, margins entire, apices usually rounded. Petals 2-3.5 cm long. Fruit 7-10 cm long, curved. Mar-Nov. Island endemic. A subspecies of special environmental concern. Common; rocky slopes and canyons, exposed ridgetops; chaparral, pine forest; 50-1200 ft (15-365 m). North side between Profile Point and Cañada del Puerto, and near Scorpion Canyon; Central Valley between upper part of Cañada Christy and Stanton Ranch airfield; higher elevations on s side between Christy Pines and Coches Prietos drainage, and on El Montañon ridge. Santa Rosa Island. Mostly restricted to igneous and metamorphic substrates. Plants from Santa Catalina and San Clemente islands, with pale green leaves and ± acute leaf apices, are called subsp. *rhamnoides* (Greene) Thorne. Subsp. *rigida* occurs on the mainland.

## **ESCHSCHOLZIA** Cham.

Annuals or perennial herbs, with clear, often orange sap. Leaves alternate, pinnate, deeply dissected, glabrous. Flowers on long pedicels. Receptacle rim ± cup-shaped, hypanthium-like. Sepals 2, fused into cap that separates and falls from flower as petals expand and corolla opens; petals 4, yellow to orange; stamens many; style short, stigmatic lobes 4-8, linear. Fruit a capsule, dehiscent from base by 2 valves. Ca. 12 species; w N. America. *E. californica*, a widely cultivated species, is the state flower of California.

1. Receptacle rim with outer, spreading to reflexed collar; petals 20-30 mm long ........ **E. californica**
1' Receptacle without outer collar; petals 5-15 mm long ........ **E. ramosa**

**E. californica** Cham. CALIFORNIA POPPY Annuals or perennial herbs, 20-60 cm tall. Leaves green, sometimes glaucous. Pedicels 6-15 cm long. Receptacle with minute, inner, hyaline rim and an outer, spreading to reflexed collar; petals 20-30 mm long, yellow to orange. Fruit 5-9 cm long. Mar-Oct. Common; coastal bluffs, rocky slopes and canyon bottoms, floodplains; coastal scrub, grasslands; 20-1050 ft (3-315 m). Widely scattered locations throughout island. All CA Channel islands except Anacapa and Santa Barbara (introduced on San Nicolas and San Clemente); Los Coronados, Todos Santos, and Guadalupe islands; s WA to n Baja CA, e to NV and NM. Although this species was widely eaten and used medicinally by Native Americans, it contains alkaloids and can be toxic to humans.

**E. ramosa** Greene ISLAND POPPY Annuals, 15-40 cm tall. Pedicels 10-40 cm long. Receptacle with minute, inner, hyaline rim, outer collar absent; petals 5-15 mm long, yellow, with orange spot near bases. Fruit 5-6 cm long. Mar-May. Island endemic. A species of special environmental concern. Scarce; coastal bluffs, rocky slopes and canyons; coastal scrub; 50-1250 ft (15-380 m). South side near Forney's Cove and Rancho Nuevo, and widely scattered locations between mouth of Laguna Canyon and slopes s of Navy Base. All CA Channel islands except San Miguel and Anacapa; all Baja CA islands except San Geronimo. Can be locally abundant in years with heavy rainfall; rarely seen in dry years.

## **MECONELLA** Nutt.

Annuals, with clear sap. Leaves mostly basal, cauline ones, when present, opposite, simple, usually entire, glabrous to pubescent. Sepals usually 3, free; petals usually 6, white to yellow; stamens 4-many; style absent, stigmatic lobes usually 3, ± ovate. Fruit a capsule, dehiscent from apex by 2-3 valves. Ca. 4 species; w N. America.

**M. denticulata** Greene Plants 3-15 cm tall. Leaves spatulate to oblanceolate, rarely linear, 0.5-3.5 cm long, entire to denticulate, glabrous, sessile or lower leaves tapered to petiolate bases. Sepals 1-2 mm long; petals 2-4 mm long, white. Fruit erect, linear, 1.5-2 cm long, spirally twisted. Mar-May. Occasional; shaded n-facing slopes, rock faces, canyon bottoms; chaparral, oak woodland, pine forest; 750-2200 ft (230-670 m). Widely scattered on n side between Cueva Valdez and Pelican Bay, and on El Montañon ridge; scattered in Central Valley between Christy Pines and Cañada de la Mina. San Francisco Bay region to s CA. Restricted to igneous and metamorphic substrates.

**PAPAVERACEAE**

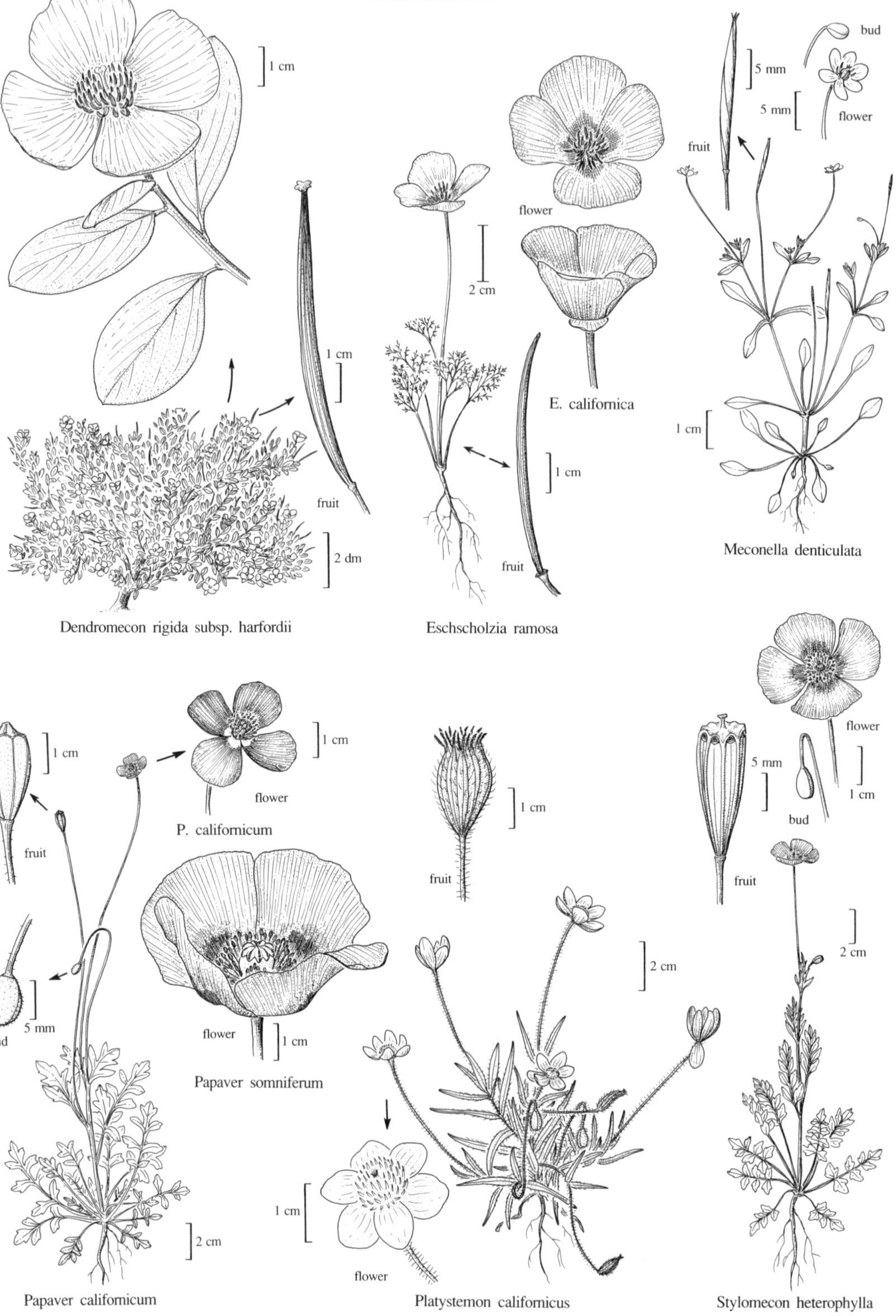

## PAPAVER L. POPPY

Annuals or perennial herbs, with opaque, white to yellow sap. Leaves glabrous to pubescent, mostly basal; cauline ones alternate, entire to pinnatifid. Flowers nodding in bud. Sepals 2, free; petals 4, white to red or purple; stamens many; style absent, stigma disk-like, sessile. Fruit a capsule, opening by pores below stigma. Ca. 50 species; temperate, mostly N. America, Europe, and Asia.

1. Petals 10-20 mm long, each with green spot near base; cauline leaves sessile; fruit narrowly obovate .................................................................. **P. californicum**

1' Petals 40-50 mm long, each with purple spot near base; cauline leaves clasping stem; fruit ± globose .................................................................. **P. somniferum**

**P. californicum** A.Gray FIRE POPPY Annuals, 1-6 dm tall. Leaves oblong to obovate in outline, 3-9 cm long, deeply lobed to divided, margins toothed. Sepals 8-10 mm long; petals 10-20 mm long, red, each with green spot at base. Fruit narrowly obovate, 10-16 mm long. Mar-May. Scarce; grassy hillsides, rocky slopes, canyon walls; chaparral, coastal scrub, grasslands, island woodland, pine forest; 100-2200 ft (30-670 m). Widely scattered on n side between Picacho Diablo and Prisoners' Harbor, and in Scorpion Canyon; Central Valley in Christy Pines and at head of Cañada de la Mina; s side in Laguna Canyon. Santa Rosa Island; San Francisco Bay region to s CA. Restricted to igneous and metamorphic substrates. Usually seen only in years with heavy rainfall or after fires.

**P. somniferum** L. OPIUM POPPY Annuals, 6-12 dm tall. Leaves oblong in outline, 10-20 cm long, coarsely toothed to lobed, upper ones with broad, clasping bases. Sepals 20-25 mm long; petals 40-50 mm long, bright red, each with deep purple spot at base. Fruit ± globose, 30-50 mm long. Sep. Introduced. Rare; rock walls, disturbed slopes; grasslands; 40-100 ft (12-30 m). North side at Pelican Bay, among ruins of Eaton resort. Santa Catalina Island; ± worldwide; native of Europe and Asia. This species is the source of raw opium, which contains morphine and other alkaloids.

## PLATYSTEMON Benth. CREAM CUPS

With only 1 species.

**P. californicus** Benth. Annuals, 1-4.5 dm tall, with clear sap. Leaves 1-9 cm long, mostly basal, cauline ones, when present, opposite, linear to lanceolate, entire, sessile, villous. Flowers on long, villous pedicels. Sepals usually 3, 6-10 mm long, free; petals usually 6, 8-15 mm long, cream to yellow; stamens many; pistil appearing as 1 but separating in fruit into 6-20 distinct segments, styles 6-10, persistent in fruit. Fruit segments linear, each with several seeds separated by constrictions, pubescence of stiff, erect, appressed trichomes. Mar-May (Aug). Common; coastal bluffs, rocky slopes, grassy hillsides; chaparral, coastal scrub, grasslands; 20-800 ft (6-245 m). Widely scattered locations throughout island. All CA Channel islands except San Clemente; Guadalupe Island; OR to Baja CA, e to UT and AZ. Leaves were eaten by Native Americans.

## STYLOMECON G.C.Taylor WIND POPPY

With only 1 species.

**S. heterophylla** (Benth.) G.C.Taylor Annuals, to 4.5 dm tall, with clear, yellow sap. Leaves 2-12 cm long, basal and cauline, alternate, pinnatifid, segments entire to deeply lobed. Sepals 2, 4-10 mm long, free; petals 4, 10-20 mm long, orange-red, each with basal purplish spot and green claw; stamens many; style short, slender, stigma capitate. Fruit an obovate or turbinate capsule, dehiscent by pores near apex. Mar-May. Rare; coastal bluffs, rocky slopes; coastal scrub; 100-130 ft (30-40 m). North side ca. 1.5 miles (2.4 km) e of West Point; s side near Smugglers' Cove (not collected there since March 1932). All CA Channel islands except San Nicolas; Los Coronados, Todos Santos, and San Martin islands; throughout CA w of Sierra Nevada, n Baja CA. Usually seen only in years with heavy rainfall or after fires.

# PLANTAGINACEAE Juss.

## PLANTAIN FAMILY

Annuals or perennial herbs, usually scapose. Leaves usually in basal rosettes, simple, entire; stipules absent. Flowers bisexual or unisexual, radial or bilateral, each subtended by 1 bract, in dense, terminal spikes. Calyx with 4 lobes, persistent in fruit; corollas cylindrical to salverform, with 4 lobes, scarious, persistent in fruit;

stamens 2-4, adnate to corolla; pistil 1, ovary superior, locules 2-4, style 1, stigma puberulent. Ca. 3 genera and 250 species; mostly temperate, ± worldwide.

## PLANTAGO L. PLANTAIN

Leaves in basal rosettes. Spikes on short to long peduncles. Sepal lobes often unequal; corolla tube cylindrical, often constricted at throat, limb usually with spreading lobes. Fruit circumscissile at or below middle. Ca. 250 species; ± worldwide.

1. Perennial herbs; peduncles usually 20-50 cm long; leaves lanceolate to ovate
    2. Corolla lobes erect, forming a beak over fruit ........ **P. subnuda**
    2' Corolla lobes spreading to reflexed, especially in fruit
        3. Leaves lanceolate to oblong; spikes cone-shaped to ovoid, becoming cylindrical in fruit; ovules and seeds 1-2 ........ **P. lanceolata**
        3' Leaves broadly elliptic to ovate; spikes narrowly cylindrical; ovules and seeds 8-18 ....... **P. major**

1' Annuals; peduncles usually 1-20 cm long; leaves linear to linear-lanceolate
    4. Corolla lobes ± lanceolate, ca. 0.5 mm long, apices acute ........ **P. elongata**
    4' Corolla lobes broadly ovate, 1.5-2.5 mm long, apices obtuse
        5. Bract ± half as long as calyx; seeds brown, dull ........ **P. erecta**
        5' Bract length ± equal to calyx length; seeds reddish yellow, shiny ........ **P. ovata**

**P. elongata** Pursh [*P. bigelovii* A.Gray subsp. *californica* (Greene) Bassett] COASTAL PLANTAIN Annuals. Leaves linear, 3-7 cm long, entire or with 2 teeth near middle, margins ciliate. Peduncles 2-10 cm long, pubescent near spike, glabrate below; spikes cylindrical, 0.5-4 cm long; bract length ± equal calyx length. Flowers bisexual, staminate, or pistillate; sepals oblong, ± 2 mm long; corolla lobes lanceolate, ± 0.5 mm long; stamens 2. Seeds 5-15, angled, irregularly and coarsely pitted, black. Apr-May. Rare; coastal flats and bluffs; coastal scrub, grasslands; 370-400 ft (110-120 m). North side near Del Mar Cove. San Miguel and Santa Rosa islands; throughout much of w N. America.

**P. erecta** E.Morris CALIFORNIA PLANTAIN Annuals. Leaves linear, 1-12 cm long, entire, appressed-pubescent. Peduncles 7-20 cm long, densely pubescent; spikes cylindrical, 0.5-2 cm long; bract length ± half calyx length. Flowers bisexual; sepals oblong to ovate, 2-3 mm long; corolla lobes 1.5-2.5 mm long, usually reflexed; stamens 4. Seeds 2, minutely pitted, dark brown. Mar-Jun. Occasional; grassy hillsides, rocky slopes, exposed ridgetops; chaparral, coastal scrub, grasslands; 50-1550 ft (15-470 m). Widely scattered locations in western and central portions of island. Santa Rosa, Anacapa, Santa Catalina, and San Clemente islands; OR to Baja CA. Leaves were used medicinally by Native Americans; seeds were used as a laxative.

**P. lanceolata** L. ENGLISH PLANTAIN Perennial herbs, with tufts of tan trichomes near leaf bases. Leaves narrowly lanceolate to oblong, 5-50 cm long, entire to denticulate, glabrous to pubescent. Peduncles 10-60 cm long, densely pubescent; spikes cone-shaped to ovoid, becoming cylindrical in fruit, 2-8 cm long; bract length ± equal calyx length. Flowers bisexual, pistillate, or staminate; sepals ± ovate, 2-2.5 mm long; corolla lobes 2-2.5 mm long, spreading to reflexed; stamens 4. Seeds 1-2, shiny, dark brown. Apr-Nov. Introduced. Occasional; disturbed sites, roadsides; grasslands; 20-1200 ft (6-365 m). North side at Prisoners' Harbor and in Cañada del Puerto; scattered locations in Central Valley between Cañada Christy and Stanton Ranch. San Clemente Island; ± worldwide; native of Europe and Asia. First collected at Prisoners' Harbor in April 1977. Young leaves can be cooked and eaten.

**P. major** L. COMMON PLANTAIN Perennial herbs, glabrous to puberulent. Leaves broadly elliptic to ovate, 5-20 cm long, entire to slightly toothed, narrowed to winged petioles. Peduncles 20-40 cm long; spikes narrowly cylindrical, 4-30 cm long; bract length ± equal calyx length, lowest ones 2-3 times longer than calyx. Flowers bisexual; sepals ± ovate, ca. 2 mm long corolla lobes ca. 1 mm long, spreading to reflexed; stamens 4. Seeds 8-18, angled, reticulate, shiny, brown. Apr-Dec. Introduced. Occasional; moist creek beds, grassy flats, disturbed sites; grasslands, riparian communities; 5-700 ft (1-215 m). Scattered locations on n side between Prisoners' Harbor and China Harbor; Central Valley at mouth of Cañada Christy; scattered locations near coast on s side between Rancho Nuevo and mouth of Laguna Canyon. San Nicolas and Santa Catalina islands; Cedros Island; ± worldwide; native of Europe and Asia. Reported by Greene (1887). Leaves have been used medicinally.

**P. ovata** Forssk. [*P. insularis* Eastw.] Annuals, villous. Leaves linear to linear-lanceolate, 6-10 cm long, entire or minutely denticulate. Peduncles 2-18 cm long; spikes cylindrical, 0.5-2 cm long; bract length ± equal calyx length. Flowers bisexual; sepals elliptic to obovate, 2-2.5 mm long; corolla lobes broadly ovate, 1.5-2.5 mm long; stamens 4. Seeds 2, shiny, reddish yellow. Mar. Possibly introduced. Rare, not seen recently; habitat,

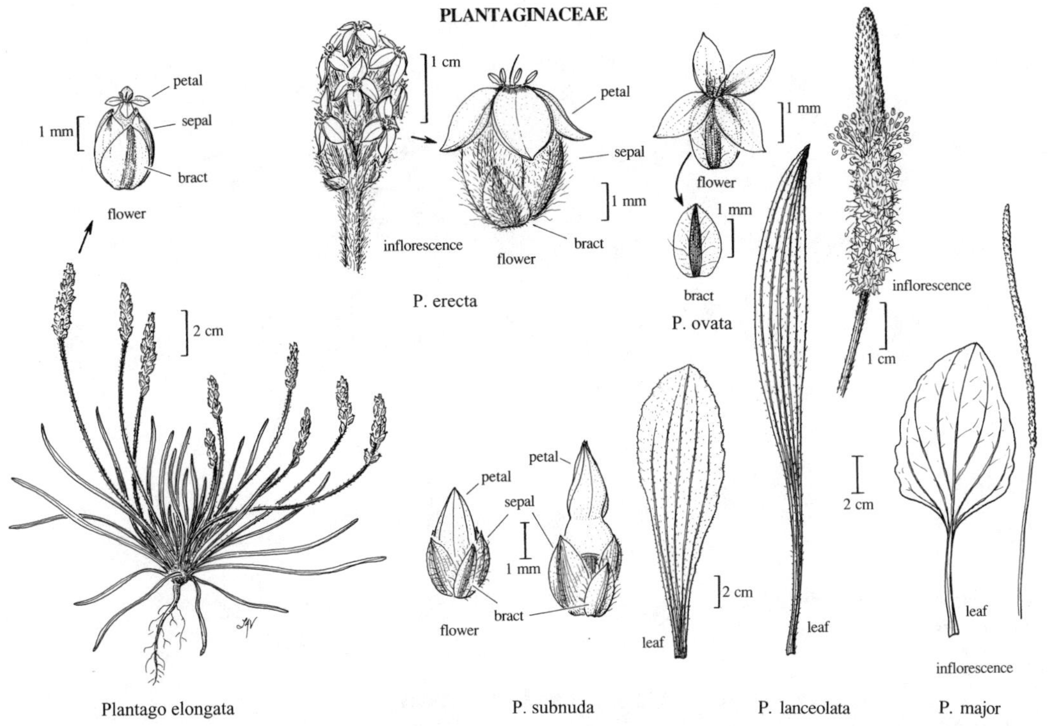


plant community, and elevation unknown. Known from a single collection without locality by R. Hoffmann in March 1925. All CA Channel islands; Guadalupe, San Benito, Cedros, and Natividad islands; throughout much of w N. America and Europe. Seeds are used to produce a laxative similar to that from *Plantago afra* L. ("psyllium").

**P. subnuda** Pilg. [*P. hirtella* Kunth var. *galeottiana* (Decne.) Pilg.] Perennial herbs, densely villous at base. Leaves elliptic, 10-25 cm long, entire to crenate-dentate near apices. Peduncles 40-60 cm long; spikes cylindrical, 5-30 cm long; bract length slightly shorter than calyx. Flowers bisexual, staminate, or pistillate; sepals ± elliptic, 2.5-3 mm long corolla lobes 2-2.5 mm long, spreading to reflexed; stamens 4. Seeds 3, dull, black. May-Sep. Rare; coastal bluffs, seeps, moist creek beds; riparian communities; 20-200 ft (6-60 m). North side in Cañada del Agua and at Aguaje Escondido; s side at mouth of Cañada de Malva Real. Santa Rosa Island; CA w of Sierra Nevada, to n Mexico.

## PLATANACEAE Dumort.

### PLANE-TREE OR SYCAMORE FAMILY

With only 1 genus. Ca. 8 species; temperate, N. America, Europe, and Asia.

### PLATANUS L.

Monoecious trees; bark often peeling in thin plates or scales. Leaves alternate, usually palmately veined and lobed, deciduous, often with stellate trichomes, petiolate; stipules often leaf-like, free or fused. Flowers unisexual, radial, subtended by bracts. Flowers in 1-6 dense, globose heads on pendant axis. Sepals minute, 3-8, minute, scale-like, free or fused; petals 3-6 or absent. Staminate flowers with 3-8 stamens. Pistillate flowers usually with 5-10 pistils, ovary superior, locule 1, style 1, persistent in fruit. Fruit an achene.

**P. racemosa** Nutt. CALIFORNIA SYCAMORE Trees, 10-25 m tall; twigs densely tomentose with stellate trichomes, glabrate and light brown. Leaf blades orbicular in outline, 10-20 cm long, palmately 5-lobed, stellate-tomentose, glabrate. Pistillate heads 3-5, 2-2.5 cm wide, sessile. Apr-May. Introduced. Scarce; floodplains, moist creek beds; riparian communities; 40-120 ft (12-35 m). North side in Cañada del Puerto; Central Valley at Stanton Ranch. Santa Catalina Island; San Francisco Bay region and Great Valley to n Baja CA. About 12 trees

PLATANACEAE

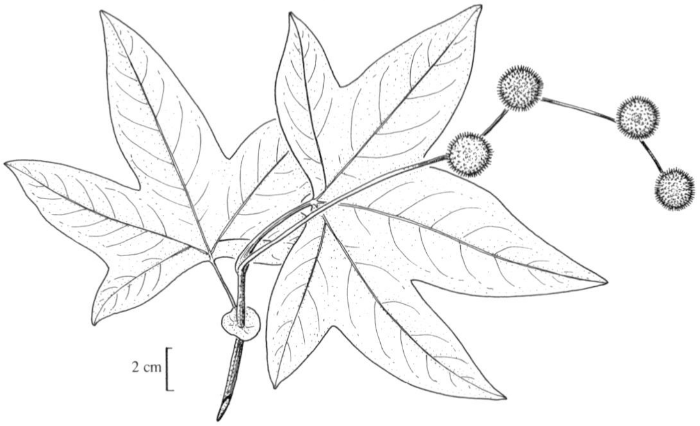


Platanus racemosa

were planted by Mrs. Edwin Stanton in Cañada del Puerto in 1937 or 1938. Bark and leaves were used medicinally by Native Americans.

## POLEMONIACEAE Juss.
### Phlox Family

Annuals, perennial herbs, or shrubs. Leaves basal or cauline, alternate or opposite, simple to compound; stipules absent. Flowers bisexual, radial to bilateral, in cymes, racemes, or panicles. Calyx with 5 lobes, tube herbaceous to translucent; corollas rotate to salverform, with 5 lobes; stamens 5, adnate to corolla, equal or unequal in length or position of insertion; pistil 1, ovary superior, locules 3, style with 3 branches or lobes. Fruit a capsule; seeds 3-many. Ca. 20 genera and 350 species; mostly temperate, N. and S. America, Europe, and Asia.

1. Leaves opposite, blades either simple and entire or palmately divided .......................................... **Linanthus**
1' Leaves alternate, blades pinnately divided
  2. Calyx lobes ± equal; bracts subtending flowers herbaceous, not stiff; flowers solitary or in ± open to moderately dense clusters .......................................................................................................... **Gilia**
  2' Calyx lobes ± unequal; bracts subtending flowers often stiff, with mucronate to spinescent apices; flowers in dense, bracteate clusters
    3. Stems, leaves and bracts lanate to woolly; leaves and bracts with mucronate apices..... **Eriastrum**
    3' Stems, leaves and bracts glandular-puberulent; leaves and bracts with spinescent apices ............. ........................................................................................................................ **Navarretia**

### ERIASTRUM Wooton & Standl.

Annuals or shrubs; stems simple to branched, glabrous to woolly. Leaves alternate, simple to pinnatifid. Flowers sessile, in dense, bracteate clusters; bract apices mucronate. Calyx lobes unequal; corollas radial to bilateral, funnelform to salverform; stamens equal to unequal in length, adnate to corolla at same level. Thirteen species; w N. America.

**E. filifolium** (Nutt.) Wooton & Standl. Annuals, to 25 cm tall; stems woolly. Leaves pinnatifid, 2-3 cm long, glabrous to woolly, segment apices s mucronate. Flowers 3-15 per cluster; bracts needle-like, to 2 cm long, woolly at bases. Calyx 5-7 mm long; corollas 6-9 mm long, tube yellow, limb blue. Fruit ca. 4 mm long. Apr-Jun. Rare; s-facing coastal bluffs, rocky slopes, loose talus slopes; coastal scrub; 150-350 ft (45-105 m). South side between mouth of Laguna Canyon and Corral Point, and in Coches Prietos drainage. Santa Catalina and San Clemente islands; Santa Barbara Co. to n Baja CA. Not collected between 1931 and 1991.

## **GILIA** Ruiz & Pav.

Annuals or perennial herbs; stems simple to branched, glabrous, glandular-puberulent, or sparsely lanate. Leaves mostly basal, cauline ones alternate, usually pinnatifid, segments usually linear. Flowers solitary or in ± open to moderately dense clusters, sessile or pedicellate; bracts simple to few-lobed, herbaceous. Calyx lobes equal; corollas radial, campanulate, funnelform, or salverform; stamens equal to unequal in length, adnate to corolla at same level. Ca. 70 species; w N. America and S. America.

1. Flowers in ± dense, capitate clusters; corolla tubes, throats, and limbs bluish; pedicels and calyces glabrous to sparsely ± white-pubescent ........ **G. capitata** subsp. **abrotanifolia**

1' Flowers solitary or in ± open clusters, each with 2-10 flowers; corolla throats or tubes yellow, limbs white to bluish or lavender; pedicels and calyces glabrous to puberulent, often glandular
  - 2. Corolla throats with purple spots; calyx tubes purplish between ribs ........ **G. clivorum**
  - 2' Corolla throats without purple spots; calyx tubes not purplish between ribs
    - 3. Corollas campanulate, 6-9 mm long; lower leaves 1-2 times pinnatifid ........ **G. angelensis**
    - 3' Corollas funnelform, 8-14 mm long; lower leaves 2-3 times pinnatifid ........ **G. nevinii**

**G. angelensis** V.E.Grant Annuals, 10-50 cm tall; stems glabrous to puberulent. Leaves 1.5-6 cm long, glabrous to puberulent, lower ones 1-2 times pinnatifid. Flowers solitary or in ± open clusters with 3-9 subsessile flowers. Calyx 3-5 mm long, glabrous to puberulent; corollas campanulate, 6-9 mm long, tube and throat yellow, limb white, bluish, or lavender; style exserted. Fruit ovoid, 4-5 mm long, shorter than calyx. Mar-May. Occasional; rocky slopes, grassy hillsides, floodplains; coastal scrub, grasslands, riparian communities; 30-1500 ft (10-460 m). Widely scattered locations throughout island, especially on n side and in Central Valley. Santa Catalina Island; Los Coronados and Todos Santos islands; San Francisco Bay region to n Baja CA.

**G. capitata** Sims subsp. **abrotanifolia** (Nutt. ex Greene) V.E.Grant BALL GILIA Annuals, 10-40 cm tall; stems glabrous to glandular-puberulent. Leaves 1.5-6 cm long, glabrous to puberulent, lower ones 1-2 times pinnatifid. Flowers in ± open to dense capitate clusters with 20-50 subsessile flowers. Calyx 4-5 mm long, glabrous to sparsely villous; corollas campanulate, 10-12 mm long, white to pale blue; style included to exserted. Fruit ovoid to subspheric, 4-5 mm long, shorter than calyx. Mar. Rare, not seen recently; s-facing slopes; coastal scrub; ca. 30 ft (10 m). Known from a single collection at Smugglers' Cove by F. R. Fosberg in March 1932. Santa Catalina Island; Monterey Co. and s Sierra Nevada to s CA. Seven other subspecies occur on the mainland.

**G. clivorum** (Jeps.) V.Grant MANY-STEMMED GILIA Annuals, 10-40 cm tall; stems simple to diffusely branched, densely glandular-puberulent above middle, pubescent below middle. Leaves 1.5-7 cm long, ± glandular-puberulent, lower ones 1-2 times pinnatifid. Flowers solitary or in ± open clusters with 2-5 subsessile flowers. Calyx 4-8 mm long, glandular-puberulent; corollas funnelform, 6-8 mm long, tube and throat yellow, throat with purple spots, limb white to lavender; style included to slightly exserted. Fruit ovoid, 5-6 mm long, shorter than calyx. Mar-Jun. Occasional; rocky slopes, grassy hillsides, exposed ridgetops; coastal scrub, grasslands; 50-2200 ft (15-670 m). Widely scattered locations throughout island, especially on n side and in Central Valley. San Miguel, Santa Rosa, Anacapa, and Santa Catalina islands; n CA coast ranges to Santa Barbara Co.

**G. nevinii** A.Gray ISLAND GILIA Annuals, 10-40 cm tall; stems simple to diffusely branched, glabrous to puberulent. Leaves 1-6 cm long, glabrous to puberulent, lower ones 2-3 times pinnatifid. Flowers solitary or in ± open clusters with 3-10 subsessile flowers. Calyx 5-6 mm long, glandular-puberulent; corollas funnelform, 8-14 mm long, tube lavender, throat yellow, limb lavender or bluish; style included to slightly exserted. Fruit ovoid, 5-6 mm long, shorter than calyx. Mar-Jun. Island endemic. A species of special environmental concern. Scarce; coastal bluffs, rocky slopes, loose talus slopes, grassy hillsides, floodplains; coastal scrub, grasslands; 20-500 ft (6-150 m). North side between Cueva Valdez and Eagle Canyon, and between Campo Grande and Scorpion Canyon; widely scattered locations in Central Valley between top of Portezuela Grade and Stanton Ranch. All CA Channel islands except San Miguel (last seen on San Nicolas Island in 1901); Guadalupe Island.

## **LINANTHUS** Benth.

Annuals or perennial herbs; stems simple to branched, glabrous to pubescent. Leaves mostly cauline, opposite, simple to palmately dissected, segments usually linear. Flowers solitary or in ± open to dense terminal clusters, sessile or pedicellate; bracts simple to palmately lobed, herbaceous. Calyx lobes equal; corollas radial, campanulate, funnelform, or salverform; stamens equal in length, adnate to corolla at same level. Ca. 40 species; w N. America and S. America.

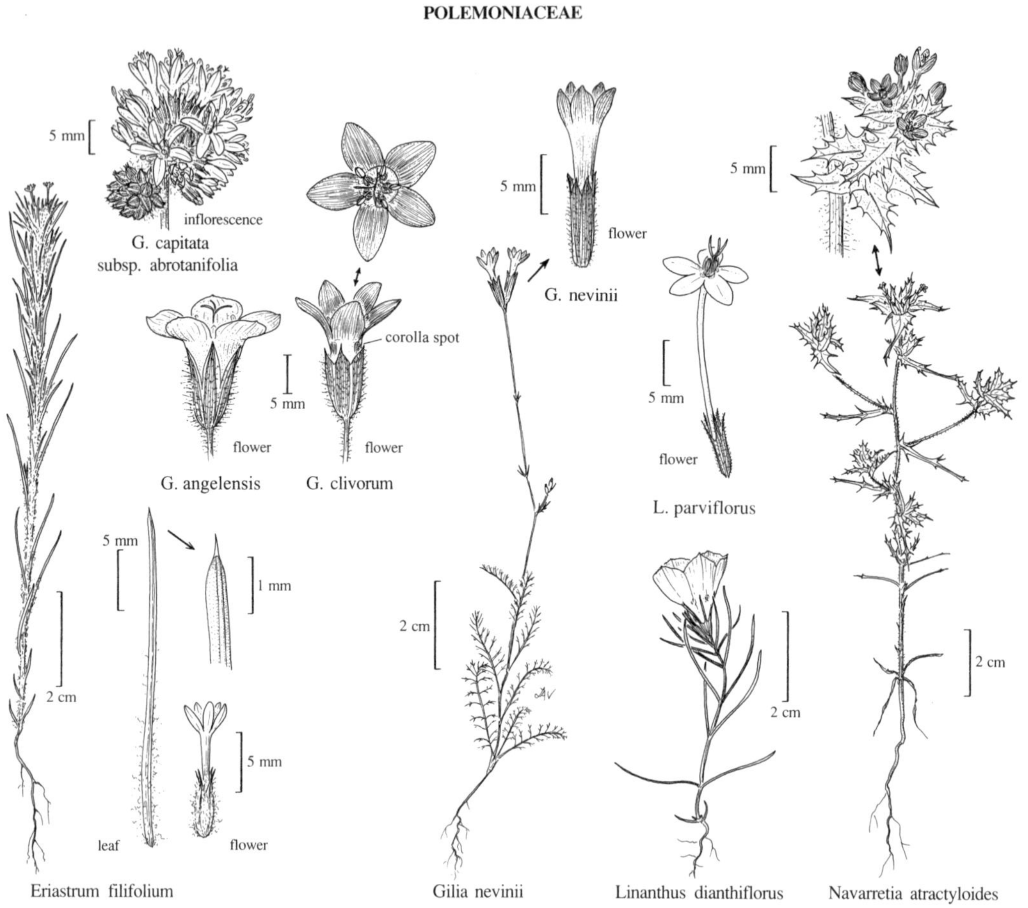


1. Corollas campanulate, tubes cylindrical, 5-7 mm long, lobe margins minutely toothed ....... **L. dianthiflorus**
1' Corollas salverform, tubes filiform, 20-35 mm long, lobe margins entire ................................ **L. parviflorus**

**L. dianthiflorus** (Benth.) Greene FRINGED LINANTHUS Annuals, to 13 cm tall; stems erect to spreading, sparsely pubescent. Leaves filiform to narrowly oblong, 5-25 mm long, entire. Flowers usually solitary, subsessile to pedicellate. Calyx 8-14 mm long; corollas campanulate, tube cylindrical, 5-7 mm long, yellow, limb white to pink, sometimes yellowish, often with purple spots, lobe margins minutely toothed; style included. Mar-Apr. Occasional; coastal flats, grassy hillsides, rocky slopes, exposed ridgetops; coastal scrub, grasslands; 20-1150 ft (6-350 m). Central Valley near Christy Ranch; widely scattered locations on s side between Black Point peninsula and Alberts Anchorage. Santa Catalina Island; Los Coronados Islands; Santa Barbara Co. to n Baja CA.

**L. parviflorus** (Benth.) Greene [*L. androsaceus* (Benth.) Greene subsp. *luteus* (Benth.) H.Mason] Annuals, to 20 cm tall; stems erect to decumbent, sparsely pubescent. Leaves 8-15 mm long, palmately divided, segments linear. Flowers in dense terminal clusters, ± sessile. Calyx 6-9 mm long; corollas salverform, tube filiform, 20-35 mm long, limb white to lavender, sometimes yellowish, lobe margins entire; style exserted. May-Jun. Scarce; rocky slopes, grassy hillsides; coastal scrub, grasslands; 50-1000 ft (15-305 m). North side near Cueva Valdez and just w of Campo del Norte; s side in upper part of Cañada de los Sauces, on Sierra Blanca ridge, and in Willows Canyon. San Miguel and Santa Rosa islands; throughout CA w of Sierra Nevada.

## **NAVARRETIA** Ruiz & Pav.

Annuals; stems usually diffusely branched, glandular-puberulent. Leaves alternate, simple to pinnatifid. Flowers sessile, in dense, bracteate clusters; bract apices spinescent. Calyx lobes unequal, spinescent; corollas radial, usually funnelform; stamens equal to unequal in length, adnate to corolla at different levels. Ca. 30 species; w N. America and S. America.

POLYGALACEAE

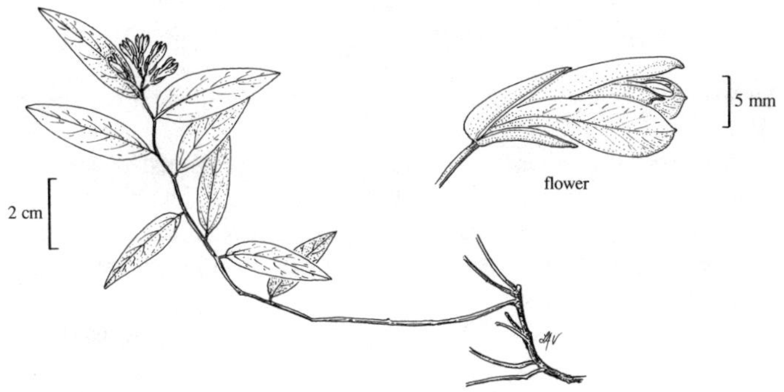


Polygala californica

**N. atractyloides** (Benth.) Hook. & Arn. Plants to 30 cm tall; stems glandular-puberulent. Leaves pinnatifid, 1-4 cm long, glandular, segment apices spinescent. Inflorescence capitate; bracts 3-lobed, glandular-puberulent, bases clasping. Calyx 6-8 mm long; corollas narrowly funnelform, 8-10 mm long, reddish purple to purple, lobes sometimes white or yellow. Fruit 4-6 mm long. May-Aug. Common; coastal bluffs, roadsides, sandy flats, floodplains, rocky slopes, grassy hillsides, exposed ridgetops; coastal scrub, grasslands; 10-1850 ft (3-565 m). Widely scattered locations throughout western and central portions of island. Santa Rosa, Santa Catalina, and San Clemente islands; OR to n Baja CA.

## POLYGALACEAE R.Br.
## Milkwort Family

Annuals or perennials. Leaves usually alternate, simple; stipules absent. Flowers bisexual, radial to bilateral; in racemes, spikes, or panicles. Sepals 5, free or fused, lateral 2 larger than other 3, petaloid; petals 3-5, lower one keel-like, apices often beak-like; stamens 3-10, partly fused into a tube, anthers opening by pores; pistil 1, ovary superior, locules 1-8, style 1, with 2 stigmatic lobes. Fruit a capsule or drupe. Ca. 18 genera and 950 species; temperate to subtropical, ± worldwide.

### POLYGALA L. MILKWORT

Annuals or perennials. Inflorescence a bracteate raceme or spike, sometimes also with cleistogamous or subterranean flowers. Stamens 6-8; locules 2. Fruit a capsule, compressed, often winged. Ca. 500 species; mostly tropical or subtropical, ± worldwide.

**P. californica** Nutt. CALIFORNIA MILKWORT Perennial herbs; stems erect to spreading, 3-35 cm tall. Leaves elliptic to broadly lanceolate, 1-4 cm long. Inflorescence a terminal raceme, small clusters of cleistogamous flowers often borne near base of stem. Lateral sepals 10-12 mm long, pink to purplish; petals pink to purplish, beak of keel 1.5-3 mm long, yellow to white at anthesis. Fruit obcordate, 5-8 mm long, with reticulate venation. Apr-Jun. Scarce; shaded n-facing slopes; oak woodland, pine forest; 300-2000 ft (90-610 m). North side on Picacho Diablo and near Pelican Bay; Central Valley in Christy Pines. Oregon to San Luis Obispo Co.

## POLYGONACEAE Juss.
## Knotweed or Buckwheat Family

Annuals, perennial herbs, shrubs, or vines, sometimes monoecious or dioecious; herbaceous stems often with enlarged nodes. Leaves basal and cauline, alternate, opposite, or whorled, entire; stipules absent or present and fused into sheath around stem. Flowers bisexual to unisexual, radial, in spikes, panicles, or umbel-like clusters, sometimes subtended by involucre of bracts. Perianth parts 5-6, if 6 then often in 2 whorls of 3, free or ± fused, persistent; stamens 3-9, in whorls of 3; pistil 1, ovary superior, locule 1, styles 2-3. Fruit a 3-angled achene. Ca. 50 genera and 1150 species; mostly temperate, worldwide.

1. Stipules present, fused into hyaline sheath surrounding stem
   2. Perianth parts 5, petaloid, greenish with white to pink margins, erect ................................ **Polygonum**
   2' Perianth parts 6, sepaloid, outer 3 spreading, inner 3 ± erect ...................................................... **Rumex**

1' Stipules absent
   3. Flowers subtended by 1 or more distinct bracts, bracts not fused into involucres (except pistillate flowers of *Pterostegia*)
      4. Leaves linear, mostly basal; inflorescence terminal, ± cylindrical, spike-like, flowers bisexual; floral bracts with hooked apices ............................................................................................ **Lastarriaea**
      4' Leaves fan-shaped, opposite; flowers 1-2 per cluster, axillary, unisexual; staminate flowers subtended by 1 bract or not, pistillate flowers subtended by 2-lobed involucre ............................ **Pterostegia**
   3' Flowers subtended by tubular to cup-shaped involucres
      5. Involucral bracts spine-tipped; flowers 1-2 per involucre; stamens 6; stems becoming brittle ........ ........................................................................................................................ **Chorizanthe**
      5' Involucral bracts acute to mucronate, not spine-tipped, flowers many per involucre; stamens 9; stems not brittle with age ........................................................................................ **Eriogonum**

## **CHORIZANTHE** R.Br. ex Benth.

Annuals, perennial herbs, or subshrubs; stems erect to spreading, becoming brittle. Leaves mostly basal, petiolate; stipules absent. Inflorescence composed of dense clusters; flowers 1-2 per involucre, bracts divergent, spinose. Perianth lobes 6; stamens 3-9. Ca. 50 species; w N. America and S. America.

**C. wheeleri** S.Watson [*C. insularis* Ralph Hoffm.] WHEELER'S SPINEFLOWER Annuals, 6-25 cm tall, pubescent. Leaves oblong to spatulate, 20-30 mm long, 2-7 mm wide, upper surfaces villous, lower surfaces tomentose. Floral bracts fused, tube cylindrical, 2.5-3 mm long. Perianth 3-3.5 mm long, pink to white; stamens 6. Fruits linear, ca. 3 mm long. Apr-Jul. Island endemic. A species of special environmental concern. Common; rocky slopes, exposed ridgetops, grassy hillsides, dry washes and ravines; chaparral, coastal scrub, grasslands; 20-1300 ft (6-395 m). Widely scattered on igneous and metamorphic substrates in western and central portions of island. Santa Rosa Island.

## **ERIOGONUM** Michx. WILD BUCKWHEAT

Annuals, perennial herbs, or shrubs; stems erect to spreading. Leaves basal or cauline, alternate, sessile to petiolate; stipules absent. Inflorescence composed of dense, often capitate clusters; flowers several-many per involucre, bracts leaf-like to scale-like. Perianth lobes 6, in 2 whorls of 3; stamens 9. Ca. 250 species; N. America.

1. Suffrutescent perennial herbs; leaves mostly near base of stem, blade margins crenate to undulate .............. ........................................................................................................................ **E. grande**

1' Much-branched shrubs; leaves cauline, in crowded clusters or evenly distributed, blade margins entire, plane or revolute
   2. Leaves linear to oblong, in crowded clusters, blade margins revolute ............................ **E. arborescens**
   2' Leaves ovate to oblong-ovate, evenly distributed, blade margins plane ........................... **E. giganteum**

**E. arborescens** Greene SANTA CRUZ ISLAND BUCKWHEAT Much-branched shrubs, 0.6-2 m tall. Leaves in crowded clusters, linear to oblong, 2-3 cm long, 2-6 mm wide, margins revolute, upper surfaces tomentulose to glabrate, lower surfaces white-tomentose, subsessile. Involucres campanulate, 2-3 mm long, tomentose. Perianth white to pink, 2-3 mm long, base villous. Fruit 2.5-3.5 mm long. Apr-Sep (Oct). Island endemic. Common; coastal bluffs, cliff faces, rocky slopes, grassy hillsides, floodplains; chaparral, coastal scrub, grasslands; 10-2000 ft (3-610 m). Widely scattered locations throughout island. Santa Rosa and Anacapa islands. Hybrids with *E. giganteum*, called *E. ×blissianum* H.Mason, were first observed at the Stanton Ranch in 1971 and along Islay Canyon and South Ridge Roads in 1980. Efforts to eradicate hybrids and eliminate gene flow from *E. giganteum* began in 1980.

**E. giganteum** S.Watson var. **giganteum** SAINT CATHERINE'S LACE Much-branched shrubs, 0.3-3.5 m tall. Leaves ovate to oblong-ovate, 3-10 cm long, 1-5 cm wide, margins plane, upper surfaces light gray-pubescent to glabrate, lower surfaces white-tomentose, petiolate. Involucres campanulate, 3-5 mm long, tomentose. Perianth white, 2.5-3.5 mm long, villous. Fruit 2-3.5 mm long. May-Aug. Introduced. Scarce; disturbed sites, roadsides, exposed ridgetops; chaparral, coastal scrub; 200-1400 ft (60-430 m). Central Valley at Stanton Ranch and near

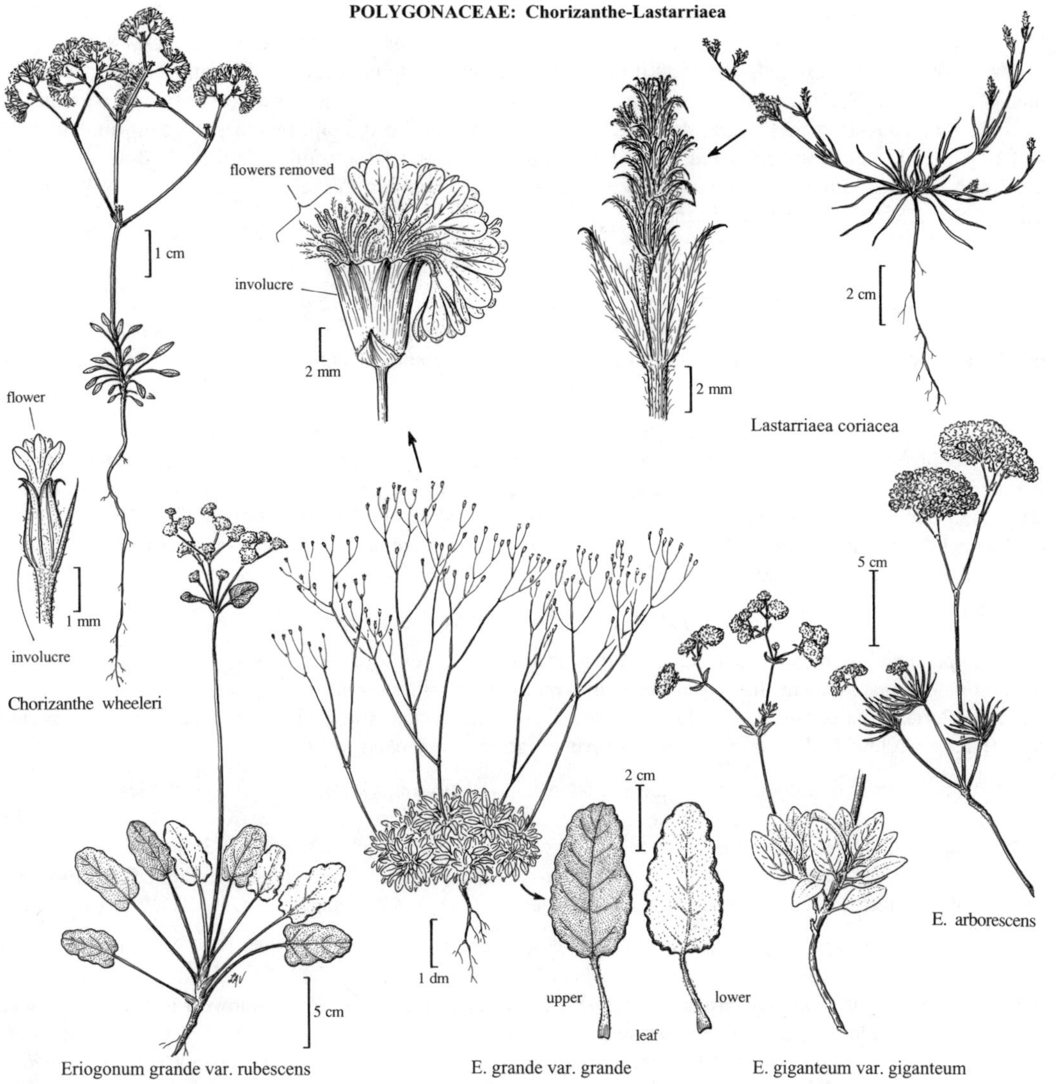


refuse dump at Stanton Ranch airfield; s side along S Ridge Road between Laguna Canyon Road and Islay Canyon Road. Endemic to Santa Catalina Island. Cuttings from cultivated plants at the Stanton Ranch were used to control erosion in culverts along Islay Canyon and South Ridge Roads, a practice that led to hybridization with *E. arborescens*. Two other varieties occur on other CA Channel islands.

**E. grande** Greene Suffrutescent perennial herbs, 0.1-2 m tall. Leaves oblong-ovate, 3-10 cm long, 1-4 cm wide, margins crenate to undulate, upper surfaces villous to glabrate, lower surfaces white-woolly, petiolate. Involucres turbinate to campanulate, 5-7 mm long, ± glabrous. Perianth white to red, 2.5-4 mm long, glabrous. Fruit 2-3 mm long. Jun-Oct (Mar). Island endemic. Common; coastal bluffs, rocky slopes, floodplains; coastal scrub, grasslands; 10-1500 ft (3-460 m). Widely scattered locations throughout island. San Miguel, Santa Rosa, Anacapa, Santa Catalina, and San Clemente islands. All CA Channel Islands except Santa Barbara. The species is composed of 4 varieties, of which 2 occur on Santa Cruz Island. Plants less than 30 cm tall, with white perianth parts, belong to var. *timorum* Reveal, which is endemic to San Nicolas Island; var. *testudinum* Reveal is endemic to nw coast of Baja CA and Todos Santos Island.

1. Plants usually 5-20 dm tall; perianth usually white ........ var. **grande**
1' Plants usually 1-5 dm tall; perianth pink to red ........ var. **rubescens**

var. **grande** ISLAND BUCKWHEAT Plants 5-20 dm tall. Involucres ± turbinate. Perianth white. Mar-Dec. A variety of special environmental concern. Jun-Oct (Mar). Common; coastal bluffs, rocky slopes, floodplains; coastal scrub, grasslands; 10-1500 ft (3-460 m). Widely scattered locations throughout island. Anacapa, Santa Catalina, and San Clemente islands. Intergrades with var. *rubescens* along nw coast of Santa Cruz Island.

var. **rubescens** (Greene) Munz RED BUCKWHEAT Plants 1-5 dm tall. Involucres ± campanulate. Perianth pink to red. Jun-Sep. Occasional; coastal bluffs, rocky slopes; coastal scrub, grasslands; 30-800 ft (10-245 m). North side between West Point and Del Mar Cove; s side between Forney's Cove and Kinton Point. San Miguel, Santa Rosa, and Anacapa islands.

## LASTARRIAEA Remy

Annuals; stems erect to decumbent, becoming brittle. Leaves mostly basal, sessile; stipules absent. Inflorescence ± spike-like or a raceme, axis brittle, composed of solitary flowers, each subtended by involucre of 5-6 bracts; bract apices attenuate, hooked. Perianth tubular, lobes 5-6; stamens 3. Three species; w N. America and S. America.

**L. coriacea** (Goodman) Hoover [*Chorizanthe c.* Goodman] Plants 5-25 cm tall; stems decumbent. Leaves linear, 1-3 cm long, margins ciliate, apices obtuse to awn-tipped. Floral bracts 3-4 mm long, reddish with age. Perianth lobes 5-6, greenish, with hooked apices. Fruit 2-2.5 mm long, smooth, dull. Apr-Jun. Scarce; coastal flats, sand dunes, floodplains; coastal beach and dune, coastal scrub; 50-1400 ft (15-430 m). North side in Cañada del Puerto; s side just e of Near Point, on Sierra Blanca ridge, at Punta Arena, and near El Tigre. Santa Rosa and Santa Catalina islands; CA w of Sierra Nevada, s to n Mexico.

## POLYGONUM L. KNOTWEED

Annuals, perennial herbs, or vines; stems erect to prostrate. Leaves basal or cauline, alternate, sessile to petiolate; stipules fused into a hyaline sheath around stem. Flowers solitary or in small clusters arranged in spikes or panicles. Perianth parts 5-6, free to fused; stamens 3-8. Fruit either 3-angled or lenticular. Ca. 300 species; temperate, ± worldwide. Leaves and stems of several species are cooked for food.

1. Stipular sheaths cylindrical, apices truncate; leaves 3.5-18 cm long; style branches 2; fruits lenticular ............ **P. lapathifolium**

1' Stipular sheaths lacerate or lobed; leaves 0.5-2.5 cm long; style branches 3; achenes 3-angled
   2. Stems prostrate to decumbent; perianth 1.5-2 mm long; achenes minutely striate, dull .. **P. arenastrum**
   2' Stems erect to ascending; perianth 2-3 mm long; achenes smooth, shiny ...................... **P. argyrocoleon**

**P. arenastrum** Boreau [*P. aviculare* L. misapplied to island plants] COMMON KNOTWEED Annuals or short-lived perennials; stems prostrate or decumbent, 1-12 dm long. Leaves oblanceolate to lanceolate-oblong, 0.5-2 cm long, 1-5 mm wide, glabrous; stipular sheaths lacerate to lobed at apices. Flowers in axillary clusters. Perianth 1.5-2 mm long, green, margins white to pink; stamens usually 8; style branches 3. Fruit 2-2.5 mm long, ± 3-angled, minutely striate, dull. May-Oct. Introduced. Occasional; gravel beaches, disturbed flats, roadsides, seasonal ponds, floodplains; grasslands, riparian communities; 5-1350 ft (1-410 m). Widely scattered locations throughout island. All CA Channel islands except San Miguel and Santa Barbara; Cedros Island; throughout N. America, native of Europe. Reported by Greene (1887).

**P. argyrocoleon** Steud. ex Kunze PERSIAN KNOTWEED Annuals; stems erect to ascending, 1-6 dm long. Leaves oblong to elliptic-lanceolate, 1-3.5 cm long, 1.5-7 mm wide, glabrous; stipular sheaths lacerate to lobed at apices. Flowers in axillary clusters. Perianth 2-3 mm long, green, margins pink; stamens usually 3-6; style branches 3. Fruit 1.5-2 mm long, 3-angled, smooth, shiny. Jun-Oct. Introduced. Scarce; disturbed sites, roadsides, seasonal ponds; grasslands, riparian communities; 250-1300 ft (75-395 m). North side near Mount Pleasant; s side near Rancho Nuevo. San Nicolas, Santa Catalina, and San Clemente islands; throughout much of CA w of Sierra Nevada, perhaps elsewhere in N. America; native of Asia. First collected near Rancho Nuevo in June 1980.

**P. lapathifolium** L. WILLOW SMARTWEED Annuals; stems erect to ascending, 3-15 dm long. Leaves lanceolate, 3.5-18 cm long, 5-50 mm wide, glabrous; stipular sheaths cylindrical, apices truncate. Flowers in axillary and terminal spikes. Perianth 1.5-2 mm long, green to pink; stamens usually 6; style branches 2. Fruit 1.5-2 mm long, lenticular, slightly shiny. Apr-Oct. Occasional; creek beds, floodplains; riparian communities; 10-350 ft (3-105 m). North side in Cañada del Puerto; Central Valley between La Cascada and Stanton Ranch. Throughout N. America.

## **PTEROSTEGIA** Fisch. & C.A.Mey.

With only 1 species.

**P. drymarioides** Fisch. & C.A.Mey. FAIRY MIST Monoecious annuals; stems prostrate, 1-4 dm long. Leaves opposite, cauline, blades broadly elliptic to fan-shaped, 5-20 mm long, entire to lobed, short-petiolate; stipules absent. Flowers unisexual, 1-2 per cluster, axillary, staminate ones not subtended by a bract, pistillate ones subtended by a 2-lobed, winged involucre, enlarged and persistent in fruit. Perianth lobes 6, ca. 1 mm long, pale yellow to pink; stamens 3 or 6. Fruit 3-angled, ca. 1 mm long, smooth, shiny. Feb-Jul. Occasional; coastal bluffs, rocky slopes, grassy flats and hillsides; coastal scrub, grasslands, oak woodland, pine forest; 50-1750 ft (15-535 m). Scattered locations throughout island. All CA Channel islands; Los Coronados, Todos Santos, San Martin, Guadalupe, and Cedros islands; OR to n Mexico, e to UT and AZ.

## **RUMEX** L. DOCK or SORREL

Monoecious or dioecious annuals or perennial herbs; stems usually erect. Leaves mostly basal, cauline ones alternate, sessile to petiolate; stipules fused into a hyaline or membranous sheath around stem. Flowers unisexual, in small clusters arranged in panicles, branches often spike-like. Perianth parts 6, ± fused at base, green, sepaloid, prominently reticulate-veined, each one sometimes with tubercle at base, inner ones convergent over fruit, often ± winged; stamens 6. Fruit 3-angled, usually shiny. Ca. 200 species; mostly temperate, N. America, Europe, and Asia. Most species contain soluble oxalates and are toxic to livestock if eaten in large quantities.

1. Plants dioecious, with slender rhizomes; lower cauline leaves hastate; inner perianth parts adherent to fruit .......... **R. acetosella**

1' Plants monoecious, with thick vertical underground stems or roots; cauline leaves cordate to tapered at bases; inner perianth parts free from fruit
  - 2. Margins of inner perianth parts toothed in fruit .......... **R. pulcher**
  - 2' Margins of inner perianth parts entire in fruit
    - 3. Inner perianth parts 4-5 mm long in fruit .......... **R. crispus**
    - 3' Inner perianth parts 2-3 mm long in fruit
      - 4. Inner perianth parts each with 1 tubercle, tubercles equal in size .......... **R. conglomeratus**
      - 4' Inner perianth parts with tubercles unequal in size, one much wider than other 2 .......... **R. salicifolius**

**R. acetosella** L. SHEEP SORREL Plants dioecious, with slender rhizomes; stems erect, 1-4 dm tall. Leaf blades linear to lanceolate, 2-6 cm long, 1-2.5 cm wide, upper cauline ones truncate to hastate at base, lower cauline ones hastate. Inflorescence a panicle composed of naked, slender, spike-like branches. Perianth parts 1.5-2.5 mm long in fruit, inner ones adherent to fruit, not much enlarged in fruit, margins entire. Fruit 1.5-2.5 mm long, reddish brown. Mar-May. Introduced. Scarce; grassy flats, rocky slopes; coastal scrub, grasslands; 100-900 ft (30-275 m). North side on Orizaba Flats; Central Valley near upper part of Cañada Christy; s side on South Ridge between Coches Prietos Anchorage and Valley Anchorage. Santa Rosa Island; throughout N. America; native of Europe. First collected at "foot of Cuesta Grade" in March 1932. Leaves were eaten by Native Americans; entire plant can be cooked and eaten.

**R. conglomeratus** Murray GREEN DOCK Plants monoecious; stems erect, 8-15 dm tall. Leaf blades oblong to ovate, 3-15 cm long, 1-6 cm wide, bases ± cordate to tapered, margins slightly undulate. Inflorescence an open panicle composed of ± leafy, spike-like branches. Perianth parts 2-3 mm long in fruit, inner ones narrowly ovate, free from fruit, each with a prominent tubercle, margins entire. Fruit 1-2 mm long, dark reddish brown. Apr-Nov. Introduced. Occasional; moist creek beds; riparian communities; 15-300 ft (5-90 m). North side in Cañada del Puerto; Central Valley in lower part of Cañada Christy and between La Cascada and Valley Anchorage; s side in canyons between Rancho Nuevo and Black Point peninsula. Santa Rosa and Santa Catalina islands; throughout CA w of Sierra Nevada; native of Europe. Reported by Greene (1887). A medicinal tea was made from the roots by the Miwok Tribe.

**R. crispus** L. CURLY DOCK Plants monoecious; stems erect, 6-12 dm tall. Leaf blades lanceolate to ± oblong, 10-30 cm long, 1-4 cm wide, bases truncate to tapered, margins strongly undulate. Inflorescence an open panicle composed of naked, spike-like branches. Perianth parts 4-6 mm long in fruit, inner ones broadly ovate to ± round, free from fruit, each with a narrow prominent tubercle, margins entire. Fruit 2-3 mm long, dark brown to black. Mar-Jun. Introduced. Common; disturbed sites, moist creek beds, floodplains; grasslands, riparian communities; 10-1400 ft (3-430 m). Widely scattered locations, primarily in Central Valley. All CA Channel islands except

POLYGONACEAE: Polygonum-Rumex

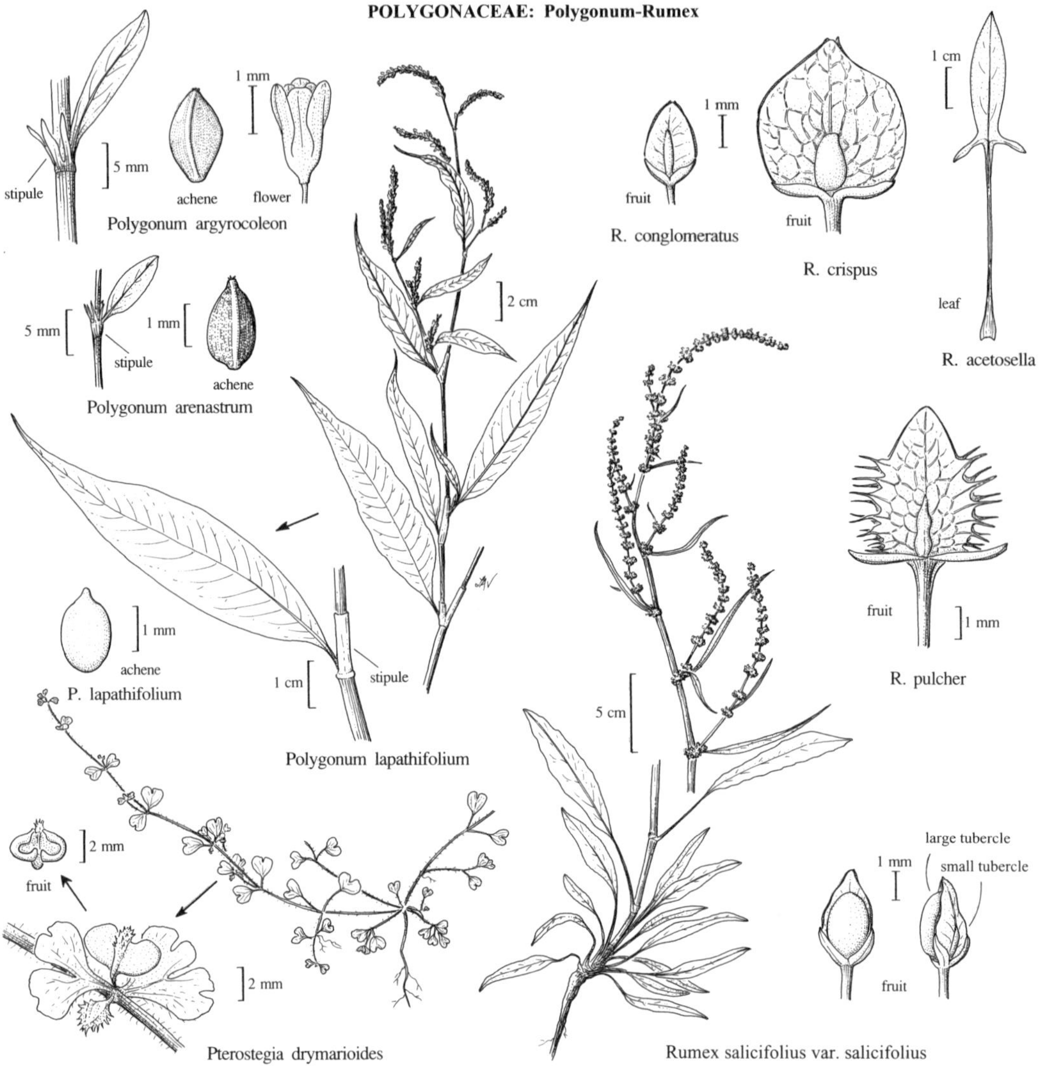


Santa Barbara; throughout N. America; native of Europe. Reported by Greene (1887). Young leaves can be boiled and eaten; seeds can be ground into flour; roots have been used medicinally.

**R. pulcher** L. FIDDLE DOCK Plants monoecious; stems erect, 3-6 dm tall. Leaf blades lanceolate, 7-30 cm long, 1-5 cm wide, bases tapered, margins slightly undulate or crenate. Inflorescence an open panicle composed of leafy, spike-like branches. Perianth parts 3-6 mm long in fruit, inner ones broadly ovate, free from fruit, each with a small tubercle, margins prominently toothed. Fruit 2.5-4 mm long, dark reddish brown. Apr-Jun. Introduced. Common; disturbed sites, open flats, floodplains; coastal scrub, grasslands, riparian communities; 10-1400 ft (3-430 m). Widely scattered locations, primarily in Central Valley. Throughout CA, native of Europe. First collected at Prisoners' Harbor in June 1930.

**R. salicifolius** Weinm. var. **salicifolius** WILLOW DOCK Plants monoecious; stems erect, 3-6 dm tall. Leaf blades lanceolate, 7-30 cm long, 1-5 cm wide, bases tapered, margins slightly undulate or crenate. Inflorescence an open panicle composed of leafy, spike-like branches. Perianth parts 3-6 mm long in fruit, inner ones broadly ovate, free from fruit, each with a small tubercle, margins prominently toothed. Fruit 2.5-4 mm long, dark reddish brown. Apr-Jun. Occasional; moist creek beds and flats, floodplains; riparian communities; 5-750 ft (1-230 m). North side in Cañada del Puerto; scattered locations in Central Valley; scattered locations on s side, especially in Coches Prietos drainage. All CA Channel islands except Anacapa and Santa Barbara; OR to n Baja

CA. Seeds were eaten by the Maidu and Miwok tribes. Five other varieties are known from mainland CA and elsewhere in N. America.

# PORTULACACEAE Juss.

## PURSLANE FAMILY

Annuals or perennial herbs, often fleshy or succulent. Leaves alternate or opposite, simple; stipules sometimes present. Flowers bisexual, radial, in racemes, panicles, or solitary. Sepals usually 2, free or fused at base; petals 3-many, free or fused at base; stamens 1-many; ovary superior or partly inferior, locule 1, placentation basal or parietal; style branches 2-8. Fruit a capsule, circumscissile or 2-3-valved. Ca. 20 genera and 400 species; temperate, America, Australia, and s Africa. Some species are edible or cultivated as ornamentals.

1. Ovary partly inferior, calyx adnate to lower half of ovary; fruit circumscissile ............................... **Portulaca**
1' Ovary superior, calyx free from ovary; fruit opening by valves
    2. Cauline leaves 2, fused and perfoliate, sessile ......................................................................... **Claytonia**
    2' Cauline leaves several, free, usually petiolate
        3. Petals fused at base, persistent and forming a cap over fruit; style branches 2; fruit opening by 2 valves ................................................................................................................ **Calyptridium**
        3' Petals free, not persistent in fruit; style branches usually 3; fruit opening by 3 valves
            4. Leaves alternate; sepals 4-8 mm long .................................................................. **Calandrinia**
            4' Leaves opposite; sepals ca. 1 mm long ........................................................................ **Montia**

## CALANDRINIA Kunth

Annuals or perennial herbs. Leaves alternate, blade tapered to petiole. Flowers in leafy racemes or panicles. Sepals persistent in fruit; petals usually 5, red, rarely white; stamens 3-18; style branches 3. Fruit opening by 3 valves. Ca. 150 species, w N. and S. America, Australia.

1. Plants glaucous; basal leaves obovate to spatulate; seeds grayish or brown; stems spreading or ascending ............................................................................................................................................. **C. maritima**
1' Plants green, not glaucous; basal leaves linear to spatulate; seeds black, ± shiny; stems ascending or prostrate
    2. Fruit 8-10 mm long, much exceeding sepals; pedicels ± reflexed in fruit ............................. **C. breweri**
    2' Fruit 4-7 mm long, barely exceeding sepals; pedicels mostly erect or ascending in fruit ........ **C. ciliata**

**C. breweri** S.Watson BREWER'S CALANDRINIA Stems prostrate to decumbent, much branched from base, 10-40 cm long. Leaves mostly basal, oblanceolate to spatulate, 3-5(15) cm long, 0.4-1.2 cm wide, green, margins minutely ciliate; cauline leaves much reduced. Inflorescence an elongate, leafy raceme. Sepals 4-6 mm long, glabrous or minutely ciliate; petals 4-5 mm long. Fruit 8-10 mm long, much exceeding sepals; pedicels reflexed; seeds ca. 1 mm wide, black, shiny, minutely tuberculate with short, black pegs. Mar-May. A species of special environmental concern. Rare; coastal bluffs, grassy hillsides, burned areas, rocky slopes, canyon bottoms; chaparral, coastal srcub, grasslands; 30-200 ft (10-60 m). North side near Pelican Bay; s side at "west end" and in lower part of Willows Canyon. Santa Rosa Island; n CA coast ranges to Baja CA. Usually seen only in years with heavy rainfall or after fires. Seeds were eaten by Native Americans.

**C. ciliata** (Ruiz & Pav.) DC. [*C. c.* var. *menziesii* (Hook.) J.F.Macbr.] RED MAIDS Stems decumbent to ascending, usually much branched from base, 5-40 cm long. Leaves mostly cauline, rarely in basal rosettes, spatulate to linear, 2-8 cm long, 0.2-1 cm wide, green, margins ciliate. Inflorescence a congested, leafy raceme. Sepals 4-8 mm long, ciliate on margins and midvein; petals 4-10(14) mm long. Fruit 4-7 mm long, barely exceeding sepals; pedicels mostly erect or ascending; seeds ca. 0.8-1 mm wide, black, shiny, faintly striate. Feb-May. Common; grassy hillsides, rocky slopes, floodplains; chaparral, coastal scrub, grasslands, oak woodland; 10-1500 ft (3-460 m). Scattered locations throughout island. All CA Channel islands; Todos Santos, San Martin, and Guadalupe islands; n CA to S. America, e to NM. Seeds, roots, and leaves were eaten by Native Americans.

**C. maritima** Nutt. SEASIDE CALANDRINIA Stems spreading or ascending, 7-25 cm long. Leaves mostly basal, spatulate-obovate, 2-6 cm long, 0.4-1.5 cm wide, glaucous, fleshy. Inflorescence a lax panicle. Sepals 4-5 cm long, glabrous; petals ca. 4 mm long. Fruit 5-6 mm long, much exceeding sepals; seeds 0.5-1 mm wide, dull brown, minutely tuberculate with short, slender, gray pegs. (Feb) Apr-Jun (Aug). A species of special environmental concern. Occasional; coastal bluffs, rocky s-facing slopes, clay flats and ridgetops; coastal scrub; 20-500 ft (6-150 m). North side at Fry's Harbor; Central Valley near Christy Ranch; scattered locations on s side

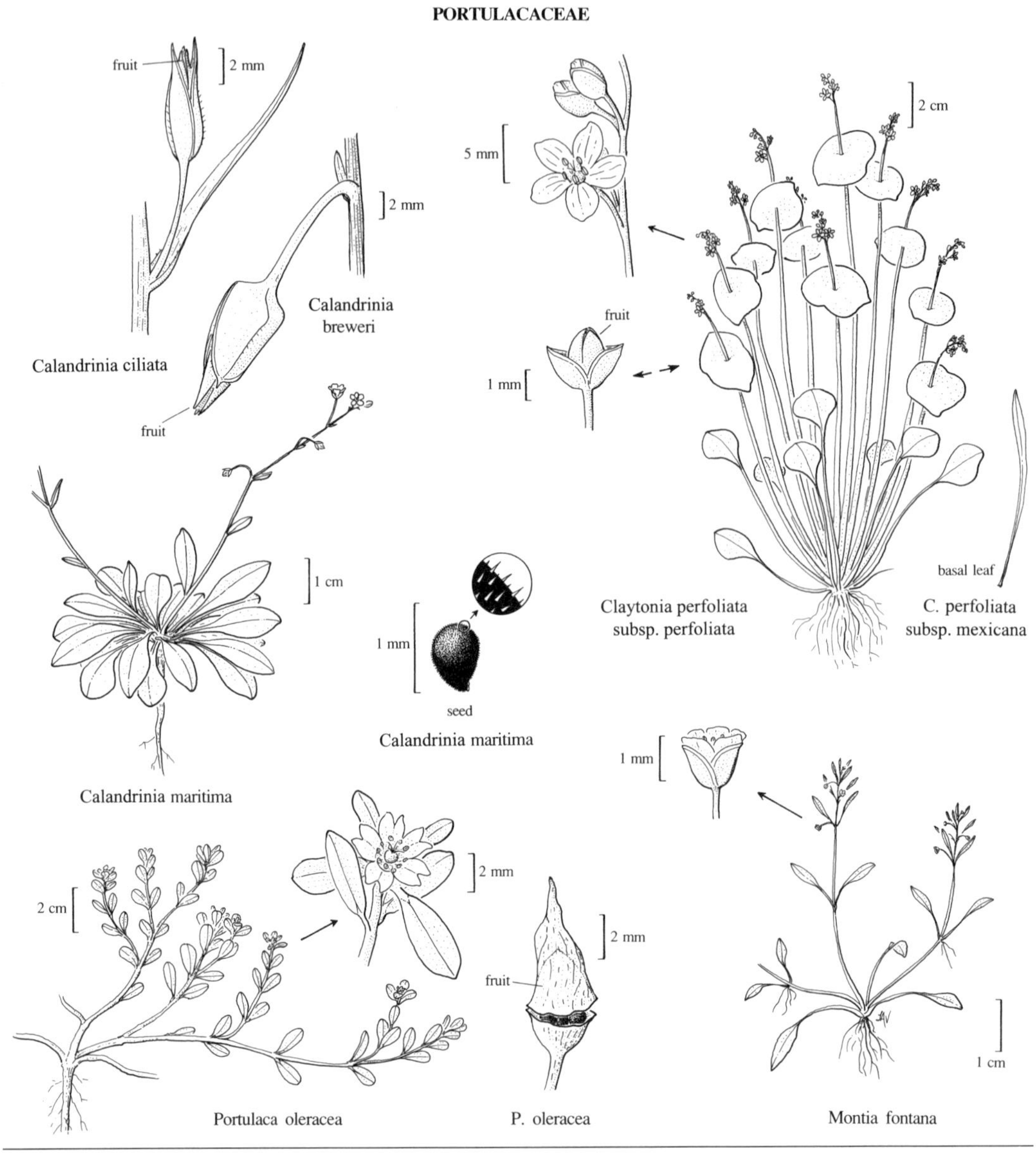


between West Point and Sandstone Point. All CA Channel islands except San Miguel and San Nicolas; all Baja CA islands except San Geronimo; Santa Barbara Co.to Baja CA. Abundance can be extremely variable; rare in most years.

## **CALYPTRIDIUM** Nutt. ex Torrey & A.Gray

Annuals or perennial herbs. Leaves basal and cauline, alternate, blades sessile or petiolate. Flowers in racemes, umbels, or panicles. Sepals persistent in fruit; petals 2-4, minute, fused at based, white to red, persistent and forming a cap over fruit; stamens 1-3; style branches 2. Fruit opening by 2 valves; seeds black, shiny. Ca. 8 species, w N. and S. America.

**C. monandrum** Nutt. Annual, 1-10 cm tall; stems spreading to decumbent. Leaves narrowly oblanceolate, basal ones 1-4 cm long, withering in flower or fruit, cauline ones 0.5-2 cm long. Inflorescence a raceme or panicle, 1-4 cm long; flowers subsessile, persistent in fruit. Sepals 1-2 mm long; petals 3, 1-3 mm long, pink to red; stamen usually 1. Fruit 3-5.5 mm long. Apr. Rare, not seen recently; habitat, plant community, and elevation unknown.

Known from a single collection without locality by T. S. Brandegee in April 1888. Monterey Co. to Baja CA, e to NV and AZ.

## CLAYTONIA L.

Annuals or perennial herbs, glabrous, ± fleshy. Leaves entire, mostly in basal rosettes, cauline ones 2, fused, usually perfoliate. Flowers in 1-sided, terminal racemes, reflexed, erect in fruit. Petals 5; ovary superior, locule 1 with basal placentation, style 1, stigmatic branches 3. Fruit opening by 3 valves. Ca. 28 species; N. America and e Asia.

1. Most basal leaves linear to narrowly oblanceolate, blades gradually tapered to bases, blade length more than 3 times width, .......... **C. parviflora**

1' Most basal leaves elliptic to ± cordate, blade bases rounded to cordate, blade length less than 3 times width .......... **C. perfoliata**

**C. parviflora** Douglas ex Hook. subsp. **parviflora** [*C. perfoliata* Willd. var. *parviflora* (Douglas ex Hook.) Torr.] Annuals, 3-15 cm tall. Basal leaves linear to narrowly oblanceolate, 1-9 cm long, 0.1-0.5 cm wide, blade bases gradually tapered to petiole, cauline pair 1-4 cm wide. Petals 2-6 mm long, white to pink. Seeds black, shiny. Feb-Apr. Scarce; shaded n-facing slopes, stream borders; chaparral, oak woodland; 200-1500 ft (60-460 m). North side near Picacho Diablo; Central Valley in Cañada de la Portezuela and near Stanton Ranch; s side in Coches Prietos drainage. Santa Barbara and San Nicolas islands; British Columbia to n Mexico, e to MT, UT, and AZ. Two other subspecies occur on the mainland, e to Rocky Mountains.

**C. perfoliata** Willd. MINER'S LETTUCE Annual, 3-30 cm tall. Basal leaves elliptic to reniform, 1.5-10 cm long, 0.1-4.5 cm wide, blade bases rounded to cordate, cauline pair 0.5-7 cm wide. Petals 2-6 mm long, white to pink. Seeds black, shiny. Feb-Jun. Common; shaded n-facing slopes, grassy hillsides, stream banks; chaparral, coastal scrub, island woodland, oak woodland; 20-1300 ft (6-400 m). Scattered locations throughout island. All CA Channel islands; Los Coronados, Todos Santos, San Martin, Guadalupe, and Cedros islands; British Columbia to C. America.

1. Basal leaf blades often wider than long, round-deltate to reniform, apices abruptly short-pointed .......... **C. perfoliata** subsp. **mexicana**

1' Basal leaf blades often longer than wide, elliptic to round-deltate, apices obtuse to acute .......... **C. perfoliata** subsp. **perfoliata**

subsp. **mexicana** (Rydb.) J.M.Miller & K.L.Chambers Plants 3-18 cm tall. Basal leaf blades round-deltate to reniform, often wider than long, apices abruptly short-pointed; cauline pair usually with acute angles. Feb-Jun. Common; shaded n-facing slopes, grassy hillsides, stream banks; chaparral, coastal scrub, island woodland, oak woodland; 20-1300 ft (6-400 m). Scattered locations throughout island. All CA Channel islands; Los Coronados, Todos Santos, San Martin, Guadalupe, and Cedros islands; throughout the range of the species.

subsp. **perfoliata** Plants 15-30 cm tall. Basal leaf blades elliptic to round-deltate, often longer than wide, apices obtuse to acute; cauline pair usually orbicular or with obtuse angles. Feb-Jun. Scarce; shaded n-facing slopes; chaparral, oak woodland; 40-300 ft (12-90 m). North side near Prisoners' Harbor and in Eagle Canyon; Central Valley between La Cascada and Stanton Ranch. Santa Rosa and San Clemente islands; throughout the range of the species.

## MONTIA L.

Annuals or perennial herbs, glabrous, fleshy. Leaves basal and cauline, entire, basal in rosettes, cauline alternate or opposite, entire, not perfoliate. Flowers in 1-sided, terminal racemes, reflexed, erect in fruit. Petals 3-5; ovary superior, locule 1, placentas basal, style 1, stigmatic branches 3. Fruit opening by 3 valves. Ca. 12 species; N. and S. America, Australia, and Siberia.

**M. fontana** L. [*M. f.* subsp. *amporitana* Sennen] WATER CHICKWEED Annual, 2-9 cm tall, often matted, stems often sprawling or prostrate. Leaves opposite, linear to oblanceolate, 0.5-2 cm long. Inflorescence axillary to terminal. Sepals ca. 1 mm long; petals 1-2 mm long, white. Seeds black, minutely tuberculate or smooth with a minutely tuberculate ridge. Feb. Rare, not seen recently; wet ledges; plant community and elevation unkown. Known from a single collection made on north side e of Orizaba in February 1932. Alaska to Mexico, e to Atlantic Coast.

## PORTULACA L. PURSLANE

Annuals, ± fleshy. Leaves alternate or opposite, entire, upper ones ± forming an involucre. Flowers solitary, terminal, sessile, mostly opening in sunshine. Sepals fused at base, adnate to ovary; petals 4-6; ovary partly inferior, locule 1, style 1, stigmatic branches 3-6. Fruit circumscissile near middle. Ca. 100 species; warm temperate, worldwide. Some species are cultivated as ornamentals or their seeds are used for food.

**P. oleracea** L. Stems prostrate, 8-20 cm long. Leaves ovate to obovate, 3-17 mm long, glabrous, fleshy. Sepals 3-4.5 mm long, keeled; petals 3-5 mm, yellow. Fruit 4-8 mm long; seeds ca. 0.5-1 mm wide, dark brown to black, minutely tuberculate. May-Oct. Introduced. Rare; disturbed sites; grasslands; 200 ft (60 m). Central Valley at Stanton Ranch. Santa Catalina Island; ± worldwide, native of Europe. First collected on the island in June 1930. Entire plant can be cooked and eaten, but plants may accumulate toxic amounts of oxalates and should not be eaten in large quantities.

# PRIMULACEAE Vent.

## PRIMROSE FAMILY

Annuals, perennial herbs, or shrubs, herbs often scapose. Leaves basal or cauline, alternate, opposite, or whorled, simple; stipules absent. Flowers bisexual, radial; flowers in racemes, umbels, or solitary. Calyx 3-5-lobed or sepals ± free; petals 4-5, free or fused; stamens 4-5, adnate to corolla, opposite corolla lobes; pistil 1, ovary superior or partly inferior, locule 1; style 1, stigma usually capitate. Fruit a capsule, circumscissile or dehiscent by valves. Ca. 25 genera and 800 species; mostly temperate, ± worldwide. Many species are cultivated as ornamentals.

1\. Leaves in basal rosettes; inflorescence an umbel ........ **Dodecatheon**
1' Leaves cauline, alternate or opposite, flowers in racemes or solitary
  2\. Annuals; leaves opposite; flowers solitary, axillary; ovary superior, free from perianth .......... **Anagallis**
  2' Perennial herbs; leaves alternate; flowers in terminal racemes; ovary partly inferior, ± adnate to calyx tube ........ **Samolus**

## ANAGALLIS L. PIMPERNEL

Annuals or perennial herbs. Leaves cauline, usually opposite. Flowers solitary, axillary. Calyx with 3-5 lobes; corolla with 5 lobes; stamens 5, free or fused, filaments pilose; ovary superior, free from perianth. Fruit circumscissile. Ca. 20 species; temperate, Europe, Africa, S. America.

**A. arvensis** L. SCARLET PIMPERNEL Annuals, 4-25 cm tall; stems ascending to prostrate, glabrous. Leaves ovate, 6-18 mm long, sessile to ± clasping. Pedicels 10-25 mm long, filiform, often recurved in fruit. Calyx lobes narrowly lanceolate, 2-5 mm long, margins scarious; corolla rotate, 8-12 mm wide, salmon-orange. Fruit globose, 3-5 mm long, glabrous. Apr-Sep. Introduced. Occasional; disturbed sites, moist streambanks, grassy hillsides; grasslands, riparian communities; 10-1000 ft (3-305 m). Widely scattered locations throughout western and central portions of island. All CA Channel islands except San Clemente; Guadalupe Island; throughout N. America; native of Europe. Reported by Hoffmann (1932b). A cleansing soap can be made from the flowers and foliage, but entire plant is toxic to humans if ingested.

## DODECATHEON L. SHOOTING STAR

Scapose perennial herbs. Leaves in basal rosettes, usually whorled. Flowers often nodding, in terminal umbels. Calyx with 5 lobes, free to near base; corolla with 5 lobes, lobes spreading to reflexed upward at anthesis; stamens 5, filaments shorter than anthers, fused; ovary superior, free from perianth. Fruit circumscissile or opening by valves. Ca. 15 species; temperate, N. America, e Asia. Roots, stems, and leaves of some species (e.g., *D. hendersonii*) were roasted and eaten by Native Americans.

**D. clevelandii** Greene subsp. **insulare** (Greene) H.J.Thomps. Plants 15-50 cm tall. Leaves oblanceolate to spatulate, 2-15 cm long, glabrous, margins entire, crenate, or serrate. Peduncles 7-45 cm tall; pedicels 1.5-6 cm long. Calyx 6-10 mm long, lobes lanceolate, ± herbaceous, glandular; corollas 10-25 mm long, lobes white to pink or purplish; filament tube dark maroon to black, anthers bright yellow. Fruit oblong to ovoid in outline, 1-2.5 cm long, opening by valves from apex. Feb-May. Occasional, grassy hillsides, rocky n-facing slopes, exposed ridgetops, and chaparral, coastal scrub; 100-1400 ft (30-430 m). Widely scattered locations throughout island. All CA Channel islands except Santa Barbara; Guadalupe Island; Monterey Co. to Santa Barbara Co. Three other subspecies occur on the mainland.

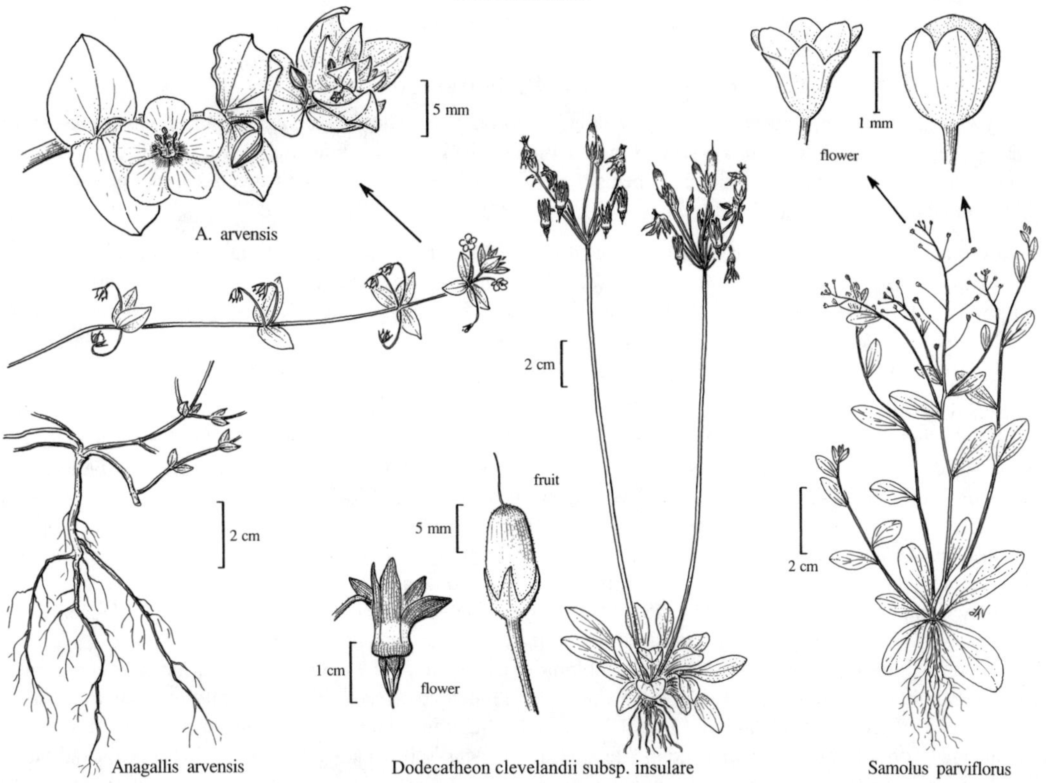


## SAMOLUS L.

Perennial herbs. Leaves cauline, alternate. Flowers in terminal racemes. Hypanthium present, fused partly to ovary; calyx with 5 lobes; corolla with 5 lobes, lobes spreading; stamens 5, free from each other, opposite petals, staminodes 5, alternate petals; ovary partly inferior. Fruit opening by 5 valves. Ca. 10 species; temperate, ± worldwide.

**S. parviflorus** (Raf.) Hulten Plants 10-50 cm tall; stems erect, glabrous. Leaves oval, obovate, or oblanceolate, 1.5-8 cm long, lower ones petiolate, upper ones sessile. Pedicels 8-15 mm long, filiform, spreading to ascending. Calyx lobes deltate, ca. 0.5 mm long, herbaceous; corolla lobes 1-2 mm long, white. Fruit globose, 1.5-3 mm long. May-Jun (Sep). Scarce; coastal bluffs, freshwater seeps, moist stream banks; coastal scrub, riparian communities; 10-250 ft (3-75 m). North side between Fry's Harbor and Orizaba Cove. Throughout much of N. and S. America. Young leaves are edible.

# RANUNCULACEAE Juss.

## Buttercup or Crowfoot Family

Annuals, perennial herbs, or woody vines. Leaves basal and cauline, alternate or opposite, simple or palmately to pinnately compound; stipules absent. Flowers bisexual to unisexual, radial to bilateral, in small clusters, racemes, panicles, or solitary. Floral parts free; sepals usually 5, sometimes petaloid; petals 4-many, sometimes absent; stamens 10-many; pistils 1-many, ovary superior, locule 1, style and stigma 1. Fruit an achene, follicle, or berry; seeds 1-many. Ca. 60 genera and 1700 species; temperate, ± worldwide. Many species, including columbines, anemones, and larkspurs, are used as ornamentals and a few are sources of medicinal drugs.

1. Woody vines; leaves opposite, pinnately compound; fruit an achene with long, pubescent styles .... **Clematis**

1' Perennial herbs; leaves basal to cauline and alternate, palmately lobed to divided; fruit a follicle or achene, persistent styles neither long nor pubescent

2. Flowers bilateral; sepals and petals purplish blue; fruit a follicle ........................................ **Delphinium**
2' Flowers radial; sepals greenish yellow, petals yellow; fruit an achene ................................ **Ranunculus**

## CLEMATIS L. VIRGIN'S BOWER or CLEMATIS

Woody vines. Leaves opposite, pinnately compound. Flowers bisexual or unisexual, radial, in small clusters or panicles. Sepals 4, petaloid; petals absent; stamens many; pistils many, styles pubescent, persistent and often plumose in fruit. Fruit an achene. Ca. 230 species; ± worldwide. Raw leaves of all species found on Santa Cruz Island can be toxic to humans if ingested.

1. Flowers many, in panicles; leaflets lanceolate to narrowly ovate; flowering from June to September ............ ............................................................ **C. ligusticifolia**
1' Flowers 1-3, in small, peduncled clusters; leaflets broadly ovate to rounded; flowering from March to May
   2. Both surfaces of sepals puberulent; achenes pubescent ........................................................ **C. lasiantha**
   2' Upper surfaces of sepals glabrous, lower surfaces puberulent; achenes glabrous ............. **C. pauciflora**

**C. lasiantha** Nutt. CHAPARRAL CLEMATIS Young stems woolly-pubescent to ± glabrous. Leaflets 3-5 per leaf, usually broadly ovate, 2-5 cm long, toothed to lobed, glabrous to puberulent; petioles 1.5-5 cm long. Peduncles 4-12 cm long; flowers mostly solitary. Sepals broadly oblong, 1.5-2.5 cm long, both surfaces puberulent. Fruits pubescent. Mar-Apr. Rare; rocky slopes; chaparral; 400-550 ft (120-170 m). Central Valley near base of Portezuela Grade; s side in upper portion of Laguna Canyon. Northern CA coast ranges to n Baja CA. Populations in Laguna Canyon are intermediate to *C. pauciflora* and may be hybrids. Steam from boiled stems or bark was used medicinally by the Shasta Tribe; tea made from the roots also had medicinal uses.

**C. ligusticifolia** Nutt. CREEK CLEMATIS Young stems ± glabrous. Leaflets 5-7 per leaf, usually lanceolate to narrowly ovate, 2-8 cm long, ± entire to lobed, glabrous to sparsely pubescent; petioles 3-7 cm long. Peduncles 3-10 cm long; flowers many, in panicles. Sepals oblong-oblanceolate, 0.8-1 cm long, both surfaces woolly tomentose. Fruits pubescent. Jun-Sep. Scarce; freshwater seeps, moist creek banks, canyon bottoms; chaparral, riparian communities; 20-200 ft (6-60 m). Central Valley near Stanton Ranch; s side in lower portions of Willows Canyon and Coches Prietos drainage, and in Aguaje Canyon near Yellowbanks Anchorage. Santa Rosa and Santa Catalina islands; British Columbia to n Mexico, e to SD and CO. Stems, bark, and leaves were used medicinally by Native Americans.

**C. pauciflora** Nutt. SMALL-FLOWERED CLEMATIS Young stems sparsely pubescent. Leaflets 3-5 per leaf, usually round-ovate, 1-3 cm long, toothed to lobed, ± glabrous; petioles 1.5-5 cm long. Peduncles 2-8 cm long; flowers 1-3 per cluster. Sepals oblong, 0.8-1.5 cm long, upper surfaces glabrous, lower surfaces puberulent. Fruits glabrous. Apr-May. Rare; rocky slopes; coastal scrub; ca. 500 ft (150 m). Known only from upper portion of Pozo Canyon. Los Coronados, Todos Santos, and Cedros islands; Los Angeles Co. to n Baja CA. Plants from the island tend to have larger leaves than those on Baja CA islands and the mainland. Bark was used medicinally by the Diegueño Tribe.

## DELPHINIUM L. LARKSPUR

Perennial herbs. Leaves basal and cauline, alternate, palmately lobed to divided. Flowers bisexual, bilateral, in racemes or spikes. Sepals 5, petaloid, upper one with a spur; petals 4, in unequal pairs; stamens many; pistils 3-5. Fruit a follicle; seeds many. Ca. 250 species; temperate, N. America, Europe, Asia. Some species are cultivated as ornamentals. Although some species were used medicinally by Native Americans, all taxa (especially young plants and seeds) can be toxic to humans and livestock if ingested. Treatment contributed by Michael Warnock.

**D. parryi** A.Gray subsp. **maritimum** (Davidson) M.J.Warnock Plants 15-45 cm tall. Leaves orbicular in outline, with 5-10 lobes, lobes ± oblong to oblong-lanceolate, 4-8 mm wide. Sepals spreading, purplish blue, lateral ones 9-20 mm long, spur 8-21 mm long; lower petals 3-11 mm long. Mar-Jun. Occasional; coastal bluffs, north-facing slopes, grassy hillsides; chaparral, coastal scrub, grasslands, pine forest; 50-800 ft (15-245 m). Widely scattered locations throughout island. San Miguel, Santa Rosa, Anacapa, and Santa Catalina islands; Los Coronados and Todos Santos islands; Monterey Co. to n Baja CA.

## RANUNCULUS L. BUTTERCUP or CROWFOOT

Annuals and perennial herbs. Leaves basal and cauline, alternate, entire to palmately divided. Flowers bisexual, radial, in small clusters or solitary, axillary or terminal. Sepals usually 5, often caducous, green to yellow; petals

**RANUNCULACEAE**

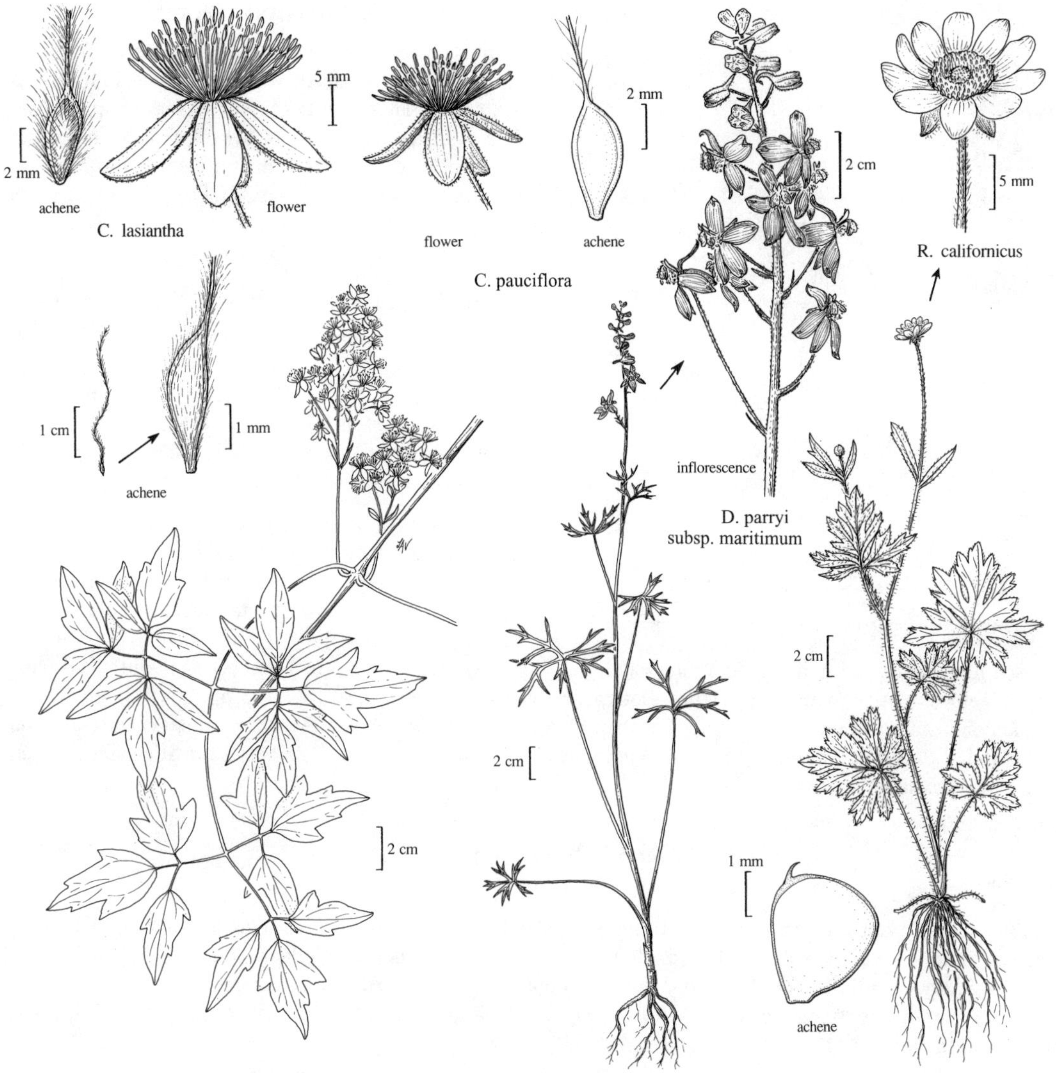

Clematis ligusticifolia Delphinium parryi subsp. maritimum Ranunculus californicus

4, in unequal pairs; stamens many; pistils 5-many; style persistent in fruit, beak-like. Fruit an achene. Ca. 250 species; temperate, worldwide. Most species have a caustic sap that can irritate the skin; can be toxic to humans and livestock if ingested. Roots and seeds of some species were cooked and eaten by Native Americans.

**R. californicus** Benth. [*R. c.* Benth. var. *cuneatus* Greene] Perennial herbs 30-70 cm tall; stems ± glabrous to hirsute. Basal leaf blades broadly ovate to orbicular in outline, 2-8 cm wide, palmately lobed to divided, usually hirsute. Sepals greenish yellow, 3.5-7 mm long; petals 8-17, 7-12 mm long, yellow. Fruits many, in a globose head. Feb-May. Occasional; coastal bluffs, rocky n-facing slopes, grassy hillsides; coastal scrub, grasslands, oak woodland, pine forest; 20-1400 ft (6-430 m). Widely scattered locations throughout island. San Miguel and Santa Rosa islands; southern OR to Baja CA.

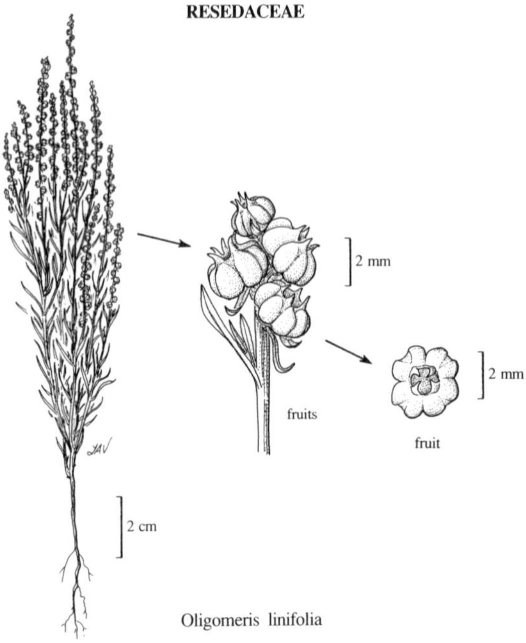


# RESEDACEAE Gray
## Mignonette Family

Annuals, perennial herbs, or shrubs. Leaves alternate, simple, entire to lobed; stipules minute. Flowers bisexual, ± bilateral, in bracteate racemes or spikes. Sepals 2-8, free; petals 2-8 or absent, free, ± unequal in shape or size; stamens 3-many; pistils 1-6, free to ± fused, ovary superior, styles 3-5. Fruit a capsule; seeds 1-many. Six genera and ca. 75 species; temperate, N. America, Europe, Africa, and Asia.

### OLIGOMERIS Cambess. OLIGOMERIS

Annuals or perennial herbs. Leaves entire, glabrous. Sepals 2-6; petals 2, persistent, sometimes absent; stamens 3-12; pistil 1, locule 1; styles 3-5. Fruit a capsule, 4-lobed, each lobe grooved. Nine species; temperate to subtropical, 8 species in Africa and s Pacific islands, 1 species in w N. America.

**O. linifolia** (M.Vahl) J.F.Macbr. Annuals 10-25 cm tall; stems erect, fleshy, glabrous. Leaves often in axillary clusters, linear, 1-4 cm long. Flowers many, in terminal spikes; bracts deltate to lanceolate. Sepals 1-3 mm long; petals oblong, white; stamens 3. Fruit depressed-globose, ca. 1.5 mm long, 2.5 mm wide. Feb-Jul (Oct). Scarce; disturbed sites, coastal bluffs and flats, clay ridgetops, loose s-facing talus slopes, floodplains; coastal scrub; 10-350 ft (3-105 m). North side between Coche Point and Scorpion Canyon; s side between Near Point and Punta Arena. All CA Channel islands; all Baja CA islands except San Martin and San Geronimo; deserts of sw N. America. Can be locally abundant in years with heavy rainfall.

# RHAMNACEAE Juss.
## Buckthorn Family

Shrubs, trees, or vines. Leaves alternate to opposite, simple, entire to toothed; stipules leaf-like or gland-like. Flowers bisexual or unisexual, radial, in small umbel-like clusters, panicles, or solitary. Sepals 4-5, partly fused; petals 4-5, often with distinct claws; stamens 4-5, opposite petals; pistil 1, ovary superior or partly inferior, locules 2-5, styles or style branches 1-3. Fruit a capsule or drupe; seeds 1-10. Ca. 50 genera and 900 species; mostly warm temperate and subtropical, ± worldwide.

1. Petals white or blue, with cup-like or hood-like blades and prominent claws; fruit a capsule, separating into 3 nut-like segments ........ **Ceanothus**

1' Petals absent or yellowish green, with flat blades tapered to base, not prominently clawed; fruit drupe-like ........ **Rhamnus**

### CEANOTHUS L. CEANOTHUS or CALIFORNIA LILAC

Shrubs, sometimes trees. Leaves alternate to opposite, evergreen or deciduous. Flowers bisexual, in panicles or umbel-like clusters. Hypanthium adnate to ovary base; calyx lobes 5, persistent in fruit; petals 5, distinct, white to blue, blades cup-like or hood-like, claws prominent; stamens 5; ovary superior, locules 3, ovules 1 per locule. Fruit a capsule, separating into 3 nut-like segments. Ca. 55 species; temperate, mostly N. America, also Europe. Many CA species are cultivated as ornamentals. Native Americans had many uses: stems of some species were used in basketry, seeds of several species were eaten, leaves and bark were used medicinally, and flowers and fruits were mashed to produce a soap.

1. Petals blue; leaves broadly ovate to elliptic, 4-7 cm long; stipules thin, early deciduous ........ **C. arboreus**

1' Petals white; leaves spatulate to obovate, 2-4.5 cm long; stipules persistent, thick, corky .... **C. megacarpus**

RHAMNACEAE

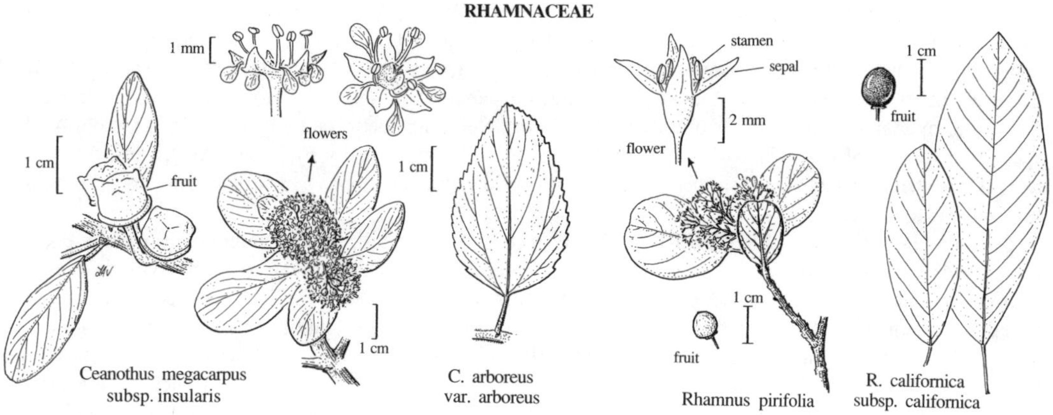


**C. arboreus** Greene ISLAND CEANOTHUS Shrubs, 1-7 m tall. Stipules thin, early deciduous; leaves alternate, semi-deciduous, broadly ovate to elliptic, 4-7 cm long, margins serrate, apices acute, upper surfaces green, lower surfaces pale green, tomentulose. Flowers in simple or compound panicles. Petals blue. Fruit 6-7 mm wide, deeply 3(4)-lobed, often with ridges or crests. Mar-Apr. Island endemic. Common; exposed ridgetops, rocky slopes (mostly n-facing), floodplains; chaparral, pine forest; 200-1900 ft (60-580 m). North side between Trident Cove and Prisoners' Harbor; slopes in Central Valley; s side between Christy Pines area and Coches Prietos drainage. Santa Rosa and Santa Catalina islands. Mostly restricted to igneous or metamorphic substrates. Plants with ± glabrous leaves have been called var. *glabra* Jeps. Putative hybrids between this species and *C. spinosus*, a mainland species not reported from Santa Cruz Island, occur in a small canyon on the s side of Central Valley near east end of Bosque Mano.

**C. megacarpus** Nutt. Shrubs, 1-6 m tall. Stipules thick, corky, persistent; leaves alternate to opposite, evergreen, spatulate to obovate, 2-4.5 cm long, margins entire, apices obtuse to slightly notched, upper surfaces green, lower surfaces pale green, canescent between veins. Flowers in umbel-like clusters or short racemes. Petals white. Fruit 8-12 mm wide, with or without horn-like appendages near tip. Dec-Apr. Common; rocky slopes, exposed ridgetops, canyon walls; chaparral; 100-1600 ft (30-490 m). Scattered locations throughout island, mostly on igneous or metamorphic substrates. San Miguel (not seen since 1886), Santa Rosa, Anacapa, Santa Catalina, and San Clemente islands; Santa Barbara Co. to San Diego Co.

1. Most leaves opposite; fruit apices with horn-like appendages absent or weakly developed .. subsp. **insularis**
1' Most leaves alternate; fruit apices with prominent horn-like appendages ........................ subsp. **megacarpus**

subsp. **insularis** (Eastw.) Raven ISLAND BIG-POD CEANOTHUS Most leaves opposite. Fruit apices with horn-like appendages absent or weakly developed. Dec-Apr. Island endemic. A subspecies of special environmental concern. Common; rocky slopes, exposed ridgetops, canyon walls; chaparral; 100-1600 ft (30-490 m). Scattered locations throughout island, mostly on igneous or metamorphic substrates. San Miguel (not seen since 1886), Santa Rosa, Anacapa, Santa Catalina, and San Clemente islands.

subsp. **megacarpus** BIG-POD CEANOTHUS Most leaves alternate. Fruit apices with prominent horn-like appendages. Mar. Rare; rocky slopes; chaparral; elevation unknown. Known from a single collection on "canyon sides" in Coches Prietos drainage in March 1929. Santa Catalina Island; Santa Barbara Co. to San Diego Co.

## RHAMNUS L. BUCKTHORN

Shrubs, sometimes trees. Leaves alternate, evergreen or deciduous. Flowers bisexual or unisexual, solitary or in umbel-like clusters. Hypanthium adnate to ovary base; calyx lobes 4-5, deciduous in fruit; petals 4-5, sometimes absent, yellowish green, blades ± flat, not cup-like or hood-like, claws short; stamens 4-5; ovary partly inferior, locules 2-4, ovules 1-2 per locule. Fruit drupe-like, seeds 2-4. Ca. 125 species; temperate and subtropical, N. and S. America, Europe, Africa.

1. Sepals 5; petals 5; terminal winter buds without scales .......................................................... **R. californica**
1' Sepals 4; petals absent; terminal winter buds covered with scales ............................................... **R. pirifolia**

**R. californica** Eschsch. subsp. **californica** CALIFORNIA COFFEE BERRY Shrubs, 1-4 m tall; bark of young twigs usually reddish; winter buds without scales. Leaves oblong to elliptic, 3-8 cm long, plane or revolute, upper surface dark green, lower surface yellowish green, margins entire to serrate. Flowers usually bisexual. Petals usually 5; style ± included. May-Jul (Sep). Scarce; canyon bottoms, rocky n-facing slopes; chaparral, oak woodland, riparian communities; 150-2000 ft (45-610 m). North side between Lady's Cove and Diablo Anchorage. Coast Ranges of CA. Although leaves and bark were used medicinally by Native Americans, they can be toxic if ingested by humans. Subsp. *occidentalis*, with yellowish green upper leaf surfaces, occurs on serpentine from sw OR to n CA.

**R. pirifolia** Greene ISLAND REDBERRY Shrubs or trees, to 10 m tall; bark of young twigs usually purplish; winter buds covered with scales. Leaves elliptic to round, 2-5(8) cm long, plane to convex-concave, margins entire to crenate, sometimes serrate. Flowers usually unisexual. Petals absent; style ± exserted. Mar-Jul. Island endemic. Common; exposed ridgetops, rocky slopes, stream banks, floodplains; chaparral, coastal scrub, oak woodland, pine forest, riparian communities; 100-1300 ft (30-395 m). North side between Platts Harbor and Eagle Canyon; Central Valley between upper portion of Cañada Christy and Stanton Ranch; s side between Laguna Canyon and Coches Prietos drainage. San Miguel (not seen since 1886), Santa Rosa, Santa Catalina, and San Clemente islands; Guadalupe Island. Mostly restricted to igneous and metamorphic substrates.

## ROSACEAE Juss.

### Rose Family

Annuals or perennials. Leaves alternate or opposite, simple to compound; stipules free or fused, persistent to deciduous. Flowers usually bisexual, radial, in racemes, panicles, umbels, or solitary. Hypanthium present, cylindrical to cup-shaped, often with minute bracts below calyx lobes; sepals usually 5; petals usually 5, free from each other, attached to hypanthium rim; stamens (1)5-many; pistils 1-many, simple or compound, ovary inferior or superior, styles 1-5. Fruit an achene, follicle, drupe, or pome; achenes and small drupes often aggregated. Ca. 110 genera and 3100 species; mostly temperate, worldwide.

1. Annuals or perennial herbs
   2. Annuals; leaves palmately lobed; flowers 1-1.5 mm long, petals absent .......... **Aphanes**
   2' Stoloniferous perennial herbs; leaves pinnately compound; flowers 10-20 mm long, petals yellow .......... **Potentilla**

1' Trees or shrubs (stems arching or reclining in *Rubus*)
   3. Most or all leaves pinnately or palmately compound
      4. Trees; stems without prickles; leaves opposite; leaflets oblong .......... **Lyonothamnus**
      4' Shrubs, stems with prickles; leaves alternate, leaflets ovate, elliptic, or oblanceolate
         5. Leaves pinnately compound; hypanthium urceolate, subglobose and leathery in fruit; petals pink to red .......... **Rosa**
         5' Leaves palmately lobed to compound; hypanthium ± cup-shaped in fruit; petals white .......... **Rubus**
   3' Leaves simple, entire, toothed, or lobed
      6. Ovary inferior .......... **Heteromeles**
      6' Ovary superior
         7. Leaf blades palmately veined and lobed; stems with prickles; pistils many per flower; fruit an aggregate of small drupes .......... **Rubus**
         7' Leaf blades pinnately veined, entire to serrate; pistil 1-5 per flower; fruit an achene or solitary drupe
            8. Leaves linear to narrowly oblong, usually in clusters, 0.4-1 cm long .......... **Adenostoma**
            8' Leaves elliptic, obovate, or ovate, not in clusters, 2-12 cm long
               9. Trees; leaf margins usually entire, sometimes dentate-serrate, lower surfaces glabrous; styles glabrous, deciduous in fruit; fruit a drupe .......... **Prunus**
               9' Shrubs, sometimes tree-like; leaf margins dentate to serrate, lower surfaces tomentulose to sparsely villosulous; styles puberulent, persistent in fruit; fruit an achene
                  10. Hypanthium tube funnelform, 6-11 mm long; petals absent; flowers few, in small axillary clusters .......... **Cercocarpus**
                  10' Hypanthium tube cup-shaped, 1.5-2 mm long; petals 5, white; flowers many, in terminal panicles .......... **Holodiscus**

## ADENOSTOMA Hook. & Arn.

Shrubs. Leaves alternate or in clusters; simple, linear to oblong, evergreen, margins entire; stipules minute or absent. Flowers many, in terminal panicles. Hypanthium tube cup-shaped; sepals 5; petals 5, white to cream; stamens 10-15; pistil 1, ovary superior, locule 1, style 1. Fruit an achene. Two species; CA and Baja CA.

**A. fasciculatum** Hook. & Arn. CHAMISE Plants to 3 m tall, openly branched or compact, mound-like, main stems prostrate to erect. Leaves in clusters, 7-14 mm long. Inflorescences 2-10 cm long; flowers many, sessile. Hypanthium tube 1-2 mm long; sepals 0.5-1 mm long; petals 1-1.5 mm long. (Feb) Apr-Jun (Oct). Common; exposed ridgetops, rocky slopes, floodplains; chaparral, pine forest; 200-1450 ft (60-440 m). North side between Twin Harbor and Prisoners' Harbor and near China Harbor; Central Valley in Cañada Christy and between La Centinela and El Sobrante; s side between Laguna Canyon and Valley Anchorage, Smugglers' Canyon, Cañada de los Sauces, near Christy Pines, and near Los Pinos del Sur. Santa Rosa, Santa Catalina, and San Clemente islands; Cedros Island; n CA coast ranges and Sierra Nevada to Baja CA. Leaves, twigs, and bark were used medicinally by Native Americans.

1. Plants erect, to 3.5 m tall, openly branched, main stems ascending to erect ....................... var. **fasciculatum**
1' Plants compact, mound-like, to 1 m tall, densely branched, main stems spreading, lower ones prostrate ...... ........................................................................................................................ var. **prostratum**

var. **fasciculatum** Plants erect, to 3.5 m tall, openly branched, main stems ascending to erect. Apr-Jun (Oct). Common; exposed ridgetops, rocky n-facing slopes, floodplains; chaparral; 200-1450 ft (60-440 m). North side between Twin Harbor and Prisoners' Harbor; Central Valley between La Centinela and El Sobrante; s side between Laguna Canyon and Valley Anchorage, and in Smugglers' Canyon. Santa Rosa, Santa Catalina, and San Clemente islands; Cedros Island; n CA coast ranges and Sierra Nevada to Baja CA.

var. **prostratum** Dunkle Plants compact, mound-like, to 1 m tall, densely branched, main stems spreading, lower ones prostrate. (Feb) May-Jun. Occasional; exposed ridgetops, rocky slopes; chaparral, pine forest; 200-1300 ft (60-440 m). North side near China Harbor; Central Valley in Cañada Christy; s side in Cañada de los Sauces, near Christy Pines, and near Los Pinos del Sur. Santa Rosa Island. May be island endemic; relationships with var. *fasciculatum* and mainland populations of low-growing forms need further study.

## APHANES L.

Annuals. Leaves alternate, palmately lobed; stipules sheathing. Flowers few, minute, usually concealed by stipules. Hypanthium tube urceolate; sepals 4, minute; petals absent; stamen usually 1; pistil 1, ovary superior, locule 1, style 1. Fruit an achene. Ca. 20 species; temperate, ± worldwide.

**A. occidentalis** (Nutt.) Rydb. [*Alchemilla o.* Nutt.] Plants 1-12 cm tall, simple to branched. Leaves broadly reniform to ± orbicular in outline, 3-7 mm long, sparsely pubescent; stipules leaf-like, 1-2 mm long. Flowers 1-1.5 mm long, 1-4 per axil. Feb-Jun. Occasional; coastal bluffs, disturbed sites, grassy hillsides, rocky n-facing slopes; coastal scrub, grasslands; 50-1500 ft (15-460 m). North side between Orizaba Cove and Prisoners' Harbor, near Mount Pleasant, and in Scorpion Canyon; scattered locations in Central Valley between upper portion of Cañada Christy and Stanton Ranch. Santa Rosa and Santa Catalina islands; Guadalupe Island; WA to Baja CA.

## CERCOCARPUS Kunth MOUNTAIN MAHOGANY

Shrubs, sometimes small trees. Leaves alternate, simple, linear to ovate, evergreen; stipules linear, deciduous. Flowers solitary or few in axillary clusters. Hypanthium tube funnelform; sepals obscure or absent; petals absent; stamens 10-many; pistil 1, ovary superior, locule 1, style 1, persistent in fruit, plumose. Fruit an achene. Ca. 13 species; w N. America.

**C. betuloides** Torr. & A.Gray var. **blancheae** (C.K.Schneider) Little Plants 1-8 m tall. Leaves elliptic to broadly obovate, 2-7 cm long, 2-6 cm wide, coriaceous, margins dentate to serrate, upper surface glabrous, lower surface tomentulose, with 6-11 pairs of lateral veins. Flowers usually 5-9 per cluster. Hypanthium tube 6-11 mm long; style 2.5-5.5 cm long in fruit. Feb-Apr (Jun). A variety of special environmental concern. Common; rocky n-facing slopes, canyon bottoms, stream banks, floodplains; chaparral, pine forest, riparian communities; 100-1800 ft (30-550 m). Scattered locations throughout island, mostly on igneous and metamorphic substrates. Santa Rosa and Santa Catalina islands; Santa Monica Mtns. Two other varieties occur on the mainland. Some plants on the island appear similar to var. *betuloides*, which usually has leaves 2-4 cm long, with 4-7 pairs of lateral veins.

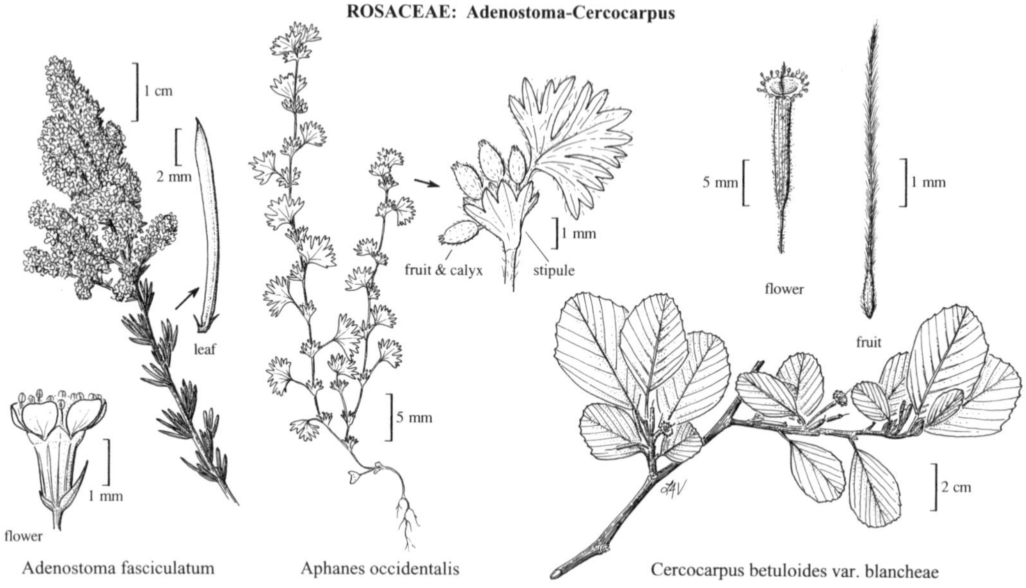


## HETEROMELES M.Roem. TOYON

With only 1 species.

**H. arbutifolia** (Lindl.) M.Roem. Shrubs or small trees, 4-10 m tall. Leaves alternate, simple, elliptic to oblong, 5.5-15 cm long, 1.5-4.5 cm wide, coriaceous, margins entire to serrate, upper surface glabrous, lower surface glaucous. Flowers many, in terminal panicles. Hypanthium tube cup-shaped, 1.5-2 mm long; sepals 5, 1-1.5 mm long; petals 5, 1.5-3 mm long, white; stamens ca. 10; pistil 1, ovary mostly inferior, locules 2-3, style 1, stigmatic lobes 2-3. Fruit a pome, globose, 6-8 mm wide, red. Jun-Oct. Abundant; n-facing coastal bluffs, rocky slopes, ridgetops, canyon bottoms; chaparral, coastal scrub, island woodland, pine forest, riparian communities; 70-1900 ft (20-580 m). Scattered locations throughout island, mostly on igneous and metamorphic substrates. All CA Channel islands except Santa Barbara (planted on San Nicolas); Los Coronados, Todos Santos, Guadalupe, and Cedros islands; n CA coast ranges to Baja CA. Plants from Santa Catalina, San Clemente, and Guadalupe islands, with fruits 8-10 mm wide, have been called var. *macrocarpa* (Munz) Munz. Although leaves and bark were used medicinally and fruits were eaten by Native Americans, this species contains cyanogenic glycosides and can be toxic to humans.

## HOLODISCUS (K.Koch) Maxim.

Shrubs. Leaves alternate, simple, elliptic to ovate, evergreen, margins usually dentate; stipules absent. Flowers many, in terminal panicles. Hypanthium tube cup-shaped; sepals 5; petals 5, usually white; stamens 15-20; pistils 5, ovary superior, locule 1. Fruit an achene. Five species; w N. America, C. and S. America. Fruits were eaten by Native Americans.

**H. discolor** (Pursh) Maxim [*H. d.* var. *franciscanus* (Rydb.) Jeps.] CREAM BUSH Plants 2-3 m tall. Leaves elliptic to broadly obovate, 3-9 cm long, 1.5-6 cm wide, margins dentate, teeth sometimes serrate, upper surface glabrous, lower surface sparsely villosulous, with 3-5 pairs of lateral veins. Hypanthium tube 1-1.5 mm long; sepals 1.5-2 mm long; petals 1.5-2 mm long, white. Fruit ciliate. (Mar) May-Jun. Rare; rocky n-facing slopes, canyon bottoms; island woodland, pine forest, riparian communities; 650-1100 ft (200-335 m). Central Valley in upper part of Cañada Christy. Santa Catalina Island; British Columbia to Mexico, e to Rocky Mtns.

## LYONOTHAMNUS A.Gray ISLAND IRONWOOD

With only 1 species.

**L. floribundus** A.Gray subsp. **aspleniifolius** (Greene) Raven SANTA CRUZ ISLAND IRONWOOD Trees to 20 m tall; bark reddish brown. Leaves opposite, evergreen, usually pinnately compound, 9-20 cm long, upper

**ROSACEAE: Heteromeles-Rubus**

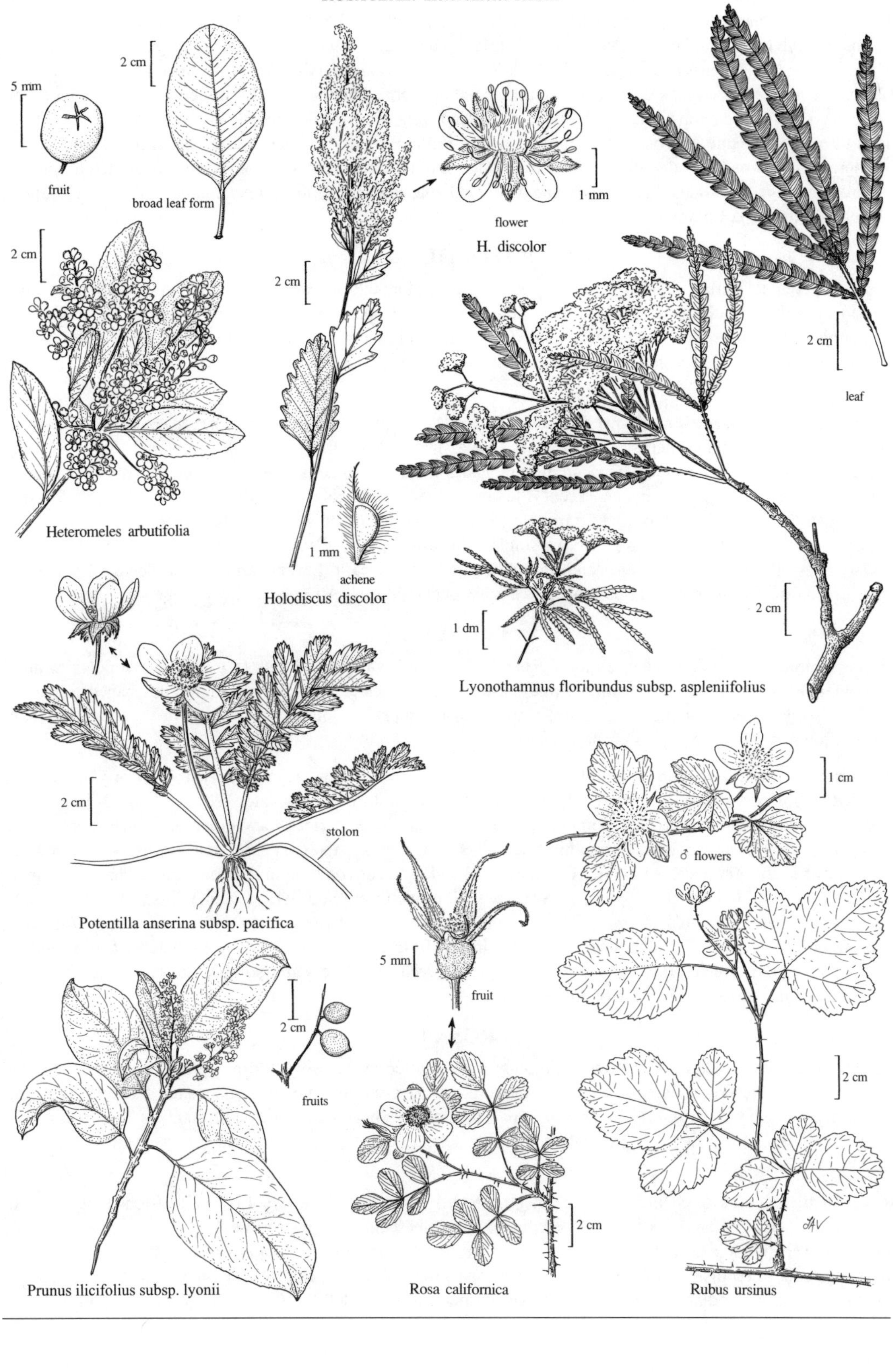

surface dark green, glabrous, coriaceous, lower surface puberulent; leaflets 3-5, narrowly oblong, 4-12 cm long, 0.8-2 cm wide, deeply toothed, teeth curved toward apex; petioles 2-3.5 cm long. Inflorescence a corymb, flowers many. Hypanthium tube cup-shaped, inner surface pubescent; sepals 1.5-2 mm long; petals 3-4 mm long, white; stamens ca. 15; pistils 2. Fruit a follicle, 5-8 mm long, woody; seeds many, minute. (Jan) May-Jul (Sep). Island endemic. A subspecies of special environmental concern. Abundant; rocky n-facing slopes, canyon bottoms, dry ravines, ridgetops; chaparral, island woodland, pine forest; 100-1900 ft (30-580 m). Scattered locations throughout n side and on slopes above Central Valley; s side between Sierra Blanca ridge and Valley Anchorage, near Los Pinos del Sur, El Montañon ridge, and in Smugglers' Canyon. Mostly restricted to igneous and metamorphic substrates. Santa Rosa and San Clemente islands. Subsp. *floribundus*, with entire leaflets, occurs only on Santa Catalina Island.

## POTENTILLA L.

Annuals, perennial herbs, or shrubs. Leaves mostly basal, pinnately to palmately compound, upper ones alternate to opposite; stipules free or fused to petiole. Flowers in open, cymose racemes or panicles or solitary. Hypanthium tube cup-shaped, usually subtended by 5 sepaloid bracts; sepals 5; petals 5, yellow; stamens ca. 20; pistils many, ovary superior, locule 1, style 1. Fruit an achene. Ca. 500 species; temperate N. America, Europe, and Asia.

**P. anserina** L. subsp. **pacifica** (Howell) Rousi [*P. egedei* Wormsk. var. *grandis* (Torr. & A.Gray) J.T.Howell] SILVERWEED Perennial herbs, with stolons. Leaves in basal rosettes, pinnately compound, 5-30 cm long; leaflets 9-21, elliptic to oblanceolate, 1-4 cm long, margins serrate, upper surface glabrous, green, lower surface tomentulose between veins, veins canescent. Flowers solitary, pedicels 5-30 cm long. Sepaloid bracts 5-9 mm long; sepals 5-8 mm long; petals 10-15 mm long, yellow. Apr. Rare; moist coastal flats; coastal beach and dune, coastal scrub; ca. 10 ft (3 m). Known from a single collection on "wet place back of beach" at mouth of Cañada Christy in April 1976. San Miguel Island; throughout n N. America, n Europe, and n Asia. Seeds were cooked and eaten by the Coast Miwok Tribe. Subsp. *anserina* occurs in e N. America and w Europe.

## PRUNUS L.

Shrubs or trees. Leaves alternate, simple, usually elliptic to ovate, deciduous to evergreen; stipules minute, deciduous. Flowers in racemes or umbel-like clusters, sometimes solitary. Hypanthium cup-shaped; sepals 5; petals 5, white to pink; stamens 15-many; pistil 1, ovary superior, locule 1, style 1. Fruit a drupe. Ca. 400 species; temperate N. America, Europe, and Asia.

**P. ilicifolia** (Nutt.) Walp. subsp. **lyonii** (Eastw.) Raven [*P. lyonii* (Eastw.) Sarg.] ISLAND or CATALINA CHERRY Shrubs or trees 2-12 m tall. Leaves broadly elliptic to broadly ovate, 5-13 cm long, 2-7 cm wide, margins usually entire, sometimes dentate-serrate, glabrous, upper surface dark green, glossy, lower surface light green. Flowers in racemes. Hypanthium tube 2-3 mm long; sepals 0.5-1 mm long; petals 1-2.5 mm long, white. Fruit 10-18 mm wide. Mar-Jun. Abundant; rocky slopes, canyon bottoms, stream banks; chaparral, riparian communities; 80-1300 ft (25-395 m). Scattered locations throughout island. Santa Rosa, Anacapa, Santa Catalina, and San Clemente islands; Baja CA. Some plants on Santa Cruz Island are similar to subsp. *ilicifolia*, with wavy, serrate leaves, which occurs on the mainland from n CA to Baja CA. Raw pulp of fruit is edible; seeds can be toxic if eaten raw but toxins (cyanogenic glycosides) are water soluble and can be leached out; processed seeds were eaten by Native Americans.

## ROSA L.

Shrubs; stems with prickles, sometimes climbing. Leaves alternate, usually odd-pinnately compound; stipules fused to petiole. Flowers solitary or few in cymose clusters. Hypanthium tube urceolate, enclosing and obscuring pistils, becoming enlarged and fleshy in fruit; sepals 5, attenuate; petals 5, white to red; stamens 20; pistils many, ovary superior, locule 1, style 1, puberulent. Fruit an achene. Ca. 100 species; temperate N. America, Europe, and Asia.

**R. californica** Cham. & Schltdl. CALIFORNIA ROSE Shrubs 1-2 m tall. Leaves 3-10 cm long, leaflets 5-7, elliptic, 1-4.5 cm long, 1-3 cm wide, subglabrous to pubescent, margins serrate. Hypanthium tube 5-10 mm long in flower, subglobose and 10-20 mm long in fruit; sepals narrowly lanceolate, 10-16 mm long; petals 12-20 mm long, pink to reddish. Mar-May (Oct). Occasional; rocky n-facing slopes, canyon bottoms, stream banks; coastal scrub, oak woodland, pine forest, riparian communities; 30-1000 ft (10-305 m). North side in Cañada del

Puerto; Central Valley between upper portion of Cañada Christy and Stanton Ranch; s side in Coches Prietos drainage. Santa Rosa and Santa Catalina islands; s OR to Baja CA. Flower buds, flowers, and fruits were eaten by Native Americans and fruits were boiled to make tea; flowers, fruits, roots, and leaves also were used medicinally.

### **RUBUS** L.

Perennial herbs or shrubs; stems prostrate to erect, often climbing, with prickles. Leaves alternate, simple and palmately lobed or palmately compound; stipules free to fused. Flowers usually few, in cymose racemes or clusters. Hypanthium tube cup-shaped; sepals 5; petals 5, white to pink; stamens many; pistils many, ovary superior, locule 1, style 1. Fruit an aggregate of small drupes, borne on a conical receptacle. Ca. 250 species; temperate, worldwide.

**R. ursinus** Cham. & Schltdl. [*R. vitifolius* Cham. & Schltdl.] CALIFORNIA BLACKBERRY Shrubs; stems often reclining or arched, 1-2 m long. Leaves 7-18 cm long, margins serrate, lower ones trifoliolate; leaflets 2-8 cm long, 1.5-6 cm wide, upper ones trifoliolate to palmately lobed. Hypanthium tube 1-3 mm long in flower, herbaceous to membranous in fruit; sepals lanceolate, 7-11 mm long; petals 10-20 mm long, white. Mar-Jun (Oct). Occasional; rocky n-facing slopes, canyon bottoms, stream banks; oak woodland, pine forest, riparian communities; 100-1200 ft (30-365 m). North side between Cueva Valdez and Cañada del Puerto, near Potato Harbor, and on El Montañon ridge; Central Valley between upper portion of Cañada Christy and Stanton Ranch; s side in Coches Prietos drainage and in Smugglers' Canyon. San Miguel, Santa Rosa, and Santa Catalina islands; San Francisco Bay region to Baja CA. Fruits are edible; leaves and roots were used medicinally by Native Americans.

## **RUBIACEAE** Juss.
## MADDER FAMILY

Annuals or perennials. Leaves opposite or appearing whorled (stipules often leaf-like, alternate leaves), simple; stipules free, often leaf-like. Flowers bisexual or unisexual, radial, in clusters, racemes, panicles, or solitary. Sepals usually 4, fused, sometimes absent; petals usually 4, fused; stamens 4, adnate to corolla, alternate corolla lobes; pistil 1, ovary inferior, locules 2 or 4, style 1-2. Fruit a berry, drupe, or capsule. Ca. 500 genera and 6000 species; temperate to tropical, worldwide. Some species are cultivated as ornamentals or for important products, including coffee and quinine.

### **GALIUM** L. BEDSTRAW or CLEAVERS

Annuals, perennial herbs, or subshrubs, often dioecious; stems often 4-angled. Leaves opposite, appearing whorled because of leaf-like stipules alternate leaves. Flowers bisexual or unisexual, in cymose clusters. Calyx lobes absent; corolla rotate, small, usually 1-3 mm wide, white to yellow or greenish; ovary 2-lobed, styles or style branches 2. Fruit a berry or, if dry, separating into 2 nutlets. Ca. 400 species; temperate, worldwide. Species with retrorse pubescence on stems have been used for stuffing matresses; several species were used medicinally by Native Americans. Cheese Rennet (*G. verum*) contains a milk-curdling enzyme and also yields a yellow dye, which was used for coloring cheese and butter.

1. Annuals; leaves apparently 5-8 per node ........ **G. aparine**

1' Subshrubs or suffrutescent perennial herbs; leaves apparently 4 per node

  2. Ovary and fruit densely hispid, trichomes long, straight, white; lower surfaces of leaves white-scabrous, especially near bases ........ **G. angustifolium** subsp. **foliosum**

  2' Ovary puberulent to glabrous, fruit puberulent to glabrous, trichomes, if present, not long, straight, and white; lower surfaces of leaves glabrous to minutely scabrous or villous

    3. Leaves linear, oblong, or broadly elliptic, 2-20 mm long, 1-3.5 mm wide, surfaces usually glabrous; stems sprawling or reclining, retrorse-hispid on angles to glabrous

      4. Stems glabrous; leaves plane, apices sharply acute, margins smooth to sparsely ciliate ........ **G. nuttallii** subsp. **insulare**

      4' Stems retrorse-hispid, scabrous; leaves often weakly revolute, apices rounded to apiculate, each with a soft, deciduous, terminal trichome, margins ciliate ........ **G. porrigens**

    3' Leaves elliptic to oblanceolate, 5-30 mm long, 2-7 mm wide, surfaces sparsely pubescent to villous, sometimes glabrous; stems erect, not sprawling or reclining, minutely scabrous or villous

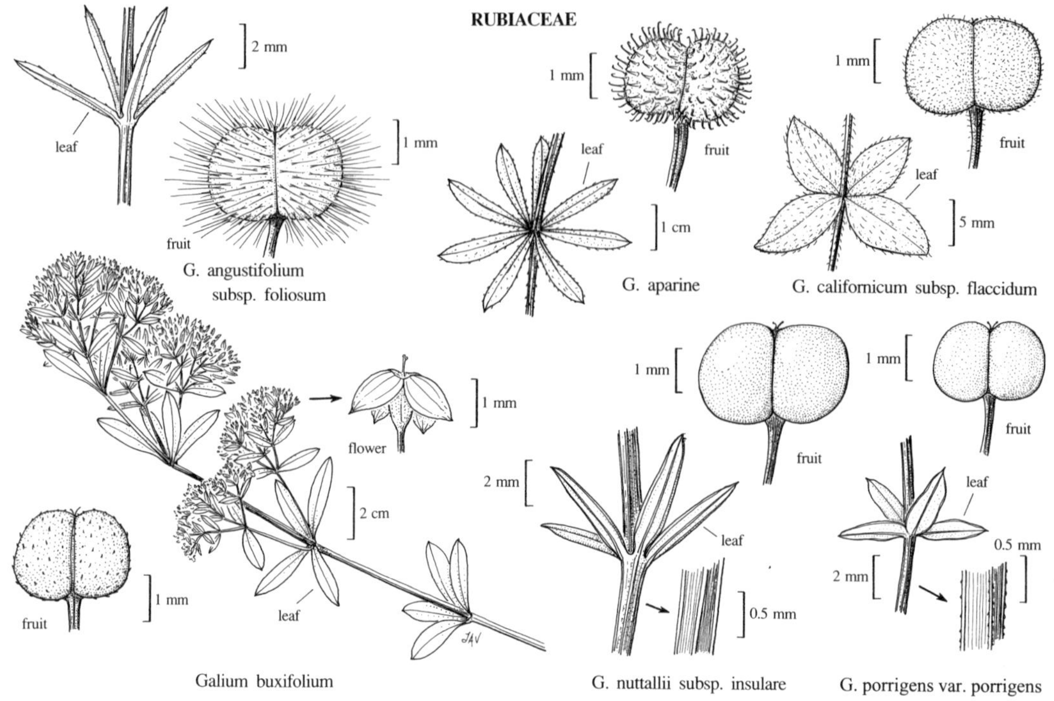


5. Leaves 15-30 mm long, sparsely pubescent to glabrous; fruit dry, separating into 2 nutlets .................. **G. buxifolium**

5' Leaves 5-20 mm long, surfaces villous; fruit fleshy, not dehiscent .. **G. californicum** subsp. **flaccidum**

**G. angustifolium** Nutt. subsp. **foliosum** (Hilend & Howell) Dempster & Stebbins NARROW-LEAVED BEDSTRAW Subshrubs, 3-6(10) dm tall, dioecious; stems glabrous. Leaves (including leaf-like stipules) 4 per node, linear, 3-16 mm long, 0.5-1 mm wide, margins ciliate, lower surfaces white-scabrous near base. Fruit dry, densely hispid, trichomes as long as fruit width. Mar-Jul (Oct). Island endemic. Common; coastal bluffs, rocky slopes, floodplains; chaparral, coastal scrub, oak woodland; 30-1900 ft (10-580 m). Widely scattered locations throughout island. Santa Rosa and Anacapa islands. Insect galls are frequently seen on stems. Seven other subspecies occur on the mainland, primarily in the deserts and mountains of s CA.

**G. aparine** L. GOOSE GRASS or CLEAVERS Annuals; stems prostrate or reclining on other plants, 3-8 dm long, often breaking easily, angles retrorsely hispid. Leaves (including leaf-like stipules) 5-8 per node, spatulate to narrowly oblanceolate, 10-30 mm long, 2-4 mm wide, margins ciliate, surfaces slightly hispid. Fruit dry, densely hispid, trichomes hooked, shorter than fruit width. Mar-Jun. Possibly introduced. Occasional; disturbed sites, rocky slopes, stream banks; coastal scrub, grasslands, oak woodland, pine forest; 20-1300 ft (6-395 m). Widely scattered locations throughout island. All CA Channel islands; Los Coronados, San Martin, Guadalupe, and Cedros islands; throughout N. America, ± worldwide. Fruits can be roasted and used as a coffee substitute.

**G. buxifolium** Greene SEA-CLIFF BEDSTRAW Subshrubs, 3-6 dm tall, usually dioecious; stems minutely scabrous, nodes often swollen. Leaves (including leaf-like stipules) 4 per node, elliptic to oblanceolate, 15-30 mm long, 3-7 mm wide, short-petiolate, margins smooth to sparsely ciliate, surfaces sparsely pubescent to glabrous. Fruit dry, puberulent or glabrate. Mar-Jul. Island endemic. A species of special environmental concern. Occasional; n-facing coastal bluffs, rocky slopes; coastal scrub, pine forest; 20-1000 ft (6-305 m). Scattered locations on n shore between Cueva Valdez and Eagle Canyon; Central Valley near Buena Vista; s side in coastal hills near Rancho Nuevo. San Miguel Island.

**G. californicum** Hook. & Arn. subsp. **flaccidum** (Greene) Dempster & Stebbins [*G. flaccidum* Greene] Subshrubs, 1-5 dm tall, dioecious; stems ± villous, trichomes ± appressed. Leaves (including leaf-like stipules) 4 per node, elliptic, 5-20 mm long, 2-6 mm wide, sessile, margins ciliate, surfaces villous. Fruit fleshy, usually puberulent. Feb-Apr. Scarce; shaded n-facing slopes, canyon bottoms; chaparral, oak woodland, pine forest; 150-2000 ft (45-610 m). North side near Picacho Diablo and Red Peak, and between Pelican Bay and Prisoners'

Harbor; Central Valley near Christy Pines and in Islay Canyon. Monterey Co. to s CA. Five other subspecies occur on the mainland of s CA; subsp. *miguelense* (Greene) Dempster & Stebbins is restricted to San Miguel and Santa Rosa islands.

**G. nuttallii** A.Gray subsp. **insulare** Ferris NUTTALL'S ISLAND BEDSTRAW Subshrubs, dioecious; stems slender, 5-15 dm long, sprawling or reclining, often densely entangled, glabrous, often red-tinged. Leaves (including leaf-like stipules) 4 per node, linear to oblong, 2-8 mm long, 1-1.5 mm wide, apices sharply acute, margins smooth to sparsely ciliate, surfaces glabrous. Fruit fleshy, glabrous. Mar-May. Island endemic. A subspecies of special environmental concern. Common; rocky n-facing slopes, exposed ridgetops, floodplains; chaparral, coastal scrub, oak woodland, pine forest, riparian communities; 200-1450 ft (60-440 m). Scattered locations on n side between Cueva Valdez and Pelican Bay, near China Harbor, and in Scorpion Canyon; Central Valley between Christy Ranch and Valley Anchorage; s side in Cañada de los Sauces, on Sierra Blanca ridge, on South Ridge between Laguna Canyon and Alberts Anchorage, and near Los Pinos del Sur. Santa Rosa and Santa Catalina islands. Subsp. *nuttallii*, with scabrous stems and leaves, occurs on the mainland of s CA and n Baja CA.

**G. porrigens** Dempster var. **porrigens** [*G. nuttallii* A.Gray var. *ovalifolium* Dempster] CLIMBING BEDSTRAW Subshrubs, dioecious; stems slender, 6-15 dm long, sprawling or reclining, scabrous, retrorse-hispid, green to red-tinged. Leaves (including leaf-like stipules) 4 per node, broadly elliptic to oblong, 2-10 mm long, 1.5-3.5 mm wide, apices rounded to apiculate, each with soft, deciduous, terminal trichome, margins ciliate, often weakly revolute, surfaces usually glabrous. Fruit fleshy, glabrous. Mar-Jun. Occasional; rocky slopes, canyon bottoms; chaparral, coastal scrub, oak woodland, pine forest; 150-950 ft (45-290 m). North side between Pelican Bay and Prisoners' Harbor, in Cañada del Puerto, and in Cañada del Agua; widely scattered locations in Central Valley between upper portion of Cañada Christy and Stanton Ranch. Santa Rosa and Santa Catalina islands; s OR to Baja CA. Var. *tenue* (Dempster) Dempster, with linear leaves, occurs on the mainland.

## SALICACEAE Mirbel
### WILLOW OR POPLAR FAMILY

Dioecious subshrubs, shrubs, or trees; bark smooth to furrowed. Leaves alternate, simple, deciduous; stipules present, deciduous. Flowers unisexual, radial, in spikes, each flower subtended by 1 scale-like bract. Perianth absent; staminate flowers with 1-many stamens; pistillate flowers with ovary superior, locule 1, placentas parietal, stigmas 2 or 4. Fruit a capsule; seeds many, minute, silky-comose. Two genera and 440 species, mostly temperate, worldwide, n hemisphere. Some species are cultivated as ornamentals or used for wood.

1. Most leaf blades deltate to broadly ovate, bases subcuneate, truncate, or cordate; inflorescences pendulous, bracts fimbriate; winter bud scales several, imbricated, resinous .......... **Populus**

1' Most leaf blades linear, lanceolate, or elliptic, bases rounded to tapered; inflorescences erect or spreading, bracts entire; winter bud scale solitary, not resinous .......... **Salix**

### POPULUS COTTONWOOD or POPLAR

Trees; bark furrowed; each winter bud enclosed by several imbricated, resinous scales. Leaves elliptic to deltate; petioles long; stipules minute. Flowers in pendulous spikes, bracts fimbriate. Stamens usually many; ovary subtended by cup-shaped disk, style short, stigmas 2 or 4, often dissected or lobed. Fruit globose to subglobose. Ca. 35 species; mostly temperate, n hemisphere.

1. Most leaf blades broadly ovate, upper surface dark green, lower surface pale green, glaucous, sometimes rust-colored; petioles slightly flattened to ± terete .......... **P. balsamifera** subsp. **trichocarpa**

1' Most leaf blades deltate, both surfaces yellowish green; petioles flattened .......... **P. fremontii**

**P. balsamifera** L. subsp. **trichocarpa** (Torr. & A.Gray ex Hook.) Brayshaw BLACK COTTONWOOD Trees, to 20 m tall. Leaf blades broadly ovate, 3-9 (20) cm long, bases truncates to cordate, margins finely crenate or serrate, apices usually broadly acute, upper surfaces dark green, lower surfaces pale green, glaucous, sometimes rust-colored; petioles slightly flattened to ± terete, 2-4 cm long. Inflorescences 4-8 cm long. Fruits 4-5 mm long. Feb-Apr. Occasional; canyon bottoms, floodplains; riparian communities; 100-1300 ft (30-395 m). North side between Trident Cove and Cañada del Puerto; Central Valley in Cottonwood Canyon; locally abundant on s side in Alamos Canyon. Santa Rosa and Santa Catalina islands; AK to n Baja CA, e to UT. Sap, roots, bark, and flower buds were used medicinally by Native Americans. Subsp. *balsamifera* occurs throughout n N. America.

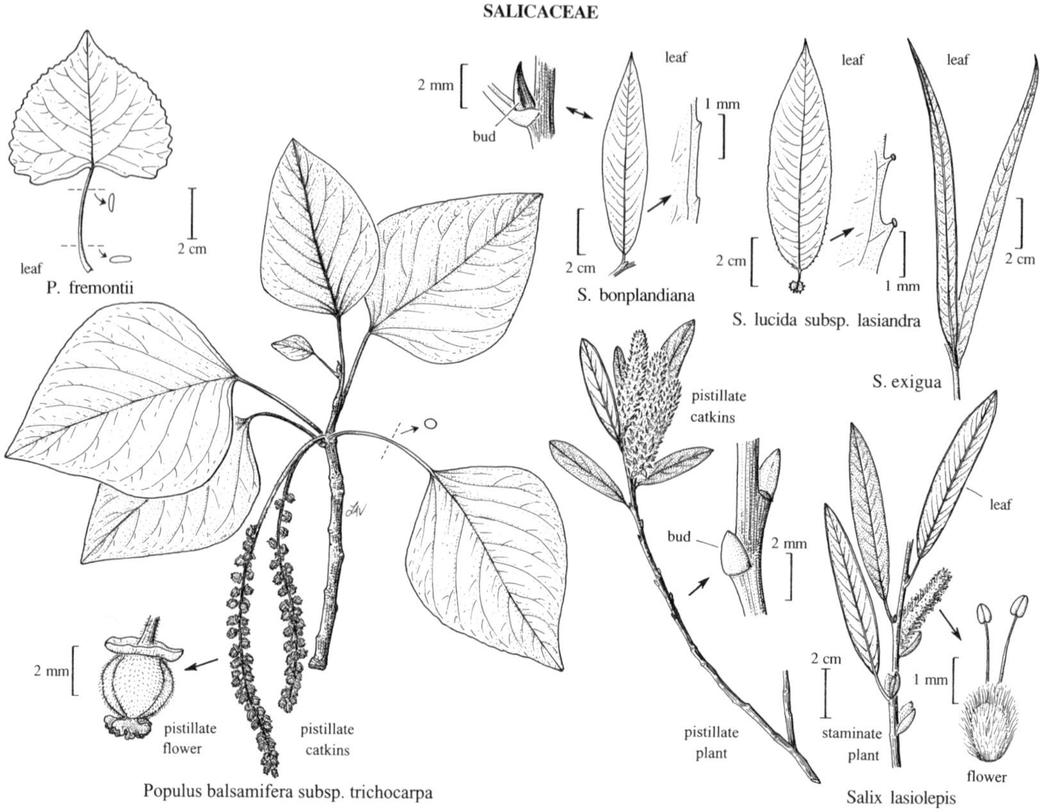


**P. fremontii** S.Watson subsp. **fremontii** FREMONT COTTONWOOD Trees, to 20 m tall. Leaf blades deltate, 4-9 cm long, bases truncate to subcuneate, margins coarsely crenate-serrate, apices usually acuminate, surfaces yellowish green; petioles flattened, 3-6 cm long. Inflorescences 4-5 cm long. Fruits 8-12 mm long. Mar-Apr. Rare; canyon bottoms; riparian communities; 250-500 ft (75-150 m). Central Valley near base of Portezuela Grade and in Cottonwood Canyon. Santa Catalina Island; throughout CA, to n Mexico, e to UT. Boiled leaves and bark were used medicinally by Native Americans. Several other subspecies occur in w N. America.

## SALIX L. WILLOW

Shrubs or trees; bark smooth or furrowed; each winter bud enclosed by a solitary, non-resinous scale. Leaves linear, lanceolate, or elliptic; petioles short or leaves subsessile; stipules often conspicuous. Flowers in erect to spreading spikes, bracts entire. Stamens 1-8; ovary or pistillate pedicel subtended by 1-2 glands, style long, stigma lobes 2. Fruit globose to ovoid. Ca. 400 species; mostly temperate, worldwide. Salicin and related compounds used in pharmaceutical products (especially aspirin) were formerly extracted from the bark of some species. Ancient Asian records indicate that willow bark has been used to reduce pain for at least 2400 years.

1. Leaves linear to linear-lanceolate, 2-11 mm wide, subsessile, both surfaces gray-tomentose or lanate .......... **S. exigua**

1' Leaves lanceolate to elliptic or oblanceolate, 10-35 mm wide, petiolate, upper surfaces glabrous, lower surfaces glabrous, glaucous, or tomentose

  2. Petioles strongly glandular near base of blade; blade margins strongly glandular-serrate .......... **S. lucida** subsp. **lasiandra**

  2' Petioles non-glandular or weakly so; blade margins entire to crenate, not glandular

    3. Margins of bud scale free, overlapping; stamens 4-5; pistillate bracts yellow, persistent .......... **S. bonplandiana**

    3' Margins of bud scale fused; stamens 2; pistillate bracts dark brown, deciduous .......... **S. lasiolepis**

**S. bonplandiana** Kunth RED WILLOW Trees, 5-15 m tall; young twigs red to yellow-brown, pubescent. Leaf blades lanceolate to broadly elliptic, 5-15 cm long, 1.5-3 cm wide, petioles 6-15 mm long, non-glandular to slightly glandular near blade base, margins entire to crenulate, upper surfaces glabrous, shiny, lower surfaces glabrous, glaucous. Spikes produced with or after leaves; bracts obovate, yellow. Stamens 4-5; filaments pilose. Fruit 3.5-7 mm long. Mar-Apr. Scarce; canyon bottoms, floodplains; riparian communities; 200-950 ft (60-290 m). North side in Cañada del Puerto; Central Valley in Cañada de la Siesta and in Hooker Canyon. Santa Catalina Island; OR to C. America, e to NV and AZ. Roots and bark were used medicinally by Native Americans. The correct name for this species may be *S. laevigata* Bebb.

**S. exigua** Nutt. [*S. hindsiana* Benth.] SAND-BAR WILLOW Shrubs, 2-6 m tall; young twigs brownish, sericeous. Leaf blades linear to linear-lanceolate, 5-16 cm long, 0.2-1.1 cm wide, subsessile, margins entire to denticulate, both surfaces gray-tomentose or lanate, especially when young. Spikes produced with or after leaves; bracts lanceolate, yellow. Stamens 2; filaments villous. Fruit 6-9 mm long. Mar-May. Occasional; canyon bottoms, floodplains; riparian communities; 100-600 ft (30-185 m). North side in Cañada del Puerto; Central Valley in Cañada Christy and between La Cascada and Bosque Mano; s side in Pozo, Horqueta, San Justiniano, and Aguaje canyons. San Nicolas and Santa Catalina islands; throughout much of N. America. Bark and stems were used medicinally by Native Americans.

**S. lasiolepis** Benth. ARROYO WILLOW Shrubs or small trees, 2-10 m tall; young twigs yellow to brown, tomentose. Leaf blades oblanceolate to oblong-lanceolate, 3.5-12 cm long, 1-3 cm wide, petioles 3-16 mm long, non-glandular, margins entire, narrowly revolute, upper surfaces pubescent, lower surfaces glabrous, glaucous, or tomentose. Spikes produced before leaves; bracts orbicular to obovate, dark brown. Stamens 2; filaments glabrous. Fruit 3-5 mm long. Feb-Mar. Common; canyon bottoms, floodplains; pine forest, riparian communities; 50-800 ft (15-245 m). Scattered locations throughout much of island. All CA Channel islands except Santa Barbara and San Clemente; WA to Mexico, e to ID and TX. Bark and leaves were used medicinally by Native Americans.

**S. lucida** Muhlenb. subsp. **lasiandra** (Benth.) E.Murray [*S. lasiandra* Benth.] YELLOW WILLOW Shrubs or trees, 4-10 m tall; young twigs brown, glabrous to puberulent. Leaf blades lanceolate, 6-17 cm long, 1.5-3.5 cm wide, petioles 5-15 mm long, glandular near blade base, margins glandular-serrate, upper surfaces glabrous, shiny, lower surfaces glabrous. Spikes produced with leaves; bracts lanceolate to ovate, yellow. Stamens 3-5; filaments puberulent near bases. Fruit 6-11 mm long. Feb-Apr. Rare, not seen recently; streambed; riparian communities; elevation unknown. Known from two collections in Valdez Canyon in September 1927 and November 1930. Alaska to s CA, e to NV. Two other subspecies occur throughout N. America.

## SAURURACEAE E.Meyer
### LIZARD'S TAIL FAMILY

Perennial herbs, with stolons or rhizomes. Leaves basal and alternate, simple; stipules adnate to petiole. Flowers bisexual, radial, in dense, terminal, bracteate spikes, often with large, petaloid bracts at base. Sepals and petals absent; stamens 6-8; pistils 1-4, ovary superior or embedded in inflorescence axis, locule 1, styles 1-4. Fruit a capsule. Ca. 5 genera and 8 species; mostly temperate, N. America and Asia.

### ANEMOPSIS Hook. & Arn. YERBA MANSA

With only 1 species.

**A. californica** (Nutt.) Hook. & Arn. Plants 10-45 cm tall, with stolons and aromatic rhizomes. Leaves mostly basal; blades elliptic to obovate, 4-24 cm long, 1.5-10 cm wide, glabrous, base obtuse to cordate, margins entire, apices rounded; petioles 3-30 cm long, glabrous to woolly. Spike narrowly conical, 1-4 cm long, bract subtending each flower 4-6 mm long, white, bracts subtending spike 5-8, whorled, 1-3 cm long, white. Pistil 1, embedded in inflorescence axis, stigmatic branches 3-4. Mar-Sep. Introduced to island. Rare; moist flats, canyon bottoms; coastal marsh and estuary, riparian communities; 5-200 ft (2-60 m). North side at Prisoners' Harbor; Central Valley near Stanton Ranch. San Nicolas, Santa Catalina, and San Clemente islands; Cedros Island; Great Valley to n Mexico, e to CO and TX. First collected in June 1930 at Prisoners' Harbor, where it was reportedly planted by a former ranch employee for medicinal uses (Hoffmann 1932a). Bark and roots were widely used medicinally by Native Americans.

SAURURACEAE

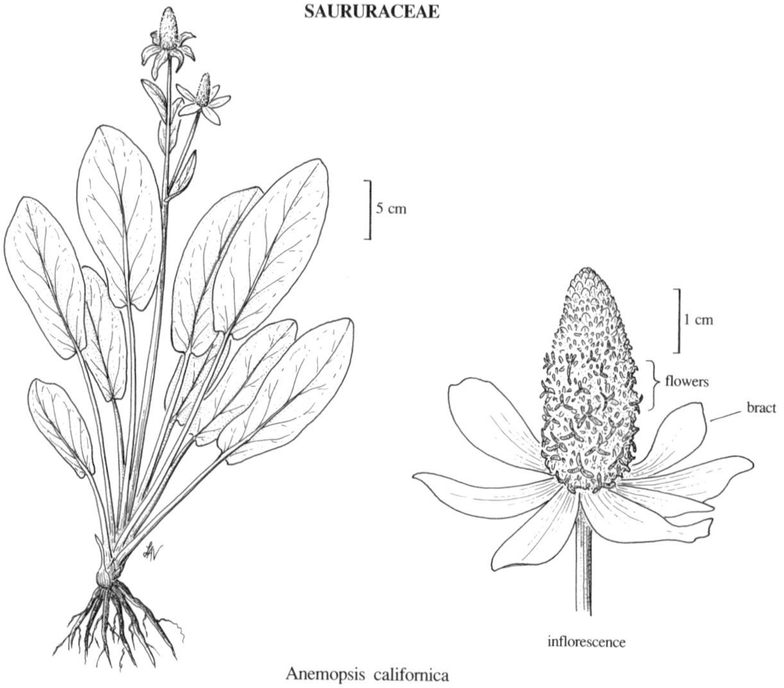


Anemopsis californica

# SAXIFRAGACEAE Juss.

## Saxifrage Family

Perennial herbs, often with rhizomes. Leaves mostly basal, sometimes cauline, alternate, usually palmately veined and lobed; stipules present. Flowers bisexual, radial, in cymes, racemes, or panicles. Hypanthium present; sepals 5; petals 5; stamens 5 or 10; pistils 1-2, simple or compound, ovary superior to partly inferior, locules 1-2, styles 1-2. Fruit a follicle or capsule. Ca. 40 genera and 600 species; arctic, temperate, mostly n hemisphere.

1. Stamens 5; petioles pilose; leaf blade margins ciliate ........ **Heuchera**
1' Stamens 10; petioles glabrous to puberulent; leaf blade margins puberulent to glabrous
   2. Leaf blades elliptic to ovate, toothed but not lobed ........ **Saxifraga**
   2' Leaf blades round-cordate, lobed, the lobes entire to toothed
      3. Leaf lobe margins toothed; plants flowering in late summer or fall; petals 3-4 mm long . **Jepsonia**
      3' Leaf lobe margins entire; plants flowering in spring; petals 5-8 mm long ........ **Lithophragma**

## HEUCHERA L. ALUMROOT

Plants with thick rhizomes, often with short underground stems. Leaves usually basal. Flowers in scapose panicles. Hypanthium cylindrical to hemispheric; stamens 5, alternate petals; ovary half-inferior, locule 1, placentas 2, parietal. Fruit a capsule. Ca. 55 species; temperate N. America. Some species are cultivated as ornamentals. Boiled or steamed leaves of some species were eaten by Native Americans; roots were used medicinally.

**H. maxima** Greene ISLAND ALUMROOT Plants 3-9 dm tall. Basal leaf blades 4-13 cm wide, margins toothed, upper surfaces sparsely strigose, lower surfaces strigose to villous; petioles 2-18 cm long. Flowers many. Hypanthium campanulate; petals acute to obtuse, white to pink. Feb-Apr. Island endemic. A species of special environmental concern. Occasional; coastal bluffs, shaded n-facing slopes, rocky canyon walls; coastal scrub, oak woodland, pine forest, riparian communities; 40-750 ft (12-230 m). North side between Del Mar Cove and Prisoners' Harbor, near Potato Harbor, and in Scorpion Canyon; scattered locations in Central Valley between upper portion of Cañada Christy and El Sobrante; s side in Smugglers' Canyon. Santa Rosa and Anacapa islands. Mostly restricted to igneous and metamorphic substrates.

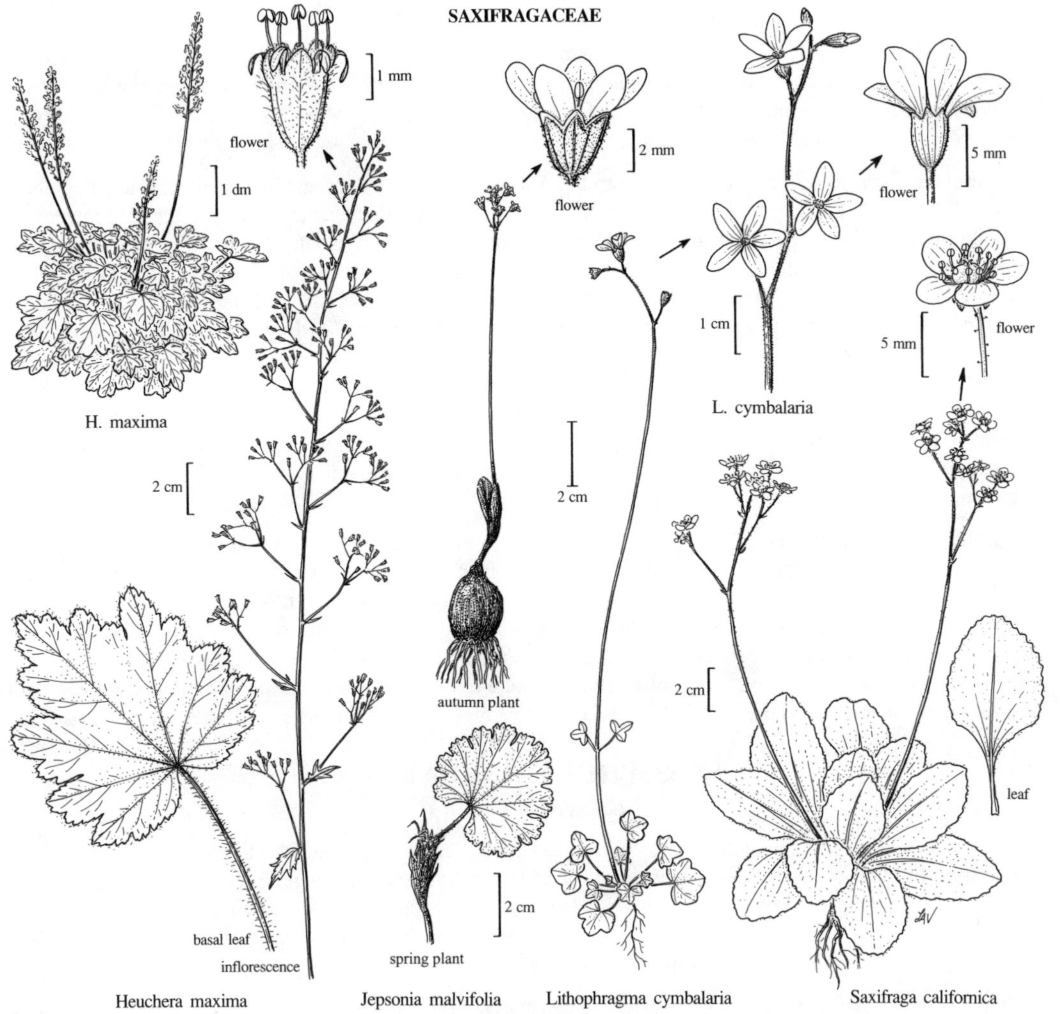


## JEPSONIA Small

Plants with ovoid to subglobose rhizomes. Leaves basal, developing with or after flowers. Flowers few, in cymes. Hypanthium campanulate; stamens 10; ovary superior, locules 2, placentas 2, axile. Fruit a follicle. 3 species; California. All species of Jepsonia are distylous, with flowers of some plants having long stamens and short styles and flowers of other plants having short stamens and long styles. The proportion of long-styled to short-styled plants in any population usually approximates a 1:1 ratio.

**J. malvifolia** (Greene) Small ISLAND JEPSONIA Plants 1-2 dm tall. Basal leaf blades 3-9 cm wide, glabrous to sparsely pubescent, margins toothed; petioles 2-7 cm long. Flowers 4-10. Petals 3-4 mm long, white to pink, with reddish veins, margins ± undulate, apices acute to obtuse. Aug-Oct. Island endemic. A species of special environmental concern. Occasional; coastal bluffs, rocky n-facing slopes; coastal scrub, chaparral, pine forest; 60-1000 ft (20-305 m). Scattered locations on n side between Cueva Valdez and Prisoners' Harbor, in Cañada del Puerto, and in Scorpion Canyon; Central Valley in upper portion of Cañada Christy and between La Cascada and El Sobrante; widely scattered locations on s side between Rancho Nuevo and Black Point peninsula, in Cañada de los Sauces, and between Sierra Blanca ridge and Valley Anchorage. Santa Rosa, San Nicolas, Santa Catalina, and San Clemente islands; Guadalupe Island. Mostly restricted to igneous and metamorphic substrates.

### LITHOPHRAGMA Nutt. WOODLAND STAR

Plants with short underground stems, slender rhizomes, often with bulblets. Leaves mostly basal, cauline few, opposite. Flowers in racemes. Hypanthium hemispheric to campanulate; stamens 10; ovary superior to ± inferior, locule 1, placentas 2-3, parietal. Fruit a capsule. Ca. 12 species; temperate, w N. America. Roots of some species were used medicinally by Native Americans.

**L. cymbalaria** Torr. & A.Gray Plants 1-5.5 dm tall. Basal leaf blades 1-3.5 cm wide, shallowly lobed, glabrous to sparsely puberulent, margins ± entire, petioles 1-9 cm long; cauline leaves 2, deeply lobed. Flowers 2-7. Hypanthium hemispheric; petals oblanceolate, 5-8 mm long, white, apices rounded. Feb-May. Occasional; canyon bottoms, grassy n-facing hillsides, shaded slopes, rocky canyon walls; chaparral, grasslands, oak woodland, pine forest; 40-2400 ft (12-730 m). Widely scattered locations on n side and in Central Valley; s side on El Montaûon ridge and in Smugglers' Canyon. Santa Rosa Island; San Francisco Bay region to San Gabriel Mtns.

### SAXIFRAGA L. SAXIFRAGE

Plants with short underground stems or rhizomes. Leaves basal. Flowers in racemes, capitate spikes, or cymose panicles. Hypanthium campanulate to hemispheric; stamens 10; pistils 1-2, ovary superior to ± inferior, locules either 1, with 1 marginal placenta, or 2, with 2 axile placentas. Fruit a capsule or follicle. Ca. 300 species; arctic, temperate, ± worldwide.

**S. californica** Greene Plants 2-3.5 dm tall. Leaves 5-9 cm long, blades ovate to elliptic, glabrous to sparsely puberulent near dentate margins, base tapered. Flowers 9-many, in cymose panicles. Petals ± elliptic, 3-5 mm long, white, apices rounded; ovary half-inferior, deeply lobed, styles 2. Feb-Apr. Scarce; stream banks, canyon bottoms, grassy n-facing hillsides, rocky n-facing slopes; chaparral, grasslands, oak woodland, pine forest; 60-1450 ft (20-440 m). North side between Platts and Prisoners' harbors, and in Scorpion Canyon; Central Valley in upper portion of Cañada Christy, just w of Bosque Mano, and between Islay Canyon and Stanton Ranch. Santa Rosa and San Clemente islands; sw OR to n Baja CA.

## SCROPHULARIACEAE Juss.
### FIGWORT FAMILY

Annuals or perennials. Leaves alternate or opposite, simple, entire to pinnatifid; stipules absent. Flowers bisexual, usually bilateral, often bilabiate, solitary or in racemes, spikes, or panicles. Sepals 4-5, free to fused; petals 4-5, fused; stamens usually 4, in 2 pairs, adnate to corolla, often with sterile stamen (staminode); ovary superior, locules 2, placentas axile, style 1, stigmatic lobes 2. Fruit a capsule, usually dehiscent by 2 valves; seeds many. Ca. 220 genera and 4500 species; mostly temperate, worldwide.

1. Stamens 5; corollas ± radial, rotate; flowers in dense cylindrical spikes .......... **Verbascum**

1' Stamens 2 or 4; corollas bilateral, often bilabiate; flowers in racemes, spikes, or panicles but inflorescence not dense and cylindrical
   2. Most upper cauline leaves alternate
      3. Upper corolla lip a galea (hood-like), entire, concealing stamens .......... **Castilleja**
      3' Upper corolla lip not hood-like, clearly lobed, not concealing stamens
         4. Corollas without spur; upper sepal wider or larger than other sepals .......... **Antirrhinum**
         4' Corollas with basal spur; sepals ± equal .......... **Linaria**
   2' Most upper cauline leaves opposite
      5. Stamens 2; corollas slightly bilateral, with 4 lobes, upper one larger than other 3; corollas pale blue to lavender .......... **Veronica**
      5' Stamens 4; strongly bilabiate, corollas with 5 lobes, corollas yellow, red, purple, or violet
         6. Sepals ± free; stigma capitate; staminode present; fruit ovoid .......... **Keckiella**
         6' Sepals fused into tube; stigmatic lobes 2, ± flattened; staminode absent; fruit cylindrical .......... **Mimulus**

### ANTIRRHINUM L. SNAPDRAGON

Annuals or perennial herbs; stems erect, twining, or reclining. Leaves linear to ovate, entire, sessile to short-petiolate, lower ones opposite, upper ones alternate. Flowers solitary and axillary or in terminal racemes. Sepals mostly free; corollas bilabiate, base swollen, white to purple, upper lip 2-lobed, lower lip 3-lobed, lower lip base often swollen, closing throat; stamens 4, in 2 unequal pairs, staminode absent; stigmas ± 2-lobed. Fruit ovoid to globose, dehiscent irregularly or by terminal slits or pores, locules equal or unequal in size. Ca. 40 species; temperate, w N. America, Mediterranean Europe.

1. Stems glabrous; pedicels 30-60 mm long, ± twining; fruit dehiscent by slits .............................. **A. kelloggii**
1' Stems glandular-puberulent; pedicels 2-15 mm long, not twining; fruit dehiscent by pores
 2. Stems erect, free-standing; leaves lanceolate, flowers in terminal racemes .................. **A. multiflorum**
 2' Stems slender, often reclining and supported by twisted or twining branches; leaves ovate or elliptic; flowers solitary and axillary ........................................................... **A. nuttallianum** subsp. **subsessile**

**A. kelloggii** Greene CLIMBING SNAPDRAGON Annuals, to 1 m tall, often supported by ± twining pedicels; stems glabrous. Lower cauline leaves narrowly lanceolate, to 4 cm long, short-petiolate. Pedicels 30-60 mm long, ± twining; corollas 10-15 mm long, blue to purplish blue. Fruit 4-5 mm long, dehiscent by apical slits, locules equal in size. Mar-May. Scarce; disturbed or burned sites, open rocky slopes, grassy hillsides; chaparral, coastal scrub; 200-700 ft (60-215 m). North side near Pelican Bay; widely scattered locations in Central Valley between Portezuela Grade and Stanton Ranch airfield; widely scattered locations on s side between Willows Canyon and Valley Anchorage. Santa Catalina Island; n CA coast ranges to n Baja CA. Usually seen only after fires or in years with heavy rainfall.

**A. multiflorum** Pennell [*A. glandulosum* Lindl.] STICKY SNAPDRAGON Annuals or short-lived perennials, to 1.5 m tall; stems glandular-puberulent, viscid. Lower cauline leaves lanceolate, to 6 cm long, sessile. Pedicels 2-4 mm long; corollas 10-15 mm long, white to pale pink. Fruit 8-11 mm long, dehiscent by apical pores, locules unequal in size. Apr-Jul. Occasional; rocky slopes, floodplains; chaparral, coastal scrub; 150-1850 ft (45-565 m). North side between Lagunitas Secas and Cañada del Puerto; Central Valley between Picacho Diablo and Cañada de la Siesta. Northern CA coast ranges to n Baja CA.

**A. nuttallianum** Benth. subsp. **subsessile** (A.Gray) D.M.Thomps. NUTTALL'S SNAPDRAGON Annuals or biennials, to 1.5 m tall; stems glandular-puberulent, viscid. Lower cauline leaves ovate or elliptic, to 3 cm long, petiolate to sessile. Pedicels 4-15 mm long; corollas 6-12 mm long, blue to purple. Fruit 9-10 mm long, dehiscent by apical pores, locules unequal in size. Feb-Jul (Sep). Common; coastal bluffs, rocky slopes, canyon bottoms, floodplains; coastal scrub, pine forest, riparian communities; 5-1400 ft (2-425 m). Widely scattered locations, especially around perimeter of island. All CA Channel islands except Santa Barbara and San Nicolas; Los Coronados, Todos Santos, San Martin, Guadalupe, and Cedros islands; Monterey Co. to Baja CA. Boiled flowers were used medicinally by the Southern Diegueño Tribe. Subspecies *nuttallianum*, with densely glandular-puberulent stems and violet-veined, white blotches on lower lips, occurs on the mainland from Los Angeles Co. to n Baja CA.

## **CASTILLEJA** Mutis ex L.f. PAINTBRUSH or OWL'S CLOVER

Annuals or perennial herbs; stems erect. Leaves alternate, linear to ovate, entire to pinnatifid, sessile to subsessile. Flowers in dense, terminal racemes or spikes, each subtended by an often colorful bract. Calyces tubular, often unequally lobed, colorful; corollas strongly bilabiate, upper lip a galea (hood-like), elongated, entire, lower lip entire to 3-toothed, often much shorter than upper lip; stamens 4, in 2 unequal pairs, staminode absent; stigmas entire to 2-lobed. Fruit ovoid to cylindrical, dehiscent by valves. Ca. 200 species; w N. America. Seeds of several species were eaten by Native Americans; entire plants of some species also were used medicinally.

1. Perennial herbs; sepal and bract apices yellow to red, sometimes greenish or gray; corollas reddish or yellowish (lower lip sometimes green)
 2. Stems and leaves ± glabrous to hispidulous; leaves narrowly lanceolate to narrowly oblong; calyx lobes 4 .......................................................... **C. affinis**
 2' Stems and leaves densely white-canescent to woolly; leaves linear to narrowly oblong; calyx lobes apparently 2, notched .......................................................... **C. lanata** subsp. **hololeuca**
1' Annuals; sepal and bract apices white, pale yellow, lavender or purple; corollas white or purplish
 3. Corollas linear, not expanded near lips; inflorescence ± oblong in outline; bract tips white to pale yellow .......................................................... **C. attenuata**
 3' Corollas abruptly expanded near lips, lower lip swollen; inflorescence ± broadly oblong to obovate in outline; bract tips purple to red-purple, sometimes white
  4. Upper lip straight, sparsely puberulent; filaments glabrous; lower leaves entire to 3-lobed .......................................................... **C. densiflora**
  4' Upper lip hooked, densely puberulent; filaments puberulent; lower leaves deeply pinnatifid .......................................................... **C. exserta**

**C. affinis** Hook. & Arn. subsp. **affinis** [*C. a.* Hook. & Arn. subsp. *insularis* (Eastw.) Munz] Perennial herbs; stems 15-50 cm tall, hispidulous, suffrutescent. Leaves narrowly lanceolate to narrowly oblong, 3-6(9) cm long, entire to pinnatifid, hispidulous. Floral bracts entire to 5-lobed, yellow to bright red. Calyces 15-25 mm long, lobes narrowly lanceolate to ovate, yellow to bright red; corollas 15-30 mm long, reddish or yellowish, lower lip 2-3 mm long, greenish. Mar-Jun(Oct). Occasional; coastal bluffs, rocky n-facing slopes; chaparral, coastal scrub, pine forest; 50-1000 ft (15-305 m). Scattered locations on north side between Platts Harbor and Cañada del Aqua; scattered in Central Valley between Christy Beach and El Sobrante; scattered on s side between Cañada de los Sauces and Coches Prietos drainage, and in Smugglers' and Aguaje canyons. San Miguel, Santa Rosa, Anacapa, and Santa Catalina islands; n CA coast to Baja CA, e to Sierra Nevada foothills. Two other subspecies occur on the mainland of northern CA.

**C. attenuata** (A.Gray) T.I.Chuang & Heckard [*Orthocarpus a.* A.Gray] NARROW-LEAVED OWL'S CLOVER Annuals, to 30 cm tall; stems puberulent. Leaves linear-lanceolate, 2-6 cm long, puberulent, lower ones entire, upper ones with 2 lateral lobes. Floral bracts 3-lobed, white to pale yellow or pale lavender. Calyces 10-20 mm long, lobes ± linear, white to pale yellow or pale lavender; corollas 10-25 mm long, white or purplish, lower lip 3-5 mm long, white or purple-tinged. Apr. Rare; rocky n-facing slopes, open grassy hillsides; chaparral, pine forest; 100-400 ft (30-120 m). North side between Pelican Bay and Prisoners' Harbor; Central Valley. Guadalupe Island; British Columbia to n Baja CA, e to Sierra Nevada foothills, disjunct in Chile. Seeds were eaten by the Miwok Tribe; this species also was used medicinally by the Costanoan Tribe.

**C. densiflora** (Benth.) T.I.Chuang & Heckard [*Orthocarpus d.* Benth.] Annuals, to 30 cm tall; stems glabrous to sparsely puberulent. Leaves linear-lanceolate, 2.5-6 cm long, sparsely glandular-puberulent, lower ones entire, upper ones often with 2 lateral lobes. Floral bracts 3-5-lobed, white to red-purple. Calyces 6-15 mm long, lobes linear-lanceolate, white to red-purple; corollas 12-25 mm long, tube expanded near lips, galea yellow to red-purple, lower lip 4-5 mm long, white or purple-tinged to purple-dotted. Mar-May. Scarce; grassy flats and hillsides; coastal scrub, grasslands; 100-400 ft (30-120 m). North side between Orizaba Flats and Prisoners' Harbor; s side in Smugglers' Canyon. San Miguel, Santa Rosa, and San Nicolas islands; n CA coast ranges to n Baja CA, e to Sierra Nevada foothills.

**C. exserta** (A.Heller) T.I.Chuang & Heckard [*Orthocarpus purpurascens* Benth.] PURPLE OWL'S CLOVER Annuals, to 40 cm tall; stems puberulent. Leaves linear to lanceolate, 1.5-6 cm long, glandular-puberulent, lower ones entire to pinnatifid, upper ones pinnatifid. Floral bracts 3-7-lobed, white to red-purple. Calyces 8-20 mm long, lobes linear-lanceolate, white to red-purple; corollas 12-25 mm long, tube expanded near lips, white to purple tinged or purple-dotted, lower lip 4-6 mm long, purple-tinged. Mar-May. Occasional; sandy flats and slopes, grassy hillsides, rocky n-facing slopes; coastal scrub, grasslands, oak woodland; 200-1200 ft (60-365 m). Central Valley between Christy and Stanton ranches; s side between Black Point peninsula and Willows Canyon (locally abundant in lower portion of Alegria Canyon and on flats just e of Near Point). San Miguel, Santa Rosa, and Santa Catalina islands; Todos Santos Island; n CA to nw Mexico, e to Sierra Nevada foothills and AZ.

**C. lanata** A.Gray subsp. **hololeuca** (Greene) T.I.Chuang & Heckard [*C. hololeuca* Greene] ISLAND PAINTBRUSH Perennial herbs, sometimes subshrubs; stems 30-100 cm tall, densely white-canescent to woolly. Leaves linear to narrowly oblong, 2-5 cm long, canescent to tomentose, lower ones entire, upper ones entire or 2-lobed. Floral bracts deeply 3-lobed, yellow, orange, or scarlet, sometimes greenish or gray. Calyces 14-18 mm long, lobes broadly ovate, notched, yellowish to reddish; corollas 20-25 mm long, reddish or yellowish, lower lip 2-3 mm long, green. Feb-Sep (Nov). Island endemic. A subspecies of special environmental concern. Common; coastal bluffs, rocky slopes; chaparral, coastal scrub, island woodland, pine forest; 200-1200 ft (60-365 m). Widely scattered locations throughout island, especially around perimeter and on n-facing slopes in Central Valley. San Miguel, Santa Rosa, and Anacapa islands. Subspecies *lanata* occurs in TX and n Mexico.

## KECKIELLA Straw

Shrubs; stems erect, spreading, or ± reclining. Leaves opposite, linear to ovate, entire to serrate, sessile to short-petiolate. Flowers in terminal racemes or panicles. Sepals ± free; corollas bilabiate, yellow to red, upper lip ± hood-like, 2-lobed, lower lip 3-lobed, lobes often reflexed; stamens 4, in 2 unequal pairs, staminode present; stigmas capitate. Fruit ovoid, dehiscent by valves. 7 species; temperate, w N. America.

**K. cordifolia** (Benth.) Straw CLIMBING PENSTEMON Plants 1-4 m tall; branches spreading or ± reclining, glabrous to puberulent. Leaves broadly lanceolate to ovate, 2-6 cm long, bases truncate to cordate, margins remotely serrate. Sepals 7-11 mm long; corollas 3-4.5 cm long, red to reddish orange. Fruits 6-9 mm long. Apr-Sep. May-Sep. Occasional; shaded n-facing slopes, canyon bottoms; chaparral, oak woodland; 40-1700 ft

**SCROPHULARIACEAE: Antirrhinum-Keckiella**

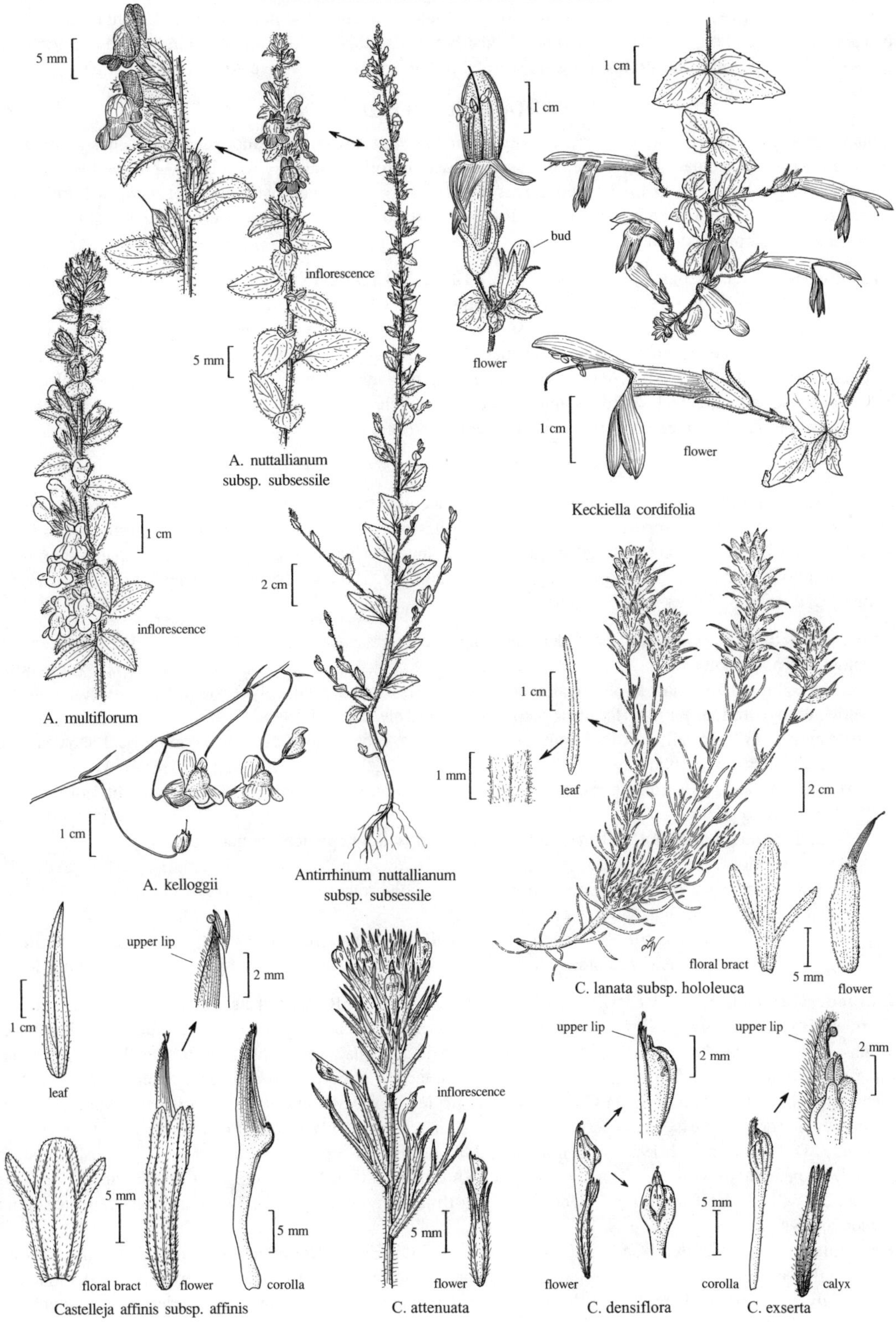

(12-520 m). North side between Peak 1848 West and Eagle Canyon; Central Valley between Centinela and El Sobrante; s side between Laguna Canyon and Coches Prietos drainage, and in Smugglers' Canyon. Santa Rosa, Anacapa, Santa Catalina, and San Clemente islands; San Luis Obispo Co. to n Baja CA. Mostly restricted to igneous and metamorphic substrates. Flowers were used medicinally by Native Americans.

## **LINARIA** Miller TOAD FLAX

Annuals or perennial herbs; stems erect. Leaves linear to ovate, lower ones opposite, upper ones alternate, ± sessile. Flowers in terminal racemes. Sepals fused at base; corollas bilabiate, with basal spur, yellow, blue, or violet, upper lip 2-lobed, lower lip 3-lobed, throat with ridges on the lower inside surface; stamens 4, in 2 pairs, staminode absent; stigma capitate or 2-lobed. Fruit globose to subglobose, dehiscent by apical slits. Ca. 100 species; mostly Europe and Asia.

**L. canadensis** (L.) Dum.Cours. [*L. texana* Scheele] BLUE TOAD FLAX Annuals or short-lived perennials, to 6 dm tall, often with basal, spreading, vegetative shoots. Leaves linear, 0.5-2 cm long, glabrous. Sepals 2-3 mm long; corollas 10-15 mm long, violet to blue, spur to 10 mm long. Fruit 2-3 mm long. Feb-May. Occasional; coastal bluffs, rocky slopes, open ridgetops; chaparral, coastal scrub; 100-1500 ft (30-460 m). Widely scattered locations throughout island. All CA Channel islands except Santa Barbara and San Nicolas; Los Coronados, Todos Santos, Guadalupe, and Cedros islands; throughout much of N. America and w S. America. This species has been used to produce medicinal teas and ointments.

## **MIMULUS** L. MONKEY FLOWER

Annuals, perennial herbs, or shrubs. Leaves opposite, lanceolate to ovate, entire to serrate, ± sessile. Flowers in axillary pairs or in racemes. Sepals fused into tube; corollas bilabiate, white, yellow, red, or purple, upper lip 2-lobed, lower lip 3-lobed, throat often with ridges on lower, inner surface; stamens 4, in 2 unequal pairs, staminode absent; stigmas 2-lobed, flat, often collapsing when touched. Ca. 100 species; mostly temperate, w N. America, S. America, e Asia, S. Africa, and Australia.

1. Leaf blades oval to orbicular, often wider than long, lower leaves petiolate; calyces inflated in fruit; corollas yellow, often red-dotted ........ **M. guttatus**

1' Leaf blades linear-lanceolate to elliptic, lower leaves subsessile to sessile; calyces not inflated in fruit; corollas yellow, orange, red, or purple, if yellow then 5-6 cm long and not red-dotted
   2. Annuals, to 5 cm tall; corolla throat and lobes purple, tube yellow; capsule asymmetrical, curved near base and attached obliquely to pedicel ........ **M. brandegei**
   2' Perennial herbs or shrubs; corollas white, orange, or red; capsule symmetrical, ± straight, not attached obliquely to pedicel
      3. Perennial herbs, not suffrutescent; stems viscid, villous; pedicel longer than calyx; corollas scarlet or yellow ........ **M. cardinalis**
      3' Shrubs or suffrutescent perennial herbs; stems glabrous to puberulent, not viscid; pedicel shorter than calyx; corollas yellow, orange, or red
         4. Corollas red, 30-40 mm long; calyces glabrous to slightly pubescent; ........ **M. flemingii**
         4' Corollas yellow to orangish, 45-60 mm long; calyces pubescent to woolly; .... **M. longiflorus**

**M. brandegei** Pennell SANTA CRUZ ISLAND MONKEY FLOWER Annuals; stems 2-5 cm long, glandular-puberulent. Leaves narrowly elliptic to broadly spatulate, 0.5-1.5 cm long, narrowed to short-petiolate base, margins entire, apices obtuse. Calyces 6-9 mm long, lobes subequal, apices obtuse; corollas 10-12 mm long, purple. Fruit ovoid, 10-15 mm long, curved near base, attached obliquely to pedicel, irregularly dehiscent or indehiscent. Mar. Endemic to Santa Cruz Island. A species of special environmental concern. Rare, not seen recently; open, rocky slopes; coastal scrub; ca. 200 ft (60 m). Known only from collections by T.S. Brandegee in 1888 and C.B. Wolf in March 1932. Label on Wolf 2800 at SBBG ("stony summit of third ridge w of Christy's Harbor") is apparently incorrect; original records at RSA cite locality as "ca. 1 mile s of Christy Ranch" for Wolf numbers 2797-2810. Wolf was on the island with R. Hoffmann and gave him the 3 specimens he collected (note in records, RSA).

**M. cardinalis** Douglas ex Benth. SCARLET MONKEY FLOWER Perennial herbs; stems erect to decumbent, 2-10 dm tall, viscid, villous. Leaves broadly oblanceolate, 3-15 cm long, glandular-puberulent, sessile, margins serrate. Calyces 25-30 mm long, lobes subequal, apices acute; corollas 35-50 mm long, scarlet red, rarely yellow. Fruit ovoid, 12-20 mm long, straight. Apr-Aug. Occasional; moist stream banks, floodplains, freshwater seeps;

**SCROPHULARIACEAE: Linaria-Veronica**

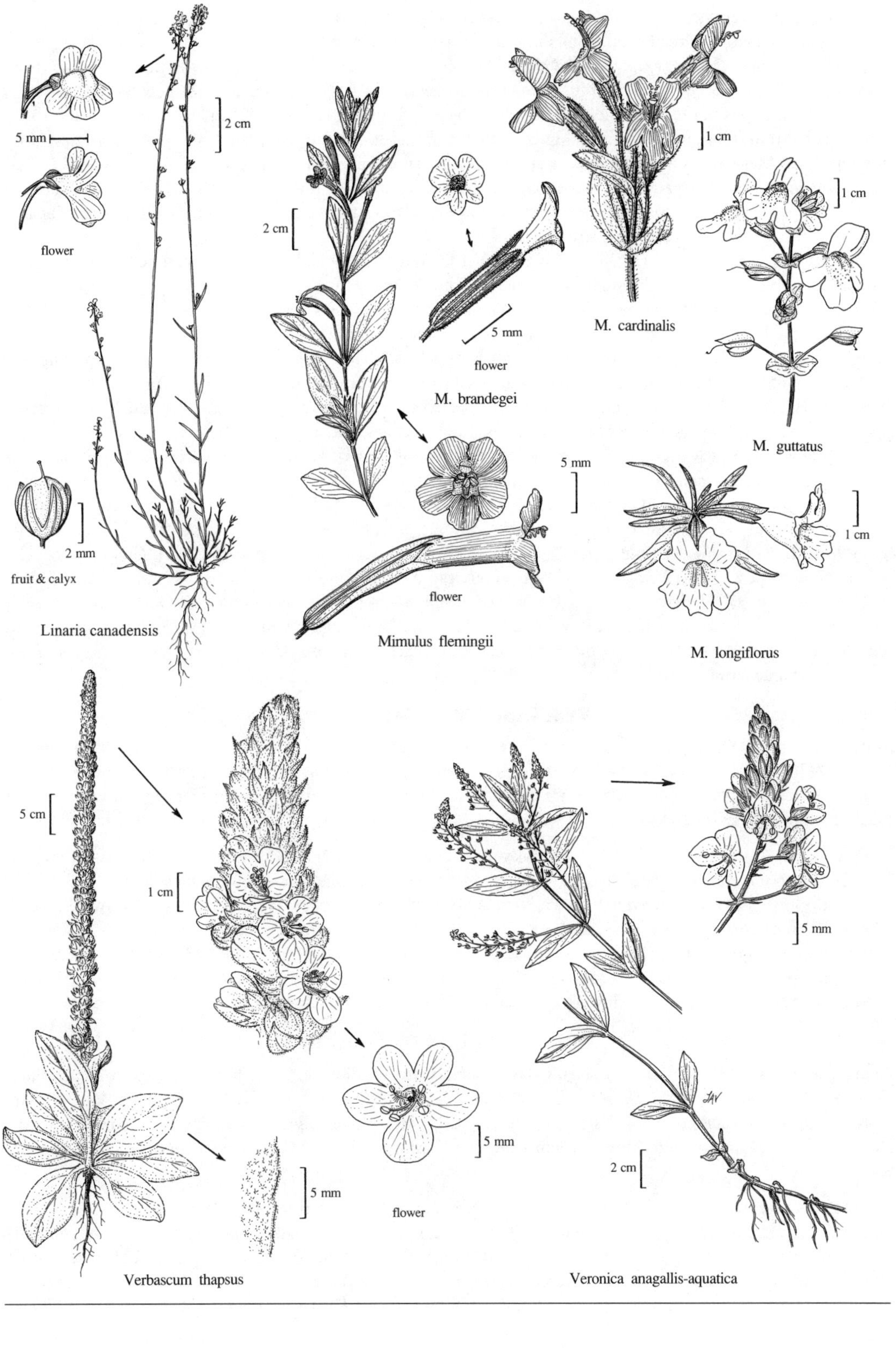

riparian communities; 10-1200 ft (3-365 m). North side between Cueva Valdez and Cañada del Puerto; Central Valley between lower portion of Cañada Christy and Rancho del Sur; s side between Pozo and Willows canyons. Santa Catalina Island; Cedros Island; southeastern OR to n Baja CA, e to UT.

**M. flemingii** Munz [*Diplacus parviflorus* Greene] ISLAND MONKEY FLOWER. Subshrubs or shrubs; stems erect to decumbent, 3-15 dm tall, glabrous to puberulent. Leaves broadly lanceolate to oblanceolate, 2.5-6(8) cm long, sessile, margins entire to denticulate, upper surface dark green, lower surface pale green. Calyces 15-20 mm long, glabrous, lobes unequal, linear, apices acute; corollas 30-40 mm long, red. Fruit ovoid, 8-16 mm long, straight. (Jan) Mar-Oct. Island endemic. A species of special environmental concern. Common; coastal bluffs, rocky n-facing slopes; chaparral, coastal scrub, island woodland, pine forest; 10-1750 ft (3-535 m). Scattered locations throughout island. Santa Rosa, Anacapa, and San Clemente islands. Hybrids between this species and *M. longiflorus* are common on Santa Cruz Island.

**M. guttatus** Fisher ex DC. COMMON MONKEY FLOWER Perennial herbs or sometimes annuals; stems erect to decumbent, 1-10 dm tall, usually glabrous. Leaf blades ovate to orbicular, 1-5 cm long, glabrous, margins denticulate to deeply dentate at base, lower leaves petiolate, upper ones sessile. Calyces 7-20 mm long, lobes unequal, linear to triangular; corollas 12-40 mm long, yellow, often red-spotted. Fruit oblong, 5-8 mm long, straight. Mar-Jun (Oct). Occasional; moist stream banks, floodplains, freshwater seeps; riparian communities; 5-1200 ft (2-365 m). Widely scattered locations on n side and in Central Valley; s side between Cañada del Agua Santa and Black Point Canyon, and in Smugglers' Canyon. Santa Rosa, Santa Catalina, and San Clemente islands; Alaska to n Mexico, e to Rocky Mountains. Young leaves and stems were eaten by Native Americans.

**M. longiflorus** (Nutt.) A.L.Grant [*Diplacus longiflorus* Nutt.] BUSH MONKEY FLOWER Subshrubs or shrubs; stems erect to ascending, 3-12 dm tall, pubescent. Leaves lanceolate to oblong, 2.5-5(8) cm long, sessile, margins entire to denticulate, often revolute, upper surface dark green, lower surface pubescent to puberulent. Calyces 20-30 mm long, pubescent to woolly, lobes strongly unequal, linear-lanceolate; corollas 45-60 mm long, cream to yellow or orangish. Fruit oblong, 10-18 mm long, straight. (Feb) Apr-Sep (Nov). Common; coastal bluffs, rocky slopes, exposed ridgetops; chaparral, coastal scrub, island woodland, oak woodland, pine forest; 10-2200 ft (3-670 m). Scattered locations throughout island. Santa Rosa Island; s San Luis Obispo Co. to n Baja CA, e to s Sierra Nevada foothills. Closely related to *M. aurantiacus* Curtis, with glabrous calyces, a widespread species throughout most of CA w of the Sierra Nevada. Hybrids between *M. longiflorus* and *M. flemingii* are common on Santa Cruz Island.

## VERBASCUM L. MULLEIN

Annuals or perennial herbs. Basal leaves in rosettes, cauline ones alternate, lanceolate to ovate, entire to dentate, ± sessile. Flowers in racemes, spikes, or panicles. Sepals ± free; corollas ± radial, white to yellow or purple; stamens 5, unequal, staminode absent; stigmas capitate. Fruit ovoid to globose, dehiscent by valves. Ca. 350 species; temperate, Europe, Asia.

**V. thapsus** L. COMMON MULLEIN Short-lived perennials; stems 2-18 dm tall, densely stellate-pubescent. Basal leaves broadly oblanceolate, oblong, or ovate, 5-40 cm long, cauline ones elliptic-lanceolate, 5-25 cm long, decurrent. Flowers in dense, terminal spikes. Sepals 6-9 mm long; corollas 12-25 mm wide, yellow; filaments glabrous to pilose. Fruit ovoid, 5-10 mm long. Jun-Oct. Introduced. Common; disturbed sites, floodplains, dry canyon bottoms, open slopes, ridgetops; grasslands, riparian communities; 10-2200 ft (3-670 m). Widely scattered locations throughout island. Throughout N. America; native of Europe. First collected at Prisoners' Harbor in August 1939. Dried leaves and flowers are used medicinally; ground capsules have been used to stupify fish.

## VERONICA L. SPEEDWELL

Annuals or perennial herbs. Leaves opposite, lanceolate, ovate, or obovate, entire to serrate, sessile to petiolate. Flowers solitary or in racemes. Sepals 4-5, ± free; corollas ± radial, rotate, usually blue to purple, lobes apparently 4, upper one (composed of 2 strongly fused petals) larger than other 3; stamens 2, staminode absent; stigmas capitate. Fruit ovoid to globose, dehiscent by valves. Ca. 250 species; temperate, N. America, Europe, and Asia.

**V. anagallis-aquatica** L. WATER SPEEDWELL Perennial herbs, with rhizomes; stems erect, 10-100 cm tall, glabrous. Leaves lanceolate to oblanceolate, 5-20 cm long, cauline ones sessile, often clasping. Flowers in axillary racemes. Sepals 3.5-4.5 mm long; corollas 5-10 mm long, pale blue to rose-pink. Fruit subglobose, apices slightly notched. Apr-Jul. Introduced. Scarce; moist stream banks, floodplains; riparian communities; 5-700 ft (2-215 m). North side at Prisoners' Harbor; Central Valley between Centinela and Cañada del Puerto; s side at mouth of Willows Canyon and in Coches Prietos drainage. Throughout N. and S. America; native of Europe. First collected at Prisoners' Harbor in April 1982; now rapidly spreading. Raw or cooked leaves are edible.

## SOLANACEAE Juss.
### NIGHTSHADE FAMILY

Annuals, perennial herbs, or shrubs. Leaves alternate, usually simple; stipules absent. Flowers bisexual, radial, solitary or in umbel-like clusters, racemes, or panicles. Sepals 5, fused; petals 5, fused, corollas rotate to salverform; stamens 5, adnate to corolla alternate lobes; pistil 1, ovary superior, locules usually 2, placentas usually axile, style 1, stigma usually capitate. Fruit a berry or capsule. Ca. 90 genera and 2600 species; temperate to tropical, ± worldwide. Several species are widely cultivated for food (including tomatoes, peppers, potatoes) and as ornamentals (petunias). Some species are toxic, including tobacco and the deadly nightshades.

1. Calyces 80-100 mm long; corollas 150-180 mm long ........ **Datura**
1' Calyces 0.5-15 mm long; corollas 3.5-40 mm long
  2. Shrubs, with spiny, woody twigs; leaves clustered at stem nodes ........ **Lycium**
  2' Annuals or perennial herbs, stems sometimes woody near base; leaves not clustered
    3. Corollas ± salverform; anthers not connivent; fruit a capsule ........ **Nicotiana**
    3' Corollas ± rotate; anthers connivent; fruit a berry
      4. Leaves compound; corollas yellow ........ **Lycopersicon**
      4' Leaves simple; corollas white, violet, purplish, or bluish ........ **Solanum**

### DATURA L. JIMSON WEED or THORN APPLE

Annuals or perennial herbs. Leaves entire to deeply lobed. Flowers solitary, on stout pedicels in forks of stem branches. Calyx tube much longer than lobes, circumscissile near base; corollas funnelform, plicate; stamens adnate to lower corolla tube; ovary with 2 or 4 locules. Fruit a capsule. Ca. 8 species; warm temperate and subtropical, N. and S. America. Some species are grown as ornamentals or used as a source of highly toxic, hallucinogenic drugs.

**D. wrightii** Regel [*D. meteloides* A.DC.] Perennial herbs, to 1 meter tall, stems ascending to spreading. Leaves ovate, 7-20 cm long, 3-10 cm wide, entire to irregularly toothed, blades glabrous, petioles puberulent. Calyces 8-10 cm long; corollas 15-18 cm long, white. Fruit subglobose, 3-5 cm wide, densely spiny. (Feb) Apr-Oct. Occasional; sandy flats, beaches, dunes, disturbed sites, stream banks, canyon bottoms; coastal scrub, grasslands, riparian communities, southern beach and dune; 5-1400 ft (2-425 m). North side in Cañada del Puerto and in Scorpion Canyon; scattered locations on s side between Laguna Canyon and Albert's Anchorage, and near Navy Base. Santa Rosa and Santa Catalina islands; San Martin Island; n CA coast ranges to Mexico, e to Great Plains and TX. Often associated with Chumash midden deposits. Although this species was used medicinally and as a ceremonial hallucinogen by Native Americans, all parts are toxic to humans and have caused many deaths.

### LYCIUM L. BOX THORN

Shrubs. Leaves often clustered at stem nodes, entire to toothed. Flowers 1-4 per cluster, in leaf axils. Calyces funnelform to campanulate, persistent; corollas rotate to funnelform; stamens adnate to corolla tube near middle; ovary with 2 locules. Fruit a berry. Ca. 100 species; warm temperate N. America and Europe. Raw or dried fruits of several species were eaten by Native Americans; dried roots also were baked and eaten by some tribes.

**L. californicum** Nutt. CALIFORNIA BOX THORN Plants intricately branched, twigs spinose. Leaves linear to oblanceolate, 4-10 mm long, glabrous. Calyces ca. 3 mm long; corollas 3.5-4 mm long, white. Fruit subglobose, 3-5 mm wide, reddish; seeds 1-2. Mar-Apr. Rare; eroded coastal bluffs, ridgetops; coastal scrub; 50-60 ft (15-18 m). Known only from s end of Christy Beach and San Pedro Point. All CA Channel islands except Santa Rosa; Los Coronados, Todos Santos, San Martin, Guadalupe, San Benito, and Natividad islands; coastal Los Angeles Co. to Baja CA.

### LYCOPERSICON L. TOMATO

Annuals or perennial herbs. Leaves deeply lobed to pinnately compound, often glandular-puberulent. Flowers in racemes. Calyx tube shorter than lobes; corollas usually rotate, lobes usually reflexed; stamens adnate to corolla tube at middle or below, anthers connivent, opening by terminal pores or slits; ovary usually with 2 locules. Fruit a berry. Ca. 7 species; S. and C. America.

**L. esculentum** L. Annuals, stems erect to spreading. Leaves compound, 5-20 cm long; leaflets unequal, 0.5-4 cm long, margins entire to lobed, upper surfaces glabrous to sparsely puberulent, lower surfaces puberulent. Calyces 5-10 mm long; corollas yellow, lobes 5-12 mm long. Fruit 2-8 cm wide, greenish to red. Jun-Sep. Introduced. Scarce; disturbed sites, gravel beaches, stream banks, floodplains; coastal scrub, grasslands, southern beach and dune; 5-40 ft (2-12 m). North side at Lady's and Prisoners' harbors; Central Valley near Christy Ranch; s side near Smugglers' Ranch. San Miguel, Anacapa, Santa Barbara, and San Nicolas islands; Los Coronados and San Geronimo islands; San Francisco Bay region and Great Valley to s CA; native of S. America. First collected at Prisoners' Harbor in June 1984. Ripe fruits are edible; unripe fruits, stems, and leaves are toxic to humans and livestock.

## **NICOTIANA** L. TOBACCO

Annuals, perennial herbs, or shrubs. Leaves entire to toothed, often viscid or glandular-puberulent. Flowers in racemes or panicles. Calyx tube longer than lobes, persistent; corollas usually salverform; stamens adnate to corolla tube at middle or below; ovary usually with 2 locules. Fruit a capsule. Ca. 750 species; N. and S. America, Australia, and s Africa. The natural source of nicotine; some species are also grown as ornamentals. Although leaves of several species were smoked and used medicinally by Native Americans, all species contain alkaloids and can be toxic to humans.

1. Annuals; leaves viscid to glandular-puberulent; calyx lobes unequal, longer than calyx tube. **N. clevelandii**
1' Shrubs, sometimes small trees; leaves glabrous, glaucous; calyx lobes equal, shorter than calyx tube .......... **N. glauca**

**N. clevelandii** A.Gray CLEVELAND'S TOBACCO Annuals, 1-4 dm tall. Leaves viscid to glandular-puberulent, lower ones ovate to oblanceolate, 4-11 cm long, upper ones ± lanceolate, 3-6 cm long. Calyces 8-11 mm long, lobes unequal, longer than tube; corollas 12-20 mm long, white to cream; stamens unequal. Mar-Sep. Scarce; coastal bluffs, disturbed sites, sandy flats; coastal scrub, southern beach and dune; 10-100 ft (3-30 m). South side near Forney's Cove, at mouth of Willows Canyon, and at Valley Anchorage. Anacapa, Santa Barbara, and Santa Catalina islands; Los Coronados, Todos Santos, San Martin, Cedros, and Natividad islands; Santa Barbara Co. to n Baja CA, e to AZ. Often associated with Chumash midden deposits.

**N. glauca** Graham TREE TOBACCO Shrubs, sometimes small trees, to 3 m tall. Leaves ovate, 7-20 cm long, glabrous, glaucous. Calyces 10-15 mm long, lobes equal, shorter than tube; corollas 30-40 mm long, yellow; stamens equal. Mar-Oct. Introduced. Occasional; coastal bluffs, disturbed sites, loose talus slopes, canyon bottoms; coastal scrub, grasslands; 10-1150 ft (3-350 m). North side near w end of China Harbor and on w side of El Montañon ridge; s side between Smugglers' Cove and Aguaje Canyon. San Nicolas and Santa Catalina islands; Guadalupe Island; n CA coast ranges to Mexico, e to Sierra Nevada foothills; native of Argentina. First collected at Smugglers' Cove in March 1971; now rapidly spreading. This species contains the alkaloid anabasine and is extremely toxic to humans and livestock.

## **SOLANUM** L. NIGHTSHADE

Annuals, perennial herbs, or shrubs. Leaves entire to deeply lobed. Flowers in cymose panicles, often umbel-like. Calyces and corollas rotate to campanulate; corollas often plicate; stamens adnate to ± middle of corolla tube, anthers connivent, opening by terminal pores or slits; ovary with 2 locules. Fruit a berry. Ca. 1400 species; ± worldwide. Some species are cultivated for food (including eggplant, potatoes, pepinos) or as ornamentals. Foliage and unripe fruits of most species are toxic to humans, especially children.

1. Leaves lanceolate to oblong, densely silvery-canescent, trichomes stellate ........................ **S. elaeagnifolium**
1' Leaves ovate, glabrous to puberulent, trichomes simple
    2. Corollas violet to purple, lobes or teeth much shorter than tube; calyces 3-4 mm long ........... **S. clokeyi**
    2' Corollas white, lobes much longer than tube; calyces 0.5-2 mm long .................................. **S. douglasii**

**S. clokeyi** Munz ISLAND NIGHTSHADE Perennial herbs or subshrubs, to 1.3 m tall; stems erect to ascending, usually woody near base. Leaves ovate, blades 2-8 cm long, entire to irregularly toothed, ± viscid, upper surfaces ± glabrous, lower surfaces puberulent, petioles 1-2 cm long. Calyces 3-4 mm long; corollas 2.5-3 cm wide, violet to purple with green dots near center, lobes or teeth much shorter than tube. Fruit 1.5-2.5 cm wide, yellowish.

**SOLANACEAE**

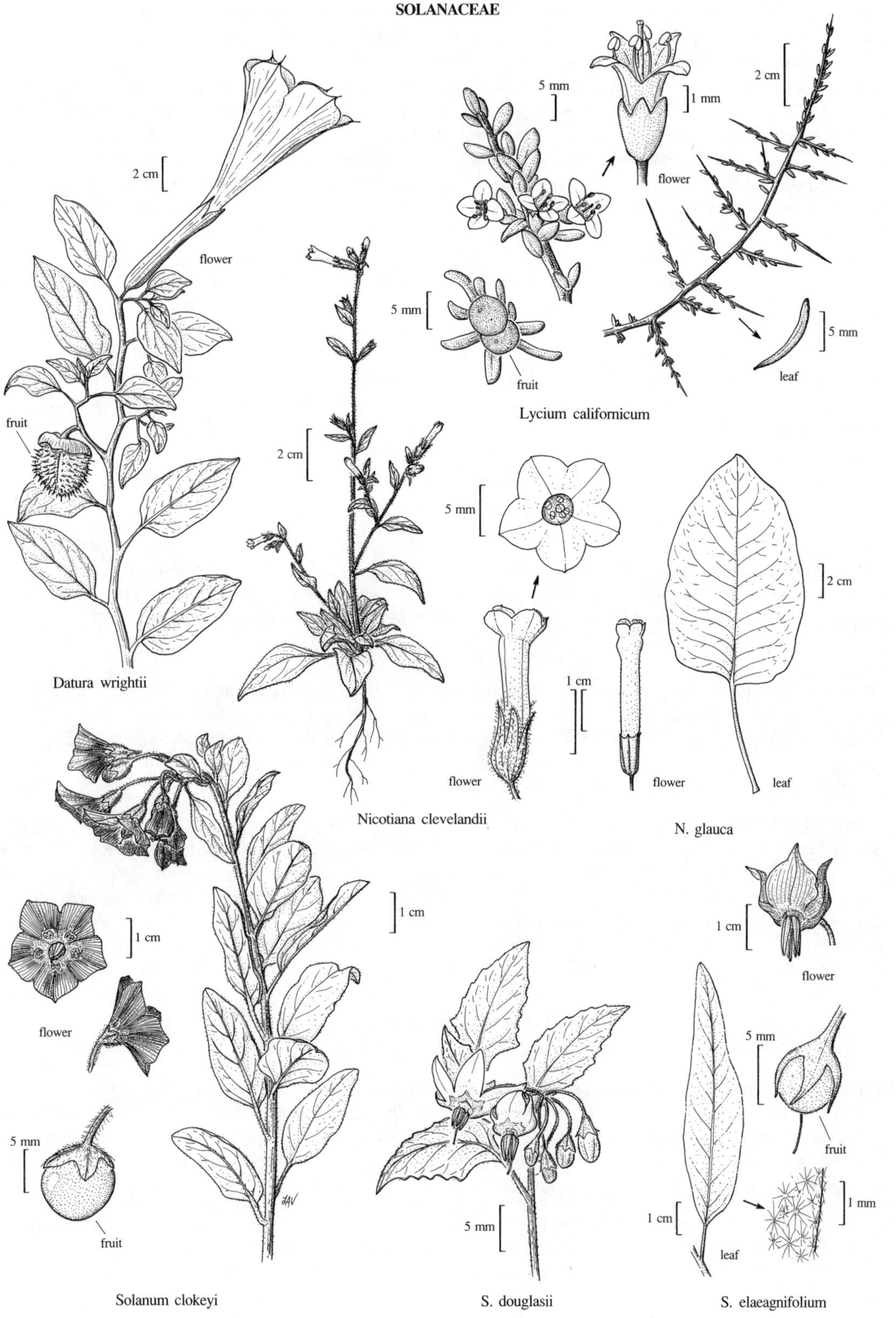
2 cm
flower
fruit
Datura wrightii
5 mm
1 mm
flower
2 cm
5 mm
fruit
5 mm
leaf
Lycium californicum
2 cm
5 mm
2 cm
1 cm
flower
flower
leaf
Nicotiana clevelandii
N. glauca
1 cm
1 cm
flower
5 mm
fruit
1 cm
flower
5 mm
fruit
1 mm
1 cm
leaf
5 mm
Solanum clokeyi
S. douglasii
S. elaeagnifolium

Mar-Jul (Sep). Island endemic. A species of special environmental concern. Common; coastal bluffs, rocky slopes, canyon bottoms; chaparral, coastal scrub, oak woodland, pine forest; 5-1300 ft (2-395 m). Scattered locations throughout island. Santa Rosa Island; Guadalupe Island.

**S. douglasii** Dunal in DC. DOUGLAS' NIGHTSHADE Perennial herbs or subshrubs, to 1 m tall; stems ascending to spreading, often woody near base. Leaves ovate, blades 2-8 cm long, entire to dentate, upper surfaces ± puberulent, lower surfaces ± canescent, petioles 0.5-1.5 cm long. Calyces 0.5-2 mm long; corollas 0.8-1.5 cm wide, white, lobes or teeth much longer than tube, often reflexed. Fruit 0.5-1.5 cm wide, dark purple to black. Apr-Oct. Occasional; stream banks, shaded slopes, canyon bottoms; chaparral, oak woodland, riparian communities; 10-1300 ft (3-395 m). Widely scattered locations throughout island. All CA Channel islands; Guadalupe Island; San Francisco Bay region to Mexico. Leaves and fruits were used medicinally by Native Americans.

**S. elaeagnifolium** Cav. SILVER-LEAVED HORSE NETTLE Perennial herbs, with rhizomes, to 8 dm tall; stems erect, often prickly. Leaves lanceolate to oblong, blades 4-10 cm long, entire to wavy, densely silvery-canescent, trichomes stellate, petioles 0.5-2 cm long. Calyces 5-10 mm long; corollas 1-2 cm wide, violet to bluish, lobes or teeth shorter than or equal to tube. Fruit 1-2 cm wide, yellow to brownish. Jul-Oct. Introduced. Rare; disturbed sites; grasslands; ca. 250 ft (75 m). Known only from corrals at Christy Barn, where first collected in October 1977. Santa Catalina Island; Sacramento Valley to Mexico; native of Great Plains of North America.

## TAMARICACEAE Link
### TAMARISK FAMILY

Shrubs or trees. Leaves alternate, simple, entire, sessile; stipules absent. Flowers bisexual, radial, solitary or in spikes or dense panicles. Sepals 4-6, free or fused at base; petals 4-5, usually attached at base of disk-like nectar-gland; stamens 4-10; ovary superior, locule 1, placentas 3-5, usually parietal, styles 2-5. Fruit a capsule, dehiscent by valves. Five genera and ca. 90 species; warm temperate, Europe, Asia, and n Africa.

### TAMARIX L. TAMARISK

Stems jointed, usually green, glabrous, twigs often drooping. Leaves scale-like, evergreen or deciduous. Flowers usually in spikes or panicles borne on lateral branches. Sepals 4-5, persistent in fruit; petals white to purple; stamens 4-5; nectar-gland entire or lobed. Fruit club-shaped; seeds comose, trichomes plumose. Ca. 15 species; Europe, Asia, and n Africa. Some species are cultivated as ornamentals or for wind breaks but are serious weeds, especially along stream banks and irrigation canals.

**T. ramosissima** Ledeb. [*T. pentandra* Pall.] Shrubs or trees, 1-5 m tall. Leaves ovate, 1.5-3 mm long, glabrous. Spikes 1.5-7 cm long. Sepals 5, 0.5-1 mm long; petals 1-2 mm long, white to rose; nectar-gland lobed; stamens alternate nectar-gland lobes. May-Oct. Introduced. Scarce; moist floodplains, canyon bottoms; riparian communities; 5-50 ft (2-15 m). North side at mouth of Cañada del Puerto; Central Valley in lower portion of Cañada Christy and Cañada Cebada; s side in a small canyon between West and Fraser points, in Cañada de los Sauces, and in Pozo Canyon (where it is rapidly spreading). Santa Rosa and San Nicolas islands; Cedros Island; Great Valley s to N Baja CA, e to Great Plains; native of e Asia. First collected at Prisoners' Harbor in August 1977.

## URTICACEAE Juss.
### NETTLE FAMILY

Annuals or perennials, monoecious or dioecious; stems and leaves usually with stinging trichomes. Leaves alternate or opposite, simple, entire to serrate; stipules usually present, minute. Flowers bisexual or unisexual, radial, in axillary clusters or pendulous racemes or panicles. Sepals 4-5; petals absent; staminate flowers with 4-5 stamens; pistillate flowers with superior ovary, locule 1, stigma 1. Fruit an achene. Ca. 50 genera and 700 species; temperate to tropical, ± worldwide.

1. Leaves alternate, without stinging trichomes, margins entire; stipules absent .............................. **Parietaria**

1' Leaves opposite, with stinging trichomes, margins toothed; stipules present

   2. Leaf margins with rounded teeth; pistillate sepals strongly fused, calyces ovoid, with 2-4 minute teeth at apex ............................................ **Hesperocnide**

   2' Leaf margins with acuminate to obtuse teeth; pistillate sepals ± free, outer ones shorter than inner ones, inner ones enfolding pistil and fruit ............................................ **Urtica**

**TAMARICACEAE**

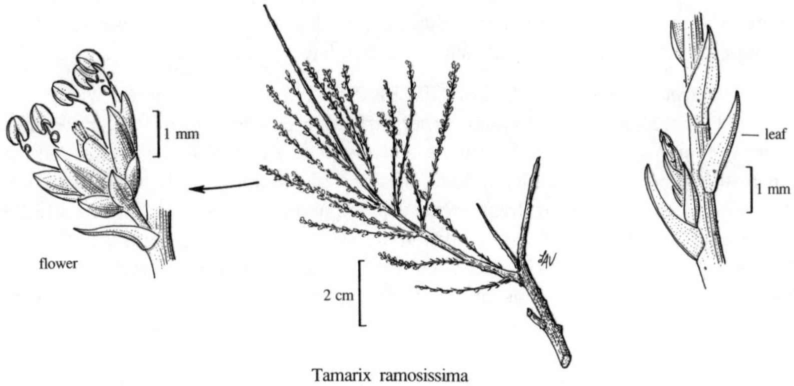


Tamarix ramosissima

## HESPEROCNIDE Torr. WESTERN NETTLE

Monoecious annuals; stems erect, with stinging trichomes. Leaves opposite, usually ovate, with stinging trichomes, margins crenate; stipules minute. Staminate and pistillate flowers mixed together in axillary clusters. Staminate flowers with 4 ± free sepals; pistillate flowers with sepals strongly fused into ovoid calyx enclosing pistil and fruit, calyces with 2-4 teeth at apex. Fruit lenticular. Two species; CA and Hawaii.

**H. tenella** Torr. Plants 2-5 dm tall; stems erect to decumbent. Leaves ovate, 5-30 mm long, 5-20 mm wide, base rounded, marginal teeth rounded. Flowers 7-25 per axil. Calyces 1-1.5 mm long, densely hispid in fruit. Mar-May. Scarce; grassy n-facing slopes, shaded hillsides, canyon bottoms; chaparral, oak woodland, riparian communities; 200-1200 ft (60-365 m). North side on El Montañon ridge and in Scorpion Canyon; Central Valley near Stanton Ranch; s side in Smugglers' Canyon. Santa Barbara, Santa Catalina, and San Clemente islands; Todos Santos, San Martin, and Guadalupe islands; n CA coast ranges to n Baja CA. Stinging trichomes on leaves and stems can cause skin irritation.

## PARIETARIA L. PELLITORY

Annuals or perennial herbs, monoecious or polygamous; stems erect to decumbent, stinging trichomes absent. Leaves alternate, lanceolate to orbicular, glabrous to hispidulous; stipules absent. Flowers mostly unisexual (sometimes bisexual), subtended by 1-3 bracts, mixed together in dense, axillary clusters. Sepals 4, ± free in bisexual and staminate flowers, fused into short tube with 4 lobes in pistillate flowers, persistent in fruit; stamens 4. Fruit ovoid. Ca. 30 species; worldwide.

**P. hespera** B.D.Hinton var. **californica** B.D.Hinton Annuals, 10-40 cm tall; stems erect to prostrate, simple or branched. Leaves broadly ovate, 3-25 mm long, 4-17 mm wide, base truncate to cordate, margins entire. Flowers 3-many per axil. Calyces 1.5-3 mm long, glabrous, lobes spreading to recurved. Mar-May. Occasional; coastal bluffs, shaded n-facing slopes, rock outcrops, canyon bottoms; chaparral, coastal scrub, oak woodland, pine forest; 20-700 ft (6-215 m). Widely scattered locations, especially on n side and around perimeter of island. All CA Channel islands; Los Coronados, Todos Santos, San Martin, Guadalupe, and Cedros islands; San Francisco Bay region to n Baja CA. Var. *hespera*, with orbicular leaves and erect sepals, occurs throughout much of sw N. America.

## URTICA L. NETTLE

Annuals or perennial herbs, monoecious or dioecious; stems erect, with stinging trichomes. Leaves opposite, lanceolate to cordate, with stinging trichomes, margins dentate to serrate, teeth acuminate to obtuse; stipules minute. Flowers unisexual, mixed together or separated, in dense clusters, elongated racemes, or panicles. Staminate flowers with 4 ± free, equal sepals, stamens 4; pistillate flowers with 4 ± free sepals, in 2 series, outer sepals smaller than inner ones, inner ones enfolding fruit. Fruit flattened, ovate to triangular in outline. Ca. 50 species; worldwide. Stinging trichomes on leaves and stems will cause skin irritation.

1. Perennial herbs, with rhizomes; inflorescences usually longer than adjacent petioles, staminate ones separate from pistillate ones; sepals hispidulous ........................................................ **U. dioica** var. **holosericea**
1' Annuals; inflorescences usually shorter than adjacent petioles; staminate and pistillate flowers mixed in same clusters; sepals glabrous, margins hispidulous to ciliolate ........................................................ **U. urens**

**U. dioica** L. var. **holosericea** (Nutt.) C. L. Hitchc. CREEK NETTLE Perennial herbs, with rhizomes; 5-10 dm tall. Leaf blades ovate to lanceolate, 3-10 cm long, upper surfaces green, lower surfaces gray-pubescent, margins coarsely serrate, teeth acuminate to acute. Staminate and pistillate flowers in separate spikes, staminate ones often longer than pistillate ones. Sepals hispidulous, margins hispidulous. Fruit ovate in outline. May-Nov. Occasional; stream banks, canyon bottoms, freshwater seeps; riparian communities; 5-1100 ft (2-335 m). Widely scattered locations throughout western and central portions of island. Anacapa and Santa Catalina islands; throughout much of w N. America. Several other varieties occur throughout N. America, Europe, and Asia. After removing the stinging trichomes, Native Americans ate young leaves either raw or cooked; the species was also used medicinally.

**U. urens** L. DWARF NETTLE Annuals, 2-6 dm tall. Leaf blades elliptic to ovate, 1-4 cm long, both surfaces dark green, margins coarsely serrate, teeth ± acute to obtuse. Staminate and pistillate flowers mixed in same clusters. Sepals ± glabrous, margins hispidulous to ciliolate. Fruit triangular in outline. Feb-Jun. Introduced. Occasional; disturbed sites, shaded slopes, canyon bottoms; chaparral, grasslands, oak woodland; 10-2200 ft (3-670 m). Widely scattered locations throughout island. San Miguel, Santa Rosa, and Santa Catalina islands; Los Coronados Islands; native of Europe. Leaves and stem tops can be cooked and eaten; leaves and flowers have also been used medicinally. Reported by Greene (1887).

## VALERIANACEAE Batsch

### VALERIAN FAMILY

Annuals or perennial herbs, often strong-smelling. Leaves in basal rosettes or opposite, simple to compound; stipules absent. Flowers bisexual or unisexual, radial to bilateral, often in hemispheric to capitate clusters or panicles. Calyces absent or lobes 5-15, inwardly coiled, uncoiled and plumose in fruit; petals 5, fused, often spurred at base; stamens 1-3. adnate to corolla; ovary inferior, locule 1, ovule 1, style 1, stigma capitate or 3-lobed. Fruit an achene. Ca. 17 genera and 400 species; mostly temperate, n hemisphere S. America.

### CENTRANTHUS DC.

Annuals or perennial herbs. Leaves mostly cauline, entire to pinnately lobed. Flowers bisexual, radial to slightly bilateral, in dense, terminal or axillary panicles. Calyx lobes 5-15, inwardly coiled, uncoiled and plumose in fruit, usually persistent; corollas slightly bilateral, spurred near base; stamen 1. Fruit slightly compressed. Ca. 12 species; temperate, Mediterranean Europe and n Africa.

**C. ruber** (L.) DC. RED VALERIAN Perennial herbs, erect, to 1 m tall. Leaves deltate to lanceolate-ovate, 3-15 cm long, 2-6.5 cm wide, entire or few-lobed. Flowers in terminal panicles. Corollas 14-18 mm long, red to pink; spur 4-6 mm long. Fruit 3-5 mm long; plumose sepals persistent, white. Mar-Oct. Introduced. Rare; canyon bottoms, shaded slopes; chaparral, coastal scrub; 950-1100 ft (290-335 m). Central Valley in ravine just w of Reparo Viejo, where it was first collected in October 1958. Santa Catalina Island; throughout coastal CA and Great Valley; native of Europe.

## VERBENACEAE J.St.-Hil.

### VERBENA FAMILY

Annuals or perennials. Leaves cauline, opposite, simple, usually toothed or lobed; stipules absent. Flowers bisexual, usually radial, in racemes, spikes, of dense panicles, each flower usually subtended by 1 bract. Sepals 5, fused; petals 4-5, fused; stamens usually 4, often in 2 unequal pairs; ovary superior, often 2- or 4-lobed, locules usually 2, placentas axile, style 1, stigmatic branches 2, unequal. Fruit a capsule or dehiscent into 2 or 4 nutlets. Ca. 90 genera and 1900 species; mostly warm temperate and tropical, worldwide.

### VERBENA L.

Annuals or perennial herbs. Leaf blade margins entire to lobed. Flowers in spikes or dense panicles. Calyx tube 5-ribbed, lobes 5; corollas radial to slightly bilateral, rotate to salverform; ovary 4-lobed, ovules 1 per locule;

## URTICACEAE

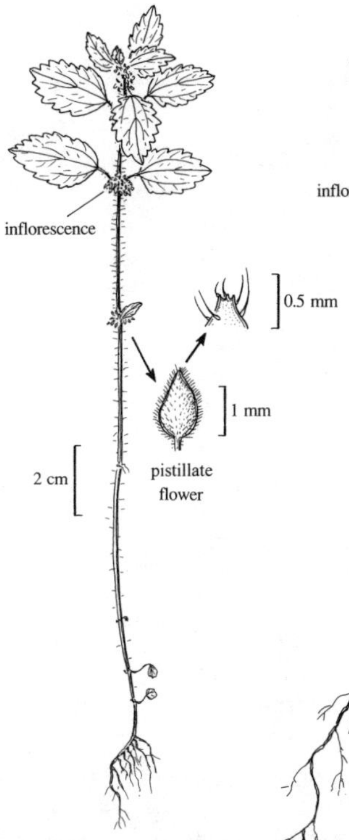


Hesperocnide tenella

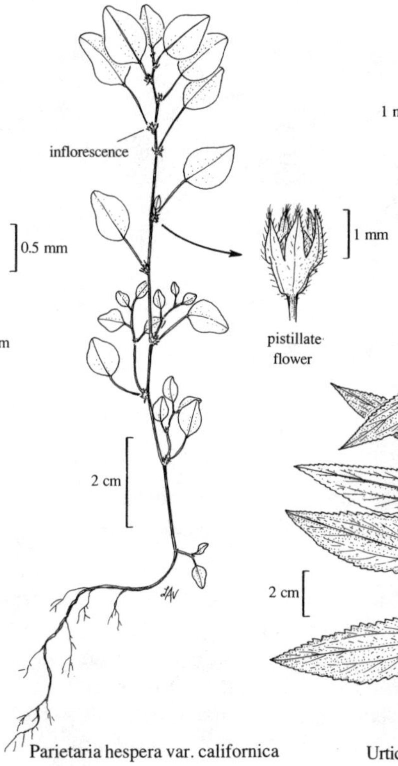


Parietaria hespera var. californica

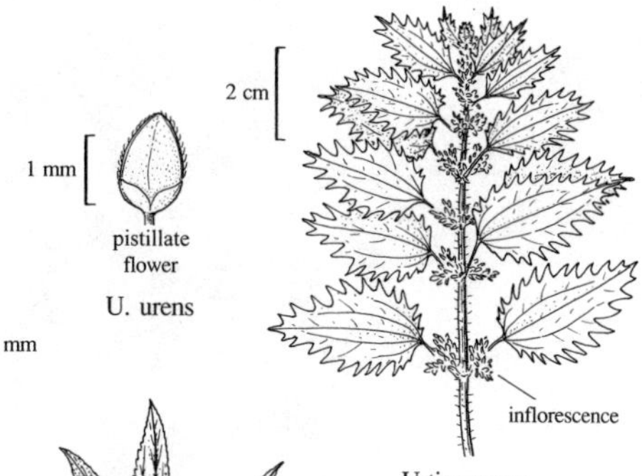


Urtica urens

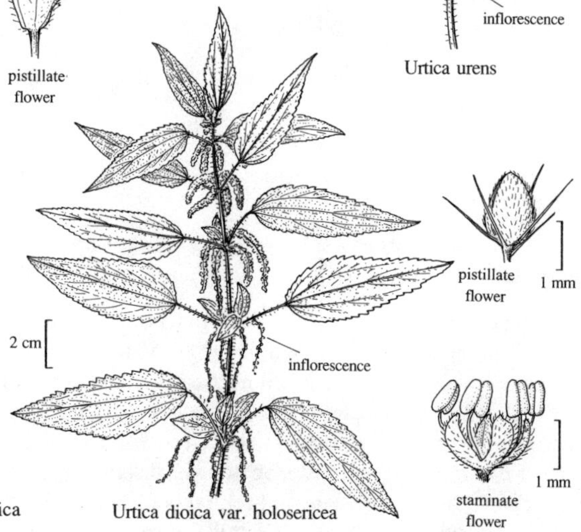


Urtica dioica var. holosericea

## VALERIANACEAE

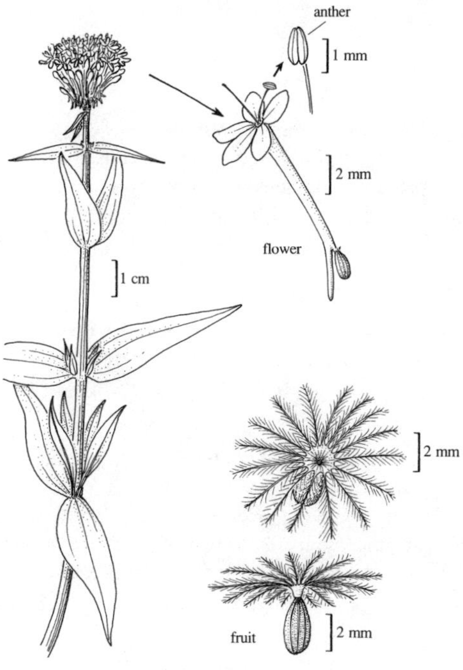


Centranthus ruber

## VERBENACEAE

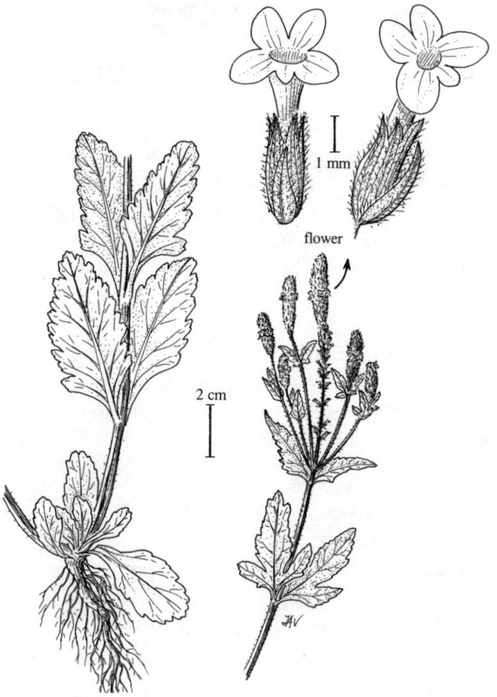


Verbena lasiostachys var. scabrida

VIOLACEAE

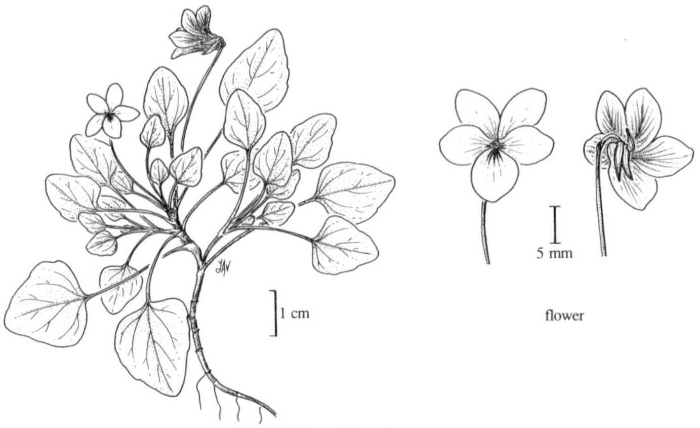


Viola pedunculata

style lobes strongly unequal, one minute, other one capitate. Fruit dehiscent into 4 nutlets. Ca. 250 species; temperate and tropical, N. and S. America, Europe, n Africa. Several species have been used medicinally.

**V. lasiostachys** Link var. **scabrida** Moldenke [*V. robusta* Greene] Perennial herbs, to 1 m tall; stems angled. Leaves oblong to ovate, 3.5-9 cm long, 1-3.5 cm wide, reduced in size upwards. lower ones petiolate, upper ones sessile, coarsely toothed to lobed near base, upper surfaces dark green, usually scabrous. Flowers in axillary or terminal spikes 3.5-15 cm long. Calyces 3-4 mm long; corollas 3-4.5 mm long, purple. Nutlets ca 1.5 mm long, curved, outer surface brown, inner surface white-papillate. May-Oct. Occasional; stream banks, canyon bottoms, freshwater seeps, moist slopes; coastal scrub, riparian communities; 15-850 ft (5-260 m). Widely scattered locations throughout western and central portions of island. Santa Rosa and Santa Catalina islands; coastal n CA to Baja CA, e to Sierra Nevada foothills. Var. *lasiostachys*, with upper surfaces of leaves grayish green and inner surfaces of nutlets rough, brownish, but not papillate, occurs on San Nicolas (introduced), Santa Catalina, and San Clemente islands and on the mainland, w of the Sierra Nevada.

## VIOLACEAE Batsch
### VIOLET FAMILY

Annuals or perennials. Leaves in basal rosettes or alternate, simple to compound; stipules present. Flowers bisexual, bilateral, solitary or in racemes or panicles. Sepals 5, free to slightly fused; petals 5, free, ± bilateral, lower one usually different in size or shape than lateral and upper ones; stamens 5; ovary superior, locule 1, placentas 3, placentas parietal, style 1. Fruit a capsule, dehiscent by 3 valves; seeds many. Ca. 25 genera and 850 species; temperate to subtropical, worldwide.

### VIOLA L. VIOLET

Annuals or perennial herbs. Leaves mostly basal, lanceolate to cordate. Sepals subequal, often with leaf-like appendages at base; lower petal usually short-spurred; stamens with filaments often shorter than anthers, lower 2 stamens with appendages extending into spur. Fruit usually explosively dehiscent. Ca. 400 species; temperate, n hemisphere and S. America.

**V. pedunculata** Torr. & A.Gray JOHNNY-JUMP-UP Perennial herbs, (5)10-30 cm tall, with short rhizomes. Leaves deltate-ovate to cordate, blades 2-5.5 cm long, 3-6 cm wide, margins crenate, petioles 3-11 cm long. Sepals 5-8 mm long, margins scarious, ciliate; petals 8-15 mm long, yellow to yellow-orange, rarely cream, lower and lateral ones with brown veins, lateral ones puberulent, upper ones brownish on back. Mar-May. Common; grassy n-facing hillsides, rocky slopes; coastal scrub, grasslands; 10-2400 ft (3-730 m). Widely scattered locations throughout island. Santa Rosa, Santa Catalina, and San Clemente islands; n CA coast ranges to Baja CA. Leaves and stems were eaten by the Luiseño and Maidu tribes.

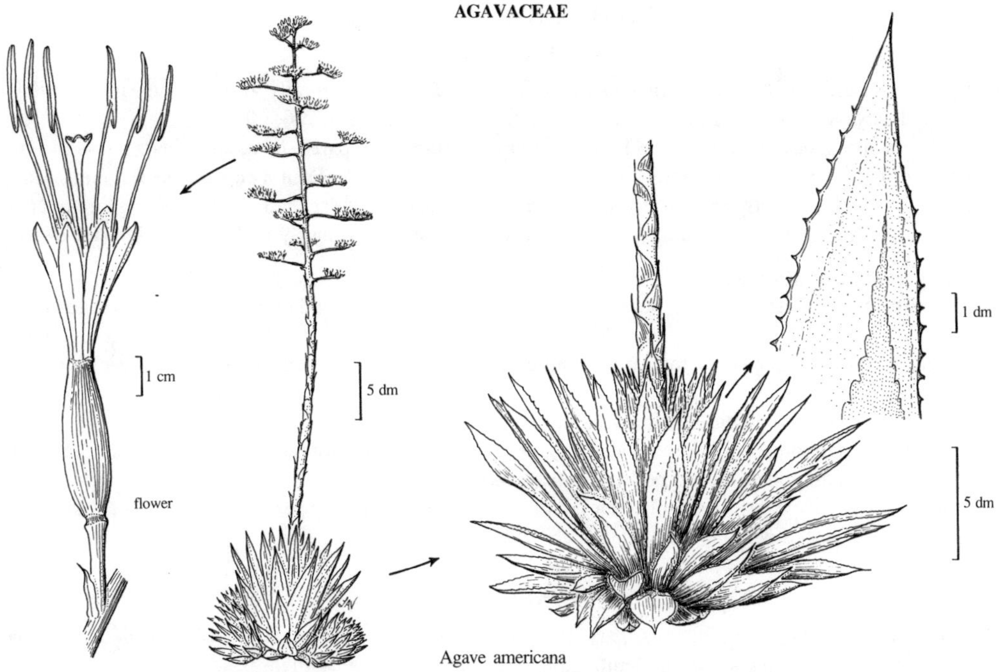


# Monocotyledonous Angiosperms

Annuals or perennial herbs, sometimes shrub-like or tree-like. Leaves usually with parallel leaf venation. Vascular bundles in stem cross-section usually scattered, not organized into a confluent ring. Perianth parts, if present, usually in multiples of 3 (4 in Potamogetonaceae); stamens often in multiples of 3.

## AGAVACEAE Endl.

### Agave Family

Perennials, shrub- or tree-like. Leaves in basal rosettes or whorled, simple, margins entire to toothed, often fibrous, thick, spinose, persistent, sessile. Flowers usually bisexual, radial, usually in terminal panicles. Sepals 3, free, petaloid; petals 3, free; stamens usually 6; ovary superior or inferior, locules 3, styles 1 or 3. Fruit a capsule or berry. Ca. 20 genera and 400 species; warm temperate, subtropical, mostly N. and S. America.

### AGAVE L.

Shrub-like perennials, usually dying after flowering once. Leaves in basal rosettes, linear to ovate, often fleshy, glabrous, margins serrate to spinose, apices spinose. Flowers in terminal, often scapose panicles. Perianth parts fused into tube; stamens 6, exserted; ovary inferior, locules 3, style 1, stigma ± 3-lobed. Fruit a capsule; seeds many, dark brown to black. Ca. 100 species; warm temperate and tropical; N. and S. America.

**A. americana** L. CENTURY PLANT Plants 4-10 m tall in flower. Basal leaf rosettes often produced vegetatively at stem base or from underground stems; leaves lanceolate, 1-2 m long, pale green to glaucous, marginal teeth 4-8 mm long. Panicle narrowly ovoid. Perianth funnelform, 5-8 cm long, yellow; stamens adnate to perianth tube; ovary 2.5-4.5 cm long. Fruit oblong, 4-6 cm long. Aug-Sep. Introduced. Scarce; disturbed sites; coastal scrub; 10-750 ft (3-230 m). Planted on n side at Pelican Bay (where naturalized), Prisoners' Harbor, Campo del Norte, and Scorpion Ranch; Central Valley at Stanton Ranch. San Nicolas and San Clemente islands; Todos Santos Island; widely cultivated and locally naturalized in sw U.S., often persisting after cultivation; native of e AZ, TX, and e Mexico. Sap from young inflorescences can be fermented to produce an alcoholic beverage; sap from leaves can cause severe contact dermatitis.

# ALLIACEAE J.Agardh

## Onion Family

Scapose perennial herbs, with bulbs, corms, or rhizomes. Leaves basal, ± spirally arranged, linear, sheathing at base. Flowers bisexual, radial; in umbels, umbels subtended by involucre of 2-6, often fused bracts. Perianth in 2 whorls of 3, petaloid, each composed of 3 parts, perianth parts free to partly fused into tube; stamens 3 or 6; ovary superior, locules 3, placentas axile, styles 1 (stigmatic lobes 3) or 3. Fruit a capsule; seeds many. Ca. 20 genera and 700 species; mostly temperate, worldwide. Many species cultivated for food or as ornamentals. Includes many genera formerly placed in Amaryllidaceae, which differ primarily by having an inferior ovary.

1. Perianth parts free to near base
   2. Perianth parts white, pink, or purple; stamen filaments not appendaged at base; leaves not keeled ........ **Allium**
   
   2' Perianth parts bright yellow to orange; stamen filaments with cup-like appendage at base; lower leaf surfaces keeled ........ **Bloomeria**

1' Perianth parts fused ± below middle into cylindrical or campanulate tube
   3. Stamens 3; staminodes 3, alternate stamens; perianth lobes ± spreading ........ **Brodiaea**
   
   3' Stamens 6; staminodes absent; perianth lobes ± erect
      4. Inflorescence dense; perianth violet to purple, tube 4-11 mm long ........ **Dichelostemma**
      
      4' Inflorescence open; perianth white, tube 2-4 mm long ........ **Triteleia**

## ALLIUM L. ONION

Plants with bulbs or rhizomes; bulb coats usually membranous. Leaves flat to terete, often with strong onion odor. Involucral bracts 2-4, free to fused, membranous. Perianth parts free; stamens 6, adnate to petals, filaments flattened near base, anthers attached at middle; ovary rounded or apex with 3-6 appendages, stigma 1 or 3-lobed. Ca. 700 species; temperate, N. America and Europe. Cultivated species include onions, chives, garlic, and leek. Bulbs, leaves, and seeds of many species were eaten by Native Americans; sap also was used medicinally. Bulbs of several species can be toxic to livestock.

1. Outer bulb coats with thick-walled, quadrangular reticulations; outer pedicels 8-15 mm long ........ **A. lacunosum**

1' Outer bulb coats with wavy, horizontally elongated reticulations; outer pedicels 15-30 mm long ........ **A. praecox**

**A. lacunosum** S.Watson var. **lacunosum** PITTED ONION Plants 1-2 dm tall; bulb coats with thick-walled, quadrangular reticulations. Leaves 1-2, terete. Involucral bracts ovate, 8-12 mm long, apices acuminate. Pedicels 5-15 mm long; perianth parts lanceolate-oblong, 5-9 mm long, white to pink, with reddish midveins. Fruit 3-4 mm long. Apr-Jun. Rare; grassy hillsides, flats; coastal scrub, grasslands; 120-1400 ft (35-425 m). Central Valley n of Centinela; s side in lower portion of Horqueta Canyon. Santa Rosa Island; San Francisco Bay region to Santa Barbara Co. Three other varieties occur in CA, from the San Francisco Bay region to s CA, e to Sierra Nevada foothills.

**A. praecox** Brandegee EARLY ONION Plants 2-5 dm tall; bulb coats with wavy, horizontally elongated reticulations. Leaves 2-4, flat. Involucral bracts lanceolate-ovate, 10-30 mm long, apices acuminate. Pedicels 15-30 mm long; perianth parts lanceolate-ovate, 9-12 mm long, pale to deep pink, with purplish midveins. Fruit 4-5 mm long. Mar-May. Occasional; n-facing slopes, grassy hillsides, rocky flats; coastal scrub, grasslands; 20-1400 ft (6-425 m). Widely scattered locations throughout island. San Miguel, Santa Rosa, Santa Catalina, and San Clemente islands; Ventura Co. to n Baja CA.

## BLOOMERIA Kellogg GOLDEN STAR

Plants with bulbs; bulb coats fibrous. Leaves ± flat, odorless, lower surface keeled. Inflorescence bracts 2-4, scarious. Perianth parts free; stamens 6, filaments filiform, each with basal appendage, anthers attached at middle; ovary rounded, stigma 3-lobed. 2 species; Monterey and San Benito cos. to n Baja CA.

**B. crocea** (Torr.) Coville Plants 1.5-4 dm tall. Leaves 1, rarely 2. Involucral bracts lanceolate, 5-12 mm long, apices attenuate. Pedicels 20-45 mm long; perianth parts linear-oblong, 8-15 mm long, yellow to gold, with greenish to reddish brown midveins. Fruit 5-6 mm long. Apr-Jun. Occasional; rocky slopes, ridgetops, dry flats; coastal scrub, grasslands; 80-1600 ft (25-490 m). Widely scattered locations throughout western and central

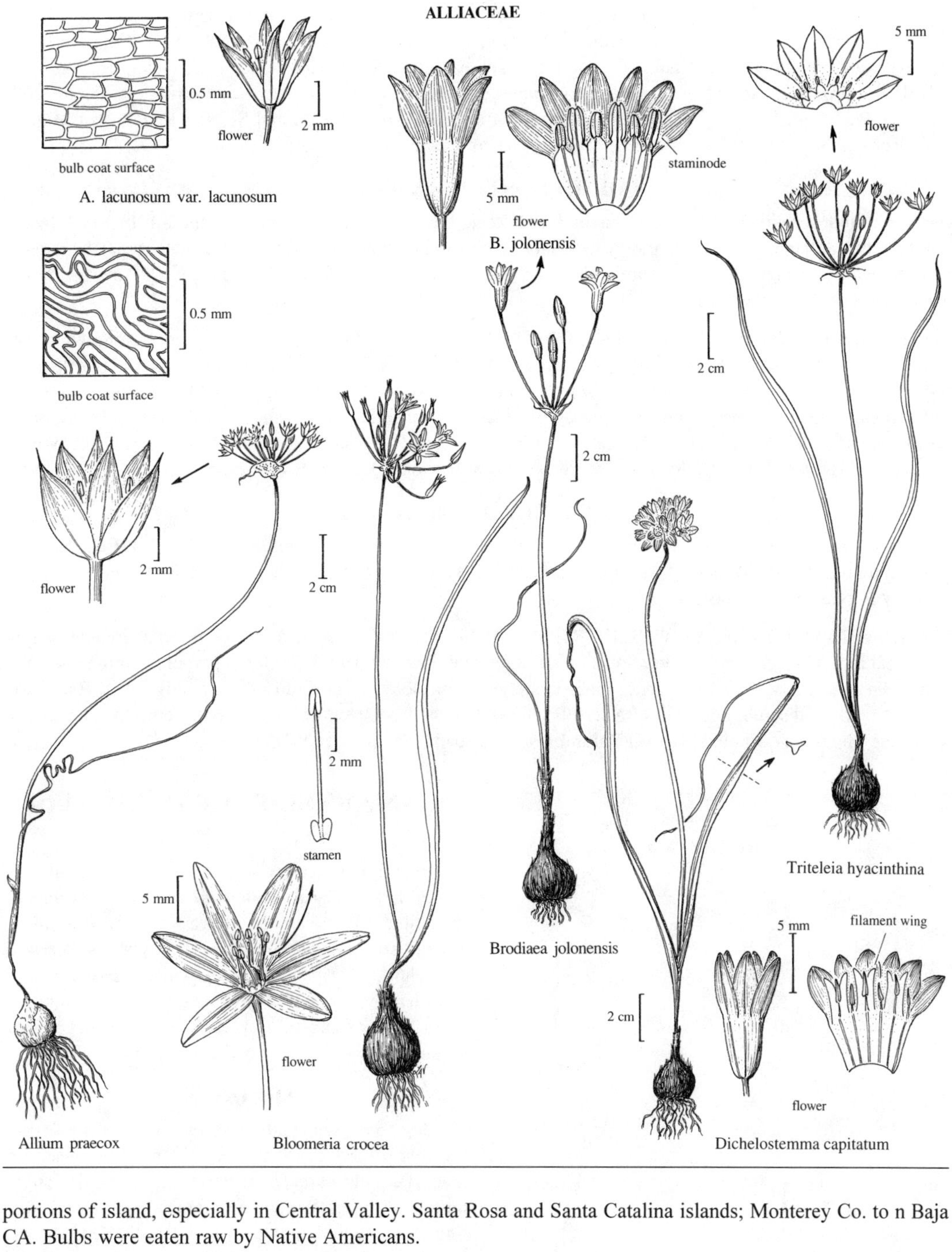


portions of island, especially in Central Valley. Santa Rosa and Santa Catalina islands; Monterey Co. to n Baja CA. Bulbs were eaten raw by Native Americans.

## BRODIAEA Sm.

Plants with bulbs; bulb coats membranous. Leaves ± flat or crescent-shaped in cross-section, odorless. Inflorescence bracts 3-4, scarious. Perianth parts fused below into tube; stamens 3, adnate to petals, anthers attached near their sagittate bases, erect, staminodes 3, alternate stamens; stigma 3-lobed. Ca. 15 species; w N. America. Bulbs of many species were eaten raw or cooked by Native Americans.

**B. jolonensis** Eastw. DWARF BRODIAEA Plants 0.7-3 dm tall. Leaves 1-2, often as long as scape. Involucral bracts lanceolate-ovate, 10-15 mm long, apices acute to acuminate. Pedicels 20-70 mm long; perianth ± violet, tube 7-12 mm long, lobes lanceolate, 10-15 mm long, spreading; staminodes 4-6 mm long, violet, tips incurved. Fruit 10-12 mm long. Apr-May. Scarce; rocky slopes, dry flats, grassy hillsides; coastal scrub, grasslands; 200-1800 ft (60-550 m). Scattered locations in Central Valley between Centinela and Stanton Ranch. San Miguel, Santa Rosa, and Santa Catalina islands; Monterey Co. to n Baja CA.

### DICHELOSTEMMA Kunth

Plants with bulbs; bulb coats fibrous. Leaves ± flat, keeled, odorless. Inflorescence bracts 2-4, free to ± fused, membranous. Perianth parts fused into cylindrical or campanulate tube; stamens 3 or 6, adnate to petals, anthers attached near their sagittate bases, staminodes 3 or absent; stigma 3-lobed. Ca. 5 species; w N. America. Some species cultivated as ornamentals.

**D. capitatum** (Benth.) A.W.Wood BLUE DICKS Plants 3-6 dm tall. Leaves 2-3, often as long as scape. Involucral bracts ovate, 8-15 mm long, apices acuminate. Pedicels 1-15 mm long; perianth violet to purple, tube 4-11 mm long, lobes ovate, 7-11 mm long, erect; stamens 6. Fruit 4-10 mm long. Mar-May. Common; coastal bluffs, rocky n-facing slopes, open flats, grassy hillsides; coastal scrub, grasslands; 10-1600 ft (3-490 m). Scattered locations throughout island. All CA Channel islands; all Baja CA islands except San Geronimo and Natividad; OR to n Mexico, e to UT and NM. Bulbs were eaten raw or cooked by Native Americans.

### TRITELEIA Douglas ex Lindl.

Plants with corms; corm coats ± fibrous. Leaves ± flat, keeled, odorless. Inflorescence bracts 2-4, free, scarious to membranous. Perianth parts fused into campanulate tube; stamens 6, anthers attached at middle; stigma minutely 3-lobed. Ca. 14 species; w N. America.

**T. hyacinthina** (Lindl.) Greene WHITE TRITELEIA Plants 3-5 dm tall. Leaves 2-3, often as long as scape. Involucral bracts lanceolate, 8-15 mm long, apices attenuate. Pedicels 10-35 mm long; perianth white, each lobe with green midvein, tube 2-4 mm long, lobes ovate, 6-12 mm long, erect. Fruit 4-6 mm long. Apr. Rare; disturbed flats; coastal scrub, grasslands; ca. 270 ft (80 m). Known from a single collection in Central Valley just w of Bosque Mano in April 1992. British Columbia to San Joaquin Valley, e to ID.

AMARYLLIDACEAE

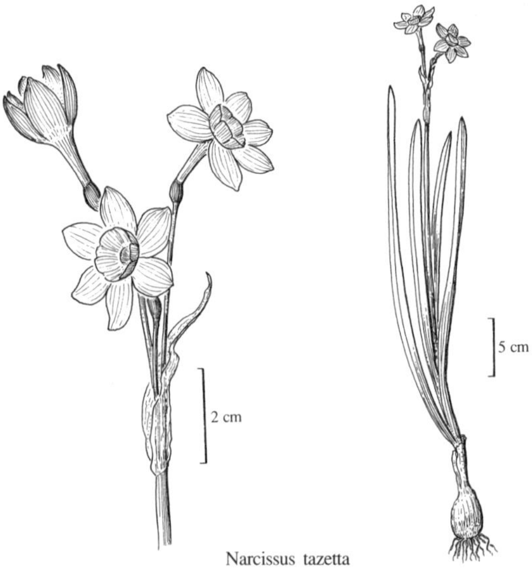


Narcissus tazetta

## AMARYLLIDACEAE J.St-Hil.

### Amaryllis Family

Perennial herbs, usually scapose, usually with bulbs or corms. Leaves basal or alternate, ± spirally arranged, sheathing at base, simple. Flowers usually in terminal umbels, subtended by involucre of 2-8 bracts. Perianth in 2 whorls of 3, petaloid, often partly fused into tube; stamens 6; ovary inferior, locules 3, style 1, stigmatic lobes 3. Fruit a capsule or berry. Ca. 50 genera and 600 species; temperate to tropical, worldwide.

### NARCISSUS L.

Plants scapose, with bulbs. Leaves linear, often ± fleshy. Flowers solitary or in umbels, subtended by sheathing bract. Perianth parts fused into tube, white to orange, tube and segments separated from stamens by tubular, ring-like, petaloid corona. Fruit a capsule; seeds black. Ca. 25 species; temperate, Europe, N. Africa.

**N. tazetta** L. Leaves 8-25 cm long, 1.5-2 cm wide. Flowers usually 3-8 per umbel. Perianth salverform, tube 2-2.5 cm long; lobes and corona white to pale yellow, shorter than perianth lobes. Jan-Feb. Rare; roadsides, disturbed sites; chaparral; 100-850 ft (30-260 m). Central Valley at Christy Ranch and along Islay Canyon Road. Widely cultivated and locally naturalized in sw U.S., often persisting after cultivation; native of Europe.

# CYPERACEAE Juss.
## SEDGE FAMILY

Annuals or perennial herbs; stems terete to triangular in cross-section, solid (filled with pith). Leaves usually in 3 ranks, alternate, usually sheathing stem; blades linear, flat to terete, sometimes much reduced or absent. Flowers bisexual or unisexual, radial, 1-many, much reduced, aggregated into spikelets, in 2 ranks or spirally arranged, each flower subtended by 1 bract; spikelets solitary or aggregated into umbels or open to dense panicles. Perianth absent or composed of scales or bristles; stamens 1-3; ovary superior, locule 1, ovule 1, styles 1-3 or style branches 2-3. Fruit an achene, lenticular to ovoid (enclosed in urn-shaped perigynium in *Carex*). Ca. 115 genera and 3600 species; temperate to tropical, worldwide.

1. Flowers unisexual; fruit enclosed in perigynium ........ **Carex**
1' Flowers bisexual; fruit not enclosed in perigynium
    2. Spikelets ± terete or ovoid; flowers and bracts spirally arranged; perianth composed of bristles; inflorescence terminal, subtended by whorl of 8 or more spreading bracts ........ **Cyperus**
    2' Spikelets ± compressed, flowers and bracts in 2 rows; perianth absent; inflorescence apparently lateral, subtended by 2-3 erect bracts ........ **Scirpus**

### CAREX L. SEDGE

Annuals or perennial herbs; stems usually triangular in cross-section. Leaves usually cauline, blades usually flat. Flowers unisexual, each staminate one subtended by 1 scale-like bract, pistillate ones enclosed in ± urn-shaped perigynium, each subtended by 1 scale-like bract; spikelets solitary or in clusters, axillary or terminal, bisexual or unisexual, if bisexual then of 2 kinds: androgynous spikelets composed of staminate flowers in distal portion, pistillate flowers in proximal portion; gynacandrous spikelets composed of pistillate flowers in distal portion, staminate flowers in proximal portion. Fruit (including perigynium) globose, lenticular (with 2 convex sides), or plano-convex (flat on side (inner) facing spikelet axis, convex on outer side); perigynium apex usually tapered to beak-like. Ca. 1000 species; mostly temperate, worldwide. Roots and stems of many species were eaten and used in basketry by Native Americans.

1. Spikelets 2.5-15 cm long, well separated, lateral or lowest ones often 1 or more cm apart
    2. Pistillate spikelets with 3-15 flowers; perigynia puberulent, triangular in cross-section; pistillate scales as wide as perigynia, base ± enfolding perigynium ........ **C. triquetra**
    2' Pistillate spikelets with many flowers; perigynia glabrous, lenticular; pistillate scales flat, narrower than perigynia
        3. Pistillate bracts lanceolate-ovate to narrowly ovate, brown except midveins, apices short-awned; perigynia elliptic, beak 0.3-0.5 mm long ........ **C. barbarae**
        3' Pistillate bracts oblong, usually dark purple except midveins, apices acute; perigynia orbicular to broadly ovate, beak less than 0.2 mm long, entire or minutely notched ........ **C. senta**
1' Spikelets 0.5-3.5 cm long, aggregated, lowest ones less than 0.5 cm apart
    4. Spikelets of 2 kinds, terminal one staminate, lateral ones mostly pistillate, sometimes androgynous; perigynia globose, puberulent ........ **C. globosa**
    4' Spikelets all alike, androgynous or gynacandrous; perigynia lenticular or plano-convex, glabrous
        5. Spikelets androgynous; perigynia lenticular, margins not winged ........ **C. tumulicola**
        5' Spikelets gynacandrous; perigynia plano-convex, margins narrowly winged
            6. Inner side of perigynium nerved; pistillate bracts as wide as perigynia; bract subtending lowest spikelet often 0.5-1 cm long, spikelets not clearly separated ........ **C. harfordii**
            6' Inner side of perigynium smooth; pistillate bracts narrower and shorter than perigynia; bract subtending lowest spikelet often less than 0.5 cm long, spikelets often somewhat separated ........ **C. subbracteata**

**C. barbarae** Dewey SANTA BARBARA SEDGE Plants 3-10 dm tall, with long, stout rhizomes; stems 1-3 per clump. Leaf blades 3.5-9 mm wide. Spikelets 5-10(15) cm long, terminal spikelet staminate (rarely androgynous),

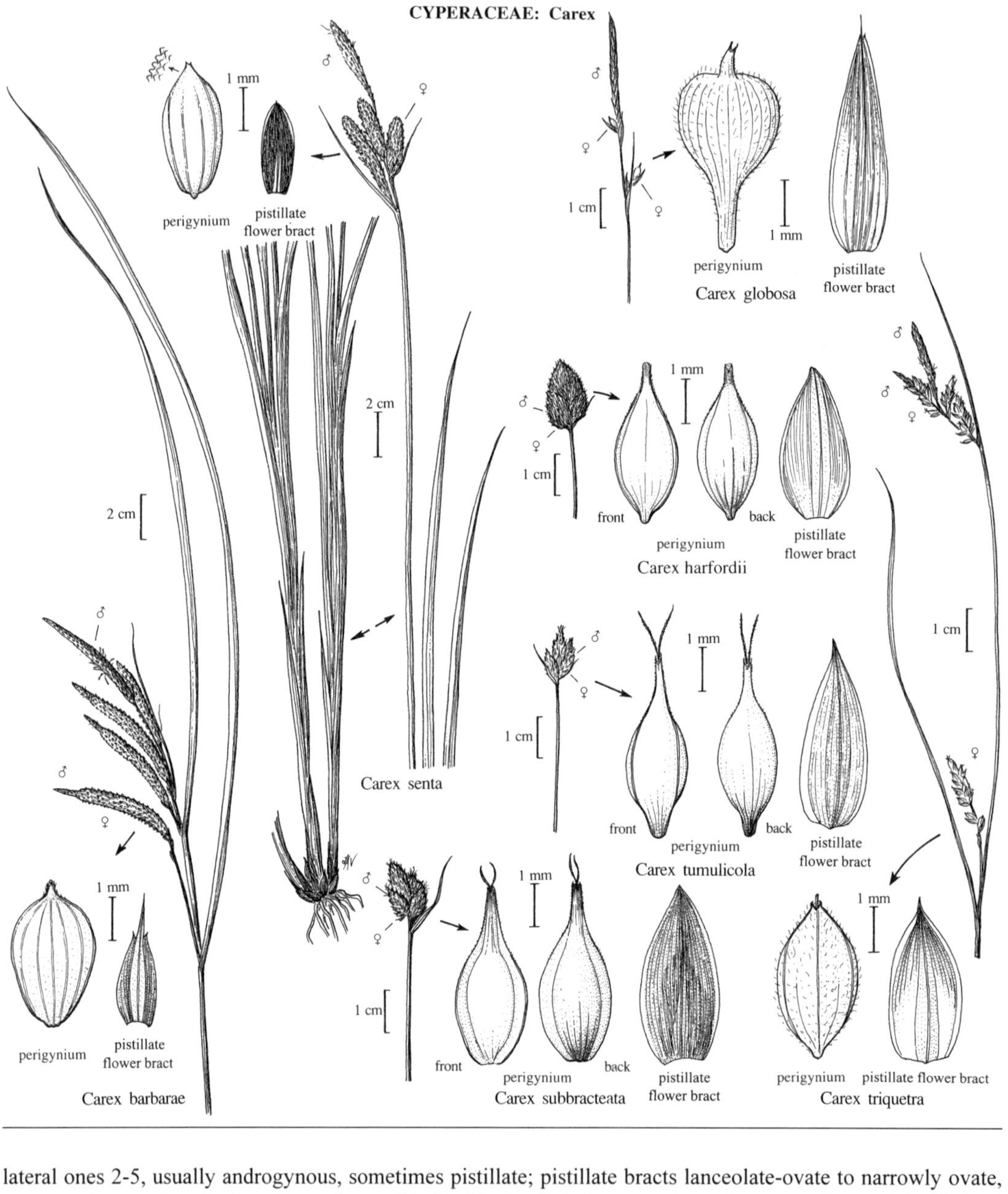


lateral ones 2-5, usually androgynous, sometimes pistillate; pistillate bracts lanceolate-ovate to narrowly ovate, narrower than perigynia, brown, purple-speckled, midveins light brown, apices short-awned, awns hispidulous. Perigynia elliptic, lenticular, glabrous, beaks 0.3-0.5 mm long, bidentate, teeth hispidulous. Mar-May. Rare; stream banks; riparian communities; 50-150 ft (15-45 m). North side in Cañada del Puerto. Southern OR to s CA, e to Cascade Ranges and Sierra Nevada foothills.

**C. globosa** Boott ROUND-FRUITED SEDGE Plants 1-4 dm tall, with rhizomes; stems many, in clumps. Leaf blades 1.5-3.5 mm wide. Spikelets 0.5-2 cm long, terminal spikelet staminate (rarely androgynous), lateral spikelets 1-3, usually pistillate, sometimes androgynous; pistillate bracts broadly lanceolate to ovate, reddish brown or purplish, bases as wide as perigynia, apices obtuse to attenuate. Perigynia globose, 2-ribbed to 2-keeled, puberulent, beaks 0.5-1.7 mm long, bidentate, teeth glabrous to sparsely puberulent. Mar-Jun. Common; shaded slopes, canyon bottoms; chaparral, oak woodland, pine forest; 100-1300 ft (30-395 m). Widely scattered locations throughout western and central portions of island. Santa Rosa and Santa Catalina islands; n CA Coast Ranges to s CA.

**C. harfordii** Mack. [*C. montereyensis* Mack.] MONTEREY SEDGE Plants 7-10 dm tall, with short, thick rhizomes; stems solitary or in clumps. Leaf blades 2.5-3 mm wide. Spikelets 0.5-1 cm long, 5-9, all gynacandrous, ± densely aggregated, staminate flowers often few; pistillate bracts elliptic to ovate, as wide as perigynia, pale green or red-tinged, apices acute to short-awned. Perigynia plano-convex, glabrous, both sides nerved, margins narrowly winged, beaks 1-2 mm long, bidentate, margins serrulate. Jul-Sep. Rare; rocky slopes, chaparral, oak woodland; 150-500 ft (45-150 m). Widely scattered locations on n side between Fry's Harbor and Pelican Bay. Northern CA coast ranges to Santa Barbara Co.

**C. senta** Boott ROUGH SEDGE Plants 4-8 dm tall, with long, stout rhizomes; stems 1-many, in dense clumps. Leaf blades 3-5 mm wide. Spikelets 2.5-6 cm long, terminal one staminate or with a few pistillate flowers near base, lateral ones 2-4, androgynous or lowest one pistillate; pistillate bracts oblong, much narrower than perigynia, usually dark purple, midveins lighter in color. Perigynia orbicular to broadly ovate, lenticular, glabrous, beaks less than 0.5 mm long, entire or notched, glabrous. Mar-May (Jul). Rare; stream banks; riparian communities; 60-500 ft (20-150 m). North side in canyon above Dick's Harbor; Central Valley in Cañada de la Siesta and near La Cascada. San Francisco Bay region to Baja CA, e to Sierra Nevada and AZ.

**C. subbracteata** Mack. Plants 3-12 dm tall, with short, thick rhizome; stems solitary or in clumps. Leaf blades 2-4 mm wide. Spikelets 0.5-1 cm long, 5-7, all gynacandrous, ± densely aggregated; pistillate bracts ovate, narrower and shorter than perigynia, reddish brown, margins narrowly white-hyaline. Perigynia plano-convex, glabrous, inner sides smooth, outer sides nerved, beaks ca. 1.5 mm long, bidentate, serrulate. Mar-May (Dec). Rare; n-facing rock ledges in canyons; chaparral; 250-500 ft (75-150 m). North side in Scorpion Canyon. Santa Rosa Island; n CA coast ranges to Santa Barbara Co.

**C. triquetra** Boott Plants 3-5.5 dm tall; stems in dense clumps. Leaf blades 2.5-6 mm wide. Spikelets 1-3.5 cm long, terminal one staminate (rarely androgynous), lateral ones usually pistillate, sometimes androgynous, lowest one usually pistillate; pistillate bracts broadly ovate, golden brown or copper-colored, base partly enfolding perigynium, as wide as perigynia, apices acute to minutely apiculate. Perigynia triangular in cross-section, puberulent, beaks less than 0.2 mm long, bidentate, glabrous. Mar. Rare, not seen recently; habitat, plant community, and elevation unknown. Known from a single collection without specific locality in March 1925. Santa Catalina Island; Monterey Co. to Baja CA.

**C. tumulicola** Mack. FOOTHILL SEDGE Plants 2-6 dm tall, with short, thick rhizomes; stems in dense clumps. Leaf blades 1.5-2.5 mm wide. Spikelets 0.5-1 cm long, 5-9, all androgynous; pistillate bracts elliptic to ovate, as wide as perigynia, reddish brown, margins often light-colored. Perigynia lenticular, glabrous, outer sides finely nerved, beaks 1-2 mm long, bidentate to notched, margins serrulate. Apr-May. Scarce; n-facing slopes; chaparral, pine forest; 500-1400 ft (150-425 m). North side near China Pines; Central Valley in Cañada Cebada and near Christy Pines; s side in Cañada de los Sauces and on Sierra Blanca ridge. Santa Rosa and San Clemente islands; WA to San Luis Obispo Co., e to Sierra Nevada.

## **CYPERUS** L. UMBRELLA SEDGE

Annuals or perennial herbs; stems round to triangular in cross-section. Leaves usually basal, usually sheathing stem, blades often absent. Flowers bisexual, each subtended by 1 bract, arranged in 2 rows; spikelets many, in terminal umbel-like or capitate clusters, clusters subtended by a whorl of leaf-like bracts. Perianth absent; style branches 3. Fruit ovoid to obovoid, 3-angled. Ca. 600 species; temperate to tropical, worldwide. Roots of several species were used medicinally and cooked and eaten by Native Americans. Many species yield strong fibers. *C. papyrus* was used for paper for over 4000 years.

**C. involucratus** Rottb. [The name *C. alternifolius* L. has been misapplied to this species] Perennial herbs, with short, thick rhizomes; stems 3-18 dm tall. Leaves reduced to bladeless, brown sheaths. Inflorescence umbel-like, involucral bracts 9-15, 8-15(30) cm long, spreading. Spikelets 5-10 mm long, flowers 10-30, spikelet axis persistent, bracts 1.5-2 mm long, usually deciduous. Fruit 0.5-0.9 mm long. Jun. Introduced. Rare, not seen recently; moist stream banks; riparian communities; elevation unknown. Known from a single collection in Valdez Canyon in June 1930. Cedros Island; Santa Barbara Co. to Los Angeles Co.; native of Africa.

## **SCIRPUS** L. BULRUSH

Annuals or perennial herbs; stems cylindrical to triangular in cross-section. Leaves basal or cauline, sheathing stem, blades present or absent. Flowers bisexual, each subtended by 1 bract, spirally arranged; spikelets many, in terminal or apparently lateral capitate clusters, clusters subtended by 1-several, ascending to erect, leaf-like bracts. Perianth composed of bristles; style branches usually 2. Fruit obovoid, lenticular, or 3-angled, usually

CYPERACEAE: Cyperus, Scirpus

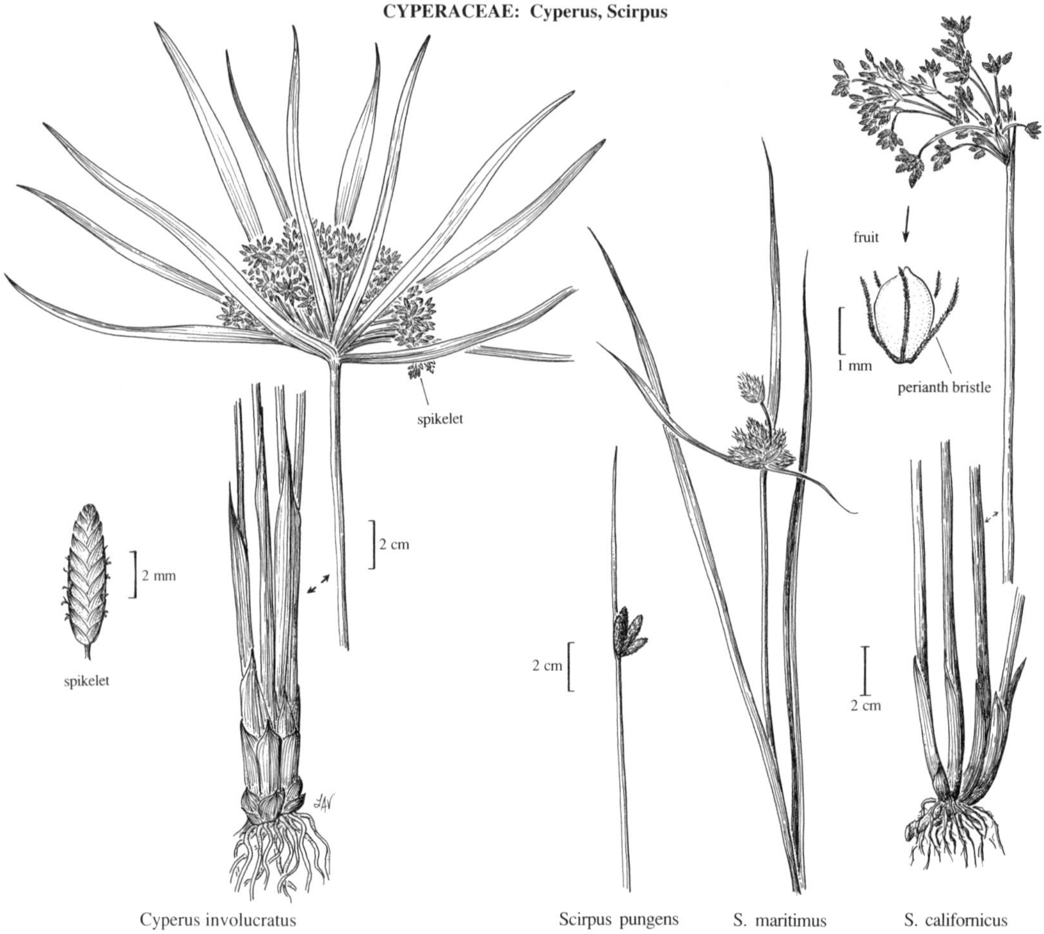


beaked or apiculate. Ca. 200 species; worldwide. Seeds, pollen, and rhizomes of several species were eaten by Native Americans; stems were used to construct water bottles, dwellings, and other items.

1. Inflorescence subtended by 2-3 ascending to spreading bracts .................. **S. maritimus**
1' Inflorescence usually subtended by 1 erect, stiff bract
   2. Spikelets many, aggregated into a flat-topped, short-pedunculate panicle; subtending bract 2-6 cm long; basal leaves without blades .................. **S. californicus**
   2' Spikelets 2-5, sessile, in capitate clusters; subtending bract 4-10 cm long, 1-3 mm wide; some or all basal leaves with blades .................. **S. pungens**

**S. californicus** (C.A.Mey.) Steud. CALIFORNIA BULRUSH Plants 1.5-4 m tall; stems terete to triangular in cross-section. Leaves reduced to ± flat, basal sheaths, blades absent. Spikelets many, 5-10 mm long, aggregated into a flat-topped, short-pedunculate panicle, entire inflorescence subtended by 1 bract; bract 2-6 cm long, erect, thick, stiff. Perianth bristles 2-4, length ± equal to fruit length. Fruit lenticular, ca. 2 mm long. Apr-Oct. Rare; moist stream banks, coastal estuaries; riparian communities; 5-100 ft (1.5-30 m). North side at Prisoners' Harbor and in Cañada del Agua. Northern CA coast to s CA, S. America, islands of Pacific Ocean.

**S. maritimus** L. Plants 0.6-0.9 m tall; stems triangular in cross-section. Leaves basal and cauline, blades flat, sheaths flat to V-shaped. Spikelets 3-many, 10-20 mm long, sessile, in 1-5 capitate clusters. Inflorescence compact, umbel-like to capitate, subtended by 2-3 bracts, bracts 3-20 cm long, longest bracts ± ascending, others ascending to spreading. Perianth bristles 5-7, length less than half fruit length. Fruit lenticular, 2-3.5 mm long, Oct-Nov. Rare; moist stream banks, edges of coastal estuaries; riparian communities; 5-10 ft (1.5-3 m). Central

IRIDACEAE

Sisyrinchium bellum

Valley at mouth of Christy Canyon; s side at mouth of Cañada de los Sauces. Throughout much of N. America, Europe, and Asia.

**S. pungens** Vahl Plants 0.6-1.5 m tall; stems triangular in cross-section. Leaves basal and cauline, sheathing, lower blades ± flat to terete, upper ones terete. Spikelets 2-5, 5-12 mm long, sessile, in 1 capitate cluster, subtended by 1-2 bracts, longest bract 4-10 cm long, ± flat, stiff, erect, other one less than 1 cm long. Perianth bristles 2-8, length half to equal to fruit length. Fruit lenticular to ± triangular in cross-section, 2-3.5 mm long. Apr-Aug. Rare; moist sandy depressions; grasslands; ca. 15 ft (5 m). South side near mouth of Pozo Canyon. Santa Rosa and Santa Catalina islands; throughout N. America, to S. America, Europe, Pacific Ocean islands.

# IRIDACEAE Juss.

## Iris Family

Perennial herbs, usually with rhizomes, bulbs, or corms. Leaves mostly basal, alternate, simple, linear, sheathing stem. Flowers bisexual, radial, solitary or in racemes or panicles, usually subtended by leaf-like bracts. Perianth parts in 2 whorls of 3, petaloid, fused into tube; stamens 3, usually adnate to or attached near base of sepals; ovary inferior, locules 3, placentas axile, styles or style branches 3, branches sometimes lobed. Fruit a capsule, dehiscent by valves. Ca. 90 genera and 1800 species; temperate to tropical, worldwide.

### SISYRINCHIUM L.

Plants with short rhizomes. Stems usually ± flat, often branched above basal rosette. Leaves folded lengthwise and flattened, overlapping, bases sheathing stem. Flowers in open to compact, often umbel-like clusters, inflorescence subtended by 2 leaf-like bracts, bracts equal or unequal in length, margins scarious to hyaline. Perianth usually rotate, all 6 segments petaloid, outer 3 often wider than inner 3; stamens free or fused into tube. Ca. 100 species; temperate, mostly N. & S. America.

**S. bellum** S.Watson BLUE-EYED GRASS Plants 15-45 cm tall; stems in clumps. Basal leaves 10-23 cm long, 2-6 mm wide. Inflorescence bracts 1.5-3 cm long, green, margins hyaline; pedicels 5-12 mm long. Perianth 10-15 mm long, blue-purple to dark blue, segments broadly elliptic, apices rounded, truncate, or notched; stamens yellow. Mar-May. Common; coastal flats, grassy n-facing slopes; coastal scrub, grasslands; 30-1400 ft (9-425 m). Widely scattered locations throughout western portion of island; scattered in Central Valley between Christy Beach and El Sobrante. San Miguel, Santa Rosa, and Santa Catalina islands; OR to s CA, e to Sierra Nevada. Leaves and roots were boiled and used medicinally by Native Americans.

# JUNCACEAE Juss.
## Rush Family

Annuals or perennial herbs; stems compressed to cylindrical, hollow or solid (filled with a spongy pith). Leaves usually basal, cauline ones few, alternate; blades linear, flat to terete; petiole sheathing stem, margins free or fused. Flowers bisexual, radial, 1-many, in small, axillary clusters, clusters aggregated into axillary or terminal panicles, each flower subtended by 1-3 minute bracts. Perianth parts in 2 whorls of 3, greenish to dark brown; stamens 3 or 6; ovary superior, locules 1 or 3, placentas axile or parietal, style branches usually 3. Fruit a capsule; seeds many, often with tail-like appendages. Ca. 10 genera and 325 species; mostly cold temperate, arctic, worldwide.

1. Sheath margins free, overlapping, glabrous; leaf blade margins glabrous; ovary with many ovules ... **Juncus**
1' Sheath margins fused except near apex, villous; leaf blade margins villous, especially near base; ovary with 3 ovules ... **Luzula**

## JUNCUS L RUSH

Annuals or perennial herbs; stems usually solid, pith spongy. Leaf blades flat to cylindrical, usually stiff, glabrous; sheath margins free, overlapping, glabrous; junction of blade and sheath often with auriculate appendages. Flowers often in axillary clusters, clusters arranged into open to compact or dense panicles, subtended by a bract. Stamens 3 or 6; locules 1 or 3, placentas parietal or axile, ovules many. Ca. 225 species; mostly temperate, worldwide. Stems of some species were roasted, boiled, or eaten raw by Native Americans; stems of several species were used in basketry.

1. Annuals; plants 5-30 cm tall; flowers usually solitary, evenly distributed along inflorescence axis ... **J. bufonius**
1' Perennial herbs, with rhizomes; plants 20-80 cm tall; flowers 5-many, in open to compact, often dense clusters
  2. Inflorescence terminal, composed of dense, capitate clusters; cauline leaves present, folded along midrib, 3-8 mm wide; basal leaf blades flat, 3.5-10 mm wide ... **J. xiphioides**
  2' Inflorescence apparently lateral, flower clusters not capitate; cauline leaves absent; basal leaf blades terete to compressed, 1-1.5 mm wide
    3. Perianth parts 3-6 mm long, filaments shorter than anthers; stems not clumped, evenly distributed along rhizome; flowers 5-30 per inflorescence ... **J. mexicanus**
    3' Perianth parts 2-3 mm long, filaments longer than anthers; stems usually densely clumped; flowers many per inflorescence ... **J. patens**

**J. bufonius** L. TOAD RUSH Annuals, 5-30 cm tall; stems tufted, erect to ascending, filiform, ca. 1 mm wide. Basal leaves 8-16 cm long; cauline leaves 1-3, 2-6 cm long, 0.5-1 mm wide, flat to involute. Inflorescence open; flowers 1 usually solitary, evenly distributed along inflorescence axis. Perianth parts 4-6 mm long; stamens 6, filaments ± equal to anthers. Fruit obovoid to oblong in outline, 3.5-5.5 mm long, shorter than perianth. Mar-Aug. Common; moist stream banks, freshwater seeps, floodplains, seasonal ponds; riparian communities; 10-1650 ft (3-505 m). Widely scattered locations throughout island. All CA Channel islands except Santa Barbara; Guadalupe Island; throughout CA, ± worldwide.

**J. mexicanus** Willd. MEXICAN RUSH Perennial herbs, 20-50 cm tall, with long, thick rhizomes; stems solitary, 1-2 mm wide, evenly distributed along rhizome, compressed, erect to twisted. Basal leaves few, 8-40 cm long, 1-1.5 mm wide, lowest ones reduced, sheath-like, brown; cauline leaves absent. Flowers 5-30, in a compact, apparently lateral panicle; bract subtending panicle 5.5-15 cm long, continuous with stem, usually exceeding inflorescence, circular in cross-section. Perianth parts 3-6 mm long; stamens 6, filaments shorter than anthers. Fruit ovoid, somewhat 3-angled, usually shorter than perianth. May-Aug. Occasional; stream banks, moist flats; grasslands, riparian communities; 200-1000 ft (60-305 m). Widely scattered locations throughout western and central portions of island. San Miguel, Santa Rosa, and Santa Catalina islands; WA to n Mexico, e to CO and TX, also in S. America.

**J. patens** E.Meyer COMMON RUSH Perennial herbs, 40-80 cm tall, with short, thick rhizomes; stems solitary, 1.5-3 mm wide, usually densely clumped, terete, erect. Basal leaves few, 10-60 cm long, 1-1.5 mm wide, lowest ones reduced, sheath-like, brown; cauline leaves absent. Flowers many, in compact to open, apparently lateral panicles; bract subtending panicle 4-21 cm long, continuous with stem, usually exceeding inflorescence, circular in cross-section. Perianth parts 2-3 mm long; stamens 6, filaments longer than anthers. Fruit subglobose, 3.5-5 mm

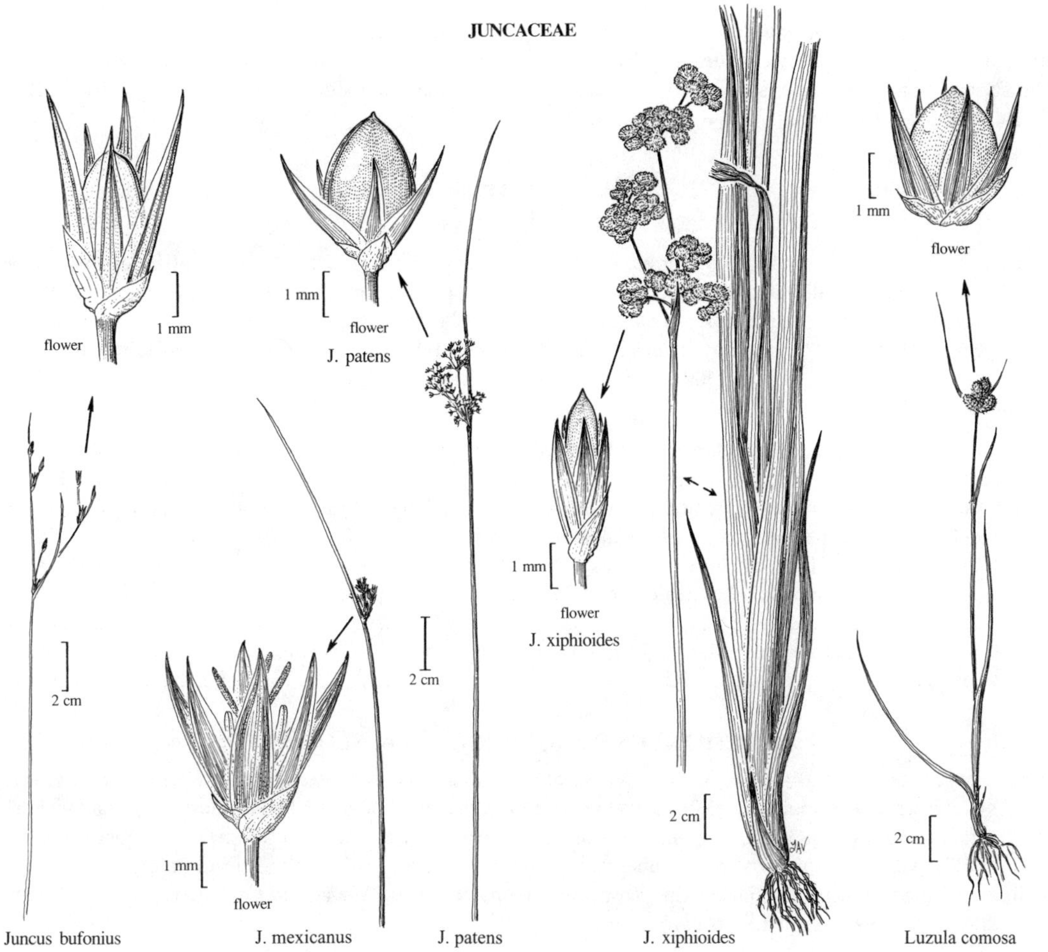


long, usually shorter than perianth. Mar-Jun. Common; grassy slopes, moist flats, stream banks, floodplains; grasslands, riparian communities; 80-1000 ft (25-305 m). Widely scattered locations throughout western and central portions of island. San Miguel, Santa Rosa, and San Clemente islands; OR to s CA.

**J. xiphioides** E.Meyer IRIS-LEAVED RUSH Perennial herbs, 35-80 cm tall, with short, thick rhizomes; stems solitary, 1.5-3 mm wide, solitary to clumped, terete, erect. Basal leaves few, equitant, 10-60 cm long, 3.5-10 mm wide, lowest ones reduced, sheath-like, brown; cauline leaves 1-3, folded along midrib, 3-8 mm wide. Flowers many, in dense, terminal, capitate clusters; bract subtending inflorescence 1-5 cm long, shorter than inflorescence, flat. Perianth parts 3-4 mm long, narrowly lanceolate; stamens 3 or 6, filaments ± equal to anthers. Fruit narrowly oblong in outline, 3.5-5 mm long, exceeding perianth. Aug-Nov. Scarce; creek banks, freshwater seeps; riparian communities; 10-900 ft (3-275 m). North side in Cañada del Puerto and in Cañada del Agua; Central Valley in Cañada Christy and Cañada Larga. Santa Catalina Island; OR to n Baja CA, e to AZ.

## LUZULA DC. HAIRY WOOD RUSH

Perennial herbs; stems hollow. Leaf blades flat, flexible, sometimes drooping. usually pubescent, sheath margins fused, free near junction with blade, usually pubescent. Flowers in open to dense panicles. Stamens 6; locule 1, placentas basal, ovules 3. Ca. 80 species; mostly arctic or temperate, n hemisphere.

**L. comosa** E.Meyer [*L. c.* E.Meyer var. *subsessilis* S.Watson] COMMON WOOD RUSH Plants 10-35 cm tall. Leaves mostly basal, blades 3-20 cm long, 3-12 mm wide, subglabrous, ciliate near junction with sheath. Flowers in dense, subcylindrical to globose panicles, panicles sessile to short-pedunculate. Perianth parts 3-5 mm long, pale to dark brown, margins hyaline. Fruit globose, 2-5 mm long. Mar-Jun. Occasional; coastal flats, grassy hillsides, rocky n-facing slopes, canyon bottoms; chaparral, coastal scrub, pine forest; 120-1250 ft (35-380 m). North side between Cueva Valdez and Prisoners' Harbor, and in Scorpion Canyon; Central Valley between Christy and Stanton ranches; s side between Cañada de los Sauces and Coches Prietos drainage. Santa Rosa Island; AK to s CA mountains, e to Rocky Mountains.

## LILIACEAE Juss.
### Lily Family

Scapose perennial herbs, with bulbs, corms, or rhizomes. Leaves linear, often sheathing at bases, basal ones usually whorled or spirally arranged, cauline ones alternate, opposite, or whorled, linear. Flowers usually bisexual, radial, in racemes, spikes, or panicles. Perianth in 2 whorls of 3, each usually composed of 3 petaloid parts; stamens 3 or 6; ovary superior, locules 3, placentas axile, styles 1 (stigmatic lobes 3) or 3. Fruit a capsule; seeds many. Ca. 160 genera and 2400 species; mostly temperate, worldwide. Many species are cultivated as ornamentals.

1. Cauline leaves whorled; perianth parts 5.5-9 cm long, orange-yellow, with many dark-purple spots, without glands ........ **Lilium**

1' Cauline leaves alternate or opposite; perianth parts 1-5 cm long, whitish, yellow, or purple, not purple-spotted, all or inner perianth parts with glands near bases
    2. Perianth whorls different, outer ones usually narrower or shorter than inner ones, inner ones 2-5 cm long; style 1, stigmatic lobes 3; flowers solitary or few in racemose to subumbellate clusters ........ **Calochortus**
    2' Outer and inner perianth whorls alike, 1-1.5 cm long; styles 3; flowers many, in terminal racemes or panicles ........ **Zigadenus**

### CALOCHORTUS Pursh MARIPOSA or SEGO LILY

Plants scapose or with leafy stems, with bulbs. Basal leaf solitary, cauline leaves few or absent, narrower than basal one. Flowers campanulate to globose, solitary or few in racemose or subumbellate clusters. Perianth whorls different, outer series sepaloid, usually narrower or shorter than inner one, inner series petaloid, upper surface of each usually with a conspicuous, often puberulent gland; style 1, stigmatic lobes 3. Fruit oblong in outline, ± sharply 3-angled, sometimes winged. Ca. 60 species; temperate, w N. America to C. America. Bulbs of many species were cooked and eaten by Native Americans.

1. Inner perianth parts yellow; flowers in subumbellate clusters ........ **C. luteus**

1' Inner perianth parts white to pink- or purple-tinged; flowers in racemose clusters
    2. Flowers ± globose, nodding; petal glands depressed, with several transverse, fringed membranes ........ **C. albus**
    2' Flowers bowl-shaped, erect; petal glands not depressed, densely puberulent ........ **C. catalinae**

**C. albus** Douglas ex Benth. FAIRY LANTERN or GLOBE LILY Plants 20-80 cm tall. Flowers 3-12, ± globose, nodding, in racemose clusters. Outer perianth parts 1-2 cm long; petals 2-2.5 cm long, white or pink-tinged, margins ciliate, basal gland depressed, with several transverse, fringed membranes; filaments 4-6 mm long. Fruit 2.5-4 cm long, angles winged. Apr-Jul. Occasional; rocky n-facing slopes; chaparral, coastal scrub, oak woodland, pine forest; 70-1200 ft (20-365 m). North side between Twin Harbors and Prisoners' Harbor; Central Valley between Cañada Christy and Stanton Ranch; s side between Cañada de los Sauces and Coches Prietos drainage. Santa Rosa and Santa Catalina islands; San Francisco Bay region and Sierra Nevada foothills to s CA.

**C. catalinae** S.Watson CATALINA MARIPOSA Plants 20-60 cm tall. Flowers 1-5, bowl-shaped, erect, in racemose clusters. Outer perianth parts 2-3.5 cm long; petals 2.5-5 cm long, white or purple-tinged, each with dark purple spot near base, margins smooth, basal gland densely puberulent; filaments 8-10 mm long. Fruit 2-5 cm long, angles acute. Mar-May. A species of special environmental concern. Occasional; grassy hillsides, rocky slopes; coastal scrub, grasslands; 50-1200 ft (15-365 m). Widely scattered locations in Central Valley between Cañada Christy and Stanton Ranch; s side between Rancho Nuevo and Cañada Christy, and in Pozo Canyon and Coches Prietos drainages. Santa Rosa and Santa Catalina islands; San Luis Obispo Co. to s CA.

**C. luteus** Douglas ex Lindl. YELLOW MARIPOSA Plants 20-50 cm tall. Flowers 1-7, campanulate, erect, in subumbellate clusters. Outer perianth parts 2-3 cm long; petals 2-3.5 cm long, deep yellow, each with red-brown lines and red-brown blotch near middle, margins smooth, basal gland puberulent, trichomes matted; filaments

LILIACEAE

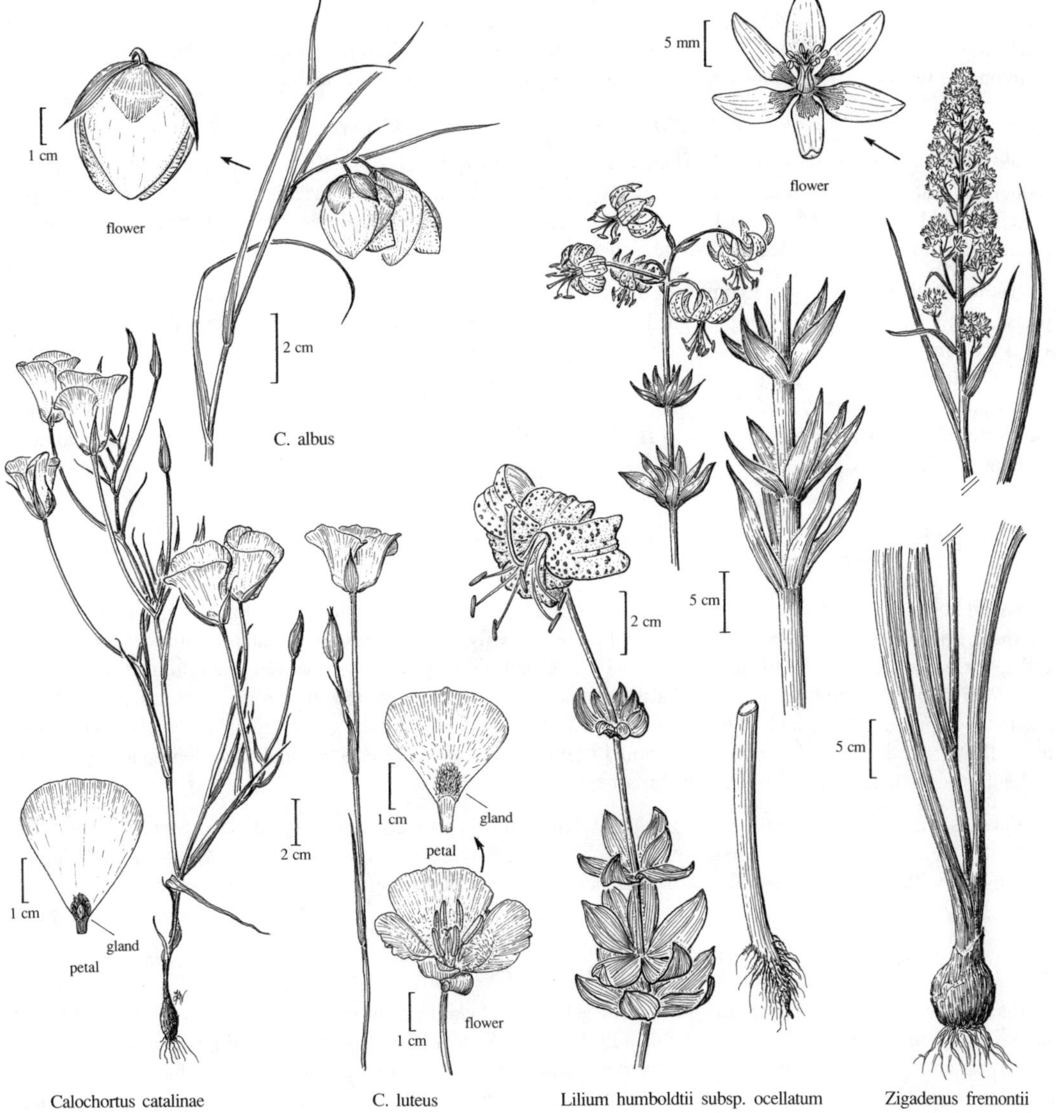


8-11 mm long. Fruit 3.5-6 cm long, angles acute. May-Jun. Scarce; grassy hillsides, rocky flats; coastal scrub, grasslands; 200-750 ft (60-230 m). Central Valley between Portezuela Grade and Rancho del Sur; s side in Coches Prietos drainage. Santa Catalina Island; nw CA to San Luis Obispo Co.

## LILIUM L. LILY

Plants scapose or with leafy stems, with short, scaly rhizomes. Basal leaves whorled, cauline leaves opposite or whorled, reduced upwards. Flowers campanulate to salverform, 1-many, often in terminal, bracteate racemes. Perianth parts alike, petaloid, without glands; style 1, stigmatic lobes 3. Fruit obovoid to oblong in outline, smooth. Ca. 100 species; temperate, n hemisphere. Rhizomes of many species were eaten by Native Americans.

**L. humboldtii** Roezel & Leichtlin subsp. **ocellatum** (Kellogg) Thorne Plants 9-20 dm tall; rhizomes subglobose to ellipsoid, 5-15 cm long. Leaves in 5-10 whorls, usually oblanceolate, 2-15 cm long. Flowers nodding, salverform. Perianth parts narrowly lanceolate, recurved, 5.5-9 cm long, orange-yellow, with many dark purple spots; stamens 3-5 cm long. Fruit 2.5-5 cm long, sharply angled. Mar-May. A subspecies of special environmental concern. Occasional; coastal bluffs, shaded n-facing slopes, creek banks; chaparral, coastal scrub, oak woodland, pine forest; 100-1400 ft (30-425 m). Widely scattered locations on n side between Cueva Valdez and Prisoners' Harbor, and in Scorpion Canyon; Central Valley between Campo Raton and El Sobrante; s side in Laguna Canyon and Coches Prietos drainage. Santa Rosa and Anacapa islands; OR to Baja CA.

### ZIGADENUS Michx. DEATH CAMAS

Plants with leafy stems, with bulbs or rhizomes. Leaves mostly basal, cauline leaves few, alternate or opposite, reduced upward. Flowers bisexual or unisexual, usually many, in terminal racemes or panicles. Perianth parts alike, petaloid, upper surfaces usually with basal glands; styles 3. Fruit cylindrical or oblong in outline, with 3 rounded angles. Ca. 20 species; temperate, n hemisphere.

**Z. fremontii** (Torr.) S.Watson Plants 3-10 dm tall; bulbs subglobose, 15-50 cm wide. Flowers bisexual. Perianth parts broadly ovate, 1-1.5 cm long, greenish white, gland obovate, greenish yellow; stamens 5-8 mm long. Fruit 1.5-3 cm long. Mar-May. Occasional; rocky slopes, grassy hillsides; chaparral, coastal scrub, grasslands; 300-1300 ft (90-395 m). Scattered locations in Central Valley between Islay Canyon and Valley Anchorage; s side near Black Point peninsula, and on South Ridge between Laguna Canyon and Willows Canyon. San Miguel, Santa Rosa, and Anacapa islands; OR to Baja CA. Entire plants, especially the bulbs, are extremely toxic to humans and livestock.

## ORCHIDACEAE Juss.
## ORCHID FAMILY

Perennial herbs, with rhizomes or corms, terrestrial in temperate regions, epiphytic in the tropics. Leaves usually alternate, linear to orbicular, sheathing at bases. Flowers usually bisexual, strongly bilateral, solitary or in racemes, spikes, or panicles. Sepals 3, usually similar in shape and color; petals 3, lower one (labellum) different from other 2 in shape, size, and colors, other 2 alike; stamens 1-2, adnate to style and stigma; ovary inferior, locule 1, placentas 3, parietal, style and stigma fused with stamens into a columnar gynandrium. Fruit a capsule, sometimes fleshy; seeds many. Ca. 800 genera and 17,500 species; temperate to tropical, worldwide. Many tropical and subtropical species are cultivated as ornamentals.

1. Flowers without spurs; sepals yellowish-green, lateral petals to reddish to purplish; labellum 15-20 mm long ........ **Epipactis**

1' Flowers with basal spurs; sepals and lateral petals green to yellowish green, labellum 2-6 mm long ........ **Piperia**

### EPIPACTIS Zinn

Stems erect, with short, creeping rhizomes. Leaves basal and cauline, narrowly lanceolate to orbicular. Flowers few to many, in bracteate racemes, bracts usually leaf-like. Sepals free, subequal; lateral petals ± like sepals but usually smaller; labellum ± equal in size to gynandrium, spreading, base saccate or pouch-like, apex enlarged and flat, entire to lobed. Fruit obovoid to ellipsoid, spreading to pendulous. Ca. 25 species; mostly temperate, n hemisphere, Africa.

**E. gigantea** Douglas ex Hook. STREAM ORCHID Plants 3-9 dm tall. Leaves 6-15 cm long, lower ones ovate, upper ones lanceolate to oblong, gradually reduced in size upwards. Flowers solitary, in upper leaf axils. Pedicels 4-8 mm long; sepals 12-18 mm long, greenish, often red-veined; lateral petals 12-15 mm long, reddish or purplish; labellum 15-20 mm long, usually yellow, red-veined, apex often 3-lobed. Fruit 2-2.5 cm long, pendulous. Apr-Jun. Scarce; stream banks, freshwater seeps; coastal scrub, riparian communities; 150-1500 ft (45-455 m). North side in Cañada del Agua and at Aguaje Escondido; Central Valley in canyon on sw side of Picacho Diablo and in Cottonwood Canyon; s side in Cañada de Malva Real. British Columbia to Mexico, e to SD and TX. Roots were used medicinally by the Maidu Tribe.

### PIPERIA Rydb. PIPERIA

Stems erect, with short tubers or bulb-like rhizomes. Leaves mostly basal, linear to broadly oblanceolate, cauline ones much reduced, bract-like. Flowers usually many, in bracteate spikes or racemes, bracts much reduced in size. Sepals and lateral petals similar, subequal; labellum with a basal spur, much longer than gynandrium, spreading to reflexed, flat, base constricted, apex acute to rounded. Fruit obovoid to ellipsoid, ± erect. Ca. 8 species; temperate, w N. America. Roots of some species were eaten by the Maidu Tribe.

ORCHIDACEAE

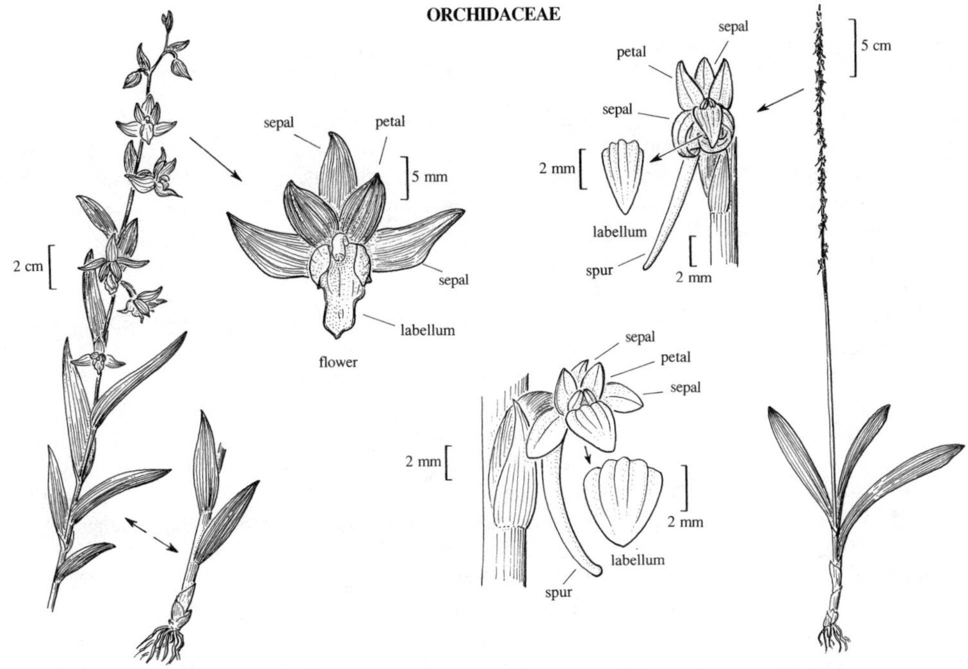


Epipactis gigantea Piperia michaelii P. elongata

1. Labellum narrowly deltate, 1-1.5 mm wide; stem width (below lowest flower) 2-5 mm; lateral petals curved, usually flat .......... **P. elongata**

1' Labellum broadly deltate, 3-4 mm wide; stem width (below lowest flower) 5-8 mm; lateral petals straight, usually concave .......... **P. michaelii**

**P. elongata** Rydb. Plants 1.5-5 dm tall; stem width (just below lowest flower) 2-5 mm. Basal leaves 8-30 cm long. Flowers in spikes, sessile, bracts 2-7 mm long. Perianth green, sepals 2-4 mm long, upper one erect, lateral ones spreading to reflexed; lateral petals usually erect, flat, curved; labellum narrowly deltate, 2-5 mm long, 1-1.5 mm wide; spur 6-12 mm long, slightly curved. Fruit 6-10 mm long, erect. Jun-Jul. Scarce; shaded slopes, canyon bottoms; chaparral, island woodland, pine forest; 300-1300 ft (90-395 m). North side in Pelican and China Pines; Central Valley near Christy Pines and in Islay Canyon; s side near Christy Pines and near Coches Prietos Anchorage. Santa Rosa Island; British Columbia to s CA, e to MT.

**P. michaelii** (Greene) Rydb. MICHAEL'S PIPERIA Plants 2-6 dm tall; stem width (just below lowest flower) 5-8 mm. Basal leaves 8-20 cm long. Flowers in spikes, sessile, bracts 2-10 mm long. Perianth green to yellowish green, sepals 2-5 mm long, upper one ascending to erect, lateral ones spreading; lateral petals usually erect, ± concave, straight; labellum broadly deltate, 3-6 mm long, 3-4 mm wide; spur 8-15 mm long, straight to slightly curved. Fruit 6-12 mm long, erect. Jun. A species of special environmental concern. Rare; coastal bluffs; coastal scrub; ca. 50 ft (15 m). Known from a single collection at extreme s end of Christy Beach in June 1968. Northern CA coast to Santa Barbara Co.

# POACEAE Juss.

## GRASS FAMILY

Annuals or perennial herbs; stems terete, hollow in cross-section. Leaves basal to cauline, usually in 2 ranks, alternate, sheathing stem; sheath margins usually free, overlapping, apices usually with small appendage (ligule) on side facing stem; leaf blades linear, flat to involute. Flowers bisexual or unisexual, radial, 1-many, much reduced, each flower subtended by 2 bracts, an outer lemma and an inner palea, the flower, lemma, and palea comprising a floret; florets 1-many, aggregated into spikelets, spikelet axis (rachilla) terminated by fertile or sterile florets, sometimes prolonged beyond upper floret as appendage, each spikelet subtended by 2 bracts (glumes). Perianth vestigial or absent; stamens 1-3; ovary superior, ovule fused to ovary wall, style branches 2.

Fruit a caryopsis, fusiform to ovoid. Ca. 640 genera and 9000 species; arctic to tropical, worldwide. Some species are critical sources of food, including rice, wheat, and corn. Caryopses of many native taxa were eaten by Native Americans.

1. Spikelets enclosed within upper leaf sheaths, 1-3; plants matted, stems stoloniferous; leaves strongly overlapping, sheaths concealing internodes .......... **Pennisetum**
1' Spikelets usually many, 1-few only on short plants, inflorescences well-developed, conspicuous, usually not concealed by leaf sheaths; leaves not usually strongly overlapping
    2. Plants usually taller than 1.5 m
        3. Inflorescence spikate or paniculate and narrow; spikelets 2-many in clusters at nodes; ligule margin entire to erose, but not pubescent .......... **Leymus**
        3' Inflorescence a plumose panicle; spikelets solitary; ligule margins pubescent
            4. Leaf blade margins entire, smooth; plants and spikelets bisexual .......... **Arundo**
            4' Leaf blade margins serrulate; plants dioecious, spikelets unisexual .......... **Cortaderia**
    2. Plants usually shorter than 1.5 m
        5. Most spikelets sessile, sometimes sunken in inflorescence axis; inflorescence either solitary and spikate or paniculate and composed of spikate branches; spikelets often in 2 rows on opposite sides of axis or borne along 1 side of inflorescence axis .......... **GROUP 1**
        5' Most spikelets short- to long-pedicellate; inflorescence usually racemose or paniculate, open, (if appearing spikate then spikelets short-pedicellate); spikelets evenly distributed around inflorescence axis, solitary or in small clusters, but not in 2 rows or along 1 side of axis
            6. Most or all spikelets with 1 fertile floret, sometimes with sterile florets much reduced in size or different from fertile ones .......... **GROUP 2**
            6' Most or all spikelets with 2-many fertile florets .......... **GROUP 3**

### GROUP 1. Most fertile spikelets sessile or sunken in inflorescence axis, often in 2 rows on opposite sides of axis or borne along 1 side of inflorescence axis.

1. Inflorescence composed of 2-7 spikate branches, each with many spikelets; spikelets evenly distributed around inflorescence axis or borne along 1 side of axis
    2. Inflorescences usually 5-15 per stem, appearing plumose, axis ± densely pubescent; spikelets evenly distributed around inflorescence axis, fertile ones sessile, each subtended by naked pedicel (vestigial spikelet); lemma awned .......... **Andropogon**
    2' Inflorescence umbellate with either 2-7 branches, axes usually glabrous; spikelets all alike, not in pairs, borne along 1 side of inflorescence axis; lemmas not awned
        3. Inflorescence umbellate, usually composed of 3-5 branches; lemma margins ciliolate; most vegetative stems prostrate .......... **Cynodon**
        3' Inflorescence composed of 2 branches; lemma margins glabrous; most vegetative stems erect .......... **Paspalum**
1' Inflorescence solitary; spikelets usually in 2 rows on opposite sides of axis or solitary
    4. Spikelets in clusters of 2-3 at each node (sometimes 1 in *Leymus*)
        5. Each spikelet with 1 floret; central spikelet bisexual, producing fruit, lateral 2 spikelets smaller than central one, usually sterile or staminate (fertile in *H. vulgare*) .......... **Hordeum**
        5' Each spikelet with 2-6 florets; spikelets bisexual, all alike
            6. Lemma awns 12-30 mm long; glumes lanceolate .......... **Elymus**
            6' Lemmas acute to short-awned, awns less than 4 mm long, glumes linear to subulate **Leymus**
    4' Spikelets solitary at each node
        7. Spikelets sunken in inflorescence axis; each spikelet with 1 bisexual floret; plants of coastal, often moist and saline habitats
            8. Inflorescence straight, 8-20 cm long; most spikelets with 1 glume .......... **Hainardia**
            8' Inflorescence curved, 3-12 cm long; most spikelets with 2 glumes, both on outer side of spikelet, their margins often overlapping .......... **Parapholis**
        7. Spikelets sessile to subsessile, not usually sunken in inflorescence axis; each spikelet with 2-many bisexual florets; plants of interior, often dry habitats

9. Spikelets in 2 rows on opposite sides of axis
   10. Flat surface of compressed spikelet facing inflorescence axis; awns 20-70 mm long; inflorescence axis breaking apart in fruit ........ **Aegilops**
   10' Edge of compressed spikelet facing inflorescence axis; longest awns 1.5-15 mm long; inflorescence axis not breaking apart in fruit; each spikelet with 1 glume (on outer edge of spikelet) ..... **Lolium**

9' Spikelets not in 2 rows
   11. Spikelets many; plants strongly rhizomatous; leaves in 2 conspicuous ranks; lemmas awnless ........ **Distichlis**
   11' Spikelet 1; annuals or perennial herbs without extensive rhizomes; leaves not conspicuously ranked; lemmas awned
      12. Lemmas apices acute, awned from tip; palea margins conspicuously stiff-ciliate ........ **Brachypodium**
      12' Lemma apices bidentate, awned from between teeth; palea margins glabrous to ciliolate, but not stiff-ciliate ........ **Bromus**

### GROUP 2. Spikelets pedicellate, evenly distributed around inflorescence axis, solitary or in small clusters; most or all spikelets with 1 fertile floret.

1. Spikelets in distinct clusters that fall together; clusters composed of 1 fertile spikelet subtended or surrounded by 1-5 sterile spikelets
   2. Spikelets spreading to nodding; fertile spikelet with lemmas awned from bidentate apices; plants 10-40 cm tall ........ **Lamarckia**
   2' Spikelets erect; fertile spikelet awnless, lemma apices acute to acuminate; plants 30-100 cm tall ........ **Phalaris paradoxa**

1' Spikelets not in distinct clusters; spikelets all alike, disarticulating either below glumes or between glumes and fertile floret
   3. Spikelets disarticulating below glumes, falling entire from their pedicels
      4. Glumes and lemmas awnless ........ **Agrostis viridis**
      4' Glumes with awns 2-8 mm long; lemmas with awns 1-4 mm long ........ **Polypogon**
   3' Spikelets usually disarticulating above glumes, floret falling from spikelet
      5. Lemma of fertile floret awnless
         6. Inflorescence spikate, dense, cylindrical; lemma of fertile floret puberulent ........ **Phalaris**
         6' Inflorescence usually paniculate, open; lemma of fertile floret glabrous to minutely scabrous
            7. Glumes 1-veined, apices acute ........ **Agrostis**
            7' Lower glume 3-5-veined, upper one 1-3-veined, apices obtuse to rounded ........ **Melica**
      5' Lemma of fertile floret awned dorsally near base, near middle, or from apex
         8. Lemma awn arising near base; lemma base densely puberulent to villosulous ........ **Calamagrostis**
         8' Lemma awn arising dorsally between middle and apex or from apex; lemma base puberulent or glabrous
            9. Glumes much shorter than lemma ........ **Muhlenbergia**
            9' Glumes as long as or longer than lemma (except awn, which often is exserted from spikelet)
               10. Glumes swollen at bases; inflorescence dense, spikate ........ **Gastridium**
               10' Glumes not swollen at bases; inflorescence open to contracted. but not dense and spikate
                  11. Awn less than 5 mm long; lemmas ± thin, often translucent
                     12. Lemmas 5-veined; palea minute; awn less than 1 mm long, arising slightly below apex ........ **Agrostis**
                     12' Lemmas obscurely veined; palea ± equal to lemma; awn 2.5-4 mm long, continuous with apex ........ **Piptatherum**
                  11' Awn 10-110 mm long; lemmas thick, becoming hard in fruit
                     13. Awn divided into 3 branches ........ **Aristida**
                     13' Awn simple, bent 1-2 times
                        14. Inflorescence open, branches spreading to nodding; lemma slightly constricted near apex; floret callus sharply acute ........ **Nassella**
                        14' Inflorescence dense, branches appressed; lemma tapered to apex; floret callus obtuse ........ **Achnatherum**

**GROUP 3. Spikelets pedicellate, evenly distributed around inflorescence axis, solitary or in small clusters; most or all spikelets with 2-many fertile florets.**

1. One or both glumes usually as long as or longer than lowest floret, sometimes as long as spikelet; lemmas dorsally awned or awnless
    2. Lemmas awned from near middle, awns 15-40 mm long; glumes 15-30 mm long; lemmas 10-25 mm long ........ **Avena**
    2' Lemmas acute to short-awned from apex, awns less than 1.5 mm long; glumes 2-12 mm long; lemmas 2-12 mm long, acute to short-awned, awn less than 1.5 mm long
        3. Plants 20-70 dm tall; leaf blades 2-5 cm wide; lemmas villous ........ **Arundo**
        3' Plants less than 6 dm; leaf blades less than 10 cm wide; lemmas glabrous or ciliate
            4. Lemmas 3-5-veined, apices acute, margins usually glabrous ........ **Koeleria**
            4' Lemmas 7-9-veined, apices notched, margins ciliate ........ **Schismus**

1' Glumes usually shorter than lowest floret; lemmas awned from apices or awnless
    5. Stems usually matted, with scaly rhizomes; inflorescences unisexual, staminate ones usually exceeding leaves, pistillate ones equal to or shorter than uppermost leaves ........ **Distichlis**
    5' Stems not matted, rhizomes if present, not scaly; inflorescences bisexual
        6. Lemma apices rounded to acute, awnless ........ **Poa**
        6' Lemmas awned from tip, awns 0.5-55 mm long (sometimes absent in *Festuca*)
            7. Spikelets strongly compressed; lemmas dorsally keeled ........ **Bromus**
            7' Spikelets not strongly compressed, lemmas dorsally rounded
                8. Lemmas lanceolate to narrowly ovate, apices bidentate, awned from between teeth ........ **Bromus**
                8' Lemmas narrowly lanceolate to attenuate, tapered to awned tip
                    9. Perennial herbs, with short rhizomes; awns 0.5-1.5 mm long, sometimes absent ........ **Festuca**
                    9' Annuals; awns 2-12 mm long
                        10. Ligules puberulent; palea margins conspicuously stiffly ciliate; spikelets 1-3 per inflorescence ........ **Brachypodium**
                        10' Ligules glabrous to minutely scabrous; palea margins glabrous to ciliolate, but not stiffly ciliate; spikelets usually many per inflorescence ........ **Vulpia**

## ACHNATHERUM Pal.

Perennial herbs; stems usually tufted. Leaves mostly basal; ligules membranous; leaf blades usually involute. Inflorescence paniculate, contracted, branches ascending to appressed; spikelets terete, pedicellate; floret 1; rachilla disarticulating above glumes, callus blunt. Glumes equal to unequal, usually longer than lemma, 1-5-veined, apices acute to attenuate; lemmas coriaceous, obscurely 5-veined, tapered to acute apex, awned apices, junction between lemma and awn distinct, awns bent 1-2 times, lower segment often twisted.

**A. diegoense** (Swallen) Barkworth [*Stipa d.* Swallen] SAN DIEGO NEEDLE GRASS. Plants 7-20 dm tall. Leaf blades 2-4 mm wide, flat to involute. Inflorescence dense, branches appressed. Glumes subequal, 7-12 mm long, attenuate; lemmas (excluding awns) 5-7 mm long, pilose, awns 20-35 mm long; palea 1.5-3 mm long. Mar-Jun. A species of special environmental concern. Common; coastal bluffs, coastal flats, rocky s-facing slopes, dry ravines; coastal scrub, grasslands; 30-1400 ft (9-425 m). North side between Pelican Bay and Prisoners' Harbor; Central Valley between Christy Ranch and Cañada Cebada, and in Islay Canyon; scattered locations on s side between Rancho Nuevo Canyon and Sandstone Point. San Miguel, Santa Rosa, Anacapa, and San Nicolas islands; Los Coronados and Todos Santos islands; San Diego Co. to Baja CA.

## AEGILOPS L. GOATGRASS

Annuals. Leaves mostly cauline; ligules membranous; junction between sheath and blade with or without auricles; leaf blades flat to slightly involute. Inflorescence spikate, axis continuous or breaking apart at nodes in fruit; spikelets in 2 opposite rows, 1 per node, sessile, strongly appressed to rachis, compressed, rachilla disarticulating above glumes; florets 2-5. Glumes oblong to obovate, subequal, coriaceous, dorsally flat to rounded, apices mucronate to awned; lemmas 5-many-veined, membranous to coriaceous, dorsally rounded, apices mucronate to awned. Ca. 20 species; temperate, Europe, w Asia, and Africa.

**A. cylindrica** Host JOINTED GOATGRASS Plants 3-6 dm tall. Auricles minute or absent; leaf blades 2-5 mm wide. Inflorescence cylindrical, rachis breaking apart at nodes in fruit; spikelets falling separately with adjacent internode. Glumes 6-11 mm long, keeled along one side, keel extended as an apical awn, awns of lowermost spikelets 0.5-2.5 cm long, those of uppermost ones 4-7 cm long; lemma apices mucronate to awned, awns, if present, 0.5-5 cm long. May-Jun. Introduced. Rare; disturbed flats, grassy swales, hillsides; coastal scrub, grasslands; 250-300 ft (75-90 m). Central Valley near Cottonwood Canyon and University of California Field Station. Throughout w N. America; native of Europe. First collected on hillside se of Field Station in June 1973.

## **AGROSTIS** L. BENT GRASS

Annuals or perennial herbs, often with rhizomes or stolons. Leaves mostly basal; ligules membranous; leaf blades flat to involute. Inflorescence paniculate to spikate, cylindrical to ovoid. Spikelets compressed to terete, pedicellate to sessile, rachilla usually disarticulating above glumes; floret 1. Glumes equal to unequal, equal to or longer than floret, 1-veined, apices acute to acuminate; lemmas 3-5-veined, thin, usually hyaline, often dorsally awned, awn straight to bent, callus glabrous to densely puberulent, apices truncate to erose; paleas absent or as long as lemma. Ca. 220 species; temperate, worldwide.

1. Disarticulation below glumes, spikelet falling entire; palea evident, slightly shorter than lemma; stems often stoloniferous .......... **A. viridis**

1' Disarticulation above glumes, floret falling from spikelet; palea minute; stems not stoloniferous, tufted, with or without rhizomes

  2. Inflorescence narrow, spikate, branches ascending to erect, with spikelets to near branch bases .......... **A. exarata**

  2' Inflorescence narrow to ± open, most branches ascending, lowest ones spreading, without spikelets at branch bases .......... **A. pallens**

**A. exarata** Trin. Perennial herbs, 1-10 dm tall; stems tufted, without rhizomes. Ligules 2-5 mm long; leaf blades 2-9 mm wide. Inflorescence narrow, spikate, branches ascending to erect, with spikelets to near their bases. Glumes 2-3.5 mm long; lemmas 1-2 mm long, awnless or dorsally short-awned near apices; paleas minute. Jun-Aug. Jun-Sep. Rare; coastal bluffs, shaded slopes, canyon bottoms, freshwater seeps; coastal scrub; 20-300 ft (6-90 m). North side near Fry's Harbor and Orizaba Cove. Santa Rosa and Santa Catalina islands; AK to Mexico, e to Great Plains.

**A. pallens** Trin. [*A. diegoensis* Vasey] Perennial herbs, 5-10 dm tall; stems solitary to tufted, with rhizomes. Ligules 1.5-3 mm wide; leaf blades 2-6 mm wide. Inflorescence narrow to ± open, most branches ascending, lowest ones spreading, without spikelets at their bases. Glumes 2-3 mm long; lemmas 1-3 mm long, awnless, sometimes dorsally short-awned near apices; paleas minute. Apr-Aug. Occasional; coastal bluffs, shaded n-facing slopes, ridgetops; chaparral, oak woodland, pine forest; 20-1300 ft (6-395 m). North side between Cueva Valdez and Prisoners' Harbor, and near China Harbor; Central Valley between Cañada Christy and Stanton Ranch; s side between Cañada de los Sauces and Coches Prietos drainage. Santa Rosa, Santa Catalina, and San Clemente islands; British Columbia to s CA, e to Rocky Mountains.

**A. viridis** Gouan [*A. semiverticillata* (Forssk.) C.Chr., *Polypogon s.* (Forssk.) H.Hyl.] Perennial herbs, sometimes flowering first year and appearing annual, 1-60 dm tall; stems decumbent, sometimes trailing and rooting at nodes. Ligules 1-5 mm long; leaf blades 2-8 mm wide. Inflorescence narrow, spikate, branches appressed to erect, with spikelets to near their bases. Glumes 1-2.5 mm long; lemmas 1-1.5 mm long, awnless; palea slightly shorter than lemma. May-Aug. Introduced. Occasional; stream banks, moist floodplains, freshwater seeps; riparian communities; 20-1000 ft (6-305 m). Widely scattered locations throughout island. Santa Rosa, Anacapa, Santa Barbara, and Santa Catalina islands; Cedros Island; throughout N. America; native of Europe. First collected in Valdez Canyon in June 1930.

## **ANDROPOGON** L. BLUESTEM

Annuals or perennial herbs. Leaves mostly cauline; ligules membranous, ciliolate; leaf blades flat to folded. Inflorescence paniculate, composed of 2-5 spikate branches. Spikelets compressed, in pairs, lower one sessile and fertile, upper one short-pedicellate and often sterile, pair falling together; florets 2, lower one minute, upper one fertile. Glumes of fertile spikelet equal, lower one often keeled, 2-5-veined, thick, longer than florets; lemmas hyaline, with or without straight to bent awn at apices; sterile spikelet modified, reduced, often pubescent, sometimes absent. Ca. 100 species; warm temperate to tropical, worldwide.

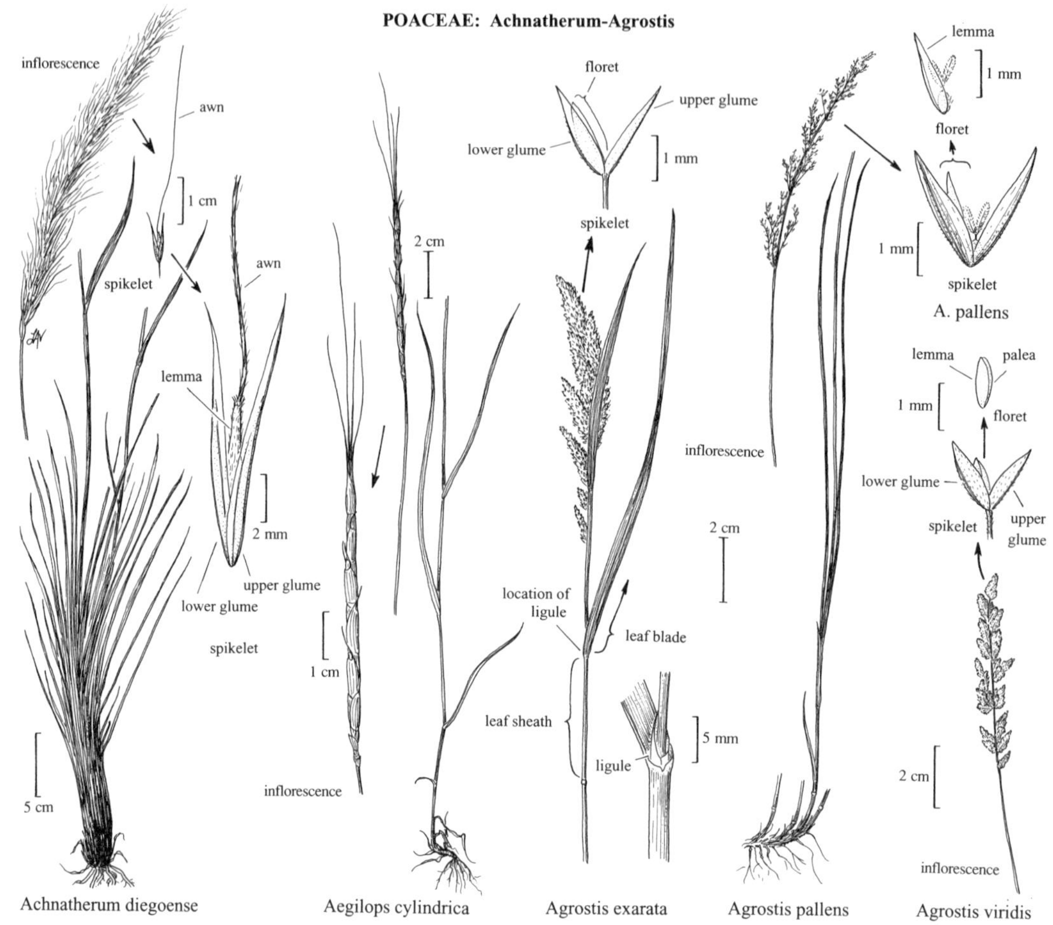


**A. glomeratus** (Walter) Britton, Sterns, & Poggenb. var. **scabriglumis** C.S.Campb. BUSHY BLUESTEM Perennial herbs, 5-10 dm tall; stems densely tufted. Leaf blades 3-8 mm wide, upper leaves each subtending and exceeding an inflorescence, often becoming reddish with age. Inflorescences 5-15 per stem, usually axillary, branches 1-3 cm long, internodes and pedicels densely pubescent, trichomes 4-6 mm long. Spikelets 3-5 mm long, fertile lemma awned, awn 0.5-1 cm long; sterile spikelet minute, reduced to pubescent pedicel 2-3 mm long. Sep-Apr. Rare; stream banks, freshwater seeps; riparian communities; 20-500 ft (6-150 m). North side in Cañada del Puerto; Central Valley in Cottonwood Canyon. Mojave Desert and Transverse Ranges (introduced elsewhere in CA), e to CO and NM. Four other varieties occur in w N. America.

## ARISTIDA L. THREE-AWN

Annuals or perennial herbs. Leaves mostly basal, ligule a ring of hairs or membranous and ciliolate; leaf blades usually involute. Inflorescence racemose or paniculate, branches spikate. Spikelets compressed to terete, short-pedicellate to sessile, rachilla disarticulating above glumes; floret 1. Glumes linear-lanceolate, unequal, equal to or longer than floret, usually 1-veined, apices acute to awn-tipped; lemmas hard, 3-veined, callus sharp-pointed, apices attenuate, tapered to 3-branched awn. Ca. 250 species; warm temperate to tropical, worldwide.

1. Annuals; inflorescences narrow, 3-10 cm long, primary branches all ascending to erect, bearing spikelets to near bases ........ **A. adscensionis**

1' Perennial herbs; inflorescences open, 15-50 cm long, primary branches widely spreading to drooping, naked ........ **A. ternipes** var. **hamulosa**

**A. adscensionis** L. Annuals, 15-30 cm tall. Leaf blades ca. 1 mm wide. Inflorescence narrow, 3-10 cm long, primary branches all ascending to erect, bearing spikelets to near bases. Glumes unequal, lower one 5-7 mm long, upper one 6-10 mm long; lemmas 6-10 mm long, awn branches equal, 6-12 mm long. Feb-Jun. Occasional; coastal bluffs, rocky s-facing slopes, exposed ridgetops, floodplains; chaparral, coastal scrub, grasslands; 100-1800 ft (30-550 m). North side in Cañada del Puerto; Central Valley between Cañada de la Portezuela and Stanton Ranch; s side between Laguna Canyon and Valley Anchorage, and in Smugglers' Canyon. Santa Rosa, Santa Catalina, and San Clemente islands; Guadalupe and Cedros islands; Monterey Co. to n Baja CA, e to Great Plains.

**A. ternipes** Cav. var. **hamulosa** (Henrard) J.S.Trent [*A. hamulosa* Henrard] Perennial herbs, 30-90 cm tall. Leaf blades ca. 1 mm wide. Inflorescence open, 15-50 cm long, primary branches widely spreading to drooping, naked. Glumes subequal, 8-14 mm long; lemmas 10-12 mm long, awn branches subequal, central one 12-22 mm long, lateral ones 8-29 mm long. Mar-Jul (Sep). Possibly introduced (Smith 1976). Scarce; rocky s-facing slopes, exposed ridgetops; chaparral, coastal scrub, grasslands; 250-950 ft (75-290 m). Central Valley between La Cascada and Islay Canyon; s side in Coches Prietos drainage. San Joaquin Valley to Mexico, e to TX.

## ARUNDO L.

Robust perennial herbs; stems cane-like. Leaves cauline; ligules membranous, ciliate; leaf blades flat to folded. Inflorescence paniculate, plumose, contracted, branches usually ascending. Spikelets bisexual, ± compressed, pedicellate; florets 4-5, rachilla disarticulating above glumes. Glumes subequal, longer than florets, linear-lanceolate, thin, 3-5-veined; lemmas 3-5-veined, villous, awnless. Ca. 3 species; warm temperate to subtropical, Europe and Asia.

**A. donax** L. GIANT REED Plants 1-2 m tall. Leaf sheaths glabrous, blades 2-5 cm wide, glabrous, grayish green, margins entire. Inflorescence 3-5 dm long, white to purplish. Glumes 10-12 mm long; lemmas 8-12 mm long, villous, trichomes silky. Flowering time unknown. Introduced. Rare; moist sandy flats, floodplains; riparian communities, southern beach and dune; 5-10 ft (1-3 m). Known only from mouth of Cañada Christy, where it was first collected in October 1992. San Miguel, Santa Rosa, and San Nicolas islands; Cedros Island; San Francisco Bay region to Baja CA, e to UT and NM; native of Europe. During heavy storms, rootstocks are washed out to sea and float across the Santa Barbara Channel; they have been found on beaches throughout the Channel Islands and occasionally establish new colonies.

## AVENA L. OAT

Annuals. Leaves basal and cauline; ligules membranous; leaf blades flat. Inflorescence racemose or paniculate, lower branches usually spreading to drooping. Spikelets compressed to terete, pedicellate, florets 2-5, rachilla disarticulating above glumes and between florets. Glumes subequal, usually longer than florets, 5-7-veined, apices acute to acuminate; lemmas coriaceous, 5-7-veined, dorsally rounded, dorsally awned near middle, awn straight to bent, apices bidentate, teeth setaceous to acute. Ca. 25 species; temperate, Europe and Asia.

1. Lemmas pubescent throughout, apical teeth setaceous, 2-5 mm long .......... **A. barbata**

1' Lemmas subglabrous to glabrous over upper half, apical teeth acute, 0.5-1 mm long

  2. Florets usually 3-4 per spikelet; lemmas usually pubescent below middle, trichomes reddish brown, awns 22-40 mm long, bent .......... **A. fatua**

  2' Florets usually 2 per spikelet; lemmas subglabrous, awns 15-25 mm long, straight or absent. **A. sativa**

**A. barbata** Link SLENDER WILD OATS Plants 2.5-6 dm tall. Leaf blades 2-5 mm wide. Glumes 17-28 mm long; florets usually 2; lemmas 11-17 mm long, pubescent throughout, trichomes reddish brown, awns bent, 22-40 mm long, apical teeth setaceous, 2-5 mm long. Mar-Jun. Introduced. Abundant; disturbed sites, rocky slopes, ravines; coastal scrub, grasslands; 10-2470 ft (3-755 m). Scattered locations throughout island. All CA Channel islands; Los Coronados, Todos Santos, Guadalupe, and Cedros islands; throughout w N. America; native of Europe. First collected (without specific locality) in March 1925.

**A. fatua** L. WILD OATS Plants 3-8 dm tall. Leaf blades 2.5-7 mm wide. Glumes 18-25 mm long; florets usually 3-4; lemmas 15-20 mm long, pubescent below middle, trichomes reddish brown, glabrous above, awns bent, 22-40 mm long, apical teeth acute, 0.5-1 mm long. Apr-Jun. Introduced. Common; disturbed sites, rocky slopes, ravines; coastal scrub, grasslands; 10-2470 ft (3-755 m). Scattered locations throughout island. All CA Channel islands; Los Coronados, Todos Santos, San Martin, Guadalupe, and Cedros islands; throughout w N. America; native of Europe. Reported by Greene (1887).

**A. sativa** L. CULTIVATED OATS Plants 3-6 dm tall. Leaf blades 4-10 mm wide. Glumes 17-28 mm long; florets usually 2; lemmas 12-23 mm long, subglabrous, awns usually straight, 15-25 mm long, sometimes absent, apical teeth acute, 0.5-1 mm long. Apr-Jun. Introduced. Rare; disturbed sites, abandoned pastures; grasslands;

**POACEAE: Andropogon-Brachypodium**

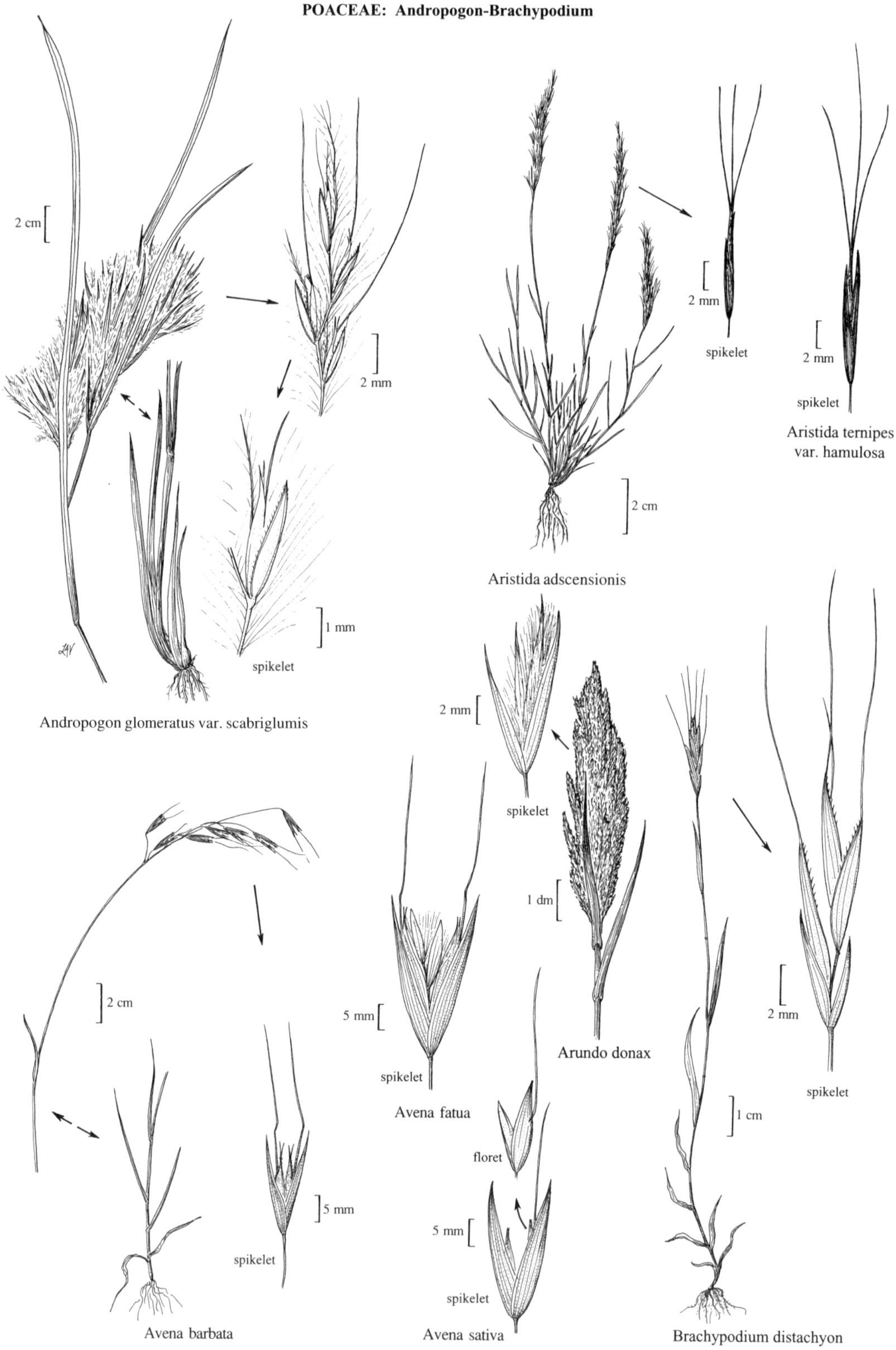

ca. 250 ft (75 m). Central Valley near University of California Field Station, where it was first collected in April 1967. Santa Rosa, Santa Catalina, and San Clemente islands; Guadalupe Island; native of Europe. This species is the source of oatmeal, rolled oats, and grains used to brew beer.

## **BRACHYPODIUM** Pal

Annuals or perennial herbs. Leaves basal and cauline; ligules membranous; leaf blades flat to involute. Inflorescence racemose or spikate. Spikelets compressed to terete, short-pedicellate to sessile; florets 7-many, rachilla disarticulating above glumes and between florets. Glumes unequal, shorter than florets, 5-9-veined, apices acute to awned; lemmas 7-9-veined, dorsally rounded, apices acute to awned, awns straight; palea margins usually stiff-ciliate. Ca. 15 species; temperate to tropical, worldwide.

**B. distachyon** (L.) P.Beauv. FALSE BROME Annuals, 10-20 cm tall. Ligules puberulent; leaf blades 2-5 mm wide, cauline leaves shorter than basal ones. Spikelets 1-3, sessile or short-pedicellate, pedicels thick, stiff. Lower glume 3-5 mm long, acute, upper one 5-7 mm long, acute to short-awned; florets 7-20; lemmas 6-9 mm long, awns 5-11 mm long; palea margins ciliate. May-Oct. Introduced. Scarce; coastal bluffs, disturbed sites, floodplains, rocky s-facing slopes, grassy flats; coastal scrub, grasslands, riparian communities; 20-1450 ft (6-440 m). North side near mouth of Scorpion Canyon; Central Valley between La Cascada and Bosque Mano; s side on South Ridge near Peak 1498, in Horqueta Canyon, and near mouth of Smugglers' Canyon. Santa Catalina Island; throughout CA; native of Europe. This species has spread rapidly since it was first collected on the South Ridge in May 1991.

## **BROMUS** L. BROME

Annuals or perennial herbs. Leaves basal and cauline; sheath margins fused along most of their length; ligules membranous; leaf blades usually flat. Inflorescence racemose to paniculate; spikelets sometimes solitary. Spikelets strongly compressed to terete, pedicellate; florets 5-many, rachilla disarticulating above glumes and between florets. Glumes unequal to subequal, shorter than lower floret, usually awnless, apices usually acute, lower glume usually 1-5-veined, upper one 3-9-veined; lemmas 5-7-veined, dorsally rounded to keeled, apices bidentate, awned from between teeth. Ca. 150 species; temperate, worldwide.

1. Spikelets strongly compressed; glumes and lemmas dorsally keeled
   2. Upper glume ± equal in length to lower lemma (excluding awn); lemma margins puberulent below middle; annuals ........ **B. arizonicus**
   2' Upper glume usually shorter than lower lemma (excluding awn); lemmas glabrous to puberulent or minutely scabrous throughout
      3. Lower inflorescence branches stiffly ascending; fruit compressed; lemma ± winged on both sides between outer veins and margin, wing ca. 1 mm wide ........ **B. stamineus**
      3' Lower inflorescence branches widely spreading to ± ascending; fruit cylindrical; lemmas not winged between outer veins and margin
         4. Lemma awns 4-15 mm long; glumes and lemmas glabrous to puberulent ........ **B. carinatus**
         4' Lemma awns 1-3 mm long; glumes and lemmas usually minutely scabrous .... **B. catharticus**

1' Spikelets terete to slightly compressed; glumes and lemmas dorsally rounded
   5. Lemma awns bent, distal part divergent ........ **B. berterianus**
   5' Lemma awns straight, ascending to erect, not divergent
      6. Inflorescence somewhat open to dense, branches ascending to erect
         7. Lemma awns 4-9 mm long, apical teeth 0.5-1.5 mm long; spikelets greenish to tan ........ **B. hordeaceus**
         7' Lemma awns 12-25 mm long, apical teeth 2-3.5 mm long; spikelets becoming reddish ........ **B. madritensis**
      6' Inflorescence open, lower branches spreading to drooping
         8. Lemma awns 25-55 mm long; lower glume 10-20 mm long; annuals ........ **B. diandrus**
         8. Lemma awns 3-5 mm long; lower glume 5-9 mm long; perennial herbs ........ **B. laevipes**

**B. arizonicus** (Shear) Stebbins ARIZONA BROME Annuals, 4-8 dm tall. Leaf blades 2-5 mm wide, glabrous to sparsely puberulent. Inflorescence 5-25 cm long, usually paniculate, ± dense, branches ascending, pedicels 5-25 mm long. Spikelet strongly compressed. Glumes dorsally keeled, glabrous, margins puberulent below middle, lower one 8-12 mm long, 3-5-veined, upper one 10-15 mm long; florets 7-20; lemmas 10-13 mm long, margins puberulent below middle, awns 7-10 mm long. Apr-May. Rare, not seen recently; habitat, plant community, and elevation unknown. Known from a single collection by T.S. Brandegee in 1888. All CA Channel islands; San Joaquin Valley to s CA, e to AZ.

**B. berterianus** Colla [*B. trinii* Desv.] CHILEAN CHESS Annuals, 2-8 dm tall. Leaf blades 2-7 mm wide, villosulous. Inflorescence 4-20 cm long, racemose to paniculate, open, lower branches spreading to drooping, pedicels 5-15 mm long. Spikelets slightly compressed. Glumes dorsally rounded, glabrous to sparsely puberulent, lower one 9-13 mm long, 1-veined, upper one 10-15 mm long; florets 3-9; lemmas 8-15 mm long, villosulous, apical teeth 3-5 mm long, awns 15-20 mm long, bent. Mar-May. Possibly Introduced. Rare, not seen recently; disturbed sites, burned areas; chaparral; elevation unknown. North side near Pelican Bay, where it was last seen in April 1932. All CA Channel islands; San Martin, Guadalupe, and Cedros islands; n CA to Baja CA, e to CO and n Mexico; believed to be native of Chile. Reported by Brandegee (1888).

**B. carinatus** Hook. & Arn. CALIFORNIA BROME. Short-lived perennial herbs, 4-20 dm tall. Leaf blades 4-10 mm wide, glabrous to villosulous. Inflorescence 5-30 cm long, usually paniculate, open to ± contracted, lower branches ascending to spreading, pedicels 2-24 mm long. Spikelet strongly compressed. Glumes dorsally keeled, glabrous to puberulent throughout, lower one 5-12 mm long, usually 3-veined, upper one 10-15 mm long; florets 7-12; lemmas 12-15 mm long, glabrous to densely puberulent, awns 4-15 mm long. Mar-Aug. Common; coastal bluffs, grassy hillsides, rocky slopes, canyon bottoms; coastal scrub, grasslands, oak woodland, pine forest; 10-1000 ft (3-305 m). Widely scattered locations throughout island. All CA Channel islands except Santa Barbara; Los Coronados, Todos Santos, and San Martin islands; AK to Baja CA, e to CO and TX.

1. Inflorescence open, lower branches widely spreading, usually longer than spikelets, pedicels 5-24 mm long; most spikelets not overlapping; plants of interior habitats ........ var. **carinatus**

1' Inflorescence ± contracted, lower branches ascending, usually shorter than spikelets, pedicels 2-8 mm long; most spikelets overlapping; plants of coastal habitats ........ var. **maritimus**

var. **carinatus** Inflorescence open, lower branches widely spreading, usually longer than spikelets, pedicels 5-24 mm long; most spikelets not overlapping. Mar-Aug. Common; coastal bluffs, grassy hillsides, rocky n-facing slopes, canyon bottoms; coastal scrub, grasslands, oak woodland, pine forest; 10-1000 ft (3-305 m). Widely scattered locations throughout island. All CA Channel islands except Santa Barbara; AK to Baja CA, e to CO and TX.

var. **maritimus** (Piper) C.L.Hitchc. [*B. maritimus* (Piper) C.L.Hitchc.] Inflorescence ± contracted, lower branches ascending, usually shorter than spikelets, pedicels 2-8 mm long; most spikelets overlapping. Jul-Aug. Scarce; coastal bluffs, rocky slopes; coastal scrub; 10-250 ft (3-75 m). South side in Coches Prietos drainage. All CA Channel islands except Santa Barbara and San Clemente; Los Coronados, Todos Santos, and San Martin islands; coastal OR to Santa Barbara Co.

**B. catharticus** Vahl [*B. willdenovii* Kunth] RESCUE GRASS Annuals or short-lived perennial herbs, 4.5-6 dm tall. Leaf blades 5-10 mm wide, minutely scabrous to puberulent. Inflorescence 12-30 cm long, usually paniculate, open, lower branches spreading, pedicels 5-20 mm long. Spikelet strongly compressed. Glumes dorsally keeled, usually minutely scabrous, lower one 8-11 mm long, usually 5-7-veined, upper one 9-12 mm long; florets 5-9; lemmas 10-12 mm long, minutely scabrous, awns 1-3 mm long. Apr. Introduced. Rare; disturbed sites; grasslands; ca. 100 ft (30 m). Known from a single collection at Christy Ranch in April 1987. San Nicolas and San Clemente islands; throughout N. America and Europe; native of S. America.

**B. diandrus** Roth [*B. rigidus* Roth] RIPGUT BROME Annuals, 2-7 dm tall. Leaf blades 3-8 mm wide, villosulous. Inflorescence 5-30 cm long, usually paniculate, open, lower branches spreading to drooping, pedicels 20-40 mm long. Spikelet compressed. Glumes dorsally rounded, glabrous to minutely scabrous, lower one 10-20 mm long, 1-veined, upper one 16-30 mm long; florets 5-9; lemmas 15-30 mm long, minutely scabrous, apical teeth 2.5-7 mm long, awns 25-55 mm long. Apr-Jun. Introduced. Common; disturbed sites, floodplains, grassy hillsides; many plant communities; 10-500 ft (3-150 m). Widely scattered locations throughout island. All CA Channel islands; Los Coronados, Todos Santos, and Guadalupe islands; British Columbia to S. America; native of Europe. First collected (without specific locality) in May 1929; reported as abundant by Hoffmann (1932a).

**B. hordeaceus** L. [The name *B. mollis* L. has been misapplied to this species.] SOFT CHESS Annuals, 1-5 dm tall. Leaf blades 1-4 mm wide, villosulous. Inflorescence 1-10 cm long, racemose to paniculate, usually dense, branches ascending to erect, sometimes composed of only 1-3 spikelets, pedicels 2-12 mm long. Spikelets compressed. Glumes dorsally rounded, glabrous to short-pubescent, lower one 5-8 mm long, usually 3-veined, upper one 5-9 mm long; florets 3-9; lemmas 6-10 mm long, glabrous to short-pubescent, apical teeth 0.5-1.5 mm long, awns 4-9 mm long. Apr-Aug. Introduced. Common; disturbed sites, grassy hillsides; coastal scrub, grasslands;

**POACEAE: Bromus**

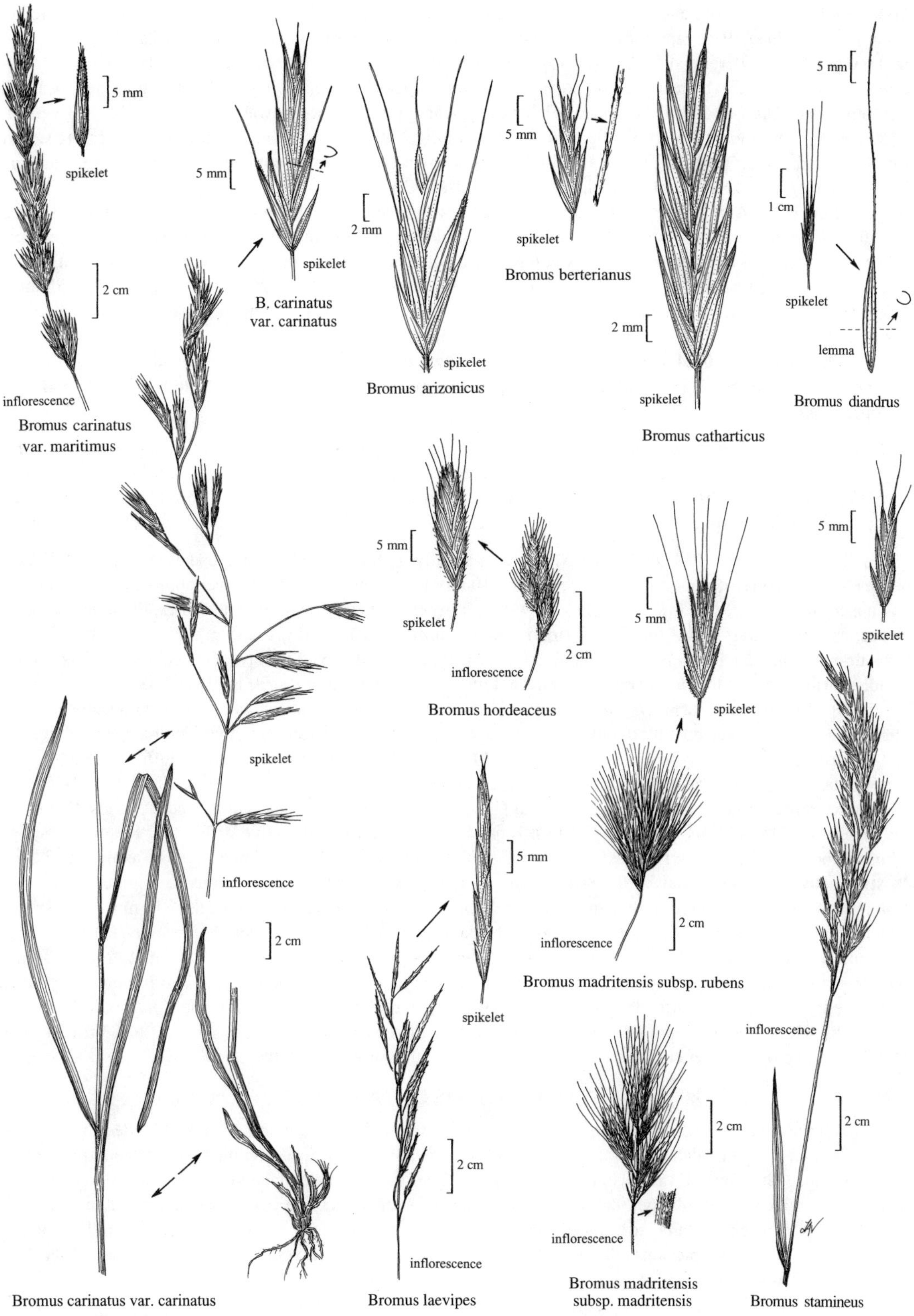

30-1250 ft (10-380 m). Widely scattered locations throughout island. All CA Channel islands; Los Coronados, Todos Santos, San Martin, and Guadalupe islands; throughout N. America; native of Europe. First collected at Main Ranch in June 1930; reported as "common and widely distributed" by Hoffmann (1932a).

**B. laevipes** Shear Perennial herbs, 4.5-6 dm. Leaf blades 2-6 mm wide, glabrous to puberulent. Inflorescence 6-10 cm long, usually racemose, open, lower branches spreading to drooping. Spikelet compressed, pedicels 3-20 mm long. Glumes dorsally rounded, glabrous to sparsely puberulent, lower one 5-9 mm long, 3-veined, upper one 5-10 mm long; florets 5-7; lemmas 8-13 mm long, awns 3-5 mm long. Jun. Rare, not seen recently; rocky wooded slopes; pine forest; elevation unknown. Known from a single collection in Christy Pines in June 1930. Santa Rosa and Santa Catalina islands; WA to n Baja CA.

**B. madritensis** L. Annuals, 4.5-6 dm tall. Leaf blades 1-3.5 mm wide, glabrous to puberulent. Inflorescence 2.5-10 cm long, racemose to ± spikate, somewhat open to dense, lower branches ascending to appressed, pedicels 1.5-10 mm long. Spikelets terete to slightly compressed. Glumes dorsally rounded, glabrous to puberulent, lower one 5-10 mm long, 1-veined, upper one 10-15 mm long; florets 3-9; lemmas 10-20 mm long, usually hispidulous, apical teeth 2-3.5 mm long, awns 12-25 mm long. Mar-Jun. Introduced. Abundant; coastal bluffs, disturbed sites, sandy flats, floodplains, grassy hillsides, rocky slopes; chaparral, coastal scrub, grasslands, pine forest, riparian communities; 10-2470 ft (3-755 m). Widely scattered locations throughout island. All CA Channel islands; Los Coronados, Todos Santos, San Martin, Guadalupe, and Cedros islands; s OR to n Baja CA; native of Europe.

1. Stems and leaf sheaths usually glabrous; spikelets not strongly overlapping, bases of upper ones usually visible ........ subsp. **madritensis**

1' Stems and leaf sheaths usually puberulent; spikelets strongly overlapping, bases of middle and upper ones obscured by lower spikelets ........ subsp. **rubens**

subsp. **madritensis** MADRID BROME Stems usually glabrous. Leaf sheaths usually glabrous. Inflorescence elongated, often oblong in outline, pedicels 3-10 mm long; spikelets strongly overlapping, bases of middle and upper ones obscured by lower spikelets. Apr-Jun. Introduced. Occasional; disturbed sites, floodplains, rocky slopes; chaparral, grasslands, pine forest, riparian communities; 40-1400 ft (12-425 m). Widely scattered locations throughtout island. San Nicolas Island; s OR to n Baja CA; native of Europe. This subspecies has spread rapidly during the last decade; it was first collected on ridge se of Stanton Ranch in May 1963.

subsp. **rubens** (L.) Husn. [*B. rubens* L.] RED BROME Stems usually puberulent; Leaf sheaths usually puberulent. Inflorescence broadly ovoid, pedicels 1.5-4 mm long; spikelets strongly overlapping, bases of middle and upper ones obscured by lower spikelets. Mar-May. Introduced. Abundant; coastal bluffs, disturbed sites, sandy flats, floodplains, grassy hillsides, rocky slopes; chaparral, coastal scrub, grasslands; 10-2470 ft (3-755 m). Widely scattered locations throughout island. All CA Channel islands; Los Coronados, Todos Santos, San Martin, Guadalupe, and Cedros islands; s OR to n Baja CA, e to Great Basin; native of Europe. First collected at China harbor in March 1928; reported as widely distributed on open and rocky slopes by Hoffmann (1932a).

**B. stamineus** Desv. Perennial herbs, 3-8 dm tall. Leaf blades 4-7 mm wide, glabrous to minutely scabrous. Inflorescence 8-15 cm long, usually paniculate, open, branches stiffly ascending, pedicels 4-15 mm long. Spikelets strongly compressed. Glumes dorsally keeled, glabrous to minutely scabrous, lower one 7-11 mm long, 5-7-veined, upper one 11-18 mm long; florets 5-7; lemmas 10-12 mm long, glabrous to minutely scabrous, winged between margins and outer veins below middle, wing ca 1 mm wide, awns 8-10 mm long. Apr-May. Introduced. Rare; coastal bluffs, disturbed sites; coastal scrub, grasslands; 10-200 ft (3-60 m). Central Valley at Stanton Ranch; s side at mouth of Willows Canyon. San Francisco Bay region to s CA, e to Sierra Nevada foothills; native of S. America. First collected at Stanton Ranch in April 1960.

## CALAMAGROSTIS Adans.

Perennial herbs, usually with rhizomes. Leaves basal and cauline; ligules membranous; leaf blades flat to involute. Inflorescence paniculate, open or contracted, branches ascending to drooping. Spikelets compressed to ± terete, pedicellate; floret 1, rachilla prolonged beyond floret as short appendage, disarticulating above glumes. Glumes ± equal, equal to or longer than floret, lower one 1-veined, upper one 1-3-veined, apices acute to acuminate; lemmas 3-5-veined, thin, dorsally awned near middle or bases, awn straight to twisted, callus densely pubescent, often villous, apices usually with 4 teeth or bristles terminating veins; paleas equal to or shorter than lemmas. Ca. 270 species; temperate, worldwide.

**C. rubescens** Buckley PINE GRASS Perennial herbs, 6-10 dm tall, with rhizomes, sometimes with stolons; stems tufted. Sheaths usually glabrous, puberulent or pubescent near junction with leaf blade; ligules 2.5-5 mm long; leaf blades 2-4.5 mm wide. Inflorescence 5-15 cm long, narrow, branches spikate, ascending to erect. Glumes 3.5-5 mm long, keel minutely scabrous; lemmas 2.5-4 mm long, callus densely puberulent to villosulous, keel minutely scabrous, awned from near bases, awns 2-4 mm long, ± straight. Jul. Rare; open sites on rocky n-facing slopes; pine forest; ca. 1300 ft (395 m). Known from a single collection in Christy Pines in July 1960. British Columbia to San Luis Obispo Co., e to Rocky Mountains.

## **CORTADERIA** Stapf PAMPAS GRASS

Robust, dioecious perennial herbs; stems densely tufted. Leaves mostly basal; ligules pubescent; leaf blades flat to folded. Inflorescence paniculate, plumose, open or contracted, branches ascending to spreading. Spikelets unisexual, ± compressed, pedicellate; florets 2-8, rachilla disarticulating above glumes. Glumes unequal, longer than lowest floret, linear-lanceolate, thin, usually hyaline, 1-veined; lemmas 3-5-veined, hyaline, glabrous to villous, awnless or awned from apices. Ca. 25 species; temperate to subtropical, S. America and New Zealand.

**C. selloana** (Schult.) Asch. & Graebn. Plants 1.5-2.5 m tall. Leaf sheaths glabrous to sparsely pubescent, blades 4-10 mm wide, bluish green, margins serrulate. Inflorescence 5-12 dm long, white to light violet. Glumes 12-18 mm long; lemma awns 3-6 mm long, lemmas of pistillate florets silky-pubescent, lemmas of staminate florets usually glabrous. Aug-Sep. Introduced. Rare; rocky s-facing slopes, canyon bottoms; coastal scrub, riparian communities; 200-500 ft (60-150 m). Central Valley at Stanton Ranch (not seen there recently); s side in Red Rock Canyon, Cañada de los Sauces, and in Laguna Canyon. San Miguel, San Nicolas, and Santa Catalina islands; coastal CA, e to s U.S.; native of S. America. First collected at Stanton Ranch in March 1932.

## **CYNODON** Rich.

Perennial herbs, with rhizomes or stolons. Leaves mostly cauline; ligules membranous, often ciliate; leaf blades flat to involute. Inflorescence usually umbellate, composed of 2-10 spikate branches. Spikelets strongly compressed, in 2 rows along 1 side of inflorescence axis, sessile, rachilla usually prolonged beyond floret as bristle-like appendage, disarticulating above glumes; floret 1. Glumes subequal, usually shorter than floret, 1-veined, thinner than lemma; lemmas 3-5-veined, keel and margins ciliolate, apices acute. Ca. 8 species; subtropical, worldwide.

**C. dactylon** (L.) Pers. BERMUDA GRASS Stems mostly stoloniferous, branches often many, spreading, flowering ones erect, 1-4 dm tall. Ligules 0.3-0.5 mm long, leaf blades 1-4 mm wide. Inflorescence branches usually 3-5, ascending to spreading, 2-6 cm long. Glumes 1-2 mm long; lemmas 2-2.5 mm long, shiny, keels glabrous to minutely scabrous, apices acute. Mar-Aug. Introduced. Occasional; sandy areas, disturbed sites, seasonal ponds, floodplains; coastal scrub, grasslands, riparian communities, southern beach and dune; 10-550 ft (3-170 m). Widely scattered locations throughout island. All CA Channel islands; Los Coronados and Cedros islands; throughout warm temperate N. America; native of Africa. First collected at Pelican Bay in June 1930. This species can cause contact dermatitis in humans and can be toxic to livestock.

## **DISTICHLIS** Raf. ALKALI GRASS

Dioecious perennial herbs, with scaly rhizomes; stems firm, internodes usually solid in cross-sectional view. Leaves mostly cauline; ligules membranous; leaf blades flat to involute. Inflorescence racemose to paniculate, dense, branches or spikelets ascending to somewhat appressed. Spikelets unisexual but ± equal in size and shape, compressed, short-pedicellate to sessile, rachilla disarticulating above glumes and between florets; florets 5-many. Glumes usually unequal, lower one obscurely 3-5-veined, upper one obscurely 5-7-veined, firm; lemmas 7-11-veined, membranous to coriaceous, usually thicker than glumes. Ca. 13 species, N. and S. America, Australia.

**D. spicata** (L.) Greene SALTGRASS Plants 1-5 dm tall, often forming extensive mats. Leaves of rhizomes scale-like, cauline ones stiffly ascending; leaf blades 2-8 cm long, 1-4 mm wide, grayish green. Inflorescences 2-6 cm long, staminate ones usually exceeding leaves, pistillate ones equal to or shorter than uppermost leaves; spikelets often purple-tinged. Glumes glabrous, keel minutely scabrous, lower one 2-5.5 mm long, upper one 2.5-8 mm long; lemmas 3-7 mm long, dorsally keeled to rounded, glabrous. May-Aug. Common; coastal bluffs, beaches, moist depressions, floodplains; coastal scrub, southern beach and dune; 10-250 ft (3-75 m). Widely scattered locations, mostly around perimeter of island. All CA Channel islands except Santa Barbara; Los Coronados, San Martin, and Cedros islands; throughout N. America. Native Americans obtained salt from burned plants; this species also was used medicinally.

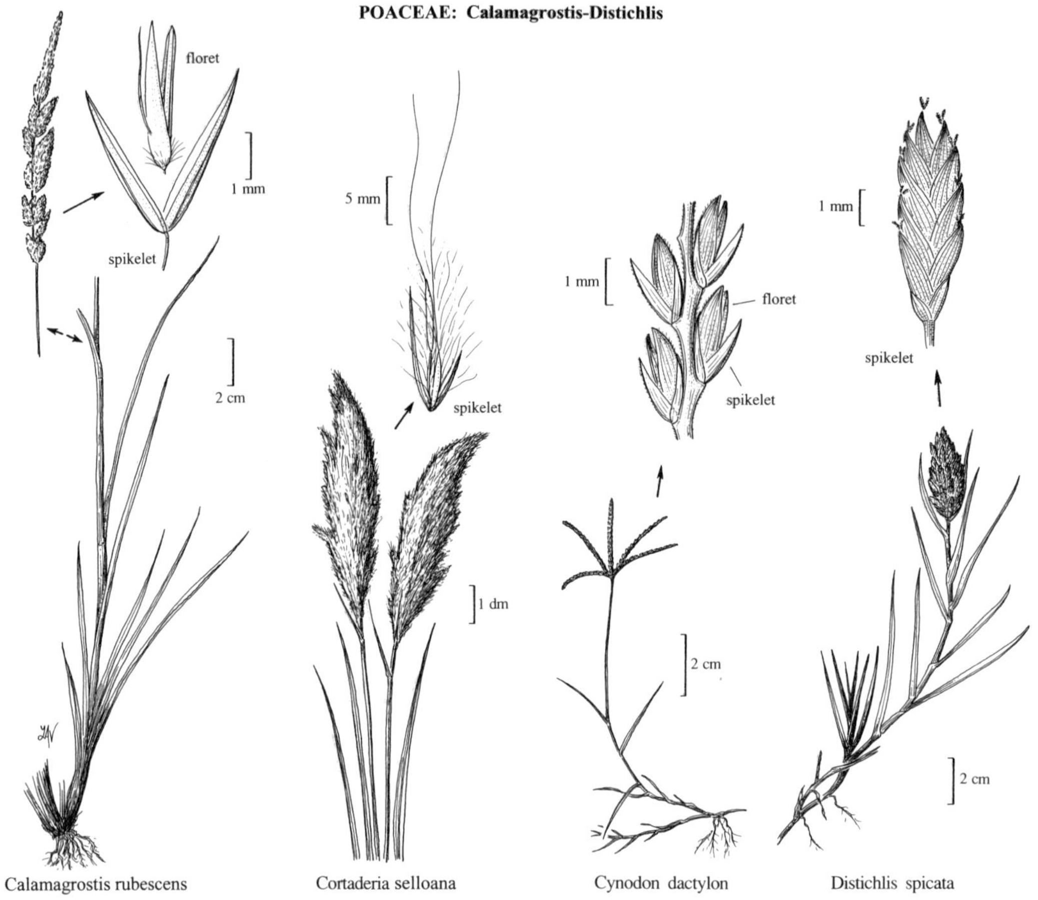


## ELYMUS L.

Perennial herbs, with or without rhizomes; stems usually tufted. Leaves mostly cauline; ligules membranous; leaf blades flat to involute. Inflorescence spikate, dense, axis usually not breaking apart at nodes in fruit. Spikelets usually in 2 ranks, 1-4 per node, compressed, subsessile to sessile, rachilla disarticulating above glumes and between florets; florets 2-7. Glumes subequal to unequal, sometimes absent, usually 1-5-veined, firm, apices awned or awnless; lemmas 3-7-veined, membranous, usually as thick as glumes, apices acute to awned. Ca. 150 species; temperate, worldwide.

**E. glaucus** Buckley subsp. **glaucus** WESTERN RYE Plants 4-12 dm tall; rhizomes absent. Leaf blades 4-10 mm wide, flat. Inflorescence 8-20 cm long; spikelets 10-18 mm long, usually 2 per node, sometimes 1 per node near bases and apices, florets 2-4. Glumes lanceolate, 6-20 mm long, longer than lowest floret, apices short-awned; lemmas 8-15 mm long, minutely scabrous, awns 12-30 mm long, straight. Apr-Jun. Common; n-facing coastal bluffs, shaded n-facing slopes, stream banks; coastal scrub, oak woodland, pine forest; 20-2000 ft (6-610 m). Scattered locations on n side between Valdez Anchorage and Cañada del Puerto, and between China Harbor and Little Scorpion Canyon; Central Valley between Cañada Christy and Valley Anchorage; s side in Cañada de los Sauces, Coches Prietos drainage, and Smugglers' Canyon. Santa Rosa and Santa Catalina islands; AK to n Mexico, e to Great Plains. Two other subspecies occur on the mainland throughout California.

## FESTUCA L. FESCUE

Perennial herbs; stems solitary or tufted. Leaves mostly basal; sheath margins free along most of their length, often overlapping; ligules membranous, apices truncate, often fimbriate or erose; leaf blades flat to involute. Inflorescence usually paniculate, branches ascending to spreading. Spikelets compressed to subterete, pedicellate; florets 2-10, rachilla disarticulating above glumes and between florets. Glumes unequal, lower one 1-3-veined, upper one 3-5-veined, apices usually awned; lemmas usually 5-veined, dorsally rounded, apices acute, usually awned. Ca. 450 species; temperate, worldwide.

**F. arundinacea** Schreb. MEADOW FESCUE Plants 3-12 dm tall, with short rhizomes. Leaves basal and cauline, blades 20-75 cm long, 4-6 mm wide. Inflorescence narrow, 15-30 cm long, lower branches ascending to spreading; florets 3-7. Lower glume 3-5 mm long, upper one 4.5-7 mm long, lemmas 5-9 mm long, minutely scabrous near apices, usually purple-tinged, awns 0.5-1.5 mm long, sometimes absent. Apr-Aug. Introduced. Rare; disturbed sites; grasslands; 100-250 ft (30-75 m). Central Valley at Christy Ranch and University of California Field Station, where it was first collected in April 1989. San Miguel, San Nicolas, and Santa Catalina islands; throughout N. America; native of Europe. Leaves and culms, especially on young plants, can be toxic to livestock.

## GASTRIDIUM Pal NIT GRASS

Annuals; stems erect, wiry, solitary or tufted. Leaves basal and cauline; ligules hyaline; leaf blades flat. Inflorescence spikate, dense, cylindrical. Spikelets short-pedicellate, compressed, disarticulating above glumes; floret 1. Glumes unequal, 1-veined, swollen at bases; lemmas much shorter than glumes, hyaline, awned from tip or awnless. Two species; temperate Europe and w Asia.

**G. ventricosum** (Gouan) Schinz & Thell. Plants 1.5-4 dm tall. Leaves few, blades 1-4 cm long, 1-2 mm wide, glabrous to minutely scabrous; ligules 1.5-4 mm long. Glumes attenuate, often hyaline between veins, keel and margins minutely scabrous, lower one 4-7 mm long, upper one 3-5.5 mm long; lemmas 0.5-1.5 mm long, awned below truncate to toothed apices, awn 3-5 mm long, straight to slightly bent. May-Sep. Introduced. Common; disturbed sites, grassy flats, exposed ridgetops, rocky slopes; 30-1600 ft (10-490 m). Widely scattered locations throughout island. Santa Rosa, Santa Catalina, and San Clemente islands; throughout N. America; native of Europe. First collected near Pelican Bay in June 1930.

## HAINARDIA Greuter THIN TAIL

With only 1 species.

**H. cylindrica** (Willd.) Greuter [*Monerma cylindrica* (Willd.) Cosson & Durieu] Annuals, 15-40 cm tall; stems erect to decumbent, tufted. Leaf blades 1-3 mm wide, flat. Inflorescence spikate, 8-17 cm long, cylindrical, straight; spikelets 1 per node, appressed, in 2 rows, sunken in inflorescence axis, axis breaking apart in fruit with 1 spikelet per segment (adjacent internode); floret 1. Most spikelets with 1 glume, glumes 5-8 mm long, usually longer than internode, thick, obscurely 9-veined, often folded or involute, apices acute, terminal spikelet often with 2 unequal glumes; lemmas 3-6 mm long, thin, hyaline, 3-veined, apices acute. May-Jul. Introduced. Rare; coastal flats, moist canyon bottoms; grasslands, riparian communities; 20-50 ft (6-15 m). Central Valley in dry ravine near n end of Christy Beach; s side near Forney's Cove and in lower portion of Cañada de los Sauces. San Miguel and San Clemente islands; San Francisco Bay region to n Baja CA; native of Europe. First collected near Forney's Cove in June 1965.

## HORDEUM L. BARLEY

Annuals or perennial herbs, usually without rhizomes. Leaves basal and cauline; ligules membranous, junction between sheath and blade with or without auricles; leaf blades flat to involute. Inflorescence spikate, axis usually breaking apart at nodes in fruit, segments often falling with glumes. Spikelets in 2 opposite rows, 3 per node, central one usually fertile, sessile, lateral ones usually sterile or staminate, short-pedicellate to subsessile, compressed to subterete, rachilla disarticulating above glumes; floret usually 1. Central spikelet with glumes unequal, awn-like to subulate, rigid, lemmas membranous, dorsally rounded, apices awned; lateral spikelets with awn-like glumes, lemmas like those of central spikelets or reduced. Ca. 40 species; temperate, N. America, Europe, and Asia.

1. All 3 spikelets alike, sessile, bisexual, lemma awns 50-150 mm long .......... **H. vulgare**
1' Central spikelet bisexual, producing fruits, sessile, lemma awns 4.5-30 mm long; lateral spikelets staminate or sterile, lemmas usually reduced, smaller than those of central spikelet
   2. Glumes conspicuously ciliate near bases .......... **H. murinum**
   2' Glumes not ciliate near bases, glabrous to minutely scabrous

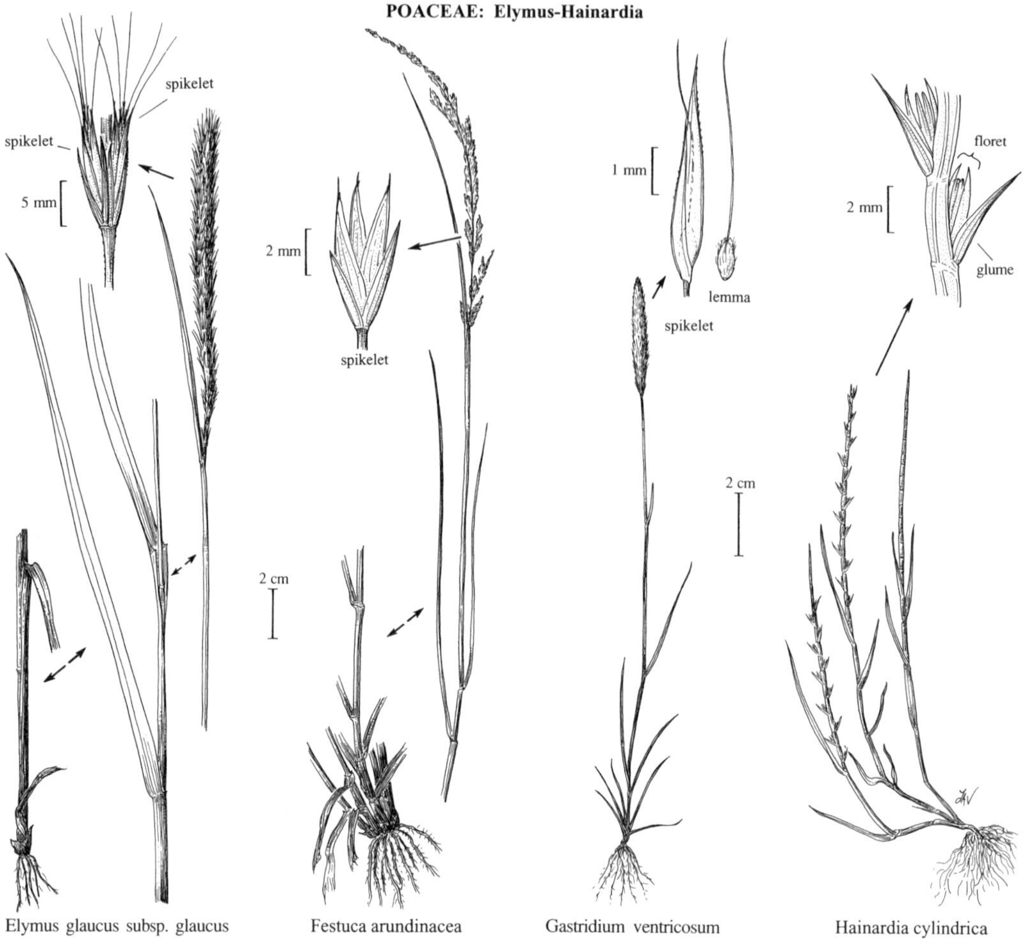


3. Perennial herb; spikelets purple-tinged ................................ **H. brachyantherum** subsp. **californicum**

3' Annual; spikelets greenish to tan, not purple-tinged

   4. Central spikelet glumes slightly wider near bases, 8-15 mm long, lemma awn 6-12 mm long; lateral spikelet lemmas with awns 3-6 mm long ................................ **H. intercedens**

   4' Central spikelet glumes linear, 12-22 mm long, lemma awn 8-23 mm long; lateral spikelet lemmas acute or with awns less than 1.5 mm long ................................ **H. marinum** subsp. **gussoneanum**

**H. brachyantherum** Nevski subsp. **californicum** (Covas & Stebbins) Bothmer, N.Jacobsen & Seberg MEADOW BARLEY Perennial herbs, 2-6.5 dm tall. Leaf sheaths with auricles to 1.5 mm long. Inflorescence 3-7 cm long, purple-tinged, breaking apart in fruit. Central spikelet with awn-like glumes 6-18 mm long, lemma 5-9 mm long, awn 4-10 mm long; lateral spikelets pedicellate, usually sterile, glumes awn-like, 10-18 mm long, lemmas 3-9 mm long, awns 1-7 mm long. Apr-Aug. Occasional; coastal bluffs, stabilized sand dunes, coastal flats, floodplains; coastal scrub, grasslands, southern beach and dune; 10-750 ft (3-230 m). Widely scattered locations on n side between West and San Pedro points; Central Valley at Christy Beach and University of California Field Station; s side between Fraser Point and mouth of Laguna Canyon. All CA Channel islands except Santa Barbara and San Clemente; OR to s CA. Subsp. *brachyantherum*, with central lemma awns shorter than 5 mm, occurs throughout much of w N. America.

**H. intercedens** Nevski [The name *H. pusillum* Nutt. has been misapplied to this species] VERNAL BARLEY Annuals, 1-3.5 dm tall. Auricles absent. Inflorescence 2-7 cm long, breaking apart in fruit, central and lateral spikelets often falling together. Central spikelet with glumes awn-like, ± subulate, slightly wider near bases, 8-15 mm long, lemma 5-9 mm long, awn 6-12 mm long; lateral spikelets short-pedicellate, staminate or sterile, glumes awn-like, slightly wider near bases, 12-18 mm long, lemmas 2-6 mm long, apices acute or with awns 0.5-1.5 mm long. Apr. A species of special environmental concern. Occasional; coastal bluffs, coastal flats, moist depressions, stabilized sand dunes; coastal scrub, grasslands, riparian communities, southern beach and dune; 30-1500 ft (10-455 m). North side on West End Flats, between Pelican Bay and Prisoners' Harbor, and between China Harbor and Little Scorpion Canyon; Central Valley in Cañada Cebada; s side between West Point and Black Point peninsula, and on Sierra Blanca ridge. All CA Channel islands; San Francisco Bay region to nw Baja CA.

**H. marinum** Hudson subsp. **gussoneanum** (Parl.) Thell. [*H. geniculatum* All.] MEDITERRANEAN BARLEY Annuals, 1-6 dm tall. Auricles absent. Inflorescence 2-5 cm long, breaking apart in fruit, central and lateral spikelets often falling together. Central spikelet with awn-like glumes 12-22 mm long, lemma 4.5-8 mm long, awn 8-23 mm long; lateral spikelets short-pedicellate, usually sterile, glumes awn-like, 10-25 mm long, lemmas 3-5 mm long, awns 3-6 mm long. Apr-Jun. Introduced. Scarce; disturbed sites, moist depressions, grassy hillsides; coastal scrub, grasslands; 150-500 ft (45-150 m). North side near Coche Point and Little Scorpion Canyon; Central Valley just w of University of California Field Station; s side in Coches Prietos drainage and near Smugglers' Ranch. Santa Rosa, Santa Catalina, and San Clemente islands; British Columbia to n Baja CA; native of Europe. First collected near Coche Point in May 1968.

**H. murinum** L. FOXTAIL Annuals, 1-5 dm tall. Leaf sheath auricles to 1 mm long. Inflorescence 4-10 cm long, breaking apart in fruit, central and lateral spikelets often falling together. Central spikelet sessile to short-pedicellate, with glumes awn-like, 12-25 mm long, margins ciliate near bases, lemma 7-15 mm long, awn 20-30 mm long; lateral spikelets short-pedicellate, staminate or sterile, glumes 10-30 mm long, lemmas 10-15 mm long, awns 20-50 mm long. Mar-Jun. Introduced. Common; disturbed sites, coastal bluffs, coastal flats, grassy hillsides, sandy areas; coastal scrub, grasslands, southern beach and dune; 10-1300 ft (3-395 m). Widely scattered locations, primarily around perimeter of island. All CA Channel islands; Los Coronados, Todos Santos, San Martin, Guadalupe, and Cedros islands; British Columbia to n Mexico; native of Europe. Reported by Brandegee (1888). Subsp. *murinum*, with central floret sessile to subsessile, apparently does not occur on Santa Cruz Island.

1. Central floret (excluding awn) slightly shorter than lateral florets; internodes of inflorescence axis 1-2 mm long with 5-8 spikelet clusters per cm of axis .......... subsp. **glaucum**

1' Central floret (excluding awn) slightly shorter than lateral florets; internodes of inflorescence axis 2-3 mm long, with 3-5 spikelet clusters per cm of axis .......... subsp. **leporinum**

subsp. **glaucum** (Steud.) Tzelev [*H. stebbinsii* Covas] Plants 1.5-4 dm tall. Inflorescence axis 3-7 cm long, internodes 1-2 mm long; central floret (excluding awn) slightly shorter than lateral florets. Mar-Jun. Introduced. Common; disturbed sites, coastal bluffs, coastal flats, grassy hillsides, sandy areas; coastal scrub, grasslands, southern beach and dune; 10-1300 ft (3-395 m). Widely scattered locations, primarily around perimeter of island. All CA Channel islands; Los Coronados, San Martin, Guadalupe, and Cedros islands; British Columbia to n Mexico; native of Europe.

subsp. **leporinum** (Link) Arcang. Plants 1-5 dm tall. Inflorescence axis 3.5-6 cm long, internodes 2-3 mm long; central floret (excluding awn) slightly shorter than lateral florets. Mar-Jun. Introduced. Common; disturbed sites, coastal bluffs, coastal flats, grassy hillsides, sandy areas; coastal scrub, grasslands, southern beach and dune; 10-1150 ft (3-350 m). Widely scattered locations, primarily around perimeter of island. All CA Channel islands; Los Coronados, Todos Santos, San Martin, and Guadalupe islands; British Columbia to n Mexico; native of Europe.

**H. vulgare** L. CULTIVATED BARLEY Annuals, 1-5 dm tall. Leaf sheaths with well developed auricles to 5 mm long. Inflorescence 3-11 cm long, usually not breaking apart in fruit. All 3 spikelets fertile, all alike, sessile, glumes 8-20 mm long; lemmas 7-15 mm long, awns 50-150 mm long. Apr-Jul. Introduced. Rare; disturbed sites, abandoned pastures; grasslands; ca. 200 ft (60 m). Central Valley just e of Stanton Ranch, where it was first collected in April 1960. Santa Rosa, San Nicolas, Santa Catalina, and San Clemente islands; cultivated throughout N. America and elsewhere; native of Europe and Asia.

## KOELERIA Pers.

Annuals or perennial herbs. Leaves mostly basal; ligules membranous; leaf blades flat to involute. Inflorescence spikate or paniculate, branches ascending to appressed. Spikelets compressed, pedicellate, florets 2-5, rachilla disarticulating above glumes and between florets, prolonged beyond upper floret as bristle-like appendage. Glumes equal to unequal, dorsally keeled, shiny, apices acute, lower one 1-veined, upper one 3-5-veined, usually

**POACEAE: Hordeum-Leymus**

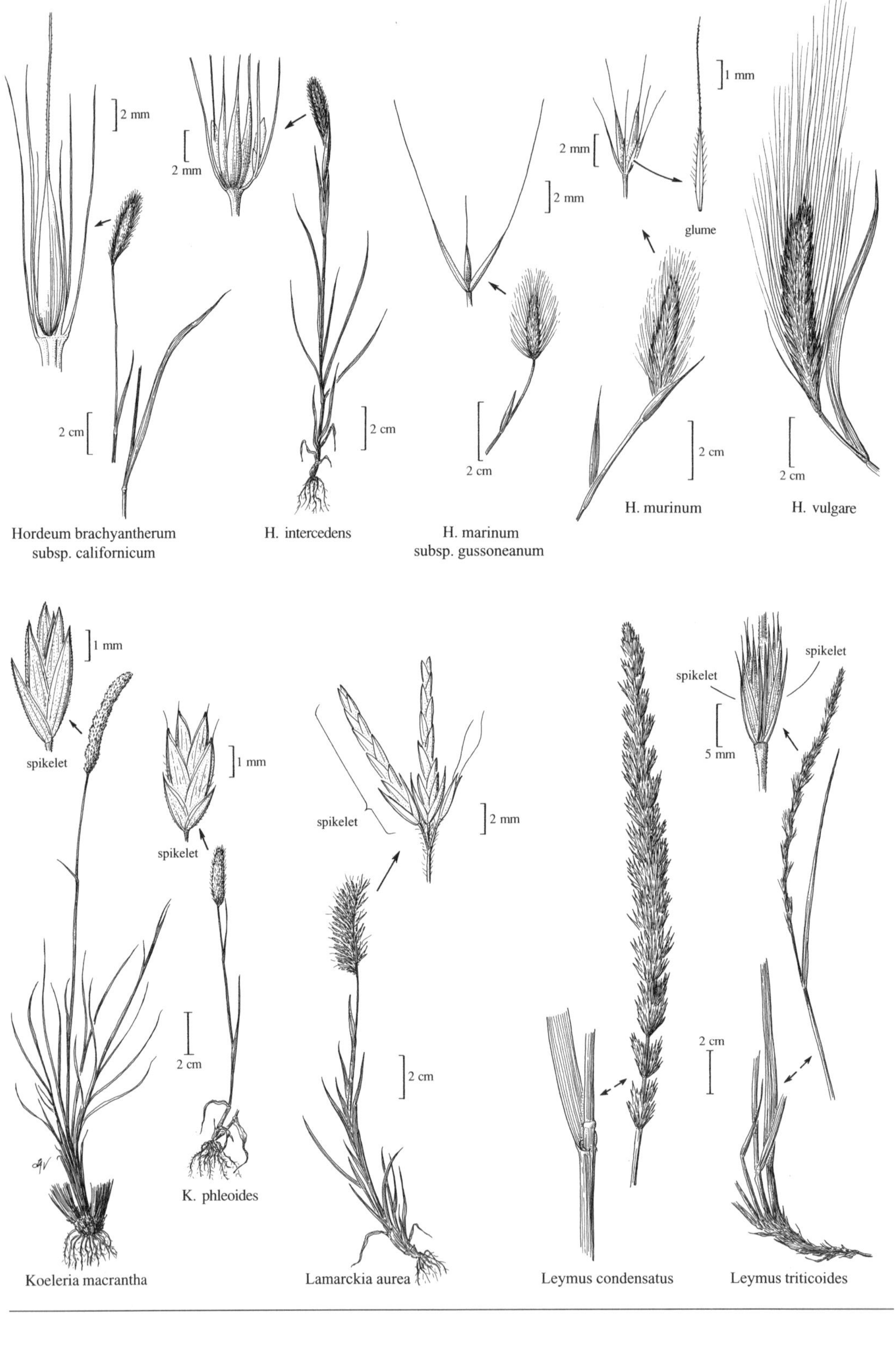

longer than lowest floret; lemmas membranous, 3-5-veined, dorsally keeled, apices acute to short-awned. Ca. 25 species; temperate, N. America, Europe, and Asia.

1. Perennial herbs; inflorescence axis puberulent; glumes glabrous to minutely scabrous, smooth .................... **K. macrantha**
1' Annuals; inflorescence axis glabrous; glumes papillose, sometimes hispidulous ...................... **K. phleoides**

**K. macrantha** (Ledeb.) Schult. JUNE GRASS Perennial herbs, 30-65 cm tall. Leaves mostly basal, blades 4-18 cm long, glabrous to puberulent. Inflorescence 6-11 cm long, axis puberulent. Glumes minutely scabrous, lower one 2-3 mm long, upper one 4-5 mm long; lemmas 2.5-4.5 mm long, lower ones minutely scabrous, apices acute to awn-tipped, awn less than 1 mm long. Apr-Jun. Occasional; coastal bluffs, rocky n-facing slopes, grassy hillsides, exposed ridgetops; chaparral, coastal scrub, pine forest; 80-600 ft (25-185 m). North side between Pelican Bay and Cañada del Puerto; Central Valley between Christy Ranch and Cañada Cebada, and between Bosque Mano and El Sobrante; s side between Cañada de los Sauces and Willows Canyon. Santa Rosa Island; AK to n Mexico, e to ne N. America.

**K. phleoides** (Vill.) Pers. Annuals, 4-35 cm tall. Leaves basal and cauline, blades 1.5-4 cm long. Inflorescence 1-20 cm long, axis glabrous. Glumes papillose, sometimes hispidulous, lower one 1.5-2 mm long, upper one 2.5-3 mm long; lemmas 2-3 mm long, lower ones papillose, apices acute to awn-tipped, awns 0.5-1.5 mm long. May. Introduced. Rare; exposed ridgetops, rocky slopes; chaparral, coastal scrub; 650-850 ft (200-260 m). North side on ridge above Valdez Anchorage; Central Valley near top of Camino de la Casa. Washington to s CA, e to s U.S.; native of Europe. First collected near Valdez Anchorage in May 1979.

## **LAMARCKIA** Moench GOLDENTOP

With only 1 species.

**L. aurea** (L.) Moench Annuals, 1-4 dm tall; stems solitary or tufted. Leaves basal and cauline, glabrous; ligules membranous, apices erose; leaf blades 2-9 cm long, 3-8 mm wide. Inflorescence paniculate, branches spreading to nodding, dense, golden yellow, often purple-tinged. Spikelets subterete, pedicellate, in distinct clusters that fall together; clusters composed of 1 fertile spikelet subtended or surrounded by 1-5 sterile spikelets; fertile spikelet with glumes 2-5 mm long, florets 1-2; lemmas 2-5 mm long, apices awned, awns 3-9 mm long; sterile spikelets with 3-10 reduced florets, glumes and lemmas thinly membranous, awnless. Mar-Jun. Introduced. Common; disturbed sites, floodplains, rocky slopes, grassy hillsides; coastal scrub, grasslands, pine forest; 10-2470 ft (3-755 m). Widely scattered locations throughout island. All CA Channel islands; Los Coronados, Todos Santos, San Martin, and Guadalupe islands; CA e to AZ; native of Europe. First collected near Scorpion Ranch in March 1929.

## **LEYMUS** Hochst.

Perennial herbs, often with rhizomes; stems solitary or tufted. Leaves mostly cauline; ligules membranous; leaf blades flat to involute. Inflorescence spikate or paniculate, dense, axis usually not breaking apart at nodes in fruit. Spikelets 2-many per node, usually in 2 ranks, ranks often obscured by densely aggregated spikelets, compressed, subsessile to sessile, rachilla disarticulating above glumes and between florets; florets 2-7. Glumes linear to subulate, subequal to unequal, obscurely 1-5-veined, thick, firm, apices acute to short-awned; lemmas obscurely 3-7-veined, membranous, usually as thick as glumes, apices acute to short-awned. Ca. 40 species; temperate, N. America, Europe, and Asia.

1. Lower inflorescence nodes with 3-many spikelets; plants usually taller than 1.5 m; rhizomes absent or short; leaf blades 10-30 mm wide .................... **L. condensatus**
1' Lower inflorescence nodes with 2-3 spikelets; plants usually shorter than 1.5 m; rhizomes well developed (plants usually colonial); leaf blades 2-8 mm wide .................... **L. triticoides**

**L. condensatus** (C.Presl) A.Löve [*Elymus condensatus* C.Presl] GIANT RYE Plants 1.5-2.5 m tall; rhizomes absent or short; stems usually densely tufted. Leaf blades 10-30 mm wide, usually flat. Inflorescence usually 15-50 cm long; spikelets 15-30 mm long, 3-many per node, florets 3-6. Glumes subulate, 5-15 mm long, longer than lowest floret; lemmas 6-15 mm long, usually glabrous, apices acute to short-awned, awns 0.5-4 mm long, straight. Jun-Aug. Common; coastal bluffs, rocky slopes, grassy flats, ridgetops; chaparral, coastal scrub, grasslands; 20-1400 ft (6-425 m). Widely scattered locations throughout island. All CA Channel islands except Santa Barbara and San Nicolas; Los Coronados and Todos Santos islands; n CA to n Mexico.

**L. triticoides** (Buckley) Pilg. [*Elymus triticoides* Buckley] ALKALI RYE Plants 0.6-1.7 m tall, usually colonial; rhizomes short to long; stems solitary or tufted. Leaf blades 2-8 mm wide, flat to involute. Inflorescence usually 5-20 cm long; spikelets 10-20 mm long, 2-3 at lower nodes, 1 per node near apices, florets 3-6. Glumes subulate, 5-15 mm long, longer than lowest floret; lemmas 5-12 mm long, glabrous to minutely scabrous, apices acute to short-awned, awns 0.5-2 mm long, straight. Jun-Jul. Occasional; coastal flats, ravines, grassy hillsides, rocky slopes; coastal scrub, grasslands; 20-1200 ft (6-365 m). Widely scattered locations throughout island. San Miguel, Santa Rosa, Anacapa, and Santa Catalina islands; WA to s CA, e to Rocky Mountains.

## LOLIUM L. RYEGRASS

Annuals or perennial herbs. Leaves basal and cauline; ligules membranous; leaf blades flat to slightly involute. Inflorescence spikate. Spikelets 1 per node, in 2 ranks, strongly compressed, one edge facing inflorescence axis, sessile to sunken in inflorescence axis; florets 2-many, rachilla disarticulating above glumes and between florets. Most spikelets with 1 glume, glume 3-9-veined, apices acute to obtuse, terminal spikelet often with 2 unequal glumes; lemmas 5-9-veined, dorsally rounded, apices acute to awned, awns straight. Ca. 8 species; temperate, Europe and Asia.

1. Glume equal to or longer than spikelet (excluding lemma awns); lower florets much thicker than upper ones, especially at their bases .......... **L. temulentum**

1' Glume usually shorter than entire spikelet (excluding lemma awns); lower florets like upper ones, not thick
   2. Lemma apices awned, awns 1.5-7 mm long .......... **L. multiflorum**
   2' Lemma apices acute, awnless .......... **L. perenne**

**L. multiflorum** Lam. ITALIAN RYEGRASS Annuals or short-lived perennials, 3-8 dm tall. Leaf blades 3-18 cm long, 2-6 mm wide, usually glabrous. Glume narrowly lanceolate, 4-8 mm long, usually shorter than entire spikelet except for lemma awns; florets 4-14; lemmas 4-6 mm long, apices awned, awns 1.5-7 mm long. Mar-Jun. Introduced. Common; coastal bluffs, disturbed sites, grassy hillsides, floodplains; coastal scrub, grasslands; 10-1500 ft (3-455 m). Widely scattered locations throughout island; locally abundant in w end of Central Valley and in sw portion of island. Santa Barbara, San Nicolas, Santa Catalina, and San Clemente islands; throughout N. America; native of Europe.

**L. perenne** L. PERENNIAL RYEGRASS Perennial herbs, 3-7 dm tall. Leaf blades 3-18 cm long, 2-6 mm wide, usually glabrous. Glume narrowly lanceolate, 4-9 mm long, usually shorter than entire spikelet except for lemma awns; florets 3-12; lemmas 4.5-6 mm long, apices acute. May-Sep. Introduced. Common; coastal bluffs, disturbed sites, grassy hillsides, floodplains; coastal scrub, grasslands; 10-500 ft (3-150 m). Widely scattered locations throughout island; locally abundant in w end of Central Valley and in sw portion of island. San Nicolas, Santa Catalina, and San Clemente islands; throughout N. America; native of Europe. First collected at Stanton Ranch in June 1930. This species can be toxic to livestock.

**L. temulentum** L. DARNEL Annuals, 3-10 dm tall. Leaf blades 3-20 cm long, 2-6 mm wide, glabrous or minutely scabrous. Glume oblong to lanceolate, 10-15 mm long, longer than entire spikelet except for lemma awns; florets 3-9; lemmas 4-7 mm long, lower ones much thicker than upper ones, especially at their bases, apices usually awned, awns 5-15 mm long. Apr-Jun. Introduced. Rare; disturbed sites, stream banks, wooded slopes; grasslands, oak woodland, riparian communities; 10-100 ft (3-30 m). North side near Pelican Bay and Prisoners' Harbor; Central Valley near Christy Ranch. Santa Rosa, Santa Catalina, and San Clemente islands; throughout N. America; native of Europe. First collected at Prisoners' Harbor in June 1930. Caryopses can be extremely toxic to livestock and humans.

## MELICA L. MELIC

Perennial herbs; stems solitary to tufted. Leaves basal and cauline; sheath margins fused along most of their length; ligules membranous, apices obtuse to truncate; leaf blades usually flat. Inflorescence racemose to paniculate. Spikelets terete, pedicellate; florets 1-7, lower ones fertile, terminal ones sterile, often much reduced; rachilla disarticulating above glumes and between florets. Glumes equal to unequal, thin, often hyaline, dorsally rounded, lower one usually 3-5-veined, upper one 1-3-veined, apices rounded; lemmas like glumes in texture, clearly 5-7-veined, dorsally rounded. Ca. 80 species; temperate, worldwide.

**M. imperfecta** Trin. COAST RANGE MELIC Plants 3-11 dm tall. Leaf blades 1-6 mm wide. Inflorescence paniculate, 10-35 cm long, lower branches ascending to spreading. Spikelets usually with 1 fertile floret, sterile ones minute, densely clustered. Glumes of fertile florets subequal, 3-6 mm long, usually purple-tinged; lemmas

POACEAE: Lolium-Muhlenbergia

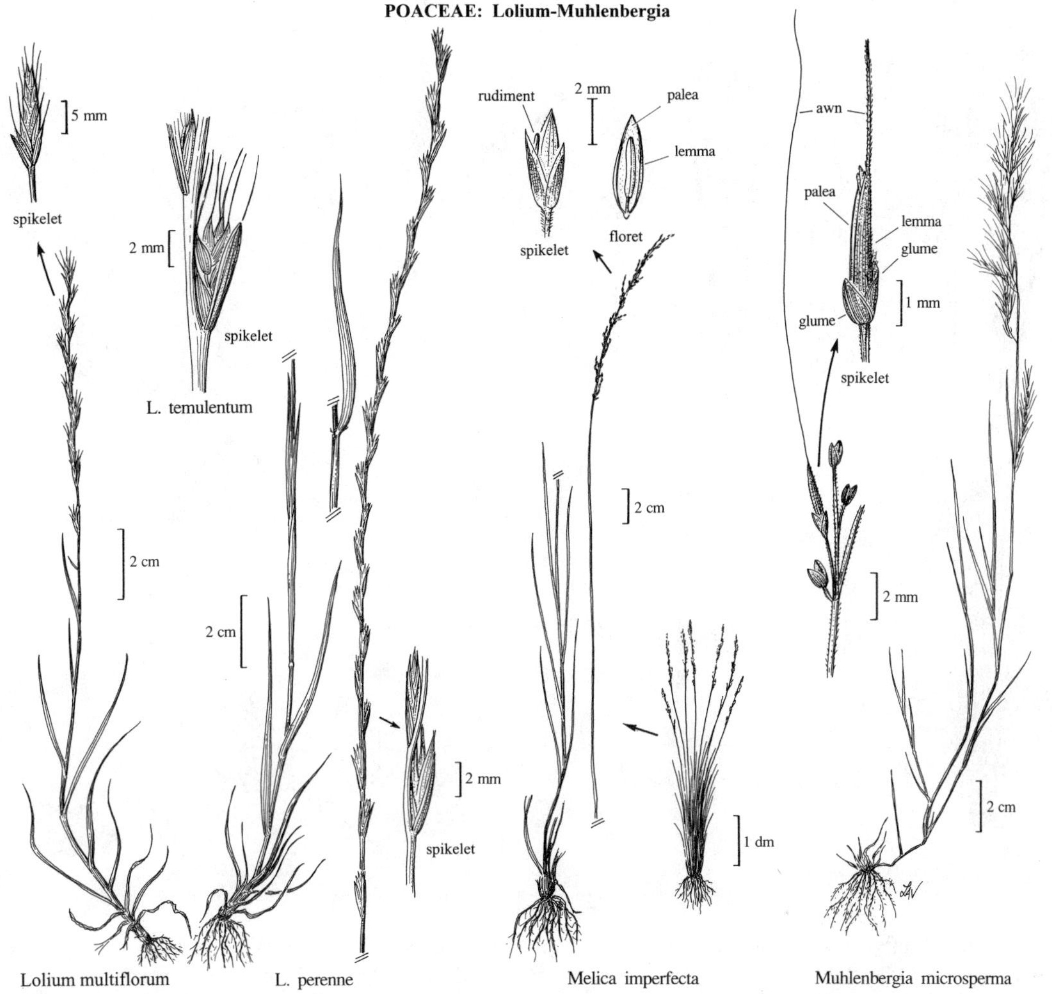


2-6 mm long, often purple-tinged. Mar-May. Common; coastal bluffs, rocky n-facing slopes, rocky canyon walls, grassy hillsides; chaparral, coastal scrub, oak woodland, pine forest; 70-1650 ft (20-505 m). Widely scattered locations throughout island. All CA Channel islands; Los Coronados, Todos Santos, San Martin, Guadalupe, and Cedros islands; Santa Cruz Mountains to Baja CA.

## MUHLENBERGIA Schreb.

Annuals or perennial herbs. Leaves basal and cauline; ligules membranous, ciliolate to fimbriate; leaf blades flat to involute. Inflorescence paniculate, cylindrical and spikate to open and ovoid. Spikelets compressed to subterete, pedicellate, rachilla disarticulating above glumes; floret 1. Glumes subequal, usually shorter than floret, apices rounded, acute, or short-awned, lower one 1-veined, upper one 1-3-veined; lemmas dorsally rounded, 3-veined, thin, obtuse to awned from apex; palea usually like lemma in texture and length. Ca. 150 species; temperate, mostly N. and S. America.

**M. microsperma** (DC.) Kunth LITTLESEED MUHLY Annuals or short-lived perennial herbs, 10-60 cm tall; stems reddish or purplish, tufted, outer ones decumbent to spreading, lowest nodes often producing much reduced spikelets with cleistogamous florets. Leaf blades 1-2 mm wide. Inflorescence branches appressed when young, ascending to spreading in flower or fruit. Spikelets usually purplish. Glumes 0.5-2 mm long, 1-veined, apices obtuse to truncate; lemmas 2-4 mm long, 3-veined, puberulent below middle, apices minutely toothed,

POACEAE: Nassella-Pennisetum

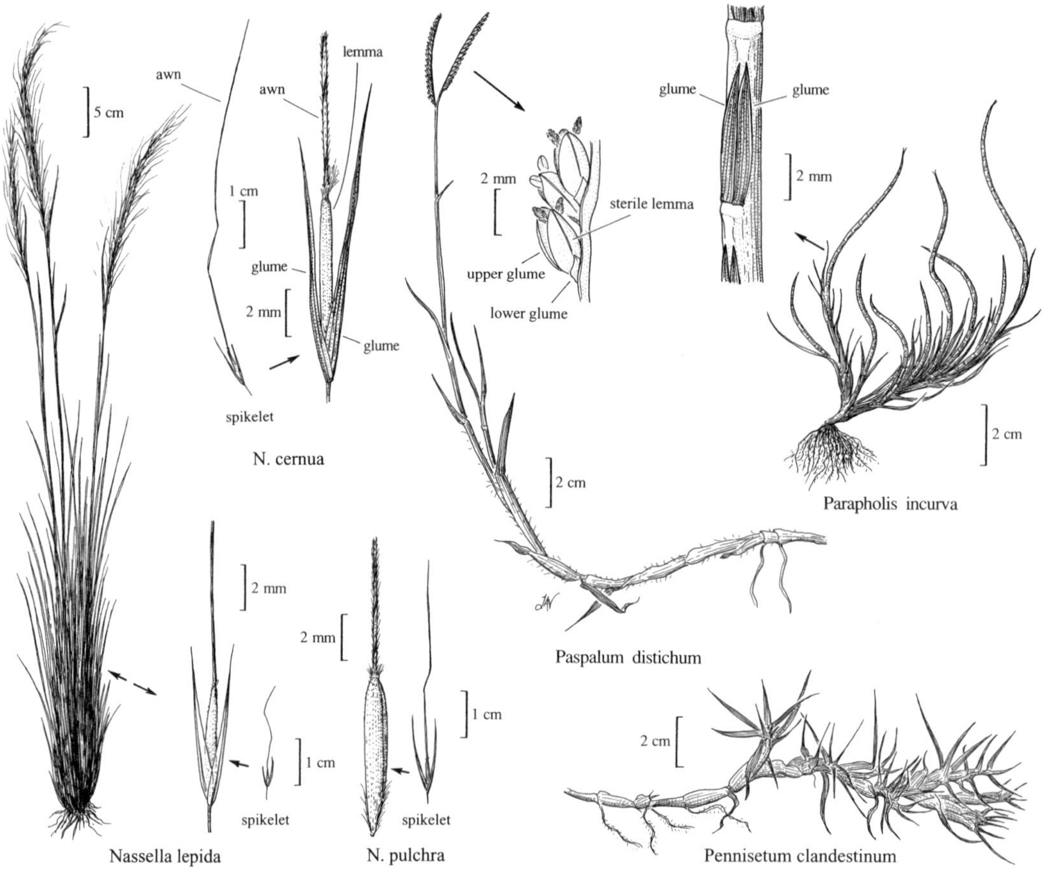


awn arising from between teeth, 10-15 mm long. Feb-May. Occasional; coastal bluffs, rocky s-facing slopes, rocky canyon walls, floodplains; chaparral, coastal scrub; 10-650 ft (3-200 m). North side between Valdez Anchorage and Cañada del Puerto; Central Valley between Cañada de la Portezuela and Valley Anchorage; widely scattered locations on s side between West Point and Smugglers' Canyon. All CA Channel islands except San Nicolas; Los Coronados, Todos Santos, San Martin, Guadalupe, and Cedros islands; San Francisco Bay region to Mexico, e to UT and AZ.

## **NASSELLA** (Trin.) E.Desv.

Perennial herbs; stems usually densely tufted. Leaves mostly basal; ligules membranous; leaf blades flat to involute. Inflorescence racemose or paniculate, open, branches spreading to drooping; spikelets terete, pedicellate; floret 1; rachilla disarticulating above glumes, callus sharply acute. Glumes equal to unequal, usually longer than lemma, 1-5-veined, apices attenuate; lemmas coriaceous, obscurely 5-veined, slightly constricted near apices, apices awned, junction between lemma and awn distinct, awns bent 1-2 times, lower segment often twisted. Ca. 18 species; temperate, w N. and S. America.

1. Terminal segment of awn straight, ± stiff ........ **N. pulchra**
1' Terminal segment of awn flexuous, soft
    2. Glumes 5-12 mm long; lemmas (excluding awns) 4-6.5 mm long; awns 20-40 mm long........ **N. lepida**
    2' Glumes 13-20 mm long; lemmas (excluding awns) 7-12 mm long; awns 45-90 mm long...... **N. cernua**

**N. cernua** (Stebbins & Love) Barkworth [*Stipa cernua* Stebbins & Love] NODDING NEEDLEGRASS Plants 4-9 dm tall. Leaf blades 0.5-2.5 mm wide, flat to involute. Inflorescence 1-8 dm long. Glumes unequal, 9-20 mm long; lemma (excluding awn) 5-10 mm long, veins pilose, awn 50-110 mm long, terminal segment flexuous, soft. Mar-May. Common; coastal bluffs, rocky slopes, grassy hillsides, exposed ridgetops; coastal scrub, grasslands;

20-1750 ft (6-535 m). Widely scattered locations throughout island. All CA Channel islands except Santa Barbara; n CA coast ranges to Baja CA.

**N. lepida** (Hitchc.) Barkworth [*Stipa lepida* Hitchc.] FOOTHILL NEEDLEGRASS Plants 6-10 dm tall. Leaf blades 1.5-4 mm wide, usually flat. Inflorescence 1-5 dm long. Glumes unequal, 5-12 mm long; lemma (excluding awn) 4-6.5 mm long, usually villosulous, sometimes sparsely pubescent between veins, awn 20-40 mm long, terminal segment flexuous, soft. Mar-May. Common; coastal bluffs, rocky slopes, grassy hillsides, exposed ridgetops; coastal scrub, grasslands, pine forest; 100-1650 ft (30-505 m). Widely scattered locations throughout island. All CA Channel islands; Los Coronados, Guadalupe, and Cedros islands; throughout CA w of Sierra Nevada to Baja CA.

**N. pulchra** (Hitchc.) Barkworth [*Stipa pulchra* Hitchc.] PURPLE NEEDLEGRASS Plants 6-10 dm tall. Leaf blades 1-5 mm wide, flat to involute. Inflorescence 1.5-6 dm long. Glumes unequal, 13-20 mm long; lemma (excluding awn) 7-12 mm long, usually villosulous, sometimes sparsely pubescent between veins, awn 45-90 mm long, terminal segment straight, ± stiff. Mar-June. Common; coastal bluffs, coastal flats, dry ravines, rocky slopes, grassy hillsides, exposed ridgetops; coastal scrub, grasslands; 10-1500 ft (3-455 m). Widely scattered locations throughout island. All CA Channel islands; Los Coronados Islands; throughout CA w of Sierra Nevada to Baja CA.

## PARAPHOLIS C.E.Hubb.

Annuals; stems decumbent to erect. Ligules membranous; leaf blades flat to involute. Inflorescence spikate, cylindrical, straight to curved; spikelets appressed to and sunken in inflorescence axis, axis breaking apart in fruit with 1 spikelet per segment (adjacent internode); floret 1, often concealed by appressed glumes. Glumes equal, margins scarious to hyaline, usually longer than floret; lemma hyaline. Ca. 5 species; temperate Europe and Asia.

**P. incurva** (L.) C.E.Hubb. SICKLE GRASS Plants 5-35 cm tall; stems usually decumbent, tufted. Leaf blades 0.5-2 mm wide, flat. Inflorescence 3-12 cm long, usually curved; spikelets 1 per node. Glumes lanceolate, 3-6 mm long, both on outer side of spikelet, margins often overlapping, apices acute; lemmas 1.5-5.5 mm long, apices acute. Mar-May. Introduced. Occasional; coastal bluffs, coastal flats, sandy beaches, rocky talus slopes, floodplains; coastal scrub, southern beach and dune; 5-300 ft (1.5-90 m). Widely scattered locations, mostly around perimeter of island. All CA Channel islands; British Columbia to Baja CA; native of Europe.

## PASPALUM L.

Annuals or perennial herbs with rhizomes or stolons. Leaves basal and cauline; ligules membranous, often ciliate; leaf blades flat to involute. Inflorescence paniculate, composed of 2-many spikate branches; spikelets strongly compressed, 1-2 per node, arranged along one side of inflorescence axis, sessile to subsessile, rachilla not prolonged beyond floret, disarticulating below glumes; florets 2, lower one sterile, much reduced, upper one fertile. Glumes usually unequal, lower one usually minute or absent, upper one equal to floret; lemmas suborbicular, thick, apices acute to rounded. Ca. 330 species; warm temperate to tropical, worldwide.

**P. distichum** L. KNOT GRASS Perennial herbs, with extensive stolons; stems erect, 2-6 dm tall. Sheaths pubescent near apex; ligules 0.5-1.5 mm long, fimbriate to ciliate, leaf blades 2-4.5 mm wide, flat. Inflorescence branches usually 2-3, ascending, 1.5-5 cm long; spikelets short-pedicellate. Glumes 2, lower one 2-2.5 mm long, upper one slightly longer; lemmas 2-2.5 mm long, margins smooth, apices acute. Jul. Rare, not seen recently; moist edge of coastal estuary; riparian communities; ca. 10 ft (3 m). Known from a single collection at Prisoners' Harbor in July 1960. Santa Catalina Island; throughout N. and S. America.

## PENNISETUM Rich. ex Pers.

Annuals or perennial herbs with stolons or rhizomes. Leaves basal or cauline; ligule membranous to ciliate; leaf blades flat to involute. Inflorescence spikate, dense; spikelets either 1-4 and enclosed in upper leaf sheaths or many, subtended by bristles and comprising a conspicuous inflorescence that usually exceeds leaves; florets 2, lower one sterile, upper one fertile. Glumes 1-2, if 2 then unequal, lower one much shorter than upper; lemmas thick, margins usually overlapping. Ca. 80 species; warm temperate to tropical, worldwide.

**P. clandestinum** Hochst. ex Chiov. KIKUYU GRASS Plants matted; stems stoloniferous. Leaves strongly overlapping, sheaths wider than blades, expanded near base, margins ciliate, upper ones concealing internodes; ligules ciliate; leaf blades spreading, 5-10 cm long, 1-4 mm wide. Spikelets 1-3, enclosed within upper leaf sheaths. Lower glume 1-2 mm long, upper glume 8-15 mm long; lemma of sterile floret longer than lemma of fertile floret; stigmas long-exserted. Apr-Jun. Introduced. Scarce; disturbed sites, ravines, edges of coastal estuaries; grasslands, riparian communities; 20-450 ft (6-140 m). North side at Prisoners' Harbor and in lower portion of Cañada del Puerto; Central Valley at Christy Ranch and at Stanton Ranch airfield; s side near Smugglers' Ranch airfield. San Miguel, Anacapa, and San Nicolas islands; coastal CA to n Baja CA; native of Africa. Can be toxic to cattle.

## PHALARIS L.

Annuals or perennial herbs; stems solitary to tufted. Leaves basal and cauline, cauline ones spreading to drooping, blades 3-12 mm wide; ligules membranous, apices obtuse to truncate; leaf blades usually flat. Inflorescence spikate, dense, often cylindrical. Spikelets compressed, short-pedicellate; florets 1-3, lower 1-2 sterile, subulate, much shorter than fertile one, sometimes absent, upper one fertile; rachilla disarticulating above glumes, florets usually falling together. Glumes equal to subequal, 3-5-veined, dorsally keeled or winged, usually longer than fertile floret; lemma or fertile floret thick, shiny. Ca. 15 species; temperate, N. America, Europe, and Asia.

1. Spikelets deciduous in clusters of 6-7, each cluster composed of 1 fertile floret surrounded by 5-6 sterile, reduced spikelets ..... **P. paradoxa**

1' Spikelets all alike, solitary, not deciduous
  2. Perennial herbs, 8-15 dm tall; keel of glumes barely winged if at all; lemmas of sterile florets unequal or sterile floret 1 ..... **P. aquatica**
  2' Annuals, 2-9 dm tall; glumes dorsally winged; lemmas of sterile florets equal or sterile floret 1 in *P. minor*
    3. Glumes not notched near apex; sterile florets 2 ..... **P. caroliniana**
    3' Glumes notched near apex; sterile floret 1 ..... **P. minor**

**P. aquatica** L. HARDING GRASS Perennial herbs, 8-15 dm tall; stems tufted, often with rhizomes. Inflorescence 5-15 cm long; spikelets solitary; sterile florets 1-2. Glumes 5-7 mm long, dorsal keel barely winged if at all, margin smooth; lemmas of sterile florets unequal, 0.5-2 mm long, lemma of fertile floret 3-4 mm long, puberulent. May-Sep. Introduced. Scarce; disturbed sites, abandoned pastures, grassy hillsides, exposed ridgetops; coastal scrub, grasslands; 200-950 ft (60-290 m). North side near Campo Grande airfield; Central Valley at e end of Cañada de la Portezuela and e of Stanton Ranch; s side on South Ridge just w of intersection with road to Coches Prietos Anchorage. San Nicolas and Santa Catalina islands; OR to s CA; native of Europe. This species has spread rapidly since it was first collected on the South Ridge in July 1982. Upper parts of plants, especially new growth, can be toxic to cattle and sheep.

**P. caroliniana** Walter CAROLINA CANARY GRASS Annuals, 2-7 dm tall; stems tufted. Leaves ascending to spreading. Inflorescence 2-7 cm long; spikelets solitary; sterile florets 2. Glumes 5-6 mm long, dorsal wing 0.2-0.5 mm wide, margin minutely scabrous near apex; lemmas of sterile florets equal, 1-2 mm long, lemma of fertile floret 3-4 mm long, puberulent. Apr-May. Probably introduced. Rare; disturbed sites, grassy hillsides; 30-250 ft (10-75 m). South side between Forney's Cove and Christy Beach. All CA Channel islands except San Miguel, Santa Rosa, and Anacapa; Guadalupe Island; CA w of Sierra Nevada; probably native of e N. America.

**P. minor** Retz. MEDITERRANEAN CANARY GRASS Annuals, 3-9 dm tall; stems tufted. Inflorescence 2-5 cm long; spikelets solitary; sterile floret 1. Glumes 4-6 mm long, dorsal wing 0.2-0.5 mm wide, margin minutely scabrous, apex notched; lemma of sterile floret 1-2 mm long, lemma of fertile floret 3-4 mm long, puberulent. Mar-Jun. Introduced. Occasional; disturbed sites, grassy hillsides, ravines, sandy areas; coastal scrub, grasslands; 100-1250 ft (30-380 m). Widely scattered locations, primarily near w and e ends of island and near Mount Pleasant. All CA Channel islands; Todos Santos, Guadalupe, and Natividad islands; w N. America; native of Europe. Upper parts of plants, especially new growth, can be toxic to cattle.

**P. paradoxa** L. HOOD CANARY GRASS Annuals, 3-10 dm tall;stems tufted. Inflorescence 3-7 cm long; spikelets in clusters composed of 1 fertile spikelet surrounded by 5-6 sterile, reduced spikelets, clusters falling from axis as 1 unit. Glumes of fertile spikelet 4-7 mm long, wing 0.1-0.5 mm wide, toothed near apex; sterile floret absent; lemma of fertile floret 2-3.5 mm long, glabrous to sparsely puberulent. May. Introduced. Rare; moist depressions; grasslands; ca. 250 ft (75 m). Known from a single collection in Central Valley ca. 1 mile w of Stanton Ranch in May 1966. San Clemente Island; throughout CA w of Sierra Nevada; native of Europe.

## PIPTATHERUM Pal

Perennial herbs. Leaves mostly basal; ligules membranous, usually truncate; leaf blades flat to involute. Inflorescence paniculate, lower branches spreading or drooping. Spikelets compressed to subterete, pedicellate, rachilla

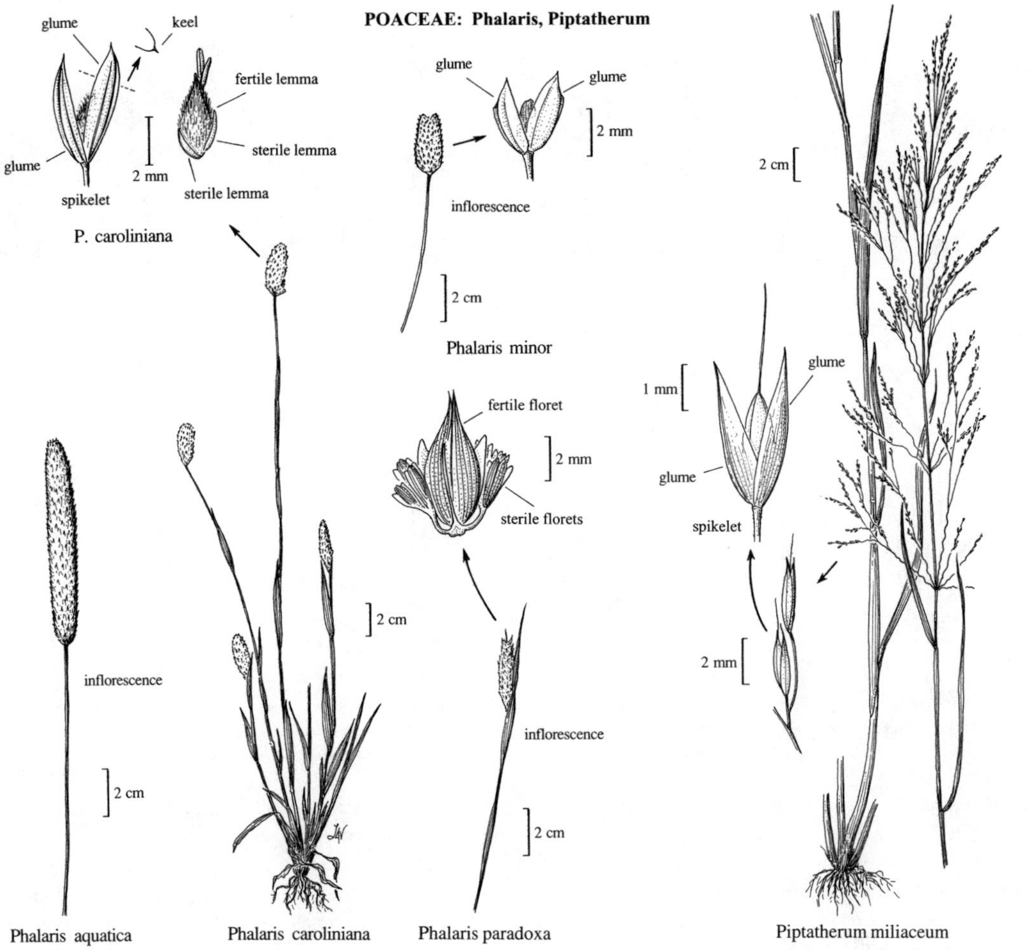


disarticulating above glumes; floret 1. Glumes subequal, longer than floret, apices rounded, acute, or short-awned, lower one 3-5-veined, upper one 5-9-veined; lemmas dorsally rounded, thin, hyaline, bases obtuse, apices awned, awn deciduous; palea usually like lemma in texture and length. Ca. 25 species; subtropical to tropical, Europe, Asia, and Africa.

**P. miliaceum** (L.) Cosson [*Oryzopsis miliacea* (L.) Asch. & Schweinf.] SMILO GRASS Plants 6-15 dm tall; stems solitary to tufted, erect or decumbent. Leaf blades 2-10 mm wide, flat. Inflorescence open, lower branches spreading. Lower glume ca. 3 mm long, apex attenuate or short-awned, upper one ca. 2 mm long, acute; lemmas 1-2 mm long, apical awns 2.5-4 mm long, straight. Apr-Sep. Introduced. Occasional; disturbed sites, moist canyon bottoms, n-facing slopes; chaparral, coastal scrub, grasslands, riparian communities; 10-1200 ft (3-365 m). North side in Cañada del Puerto and Scorpion Canyon; Central Valley between Cañada Christy and Stanton Ranch. All CA Channel islands except Santa Barbara; Great Valley to s CA; native of Europe and Asia. This species has spread rapidly since it was first collected in Cañada de la Portezuela in April 1967.

## **POA** L. BLUEGRASS

Annuals or perennial herbs with rhizomes; stems solitary or tufted. Leaves mostly basal; sheath margins usually free along most of their length, often overlapping; ligules membranous; leaf blades flat to involute. Inflorescence usually paniculate, branches ascending to drooping. Spikelets compressed to subterete, pedicellate; florets 3-12, rachilla disarticulating above glumes and between florets. Glumes subequal, lower one 1-3-veined, upper one 3-5-veined, apices acute to obtuse, dorsally keeled to rounded; lemmas usually 3-5-veined, dorsally rounded, apices acute to rounded. Ca. 500 species; temperate, worldwide.

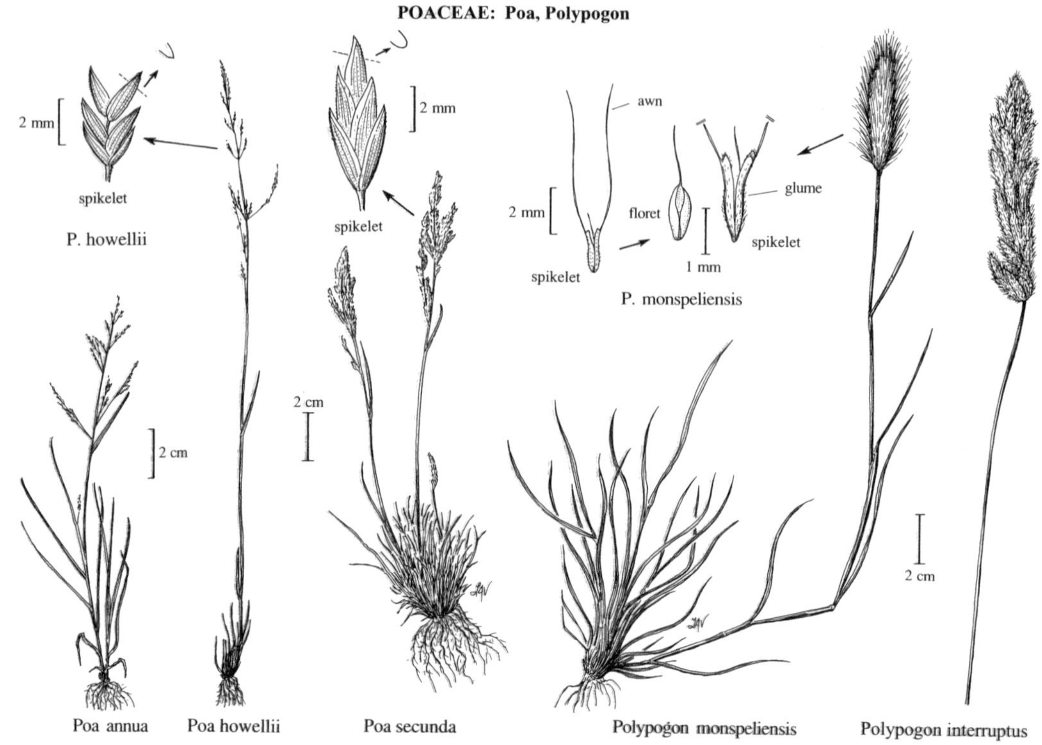


1. Perennial herbs; spikelets subterete, 5-10 mm long, lemmas dorsally rounded .............................. **P. secunda**

1' Annuals; spikelets compressed, 3-5 mm long, lemmas dorsally keeled or ridged

  2. Plants 6-30 cm tall; stems and leaf sheaths glabrous; lemmas glabrous to villosulous on veins, especially near base, callus glabrous .............................. **P. annua**

  2' Plants 30-85 cm tall; stems and leaf sheaths minutely scabrous; lemmas minutely scabrous to puberulent, callus with cobwebby trichomes .............................. **P. howellii**

**P. annua** L. ANNUAL BLUEGRASS Annuals, 6-30 cm tall; stems tufted, glabrous. Leaves mostly basal, blades 1-4 cm long, 1-3 mm wide; sheaths glabrous; ligules 1-5 mm long, apex rounded. Inflorescence 1-7 cm long, open, lower branches ascending to spreading; spikelets compressed, 3.5-4.5 mm long, florets 3-7. Glumes 0.5-2.5 mm long; lemmas 2-3.5 mm long, glabrous to villosulous on veins near base, callus glabrous. Jan-May. Introduced. Occasional; disturbed sites, moist depressions, sandy flats; coastal scrub, grasslands; 10-850 ft (3-260 m). North side at Lady's Harbor, Pelican Bay, Prisoners' Harbor, and Scorpion Ranch; scattered locations in Central Valley between Christy Beach and Stanton Ranch; s side in Cañada de los Sauces, Willows Canyon, and Coches Prietos drainage. All CA Channel islands; Guadalupe Island; throughout N. America; native of Europe. Reported by Brandegee (1888).

**P. howellii** Vasey & Scribn. [*P. bolanderi* Vasey subsp. *howellii* Vasey & Scribn.) Keck] Annuals, 30-85 cm tall; stems solitary to loosely tufted, minutely scabrous. Leaves mostly basal, blades 3-15 cm long, 2-6 mm wide; sheaths minutely scabrous; ligules 2-4 mm long, apices acute or truncate. Inflorescence 6-25 cm long, open, lower branches spreading to drooping; spikelets compressed, 3-5 mm long, florets 3-5. Glumes 1-3 mm long; lemmas 2.5-3.5 mm long, minutely scabrous to puberulent, callus with cobwebby trichomes. Apr-Jun. Rare; rocky n-facing slopes; pine forest; elevation unknown. Central Valley in Christy Pines. British Columbia to s CA.

**P. secunda** J.Presl [*P. scabrella* (Thurb.) Vasey] ONE-SIDED BLUEGRASS Perennial herbs, 30-80 cm tall; stems densely tufted, minutely scabrous. Leaves mostly basal, blades 3-16 cm long, 0.5-3 mm wide; sheaths glabrous to scabrous near margins; ligules 1-5 mm long, acute to truncate. Inflorescence 3-15 cm long, contracted, lower branches ascending to erect; spikelets terete, 5-10 mm long, florets 3-7. Glumes 2-5 mm long; lemmas 2.5-5 mm long, usually minutely scabrous, sometimes glabrous, callus glabrous or puberulent but trichomes not cobwebby. Mar-May. Common; coastal bluffs, rocky n-facing slopes, grassy hillsides; chaparral, coastal scrub, grasslands; 10-2300 ft (3-700 m). Widely scattered locations throughout island. All CA Channel islands except Santa Barbara; Todos Santos and Guadalupe islands; throughout w N. America.

## **POLYPOGON** Desf. BEARD GRASS

Annuals or perennial herbs. Leaves mostly basal; ligules membranous, thin, apices ciliolate to denticulate; leaf blades flat. Inflorescence paniculate to dense and ± spikate, ovoid to cylindrical. Spikelets subterete, short-pedicellate, falling with part of pedicel; floret 1. Glumes equal to subequal, longer than floret, 1-veined, apices acute to obtuse, apices awned, awn straight; lemmas faintly 5-veined, thin, hyaline, apices entire to bidentate, usually awned, awn straight, callus glabrous to densely puberulent, apices truncate to erose; paleas as long as or slightly shorter than lemmas. Ca. 20 species; warm temperate, Europe, Asia, Africa, and S. America.

1. Glume awns 2-4 mm long; spikelets purplish; longer inflorescences interrupted, branches ascending .......... **P. interruptus**
1. Glume awns 4-8 mm long; spikelets greenish to yellowish; inflorescence spice-like throughout, appearing plumose .......... **P. monspeliensis**

**P. interruptus** Kunth DITCH BEARD GRASS Perennial herbs, often short-lived, 1.5-8 dm tall. Inflorescence paniculate, 2-15 cm long, longer ones often interrupted below. Glumes 2-3 mm long, usually purplish, awns 2-4 mm long; lemmas 1-2 mm long, awns 1-3 mm long. May-Sep. Introduced. Occasional; stream banks, moist floodplains, freshwater seeps; coastal scrub, riparian communities; 10-1300 ft (3-395 m). Widely scattered locations throughout western and central portions of island. San Miguel, Santa Rosa, Santa Catalina, and San Clemente islands; throughout N. America; native of S. America. First collected (without exact locality) in March 1927.

**P. monspeliensis** (L.) Desf. RABBITSFOOT GRASS Annuals, 1.5-5 dm tall. Inflorescence paniculate, dense, spike-like, 2-14 cm long. Glumes 1-2.5 mm long, greenish to yellowish, awns 4-8 mm long; lemmas 0.5-1 mm long, awns 1-4 mm long. Jun-Sep. Introduced. Occasional; stream banks, moist floodplains, freshwater seeps, sandy flats; coastal scrub, riparian communities; 10-500 ft (3-150 m). Widely scattered locations throughout island. All CA Channel islands; Todos Santos, Guadalupe, and Cedros islands; throughout N. America; native of Europe. Reported by Greene (1887).

## **SCHISMUS** P.Beauv. MEDITERRANEAN GRASS

Annuals; stems tufted. Leaves mostly basal; sheath margins free along most of their length; ligules mostly ciliate; leaf blades flat to involute. Inflorescence usually paniculate, dense. Spikelets compressed to subterete, pedicellate; florets 3-10, rachilla disarticulating above glumes and between florets. Glumes equal to subequal, 3-7-veined, apices acute; lemmas 7-9-veined, dorsally rounded, apices bidentate. Ca. 5 species; temperate to tropical, s Europe, Africa, and Asia.

1. Lemmas of lower florets 2.5-4 mm long; palea of lowest floret 2-3 mm long, apex acute .......... **S. arabicus**
1' Lemmas of lower florets 2-2.5 mm long; palea of lowest floret 2-2.5 mm long, apex obtuse ..... **S. barbatus**

**S. arabicus** Nees Plants 3-20 cm tall. Leaf blades 0.5-1.5 mm wide, involute, glabrous to sparsely pubescent. Inflorescence 1-3.5 cm long; spikelets tan to brownish; florets 5-7. Glumes 4.5-6 mm long; lemmas obovate, 2.5-4 mm long, puberulent, margins ciliate, apices bidentate; paleas 1.5-3 mm long, apices acute. Apr-May. Introduced. Rare; disturbed sites, floodplains; riparian communities; ca. 300 ft (90 m). Known from a single collection in Central Valley near La Cascada in May 1990. Santa Barbara, Santa Catalina, and San Clemente islands; San Joaquin Valley to n Mexico, e to UT and TX; native of Europe and Asia.

**S. barbatus** (L.) Thell. Plants 3-20 cm tall. Leaf blades 0.5-2 mm wide, involute, glabrous to sparsely pubescent. Inflorescence 1-3.5 cm long; spikelets often purplish; florets 5-7. Glumes 3.5-5 mm long; lemmas obovate, 2-2.5 mm long, puberulent below middle, margins ciliate, apices bidentate, often mucronate; paleas 1-2.5 mm long, apices obtuse. Feb-May. Introduced. Scarce; coastal bluffs, disturbed sites, roadbeds, floodplains, rocky s-facing slopes, exposed ridgetops; coastal scrub, riparian communities; 60-1750 ft (20-535 m). North side in Cañada del Puerto; scattered locations in Central Valley between Christy and Stanton ranches; s side near Coches Prietos Anchorage. Santa Catalina Island; s CA to n Mexico, e to TX; native of Europe and Africa. This species has spread rapidly since it was first collected in Cañada del Puerto in August 1977.

## **VULPIA** C.C.Gmel.

Annuals; stems solitary or few and loosely tufted. Leaves usually cauline; sheath margins free along most of their length, often overlapping; ligules membranous, apices minutely fimbriate; leaf blades narrow, usually involute.

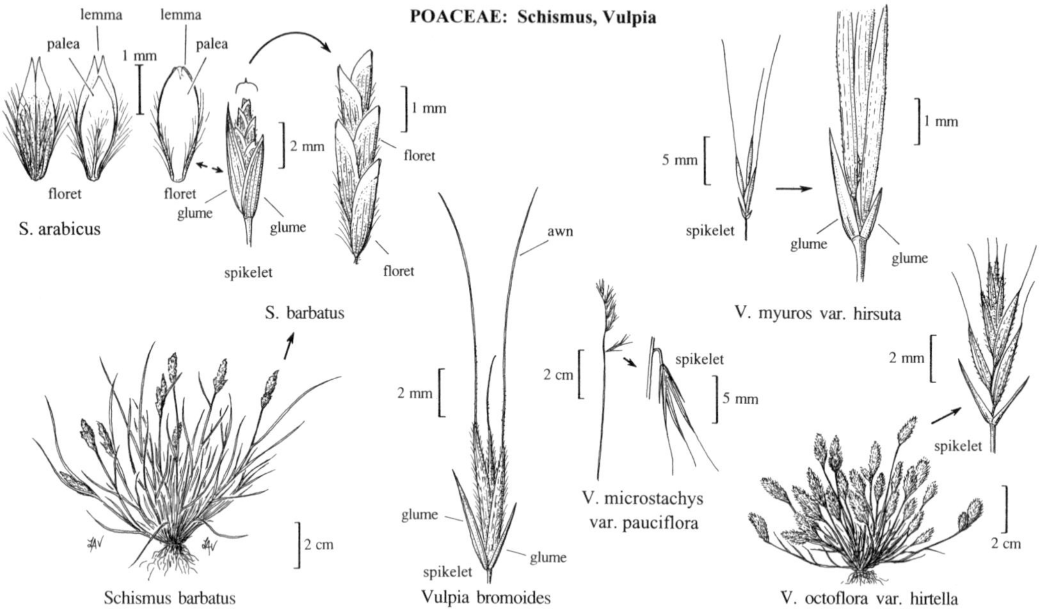


Inflorescence usually paniculate, open to dense, branches appressed to spreading. Spikelets compressed to subterete, pedicellate; florets 2-10, rachilla disarticulating above glumes and between florets. Glumes unequal, usually longer than lowermost florets, lower one 1-3-veined, upper one 3-veined, apices acute to short-awned; lemmas 3-5-veined, dorsally rounded, apices awned. Ca. 20 species; temperate, Europe and w N. America.

1. Lower glume 0.5-2 mm long, less than half the length of upper glume ...................... **V. myuros** var. **hirsuta**
1' Lower glume 2-5 mm long, more than half the length of upper glume
  2. Florets 5-15 per spikelet; awns of lower florets 2-5 mm long .......................... **V. octoflora** var. **hirtella**
  2' Florets 2-7 per spikelet; awns of lower florets 3-12 mm long
    3. Lowest inflorescence branches appressed to erect ....................................................... **V. bromoides**
    3' Lowest inflorescence branches spreading ..................................... **V. microstachys** var. **pauciflora**

**V. bromoides** (L.) A.Gray [*Festuca dertonensis* (All.) Asch. & Graebn.] BROME FESCUE Plants 6-40 cm tall. Leaf blades 3-12 cm long, Inflorescence dense, 3-11 cm long, branches 1-3 per node, appressed; florets 3-7, not densely aggregated, rachilla between lower florets visible in fruit. Glumes subequal, 3-7 mm long; lemma of lowermost floret 5-8 mm long; awns of lower florets 3-12 mm long. Apr-Jun. Introduced. Common; disturbed sites, grassy flats, hillsides, rocky slopes, exposed ridgetops; chaparral, coastal scrub, grasslands, pine forest; 50-2470 ft (15-755 m). Widely scattered locations throughout island. All CA Channel islands except Santa Barbara and San Nicolas; Guadalupe Island; throughout N. America; native of Europe. First collected at Coches Prietos Anchorage in March 1929; reported as "common and widely distributed" by Hoffmann (1932a).

**V. microstachys** (Nutt.) Benth. var. **pauciflora** (Beal) Lonard & Gould PACIFIC FESCUE Plants 7-17 cm tall. Leaf blades 3.5-7 cm long, Inflorescence usually open, 3-8 cm long, branches 1 per node, usually spreading; florets 2-5, not densely aggregated, rachilla between lower florets visible in fruit. Glumes unequal, lower one 2-5 mm long, upper one 3-7.5 mm long; lemma of lowermost floret 4-9 mm long; awns of lower florets 3-10 mm long. Apr. Scarce; disturbed sites, rocky n-facing slopes, moist depressions; chaparral, coastal scrub; 1000-1950 ft (305-595 m). North side between Picacho Diablo and Red Peak; Central Valley near Red Peak. San Miguel, Anacapa, Santa Catalina, and San Clemente islands; San Martin, Guadalupe, and Cedros islands; WA to Baja CA. Three other varieties occur on the mainland of w N. America.

**V. myuros** (L.) C.C.Gmel. var. **hirsuta** (Hack.) Asch. & Graebn. [*Festuca megalura* Nutt.] FOXTAIL FESCUE. Plants 10-60 cm tall. Leaf blades 2.5-12 cm long, Inflorescence dense, usually 5-20 cm long, branches appressed to ascending, sometimes drooping in fruit; florets 3-7, not densely aggregated, rachilla between lower florets visible in fruit. Glumes strongly unequal, lower one 0.5-2 mm long, upper one 2.5-5.5 mm long; lemma of lowermost floret 4.5-7 mm long; awns of lower florets 4.5-7 mm long; lemma margins ciliate near apices. Mar-

Jun. Introduced. Common; disturbed sites, grassy flats, hillsides, rocky slopes, exposed ridgetops; chaparral, coastal scrub, grasslands; 50-2470 ft (15-755 m). Widely scattered locations throughout island. All CA Channel islands; Los Coronados, Todos Santos, and Guadalupe islands; worldwide; native of Europe. Var. *myuros*, with glabrous lemma margins, occurs on the adjacent mainland. First collected (without exact locality) by T. Brandegee in 1888.

**V. octoflora** (Walter) Rydb. var. **hirtella** (Piper) Henrard SIX-WEEKS FESCUE Plants 3-20 cm tall. Leaf blades 1-4 cm long, Inflorescence dense, 1.5-3 cm long, branches appressed to erect; florets 5-15, densely aggregated, rachilla not visible. Glumes subequal, 2-6 mm long; lemma of lowermost floret 2.5-6 mm long, scabrous near apex; awns of lower florets 2-5 mm long. Mar-Jun. Scarce; coastal flats, rocky slopes; coastal scrub, grasslands, pine forest; 100-750 ft (30-230 m). North side near Pelican Bay; Central Valley in Cañada Christy; s side at Pozo Beach. All CA Channel islands; San Martin, Guadalupe, and Cedros islands; throughout w N. America. Var. *octoflora*, with glabrous to minutely scabrous lemmas, occurs on the adjacent mainland.

## POTAMOGETONACEAE Dumort
### Pondweed Family

Annuals or perennial herbs, aquatic, usually submersed, with rhizomes or tuber-like buds. Leaves cauline, alternate or opposite; blades simple, flat to filiform, submersed or floating; stipules free or fused, usually sheathing stem. Flowers bisexual to unisexual, radial, 2-many, whorled or in small, axillary to terminal spikes. Perianth parts 4 or absent, hyaline, green, or brownish; stamens 2 or 4, sessile; pistils 4, locule 1, ovules 1 per locule. Fruit drupe-like or an achene. Ca. 3 genera and 9 species; worldwide.

1. Perianth parts 4; stamens 4; inflorescence a spike, composed of 2-6 whorls of sessile flowers, pistils and fruits sessile ........ **Potamogeton**

1' Perianth parts absent; stamens 2; inflorescence composed of 2 sessile flowers, pistils sessile in flower, long-stalked in fruit ........ **Ruppia**

### POTAMOGETON L. PONDWEED

Perennial herbs, with rhizomes or tuber-like buds. Leaves cauline, usually alternate; floating blades, if present, usually elliptic to ovate, submersed ones narrowly elliptic to filiform. Flowers bisexual, radial, in small, axillary to terminal spikes. Perianth parts 4, hyaline to brownish; stamens 4; pistils 4, locule 1, ovules 1 per locule. Fruit an achene, slightly compressed. Ca. 90 species; temperate, Europe, Asia, and N. America.

**P. pectinatus** L. FENNEL-LEAF PONDWEED Leaves submerged, filiform, 3-15 cm long, 1-1.5 mm wide, apices sharply acute; stipules adnate to petiole, 2-5 cm long. Inflorescence a spike, composed of 2-6 whorls of sessile flowers; peduncles 5-25 cm long. Perianth parts 1-2 mm long, green to brownish. Fruit 2-4 mm long, dorsally keeled. Apr-Jun. Rare, not seen recently; stream beds, shallow ponds; riparian communities; ca. 10 ft (3 m). Known only from three collections at Prisoners' Harbor in April, May, and June 1930. Santa Rosa and Santa Catalina islands; throughout N. America, Europe and Asia. Entire plant can be eaten.

### RUPPIA L. DITCHGRASS

Perennial herbs, with slender rhizomes. Stems filiform, widely branching. Leaves submerged, cauline, usually alternate, linear to filiform; stipules adnate to petioles. Flowers bisexual, radial, usually 2, in terminal, sessile clusters. Perianth parts absent; stamens 2; pistils 4, locule 1, ovules 1 per locule. Fruit drupe-like, usually ovoid, usually stalked in fruit. Ca. 7 species; temperate to subtropical, worldwide.

**R. maritima** L. Stems 20-50 cm long. Leaves 2-10 cm long, ca. 0.5 mm wide; stipules sheathing, 2-12 mm long. Peduncles 2-10 mm long, straight. Fruit ovoid, 2-4 mm long, stalks 5-15 mm long. Apr-Nov. Rare; stream beds, shallow ponds; riparian communities; 10-70 ft (3-20 m). North side at Prisoners' Harbor; s side in Rancho Nuevo Canyon. Santa Rosa, San Nicolas, Santa Catalina, and San Clemente islands; ± worldwide.

## TYPHACEAE Juss.
### Cattail Family

Monoecious perennial herbs, usually aquatic, lower stems and rhizomes usually submersed; stems usually cylindrical, solid (filled with a spongy pith), often with thick rhizomes. Leaves cauline, alternate; blades linear, flat or

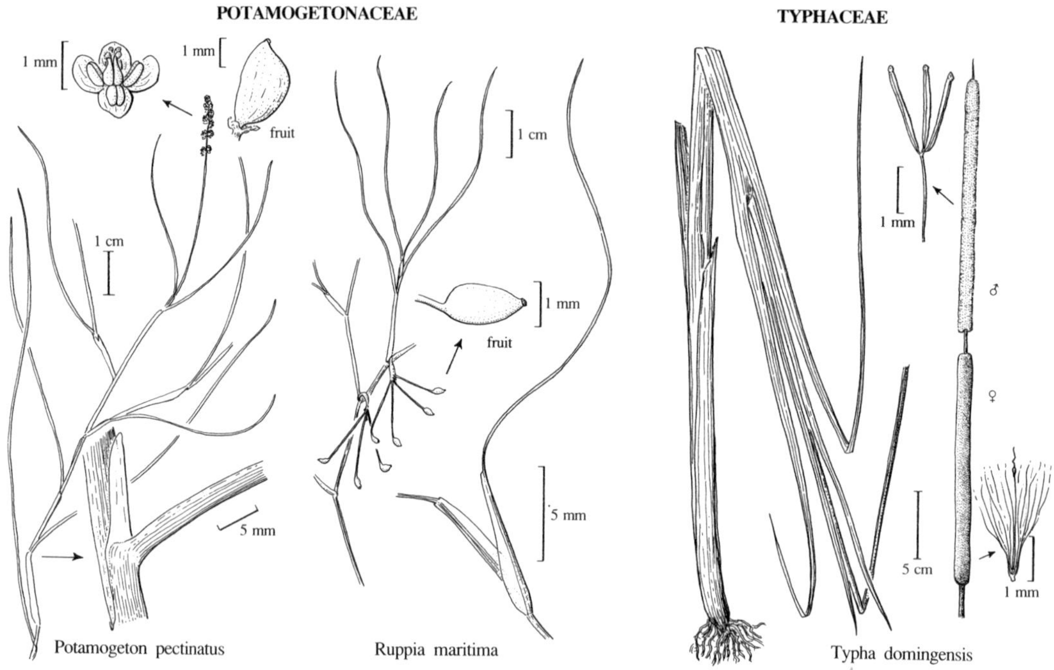


folded, sheathing stem. Flowers unisexual, radial, many, in dense, cylindrical or globose spikes. Perianth parts few or absent, minute, scale-like, greenish to brownish; staminate flowers with 1-8 stamens; pistillate flowers with 1 pistil, 1-2 locules, style 1. Fruit an achene, sessile or pedicellate; seeds 1-2. Two genera and ca. 25 species; worldwide.

## TYPHA L. CATTAIL

Stems erect, stiff, thick. Leaf sheath margins overlapping, blades linear, folded below, flat above middle. Flowers in dense, cylindrical spikes, staminate flowers usually borne above pistillate ones; perianth absent. Staminate flower pedicellate, subtended by minute, hyaline to membranous scales, stamens 2-7; pistillate flowers sterile or fertile, borne on persistent, long-villous pedicels, style filiform, stigmas linear to lanceolate. Fruit fusiform, deciduous with pedicel. Ca. 12 species; mostly temperate, worldwide. Flower buds, pollen, seeds, and rhizomes were eaten by Native Americans; flowers and rhizomes also were used medicinally.

1. Staminate segment of spike separated from pistillate segment by naked axis; pistillate segment 15-25 mm wide in fruit; pistillate pedicel subtended by minute, linear to narrowly spatulate bract (often obscured by villous pubescence) .......... **T. domingensis**

1' Staminate segment of spike confluent with pistillate segment; pistillate segment 25-35 mm wide in fruit; pistillate pedicel bractless .......... **T. latifolia**

**T. domingensis** Pers. NARROW-LEAVED CATTAIL Plants 1-3.5 m tall. Leaves 5-15 mm wide, upper surface glandular near junction with sheath. Axis between staminate segment of spike and pistillate segment naked, 0.5-3 cm long; pistillate segment 15-25 cm wide in fruit. Staminate bracts simple or branched, straw-colored to reddish brown. Pistillate pedicel ca. 0.5 mm long, subtended by minute, linear to narrowly spatulate bract; stigma linear, usually deciduous in fruit; sterile ovaries rounded to truncate, fertile ones fusiform. Jul-Oct. Occasional; stream beds, shallow ponds, freshwater seeps, floodplains; riparian communities; 10-500 ft (3-150 m). Widely scattered locations throughout island. San Miguel, Santa Rosa, San Nicolas, and Santa Catalina islands; Cedros Island; throughout CA; ± worldwide.

**T. latifolia** L. BROAD-LEAVED CATTAIL Plants 1-3 m tall. Leaves 10-20 mm wide, upper surface glabrous near junction with sheath. Staminate and pistillate segments of inflorescence confluent; pistillate segment 25-35 cm wide in fruit. Staminate bracts simple, whitish. Pistillate pedicel 1.5-3 mm long, bractless (except for villous trichomes); stigma lanceolate, usually persistent in fruit; sterile ovaries rounded, fertile ones fusiform. Aug. Rare; floodplains, stream beds; riparian communities; 10-120 ft (3-35 m). North side at Prisoners' Harbor; s side in Cañada de los Sauces. San Nicolas, Santa Catalina, and San Clemente islands; N. and C. America, Europe, Asia, and Africa.

## ZANNICHELLIACEAE Dumort
### HORNED PONDWEED FAMILY

Monoecious or dioecious perennial herbs, with rhizomes, aquatic, submersed. Leaves cauline, lower ones alternate, upper ones opposite or whorled; blades simple, flat to filiform; stipules free or fused, usually sheathing stem. Flowers unisexual, radial, few, axillary, staminate ones usually below pistillate ones. Staminate flowers with perianth parts 3 or absent, stamens 1-3. Pistillate flowers without perianth, pistil 1, locule 1, style simple or 3-lobed. Fruit an achene or nutlet. Ca. 4 genera and 10 species; ± worldwide.

### ZANNICHELLIA L. HORNED PONDWEED

With only 1 species.

**Z. palustris** L. Plants monoecious. Stems filiform, simple to branched. Leaves subopposite, upper ones ± whorled, 2-10 cm long, blade 0.5-1 mm wide, apex acute; stipules free, scarious. Staminate flowers solitary, without perianth, longer than pistillate flowers; stamen 1; Pistillate flowers 3-5, sessile, subtended by hyaline bract; stigma peltate. Fruit a nutlet, pedicellate, dorsally ridged, denticulate, style persistent. Sep. Rare, not seen recently; stream beds; riparian communities; ca. 200 ft (60 m). Known from a single collection in Cañada del Puerto in September 1932. Throughout CA; worldwide.

## ZOSTERACEAE Dumort
### EELGRASS FAMILY

Aquatic, marine perennial herbs, with rhizomes. Leaves cauline, alternate, usually in 2 ranks or tufted; sheath blades simple, flat, ribbon-like, submersed or floating; stipules absent. Flowers unisexual, radial, many, usually in 2 rows, sessile, in axillary flattened spikes, each subtended by one thin, leaf-like bract. Perianth parts absent; staminate flowers with 1 stamen, anthers usually sessile, pollen filiform; pistillate flowers with 1 pistil, locule 1, ovule 1 per locule. Fruit drupe-like or an achene. Ca. 3 genera and 20 species; seacoasts, ± worldwide.

1. Plants with short, thick rhizomes, occurring in tidal zone, often on wave-swept rocks; leaves tufted, 1-5 mm wide, blades thick to subterete; inflorescence subtended by leaf-like bract .............................. **Phyllospadix**

1' Plants with long, slender rhizomes, usually occurring in deep waters below tidal zone, often found stranded on beaches; leaves cauline, well-separated, 2-18 mm wide, blades thin; inflorescence axillary, enclosed by sheath-like leaf base .............................. **Zostera**

### PHYLLOSPADIX Hook. SURF-GRASS

Plants dioecious. Rhizome thick, internodes short. Leaves usually tufted, thick to subterete, 0.2-1 m long, 1-5 mm wide, lower segment of sheath fibrous with age. Staminate and pistillate inflorescences each subtended or enclosed by thin, leaf-like bract. Fruit a nutlet, crescent-shaped, base with 2 horn-like lobes. Ca. 5 species; n Pacific coast. Leaves of both species were woven into mats by Native Americans. Insular distributions of both species in CA need further study.

1. Leaves thin, flat, 1.5-4 mm wide; pistillate inflorescences near base, usually 1 per stem near base .............................. **P. scouleri**

1' Leaves thick, cylindrical to elliptic in cross-section, 0.5-2 mm wide; pistillate inflorescences cauline, 3-7 per stem .............................. **P. torreyi**

**P. scouleri** Hook. Leaves 4-10 dm long, 1.5-4 mm wide, flat, thin. Pistillate inflorescences usually solitary, borne near base of stem; peduncles 1-5 cm long. Apr-May. Common; rocky shorelines; intertidal and subtidal marine community; just below mean sea level. Around perimeter of island. Anacapa, Santa Barbara, San Nicolas, Santa Catalina, and San Clemente islands; all Baja CA islands except Todos Santos and Guadalupe; AK to Baja CA.

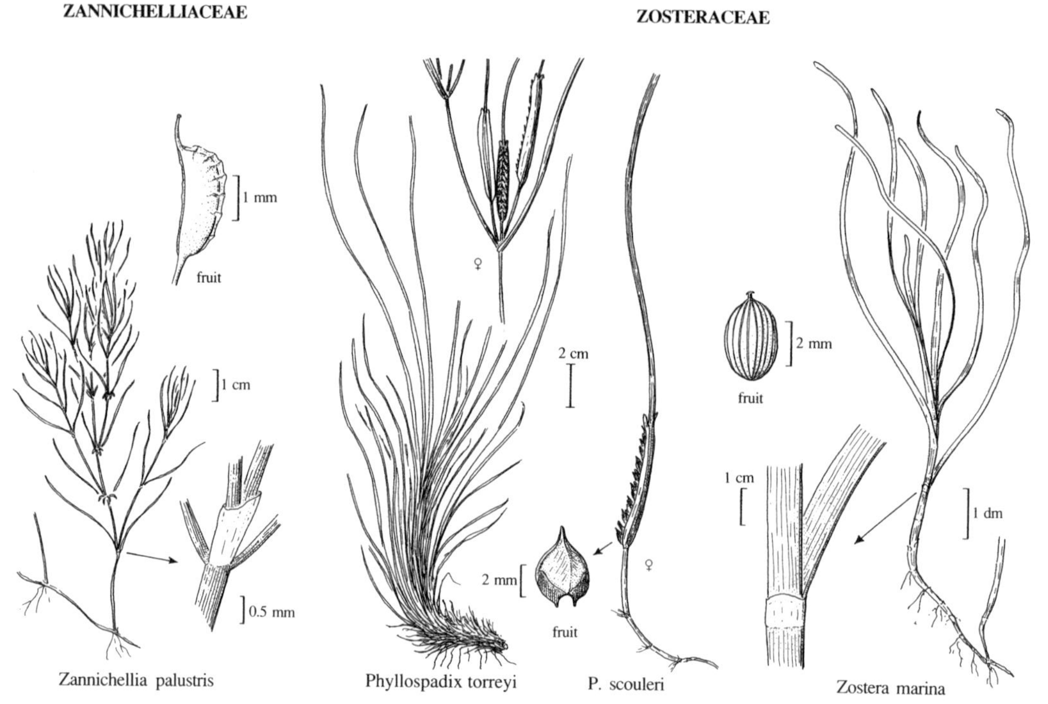


**P. torreyi** S.Watson Leaves 3-6.5 dm long, 0.5-2 mm wide, folded, elliptic in cross-section, thick. Pistillate inflorescences usually in pairs, 3-7 per stem, cauline; peduncles 6-25 cm long. May-Jun. Common; rocky shorelines; intertidal and subtidal marine community; just below mean sea level. Around perimeter of island. All CA Channel islands; Todos Santos, San Martin, and Guadalupe islands; OR to Baja CA.

## ZOSTERA L. EEL-GRASS

Plants monoecious. Rhizome slender, internodes long. Leaves cauline, well-separated, thin, often translucent, 0.5-1.5 m long, 2-18 mm wide, lower segment of sheath scaly or fragmented with age. Staminate and pistillate flowers alternating in a spike, subtended or enclosed by sheath-like leaf base. Fruit a nutlet, ovoid to cylindrical, base truncate. Ca. 12 species; cold to temperate marine waters, ± worldwide. Insular distributions of both species in CA need further study.

1. Leaves 5-12 mm wide, apices usually obtuse; seed surface with 20-25 longitudinal ridges; rooted plants usually submersed no more than 2-4 m below mean tide .......... **Z. marina**

1' Leaves 12-18 mm wide, apices usually notched; seed surface smooth to faintly ridged; rooted plants usually in deep waters, often as deep as 15 m below mean tide .......... **Z. pacifica**

**Z. marina** L. Leaves 0.3-1.5 m long, 5-12 mm wide, apices usually obtuse. Inflorescence 2.5-6 cm long. Fruit ca. 3 mm long; seed surface with 20-25 longitudinal ridges. Flowering in Spring. Scarce; sandy sites; subtidal marine community; 0-12 ft (0-4 m) below mean sea level. North side in Prisoners' Harbor and at Scorpion Anchorage; s side between Forney's Cove and Malva Real Anchorage. Santa Rosa, Anacapa, and Santa Catalina islands; Los Coronados, San Martin, San Geronimo, Guadalupe, and Cedros islands; AK to Baja CA and Gulf of California. Rhizomes and leaf bases can be eaten.

**Z. pacifica** L. Leaves 0.5-1.5 m long, 12-18 mm wide, apices usually notched. Inflorescence 2-5 cm long. Fruit ca. 3 mm long; seed surface smooth to faintly ridged. Flowering in late summer. Scarce; sandy sites; subtidal marine community; 15-50 ft (4.5-15 m) below mean sea level. North side at China Harbor; s side off Cañada de los Sauces, Pozo Beach, and Smugglers' Cove. British Columbia to s CA.

# LITERATURE CITED

Abrams, L. 1940-1951. Illustrated flora of the Pacific states. Volumes 1-3. Stanford University Press, Stanford, California. 2037 pp.

_____. and R. Ferris. 1960. Illustrated flora of the Pacific states. Volume 4. Stanford University Press, Stanford, California. 732 pp.

Apt, K., C. D'Antonio, J. Crisp, and J. Gauvain. 1988. Intertidal macrophytes of Santa Cruz Island, California. Publication No. 6. Herbarium, University of California at Santa Barbara. 87 pp.

Axelrod, D. 1980. History of the maritime closed-cone pines, Alta and Baja California. University of California Publications in Geological Sciences 120: 1-143.

_____. 1983. New Pleistocene conifer records, coastal California. University of California Publications in Geological Sciences 127: 1-108.

Baber, D. 1982. Report on a survey of feral pigs on Santa Cruz Island, California: Ecological implications and management recommendations. Unpublished report submitted to The Nature Conservancy, Santa Barbara, California. 19 pp.

Bancroft, H. 1884. The works of Hubert Howe Bancroft. Volume 18. History of California. Volume 1: 1542-1800. The History Company, San Francisco, California. 744 pp.

_____. 1886. The works of Hubert Howe Bancroft. Volume 20. History of California. Volume 3: 1825-1840. The History Company, San Francisco, California. 792 pp.

Barrett, L. 1935. A record of forest and field fires in California from the days of the early explorers to the creation of the Forest Reserves. Unpublished manuscript submitted to U.S. Forest Service, San Francisco, California. On file in Library, University of California at Berkeley.

Bean, J. and K. Saubel. 1972. *Temalpakh*: Cahuilla Indian knowledge and usage of plants. Malki Museum Press, Banning, California. 225 pp.

Beatty, S. and D. Licari. 1992. Invasion of fennel (*Foeniculum vulgare* Mill.) into shrub communities on Santa Cruz Island, California. Madroño 39: 54-66.

Bjørndalen, J. 1978. The chaparral vegetation of Santa Cruz Island, California. Norwegian Journal of Botany 25: 255-269.

Blake, E. 1887. The coast fox. West American Scientist 3: 49-52.

Brandegee, T. 1888. Flora of the Santa Barbara Islands. Proceedings of the California Academy of Sciences (2nd series) 1: 201-226.

_____. 1890. Flora of the California islands. Zoe 1: 129-148.

Bremner, C. 1932. Geology of Santa Cruz Island, Santa Barbara County, California. Santa Barbara Museum of Natural History Occasional Papers 1: 1-33.

Brumbaugh, R. 1980a. Erosion and soils, Santa Cruz Island, California. Unpublished report submitted to The Nature Conservancy, Santa Barbara, California. 110 pp.

_____. 1980b. Recent geomorphic and vegetal dynamics on Santa Cruz Island, California. Pp. 139-158 *in* D. M. Power (ed.) The California Islands: Proceedings of a multidisciplinary symposium. Santa Barbara Museum of Natural History, Santa Barbara, California.

_____. 1983. Hillslope gullying and related changes, Santa Cruz Island, California. Unpublished Ph.D. dissertation, University of California at Los Angeles. 192 pp.

Brummitt, R. and C. Powell (eds.) 1992. Authors of plant names. Royal Botanic Gardens, Kew, England. 732 pp.

Caire, D. 1933. First visit to Santa Cruz Island (1880), an excerpt from the memoirs of Delphine Adelaide Caire. Unpublished manuscript on file at Santa Cruz Island Foundation, Santa Barbara, California. 7 pp.

Caire, H. 1993. Santa Cruz Island: A history and recollections of an old California rancho. Arthur H. Clark Company, Spokane, WA. 191 pp.

California State Water Resources Control Board. 1979. California's marine waters areas of special biological significance: Reconnaissance survey report for Santa Cruz Island. Water Quality Monitoring Report No. 79-8: 1-109.

Camp, C. 1963. Lorenzo Gordin Yates (1837-1909). Journal of the Society for the Bibliography of Natural History 4: 178-193.

Cantelow, E. and H. Cantelow. 1957. Biographical notes on persons in whose honor Alice Eastwood named native plants. Leaflets of Western Botany 8: 83-101.

Carlquist, S. 1974. Island biology. Columbia University Press, New York, N.Y. 660 pp.

Carroll, M., L. Laughrin, and A. Bromfield. 1993. Fire on the California Islands: Does it play a role in chaparral and closed-cone pine forest habitats? Pp. 73-88 *in* F. Hochberg (ed.) Third California Islands Symposium: Recent advances in research on the California Islands. Santa Barbara Museum of Natural History, Santa Barbara, California.

Chaney, R. and H. Mason. 1930. A pleistocene flora from Santa Cruz Island, California. Carnegie Institute of Washington Publication No. 415: 1-24.

Cheatham, N. and J. Haller. 1975. An annotated list of California habitat types. Unpublished manuscript prepared for University of California Natural Land and Water Reserve System. 81 pp.

Clark, R., W. Halvorson, A. Sawdo, and K. Danielsen. 1990. Plant communities of Santa Rosa Island, Channel Islands National Park. Technical Report No. 42. Cooperative National Park Resources Studies Unit, University of California at Davis. 93 pp.

Clark, W. 1979. The taxonomy of *Hazardia* (Compositae: Astereae). Madroño 26: 105-127.

Clarke, C. 1977. Edible and useful plants of California. University of California Press, Berkeley. 280 pp.

Clokey, I. 1931. Notes on the flora of Santa Cruz Islands, California. Bulletin of the Southern California Academy of Sciences 30: 60-61.

Cole, K. and G. Liu. 1994. Holocene paleoecology of an estuary on Santa Rosa Island, California. Quaternary Research 41: 326-335.

Daily, M. and C. Stanton. 1983. Historical highlights of Santa Cruz Island. La Reata 5: 14-19.

Dibblee, T. 1976. Geology of Santa Barbara County and its influence on vegetation. Pp. 7-13 *in* C. Smith. A Flora of the Santa Barbara Region, California. Santa Barbara Museum of Natural History, Santa Barbara, California.

_____. 1982. Geology of the Channel Islands, southern California. Pp. 27-39 *in* D. Fife and J. Minch (eds.) Geology and mineral wealth of the California Transverse Ranges. Annual Symposium and Guidebook Number 10. South Coast Geological Society, Santa Ana, California.

Doerner, D. 1969. Lower tertiary biostratigraphy of southwestern Santa Cruz Island. Pp. 17-29 *in* D. Weaver (ed.) Geology of the Northern Channel Islands. Special publication. American Association of Petroleum Geologists (Pacific Section) and Society of Economic Paleontologists and Mineralogists.

Dunkle, M. 1950. Plant ecology of the Channel Islands of California. Allan Hancock Pacific Expeditions 13 (3): 247-386. University of Southern California Press, Los Angeles.

Eastwood, A. 1933. New manzanitas from Santa Cruz Island, California. Leaflets of Western Botany 1: 61-63.

_____. 1941. The islands of southern California and a list of the recorded plants. Leaflets of Western Botany 3: 27-36, 54-78.

Eaton, M. 1980. Diary of a sea captain's wife: Tales of Santa Cruz Island. McNally and Loftin, Santa Barbara, California. 256 pp.

Ellison, W. 1937. History of the Santa Cruz Island grant. Pacific Historical Review 7: 270-283.

Erlandson, J. 1994. Early Hunter-Gatherers of the California coast. Plenum Press, New York. 336 pp.

Ewan, J. 1955. San Francisco as a mecca for nineteenth century naturalists, with a roster of biographical references to visitors and residents. Pp. 1-63 *in* A century of progress in the natural sciences, 1853-1953. California Academy of Sciences, San Francisco.

_____. and N. Ewan. 1981. Biographical dictionary of Rocky Mountain naturalists: a guide to the writings and collections of botanists, zoologists, geologists, artists, and photographers, 1682-1932. Regnum Vegetabile 107: 1-253.

Fenzi, E. 1976. Doctor Franceschi and the Catalina ironwood. Pacific Horticulture 37: 15-16.

Ford, H. 1890. The indigenous shrubs of Santa Barbara County. Bulletin of the Santa Barbara Society of Natural History 1: 29-31.

Foreman, R. 1967. Observations on the flora and ecology of San Nicolas Island. Report USNRDL-TR-67-8. U.S. Naval Radiological Defense Laboratory, San Francisco, California. 79 pp.

Forney, S. 1875. Progress report for 1 November 1874 to 30 June 1875. Unpublished manuscript on file at U.S. National Archives, Washington, D.C. Record Group 23 (records of the U.S. Coast and Geodetic Survey), Bache Records, Entry 22, Correspondence from assistants (1866-1875).

Fuller, T. and E. McClintock. 1986. Poisonous plants of California. University of California Press, Berkeley. 433 pp.

Gentry, H. 1949. Land plants collected by the *Velero III*, Allan Hancock Pacific Expeditions 1937-1941. Allan Hancock Pacific Expeditions 13 (2): 7-245. University of Southern California Press, Los Angeles.

Gherini, P. 1966. Island rancho. Noticias 12: 14-20.

Gillett, G. 1955. Variation and genetic relationships in the Whitlavia and Gymnobythus Phacelias. University of California Publications in Botany 28: 19-77.

Glassow, M. 1977. Archaeological overview of the Northern Channel Islands, including Santa Barbara Island. Western Archaeological Center, National Park Service, Tucson, Arizona.

_____. 1980. Recent developments in the archaeology of the Channel Islands. Pp. 79-99 *in* D. Power (ed.) The California Islands: Proceedings of a multidisciplinary symposium. Santa Barbara Museum of Natural History, Santa Barbara, California.

Goeden, R., C. Fleschner, and D. Ricker. 1967. Biological control of prickly pear cacti on Santa Cruz Island, California. Hilgardia 38: 579-606.

_____. and D. Ricker. 1981. Santa Cruz Island revisited. Sequential photography records the causation, rates of progress, and lasting benefits of successful biological weed control. Pp. 355-365 *in* E. Del Fosse (ed.) *?,*Proceedings of Fifth International Symposium on the Biological Control of Weeds. Commonwealth Scientific and Industrial Research Organization, Brisbane, Australia.

Goodyear, W. 1890. Santa Cruz Island. Pp. 155-170 *in* Ninth Annual Report of the State Mineralogist. California State Mining Bureau, Sacramento.

Grant, A. 1924. A monograph of the genus *Mimulus*. Annals of the Missouri Botanical Garden 11: 99-388.

Greene, E. 1887. Studies in the botany of California and parts adjacent. VI. Notes on the botany of Santa Cruz Island. Bulletin of the California Academy of Sciences 2: 377-418.

Greenwell, W. 1856-1857. Description of signals, Santa Cruz Island, Santa Barbara Channel, Section 10. Unpublished field notes on file at U.S. National Archives, Washington, D.C. Record Group 23 (records of the U.S. Coast and Geodetic Survey), Box 106. 47 pp.

Greuter, W. 1988. International code of botanical nomenclature. Regnum Vegetabile 118: 1-328.

Hall, H. 1907. Compositae of southern California. University of California Publications in Botany 3 (1): 1-302.

Haller, J. 1986. Taxonomy and relationships of the mainland and island populations of *Pinus torreyana* (Pinaceae). Systematic Botany 11: 39-50.

Helen, R. 1979. A brief history of the Fry's Harbor quarry, Santa Cruz Island, California. Unpublished manuscript on file at Santa Cruz Island Foundation, Santa Barbara, California. 6 pp.

Heusser, L. 1978. Pollen in Santa Barbara Basin, California: A 12,000 year record. Geological Society of America Bulletin 89: 673-678.

Hevly, R. and J. Hill. 1970. Pollen in archaeological middens from Santa Cruz Island, California. University of California Archaeological Survey Annual Report 12: 108-118.

Hickman, J. (ed.). 1993. The Jepson manual: higher plants of California. University of California Press, Berkeley. 1400 pp.

Hill, D. 1976. Geology of the Jurassic basement rocks, Santa Cruz Island, California, and correlation with other Mesozoic basement terranes in California. Pp. 16-46 *in* D. Howell (ed.) Aspects of the geologic history of the California continental borderland. Miscellaneous Publication No. 24. Pacific Section, American Association of Petroleum Geologists.

Hobbs, E. 1978. The effects of grazing on bishop pine (*Pinus muricata*) forests, Santa Cruz Island, California. Unpublished M.A. thesis, University of California at Los Angeles.

_____. 1980. Effects of grazing on the northern population of *Pinus muricata* on Santa Cruz Island, California. Pp. 159-165 *in* D. M. Power (ed.) The California Islands: Proceedings of a multidisciplinary symposium. Santa Barbara Museum of Natural History, Santa Barbara, California.

_____. 1983. Factors controlling the form and location of the boundary between coastal sage scrub and grassland in southern California. Unpublished Ph.D. dissertation, University of California at Los Angeles. 307 pp.

Hochberg, M. 1980. Factors affecting leaf size of chaparral shrubs on the California islands. Pp. 189-206 *in* D. M. Power (ed.) The California Islands: Proceedings of a multidisciplinary symposium. Santa Barbara Museum of Natural History, Santa Barbara, California.

_____, S. Junak, and R. Philbrick. 1980. Botanical study of Santa Cruz Island for The Nature Conservancy. Volume 1. Unpublished report prepared by Santa Barbara Botanic Garden, Santa Barbara, California. 90 pp.

Hoffmann, R. 1931. The oaks of Santa Cruz Island. Santa Barbara Museum of Natural History Leaflets 6: 8-11.

_____. 1932a. Notes on the flora of the Channel Islands off Santa Barbara, California. Bulletin of the Southern California Academy of Sciences 31: 46-60.

_____. 1932b. Notes on the flora of the Channel Islands off Santa Barbara, California II. Bulletin Southern California Academy of Sciences 31: 101-120.

Holder, C. 1910. The Channel Islands of California. A. C. McClurg and Company, Chicago, Illinois. 397 pp.
Holmgren, P., N. Holmgren, and L. Barnett (eds.) 1990. Index Herbariorum. Part I: The herbaria of the world. Regnum Vegetabile 120: 1-693.
Holland, R. 1986. Preliminary descriptions of the terrestrial natural communities of California. California Department of Fish and Game, Sacramento. 156 pp.
Howell D. (ed.) 1976. Aspects of the geologic history of the California continental borderland. Miscellaneous Publication 24. American Association of Geologists (Pacific Section).
Hussey, J. 1961. A brief sketch of the history of Santa Cruz Island, California. Western Explorer 1: 1-8.
James, J. 1982. Channel Islands climate: Addenda to a modified Koppen classification of California's climate. California Geographer 22: 33-49.
Jepson, W. 1927. Field notes for 27-28 May 1927. Jepson field book 45: 183-200. Unpublished manuscript on file at Jepson Herbarium, University of California at Berkeley.
_____. 1933. The botanical explorers of California VIII. Madroño 2: 83-88.
_____. 1909-1943. A flora of California. Associated Students Store, University of California at Berkeley. 3 volumes. 1726 pp.
Johnson, W. 1858. Letter of 20 June 1858 to W.E. Greenwell. Unpublished manuscript on file at U.S. National Archives, Washington, D.C. Record Group 23 (records of the U.S. Coast and Geodetic Survey), Entry 5, Bache Records, 16: 281-283.
Johnson, D. 1972. Landscape evolution on San Miguel Island, California. Unpublished Ph.D. dissertation, University of Kansas, Lawrence. 390 pp.
_____. 1977. The late Quaternary climate of coastal California: Evidence for an Ice Age refugium. Quaternary Research 8: 154-179.
_____. 1980. Episodic vegetation stripping, soil erosion, and landscape modification in prehistoric and recent historic time, San Miguel Island, California. Pp. 103-121 *in* D. M. Power (ed.) The California Islands: Proceedings of a multidisciplinary symposium. Santa Barbara Museum of Natural History, Santa Barbara, California.
Johnson, J. 1982. An ethnohistoric study of the Island Chumash. Unpublished M.A. thesis, University of California at Santa Barbara. 323 pp.
Jones, J., S. Junak, and R. Paul. 1993. Progress in mapping vegetation on Santa Cruz Island and a preliminary analysis of relationships with environmental factors. Pp. 97-104 *in* F. Hochberg (ed.) Third California Islands Symposium: Recent advances in research on the California Islands. Santa Barbara Museum of Natural History, Santa Barbara, California.
Junak, S. 1987. Environmental factors correlated with the distribution of island ironwood (*Lyonothamnus floribundus* subsp. *aspleniifolius*) on Santa Cruz Island, California. Unpublished M.A. thesis, University of California at Santa Barbara. 162 pp.
Junger, A. and D. Johnson. 1980. Was there a Quaternary land bridge to the Northern Channel Islands? Pp. 33-39 *in* D. Power (ed.) The California Islands: Proceedings of a multidisciplibary symposium. Santa Barbara Museum of Natural History, Santa Barbara, California.
King, C. 1976. Chumash inter-village economic exchange. Indian Historian 4: 30-43.
Klinger, R., P. Schuyler, and J. Sterner. 1994. Vegetation response to the removal of feral sheep from Santa Cruz Island. Pp. 341-350 *in* W. Halvorson and G. Maender (eds.) The fourth California Islands symposium: Update on the status of resources. Santa Barbara Museum of Natural History, Santa Barbara, California.
Krythe, M. 1958. The *Senator*, favorite coastwise steamer. Historical Society of Southern California Quarterly 40: 33-57.
Leishman, N. 1981. Effects of feral animals on woody vegetation, Santa Cruz Island, California. Unpublished M.A. thesis, University of California at Los Angeles. 71 pp.
Lenz, L. 1986. Marcus E. Jones: Western geologist, mining engineer, and botanist. Rancho Santa Ana Botanic Garden, Claremont, California. 486 pp.
Loew, O. 1876. Report on the physical and agricultural features of southern California, and especially of the Mohave Desert. Appendix H6 *in* G. Wheeler. Annual report upon the geographical surveys west of the one hundreth meridian in California, Nevada, Utah, Colorado, New Mexico, Arizona, and Montana. Appendix JJ of the Annual Report of the Chief of Engineers for 1876. U.S. Government Printing Office, Washington, D.C.
Mallis, A. 1971. American entomologists. Rutgers University Press, New Brunswick, New Jersey. 549 pp.
Mason, H. 1927. Field notes for 23-25 September 1927. Unpublished manusript on file at University Herbarium, University of California at Berkeley.

McElrath, C. 1967. On Santa Cruz Island: The ranching recollections of Clifford McElrath. Published for Santa Barbara Historical Society by Dawson's Book Shop, Los Angeles, California. 125 pp.
Millspaugh, C. and L. Nuttall. 1923. Flora of Santa Catalina Island, California. Field Museum of Natural History Publication No. 212: 1-413.
Minnich, R. 1980. Vegetation of Santa Cruz and Santa Catalina Islands. Pp. 123-137 *in* D. Power (ed.) The California Islands: Proceedings of a multidisciplibary symposium. Santa Barbara Museum of Natural History, Santa Barbara, California.
Munz, P. 1935. A manual of southern California botany. Claremont Colleges, Claremont, California. 642 pp.
_____. 1968. Supplement to A California Flora. University of California Press, Berkeley. 224 pp.
_____. 1974. A flora of southern California. University of California Press, Berkeley. 1086 pp.
_____. and D. Keck. 1963. A California flora. University of California Press, Berkeley. 1681 pp.
Newmark, M.H. and M.R. Newmark (eds.) 1930. Sixty years in southern California, 1853-1913, containing the reminiscences of Harris Newmark. Houghton Mifflin Company, Boston, Massachusetts. 744 pp.
Nolf, B. and P. Nolf. 1969. Santa Cruz Island volcanics. Pp. 91-94 *in* D. Weaver (ed.) Geology of the Northern Channel Islands. Special publication. American Association of Petroleum Geologists (Pacific Section) and Society of Economic Paleontologists and Mineralogists.
Norris, R. and R. Webb. 1976. Geology of California. John Wiley and Sons, New York, New York. 365 pp.
Ord, A. 1956. Occurrences in hispanic California. Academy of American Franciscan History, Washington, D.C. 98 pp.
Orr, P. 1968. Prehistory of Santa Rosa Island. Santa Barbara Museum of Natural History, Santa Barbara, California. 253 pp.
Panetsos, C. and H. Baker. 1967. The origin of variation in "wild" *Raphanus sativus* (Cruciferae) in California. Genetica 38: 243-274.
Peart, D., D. Patten, and S. Lohr. 1994. Feral pig disturbance and woody species seedling regeneration and abundance beneath coast live oaks (*Quercus agrifolia*) on Santa Cruz Island, California. Pp. 313-322 *in* W. Halvorson and G. Maender (eds.) The fourth California Islands symposium: Update on the status of resources. Santa Barbara Museum of Natural History, Santa Barbara, California.
Philbrick, R. 1963. Biosystematic studies of two Pacific coast opuntias. Unpublished Ph.D. dissertation, Cornell University, Ithaca, New York. 177 pp.
_____. 1978. A botanist's view of Santa Cruz Island. Fremontia 6: 6-10, 23.
_____. 1980. Distribution and evolution of endemic plants of the California Islands. Pp. 173-187 *in* D. M. Power (ed.) The California Islands: Proceedings of a multidisciplinary symposium. Santa Barbara Museum of Natural History, Santa Barbara, California.
_____. and J. Haller. 1977. The southern California islands. Pp. 893-906 *in* M. Barbour and J. Major (eds.) Terrestrial vegetation of California. John Wiley and Sons, New York.
Phillips, M. 1927. History of Santa Barbara County, California, from its earliest settlement to the present time. Volume 1. S.J. Clarke Publishers, Los Angeles, California. 464 pp.
_____. 1988. Fifty years and more in Santa Barbara (by Jane Merritt Kimberly as told to Michael J. Phillips). Noticias 34: 50-62.
Pinney, T. 1994. The wine of Santa Cruz Island. Santa Cruz Island Foundation, Santa Barbara, and Zamorano Club, Los Angeles, California. 99 pp.
Rand, W. 1933. The geology of Santa Cruz Island, California. Unpublished Ph.D. dissertation, University of California at Berkeley. 192 pp.
Raven, P. 1967. The floristics of the California islands. Pp. 57-67 *in* R. Philbrick (ed.) Proceedings of the symposium on the biology of the California islands. Santa Barbara Botanic Garden, Santa Barbara, California.
Reifschneider, O. 1964. Biographies of Nevada botanists, 1844-1963. University of Nevada Press, Reno. 165 pp.
Rothrock, J. 1876. Report upon the operations of a special natural history party and Main Field Party No. 1, California section, field season of 1875, being the results of observations upon the economic botany and agriculture of portions of southern California. Appendix H5 *in* G. Wheeler. Annual report upon the geographical surveys west of the one hundreth meridian in California, Nevada, Utah, Colorado, New Mexico, Arizona, and Montana. Appendix JJ of the Annual Report of the Chief of Engineers for 1876. U.S. Government Printing Office, Washington, D.C.
Rouse, S. 1980. In old Santa Barbara (50 years ago), 15 Oct 1980. Santa Barbara News-Press, Santa Barbara, California.

_____. 1983a. In old Santa Barbara (75 years ago), 5 December 1983. Santa Barbara News-Press, Santa Barbara, California.

_____. 1983b. In old Santa Barbara (50 years ago), 14 June 1983. Santa Barbara News-Press, Santa Barbara, California.

Sauer, J. 1988. Plant Migration: The dynamics of geographic patterning in seed plant species. University of California Press, Berkeley. 282 pp.

Schumacher, P. 1877. Researches in the kjökkenmöddings and graves of a former population of the Santa Barbara Islands and the adjacent mainland. Bulletin U.S. Geological and Geographical Surveys of the Territories 3: 37-56.

Schuyler, P. 1988. The status of the Santa Cruz Island Preserve. Santa Cruz Island Preserve News No. 27: 1. The Nature Conservancy, Santa Barbara, California.

_____. 1993. Control of feral sheep (*Ovis aries*) on Santa Cruz Island, California. Pp. 443-452 *in* F. Hochberg (ed.) Third California Islands Symposium: Recent advances in research on the California Islands. Santa Barbara Museum of Natural History, Santa Barbara, California.

Shipman, G. 1972. Soil survey of northern Santa Barbara area, California. U.S. Department of Agriculture, Soil Conservation Service, in cooperation with University of California Agricultural Experiment Station. 182 pp.

Skinner, M. and B. Pavlik (eds.) 1994. Inventory of rare and endangered plants of California. Special Publication No. 1, 5th edition. California Native Plant Society, Sacramento. 338 pp.

Smith, C. 1952. A flora of Santa Barbara: An annotated catalog of the native and naturalized plants of Santa Barbara, California, and vicinity. Santa Barbara Botanic Garden, Santa Barbara, California. 100 pp.

_____. 1976. A flora of the Santa Barbara region, California. Santa Barbara Museum of Natural History, Santa Barbara, California. 331 pp.

Spaulding, E. 1964. A brief history of Santa Barbara. Santa Barbara Historical Society, Santa Barbara, California. 78 pp.

Stanton, C. 1984. An island memoir. Santa Cruz Island Company, Los Angeles, California. 38 pp.

Stenzel, F. 1975. James Madison Alden: Yankee artist of the Pacific coast, 1854-1860. Amon Carter Museum, Fort Worth, Texas. 209 pp.

Strike, S. 1994. Ehtnobotany of the California Indians. Volume 2. Aboriginal uses of California's indigenous plants. Koeltz Scientific Books, Champaign, Illinois. 210 pp.

Symmes and Associates. 1922. Report on Santa Cruz Island, Santa Barbara County, California. Symmes and Associates, San Francisco, California. 93 pp.

Tanaka, T. 1976. Tanaka's cyclopedia of edible plants of the world. Keigaku Publishing Company, Tokyo. 924 pp.

Tays, G. 1935. Captain Andres Castillero, diplomat. California Historical Society Quarterly 14: 230-268.

Tharp, B. and M. Johnston. 1961. Recharacterization of *Dichondra* (Convolvulaceae) and a revision of the North American species. Brittonia 13: 346-360.

Thomas, J. 1959. Herman Knoche, 1870-1945. Contributions of the Dudley Herbarium 5: 123-127.

Thompson, T. and A. West. 1883. History of Santa Barbara County, California. Thompson and West, Oakland, California. 477 pp.

Thorne, R. 1969. The California Islands. Annals of the Missouri Botanical Garden 56: 391-408.

Timbrook, J. 1984. Chumash ethnobotany: A preliminary report. Journal of Ethnobiology 4: 141-169.

_____. 1993. Island Chumash ethnobotany. Pp. 47-62 *in* M. Glassow (ed.) Archaeology on the Northern Channel Islands of California: Studies of subsistence, economics, and social organization. Coyote Press Archives of California Prehistory No. 34. Coyote Press, Salinas, California.

Towne, C.and E. Wentworth. 1945. Shepherd's empire. University of Oklahoma Press, Norman. 364 pp.

Tucker, J. 1945. Francesco Franceschi, botanist and horticulturist. Santa Barbara Botanic Garden Leaflets 1: 1-10.

U.S. Bureau of Census. 1870. Ninth census of U. S. Volume 2. Schedule of agricultural recapitulation, Santa Barbara County. California State Library, Sacramento.

U. S. District Court (San Francisco). 1857. Transcripts of proceedings for Case No. 176. Petition of Andres Castillero for the island of Santa Cruz (No. 340SD).

Uphof, J. 1968. Dictionary of economic plants. 2nd Edition. J. Cramer, Lehre. 591 pp.

Van Vuren, D. and B. Coblentz. 1987. Some ecological effects of feral sheep on Santa Cruz Island, California. Biological Conservation 41: 253-268.

Vedder, J. 1976. Precursors and evolution of the name California Continental Borderland. Pp. 6-11 *in* D. Howell (ed.) Aspects of the geologic history of the California continental borderland. Miscellaneous Publication 24. American Association of Geologists (Pacific Section).

_____. and D. Howell. 1980. Topographic evolution of the southern California borderland during late Cenozoic time. Pp. 7-31 *in* D. Power (ed.) The California Islands: Proceedings of a multidisciplibary symposium. Santa Barbara Museum of Natural History.

Wagner, H. 1966. Spanish voyages to the northwest coast of America in the Sixteenth Century. California Historical Society, San Francisco. 571 pp.

Wallace, G. 1985. Vascular plants of the Channel Islands of southern California and Guadalupe Island, Baja California, Mexico. Natural History Museum of Los Angeles County, Contributions in Science No. 365: 1-136.

Warren, E., Jr. ca. 1954. The agriculture of Santa Cruz Island. Unpublished thesis, University of California at Davis. 50 pp.

Weaver, D. (ed.) 1969a. Geology of the Northern Channel Islands. Special publication. American Association of Petroleum Geologists (Pacific Section) and Society of Economic Paleontologists and Mineralogists.

_____. 1969b. The pre-tertiary rocks. Pp. 11-13 *in* D. Weaver (ed.) Geology of the Northern Channel Islands. Special publication. American Association of Petroleum Geologists (Pacific Section) and Society of Economic Paleontologists and Mineralogists.

_____. and G. Meyer. 1969. Stratigraphy of northeastern Santa Cruz Island. Pp. 95-104 *in* D. Weaver (ed.) Geology of the Northern Channel Islands. Special publication. American Association of Petroleum Geologists (Pacific Section) and Society of Economic Paleontologists and Mineralogists.

_____., G. Griggs, D. McClure, and P. Nolf. 1969b. Volcaniclastic sequence, south central Santa Cruz Island. Pp. 91-94 *in* D. Weaver (ed.) Geology of the Northern Channel Islands. Special publication. American Association of Petroleum Geologists (Pacific Section) and Society of Economic Paleontologists and Mineralogists.

Wells, P. 1962. Vegetation in relation to geological substratum and fire in the San Luis Obispo quadrangle, California. Ecological Monographs 32: 79-103.

Wehtje, W. 1994. Response of a bishop pine (*Pinus muricata*) population to removal of feral sheep on Santa Cruz Island, California. Pp. 331-340 *in* W. Halvorson and G. Maender (eds.) The fourth California Islands symposium: Update on the status of resources. Santa Barbara Museum of Natural History, Santa Barbara, California.

Wenner, A. and D. Johnson. 1980. Land vertebrates on the California Channel Islands: sweepstakes or bridges? Pp. 497-530 *in* D. Power (ed.) The California Islands: Proceeedings of a multidisciplinary symposium. Santa Barbara Museum of Natural History, Santa Barbara, California.

Westman, W. 1983. Island biogeography: studies on the xeric shrublands of the inner Channel Islands, California. Journal of Biogeography 10: 97-118.

Wheeler, S. 1944. California's little-known Channel Islands. U.S. Naval Institute Proceedings 70: 257-267.

Wheeler, G. 1876. Annual report upon the geographical surveys west of the one hundreth meridian in California, Nevada, Utah, Colorado, New Mexico, Arizona, and Montana. Appendix JJ of the Annual Report of the Chief of Engineers for 1876. U.S. Government Printing Office, Washington, D.C.

Williams, M. 1954a. California's offshore islands. Pacific Discovery 7: 16-17.

_____. 1954b. Santa Cruz: An island museum. Pacific Discovery 7: 18-21.

Wolf, C. 1935. California plant notes I. Occasional Papers Rancho Santa Ana Botanic Garden (Series I) No. 1: 31-43.

Woodring, W. and M. Bramlette. 1950. Geology and paleontology of the Santa Maria district, California. U.S. Geological Survey Professional Paper No. 222: 1-142.

Worts, G., Jr., 1951 Geology and ground water resources of the Santa Maria Valley area, California. U.S. Geological Survey Watyer Supply Paper 1000: 1-169.

Yates, L. 1890a. Notes on the ferns of the Channel Islands off the coast of California. Bulletin of the Santa Barbara Society of Natural History 1: 8-10.

_____. 1890b. Insular floras. Pp. 179-188 *in* Ninth Annual Report of the State Mineralogist. California State Mining Bureau, Sacramento.

# APPENDIX I. History of Collecting Trips on Santa Cruz Island.

Collectors, their field numbers, and general localities are listed by year and dates (when known). Dates or field numbers preceded by "ca." are approximate and estimated from known herbarium specimens. Data from specimens were clarified, in some cases, by examination of original field notes.

1874: Nov, A.Kellogg & W.Harford ca. 382.

1875: late spring, J.Rothrock, Prisoners' Harbor, Forney's Cove.

1876: C.C.Parry.

1885: B.Hazard.

1886: Jul-Aug, E.L.Greene, west end, north shore, North Ridge, Prisoners' Harbor, Cañada del Puerto, Cañada Christy, Central Valley, Laguna Canyon, south side.

1887: Apr, H.C.Ford.

1888: 26 Mar-1 May, Jun, T.S.Brandegee.

1891: L.Yates.

1893: Aug, L.Yates.

1894: F.Franceschi.

1900: Oct, B.Trask.

1901: Jul, R.E.Snodgrass, Cueva Valdez, east end.

1903: 3 Sep, J.Grinnell, Fry's Harbor.

1908: 31 May-2 Jun, H.M.Hall 8175-8207, Fry's Harbor, Prisoners' Harbor; 31 May, F.Niedermuller, Fry's Harbor.

1912: Sep, H.J.Keech.

1916: 11 Aug, Mrs. C.E.Miller.

1917: R.Kuhn; Mrs. A.Swain; 16-17 Jul, A.Eastwood 6368-6441d.

1918: 24 May, C.E.Miller; 6 Jun, Mrs. C.E.Miller.

1919: Mrs. A.Swain; 13 Sep, H.Knoche ca. A2117-A2122, Pelican Bay, North Ridge.

1920: R.Kuhn; F. Sexuar.

1921: 28-30 Mar, R.C. & J.Robbins; May, O.Thacher.

1922: 28 Mar, R.C. & J.Robbins.

1923: 28-29 Jul, A.L.Grant ca. 1696-1734, Valdez Anchorage, Fry's Harbor.

1924: 13-14 Apr, M.E.Jones.

1925: 28 Mar, R.Hoffmann, west end, Christy Ranch, Christy Pines, Central Valley, Pelican Bay.

1927: 28 Mar, R.Hoffmann; 12-14 Apr, M.B.Peirson 7122-7152, Lady's Harbor, Prisoners' Harbor; 12 Apr, M.Stason; 14 May, R.Hoffmann, Smugglers' Cove; 27-28 May, W.L.Jepson 12053-12126, Pelican Bay; 23 Jun, R.Hoffmann, west end; 4 Sep, G.R.Johnstone; 5 Sep, M.E.Jones, Valdez Anchorage; 23-25 Sep, H.L.Mason 4082-4110, Pelican Bay, Prisoners' Harbor, Central Valley, Cañada de los Sauces.

1928: 10-11 Mar, R.Hoffmann, China Harbor, Scorpion Ranch; 31 Aug-1 Sep, R.Hoffmann, Pelican Bay; 1 Sep, S.W.Hutchinson; 2-3 Sep, M.N.Ackley & S.W.Hutchinson ca. 957, Pelican Bay, Prisoners' Harbor; 2 Sep, O.Thacher; Dec, G.R.Johnstone.

1929: 17 Mar, R.Hoffmann, Scorpion Anchorage, Scorpion Canyon; 22-24 Mar, R.Hoffmann, Coches Prietos Anchorage, Valley Anchorage, Scorpion Anchorage; 25 Mar, M.E.Jones; 26 Mar, F.Fosberg, S953, Pelican Bay; 29 Mar, R.Hoffmann, Coches Prietos; 15-16 Apr, R.Hoffmann, Baby's Cove; 12-15 May, L.Ellison; 14 May, R.Hoffmann; 25 May, A.Blake.

1930: 21 Feb, R.Hoffmann, Pelican Bay; 7 Apr, R.Hoffmann, Hazard's Canyon; 7 Apr, P.A.Munz & E.Crow 11501-11559, Hazard's Canyon; 10-12 Apr, R.Hoffmann, Lady's Cove, Lady's Canyon, Main Ranch; 11 Apr, P.A.Munz & E.Crow 11831-11871, 11885-11897, Lady's Harbor; 25-28 Apr, L.Abrams & I.L.Wiggins 1-173, Fry's Harbor, Pelican Bay, Prisoners' Harbor; ca. 23 May-29 Jul, I.W.Clokey ca. 4909-5117, Dick's Canyon, Lady's Canyon, Twin Harbor, Pelican Bay, Prisoners' Harbor, China Harbor, Potato Harbor, Scorpion Anchorage; 12 Jun-13 Aug, G.S.Rand ca. 11-80, Central Valley, Yellowbanks, Scorpion Anchorage; 13-15 Jun, R.Hoffmann, Valdez Anchorage, Pelican Bay, Pelican Pines, North Ridge, Prisoners' Harbor, Central Valley, Main Ranch, Smugglers' Cove; ca. 15-19 Jun, L.Rowntree, Prisoners' Harbor; 28 Jun-1 Jul, R.Hoffmann, Forney's Cove, Christy Ranch, Christy Pines, Canyon west of Gull Rock, near Gull Rock, Willows Canyon, Prisoners' Harbor, Main Ranch, Coches Prietos Anchorage; 7 Jul, R.Hoffmann, Lady's Cove; 11 Jul, M.Dearing, Scorpion Anchorage; 6-7 Aug, R.Hoffmann, Lady's Cove; 20 Sep, I.Cooke, China Harbor, China Water Hole; 20-22 Sep, R.Hoffmann, Fry's Harbor, Dick's Harbor, Twin Harbor, Prisoners' Harbor, Main Ranch, South Ridge, Valley Anchorage, China Harbor, Scorpion Anchorage, Smugglers' Cove; 28 Sep, R.Hoffmann, China Harbor; 8-10 Nov, R.Hoffmann, Valdez Harbor, Valdez Canyon, Hazard's Anchorage, Alamo Canyon, Fry's Harbor; 7 Dec, R.Hoffmann, Platts Harbor, Dick's Canyon.

1931: 9-15 Apr, J.T.Howell 6183-6408, Pelican Bay, Prisoners' Harbor, Cañada del Puerto, Central Valley, Valley Anchorage, Coches Prietos Canyon, South Ridge, Christy Pines, China Pines, China Harbor; 10-13 Apr, R.Hoffmann, Christy Pines, Pelican Bay, Prisoners' Harbor, Central Valley, Main Ranch, South Ridge, Coches Prietos, China Harbor, China Pines; ca. 18 Apr-3 May, I.W.Clokey ca. 5142-5226, Fry's Canyon, Dick's Canyon, Pelican Bay; 1 Jun, H.H.Sheldon, Picacho Diablo; 29 Jun, L.Abrams and I.L.Wiggins 181-222, Smugglers' Cove; 8-10 Sep, R.Hoffmann, Cañada Christy, Buena Vista, Picacho Diablo, Fry's Harbor, Fry's Canyon, Dick's Canyon; 21 Sep, R.Hoffmann, Valley Anchorage; 6 Oct, F.Caire, Central Valley.

1932: 20-21 Feb, R.Hoffmann, Pelican Bay, Pelican Pines, Prisoners' Harbor; 27-28 Feb, R.Hoffmann, Platts Harbor, Orizaba Canyon, Pelican Pines; 2 Mar, R.Hoffmann, Laguna Canyon; 5-6 Mar, R.Hoffmann, Twin Harbor, Pelican Bay, Pelican Pines, Prisoners' Harbor; 19-27 Mar, R.Hoffmann, Christy Ranch, Central Valley, Laguna Canyon, Cañada de los Sauces, Portezuelo cabin, S Ranch, La Cuesta, Dick's Canyon, Orizaba Cove, Baby's Canyon; 19-27 Mar, C.B.Wolf 2716-2917, Orizaba Canyon, Prisoners' Harbor, Cañada del Puerto, Central Valley, Christy Ranch, Cañada de los Sauces, Laguna Canyon, Willows Canyon; 20-26 Mar, F.R.Fosberg ca. 7512-7709, Twin Harbors, Prisoners' Harbor, Smugglers' Cove, Hungryman Gulch; 21-22 Apr, R.Hoffmann, Diablo Canyon, Fry's Harbor, Pelican Bay, Pelican Pines; 21-22 Apr, H.McMinn ca. 2815-2835; 23 May, R.Hoffmann, Dick's Canyon, Platts Harbor, Orizaba Canyon, Twin Harbor Canyon, Pelican Bay; 14 Jun, R.Hoffmann, Pelican Pines; 20-24 Sept, C.B.Wolf 4125-4171, Cañada del Puerto, Central Valley.

1934: 25-28 Mar, F.W.Peirson 11022-11095, 11103, Diablo Canyon, Fry's Harbor, North Ridge; 26 Mar, M.B.Peirson 11096-11102, Prisoners' Harbor.

1936: 18 Apr, F.Elmore, Valdez Anchorage; 18 Apr, O.A.Plunkett; 18 Apr, R.S.Woglum ca. 116-139, Cueva Valdez; 20 Apr, Harvey; 20 Jun, M.Van Rensselaer, Fry's Harbor; 31 Jul-1 Aug, R.Bacigalupi ca. 2378-2385, Fry's Canyon; 1 Aug, M.Van Rensselaer, Fry's Harbor; 2 Sep, M.Van Rensselaer, North Ridge, Fry's Harbor.

1938: 5 May, N.E.Bilderback; 2 Aug-14 Sep, F.Elmore.

1939: Jun-Aug, M.W.Williams ca. 10-104, Pelican Bay, Prisoners' Harbor, Cañada del Puerto, Cavern Point, Scorpion Harbor, Smugglers' Cove, Willows Anchorage, Pozo Anchorage; 11-17 Aug, M.B.Dunkle 8531-8657, Pelican Bay, Prisoners' Harbor, Central Valley; 15 Oct, R.Hartwell 427-431, Pelican Bay.

1941: 23-29 Mar, R.V.Moran 745-776, Pelican Bay, North Ridge; 17 Apr, F.H.Elmore 254-293, 440-468, Pelican Bay, Prisoners' Harbor; 5 Jul, H.Baer, Fry's Harbor.

1947: 26 Apr, C.H.Muller ca. 8838, Christy Canyon.

1948: 8 Aug, Fisher.

1950: 5-7 Mar, R.V.Moran 3319-3334, Prisoners' Harbor, Central Valley, Fraser Point; 11-12 Mar, R.V.Moran 3354-3362, Fraser Point, Willows Anchorage.

1951: ca. 25 Jun, C.H.Muller ca. 8939-8949, Pelican Bay.

1958: 21-22 Apr, C.F.Smith 5731-5739, Cañada de los Sauces, Central Valley, China Pines; 10-17 Oct, E.K.Balls & E.R.Blakley ca. 23620-23745, Prisoners' Harbor, Cañada del Puerto, Central Valley; 10-17 Oct, E.R.Blakley & E.K.Balls B2474-B2640, Prisoners' Harbor, Cañada del Puerto, Central Valley, Valley Anchorage, Willows Canyon, Christy Canyon, Cañada de los Sauces, Valdez Canyon, China Pines, Laguna Canyon.

1959: 25-28 May, C.F.Smith 5945-5975, Prisoners' Harbor, Central Valley, Valley Anchorage, Fraser Point, North Ridge, Christy Ranch, Isthmus, China Pines.

1960: 25-29 Apr, P.H.Raven & C.F.Smith 15127-15317, Fraser Point, China Pines, Christy Pines, Central Valley, South Ridge; 25-29 Apr, C.F.Smith & P.H.Raven 6126-6138, West Point, Christy Ranch; 29 Apr-1 May, E.R.Blakley B3281-B3395, Coches Prietos drainage; 22-26 Jul, E.R.Blakley & C.H.Muller B3722-B3788, Pelican Bay, Prisoners' Harbor, Central Valley, Christy Canyon, Cañada de los Sauces, Punta Arena, North Ridge; 20 Dec, C.F.Smith 6338-6343, Christy Ranch, Christy Pines.

1961: 11-14 Mar, E.R.Blakley B3872-B4009, Prisoners' Harbor, Cañada del Puerto, Central Valley, Rancho del Sur, Fraser Point, West End Ridge, Cañada Christy; 6-9 Jun, C.F.Smith 6432-6475, Fraser Point, Christy Ranch, Central Valley, Isthmus; 2-3 Dec, E.R.Blakley B4825-B4832, South Ridge, Pinos del Sur.

1962: 15-17 May, D.E.Breedlove 2785-2884, Prisoners' Harbor, Eagle Canyon, Cañada del Puerto, Christy Beach, Christy Pines, Central Valley, China Pines; 15-18 May, M.Piehl ca. 62-280, Central Valley; 29-30 Jun, E.R.Blakley 5270a-5291, Prisoners' Harbor, Isthmus, Central Valley.

1963: 6-9 May, C.F.Smith 8140-8166, Fraser Point, Christy Pines, Central Valley, South Ridge.

1964: 16-19 Jun, R.N.Philbrick B64-20 - B64-204, Christy Beach, Christy Ranch, Christy Pines, Campo Raton, canyon north of Cañada de los Sauces, Laguna Canyon, Central Valley; 13-14 Jul, R.N.Philbrick B64-282 - B64-302, West Point, Fraser Point, North Ridge, Central Valley, Cascada, Prisoners' Harbor, Cañada del Puerto.

1965: 1-2 Feb, R.N.Philbrick, D.Smith, & A.Flinck B65-70 - B65-172, Prisoners' Harbor, Cañada del Puerto, Isthmus; 1-2 Mar, R.N.Philbrick & M.R.Benedict B65-443 - B65-473, North Ridge, Central Valley; 20-23 Apr, R.N.Philbrick & M.R.Benedict B65-700 - B65-852, Fraser Point, Christy Ranch, Central Valley, South Ridge, Isthmus; 20-23 Apr, C.F.Smith 8659-8673, Christy Beach, Christy Ranch, Central Valley, South Ridge; 11-14 Jun,

R.N.Philbrick & M.R.Benedict B65-1301 - B65-1421, Ragged Mountain, Profile Point, Pozo Beach, Punta Arena, Central Valley.

1966: 3-6 May, R.N.Philbrick, O.Clarke, & D.W.Ricker B66-152 - B66-216, North Ridge, Pelican Bay, Cañada Christy, Central Valley, Black Point peninsula, South Ridge, Laguna Canyon; 3-7 Jun, R.Gankin & W.Hildreth 754-805, Pelican Bay, Prisoners' Harbor, Navy Road, China Pines, Central Valley, South Ridge, Willows Canyon; 20-23 Sep, R.Gankin, G.Webster & E.Gedling 841-868, Pelican Bay, Prisoners' Harbor, Navy Road, Cañada Christy, Central Valley, Cañada de los Sauces, Pozo Canyon, South Ridge, Willows Canyon; 20-22 Sep, G.L.Webster, R.Gankin, & E.Gedling 7426-7445, Prisoners' Harbor, Christy Beach, Cañada Christy, Central Valley, Pozo Beach, South Ridge.

1967: 14-18 Apr, R.N.Philbrick, R.F.Thorne, & P.C.Everett B67-52 - B67-164, Pelican Bay, Prisoners' Harbor, Christy Beach, Central Valley, Cañada de los Sauces, Sierra Blanca ridge, Ragged Mountain, Willows Canyon, Coches Prietos; 14-18 Apr, R.F.Thorne & P.C.Everett ca. 36772; 15 Apr, C.B.Hardham, South Ridge; May-Jun, M.R.Benedict, Christy Beach, Cañada del Agua.

1968: ca. 28 Mar-26 Apr, M.R.Benedict, Cueva Valdez, Horqueta Canyon, Alamos Canyon, Cañada del Puerto; 18-19 Apr, C.F.Smith 9895-9906, Pozo Beach, Pozo Creek, Prisoners' Harbor, Cañada del Puerto; 7-10 May, R.N.Philbrick & M.R.Benedict B68-268 - B68-313, Prisoners' Harbor, Eagle Canyon, Mount Pleasant, El Montañon, High Mount, Potato Harbor, Scorpion Canyon, Central Valley; 3-14 Jul, R.L.Hoover ca. 88, Forney's Cove, Willows Anchorage; 7 Sep, R.Gankin 1263-1274, Navy Road, Central Valley, South Ridge.

1969: 7-9 May, R.N.Philbrick & M.R.Benedict B69-114 - B69-133, Central Valley, Alamos Canyon, Horqueta Canyon, Willows Canyon.

1970: 8-10 May, J.Sauer 4768-4805; 14-16 Aug, D.Wilken 8805-8840, Coches Prietos.

1971: 23-26 Mar, R.N.Philbrick & M.R.Benedict B71-1 - B71-62, Cañada del Puerto, Cañada del Agua, El Montañon, Central Valley, Islay Canyon, South Ridge, El Tigre, Smugglers' Canyon, Aguaje Canyon; 29 Apr, R.N.Philbrick, P.W.Shaw, & M.R.Benedict B71-63 - B71-104, Scorpion Canyon, Smugglers' Field, Hungryman Gulch; 10 Jul, D.Weissman, ca. 312.

1972: 21-24 Apr, S.L.Timbrook 195-206; 2-5 Jun, J.Sauer & Ware 5079-5110, Cañada de los Sauces.

1973: ca. 30 Apr, D.Martin, ca. LASCA 1021; 6 May, J.Sauer 5453; 8-11 May, L.Heckard & R.N.Philbrick 3350-3392, Christy Pines, Central Valley, South Ridge, Willows Canyon, Coches Prietos; 8-11 May, R.N.Philbrick & L.Heckard B73-289 - B73-327, Cañada del Agua, Navy Road, Christy Pines, Central Valley, Willows Canyon, Coches Prietos; 22-24 May, J.Sauer 5493-5525; 2 Jun, G.D.Barbe ca. 1566, Central Valley; 2 Jun, B.Kasapligil, ca. 4584, Central Valley; 10-13 Nov, R.N.Philbrick & D.W.Ricker B73-288 - B73-294, Fry's Harbor, Prisoners' Harbor, Cañada del Puerto, Central Valley.

1974: 2-4 Nov, R.N.Philbrick B74-151 - B74-154. Cañada del Puerto, Central Valley, Fraser Point.

1975: 24-28 Mar, G.L.Webster, D.I.Axelrod, et al. 1-132, Pelican Bay, Prisoners' Harbor, North Ridge, Red Peak, Christy Pines, Central Valley, Fraser Point, Cañada de los Sauces, South Ridge; 16-19 May, F.Almeda, M.Mathias, & J.Sauer 2550-2564, west end, China Pines; 16-19 May, J.Sauer & M.Mathias 5672-5692; 13 Sep, B.Bartholomew, ca. 310, Central Valley; 20 Nov, W.Cox.

1976: 10-11 Apr, D.Daum ca. 34-42, Islay Canyon.

1977: 21-23 Aug, R.N.Philbrick & M.J.Cummings B77-34 - B77-48, Valdez Anchorage, Dick's Harbor, Cañada del Puerto, Big Scorpion Rock, Central Valley; 25-27 Nov, R.N.Philbrick & S.A.Junak B77-45 - B77-75, Pelican Bay, Prisoners' Harbor, Cañada del Puerto, Christy Pines.

1978: 5-7 Apr, R.N.Philbrick, S.A.Junak, & M.R.Benedict B78-1 - B78-47, North Ridge, Peak 1941, Cañada del Puerto, Central Valley; 29-30 Apr, K.Nixon ca. 1547-1577, Lagunitas Secas, China Pines; 19-22 May, J.Sauer 5019-5035; 22 May, R.N.Philbrick B78-176 - B78-184, Fry's Harbor; 3-4 Jun, E.Hobbs ca. 532-536, Pelican Pines, China Pines, Christy Pines; 27-28 Jun, G.Gorelick, La Cascada, Central Valley; 27 Jul, M.Emerson, Fraser Point, Christy Ranch.

1979: 10 Feb, K.Nixon ca. 1923-1931, Picacho Diablo, Central Valley; 15-18 Feb, S.A.Junak & M.Hochberg SC9-SC19b, Christy Ranch, Cañada Christy, Central Valley, South Ridge; 25 Mar, J.Leishman, ca. 603-604, Willows Canyon; 23-25 Apr, R.F.Thorne, J.Dourley, & W.Wisura 52453-52639, Fraser Point, Christy Ranch, Christy Pines, North Ridge, Central Valley, Prisoners' Harbor; 7-11 May, S.A.Junak, M.Hochberg, & R.N.Philbrick SC20-SC84, Prisoners' Harbor, China Pines, Christy Pines, Cañada Christy, Central Valley, Mission Spring Canyon, Gallina Canyon, Stanton Ranch, Willows Canyon, South Ridge, Coches Prietos; 29 May-1 Jun, S.L.Timbrook; 2 Jun, J.Sauer 6081-6082; 9-13 Jul, S.A.Junak, M.Hochberg, & R.N.Philbrick, SC125-SC150, China Pines, Christy Beach, Cañada Christy, Central Valley, Fraser Point, Morse Point, Johnson Canyon, Punta Arena, Cañada de Malva Real, Los Pinos del Sur; 24-28 Sep, S.A.Junak, M.Hochberg, & R.N. Philbrick SC175-SC213, Fry's Harbor, Dick's Canyon, Twin Harbor, Orizaba Cove, Pelican Bay, Cañada del Muro, Campo del Norte, China Harbor, Cañada Christy, Christy Pines, Central Valley.

1980: 7-11 Jan, S.L.Timbrook 662-666; 6-14 Mar, S.A.Junak, M.Hochberg, & R.N.Philbrick SC250-SC292, Hazard's Anchorage, North Ridge, Peak 1848, Picacho Diablo, Lady's Canyon, Diablo Anchorage, Central Valley, Black Point; 25-27 Mar, UCSB Botanical Society; 12-16 May, S.A.Junak & M.Hochberg SC293-SC303, North Ridge, Picacho Diablo, Cañada del Puerto, Christy Pines, Lagunitas Secas, Cottonwood Canyon, Central Valley, Cañada de los Sauces, South Ridge; 16-20 Jun, S.A.Junak & M.Hochberg SC304-SC316, West Point ridge, Profile Point, Fraser Point, Rancho Nuevo, Black Point; 25-29 Aug, S.A.Junak & M.Hochberg SC317-SC336, North Ridge, Alta 2, Christy Pines, Embudo Canyon, Rancho Nuevo, South Ridge, El Tigre; 17-21 Nov, S.A.Junak, M.Hochberg, &. L.Laughrin SC337-SC352, Lady's Canyon, Baby's Canyon, Prisoners' Harbor, China Pines, Cottonwood Canyon, Cañada de la Siesta; 15-19 Dec, S.A.Junak, M.Hochberg, & S.Timbrook SC353-SC363, Prisoners' Harbor, No Man's Land, Christy Pines, Los Pinos del Sur.

1981: 23-26 Jan, S.L.Timbrook 680-687; 23-30 Mar, UCSB Botanical Society; 8-12 May, S.A.Junak SC370-SC399, Cañada del Puerto, Central Valley, Fraser Point, Willows Canyon, Coches Prietos; 29-31 May, J.Sauer 6140-6145; 5-8 Jun, S.A.Junak, M.Daily & C.Stanton SC400-SC419, Fry's Harbor, Platts Harbor, Orizaba Canyon, Olsen's Cave, Cañada del Puerto, Central Valley, Coches Prietos; 15 Jun, S.A.Junak & R.N.Philbrick SC420-SC437, Potato Harbor, Smugglers' Cove.

1982: 12-15 Feb, S.A.Junak SC459-SC470, Cañada del Puerto, Christy Pines, Central Valley, Willows Canyon; 22-26 Mar, S.A.Junak SC470a-SC497, Prisoners' Harbor, Navy Road, North Ridge, Picacho Diablo, Christy Beach, Christy Pines, Central Valley, Forney's Cove, South Ridge, Peak 777; 22-26 Mar, UCSB Botanical Society; ca. 15 Apr, T.Ledwith & R.Collett, Valley Anchorage; 30 Apr-3 May, S.A.Junak SC510-SC533, Cañada del Puerto, China Harbor, Central Valley, Cañada de los Sauces; 7-9 May, S.A.Junak SC534-SC553, North Ridge, Picacho Diablo, Lagunitas Secas, Islay Canyon, Central Valley; 5-9 Jul, S.A.Junak SC555-SC567, China Harbor, Islay Canyon, Central Valley, Johnson Canyon; 9 Aug, J.Vanderwier, Navy Base; 10-13 Sep, S.A.Junak SC570-SC593, Cañada del Puerto, Navy Road, Christy Beach, Central Valley.

1983: 21-28 Mar, UCSB Botanical Society; 25-28 Mar, S.A.Junak, SC610-SC663, Prisoners' Harbor, Christy Barn, Central Valley, Islay Canyon, Willows Canyon, South Ridge, El Tigre, Coches Prietos; 7 Apr, C.A.Drost 320-332, Scorpion Anchorage, Smugglers' Canyon; 15-20 Apr, S.A.Junak SC664-SC675, China Harbor, Cañada de la Siesta, Central Valley; 10 May, J.Vanderwier, Navy Base; 20-23 May, S.A.Junak SC676-SC704, Pelican Bay, Prisoners' Harbor, Islay Canyon, Forney's Cove, South Ridge; 3-6 Jun, S.A.Junak SC705-SC734, North Ridge, Navy Road, China Pines, Christy Pines, Cañada Larga, Islay Canyon, Central Valley, Cañada de los Sauces; 3-5 Aug, S.A.Junak SC757-SC777, North Ridge, Peak 2021, Islay Canyon, 2-Fork Canyon, Central Valley; 28 Aug-2 Sep, J.Sauer 6257-6269; 2-6 Sep, S.A.Junak SC778-794, Fry's Harbor, North Ridge, South Ridge; 23-26 Sep, S.A.Junak SC795-SC802, Campo del Norte, El Montañon, Central Valley, South Ridge.

1984: 27-30 Jan, M.Bourell 1925-1991, Cañada del Puerto, Christy Pines, Fraser Point, Coches Prietos; 27-30 Jan, S.A.Junak SC804-SC820, Eagle Canyon, Navy Road, Islay Canyon; 24-27 Feb, S.A.Junak SC821-SC842, Pelican Bay, Cañada del Puerto, Navy Road, Central Valley, South Ridge, Peak 777; 16-19 Mar, S.A.Junak SC844-SC861, Cañada del Puerto, Navy Road, China Harbor, Cañada de la Siesta, Central Valley, Willows Canyon, Coches Prietos; 4-7 May, R.F.Thorne 58324-58453, La Cascada, Central Valley, Cañada del Puerto, Prisoners' Harbor; 15-18 Jun, S.A.Junak, Pelican Bay, Cañada del Puerto, China Harbor; 18-22 Jun, J.Sauer 6281-6283; Jul, S.Unser, Smugglers' Canyon, Aguaje Canyon; 28 Sep-1 Oct, S.A.Junak SC885-SC915, Cañada del Puerto, Navy Road, Christy Ranch, Central Valley, Willows Canyon, South Ridge, Coches Prietos; 5 Oct, S.Unser, Smugglers' Canyon; 2-5 Nov, S.A.Junak SC916-SC941, North Ridge, Peak 1941, Peak 2021, Cañada del Agua, Campo del Norte, Hooker Canyon.

1985: 1-4 Feb, S.L.Timbrook 774-787; 9-11 Mar, S.Unser, Smugglers' Canyon; 22-25 Mar, S.A.Junak SC942-1013, North Ridge, Orizaba Canyon, Prisoners' Harbor, Central Valley, Centinela, Canyon S of Stanton Ranch, Laguna Canyon; 12-15 Apr, S.A.Junak SC1014-SC1069, West Point ridge, North Ridge, Orizaba Canyon, Prisoners' Harbor, Cañada del Puerto, Central Valley, Centinela, Forney's Cove; 10-12 May, S.A.Junak SC1070-SC1096, Navy Road, China Pines, Christy Beach, Central Valley, Centinela, Black Point, South Ridge; 19-23 Jul, S.A.Junak SC1097-1102, Central Valley, Cottonwood Canyon; 23-26 Aug, S.A.Junak SC1103-SC1115, Cañada del Puerto, Central Valley, South Ridge, Willows Canyon.

1986: 18-21 Apr, S.A.Junak SC1129-SC1157, Pelican Bay, Central Valley, Islay Canyon, Fraser Point, Coches Prietos.

1987: 1-2 Aug, T.Elias & W.Wisura, ca. 10537-10586, Prisoners' Harbor, Fraser Point; 6-9 Oct, S.A.Junak SC1160-SC1193, Pelican Bay, Prisoners' Harbor, Cañada del Puerto, Central Valley, South Ridge.

1988: 24-26 Aug, S.A.Junak SC1194-SC1209, Valdez Canyon, Lady's Harbor, Diablo Canyon, Cañada Christy.

1989: 17-19 Mar, S.A.Junak SC1211-SC1226, Scorpion Canyon, Smugglers' Cove; 27-29 Mar, S.A.Junak, T.Ayers, & R.Scott, SC1227-SC1264, Mount Pleasant, China Pines, Christy Beach, Central Valley, Fraser Point, Black Point peninsula, South Ridge, Laguna Canyon, Willows Canyon; 27-29 Mar, D.Young, S.A.Junak, & T.Ayers, S1854-S1889, Mount Pleasant, China Pines, Christy Beach, Central Valley, Fraser Point, Black Point peninsula, South Ridge, Laguna Canyon, Willows Canyon; 17-21 Apr, T.Ayers, R. Scott, & D.Young 619-711, North Ridge, Picacho Diablo, Platts Harbor, Twin Harbor, Cañada del Puerto, Central Valley, Cañada de los Sauces, Pozo Canyon, South Ridge, Sierra Blanca ridge, Coches Prietos; 17-21 Apr, S.A.Junak, T.Ayers, & R.Scott, SC1264a-SC1393, North Ridge, Picacho Diablo, Platts Harbor, burn west of Pelican Bay, Central Valley, Cañada de los Sauces, Pozo Canyon, South Ridge, Sierra Blanca ridge, Coches Prietos; 17-19 Apr, D.Young, T.Ayers, & S.A.Junak, S1890-S1939, Cañada del Puerto, Central Valley, Cañada de los Sauces, Pozo Canyon, South Ridge, Coches Prietos; 11-15 Sep, T.Ayers, S.A.Junak, & R.Scott 777-813, Profile Point, Peak 1848 West, North Ridge, Lady's Canyon, Baby's Canyon, Prisoners' Harbor, Cañada Christy; 11-15 Sep, S.A.Junak SC1394-SC1449, Profile Point, Peak 1848, North Ridge, Little Valdez Canyon, Trident Canyon, Cañada del Puerto, Cañada del Agua, Campo del Norte, Cañada Christy.

1990: 9-13 Apr, S.A.Junak SC1451-SC1571, West Point ridge, Platts Harbor, Navy Base, China Pines, Cañada Christy, Central Valley, Valley Anchorage, Forney's Cove, Black Point peninsula, South Ridge, Willows Canyon, San Justiniano, Coches Prietos, 3-Fork Canyon; 9-13 Apr, S. McCabe, Platts Harbor; 20-23 Apr, S.A.Junak SC1572-SC1643, Pelican Bay, Prisoners' Harbor, Cañada del Puerto, Navy Road, China Pines, Christy Beach, Campo Raton, Central Valley, Islay Canyon, Fraser Point, South Ridge; 7-11 May, S.A.Junak SC1644-SC1836, Cañada Christy, Embudo Canyon, Campo Raton, Centinela, Cañada de la Portezuela, Central Valley, Cañada Larga, Cañada de la Mina, Cañada de los Sauces, South Ridge; 19 May, S.A.Junak SC1837-SC1842, Christy Barn, Christy Pines; 29-31 May, S.A.Junak SC1843-SC1971, Cañada del Puerto, Cañada Christy, canyon SW of Picacho Diablo, Reparo Viejo, Central Valley, Cañada de la Siesta, Valley Peak, El Sobrante, Cañada de los Sauces, South Ridge; 19-21 Sep, S.A.Junak SC1972-SC2022, Diablo Canyon, ridge between Diablo and Fry's canyons, North Ridge, Cañada del Puerto, Navy Road, Central Valley, Cañada Larga, Islay Canyon, Laguna Beach, Corral Point; 15-19 Oct, S.A.Junak SC2023-SC2109, China Harbor, Central Valley, Valley Peak, Cañada de los Sauces, Alegria Canyon, South Ridge, Pozo Beach, Playa Larga, Laguna Canyon, Corral Point, ridge above Alberts Anchorage.

1991: 1-4 Apr, S.A.Junak SC2111-SC2198, Navy Road, Cañada Christy, Central Valley, Valley Anchorage, Cañada de los Sauces, Christy Pines, South Ridge, Sierra Blanca ridge, Ragged Mountain; 15-18 Apr, S.A.Junak SC2199-SC2335, Christy Beach, Central Valley, Islay Canyon, Cañada de los Sauces, South Ridge, Laguna Canyon, Laguna Beach, San Justiniano, Coches Prietos; 6-10 May, S.A.Junak SC2336-SC2549, West Point Ridge, Cañada Christy, Central Valley, Rancho Nuevo, Cañada de los Sauces, South Ridge, Pozo Beach, Willows Beach, San Justiniano, Coches Prietos; 20-22 May, S.A.Junak SC2550-SC2615, West Point Ridge, Prisoners' Harbor, Central Valley, Gallina Canyon, Valley Peak, Forney's Cove; 27-30 May, S.A.Junak SC2616-SC2732, Black Point, Christy Beach, Christy Pines, Central Valley, South Ridge, Laguna Canyon, Corral Point, San Justiniano Canyon, Los Torres, Coches Prietos; 22 Jun, S.A.Junak SC2733-SC2737, Forney's Cove, Christy Ranch; 26-28 Jun, S.A.Junak SC2738-SC2770, Christy Ranch, Central Valley, Christy Pines, South Ridge, Pozo Canyon, Pozo Beach, Morse Point, Laguna Canyon, Corral Point; 31 Jul-2 Aug, S.A.Junak SC2771-SC2816, China Harbor, Central Valley, Cañada de los Sauces, Playa Larga, Johnson Canyon, Punta Arena, Coches Prietos; 13-14 Sep, S.A.Junak SC2817-SC2885, Pelican Bay, Coche Point Scorpion Ranch, Scorpion Canyon, Smugglers' Ranch, Smugglers' Canyon; 7-12 Oct, S.A.Junak SC2886-SC2944, Prisoners' Harbor, Navy Road, Coche Point, El Montañon, Christy Beach, Central Valley, Islay Canyon, Forney's Cove, Cañada de los Sauces, Sierra Blanca, Coches Prietos, Valley Anchorage; 18-19 Oct, S.A.Junak SC2945-SC2979, Campo Grande, Cavern Point, Scorpion Beach, Scorpion Canyon, Little Scorpion Canyon.

1992: 2-6 Mar, S.A.Junak SC2982-SC3044, Pelican Bay, Cañada del Puerto, Christy Ranch, Central Valley, ridge north of Cañada de los Sauces; 16-21 Mar, S.A.Junak SC3045-SC3095, Cañada del Puerto, Triangulation Point "John", Campo Raton, Valley Peak Road, South Ridge, Cañada de Malva Real, Laguna Canyon, Coches Prietos, south side, west of Navy Base; 28 Mar, S.A.Junak SC3096-SC3114, Cavern Point, Scorpion Ranch, Scorpion Canyon; 13-16 Apr, S.A.Junak SC3115-SC3226, Navy Road, Campo Raton, Christy Pines, Triangulation Point "John", Central Valley, Cañada Larga, Cañada de la Siesta, La Selva, South Ridge, El Tigre, south side west of Navy Base; 17-18 Apr, S.A.Junak SC3227-SC3286, Scorpion Ranch, Scorpion Canyon, Smugglers' Canyon; 25-27 Apr, S.A.Junak SC3287-SC3351, West Point Ridge, Christy Pines, Central Valley, Rancho Nuevo, South Ridge, Willows Canyon; 18-21 May, S.A.Junak SC3352-SC3411, North Ridge, Peak 1941, Red Mountain, Cañada del Agua, Navy Road, Christy Pines, Cottonwood Canyon, Islay Canyon, Alegria Canyon, Montañon Canyon; 29 May-1 Jun, S.A.Junak SC3412-SC3469, Pelican Bay, Prisoners' Harbor, China Harbor, China Pines, Central Valley, Cottonwood Canyon, Islay Canyon; 17-18 Sep, S.A.Junak SC3470-SC3486, Cañada del Puerto, Central Valley, Cañada de Malva Real, Laguna Canyon, South Ridge; 17-18 Sep, T.Ross, Cañada del Puerto, Central Valley, Cañada de Malva Real, Laguna Canyon, South Ridge; 2-4 Oct, S.A.Junak SC3487-SC3513, Christy Beach, Christy Ranch, Near Point, Pozo Beach, west of Cañada de Malva Real, Laguna Canyon; 6-7 Nov, S.A.Junak SC3515-SC3531, Cavern Point, Scorpion Ranch, Smugglers' Canyon, Smugglers' Cove, Yellowbanks Beach, Aguaje Canyon.

1993: 1-2 Jan, S.A.Junak SC3532-SC3538, Scorpion Canyon; 1-4 Apr, S.A.Junak SC3539-SC3594, Christy Beach, Christy Ranch, Cañada Christy, Near Point, Pozo Beach; 16-18 Apr, S.A.Junak SC3595-SC3642, Christy Ranch, Rancho Nuevo Canyon, Laguna Canyon; 20-23 Apr, S.A.Junak SC3643-SC3729, Potato Harbor, Cavern Point, Scorpion Ranch, Scorpion Canyon, Little Scorpion Canyon, Smugglers' Canyon; 2-4 Jun, S.A.Junak SC3730-SC3780, Road NW of Centinela, Central Valley, Horqueta Canyon, Los Pinos del Sur; 6-7 Aug, S.A.Junak SC3781-SC3819, West Point Ridge, Christy Beach, Christy Ranch, Christy Pines, Cañada de los Sauces, Pozo Canyon.

1994: 11-12 Mar, S.A.Junak SC3820-SC3831, Scorpion Canyon, San Pedro Point; 11-14 Apr, S.A.Junak SC3832-SC3923, Prisoners' Harbor, Campo del Norte, Christy Ranch, Central Valley, Pozo Canyon, Alamos Canyon, Horqueta Canyon, Los Pinos del Sur, Montañon Canyon; 9-12 May, S.A.Junak SC3924-SC3967, West Point Ridge, Fraser Point, Pelican Bay, Cañada del Puerto, China Pines, Christy Pines, Rancho del Sur, Coches Prietos; 20-22 May, S.A.Junak SC3968-SC4008, Fraser Point, Christy Beach, Rancho Nuevo, Black Point Canyon, Pozo Canyon, Pozo Beach; 24-25 Jun, S.A.Junak SC4009-SC4029, Prisoners' Harbor, Cañada del Puerto, Stanton Ranch, Valley Anchorage, South Ridge; 22-24 Jul, S.A.Junak SC4030-SC4053, Fraser Point, Christy Beach, Christy Ranch, Christy Pines, Cañada de los Sauces; 5-7 Aug, S.A.Junak SC4054-SC4075, North Ridge, Christy Beach, Christy Ranch, Road to Triangulation Point "John", Cañada de los Sauces, Near Point, Pozo Canyon; 26-30 Sep, D.Wilken 15132-15172, Prisoners' Harbor, Cañada del Puerto, Cañada Cebada, Cañada de los Sauces, South Ridge.

# APPENDIX II. Representative Herbarium Specimens.

A maximum of three specimens are cited for each taxon. Specimens are listed in the same geographic order used in the taxonomic treatments and are deposited at the Santa Barbara Botanic Garden unless otherwise indicated. Standard herbarium abbreviations (Holmgren et al. 1990) are used, except for the Santa Cruz Island Reserve Herbarium, which is abbreviated SCIR. Herbarium accession numbers are only cited for critical specimens (e.g., *Quercus dumosa*). Label data on some specimens have been edited for consistency (e.g., punctuation and spelling of some place names). Most place names used by collectors can be found on the supplemental map of Santa Cruz Island.

## Pteridophytes

### BLECHNACEAE

**Woodwardia fimbriata** Smith: w fork of first canyon w of Profile Point, 12 Sep 1989, Junak & Scott SC-1411; Central Valley, in upper e fork of Hooker Canyon, 4 Nov 1984, Junak SC-936; West Aguaje Spring, n of Sandstone Point, 24 Mar 1971, Philbrick et al. B71-29.

### DENNSTAEDTIACEAE

**Pteridium aquilinum** L. var. **pubescens** Underw.: Lagunitas Secas, 1 Mar 1965, Philbrick & Benedict B65-463; head of easternmost fork of Lady's Canyon, 19 Nov 1980, Junak et al. SC-341; Pelican Bay, 14 Aug 1939, Dunkle 8603 (RSA-POM,SBBG).

### DRYOPTERIDACEAE

**Athyrium filix-femina** (L.) Mertens var. **cyclosorum** Rupr.: canyon above Lady's Harbor, 20 Aug 1967, Philbrick s.n.; w fork of drainage to Baby's Cove, 14 Sep 1989, Ayers & Scott 803; Pelican Bay, 26 Jul 1960, Blakley 3778.
**Cystopteris fragilis** (L.) Bernh.: 0.75 mi nw of Picacho Diablo, in canyon which drains w of Arch Rock, 24 Mar 1982, Junak SC-484; w fork of drainage to Baby's Cove, 14 Sep 1989, Ayers & Scott 804; upper Orizaba Canyon, 23 May 1932, Hoffmann s.n. (RSA-POM,SBBG).
**Dryopteris arguta** (Kaulf.) Watt.: upper e fork of Diablo Canyon, 20 Sep 1990, Junak & Bornstein SC-1994; between Cañada del Portezuelo and La Centinela, 25 Jul 1960, Blakley & Muller 3746; w side of Coches Prietos Canyon, in small canyon on n side of ridge, 1 May 1960, Blakley 3379 (RSA-POM,SBBG).
**Polystichum munitum** (Kaulf.) C. Presl: upper e fork of Baby's Canyon, 19 Nov 1980, Junak et al. SC-343; n-facing slope of Christi Pines, ca. 0.2 mi w of Centinela, 21 Dec 1967, Benedict s.n.; upper Laguna Canyon, 25 Nov 1977, Philbrick & Junak B77-46.

### EQUISETACEAE

**Equisetum hyemale** L. subsp. **affine** (Engelm.) Calder & R.L.Taylor: Water Canyon, just above beach, 11 Sep 1989, Ayers et al. 781; water hole near China Harbor, 12 Apr 1931, Hoffmann s.n.
**Equisetum laevigatum** A.Braun: near mouth of Point Diablo Canyon, 21 Apr 1932, Hoffmann s.n.; Cañada del Puerto, 6 Jun 1961, Smith 6438; upper Cañada Christy at Campo Raton, 21 Apr 1990, Junak SC-1585.
**Equisetum telmateia** Ehrh. subsp. **braunii** (J.Milde) Hauke: lower portion of Diablo Canyon, 13 Mar 1980, Junak SC-286; Fry's Harbor, 21 Sep 1930, Hoffmann s.n.; water hole near China Harbor, 20 Sep 1930, Hoffmann s.n.

### POLYPODIACEAE

**Polypodium californicum** Kaulf.: e side of Prisoners' Harbor, 13 Mar 1961, Blakley 3946; upper w fork of Scorpion Canyon, 23 Apr 1993, Junak SC-3723; small canyon s of Stanton Ranch, 14 Mar 1961, Blakley 4006.
**Polypodium scouleri** Hook. & Grev.: Santa Cruz Island, Aug 1889, Yates s.n.; Santa Cruz Island, Aug 1892, Yates s.n.; Valdez Harbor, 10 Nov 1930, Hoffmann s.n.

### PTERIDACEAE

**Adiantum aleuticum** (Rupr.) Paris: canyon s of Cueva Valdez, 13 Sep 1989, Junak et al. SC-1424; Lady's Harbor, 11 Apr 1930, Hoffmann s.n.; lower Tinker's Canyon, 26 Nov 1977, Philbrick et al. B77-52.

**Adiantum capillus-veneris** L.: ca. 0.25 mi W of West Twin Harbor, 11 Oct 1967, Benedict s.n.; upper part of n tributary of Griffith Canyon, directly s of Picacho Diablo, 5 Oct 1967, Benedict s.n.; in small stream at e side of mouth of Laguna Canyon, 15 Oct 1958, Blakley & Balls B-2606.
**Adiantum jordanii** C. Mueller: canyon draining n from Triangulation Point "Coche Point" to Potato Harbor, 8 May 1968, Philbrick et al. B68-281; upper Cañada Christy, at bottom of Centinela Grade, 10 May 1990, Junak SC-1798; canyon draining into Coches Prietos Beach from w, 14 Jun 1967, Benedict s.n.
**Aspidotis californica** (Hook) Nutt. ex Copel.: above Lady's Harbor, 12 Apr 1930, Hoffmann s.n.; ca. 1.25 mi n of Peak 1941, 22 Feb 1969, Benedict s.n.; s of Pelican Bay, 21 Feb 1932, Hoffmann s.n.
**Cheilanthes clevelandii** D.C.Eaton: lower portion of Eagle Canyon, 10 May 1968, Philbrick & McPherson B68-311; Central Valley, about 0.5 mi e of Centinela, 1 Mar 1965, Philbrick & Benedict B65-455; head of Aguaje Canyon, w of Yellowbanks Anchorage, 25 Mar 1971, Benedict s.n.
**Cheilanthes cooperae** D.C.Eaton: Santa Cruz Island, ca. 1893, Yates s.n.
**Pellaea andromedifolia** (Kaulf.) Fee: Cañada del Puerto, near second creek crossing below Stanton Ranch, 1 Feb 1965, Philbrick et al. B65-89; Cañada Islay, 16 Jun 1964, Philbrick et al. B64-22; upper portion of Pozo Canyon Road, 0.4 mi from top of road, 7 May 1991, Junak SC-2412.
**Pellaea mucronata** (D.C.Eaton) D.C.Eaton: near Prisoners' Harbor, 2 Feb 1965, Philbrick & Flinck B65-147; Willows Canyon, at s point of ridge which Willows Road follows, 26 Mar 1983, Junak SC-642; 200 yards w of main summit of El Tigre, 1 Feb 1968, Benedict s.n.
**Pentagramma triangularis** (Kaulf.) Yatsk., Windham, & E.Wollenw. subsp. **triangularis**: Pelican Harbor, 12 Apr 1931, Hoffmann 235; Central Valley, at bottom of Portezuelo Grade, 11 Mar 1961, Blakley 3891; n side of Ridge Road near head of Willows Canyon, 19 Jun 1967, Benedict s.n.
**Pentagramma triangularis** subsp. **viscosa** (Nutt. ex D.C.Eaton) Yatsk., Windham, & E.Wollenw.: "Alamo" Canyon, 2 mi e of Valdez Harbor, 9 Nov 1930, Hoffmann s.n.; Prisoners' Harbor, 13 Apr 1931, Hoffmann 236; interior of island, 10 Apr 1931, Hoffmann 231.

### SELAGINELLACEAE

**Selaginella bigelovii** Underw.: Pelican Bay, 22 Apr 1990, Junak SC-1616; South Ridge Road, just s of turn into Cañada Cebada, 4 Apr 1991, Junak SC-2195; canyon n of Stanton Ranch, 27 Apr 1960, Raven & Smith 15270 (RSA-POM,SBBG).

## Gymnosperms

### CUPRESSACEAE

**Cupressus macrocarpa** Hartw. ex Gord: sw of mouth of Scorpion Canyon, 29 Apr 1971, Philbrick B71-90; Sauces Canyon, 13 Feb 1982, Philbrick et al. B82-5.

### PINACEAE

**Pinus muricata** D.Don forma **muricata**: Orizaba Flats, 8 Aug 1968, Benedict s.n.; pine grove above China Harbor, 24 Jul 1960, Blakley & Muller 3740.
**Pinus muricata** forma **remorata** (H.Mason) Hoover: Fry's Harbor, 3 Aug 1936, Van Rensselaer s.n.; between Navy Base and China Harbor, 3 Dec 1961, Blakley 4832; China Harbor, 14 Apr 1931, Howell 6371 (CAS,SBBG).
**Pinus pinea** L.: Pelican Bay, 14 Sep 1991, Junak SC-2877; Prisoners' Harbor, n.d., Van Rensselaer s.n.

## Dicotyledonous Angiosperms

### ACERACEAE

**Acer macrophyllum** Pursh: upper e fork of Diablo Canyon, 20 Sep 1990, Junak & Bornstein SC-1998; mouth of Dick's Canyon, 12 Mar 1981, Junak et al. SC-374; upper w fork of Water Canyon, 23 Mar 1971, Philbrick et al. B71-3.

### ADOXACEAE

**Sambucus mexicana** C.Presl: e side of Prisoners' Harbor, 13 Mar 1961, Blakley 3940; Cañada de la Portezuela, ca. 0.25 mi w of Rancho Portezuelo ruins, 9 May 1990, Junak SC-1729; upper Pozo wash, 18 Apr 1989, Young et al. 1928.

## AIZOACEAE

**Carpobrotus chilensis** (Molina) N.E.Br.: trailing over cliffs, 28 Mar 1925, Hoffmann s.n.; Christy Beach, 11 Apr 1990, Junak SC-1511; n side of mouth of Cañada de los Sauces, 8 May 1990, Junak SC-1684.

**Carpobrotus edulis** (L.) N.E.Br.: Pelican Bay, 18 Apr 1986, Junak SC-1137; General Motors facility at Valley Anchorage, 7 May 1979, Junak et al. SC-33.

**Mesembryanthemum crystallinum** L.: beach w of Christi Ranch, 18 Jun 1964, Philbrick et al. B64-131; along shore near Christi Ranch, 16 May 1962, Piehl 62-214; Fraser Point, 12 Mar 1961, Blakley 3929.

**Mesembryanthemum nodiflorum** L.: beach w of Christi Ranch, 18 Jun 1964, Philbrick et al. B64-128; Fraser Point, 7 Mar 1950, Moran 3331 (UC); flats above Forney's Cove, 26 May 1959, Smith 5959.

**Tetragonia tetragonioides** (Pallas) Kuntze: Scorpion Canyon, just n of Scorpion Ranch, 19 Oct 1991, Junak SC-2963; s end of Christy Beach, 28 Mar 1989, Young et al. 1871; sw along beach from mouth of Cañada de los Sauces toward Kinton Point, 15 Oct 1958, Blakley & Balls B-2616.

## AMARANTHACEAE

**Amaranthus albus** L.: Pelican Bay, 14 Jun 1930, Hoffmann s.n.; ca. 0.75 mi w of Campo del Norte, 24 Sep 1979, Junak et al. SC-179; Main Ranch, 15 Aug 1939, Dunkle 8644 (RSA-POM,SBBG).

**Amaranthus blitoides** S.Watson: Prisoners' Harbor, 18 Jun 1984, Junak SC-883; Central Valley, in wash behind University of California Field Station, 27 Sep 1968, Benedict s.n.

**Amaranthus deflexus** L.: stockpen on s side of Christy Ranch airfield, 10 Jul 1979, Junak et al. SC-130; Stanton Ranch, 25 Apr 1960, Raven & Smith 15162 (RSA-POM,SBBG).

**Amaranthus powellii** S.Watson: head of w fork of Cañada del Agua, 3 Nov 1984, Junak SC-916; ca. 0.75 mi w of Campo del Norte, 24 Sep 1979, Junak et al. SC-180; vineyard, Main Ranch, 21 Sep 1930, Hoffmann s.n.

## ANACARDIACEAE

**Rhus integrifolia** (Nutt. in Torr. & A.Gray) Brewer & S.Watson: near Prisoners' Harbor, 2 Feb 1965, Philbrick & Flinck B65-150; Cañada del Puerto, 0.5 mi from Prisoners' Harbor, 13 Mar 1961, Blakley 3948; mouth of Laguna Canyon, 17 Jun 1964, Philbrick et al. B64-66.

**Rhus ovata** S.Watson: Central Valley at Portezuelo Grade, 11 Mar 1961, Blakley 3896; 1 mi down Willows Canyon Road, 26 Mar 1980, Millett-UCSB Botanical Society 2004 (UCSB,SBBG); canyon, Cochies Prietos, 22 Mar 1929, Hoffmann s.n.

**Schinus molle** L.: sw of mouth of Scorpion Canyon, 29 Apr 1971, Philbrick B71-89; nw of Stanton Ranch, 27 Nov 1977, Philbrick & Junak B77-72; ridge between Coches Prietos and Canyon del Medio, 30 Apr 1960, Blakley 3337 (RSA-POM,SBBG).

**Toxicodendron diversilobum** (Torr. & A.Gray) Greene: Hazard's, 7 Apr 1930, Hoffmann s.n.; s side of Islay Pasture, 11 Mar 1961, Blakley 3887.

## APIACEAE

**Apiastrum angustifolium** Nutt.: just w of Prisoners' Harbor, 22 May 1983, Junak SC-682; in first canyon e of winery buildings at Stanton Ranch, 22 Mar 1985, Junak SC-954; at top of road to Los Pinos del Sur, 23 Mar 1982, Junak SC-478.

**Apium graveolens** L.: Prisoners' Harbor, 15-18 May 1962, Piehl 62-248; Prisoners' Harbor, 18 Jun 1970, Benedict s.n.; near mouth of creek at Valley Anchorage, 7 May 1979, Junak et al. SC-26.

**Berula erecta** (Huds.) Coville: Prisoners' Harbor, 25 Apr 1979, Thorne et al. 52600 (RSA-POM); mouth of Cañada Christy, 3 Oct 1992, Junak SC-3494; Cañada de Malva Real, 0.1 mi inland from beach, 17 Sep 1992, Junak SC-3476.

**Bowlesia incana** Ruiz & Pav.: n side of Picacho Diablo near summit, 19 Apr 1989, Junak et al. SC-1350; Smugglers' Field, s of Scorpion Rocks and w of mouth of Hungryman's Gulch, 29 Apr 1971, Philbrick et al. B71-68; lower portion of Cañada de la Mina, 9 May 1990, Junak SC-1718.

**Conium maculatum** L.: Cañada del Puerto, ca. 0.5 mi from Stanton Ranch, 15-17 May 1962, Piehl 62-161; Central Valley, ca. 0.1 mi w of top of Portezuela Grade, 8 May 1990, Junak SC-1669; Central Valley, ca. 0.2 mi w of University of California Field Station, 10 May 1990, Junak SC-1773.

**Daucus pusillus** Michx.: first major canyon e of Profile Point, 12 Jun 1965, Philbrick & Benedict B65-1365; small canyon on crest of ridge on s side of Central Valley, ca. 3 mi w of Stanton Ranch, 16 May 1962, Breedlove 2827 (CAS,SBBG); Coches Prietos Canyon, 29 Apr 1960, Blakley 3283.

**Foeniculum vulgare** Mill.: Prisoners' Harbor, 12 Aug 1939, Dunkle 8558 (RSA-POM,SBBG); just w of mouth of Scorpion Canyon, 19 Oct 1991, Junak SC-2962; Main Ranch, 1 Jul 1930, Hoffmann s.n.
**Lomatium caruifolium** (Hook. & Arn.) J.M.Coult. & Rose: Central Valley just e of Christy Ranch, 16 Apr 1991, Junak SC-2248; ridge on s side of Central Valley, 200 yards w of El Sobranti, 13 Mar 1961, Blakley 3972; lower portion of Cañada de los Sauces, 16 Apr 1991, Junak SC-2260.
**Sanicula arguta** Greene ex J.M.Coult. & Rose: Navy Road near Lake Pasture, ca. 5 mi e of Stanton Ranch, 15-18 May 1962, Piehl 62-117; Central Valley ca. 0.1 mi e of Reparo Viejo, 29 May 1990, Junak SC-1854; Willows Canyon, ca. 3 km n of Willows Anchorage, 29 Mar 1989, Young et al. 1885.
**Sanicula hoffmannii** (Munz) C.R.Bell: near ridge crest in Christi Pines, Apr-May 1967, Benedict s.n.; Arroyo del Medio, 1.1 mi w of Stanton Ranch, 26 Apr 1960, Raven & Smith 15174 (CAS-DS,RSA-POM,SBBG); lower portion of Cañada de los Sauces, 8 May 1990, Junak SC-1704.
**Torilis nodosa** (L.) Gaertn.: Prisoners' Harbor, 15 Jun 1930, Hoffmann s.n.; upper Cañada Christy, at bottom of Centinela Grade, 10 May 1990, Junak SC-1795; second major canyon nw of Near Point, nw of Morse Point, 18 Apr 1968, Benedict s.n.
**Yabea microcarpa** (Hook. & Arn.) Koso-Pol.: n side of Picacho Diablo near summit, 19 Apr 1989, Junak et al. SC-1349; Central Valley, ca. 0.3 mi e of Buena Vista, 10 May 1990, Junak SC-1875; n side of Las Torres ridge, 1 mi w of intersection of San Justiniano and Coches Prietos roads, 12 Apr 1990, Junak SC-1547.

## APOCYNACEAE

**Vinca major** L.: Pelican Harbor, 20 Aug 1967, Philbrick s.n.; Cañada del Puerto, 27 Nov 1977, Philbrick & Junak B77-75; established on bank of stream, Main Ranch, 25 Mar 1932, Hoffmann s.n.

## ASCLEPIADACEAE

**Asclepias fascicularis** Decne. in A.DC.: ridgetop above e fork of upper Fry's Canyon, 20 Sep 1990, Junak SC-1993; lower part of n tributary of Griffith Canyon, directly s of Picacho Diablo, 5 Oct 1967, Benedict s.n.; near mouth of Laguna Canyon, 17 Jun 1964, Philbrick et al. B64-69.

## ASTERACEAE

**Achillea millefolium** L.: head of westernmost canyon emptying into China Harbor, e of Mount Pleasant, 8 May 1968, Philbrick et al. B68-269; Central Valley, 0.3 mi e of Buena Vista, 10 May 1990, Junak SC-1784; ridge between Coches Prietos Canyon and Alberts Anchorage, 29 Apr 1960, Blakley 3289.
**Achyrachaena mollis** Schauer: on n flanks of Peak 1450, just se of Mount Pleasant, 13 Apr 1992, Junak SC-3123; near Main Ranch, 10 Apr 1931, Hoffmann 219; South Ridge Road, just e of road to Coches Prietos Anchorage, 17 Apr 1989, Ayers et al. 643.
**Acourtia microcephala** DC.: near bottom of road leading to w end of China Harbor, 17 Jun 1984, Junak SC-874; at spring 0.5 mi n of ranch house and w of Cañada del Puerto, 29 Jun 1962, Blakley 5284; at base of Johnson Canyon Road, 1 Aug 1991, Junak SC-2799.
**Agoseris grandiflora** (Nutt.) Greene: lower Embudo, ca. 6 mi w of top of Portezuela Grade, 4 May 1966, Philbrick B66-189; Central Valley, ca. 50 yards e of Reparo Viejo overlook, 8 May 1990, Junak SC-1675; ridge between Posa Canyon and Sauces Canyon, 5 May 1968, Benedict s.n.
**Amblyopappus pusillus** Hook. & Arn.: field near San Pedro Point, 29 Apr 1971, Philbrick et al. B71-97; Fraser Point, 29 Apr 1960, Raven & Smith 15306 (RSA-POM,SBBG); mouth of Laguna Canyon, 17 Jun 1964, Philbrick et al. B64-65.
**Ambrosia chamissonis** (Less.) Greene: Scorpion Anchorage, 19 Oct 1991, Junak SC-2979; beach w of Christy Ranch, 18 Jun 1964, Philbrick et al. B64-126; dunes at Coches Prietos, 17 Apr 1989, Ayers et al. 638.
**Anthemis cotula** L.: Prisoners' Harbor, 15 Jun 1930, Hoffmann s.n.; Prisoners' Harbor, 28 Jul 1930, Clokey 5113 (RSA-POM); Prisoners' Harbor, 13 Oct 1958, Blakley B-2550.
**Artemisia californica** Less.: e side of Scorpion Canyon, 18 Oct 1991, Junak SC-2949; e of Christi Ranch, 18 Jun 1964, Philbrick et al. B64-159; Coches Prietos, 10 May 1973, Philbrick & Philbrick B73-309.
**Artemisia douglasiana** Besser in Hook.: Cañada del Puerto, 1.3 mi inland from Prisoners' Harbor, 21 Sep 1990, Junak & Bornstein SC-2014; upper Cañada Christy, just e of Embudo Trough, 27 Aug 1980, Hochberg & Junak 553; 4 mi w of Stanton Ranch, 14 Oct 1958, Blakley & Balls B-2597.
**Aster chilensis** Nees: water hole near China Harbor, 20 Sep 1930, Hoffmann s.n.; Christy Canyon, 100 m downstream from bridge at Christy Ranch, 29 Sep 1984, Junak SC-899.
**Aster radulinus** A.Gray: Buena Vista, 26 Sep 1968, Benedict s.n.; 0.3 mi e of Reparo Viejo, 8 Oct 1991, Junak SC-2901; Cañada de los Sauces, on ridge between 2 forks of canyon, 8 Aug 1993, Junak SC-3810.

**Aster subulatus** Michx. var. **ligulatus** Shinners: water hole near China Harbor, 20 Sep 1930, Hoffmann s.n.; near mouth of Christy Canyon, 12 Sep 1989, Junak et al. SC-1419; creek crossing along road to Fraser Point, on n side of Black Point peninsula, 11 Oct 1991, Junak SC-2941.
**Baccharis douglasii** DC.: spring at e edge of Potato Harbor, 15 Jun 1981, Philbrick s.n.; Christy Canyon, just downstream from bridge at Christy Ranch, 29 Sep 1984, Junak SC-892; 1.5 mi se of Forney's Cove on road to Christi Beach, 23 Jul 1960, Blakley & Muller 3730.
**Baccharis pilularis** DC.: hills above Valdez Harbor, 8 Nov 1930, Hoffmann s.n.; 1.5 mi e of Stanton Ranch near El Sobrante, 12 Oct 1958, Blakley B-2531; lower portion of Laguna Canyon, 17 Oct 1990, Junak SC-2080.
**Baccharis plummerae** A.Gray subsp. **plummerae**: Prisoners' Harbor, 13 Aug 1939, Dunkle 8579 (NY,RSA-POM,SBBG); Islay Canyon Road, just n of intersection with South Ridge Road, 17 Oct 1990, Junak SC-2059; just inland from mouth of Laguna Canyon, 17 Oct 1990, Junak SC-2068.
**Baccharis salicifolia** (Ruiz & Pav.) Pers.: Valdez, 15 Jun 1930, Hoffmann s.n.; ca. 1 mi w of Stanton Ranch, 16 Jun 1964, Philbrick et al. B64-55; Coches Prietos Canyon, 30 Apr 1960, Blakley 3341.
**Blennosperma nanum** (Hook.) S.F.Blake var. **nanum**: road just n of summit of saddle to n of Point, w end of island, 5 Apr 1970, Benedict s.n.; on ridge e of West Point, 12 Mar 1961, Blakley 3899; saddle on ridge, 1.2 mi ne of Forney's Cove, 13 Feb 1967, Benedict s.n.
**Brickellia californica** (Torr. & A.Gray) A.Gray: middle portion of Little Scorpion Canyon, 18 Oct 1991, Junak SC-2957; Central Valley, ca. 1 mi w of Stanton Ranch, 22 Sep 1966, Webster 7445 (DAV,SBBG); Laguna Canyon Road, 3.5 mi inland from beach, 19 Sep 1990, Junak & Bornstein SC-1987.
**Centaurea melitensis** L: Valdez, 14 Jun 1930, Hoffmann s.n.; lower Embudo, ca. 6 mi w of top of Portezuela Grade, 4 May 1966, Philbrick et al. B66-188; Alberts Anchorage, 30 Apr 1960, Blakley 3349.
**Centaurea solstitialis** L.: Prisoners' Harbor, 12 Aug 1939, Dunkle 8546 (NY,RSA-POM,SBBG); Main Ranch, 1 Jul 1930, Hoffmann s.n.; upper portion of Laguna Canyon Road, ca. 1 mi s of intersection with South Ridge Road, 17 Oct 1990, Junak SC-2064.
**Chamomilla suaveolens** (Pursh) Rydb.: Prisoners' Harbor, 23 Apr 1990, Junak SC-1642; Main Ranch, 15 Jun 1930, Hoffmann s.n.; middle portion of Cañada de los Sauces, 16 Apr 1991, Junak SC-2271.
**Chrysanthemum coronarium** L.: Pelican Bay, 14 Aug 1939, Dunkle 8622 (RSA-POM,SBBG).
**Cichorium intybus** L.: road above Christy's, 29 Jun 1930, Hoffmann s.n.; 0.25 mi e of Christi Ranch, 25 Jul 1960, Blakley & Muller 3752; streambed at Christi Ranch, 16 May 1962, Piehl 62-218.
**Cirsium brevistylum** Cronquist: Central Valley just e of Campo Raton, 31 May 1979, Junak et al. SC-119; ridge n of Ragged Mountain, 11 Jun 1965, Philbrick & Benedict B65-1303; e-facing slope of Sierra Blanca summit, 16 Apr 1967, Philbrick B67-133.
**Cirsium occidentale** var. **californicum** (A.Gray) D.J.Keil & Turner: Scorpion Ranch, 10 Mar 1928, Hoffmann s.n.; hillsides above Eucalyptus grove e of Christi, 19 Apr 1969, Benedict s.n.; near Cochies Prietos, 12 Apr 1931, Hoffmann s.n.
**Cirsium occidentale** (Nutt.) Jeps. var. **occidentale**: lower part of first major canyon e of Profile Point, 12 Jun 1965, Philbrick & Benedict B65-1378; Central Valley e of El Sobranti, 13 Mar 1961, Blakley 3966; top of ridge between Coches Prietos Canyon and Alberts Anchorage, 29 Apr 1960, Blakley 3319.
**Cirsium vulgare** (Savi) Ten.: first major drainage w of Profile Point, 12 Sep 1989, Ayers et al. 794; middle portion of canyon between Cueva Valdez and Arch Rock, 14 Sep 1989, Junak SC-1444.
**Cnicus benedictus** L.: Central Valley, 2 mi w of Stanton Ranch, 16 May 1962, Breedlove 2862 (CAS-DS,SBBG).
**Conyza bonariensis** (L.) Cronquist: Pelican Bay, 14 Jun 1930, Hoffmann s.n.; South Ridge Road, just e of intersection with road to Centinela Grade, 11 Oct 1991, Junak SC-2932; Smugglers' Ranch, 13 Sep 1991, Junak & Meadows SC-2839.
**Conyza canadensis** (L.) Cronquist: Fry's Harbor, 21 Sep 1930, Hoffmann s.n.; Christy Canyon, just downstream from bridge at Christy Ranch, 29 Sep 1984, Junak SC-893; upper Laguna Canyon, 17 Oct 1990, Junak SC-2066.
**Coreopsis gigantea** (Kellogg) H.M.Hall: Prisoners' Cove, 2 Sep 1928, Ackley 1191; 1.25 mi sw of Devil's Peak, 26 Mar 1980, Steuart-UCSB Botanical Society 341 (UCSB,SBBG); Alberts Anchorage, 30 Apr 1960, Blakley 3348.
**Cotula australis** (Sieber) Hook.f.: Santa Cruz Island, 28 Mar 1925, Hoffmann s.n.; burn, top of ridge s of Prisoners' Harbor, 21 Feb 1932, Hoffmann s.n.; mouth of Scorpion Canyon, 28 Mar 1992, Junak SC-3114.
**Cotula coronopifolia** L.: Prisoner's, 17 Jun 1930, Hoffmann s.n.; upper reaches of Cañada Christy, on Centinela Grade, 10 May 1990, Junak SC-1821; beach at mouth of Laguna Canyon, 17 Jun 1964, Philbrick et al. B64-63.
**Encelia californica** Nutt.: mouth of Scorpion Canyon, 14 Sep 1991, Junak SC-2849; Valley Anchorage, 24 Mar 1929, Hoffmann s.n.; grassy mesa 1 mi w of Sandstone Point, 17 May 1962, Breedlove 2875 (CAS,SBBG).

**Erechtites glomerata** (Poir.) DC.: Cañada de los Sauces, ca. 0.2 mi inland from canyon mouth, 8 May 1990, Junak SC-1695; ridge s of Sauces Canyon Road, on ne extension of Ragged Mountain, 18 Apr 1989, Ayers et al. 664; e side of Sierra Blanca ridge, at saddle n of Ragged Mountain, 8 Oct 1991, Junak SC-2911.

**Erigeron foliosus** Nutt. var. **foliosus**: rocky ridge near Pelican Pines, 14 Jun 1930, Hoffmann s.n.; wooded bank above Christy's, 29 Jun 1930, Hoffmann s.n.

**Erigeron foliosus** var. **stenophyllus** (Nutt.) A.Gray: Mount Diablo, 9 Sep 1931, Hoffmann s.n.; Cañada Larga Road, 0.2 mi n of Valley Road, 30 May 1990, Junak SC-1884; coastal ridge on w side of Coches Prietos Beach, 30 May 1991, Junak SC-2709.

**Erigeron glaucus** Ker Gawl.: first cove e of Twin Harbors, 20 Apr 1989, Ayers et al. 698; Prisoners' Harbor, 13 Aug 1939, Dunkle 8580 (RSA-POM,SBBG); ca. halfway between Scorpion Canyon and Cavern Point, 14 Sep 1991, Junak & Meadows SC-2853.

**Erigeron sanctarum** S.Watson: small ravine leading to summit of Ragged Mountain, ca. 200 yards n of summit, 11 Jun 1965, Philbrick et al. B65-1335.

**Eriophyllum confertiflorum** (DC.) A.Gray var. **confertiflorum**: on ridge above Chinese Harbor, 15 May 1962, Breedlove 2789 (CAS,SBBG); brushy hill top s of Main Ranch, 11 Apr 1931, Hoffmann s.n.; Coches Prietos Beach, 17 Apr 1989, Young et al. 1904.

**Eriophyllum staechadifolium** Lagasca: West Point, 15 Jun 1967, Benedict s.n.; Valdez, 9 Nov 1930, Hoffmann s.n.; mouth of Potato Canyon, 8 May 1968, Philbrick et al. B68-293.

**Filago californica** Nutt.: Fry's Harbor, 22 May 1978, Philbrick B78-180; Islay Canyon Road 1.3 mi from intersection with South Ridge Road, 15 Apr 1991, Junak SC-2210; upper portion of Pozo Canyon Road, 0.4 mi from top of road, 7 May 1991, Junak SC-2411.

**Filago gallica** L.: on ridge above Cueva Valdez, 30 May 1979, Junak et al. SC-105; Christy Pines, along South Ridge Road ca. 2.3 mi e of Christy Barn, 14 May 1980, Junak & Hochberg SC-298; El Tigre, 6 May 1969, Benedict s.n.

**Gnaphalium bicolor** Bioletti: Fry's Harbor, 21 Sep 1930, Hoffmann s.n.; 1 mi nw of Centinela, along road toward Cueva Valdez, 1 Mar 1965, Philbrick & Benedict B65-460; e end of South Ridge Road, 24 Feb 1984, Junak SC-827.

**Gnaphalium californicum** DC.: pine grove above Chinese Harbor, 15-18 May 1962, Piehl 62-139; Central Valley, 3 mi w of top of Portezuela Grade, 4 May 1966, Philbrick B66-174; above beach at Coches Prietos, 17 Apr 1989, Ayers et al. 628.

**Gnaphalium canescens** DC. subsp. **beneolens** (Davidson) Stebbins & D.J.Keil: upper portion of Sauces Canyon road, 7 Aug 1993, Junak SC-3800.

**Gnaphalium canescens** DC. subsp. **microcephalum** (Nutt.) Stebbins & D.J.Keil: Cañada del Puerto, 0.8 mi inland from Prisoners' Harbor, 21 Sep 1990, Junak & Bornstein SC-2007; vineyard, Main Ranch, 21 Sep 1930, Hoffmann s.n.; Smugglers' Canyon, 25 Mar 1971, Philbrick & Ricker B71-34.

**Gnaphalium luteo-album** L.: Prisoners' Harbor, 13 Oct 1958, Blakley & Balls B-2555; lower Embudo, ca. 6 mi w of top of Portezuela Grade, 4 May 1966, Philbrick B66-184; near mouth of Willows Canyon, 12 Apr 1990, Junak SC-1555.

**Gnaphalium palustre** Nutt.: Prisoners' Harbor, 18 Jun 1984, Junak SC-878; near Mount Pleasant, 24 Jul 1960, Blakley & Muller 3741; Main Ranch, 15 Aug 1939, Dunkle 8642 (RSA-POM,SBBG).

**Gnaphalium purpureum** L.: w end of Platt's Harbor, 20 Apr 1989, Ayers et al. 705; upper Cañada Christy at Campo Raton, 19 May 1990, Junak SC-1838; ridge n of Ragged Mountain, 11 Jun 1965, Philbrick & Benedict B65-1315.

**Gnaphalium ramosissimum** Nutt.: first major drainage w of Profile Point, 12 Sep 1989, Ayers et al. 791; Tinker's Canyon, 14 Sep 1991, Junak SC-2881; lower Willows Canyon, 28 Sep 1984, Junak SC-886.

**Gnaphalium stramineum** Kunth: ne flank of Picacho Diablo, 19 Apr 1989, Junak SC-1357; Prisoners' Harbor, 13 Aug 1939, Dunkle 8588 (RSA-POM,SBBG); lower Willows Canyon, ca. 0.2 mi inland from beach, 28 Sep 1984, Junak SC-885.

**Grindelia camporum** Greene var. **bracteosum** (J.T.Howell) M.A.Lane: Prisoners' Harbor, 20 Nov 1980, Junak & Hochberg SC-345; Central Valley, just s of University of California Field Station, 23 Jul 1985, Junak SC-1100; upper portion of Laguna Canyon Road, 17 Oct 1990, Junak SC-2062.

**Hazardia detonsa** (Greene) Greene: Valdez Harbor, 8 Nov 1930, Hoffmann s.n.; upper Christi Canyon, 21 Sep 1966, Webster 7431 (DAV,SBBG); upper w side of Albert's Anchorage drainage, 18 Oct 1990, Junak SC-2086.

**Hazardia squarrosa** (Hook. & Arn.) Greene var. **grindelioides** (DC.) W.D.Clark: Prisoners' Cove, 2 Sep 1928, Ackley 426; Central Valley, at w end of Bosque Mano, 16 Oct 1990, Junak SC-2031; bluffs w of Coches Prietos, 10 Aug 1968, Benedict s.n.

**Helianthus annuus** L.: seasonal pond just s of top of steep portion of road to China Harbor beach, 2 Aug 1991, Junak SC-2815; Central Valley, in canyon below Mission Spring, 24 Jul 1978, Laughrin s.n.
**Hemizonia clementina** Brandegee: coastal flats w of Sandstone Point, ca. 0.25 mi sw of Peak 712, 13 Apr 1994, Junak SC-3898.
**Hemizonia fasciculata** (DC.) Torr. & A.Gray: road e of Scorpion Canyon, 13 Sep 1991, Junak & Meadows SC-2823; Central Valley, 0.1 mi w of The Nature Conservancy cabins, 29 May 1990, Junak SC-1843; flats n of Forney's Cove, 22 Jun 1991, Junak SC-2737.
**Hemizonia fitchii** A.Gray: Islay Pasture, 0.75 mi w of Stanton Ranch, 14 Oct 1958, Blakley & Balls 2600; ridges directly s of University of California Field Station, 6 Jun 1969, Benedict s.n.; near top of road to Alameda de los Coches, 0.1 mi s of intersection with South Ridge Road, 8 Oct 1987, Junak SC-1170.
**Hemizonia increscens** (H.M.Hall ex Keck) Tanowitz subsp. **increscens**: open fields above Christy's, 29 Jun 1930, Hoffmann s.n.; lower Cañada Christy, 0.3 mi e of Christy Ranch, 27 Jun 1991, Junak SC-2757.
**Heterotheca grandiflora** Nutt.: Cañada del Puerto, 1.5 mi from Prisoners' Harbor, 11 Oct 1958, Blakley B-2501; stream bed near Christi Ranch, 16 May 1962, Piehl 62-221; Coches Prietos Beach, 31 Jul 1991, Junak SC-2777.
**Hieracium argutum** Nutt.: rocky ridge near Pelican, 14 Jun 1930, Hoffmann s.n.; Central Valley, between Canyon del Portezuelo and La Centinela, 25 Jul 1960, Blakley & Muller 3747; road to Coches Prietos Beach, 0.2 mi e of Alameda de los Coches Prietos, 31 Jul 1991, Junak SC-2775.
**Hypochaeris glabra** L.: n face of Picacho Diablo, 26 Apr 1968, Benedict s.n.; Stanton Ranch, 29 Jun 1962, Blakley 5275; Smugglers' Canyon, ca. 0.25 mi inland from beach, 9 Mar 1985, Unser s.n.
**Isocoma menziesii** (Hook. & Arn.) G.L.Nesom var. **sedoides** (Greene) G.L.Nesom: beach w of Christi Ranch, 18 Jun 1964, Philbrick et al. B64-125; s end of Christy Beach, 12 Sep 1982, Junak SC-589; near Gull Rock, 30 Jun 1930, Hoffmann s.n.
**Isocoma menziesii** var. **vernonioides** (Nutt.) G.L.Nesom: just w of Triangulation Point "Santa Cruz East" above Cavern Point, 19 Oct 1991, Junak SC-2972; Cuesta Road, 22 Sep 1966, Webster 7438 (DAV,SBBG); Navy Road, 0.1 mi w of Navy Base, 30 Sep 1984, Junak SC-908.
**Jaumea carnosa** (Less.) A.Gray: mouth of Sauces Canyon, 28 Sep 1969, Benedict s.n.; mouth of small canyon ca. 0.25 mi w of mouth of Poso Canyon, 12 Sep 1968, Benedict s.n.
**Lactuca saligna** L.: lower portion of canyon at Valdez Anchorage, 25 Aug 1988, Junak SC-1200; n side of Central Valley, ca. 1 mi from Christy Ranch on road leading to Triangulation Point "John", 18 Mar 1992, Junak SC-3090; w fork of Laguna Canyon, n of Peak 869, 17 Oct 1990, Junak SC-2081.
**Lactuca serriola** L.: saddle at extreme e end of Navy Road, 9 Oct 1991, Junak & Young SC-2929; corrals n of Christy Ranch, 9 Sep 1981, Junak SC-446; Stanton Ranch, 15-18 May 1962, Piehl 62-288.
**Lasthenia californica** Lindl.: first major canyon e of Profile Point, 12 Jun 1965, Philbrick & Benedict B65-1370; lower portion of Cañada Christy, just w of Christy Barn, 16 Apr 1991, Junak SC-2245; Hungryman's Gulch, 20 Mar 1932, Fosberg 7523.
**Layia platyglossa** (Fisch. & C.A.Mey.) A.Gray: pine grove above Chinese Harbor, 15-18 May 1962, Piehl 62-128; Valley Anchorage, 11 Apr 1931, Hoffmann s.n.; Near Point, 13 May 1967, Benedict s.n.
**Lepidospartum squamatum** (A.Gray) A.Gray: Cañada del Puerto, 0.8 mi inland from Prisoners' Harbor, 21 Sep 1990, Junak & Bornstein SC-2006; Central Valley, 1.1 mi w of Stanton Ranch, 22 Sep 1966, Webster 7437 (DAV,SBBG).
**Lessingia filaginifolia** (Hook. & Arn.) M.A.Lane var. **filaginifolia**: bluffs ca. 2 mi e of West Point, 12 Sep 1989, Ayers et al. 786; Christy Canyon, just downstream from bridge at Christy Ranch, 29 Sep 1984, Junak SC-898; ridge n of Ragged Mountain, 0.75 mi sw of upper road to Christi Ranch, 11 Jun 1965, Philbrick & Benedict B65-1304.
**Madia exigua** (Smith) A.Gray: summit of High Mount, 7 May 1968, Philbrick & McPherson B68-251; n side of Central Valley, on ridge along e side of Cañada Larga, 4 Jun 1983, Junak SC-722; San Justiniano Road, 1.3 mi w of intersection with Coches Prietos Road, 30 May 1991, Junak SC-2718.
**Madia gracilis** (Smith) Keck: Prisoners' Harbor, 13 Apr 1931, Hoffmann 222; Central Valley, in field just e of University of California Field Station, 6 Jun 1983, Junak SC-756.
**Madia sativa** Molina: Prisoners' Harbor, 30 Jun 1962, Blakley 5288; Central Valley, 0.2 mi e of Campo Raton, 27 Aug 1980, Junak SC-318.
**Malacothrix coulteri** A.Gray: Santa Cruz Island, 1888, Brandegee s.n. (UC); Valley Anchorage, 24 Mar 1929, Hoffmann s.n.; Smugglers' Cove, 21 Mar 1932, Fosberg 7535 (RSA-POM).
**Malacothrix incana** (Nutt.) Torr. & A.Gray: Santa Cruz Island, Jul & Aug 1886, Greene s.n. (ND-G).

**Malacothrix indecora** Greene: Twin Harbor, 18 Jul 1939, Williams 87 (RSA-POM); Black Point, 19 Jun 1980, Junak SC-312 (DHL,SBBG).
**Malacothrix saxatilis** (Nutt.) Torr. & A.Gray var. **implicata** (Eastw.) H.M.Hall: lower portion of canyon at Valdez Anchorage, 25 Aug 1988, Junak SC-1195; Christy's, 28 Jun 1930, Hoffmann s.n.; Pozo Anchorage, 18 Apr 1989, Ayers et al. 675.
**Malacothrix similis** W.S.Davis & Raven: Santa Cruz Island, Apr 1888, Brandegee (UC).
**Malacothrix squalida** Greene: Santa Cruz Island, 1888, Brandegee s.n. (UC); rocky promontory above Prisoners' Harbor, Jul & Aug 1886, Greene s.n. (CAS,UC,US); canyon draining n from Triangulation Point "Coche Point" to Potato Harbor, 8 May 1968, Philbrick et al. B68-292.
**Micropus californicus** Fisch. & C.A.Mey. var. **californicus**: Fry's Harbor, 22 May 1978, Philbrick B78-182; n side of Central Valley, on slopes above e fork of Cañada Larga, 24 Mar 1985, Junak SC-994; ridge between Coches Prietos Canyon and Alberts Anchorage, 29 Apr 1960, Blakley 3295.
**Microseris douglasii** (DC.) Sch.Bip. subsp. **tenella** (A.Gray) K.L.Chambers: grassy mesa s of Christy's, 21 Mar 1932, Hoffmann s.n.
**Microseris elegans** Greene ex A.Gray: n shore ca. 0.9 mi e of West Point, 13 Apr 1985, Junak SC-1043; pine forest, w end, 25 Mar 1932, Hoffmann s.n.; field at head of canyon nw of Smugglers' Field airstrip, 29 Apr 1971, Philbrick et al. B71-69.
**Osteospermum fruticosum** L.: s end of Christy Beach, 28 Mar 1989, Junak et al. SC-1246.
**Perityle emoryi** Torr.: first cove e of Pelican Bay, 25 Feb 1984, Padgett 15; mouth of Laguna Canyon, 17 Jun 1964, Philbrick et al. B64-64; canyon w of Los Pinos del Sur, 25 Mar 1981, Roberts-UCSB Botanical Society 2128 (UCSB,SBBG).
**Picris echioides** L.: Alegria Canyon, just inland from mouth, 19 May 1992, Junak SC-3387.
**Pluchea odorata** (L.) Cass.: Prisoners' Harbor, 26 Sep 1979, Junak SC-203; Malva Real Anchorage, 25 Jul 1960, Blakley & Muller 3754; near Valley Anchorage, 8 Nov 1981, Hochberg 718.
**Psilocarphus tenellus** Nutt. var. **tenellus**: n slope of island, ca. 0.75 m w of Picacho Diablo, 8 May 1982, Junak & Smith SC-540; Cañada del Puerto, ca. 0.5 mi from Stanton Ranch, 15-17 May 1962, Piehl 62-140; Willows Canyon, 26 Mar 1983, Junak SC-640.
**Rafinesquia californica** Nutt.: canyon e of Prisoners' Harbor, 17 May 1962, Breedlove 2869 (CAS,SBBG); Central Valley in second large canyon w of Rancho Portezuelo ruins, 9 May 1990, Junak SC-1731; ca. 0.4 mi ese of Peak 1387, s of Navy Road, 10 May 1985, Junak SC-1077.
**Senecio aphanactis** Greene: near Fry's Harbor, 25 Mar 1934, Peirson 11033 (RSA-POM); ridge n of Sauces Canyon, 2 Apr 1991, Junak SC-2133; w-facing slope between Ragged Mountain and Sierra Blanca Peak, 16 Apr 1967, Philbrick B67-141.
**Senecio flaccidus** Less. var. **douglasii** (DC.) B.L.Turner & T.M.Barkley: Cañada del Puerto, 1.5 mi from Prisoners' Harbor, 11 Oct 1958, Blakley & Balls B-2503; Central Valley just e of Christy Barn, 12 Sep 1982, Junak SC-586; mouth of Laguna Canyon, 17 Jun 1964, Philbrick et al. B64-67.
**Senecio mikanioides** Walp.: Ridge Road, 0.3 mi e of pine forest, 4 Jan 1977, Daily 172 (SCIR).
**Senecio vulgaris** L.: ridge e of West Point, 12 Mar 1961, Blakley 3902; near Mount Pleasant, 1 Feb 1965, Philbrick et al. B65-124a; Central Valley, ca. 0.25 mi e of Buena Vista, 7 May 1990, Junak SC-1663.
**Silybum marianum** (L.) Gaertn.: hill above Valdez, 15 Jun 1930, Hoffmann s.n.; Scorpion Ranch, 29 Apr 1971, Philbrick B71-85; ridge road ca. 1 mi wsw of Stanton Ranch, 13 Jul 1964, Philbrick B64-296.
**Solidago californica** Nutt.: between China and Scorpion, 30 Sep 1930, Hoffmann s.n.; Central Valley, 1.2 mi e of Stanton Ranch, Junak SC-1939; Laguna Canyon, 25 Nov 1977, Philbrick & Junak B77-48.
**Sonchus asper** (L.) Hill: headland on e side of Dick's Canyon, 10 Apr 1990, Junak & Scott SC-1494; Central Valley ca. 1.5 mi w of Stanton Ranch, 15 Apr 1967, Philbrick B67-71; Coches Prietos, 20 Mar 1967, Benedict s.n.
**Sonchus oleraceus** L.: Santa Cruz Island, ca. 1893, Yates s.n.; Prisoners' Harbor, 13 Aug 1939, Dunkle 8589 (RSA-POM,SBBG); Willows Canyon, 29 Mar 1989, Young et al. 1874.
**Stebbinsoseris heterocarpa** (Nutt.) K.L.Chambers: South Ridge Road, 0.8 mi e of Christy Barn, 31 May 1979, Junak et al. SC-114; hillside nw of Christy Ranch, on way to West Point, 7 Apr 1963, Smith 8151; upper portion of Pozo Canyon Road, 0.4 mi from top of road, 7 May 1991, Junak SC-2413.
**Stephanomeria cichoriacea** A.Gray: Prisoners' Harbor, 11 Aug 1939, Dunkle 8531 (RSA-POM,SBBG); Central Valley at w end of Bosque Mano, 16 Oct 1990, Junak SC-2030.
**Stephanomeria diegensis** Gottlieb: lower portion of canyon above Lady's Harbor, 25 Aug 1988, Junak SC-1205; Coches Prietos, 1 Jul 1930, Hoffmann s.n.

**Stephanomeria exigua** Nutt. subsp. **coronaria** (Greene) Gottlieb: Navy Road, 0.6 mi w of Navy Base, 10 Sep 1982, Junak SC-571; Central Valley, ca. 1 mi w of Stanton Ranch, 22 Sep 1966, Webster 7444 (DAV,SBBG); Main Ranch, 15 Aug 1939, Dunkle 8625 (RSA-POM,SBBG).

**Stephanomeria virgata** Benth. subsp. **virgata**: near head of Water Canyon, ca. 1 mi se of Prisoners' Harbor, 26 Sep 1967, Benedict s.n.; 0.25 mi w of Mount Pleasant, 6 Oct 1969, Benedict s.n.

**Stylocline gnaphaloides** Nutt.: upper e fork of Orizaba Canyon, 24 Mar 1985, Junak SC-1010; Central Valley, 3 mi w of top of Portezuela Grade, 4 May 1966, Philbrick B66-173; near mouth of Laguna Canyon, 17 Jun 1964, Philbrick et al. B64-75.

**Taraxacum officinale** F.H.Wigg.: Prisoners' Harbor, 15-18 May 1962, Piehl 62-251; middle portion of canyon between Cueva Valdez and Arch Rock, 14 Sep 1989, Junak SC-1441.

**Tragopogon porrifolius** L.: Cañada del Puerto, ca. 0.2 mi inland from Prisoners' Harbor, 1 Jun 1979, Junak et al. SC-123; Main Ranch, 15 Jun 1930, Hoffmann s.n.; Central Valley, 0.2 mi w of Stanton Ranch airfield, 3 Jun 1983, Junak SC-718.

**Uropappus lindleyi** (DC.) Nutt.: n face of Picacho Diablo, 26 Apr 1968, Benedict s.n.; Islay Canyon Road, 15 Apr 1985, Junak SC-1065; ridge between Coches Prietos Canyon and Alberts Anchorage, 30 Apr 1960, Blakley 3356.

**Venegasia carpesioides** DC.: W Twin Canyon, 20 Apr 1989, Ayers et al. 707; bottom of Portezuelo Grade, 11 Mar 1961, Blakley 3898; Smugglers' Canyon, 25 Mar 1971, Philbrick & Ricker B71-37.

**Xanthium spinosum** L.: Prisoners' Harbor, 13 Oct 1958, Blakley & Balls B-2567; Central Valley, at foot of Portezuela Grade, 13 May 1980, Junak & Hochberg SC-296; mouth of Posa Canyon, 13 Jun 1965, Philbrick & Benedict B65-1393.

**Xanthium strumarium** L.: estuary, Christi Harbor, 26 Sep 1968, Benedict s.n.; 1.5 mi e of Stanton Ranch, near El Sobrante, 12 Oct 1958, Blakley & Balls B-2532.

## BERBERIDACEAE

**Berberis pinnata** Lagasca subsp. **insularis** Munz: middle fork of Hazard's Canyon, 19 Nov 1983, Schuyler 503; Campo Raton, 23 Mar 1982, Junak SC-479; Christy Pines, 27 Sep 1979, Junak et al. SC-208.

## BORAGINACEAE

**Amsinckia menziesii** var. **intermedia** (Fisch. & C.A.Mey.) Ganders: bluffs e of Campo Grande airstrip, 28 Mar 1992, Junak SC-3107; Central Valley, 0.5 mi n of top of Centinela Grade, 24 Mar 1982, Junak SC-481; Fraser Point, 27 Mar 1980, Millett-UCSB Botanical Society 2036.

**Amsinckia menziesii** (Lehm.) A.Nelson & J.F.Macbr. var. **menziesii**: e end of Lagunitas Secas area, 8 May 1982, Junak SC-545; ca. 0.25 mi w of Cascada, 3 Apr 1982, Junak SC-498; just e of Stanton Ranch, 27 Apr 1960, Raven & Smith 15261 (RSA-POM,SBBG).

**Amsinckia spectabilis** Fisch. & C.A.Mey. var. **spectabilis**: Fraser Point, 12 Mar 1961, Blakley 3925; mouth of Posa Creek, 18 Apr 1968, Smith 9896; mouth of Willow Creek, 21 Mar 1932, Hoffmann 668.

**Cryptantha clevelandii** Greene: tip of San Pedro Point, 29 Apr 1971, Philbrick et al. B71-95; Valley Anchorage, 24 Mar 1929, Hoffmann s.n.; Fraser Point, 29 Apr 1960, Raven & Smith 15314 (RSA-POM,SBBG).

**Cryptantha micromeres** (A.Gray) Greene: Pelican Bay, 22 May 1983, Junak SC-693; South Ridge Road, 26 Mar 1983, Junak SC-643; Coches Prietos, 23 Mar 1929, Hoffmann s.n.

**Cryptantha muricata** (Hook. & Arn.) A.Nelson & J.F.Macbr.: rocky ridge above Pelican Harbor, 22 Apr 1932, Hoffmann s.n.

**Heliotropium curassavicum** L.: Prisoners' Harbor, 13 Aug 1939, Dunkle 8593 (RSA-POM,SBBG); beach w of Christi Ranch, 18 Jun 1964, Philbrick et al. B64-122; Punta Arena, 25 Jul 1960, Blakley & Muller 3757.

**Pectocarya linearis** (Ruiz & Pav.) DC. subsp. **ferocula** (I.M.Johnst.) Thorne: saddle w of third peak w of Peak 2021, 24 Mar 1985, Junak SC-998; slopes above e fork of Cañada Larga, 24 Mar 1985, Junak SC-992; first saddle n of Ragged Mountain, 3 Apr 1991, Junak SC-2171.

**Pectocarya penicillata** (Hook. & Arn.) A.DC.: dunes at Fraser Point, 29 Apr 1960, Smith 6129.

**Plagiobothrys acanthocarpus** (Piper) I.M.Johnst.: flats w of Pelican Bay, 11 Mar 1979, Daily 568 (SCIR); 0.5 mi s of Twin Harbors, 27 Mar 1987, Dunn-UCSB Botanical Society 2574 (UCSB).

**Plagiobothrys canescens** Benth.: Mount Pleasant, 29 Mar 1989, Junak et al. SC-1259; just e of Stanton Ranch, 27 Apr 1960, Raven & Smith 15259 (RSA-POM,SBBG); ridgetop nw of Smugglers' Ranch airfield, 18 Mar 1989, Junak SC-1219.

**Plagiobothrys collinus** (Phil.) I.M.Johnst. var. **gracilis** (I.M.Johnst.) L.C.Higgins: saddle w of third peak w of Peak 2021, 24 Mar 1985, Junak SC-997; ridge above e fork of Cañada Larga, 4 Jun 1983, Junak SC-724; lower Cañada de Malva Real, 17 Mar 1992, Junak SC-3074.

## BRASSICACEAE

**Arabis glabra** (L.) Bernh.: Santa Cruz Island, Apr 1888, Brandegee s.n. (CAS-DS,SBBG); s of Main Ranch, 11 Apr 1931, Hoffmann s.n.; 1.5 mi e of Main Ranch, near El Sobrante, 24 Mar 1932, Wolf 2857 (RSA-POM).

**Arabis hoffmannii** (Munz) Rollins: sea cliff e of Dick's, 28 Feb 1932, Hoffmann 653; 100 yards n of road intersection at Centinela, 15 Apr 1967, McPherson s.n.; 200 yards n of Buena Vista summit, 23 Mar 1985, Junak SC-973.

**Athysanus pusillus** (Hook.) Greene: n flanks of Picacho Diablo, 19 Apr 1989, Junak et al. SC-1352; Arroyo del Medio, 1.7 mi w of Stanton Ranch, 26 Apr 1960, Raven & Smith 15193 (RSA-POM,SBBG); road to Coches Prietos Anchorage, 6 May 1991, Junak SC-2358.

**Brassica nigra** (L.) Koch: field ca. 0.5 mi w of Main Ranch, 19 Mar 1932, Hoffmann s.n.; road between Rancho del Sur and Valley Anchorage, 3 Apr 1991, Junak SC-2148; Smugglers', 15 Jun 1930, Hoffmann s.n.

**Brassica rapa** L.: Prisoners' Harbor, 5 Mar 1932, Hoffmann s.n.; Central Valley, ca. 0.25 mi w of Bosque Mano, 7 Mar 1980, Junak & Hochberg SC-251; Main Ranch, 15 Jun 1930, Hoffmann s.n.

**Cakile maritima** Scop.: Prisoners' Harbor, 13 Mar 1961, Blakley 3947; Fraser Point, 29 Apr 1960, Raven & Smith 15304 (RSA-POM,SBBG); just n of mouth of Cañada de los Sauces, 13 Oct 1958, Blakley & Balls B-2576.

**Capsella bursa-pastoris** (L.) Medik.: bluffs e of Campo Grande airstrip, 28 Mar 1992, Junak SC-3106; Main Ranch, 15 Aug 1939, Dunkle 8633 (RSA-POM,SBBG); Coches Prietos Canyon, 1 May 1960, Blakley 3368.

**Cardamine californica** (Torr. & A.Gray) Greene var. **californica**: Dick's Canyon, 10 Apr 1990, Junak SC-1498; upper Cañada del Puerto, 3 Feb 1968, Benedict s.n.; small canyon s of Stanton Ranch, 14 Mar 1961, Blakley 4002.

**Cardamine oligosperma** Torr. & A.Gray: Santa Cruz Island, 28 Mar 1925, Hoffmann s.n.

**Cardaria draba** (L.) Desv.: Prisoners' Harbor, 29 Apr 1969, Benedict s.n.; Cañada Christy, at foot of Portezuela Grade, 22 Mar 1986, Junak SC-1128; s of Main Ranch, 11 Apr 1931, Hoffmann s.n.

**Descurainia pinnata** (Walter) Britton subsp. **menziesii** (DC.) Detling: mouth of Scorpion Canyon, 18 Mar 1989, Junak SC-1215; Valley Anchorage, 7 May 1979, Junak et al. SC-29; s end of Smugglers' Cove, 18 Mar 1989, Junak SC-1220.

**Draba cuneifolia** Torr. & A.Gray var. **integrifolia** S.Watson: e face of Picacho Diablo, 22 Feb 1970, Benedict s.n.; canyon just e of Christi Beach, 15 Apr 1967, Philbrick B67-93.

**Erysimum insulare** Greene: Santa Cruz Island, 1888, Brandegee s.n. (UC); outer rock, Willows Anchorage, 11 Jan 1966, Benedict s.n.

**Guillenia lasiophylla** (Hook. & Arn.) Greene: just w of first canyon w of Prisoners' Harbor, 22 Apr 1990, Junak SC-1597; Central Valley, in first canyon e of winery buildings at Stanton Ranch, 22 Mar 1985, Junak SC-959; road to Coches Prietos Anchorage, 1 mi inland from beach, 6 May 1991, Junak SC-2361.

**Hirschfeldia incana** (L.) Lagr.-Foss.: top of Peak 1941, 20 May 1992, Junak SC-3399; creek bed at Christy's, 20 Dec 1960, Smith 6339; South Ridge Road, 0.1 mi e of junction with road to Centinela Grade, 11 Apr 1990, Junak et al. SC-1504.

**Hutchinsia procumbens** (L.) Desv.: nw tip of West Point, 30 Apr 1967, Benedict s.n.; rocky cliff just n of Los Sauces, 21 Mar 1932, Wolf 2795 (CAS,RSA-POM); mouth of Willow Creek [Cañada de los Sauces], 21 Mar 1932, Hoffmann 667.

**Lepidium latipes** Hook. var. **latipes**: ca. 0.25 mi s of Scorpion Ranch, 17 Mar 1989, Junak SC-1211; head of valley w of Portezuela Grade, 28 Apr 1960, Raven & Smith 15286 (RSA-POM,SBBG); Forney's Cove, 6 Apr 1969, Benedict s.n.

**Lepidium nitidum** Torr. & A.Gray var. **nitidum**: Lake Pasture between Prisoners' Harbor and Mount Pleasant, 1 Feb 1965, Philbrick et al. B65-117; Central Valley between Rancho del Sur and Valley Anchorage, 3 Apr 1991, Junak SC-2149; lower portion of Laguna Canyon, 15 Apr 1991, Junak SC-2218.

**Lepidium oblongum** Small var. **insulare** C.L.Hitchc.: Christy Beach, 11 Apr 1990, Junak SC-1514; n end of Sierra Blanca ridge, 3 Apr 1991, Junak SC-2175; bluffs just w of mouth of Laguna Canyon, 17 Mar 1992, Junak SC-3061a.

**Lepidium virginicum** L. var. **robinsonii** (Thell.) C.L.Hitchc.: rocky slope above Pelican Harbor, 5 Mar 1932, Hoffmann 654; bluffs just w of mouth of Laguna Canyon, 17 Mar 1992, Junak SC-3061; e of Alameda de los Coches Prietos, 1.1 mi inland from beach, 17 Apr 1991, Junak SC-2298a.

**Raphanus raphanistrum** L.: Prisoners' Harbor, 10 Apr 1990, Junak SC-1502; s side of Christy Ranch airfield, 10 Jul 1979, Junak et al. SC-131; Cañada Cebada, 4 Apr 1991, Junak SC-2193.

**Raphanus sativus** L.: Scorpion Harbor, 11 Jul 1930, Dearing s.n., Main Ranch, 15 Aug 1939, Dunkle 8630 (RSA-POM,SBBG); Smugglers, 15 Jun 1930, Hoffmann s.n.

**Rorippa nasturtium-aquaticum** (L.) Hayek: Prisoners, 1 Jul 1930, Hoffmann s.n.; Prisoners' Harbor, 12 Jul 1939, Williams 45 (CAS,SBBG).

**Sibara filifolia** (Greene) Greene: stony ridge above Pelican Bay, 15 Jun 1930, Hoffmann s.n.; stony ridge above Laguna Canyon, 21 Mar 1932, Hoffmann s.n.

**Sinapis arvensis** L.: Central Valley, along roadside directly across from the Field Station, 15 Apr 1979, Daily s.n.; near Main Ranch, 10 Apr 1931, Hoffmann s.n.; Valley Anchorage, 24 Mar 1929, Hoffmann s.n.

**Sisymbrium officinale** L.: Scorpion Canyon, ca. 1.2 mi inland from Scorpion Anchorage, 22 Apr 1993, Junak SC-3709; Main Ranch, 15 Jun 1930, Hoffmann s.n.; near Alberts Anchorage, 30 Apr 1960, Blakley 3355.

**Thysanocarpus conchuliferus** Greene: upper w fork of Orizaba Canyon, 14 Apr 1985, Junak SC-1053; stony ridge above Pelican Harbor, 5 Mar 1932, Hoffmann s.n.; Central Valley, in second large canyon w of Rancho Portezuelo ruins, 9 May 1990, Junak SC-1734.

**Thysanocarpus curvipes** Hook.: above upper Twin Harbor Canyon, 5 Mar 1932, Hoffmann 654; near Pelican Bay, 24 Mar 1941, Moran 750 (in part); ridge road 0.5 mi w of Eastern Pines, 21 Mar 1967, Benedict s.n.

**Thysanocarpus laciniatus** Nutt. ex Torr. & A.Gray: woodland above China Harbor, 25 Apr 1960, Raven & Smith 15152 (RSA-POM,SBBG); Arroyo del Medio, 1.7 mi w of Stanton Ranch, 26 Apr 1960, Raven & Smith 15194 (RSA-POM,SBBG); Cochies Prietos, 22 Mar 1929, Hoffmann s.n.

## CACTACEAE

**Opuntia ficus-indica** (L.) J.Mill.: tip of peninsula between Pelican Bay and Tinker's Cove, 16 Jun 1984, Junak SC-873.

**Opuntia littoralis** (Engelm.) Cockerell var. **littoralis**: 1.5 mi e of Stanton Ranch, near El Sobranti, 12 Oct 1958, Blakley & Balls B-2539; Smugglers', 15 Jun 1930, Hoffmann s.n.

**Opuntia oricola** Philbrick: "No Man's Land" on w side of Montañon Ridge, 9 Oct 1991, Junak & Young SC-2928.

**Opuntia prolifera** Engelm.: just e of mouth of Malva Real Canyon, 17 Oct 1990, Junak SC-2070; coastal headland on e side of Laguna Canyon, 19 Sep 1990, Junak SC-1983.

## CAMPANULACEAE

**Githopsis diffusa** A.Gray subsp. **diffusa**: w of Fry's Harbor, 21 Apr 1932, Hoffmann s.n.; Central Valley, 0.25 mi w of top of Portezuela Grade, 11 May 1985, Junak SC-1080; Islay Canyon, n of Triangulation Station "Center 2", 21 Apr 1986, Junak SC-1152.

**Triodanis biflora** (Ruiz & Pav.) Greene: burn above Pelican Harbor, 22 Apr 1932, Hoffmann s.n.; Central Valley, at foot of Centinela Grade, 8 May 1982, Junak SC-551; upper portion of Cañada de los Sauces, 5 Jun 1983, Junak SC-732.

## CAPRIFOLIACEAE

**Lonicera hispidula** Douglas ex Torr. & A.Gray var. **vacillans** A.Gray: Valdez, 15 Jun 1930, Hoffmann s.n.; 0.5 mi above W Twin Harbor, 11 Oct 1967, Benedict s.n.; small canyon on n side of ridge on w side of Coches Prietos Canyon, 1 May 1960, Blakley 3378.

**Lonicera interrupta** Benth.: Central Valley at w end of Bosque Mano, 1 Aug 1991, Junak SC-2783.

**Lonicera subspicata** Hook. & Arn.: bank above Baby's Canyon, 26 Mar 1931, Hoffmann s.n.; small canyon on n side of ridge on w side of Coches Prietos Canyon, 1 May 1960, Blakley 3377; 0.5 mi above beach at Coches Prietos, 18 Oct 1967, Benedict s.n.

**Symphoricarpos mollis** Nutt. in Torr. & A.Gray: "Alamo Canyon", 2 mi e of Valdez Harbor, 9 Nov 1930, Hoffmann s.n.; upper Cañada Christy at bottom of Centinela Grade, 10 May 1990, Junak SC-1799; Smugglers' Cove, 14 May 1927, Hoffmann s.n.

## CARYOPHYLLACEAE

**Cardionema ramosissimum** (Weinm.) A.Nelson & J.F.McBride: first major canyon e of Profile Point, 12 Jun 1965, Philbrick & Benedict B65-1385; Central Valley, ca. 1 mi w of Stanton Ranch, 16 Jun 1964, Philbrick et al. B64-59; ridge w of Coches Prietos Canyon, 1 May 1960, Blakley 3376.

**Cerastium glomeratum** Thuill.: Lagunitas Secas, 19 Apr 1968, Benedict s.n.; between Prisoners' Harbor and Christy Ranch, 28 Mar 1925, Hoffmann s.n.; Coches Prietos Canyon, 29 Apr 1960, Blakley 3282.
**Minuartia douglasii** (Torr. & A.Gray) Mattf.: upper w fork of Orizaba Canyon, 14 Apr 1985, Junak SC-1063; head of Twin Harbor Canyon, 23 May 1932, Hoffmann s.n.; n of Red Peak, 23 Apr 1979, Thorne et al. 52474 (RSA-POM).
**Polycarpon depressum** Nutt.: stony wash, Willows, 30 Jun 1930, Hoffmann s.n.
**Polycarpon tetraphyllum** (L.) L.: Stanton Ranch, 25 Jun 1994, Junak SC-4020.
**Sagina apetala** Ard.: Pelican Bay, 18 Apr 1986, Junak SC-1136.
**Sagina decumbens** (Elliott) Torr. & A.Gray subsp. **occidentalis** (S.Watson) G.Crow: w fork of major canyon above E Twin Harbor, 5 May 1966, Philbrick B66-200; between Pelican Bay and Prisoners' Harbor, 14 Apr 1967, Philbrick & Hardham B67-51; Central Valley, in upper e fork of first canyon e of Cascada, 4 Nov 1984, Junak SC-939.
**Silene antirrhina** L.: rocky ridge near Pelican, 14 Jun 1930, Hoffmann s.n.; w branch of Laguna Canyon, 21 Mar 1932, Hoffmann s.n.; ridge on e side of Willows Canyon, 9 May 1991, Junak SC-2525.
**Silene gallica** L.: Twin Harbor, 21 Sep 1930, Hoffmann s.n.; Mount Pleasant reservoir, 7 May 1968, Philbrick & McPherson B68-249; Cochies Prietos, 1 Jul 1930, Hoffmann s.n.
**Silene laciniata** Cav. subsp. **major** C.L.Hitchc. & Maguire: bluffs ca. halfway between Scorpion Canyon and Cavern Point, 14 Sep 1991, Junak & Meadows SC-2855; Central Valley, 0.1 mi e of Reparo Viejo, 29 May 1990, Junak SC-1852; butte at w end of Coches Prietos Beach, 30 May 1991, Junak SC-2710.
**Silene multinervia** S.Watson: burn above Pelican Harbor, 22 Apr 1932, Hoffmann s.n.
**Spergularia bocconii** (Scheele) Merino: vernal pool along Navy Road, 2 mi e of Prisoners' Harbor, 16 Oct 1958, Blakley & Balls B-2621; Cañada Christy, ca. 0.5 mi e of Christy Barn, 22 Mar 1986, Junak SC-1127; Smugglers' Field, w of mouth of Hungryman's Gulch, 29 Apr 1971, Philbrick et al. B71-73.
**Spergularia macrotheca** (Hornem.) Heyn var. **macrotheca**: Prisoners' Harbor, 13 Aug 1939, Dunkle 8576 (RSA-POM,SBBG); Valley Anchorage, 11 Apr 1931, Hoffmann s.n.; near mouth of Poso Canyon, 22 Sep 1966, Webster 7442 (DAV,SBBG).
**Spergularia marina** (L.) Griseb.: Dick's Harbor, 22 Sep 1930, Hoffmann s.n.; Prisoners' Harbor, 7 Oct 1991, Junak SC-2888; mouth of Scorpion Canyon, 14 Sep 1991, Junak SC-2860.
**Stellaria media** (L.) Vill.: Prisoners' Harbor, 1 Feb 1965, Philbrick et al. B65-116; Main Ranch, 15 Aug 1939, Dunkle 8624 (RSA-POM,SBBG); Coches Prietos, 22 Mar 1929, Hoffmann s.n.
**Stellaria nitens** Nutt.: Orizaba, 28 Feb 1932, Hoffmann s.n.; main crest of island, 4.7 mi e of Stanton Ranch, 25 Apr 1960, Raven & Smith 15132 (RSA-POM,SBBG); Central Valley, in first canyon e of winery buildings at Stanton Ranch, 22 Mar 1985, Junak SC-949.

## CHENOPODIACEAE

**Aphanisma blitoides** Nutt. ex Moq.: cliffs in cactus, Christy's, 28 Jun 1930, Hoffmann s.n.; Christy Rnach, along s stream bank between house and beach, 23 Jul 1939, Williams 54; edge of sea cliff, Willow Creek, w of Christy's Harbor, 21 Mar 1932, Hoffmann s.n.
**Atriplex argentea** Nutt. subsp. **expansa** (S.Watson) H.M.Hall & Clem.: near beach, Smugglers', 15 Jun 1930, Hoffmann s.n.; mouth of creek near beach, Smugglers' Harbor, 20 Sep 1930, Hoffmann s.n.; mouth of small canyon near n end of Smugglers' Beach, 7 Nov 1992, Junak SC-3523.
**Atriplex californica** Moq. in DC.: Little Scorpion Rock, 21 Aug 1977, Philbrick & Cummings B77-37; Christy Beach, 11 Apr 1990, Junak SC-1508; sandy fields behind Fraser Point, 29 Apr 1960, Raven & Smith 15315 (RSA-POM,SBBG).
**Atriplex coulteri** (Moq.) D.Dietr.: mouth of Scorpion Canyon, 14 Sep 1991, Junak SC-2866; open fields above Christy's, 29 Jun 1930, Hoffmann s.n.; South Ridge Road, 0.1 mi w of junction with road to Centinela Grade, 11 Apr 1990, Junak et al. SC-1505.
**Atriplex lentiformis** Nutt. subsp. **breweri** (S.Watson) H.M.Hall & Clem.: mouth of Scorpion Canyon, 14 Sep 1991, Junak SC-2865; mouth of Willows Canyon, 14 Feb 1982, Junak SC-466; Valley Anchorage, 24 Jul 1960, Blakley & Muller 3743.
**Atriplex leucophylla** (Moq.) D.Dietr.: mouth of Scorpion Canyon, 14 Sep 1991, Junak SC-2863; beach w of Christi Ranch, 18 Jun 1964, Philbrick et al. B64-129; mouth of Laguna Canyon, 15 Apr 1991, Junak SC-2224.
**Atriplex pacifica** A.Nelson: headland just w of Coches Prietos Beach, 31 Jul 1991, Junak SC-2781; just w of Valley Anchorage, 7 May 1979, Junak et al. SC-28.
**Atriplex semibaccata** R.Br.: Prisoners' Harbor, 11 Oct 1958, Blakley & Balls B-2512; Christy Beach, 6 Jun 1968, Benedict s.n.; vicinity of Near Point, 18 Apr 1968, Philbrick s.n.

**Atriplex serenana** A.Nelson var. **davidsonii** (Standl.) Munz: Pelican Bay, 14 Jun 1930, Hoffmann s.n.
**Atriplex triangularis** Willd.: Prisoners' Harbor, 11 Oct 1958, Blakley & Balls B-2509; Scorpion Canyon, 19 Oct 1991, Junak SC-2964; mouth of Cañada de los Sauces, 8 Oct 1991, Junak SC-2915.
**Atriplex watsonii** A.Nelson: s slope of West Point, 13 Jan 1971, Benedict s.n.; 0.7 km e of Frasier Point, 19 May 1979, Ferren 1907; n end of Sauces Beach, 14 May 1980, Junak & Hochberg SC-299.
**Beta vulgaris** L.: just s of Cavern Point, 28 Mar 1992, Junak SC-3111; Scorpion Ranch, 28 Mar 1992, Junak SC-3096.
**Chenopodium ambrosioides** L.: mouth of canyon at Valdez Anchorage, 25 Aug 1988, Junak SC-1202; Main Ranch, 15 Aug 1939, Dunkle 8645 (RSA-POM,SBBG); Coches Prietos, 23 Mar 1929, Hoffmann s.n.
**Chenopodium berlandieri** Moq.: Pelican, 14 Jun 1930, Hoffmann s.n.; Coches Prietos Beach, 30 May 1991, Junak SC-2706; Valley Anchorage, 30 May 1991, Junak SC-2732.
**Chenopodium californicum** (S.Watson) S.Watson: near mouth of Scorpion Canyon, 14 Sep 1991, Junak SC-2868; top of first summit n of Centinela Grade, 4 Apr 1991, Junak SC-2179; Cochies Prietos, 23 Mar 1929, Hoffmann s.n.
**Chenopodium murale** L.: Prisoners' Harbor, 13 Aug 1939, Dunkle 8592 (RSA-POM,SBBG); Valley Anchorage, 24 Mar 1929, Hoffmann s.n.; mouth of Laguna Canyon, 15 Apr 1991, Junak SC-2225.
**Monolepis nuttalliana** (Schult.) Greene: Santa Cruz Island, 28 Mar 1925, Hoffmann s.n.; base of bluff, beach at Christy Ranch, 23 Apr 1965, Smith 8669; Fraser Point Road, on n side of first ridge n of Black Point peninsula, 8 May 1991, Junak SC-2470.
**Salicornia subterminalis** Parish: w side of Potato Harbor, 15 Jun 1981, Junak SC-437; West Point, 14 Jul 1964, Philbrick B64-302; Fraser Point, 29 Apr 1960, Raven & Smith 15308 (RSA-POM,SBBG).
**Salicornia virginica** L.: mouth of Cañada de los Sauces, 13 Oct 1958, Blakley & Balls B-2574; mouth of Sauces Canyon, 28 Sep 1969, Benedict s.n.; mouth of Cañada de los Sauces, 8 May 1990, Junak SC-1686.
**Salsola tragus** L.: Prisoners' Harbor, 16 Aug 1939, Dunkle 8653 (RSA-POM,SBBG); Central Valley ca. 0.3 mi w of Valley Anchorage, 10 Oct 1991, Junak SC-2930; ca. 0.5 mi w of Navy Base, 16 Mar 1992, Junak SC-3054.
**Suaeda taxifolia** (Standl.) Standl.: bluffs just e of third canyon e of Scorpion Canyon, 18 Oct 1991, Junak SC-2955; Valley Anchorage, 24 Jul 1960, Blakley & Muller 3742; Smugglers' Cove, 21 Mar 1932, Fosberg 7706.

## CISTACEAE

**Helianthemum greenei** B.L.Rob.: pine forest above China Harbor, 9 May 1979, Junak et al. SC-58; South Ridge Road just w of Christy Pines, 21 May 1983, Junak SC-676; Cañada Horqueta, 28 Mar 1968, Benedict s.n.
**Helianthemum scoparium** Nutt.: Cañada del Puerto, 1.2 mi inland from Prisoners' Harbor, 21 Sep 1990, Junak & Bornstein SC-2012; ridge s of Stanton Ranch, 27 Apr 1960, Raven & Smith 15240 (RSA-POM,SBBG); e side of Sierra Blanca ridge, 18 Apr 1989, Junak & Ayers SC-1312.

## CONVOLVULACEAE

**Calystegia macrostegia** (Greene) Brummitt subsp. **macrostegia**: Prisoners' Harbor, 13 Mar 1961, Blakley 3951; 1.5 mi e of Stanton Ranch, near El Sobrante, 12 Oct 1958, Blakley & Balls B-2530; 1 mi up Coches Prietos Canyon from mouth, 30 Apr 1960, Blakley 3324.
**Calystegia soldanella** (L.) R.Br.: beach at mouth of Water Canyon, 2 May 1969, Benedict s.n.; Christy Beach, 21 Apr 1990, Junak SC-1581; s third of Sauces Beach, 23 Jul 1994, Junak SC-4042.
**Convolvulus arvensis** L.: Main Ranch, 15 Jun 1930, Hoffmann s.n.; road to Coches Prietos beach, 0.8 mi s of intersection with South Ridge Road, 12 Apr 1990, Junak et al. SC-1540; Smuggler's Cove Ranch, 26 Jul 1984, Unser s.n.
**Convolvulus simulans** L.M.Perry: road to Fraser Point, on n side of Black Point peninsula, 21 May 1991, Junak SC-2575; n fork of Black Point Canyon, 21 May 1994, Junak SC-3981.
**Cressa truxillensis** Kunth: Scorpion Harbor, 21 Sep 1930, Hoffmann s.n.; Playa Larga, ca. 0.25 mi e of Johnson Canyon mouth, 16 Oct 1990, Junak SC-2057.
**Dichondra occidentalis** House: s of West Point, just w of Triangulation Point 717, 10 Jul 1979, Junak SC-136; ridge n of Ragged Mountain, 0.75 mi sw of upper road to Christi Ranch, 11 Jun 1965, Philbrick & Benedict B65-1305; Sierra Blanca ridge, at second saddle n of Ragged Mountain, 3 Apr 1991, Junak SC-2152.

## CRASSULACEAE

**Crassula connata** (Ruiz & Pav.) A.Berger: mouth of Orizaba Canyon, 26 Mar 1932, Wolf 2897 (RSA-POM,SBBG); Christi Ranch, 6 Apr 1969, Benedict s.n.; upper portion of Pozo Canyon Road, 0.4 mi from top of road, 7 May 1991, Junak SC-2409.

**Dudleya candelabrum** Rose: first major canyon e of Profile Point, 12 Jun 1965, Philbrick & Benedict B65-1356; Cañada de la Portezuela at second canyon crossing w of Reparo Viejo, 26 Jun 1991, Junak SC-2740; mouth of Laguna Canyon, 26 Jun 1991, Junak SC-2747.
**Dudleya greenei** Rose: just w of Cueva Valdez, 8 May 1969, Philbrick & Benedict B69-131; Pelican Harbor, 5 May 1966, Philbrick B66-212; Smugglers' Canyon, 25 Mar 1971, Philbrick & Ricker B71-41.
**Dudleya nesiotica** Moran: flats ne of Fraser Point, 26 May 1959, Smith 5955; Fraser Point, 23 Apr 1965, Philbrick & Benedict B65-849; e end of Forney's Cove, 21 May 1991, Junak SC-2582.

## CUCURBITACEAE

**Cucurbita foetidissima** Kunth: beach at Fry's, 21 Sep 1930, Hoffmann s.n.; near shore below Christi Ranch, 16 May 1962, Piehl 62-213; Christy Beach, 12 Sep 1989, Junak et al. SC-1416.
**Marah fabaceus** (Naudin) Greene: Bosque Cabrillo Eucalyptus grove, Cañada del Puerto, 22 Apr 1968, Benedict s.n.; Central Valley between Centinela and Portezuela, 16 Apr 1967, Philbrick B67-125; Arroyo del Medio, 1.1 mi w of Stanton Ranch, 26 Apr 1960, Raven & Smith 15171 (RSA-POM,SBBG).
**Marah macrocarpus** (Greene) Greene var. **major** (Dunn) Stocking: Bosque Cabrillo Eucalyptus grove, Cañada del Puerto, 13 Apr 1968, Benedict s.n.; mouth of Cañada Islay, 8 May 1969, Philbrick & Benedict B69-122; ridge s of Stanton Ranch, 27 Apr 1960, Raven & Smith 15244 (RSA-POM,SBBG).

## CUSCUTACEAE

**Cuscuta salina** Engelm. var. **major** Yunck.: tip of Fraser Point, 30 Apr 1967, Benedict s.n.; ca. 0.5 mi e of Fraser Point, 23 Apr 1965, Philbrick & Benedict B65-846.
**Cuscuta subinclusa** Durand & Hilg.: Santa Cruz Island, July & Aug 1886, Greene s.n. (ND-G); n of old chapel, Stanton Ranch, 30 Sep 1969, Benedict s.n.

## ERICACEAE

**Arbutus menziesii** Pursh: near head of Little Valdez Canyon, 2 Apr 1969, Benedict s.n.; head of side canyon sw of Fry's Harbor, 9 Sep 1931, Hoffmann 599; ca. 0.5 mi n of Peak 1848, nnw of Red Peak, 28 Jan 1979, Junak SC-8.
**Arctostaphylos insularis** Greene: Fry's Harbor, 30 Jun 1936, Van Rensselaer s.n.; Central Valley, ca. 0.5 mi e of Centinela, 1 Mar 1965, Philbrick & Benedict B65-454; canyon above Cochies, 1 Jul 1930, Hoffmann s.n.
**Arctostaphylos tomentosa** (Pursh) Lindl. subsp. **insulicola** P.V.Wells: first major canyon e of Twin Harbors, 5 May 1966, Philbrick B66-209; Central Valley, ca. 1 mi sw of Picacho Diablo, 27 Sep 1979, Junak et al. SC-211; ca. 0.25 mi sw of El Tigre summit, 1 Feb 1968, Benedict s.n.
**Arctostaphylos tomentosa** subsp. **subcordata** (Eastw.) P.V.Wells: n side of saddle between Peak 1941 and Peak 2021, 4 Nov 1984, Junak SC-934; Buena Vista, 27 Oct 1968, Benedict s.n.; East Ridge Road, n of El Tigre, 26 Mar 1971, Philbrick et al. B71-58.
**Arctostaphylos viridissima** (Eastw.) McMinn: Eastern Pines, 7 Aug 1966, Benedict s.n.; extreme upper Scorpion Canyon, 29 Apr 1971, Benedict s.n.; head of Aguaje Canyon, w of Yellowbanks Anchorage, 25 Mar 1971, Benedict s.n.
**Comarostaphylis diversifolia** (Parry) Greene subsp. **planifolia** (Jeps.) G.D.Wallace: Cueva Valdez, 18 Apr 1936, Woglum 139; large pine grove e of Christy's, 26 Apr 1960, Raven & Smith 15216 (RSA-POM,SBBG); small canyon on ridge, w side of Coches Prietos Canyon, 1 May 1960, Blakley 3372.
**Vaccinium ovatum** Pursh: pine forest near head of Laguna Canyon, just e of Centinela and just n of Ridge Road, 17 Jun 1964, Philbrick et al. B64-91; ridge n of Ragged Mountain, 2.5 mi n of Punta Arena, 11 Jun 1965, Philbrick & Benedict B65-1322.

## EUPHORBIACEAE

**Chamaesyce maculata** (L.) Small: Stanton Ranch, 25 Jun 1994, Junak SC-4019.
**Eremocarpus setigerus** (Hook.) Benth.: ephemeral pond on road to Water Canyon, 0.5 mi n of Navy Road, 11 Sep 1989, Ayers et al. 780; Cañada de la Portezuela, 0.1 mi w of top of Portezuela Grade, 21 Sep 1990, Junak & Bornstein SC-2022; Laguna Canyon Road, 2.5 mi inland from beach, 19 Sep 1990, Junak & Bornstein SC-1986.
**Euphorbia misera** Benth.: e side of Laguna Canyon, in small bowl-shaped cove just w of Corral Point, 26 Jun 1991, Junak SC-2746.
**Euphorbia peplus** L.: Fry's Harbor, 22 May 1978, Philbrick B78-184; Pelican Bay, 25 Feb 1984, Junak SC-836; Main Ranch, 20 May 1978, Daily 522.

**Stillingia linearifolia** S.Watson: bluffs ca. midway between e end of Laguna Beach and Corral Point, 29 May 1991, Junak SC-2682; e side of spine leading down to Corral Point, 29 May 1991, Junak SC-2695.

## FABACEAE

**Acacia melanoxylon** R.Br.: Cañada del Puerto, 26 Jul 1960, Blakley 3775.

**Albizia lophantha** (Willd.) Benth.: Prisoners' Harbor, 13 Oct 1958, Blakley & Balls B-2549; Prisoners' Harbor, 26 Nov 1977, Philbrick & Junak B77-49; mouth of Cañada del Puerto, 20 Nov 1980, Junak SC-348.

**Astragalus didymocarpus** Hook. & Arn. var. **didymocarpus**: ca. 0.9 mi e of West Point, 13 Apr 1985, Junak et al. SC-1042; Fraser Point, 30 Apr 1967, Benedict s.n.; hill just w of Navy Compound, 20 Apr 1965, Philbrick et al. B65-750.

**Astragalus gambelianus** E.Shield.: 1.2 mi w of Stanton Ranch, 26 Apr 1960, Raven & Smith 15185 (RSA-POM,SBBG); Forney's Cove, 6 Apr 1969, Benedict s.n.

**Astragalus miguelensis** Greene: w end of Potato Harbor, 15 Jun 1981, Junak SC-436; near Cavern Point, 19 Oct 1991, Junak SC-2970; Fraser Point, 29 Apr 1960, Raven & Smith 15313 (RSA-POM,SBBG).

**Astragalus trichopodus** (Nutt.) A.Gray var. **lonchus** (M.E.Jones) Barneby: bluffs e of Coche Point peninsula, 9 Oct 1991, Junak & Young SC-2924; Cañada Christy near Christy airfield, 8 May 1990, Junak SC-1677a; canyon w of Gull Rock, 30 Jun 1930, Hoffmann s.n.

**Lathyrus vestitus** Nutt. var. **vestitus**: Lady's Harbor, 12 Apr 1930, Hoffmann s.n.; 0.2 mi w of Portezuela Grade summit, 18 Apr 1989, Ayers et al. 648; Smugglers' Cove, 21 Mar 1932, Fosberg 7708.

**Lotus argophyllus** (A.Gray) Greene var. **niveus** (Greene) Ottley: mouth of Dick's Canyon, 10 Apr 1990, Junak SC-1501; first canyon w of Reparo Viejo Canyon, 30 May 1990, Junak SC-1915; near mouth of Smugglers' Canyon, 13 Sep 1991, Junak & Meadows SC-2828.

**Lotus corniculatus** L.: Prisoners' Harbor, 26 Jul 1960, Blakley 3782; lower portion of Cañada Christy, downstream from Christy Ranch, 22 Jul 1994, Junak SC-4031.

**Lotus dendroideus** (Greene) Greene var. **dendroideus**: w of Pelican Bay, 25 Mar 1941, Moran 768 (RSA-POM,SBBG; Islay Canyon Road, 29 Jan 1984, Junak SC-813; Smugglers' Canyon, ca. 0.25 mi inland from beach, 9 Mar 1985, Unser s.n.

**Lotus grandiflorus** (Benth.) Greene var. **grandiflorus**: flats above first cove e of Twin Harbor, 8 Oct 1987, Junak SC-1185; 1 mi w of Stanton Ranch, 16 May 1962, Breedlove 2824; e side of Sierra Blanca ridge, ne of Peak 1213, 18 Apr 1989, Junak & Ayers SC-1311.

**Lotus hamatus** Greene: Islay Canyon Road, 15 Apr 1991, Junak SC-2205; slopes at e end of Laguna Beach, 29 May 1991, Junak SC-2674; ridge bordering nw side of Coches Prietos Anchorage, 23 Mar 1967, Benedict s.n.

**Lotus micranthus** Benth.: flats w of second canyon w of Prisoners' Harbor, 25 Feb 1984, Junak SC-835; lower portion of Cañada Islay, 22 Apr 1990, Junak SC-1590; knoll 0.5 mi n of S Ranch, 24 Mar 1932, Hoffmann s.n.

**Lotus purshianus** (Benth.) Clem. & E.G.Clem.: Cañada del Puerto, ca. 0.25 mi from Stanton Ranch, 24 Aug 1985, Junak SC-1110; Cañada de la Portezuela s of Picacho Diablo, 13 Sep 1989, Junak et al. SC-1421; wash n of University of California Field Station, 16 Jun 1967, Benedict s.n.

**Lotus salsuginosus** Greene var. **salsuginosus**: w side of Potato Harbor, 15 Jun 1981, Junak & Bornstein SC-423; Cañada Christy near Christy airfield, 8 May 1990, Junak SC-1678a; lower Smugglers' Canyon, 13 Sep 1991, Junak SC-2837.

**Lotus strigosus** (Nutt. in Torr. & A.Gray) Greene: Prisoners' Harbor, 14 Apr 1967, Philbrick B67-54; ridge ca. 0.25 mi w of La Cascada, 3 Apr 1982, Junak SC-499; Smugglers' Cove, ca. 0.25 mi inland from canyon mouth, 9 Mar 1985, Unser s.n.

**Lotus wrangelianus** Fisch. & C.A.Mey.: first canyon w of Reparo Viejo Canyon, 30 May 1990, Junak SC-1904; slope above Main Ranch, 12 Apr 1931, Hoffmann s.n.; e ridge above Willows Canyon, 10 May 1973, Philbrick & Philbrick B73-317;

**Lupinus albifrons** Benth. var. **douglasii** (J.Agardh) C.P.Sm.: Lake Pasture, between Prisoners' Harbor and Mount Pleasant, 1 Feb 1965, Philbrick et al. B65-119; upper Cañada Christy, 0.4 mi w of Embudo Trough, 11 Apr 1990, Junak SC-1536; base of beach bluff w of Pozo Creek, 19 Apr 1968, Smith 9898.

**Lupinus bicolor** Lindl.: w end of China Harbor, 16 Apr 1983, Junak & Turner SC-673; first canyon w of Reparo Viejo Canyon, 30 May 1990, Junak SC-1916; Fraser Point, 12 Mar 1961, Blakley 3916.

**Lupinus concinnus** J.Agardh: near mouth of Tinker's Canyon, 22 Apr 1990, Junak SC-1604; canyon due n of University of California field station, 11 May 1981, Junak & Timbrook SC-392; road to Coches Prietos Anchorage, ca. 0.8 mi inland from beach, 6 May 1991, Junak SC-2346.

**Lupinus hirsutissimus** Benth.: near Prisoners' Harbor, 19 Mar 1932, Hoffmann s.n.; just w of Stanton Ranch airfield, 12 May 1981, Junak SC-399; bluffs just w of mouth of Laguna Canyon, 17 Mar 1992, Junak SC-3066.
**Lupinus microcarpus** Sims var. **microcarpus**: w end of Bosque Mano, 12 May 1981, Hochberg 701; Fraser Point, 12 Mar 1961, Blakley 3918; s end of Sauces Beach, 14 May 1980, Junak & Hochberg SC-300.
**Lupinus succulentus** Koch: road to Los Pinos de la Calera, 29 Mar 1989, Junak et al. SC-1264; s end of Christy Beach, 28 Mar 1989, Junak et al. SC-1247; Posa Canyon, ca. 1.8 mi inland from beach, 13 Jun 1965, Philbrick & Benedict B65-1046.
**Lupinus truncatus** Hook. & Arn.: Prisoners' Harbor, 25 Mar 1932, Fosberg 7673; hill ne of Stanton Ranch, 27 Apr 1960, Raven & Smith 15256 (RSA-POM,SBBG); Laguna Canyon, ca. 4 km n of Laguna Harbor, 27 Mar 1989, Young et al. 1854.
**Medicago polymorpha** L.: Valdez Cove, 18 Apr 1936, Elmore 253 (RSA-POM,SBBG); creek near Christy Ranch, 7 Jun 1961, Smith 6466; Coches Prietos, 22 Mar 1929, Hoffmann s.n.
**Medicago sativa** L.: Prisoners' Harbor, 13 Aug 1939, Dunkle 8572 (RSA-POM,SBBG); upper e fork of Wall Canyon, ca. 0.4 mi sw of Campo del Norte, 25 Sep 1979, Junak et al. SC-183; Stanton Ranch, 8 Nov 1991, Junak SC-2981.
**Melilotus albus** Medik.: Prisoners' Harbor, 15-18 May 1962, Piehl 62-256; Cañada del Puerto, ca. 0.3 mi inland from Prisoners' Harbor, 24 Jun 1994, Junak SC-4011.
**Melilotus indicus** (L.) All.: Prisoners' Harbor, 13 Aug 1939, Dunkle 8571 (RSA-POM,SBBG); La Cascada, 12 Apr 1990, Junak SC-1569; Cochies Prietos, 22 Mar 1929, Hoffmann s.n.
**Pickeringia montana** Nutt. ex Torr. & A.Gray: ridge above Pelican Harbor, 23 May 1932, Hoffmann s.n.; Pelican Pines, 20 Apr 1969, Benedict s.n.
**Pisum sativum** L.: Islay Canyon, 100 ft s of Valley Road, 9 May 1982, Junak SC-552.
**Robinia pseudoacacia** L.: Prisoners' Harbor, 26 Nov 1977, Philbrick & Junak B77-64; Scorpion Ranch, 23 Apr 1993, Junak SC-3726; Cañada del Puerto, 0.5 mi from Stanton Ranch, 19 Apr 1989, Junak et al. SC-1324.
**Trifolium albopurpureum** Torr. & A.Gray: ridgetop just e of Lagunitas Secas, 19 Apr 1989, Junak et al. SC-1359; head of e fork of Cañada Larga, 24 Mar 1985, Junak SC-995; summit of Peak 777, ca. 0.8 mi wsw of Valley Anchorage, 22 Mar 1982, Junak SC-472.
**Trifolium ciliolatum** Benth.: Hazard's, 7 Apr 1930, Hoffmann s.n.; ca. 1 mi w. of Main Ranch, 8 May 1973, Philbrick & Philbrick B73-289; lower Willows Canyon, 26 Mar 1983, Junak SC-635.
**Trifolium depauperatum** Desv. var. **truncatum** (Greene) Isely: N Ridge between Picacho Diablo and Lagunitas Secas, 19 Apr 1989, Junak et al. SC-1334; ridge above Christy's, 21 Mar 1932, Hoffmann s.n.; Smugglers' Field, w of mouth of Hungryman's Gulch, 29 Apr 1971, Philbrick et al. B71-74.
**Trifolium fucatum** Lindl.: first arroyo s of mouth of Cañada de Christi, 30 Apr 1967, Benedict s.n.; Fraser Point, 22 Apr 1979, Daily s.n.; just n of lower portion of Cañada de los Sauces, 16 Apr 1991, Junak SC-2267.
**Trifolium gracilentum** Torr. & A.Gray var. **gracilentum**: Fry's Harbor, 22 May 1978, Philbrick B78-178; Valley Anchorage, 24 Mar 1929, Hoffmann s.n.; bluffs above e end of Willows Beach, 9 May 1991, Junak SC-2520.
**Trifolium macraei** Hook. & Arn.: hilltop sw of Fry's, 21 Apr 1932, Hoffmann s.n.; road to Lagunitas Secas, 19 Apr 1989, Junak et al. SC-1331; upper portion of Cañada de los Sauces road, ca. 0.4 mi s of South Ridge Road, 18 Apr 1989, Junak et al. SC-1306.
**Trifolium microcephalum** Pursh: just e of mouth of Dick's Canyon, 10 Apr 1990, Junak & Scott SC-1484; w end pine forest, 25 Mar 1932, Hoffmann 673; lower portion of Laguna Canyon, 15 Apr 1991, Junak SC-2220.
**Trifolium variegatum** Nutt.: Cañada del Puerto, 12 Oct 1958, Blakley & Balls B-2529; first canyon w of Reparo Viejo Canyon, 30 May 1990, Junak SC-1898; Cañada Larga, 29 May 1979, Junak et al. SC-103.
**Trifolium willdenovii** Spreng.: Baby's Cove, 15 Apr 1929, Hoffmann s.n.; upper Cañada Christy, 0.1 mi e of Campo Raton, 10 May 1990, Junak SC-1801; lower Willows Canyon, 26 Mar 1983, Junak SC-634.
**Vicia americana** Willd.: Navy Road, ca. 0.5 mi e of Navy Base, 9 Apr 1990, Junak et al. SC-1467; ca. 0.1 mi w of Rancho Portezuelo ruins, 9 May 1990, Junak SC-1728; 1.5 mi down Willows Canyon Road, 26 Mar 1980, Mason-UCSB Botanical Society 2005.
**Vicia benghalensis** L.: Central Valley, ca. 0.4 mi w of The Nature Conservancy cabins at w end of Bosque Mano, 29 May 1992, Junak SC-3415.
**Vicia hassei** S.Watson: upper Tinker's Canyon, 3 May 1969, Benedict s.n.; flats ca. 0.75 mi n of Black Point, 9 Mar 1980, Junak SC-267; Cochies Prietos, 22 Mar 1929, Hoffmann s.n.
**Vicia ludoviciana** Nutt. subsp. **ludoviciana**: Central Valley, 3 mi w of top of Portezuela Grade, 4 May 1966, Philbrick B66-178; road to Punta Arena, ca. 2.3 mi e of Posa Canyon, 13 Jun 1956, Philbrick & Benedict B65-1399; lower Willows Canyon, 26 Mar 1983, Junak SC-637.

**Vicia sativa** L.: Central Valley, 0.1 mi w of The Nature Conservancy cabins at w end of Bosque Mano, 14 Apr 1992, Junak SC-3137; at La Selva Eucalyptus grove, ca. 0.6 mi e of Stanton Ranch, 13 Apr 1992, Junak SC-3116.

## FAGACEAE

**Quercus agrifolia** Nee var. **agrifolia**: Picacho Diablo, 19 Feb 1968, Benedict s.n.; Portezuela Grade, 26 Apr 1960, Raven & Smith 15205 (RSA-POM,SBBG); Coches Prietos Canyon, 0.5 mi from mouth, 1 May 1960, Blakley 3374.

**Quercus chrysolepis** Liebm. in Benth.: n face of Peak 1848, ca. 0.5 mi w of Lagunitas Secas, 27 Apr 1968, Benedict s.n.; Montañon ridge, on Peak 1808, 29 Mar 1969, Benedict s.n.; n side of El Tigre ridge, just n of Peak 1367, 26 Aug 1980, Junak & Hochberg SC-317.

**Quercus douglasii** Hook. & Arn.: ridge w of Alta 2, 22 Jul 1960, Muller 10865; Cañada del Puerto, half way between Prisoners' Harbor and Main Ranch, 14 Oct 1968, Benedict s.n.; Cañada de la Mina, 0.5 mi s of Central Valley Road, 6 Apr 1968, Benedict s.n.

**Quercus dumosa** Nutt.: Santa Cruz Island, Aug 1893, Yates s.n. (SBBG 77407).

**Quercus lobata** Nee: n face of Peak 1848, ca. 0.5 mi w of Lagunitas Secas, 27 Apr 1968, Benedict s.n.; above China Harbor, 25 Apr 1960, Raven & Smith 15156 (RSA-POM,SBBG); Canon Medio, ca. 1.5 mi above Stanton Ranch, 22 Jul 1960, Muller 10850.

**Quercus ×macdonaldii** Greene: ca. 2 mi sw of Lady's Harbor, n of Lagunitas Secas, 28 Aug 1967, Benedict s.n.; Cascada Canyon, n.d., Daily 416; lower Aguaje Canyon, downstream from East Aguaje Spring, 25 Mar 1971, Benedict s.n.

**Quercus pacifica** Nixon & C.H.Mull.: middle portion of Well Canyon, 4 Oct 1969, Benedict s.n.; Central Valley, n of base of Portezuela Grade, 5 Mar 1992, Junak SC-3041; Alameda de los Coches Prietos, 10 May 1979, Junak et al. SC-73.

**Quercus parvula** Greene var. **parvula**: just w of Pelican Bay, 16 Jun 1984, Junak SC-872; Christy Canyon, 26 Apr 1947, Muller 8842; offshore side of island, opposite China Harbor, 17 Dec 1980, Junak & Hochberg SC-356.

**Quercus parvula** Greene × **Q. kelloggii** Newb.: w of Puerta Zuela, 12 Apr 1931, Hoffmann s.n.; Buena Vista Grade, 20 Mar 1932, Hoffmann s.n.

**Quercus tomentella** Engelm.: between Pelican Bay and Prisoners' Harbor, 6 Jun 1951, Muller 8952; head of Canon Medio, ca. midway between Reparo del Viejo and Buena Vista, 22 Jul 1960, Muller 10851; El Tigre ridge, 0.25 mi sw of Peak 1484, 1 Feb 1968, Benedict s.n.

## FRANKENIACEAE

**Frankenia salina** (Molina) I.M.Johnst.: Big Scorpion Rock, 21 Aug 1977, Philbrick & Cummings B77-36; Forney's Cove, 22 Jun 1991, Junak SC-2735; mouth of small canyon near n end of Smugglers' Beach, 7 Nov 1992, Junak SC-3521.

## GARRYACEAE

**Garrya veatchii** Kellogg: Peak 1941, 14 Mar 1970, Benedict s.n.; ca. 0.5 mi n of Peak 1848, nnw of Red Peak, 28 Jan 1979, Junak SC-1; middle portion of Willows Canyon, ca. 0.4 mi n of Peak 678, 14 Feb 1982, Junak & Laughrin SC-464.

## GENTIANACEAE

**Centaurium davyi** (Jeps.) Abrams: w end of Prisoners' Harbor, 22 May 1983, Junak SC-678; Christy Pines, 5 Jun 1983, Junak SC-728; w side of Coches Prietos, 10 May 1973, Philbrick & Philbrick B73-315.

## GERANIACEAE

**Erodium botrys** (Cav.) Bertol.: n side of Picacho Diablo near summit, 19 Apr 1989, Junak et al. SC-1346; Arroyo del Medio, 1.1 mi w of Stanton Ranch, 26 Apr 1960, Raven & Smith 15172 (RSA-POM,SBBG); e side of Sierra Blanca ridge, ne of Peak 1213, 18 Apr 1989, Junak & Ayers SC-1309.

**Erodium cicutarium** (L.) L.Her.: Orizaba, 28 Feb 1932, Hoffmann s.n.; lower Cascada Canyon, ca. 2 mi w of Stanton Ranch, 3 May 1966, Philbrick B66-157; East Aguaje Spring, w of Yellowbanks Anchorage, 24 Mar 1971, Philbrick et al. B71-12.

**Erodium macrophyllum** Hook. & Arn.: Santa Cruz Island, Apr 1888, Brandegee s.n. (UC).

**Erodium moschatum** (L.) L'Her.: near Prisoners' Harbor, 5 Mar 1932, Hoffmann s.n.; Main Ranch, 15 Aug 1939, Dunkle 8627 (RSA-POM,SBBG); Valley Anchorage, 24 Mar 1929, Hoffmann s.n.

**Geranium carolinianum** L.: lower portion of Cañada de la Mina, 9 May 1990, Junak SC-1714; Cochies Prietos, 23 Mar 1929, Hoffmann s.n.

**Pelargonium** ×**hortorum** L.H.Bailey: Frey's Harbor, 12 Nov 1973, Philbrick & Ricker B73-294; lower Tinker's Canyon, 26 Sep 1979, Junak et al. SC-189.

## GROSSULARIACEAE

**Ribes malvaceum** Sm. var. **malvaceum**: just e of mouth of Eagle Canyon, 29 Jan 1984, Junak SC-818; base of w end pine forest, 12 Apr 1931, Hoffmann 190; Ragged Mountain, in ravine ca. 200 yards n of summit, 11 Jun 1965, Philbrick & Benedict B65-1330.

**Ribes thacherianum** (Jeps.) Munz: Valdez Canyon, 14 Jun 1930, Hoffmann s.n.; pine grove e of Christy Ranch, 28 May 1959, Smith 5972; head of Cañada de los Sauces, 22 Apr 1965, Philbrick et al. B65-796h.

## HYDROPHYLLACEAE

**Emmenanthe penduliflora** Benth. var. **penduliflora**: Pelican Harbor, 22 Apr 1932, Hoffmann s.n.; S Ranch, 24 Mar 1932, Hoffmann s.n.; bluffs at e end of Laguna Beach, 29 May 1991, Junak SC-2679.

**Eucrypta chrysanthemifolia** (Benth.) Greene var. **chrysanthemifolia**: pine grove above Chinese Harbor, 15-18 May 1962, Piehl 62-131; first canyon s of bunkhouse at Christy Ranch, 22 Mar 1986, Junak SC-1125; Smugglers' Cove, ca. 0.25 mi inland from beach, 9 Mar 1985, Unser s.n.

**Nemophila pedunculata** Douglas ex Benth.: n flanks of Picacho Diablo, 19 Apr 1989, Junak et al. SC-1351; upper Cañada Christy, just w of base of Centinela Grade, 2 Apr 1991, Junak SC-2124; Cañada del Medio, 0.5 mi w of top of Portezuela Grade, 12 Apr 1985, Junak SC-1022.

**Phacelia cicutaria** Greene var. **hispida** (A.Gray) J.T.Howell: Cañada del Puerto, 0.6 mi from Prisoners' Harbor, 15 Apr 1985, Junak SC-1066; hillside n of Stanton Ranch, 27 Apr 1960, Raven & Smith 15263 (RSA-POM,SBBG); San Justiniano Canyon, along road to Willows Anchorage, 30 May 1991, Junak SC-2723.

**Phacelia cicutaria** var. **hubbyi** (J.F.Macbr.) J.T.Howell: w portion of China Harbor, 1 May 1982, Junak SC-527; e edge of Coches Prietos Beach, 10 May 1973, Philbrick & Philbrick B73-308; Smugglers' Cove, 14 May 1927, Hoffmann s.n.

**Phacelia distans** Benth.: first major canyon e of Profile Point, 12 Jun 1965, Philbrick & Benedict B65-1371; canyon draining n from Triangulation Point "Coche Point" to Potato Harbor, 8 May 1968, Philbrick et al. B68-289; mouth of Willow Creek, w of Christy's Harbor, 21 Mar 1932, Hoffmann 666.

**Phacelia ramosissima** Lehm. var. **austrolitoralis** Munz: Valdez Canyon, 8 Nov 1930, Hoffmann s.n.; n side of Fraser Point, 27 Mar 1980, Simmons-UCSB Botanical Society 626; Smugglers' Cove, 14 May 1927, Hoffmann s.n.

**Phacelia viscida** (Benth.) Torr.: w side of Potato Harbor, 15 Jun 1981, Junak & Bornstein SC-424; canyon due n of University of California Field Station, 11 May 1981, Junak & Timbrook SC-389; 1 mi up Coches Prietos Canyon from mouth, 30 Apr 1960, Blakley 3323.

**Pholistoma auritum** (Lindl.) Lilja var. **auritum**: Prisoners' Harbor, 25 Apr 1979, Thorne et al. 52622 (RSA-POM,SBBG,SD); Cañada del Puerto, 0.25 mi n of Stanton Ranch, 14 Mar 1961, Blakley 3992; China Harbor, 10 Mar 1928, Hoffmann s.n.

**Pholistoma racemosum** (Nutt.) Constance: canyon above Lady's Harbor, 13 Mar 1980, Junak SC-283c; upper w fork of Orizaba Canyon, 14 Apr 1985, Junak SC-1055; Mountain Pasture, in Scorpion Canyon, 9 May 1968, Philbrick & McPherson B68-298.

## LAMIACEAE

**Lamium amplexicaule** L.: Portezuela Cabin, 23 Mar 1932, Hoffmann s.n.; near Puerto Zuelo Ranch cabin, 25 Mar 1932, Wolf 2886 (RSA-POM); upper part of Cañada de la Mina, 6 Apr 1968, Benedict s.n.

**Lepechinia fragrans** (Greene) Epling: ridge w of summit of El Tigre, 5 Jan 1969, Benedict s.n.; e of El Tigre ridge, 0.2 mi ne of Benchmark 1484, 3 Apr 1982, Junak SC-500.

**Marrubium vulgare** L.: ridge above Pelican Bay, 15 Jun 1930, Hoffmann s.n.; Rancho del Sur, 9 Apr 1990, Junak et al. SC-1456; canyon w of Gull Rock, 30 Jun 1930, Hoffmann s.n.

**Mentha spicata** L.: Prisoners' Harbor, 15 Aug 1976, Daily 89 (SCIR); Prisoners' Harbor, 26 Nov 1977, Philbrick & Junak B77-61; Prisoners' Harbor, 21 Sep 1990, Junak & Bornstein SC-2001.

**Salvia columbariae** Benth.: first canyon w of Reparo Viejo Canyon, 30 May 1990, Junak SC-1905; mouth of Cañada Islay, Apr 1969, Benedict s.n.; road to Fraser Point, on s side of Black Point peninsula, 11 Apr 1990, Junak SC-1517.

**Salvia leucophylla** Greene: sea cliffs just w of mouth of Scorpion Canyon, 14 Sep 1991, Junak & Meadows SC-2858.

**Salvia mellifera** Greene: Cañada del Puerto, near Prisoners' Harbor, 2 Feb 1965, Philbrick & Flinck B65-148; 0.1 mi e of top of Portezuela Grade, 4 May 1966, Philbrick B66-172; bluffs at e end of Laguna Beach, 29 May 1991, Junak SC-2680.

**Satureja douglasii** (Benth.) Briq.: Valley Anchorage, in small creekbed adjacent to General Motors installation, 15 Apr 1982, Ledwith & Collett s.n. (SBBG,SCIR).

**Scutellaria tuberosa** Benth.: near Pelican Bay, 24 Mar 1941, Moran 749 (RSA-POM,SBBG); first canyon e of winery buildings at Stanton Ranch, 22 Mar 1985, Junak SC-960; lower Willows Canyon Road, 26 Mar 1983, Junak SC-630.

**Stachys bullata** Benth.: near mouth of second major canyon w of Prisoners' Harbor, 18 Apr 1986, Junak SC-1133; bottom of Portezuelo Grade, 11 Mar 1961, Blakley 3888; Alberts Anchorage, 30 Apr 1960, Blakley 3350.

## LOASACEAE

**Mentzelia affinis** Greene: e side of canyon sw of Los Pinos del Sur, 2 Jun 1993, Junak SC-3738; Smugglers' Cove, 21 Mar 1932, Fosberg 7536.

**Mentzelia micrantha** (Hook. & Arn.) Torr. & A.Gray: talus slope near Prisoners', 15 Jun 1930, Hoffmann s.n.; w side of Cañada del Puerto, ca. 0.25 mi n of Stanton Ranch, 9 May 1979, Junak SC-55; e side of canyon sw of Los Pinos del Sur, 2 Jun 1993, Junak SC-3739.

## LYTHRACEAE

**Lythrum californicum** Torr. & A.Gray: second canyon w of The Nature Conservancy cabins at w end of Bosque Mano, 19 Jul 1985, Junak & Bornstein SC-1098.

**Lythrum hyssopifolium** L.: Prisoners' Harbor, 1 Jun 1979, Junak SC-122; just downstream from La Cascada, 11 Sep 1982, Junak SC-576; mouth of Cañada Pomona at Valley Anchorage, 10 Jul 1967, Benedict s.n.

## MALVACEAE

**Alcea rosea** L.: Cañada del Puerto, ca. 1 mi from Prisoners' Harbor, 8 Jun 1981, Junak & Daily SC-419.

**Eremalche exilis** (A.Gray) Greene: Santa Cruz Island, Apr 1888, Brandegee s.n. (UC).

**Lavatera assurgentiflora** Kellogg subsp. **assurgentiflora**: Islay Canyon Road, 31 Dec 1981, Hochberg 716; road to Central Valley from e end of South Ridge Road, 0.3 mi from Valley Road, 10 May 1985, Junak SC-1074.

**Lavatera assurgentiflora** subsp. **glabra** Philbrick: Valley Road just w of Cascada Creek crossing, 12 Apr 1990, Junak SC-1565.

**Malacothamnus fasciculatus** (Torr. & A.Gray) Greene var. **nesioticus** (B.L.Rob.) Kearney: Christy's, 28 Jun 1930, Hoffmann s.n.; Christy Ranch, 27 Jun 1991, Junak SC-2761; Central Valley, in small canyon directly n of University of California Field Station, 5 May 1993, Gleissman & Wenner s.n. (RSA-POM,SBBG,SCIR).

**Malva nicaeensis** All.: intersection of Navy Road and Valley Peak Road, 20 May 1991, Junak SC-2558; Main Ranch, 15 Jun 1930, Hoffmann s.n.; 0.8 mi e of Stanton Ranch, 31 May 1990, Junak SC-1933.

**Malva parviflora** L.: low ground, 14 May 1927, Hoffmann s.n.; Scorpion Harbor, 21 Sep 1930, Hoffmann s.n.; Cochies Prietos, 22 Mar 1929, Hoffmann s.n.

**Sidalcea malviflora** (DC.) A.Gray ex Benth. subsp. **malviflora**: arroyo 0.25 mi n of mouth of Cañada de Christi, 7 May 1967, Benedict s.n.; slope back of Christy Ranch, 7 Jun 1961, Smith 6463; Fraser Point, 30 Apr 1967, Benedict s.n.

## MORACEAE

**Ficus carica** L.: Pelican Harbor, 26 Nov 1977, Philbrick & Junak B77-59; Pelican Bay, 22 Apr 1990, Junak SC-1609; Christi Ranch, 5 Jun 1977, Daily 399 (SCIR).

## MYRTACEAE

**Eucalyptus camaldulensis** Dehnh.: Central Valley at Rialto Grove, 9 Apr 1990, Junak et al. SC-1453; Navy Base, 25 Sep 1979, Junak et al. SC-186.

**Eucalyptus globulus** Labill.: w end of China Harbor beach, 18 Oct 1990, Junak SC-2105; lower portion of Scorpion Canyon, 19 Oct 1991, Junak SC-2977; lower portion of Cañada de los Sauces, 8 May 1990, Junak SC-1696.

**Eucalyptus sideroxylon** A.Cunn. ex Woolls: Central Valley, ca. 0.5 mi e of Stanton Ranch, 5 Apr 1978, Philbrick B78-24.

**Eucalyptus tereticornis** Smith: Pelican Bay, 8 Oct 1987, Junak SC-1175; just s of Prisoners' Harbor, 19 Apr 1968, Smith 9904; Central Valley at Rialto Grove, 9 Apr 1990, Junak et al. SC-1452.

## NYCTAGINACEAE

**Abronia maritima** S.Watson: Prisoners' Harbor, 13 Aug 1939, Dunkle 8583 (RSA-POM,SBBG); beach w of Christi Ranch, 18 Jun 1964, Philbrick et al. B64-127; Punta Arena, 25 Jul 1960, Blakley & Muller 3756.

**Abronia umbellata** Lam.: Fraser Point, 29 Apr 1960, Raven & Smith 15301 (RSA-POM,SBBG); ca. 0.25 mi ne of top of Punta Arena, 9 Apr 1967, Benedict s.n.; mouth of Laguna Canyon, 17 Jun 1964, Philbrick et al. B64-60.

**Mirabilis californica** A.Gray: Cañada del Puerto, ca. 2 mi from Stanton Ranch, 15-17 May 1962, Piehl 62-167; hillside n of Stanton Ranch, 27 Apr 1960, Raven & Smith 15264 (RSA-POM,SBBG); Coches Prietos Canyon, 29 Apr 1960, Blakley 3281.

## OLEACEAE

**Olea europaea** L.: pine woodland, ca. 1 mi above Pelican Bay, 20 Apr 1969, Benedict s.n.; Scorpion Canyon, 9 May 1968, Philbrick & McPherson B68-308; Smugglers' Canyon, e of High Mount, 25 Mar 1971, Philbrick & Ricker B71-33.

## ONAGRACEAE

**Camissonia californica** (Torr. & A.Gray) Raven: Prisoners', 1 Jul 1930, Hoffmann s.n.; Portezuela Grade, 7 Jun 1961, Smith 6447; e side of spine leading down to Corral Point, 29 May 1991, Junak SC-2692.

**Camissonia cheiranthifolia** (Hornem. ex Spreng.) Raim. subsp. **cheiranthifolia**: beach w of Christi Ranch, 18 Jun 1964, Philbrick et al. B64-132; Fraser Point, 27 Mar 1980, Millett-UCSB Botanical Society 2042 (UCSB,SBBG); point near Malva Real Anchorage, 7 Jun 1961, Smith 6467.

**Camissonia hirtella** (Greene) Raven: Pelican Bay, 7 Jun 1930, Clokey 5010 (CAS,GH,NY,RSA-POM,SD,US); lower Cañada Islay, 17 Jun 1969, Benedict s.n.

**Camissonia ignota** (Jeps.) Raven: road to Coches Prietos Anchorage, 1.1 mi inland from beach, 17 Apr 1991, Junak SC-2293.

**Camissonia intermedia** Raven: Cañada del Puerto, ca. 1 mi inland from Prisoners' Harbor, 30 Apr 1982, Junak SC-515; Mission Spring Canyon, 11 May 1979, Junak SC-81; ravine ca. 200 yards n of summit of Ragged Mountain, 11 Jun 1965, Philbrick & Benedict B65-1336.

**Camissonia micrantha** (Hornem. ex Spreng.) Raven: Scorpion Harbor, 24 Mar 1929, Hoffmann s.n.; Valley Anchorage, 24 Mar 1929, Hoffmann s.n.; bluffs just w of Coches Prietos Beach, 30 May 1991, Junak SC-2716.

**Camissonia robusta** Raven: Fry's Harbor, 22 May 1978, Philbrick B78-176.

**Clarkia epilobioides** (Nutt.) A.Nelson & J.F.Macbr.: Scorpion Canyon, ca. 1.6 mi inland from Scorpion Anchorage, 23 Apr 1993, Junak SC-3719; upper Cañada Christy, in lower w fork of Embudo Canyon, 10 May 1990, Junak SC-1811; road to Coches Prietos Anchorage, 1 mi inland from beach, 6 May 1991, Junak SC-2360.

**Clarkia purpurea** (Curtis) A.Nelson & J.F.Macbr. subsp. **quadrivulnera** (Douglas) F.H.Lewis & M.R.Lewis: ca. 200 yards w of University of California Field Station, 14 May 1970, Benedict s.n.; just w of University of California Field Station, 12 May 1981, Hochberg 698.

**Clarkia unguiculata** Lindl.: small canyon on w side of Scorpion Canyon, ca. 0.5 mi upstream from ranch, 31 May 1986, Junak SC-1158; canyon upstream from La Cascada, 24 Mar 1985, Junak et al. SC-101; just w of El Sobrante, 22 Jun 1967, Benedict s.n.

**Epilobium brachycarpum** C.Presl: e fork of drainage to Lady's Harbor, 14 Sep 1989, Ayers & Scott 811.

**Epilobium canum** subsp. **angustifolium** (D.D.Keck) Raven: first major drainage w of Profile Point, 12 Sep 1989, Ayers et al. 793; Cañada del Puerto, 1.8 mi inland from Prisoners' Harbor, 21 Sep 1990, Junak & Bornstein SC-2016.

**Epilobium canum** (Greene) Raven subsp. **canum**: Cañada del Puerto, 0.5 mi ne of Stanton Ranch, 10 Oct 1958, Blakley & Balls B-2480; just e of first large canyon e of Buena Vista, 16 Oct 1990, Junak SC-2034; s fork of Smugglers' Canyon, ca. 0.25 mi e of Peak 511, 7 Nov 1992, Junak SC-3529.

**Epilobium canum** subsp. **mexicanum** (C.Presl) Raven: Cañada Cascada, 3 Aug 1969, Benedict s.n.; 1.5 mi w of Stanton Ranch, 14 Oct 1958, Blakley & Balls B-2592.

**Epilobium ciliatum** Raf. subsp. **ciliatum**: lower portion of canyon at Valdez Anchorage, 25 Aug 1988, Junak SC-1199; just e of Campo Raton, 31 May 1979, Junak et al. SC-118; Alameda de los Coches Prietos, 10 May 1979, Junak et al. SC-72.

**Ludwigia peploides** (Kunth) Raven subsp. **peploides**: Prisoners' Harbor, 13 Aug 1939, Dunkle 8569 (RSA-POM,SBBG); Prisoners' Harbor, 13 Oct 1958, Blakley & Balls B-2558; mouth of Cañada Christy, 3 Oct 1992, Junak SC-3495.
**Oenothera elata** Kunth subsp. **hirsutissima** (A.Gray ex S.Watson) W.Dietr.: Prisoners' Harbor, 15 Sep 1989, Junak SC-1450; Griffith Canyon, Buena Vista Pasture, 18 Nov 1970, Benedict s.n.; just downstream from University of California Field Station, 7 Jul 1982, Junak SC-560.

## OROBANCHACEAE

**Orobanche bulbosa** Beck: nw flank of Peak 1848, w of Lagunitas Secas, 10 Sep 1981, Ferguson et al. s.n.; side canyon se of Cañada del Puerto, ca. 2 mi from Prisoners' Harbor, 19 Jun 1964, Philbrick et al. B64-198; Christy Pines, along South Ridge Road 0.7 mi w of road to Cañada de los Sauces, 5 Jun 1983, Junak & Hawks SC-734.
**Orobanche fasciculata** Nutt.: Central Valley, on Centinela Grade, 5 Jun 1977, Daily 376 (SCIR); ridge above Albert's Anchorage, ca. 0.6 mi nw of Triangulation Station "Albert", 26 Jun 1981, Hochberg & Hill 703; road to Los Pinos del Sur, ca. 0.5 mi s of Navy Road, 2 Jun 1993, Junak SC-3732.
**Orobanche parishii** (Jeps.) Heckard subsp. **brachyloba** Heckard: "No Man's Land", w of Montañon ridge, 9 Oct 1991, Junak SC-2922; s end of Christy Beach, 11 Oct 1991, Junak SC-2938; n face of Kinton Point, 28 Sep 1968, Benedict s.n.
**Orobanche uniflora** L.: upper w fork of Orizaba Canyon, 14 Apr 1985, Junak & Padgett SC-1056; canyon just back of Pelican Bay, 25 Apr 1930, Abrams & Wiggins 12 (CAS-DS,SBBG); first canyon e of winery buildings at Stanton Ranch, 22 Mar 1985, Junak SC-957.

## OXALIDACEAE

**Oxalis albicans** Kunth subsp. **californica** (Abrams) Eiten: Prisoners' Harbor, 25 Mar 1932, Fosberg 7672; Cañada del Puerto, 0.25 mi n of Stanton Ranch, 14 Mar 1961, Blakley 3993; lower Smugglers' Canyon, 18 Mar 1989, Junak SC-1221.
**Oxalis corniculata** L.: Christy Ranch, 31 May 1979, Junak et al. SC-115; Christy Ranch, 4 Mar 1992, Junak SC-3020; Main Ranch, 14 Jun 1930, Hoffmann s.n.
**Oxalis pes-caprae** L.: Pozo Beach, just e of first large canyon w of mouth of Pozo Canyon, 3 Apr 1993, Junak & Daily SC-3575.

## PAPAVERACEAE

**Dendromecon rigida** Benth. subsp. **harfordii** (Kellogg) Raven: sw fork of upper Scorpion Canyon, 9 May 1968, Philbrick & McPherson B68-300; ca. 0.5 mi e of Centinela, 1 Mar 1965, Philbrick & Benedict B65-457; e end of Ridge Road, 22 Sep 1966, Webster 7441 (DAV,SBBG).
**Eschscholzia californica** Cham.: Scorpion Canyon, just inland from mouth, 14 Sep 1991, Junak SC-2859; El Pato water trough, 3 Apr 1970, Benedict s.n.; ridge between Coches Prietos Canyon and Alberts Anchorage, 29 Apr 1960, Blakley 3298.
**Eschscholzia ramosa** Greene: bluffs w of mouth of Laguna Canyon, 17 Mar 1992, Junak SC-3064; ridge between Alberts and Coches, 28 Mar 1978, Hochberg 464 (SCIR).
**Meconella denticulata** Greene: upper e fork of Orizaba Canyon, 24 Mar 1985, Junak SC-1003; e fork of Cañada de la Mina, 9 May 1990, Junak SC-1722; Montañon ridge, between High Mount and Peak 1808, 23 Mar 1971, Philbrick et al. B71-5.
**Papaver californicum** A.Gray: upper w fork of Orizaba Canyon, 14 Apr 1985, Junak SC-1050; upper w fork of Scorpion Canyon, 23 Apr 1993, Junak SC-3724; lower Laguna Canyon, 29 May 1991, Junak SC-2700.
**Papaver somniferum** L.: Pelican Bay, 26 Sep 1979, Junak et al. SC-192; Pelican Bay, 22 May 1983, Junak SC-690.
**Platystemon californicus** Benth.: just e of cove on e side of Orizaba Canyon, 20 Apr 1989, Junak et al. SC-1378; 0.1 mi w of Portezuela Grade, 28 Mar 1989, Young et al. 1859; bluffs just n of Fraser Point peninsula, 25 Mar 1982, Junak SC-494.
**Stylomecon heterophylla** (Benth.) G.C.Taylor: sea cliff ca. 1.5 mi e of West Point, 5 Apr 1970, Benedict s.n.; Smugglers' Cove, 21 Mar 1932, Fosberg 7620 (RSA-POM).

## PLANTAGINACEAE

**Plantago elongata** Pursh: coastal flats ca. 0.9 mi e of West Point, 13 Apr 1985, Junak SC-1040.

**Plantago erecta** E.Morris: first major headland w of Prisoners' Harbor, 18 Apr 1986, Junak SC-1130; ridge s of Stanton Ranch, 27 Apr 1960, Raven & Smith 15241 (RSA-POM,SBBG); e side of Sierra Blanca ridge, ne of Peak 1213, 18 Apr 1989, Junak & Ayers SC-1315.

**Plantago lanceolata** L.: Prisoners' Harbor, 27 Apr 1977, Daily 316 (SCIR); foot of Portezuela Grade, 7 Oct 1987, Junak SC-1168; n side of Central Valley, across from generator building at Stanton Ranch, 9 Jan 1989, Junak SC-1210.

**Plantago major** L.: Prisoners' Harbor, 13 Oct 1958, Blakley & Balls B-2552; Christy Canyon, just downstream from bridge at Christy Ranch, 29 Sep 1984, Junak SC-890; mouth of Laguna Canyon, 29 May 1991, Junak SC-2673.

**Plantago ovata** Forssk.: Santa Cruz Island, 28 Mar 1925, Hoffmann s.n.

**Plantago subnuda** Pilg.: mouth of Cañada del Agua, 3 Nov 1984, Junak SC-926; second beach e of Prisoners' Harbor, near mouth of Water Canyon, 3 Apr 1970, Benedict s.n.; lower portion of Cañada de Malva Real, 17 Mar 1992, Junak SC-3071.

## PLATANACEAE

**Platanus racemosa** Nutt.: Cañada del Puerto, ca. 2 mi from Stanton Ranch, 15-17 May 1962, Piehl 62-173; Stanton Ranch, 19 Apr 1968, Smith 9902.

## POLEMONIACEAE

**Eriastrum filifolium** (Nutt.) Wooton & Standl.: bluffs just w of Corral Point, 29 May 1991, Junak SC-2689; sea cliff, Cochies Prietos, 12 Apr 1931, Hoffmann s.n.; lower Coches Prietos Canyon, on third ridge inland from beach, 6 May 1991, Junak SC-2352.

**Gilia angelensis** V.E.Grant: Prisoners' Harbor, 25 Mar 1932, Fosberg 7674; slope near S Ranch, 24 Mar 1932, Hoffmann s.n.; ridge between Horqueta and Alamos canyons, 3 mi n of beach, 28 Mar 1968, Benedict s.n.

**Gilia capitata** Sims subsp. **abrotanifolia** (Nutt. ex Greene) V.E.Grant: Smugglers' Cove, 21 Mar 1932, Fosberg 7700.

**Gilia clivorum** (Jeps.) V.Grant: Hazard's, 7 Apr 1930, Hoffmann s.n.; Navy Road, in seasonal pond just w of Mount Pleasant, 23 Mar 1982, Junak SC-477; Sierra Blanca ridge, in saddle n of Ragged Mountain, 11 Mar 1980, Junak et al. SC-272.

**Gilia nevinii** A.Gray: just s of Pelican Harbor, 14 Apr 1967, Philbrick B67-59; top of bluffs between Cavern Point and large offshore rock e of Campo Grande, 22 Apr 1993, Junak SC-3697; upper Cascada, ca. 2 mi w of Stanton Ranch, 18 May 1967, Benedict s.n.

**Linanthus dianthiflorus** (Benth.) Greene: 0.5 mi e of Christi Ranch, 5 Mar 1967, Benedict s.n.; n end of Sierra Blanca ridge, ca. 0.1 mi s of Sauces Road; ridge between Coches Prietos Canyon and Alberts Anchorage, 29 Apr 1960, Blakley 3310.

**Linanthus parviflorus** (Benth.) Greene: middle portion of Valdez Canyon, 30 May 1979, Junak et al. SC-107; just w of Campo del Norte, 4 May 1982, Hochberg & Bromfield 717; upper portion of Cañada de los Sauces, 1.2 mi w of South Ridge Road, 5 Jun 1983, Junak SC-731.

**Navarretia atractyloides** (Benth.) Hook. & Arn.: N Ridge ca. 0.25 mi w of Peak 2021, 4 Jun 1983, Junak SC-726; above Portezuela, 6 Jun 1961, Smith 6441; road to Los Pinos del Sur, 0.5 mi s of Navy Road, 2 Jun 1993, Junak SC-3730.

## POLYGALACEAE

**Polygala californica** Nutt.: directly n of Picacho Diablo, 25 Apr 1968, Benedict s.n.; s of Pelican Harbor, 21 Apr 1932, Hoffmann s.n.; Christy Pines, 27 Jun 1991, Junak SC-2750.

## POLYGONACEAE

**Chorizanthe wheeleri** S.Watson: above Dick's Harbor, 20 Apr 1989, Ayers et al. 703; between base of Cuesta Grade and Christy Ranch barn, 7 Jun 1961, Smith 6452; mouth of Laguna Canyon, 17 Apr 1993, Junak SC-3624.

**Eriogonum arborescens** Greene: middle portion of Little Scorpion Canyon, 18 Oct 1991, Junak SC-2958; Islay Canyon Road, 19 Sep 1990, Bornstein & Junak s.n.; road to Coches Prietos, 17 Apr 1989, Ayers et al. 619.

**Eriogonum giganteum** S.Watson var. **giganteum**: junction of E Ridge Road and Camino Viejo at head of Cañada Islay, 26 Mar 1971, Philbrick et al. B71-54; dumpsite at e end of Stanton Ranch airstrip, 9 Apr 1990, Junak et al. SC-1470; South Ridge Road, ca. 2.5 mi e of Laguna Canyon Road, 7 Oct 1987, Junak SC-1161.

**Eriogonum grande** Greene var. **grande**: Cañada del Puerto, 0.25 mi n of Stanton Ranch, 10 Oct 1958, Blakley & Balls B-2479; Centinela Grade, 0.3 mi w of summit, 12 Sep 1982, Junak SC-582; Smugglers' Cove, 14 May 1927, Hoffmann s.n.

**Eriogonum grande** var. **rubescens** (Greene) Munz: s Christi Beach, 21 Sep 1966, Benedict s.n.; sea cliffs between Christi Ranch and Cañada de los Sauces, 18 Jun 1964, Philbrick et al. B64-133.

**Lastarriaea coriacea** (Goodman) Hoover: Cañada del Puerto, ca. 0.5 mi from Stanton Ranch, 15 May 1962, Piehl 62-143; w-facing slope between Ragged Mountain and Sierra Blanca Peak, 16 Apr 1967, Philbrick B67-142; El Tigre, 6 May 1969, Benedict s.n.

**Polygonum arenastrum** Boreau: Pelican Bay, 14 Jun 1930, Hoffmann s.n.; Scorpion, 20 Sep 1930, Hoffmann s.n.; s end of Christy Beach, 9 Sep 1981, Junak & Hochberg SC-445.

**Polygonum argyrocoleon** Steud. ex Kunze: near ruins of Rancho Nuevo, 16 Jun 1980, Junak & Hochberg SC-304.

**Polygonum lapathifolium** L.: Cañada del Puerto, 0.6 mi from Stanton Ranch, 7 Oct 1991, Junak SC-2894; just downstream from La Cascada, 7 Oct 1991, Junak SC-2898; ca. 0.25 mi w of The Nature Conservancy cabins at w end of Bosque Mano, 20 Apr 1989, Junak SC-1360.

**Pterostegia drymarioides** Fisch. & C.Mey.: Montañon ridge between High Mount and Peak 1808, 30 Mar 1969, Benedict s.n.; Arroyo del Medio, 1.2 mi w of Stanton Ranch, 26 Apr 1960, Raven & Smith 15190 (RSA-POM,SBBG); lower Horqueta Canyon, 28 Mar 1968, Benedict s.n.

**Rumex acetosella** L.: Orizaba Flats, 20 Apr 1989, Junak & Ayers SC-1380; lower La Questa grade, 2.7 road mi e of Christy Barn, 9 May 1973, Philbrick & Philbrick B73-299; South Ridge Road, ca. 0.8 mi e of road to Alameda de los Coches, 11 May 1979, Junak et al. SC-77.

**Rumex conglomeratus** Murray: Prisoners' Harbor, 21 Nov 1980, Junak & Hochberg SC-351; Mission Spring Canyon, 11 May 1979, Junak SC-80; ca. 0.7 mi w of Stanton Ranch airstrip, 7 May 1979, Junak et al. SC-20.

**Rumex crispus** L.: road to Navy Compound, ca. 1 mi e of Prisoners' Harbor pier, 20 Apr 1965, Philbrick et al. B65-709; creek w of Christi Ranch, 18 Jun 1964, Philbrick et al. B64-151; Valley Anchorage, 24 Mar 1929, Hoffmann s.n.

**Rumex pulcher** L.: Prisoners', 15 Jun 1930, Hoffmann s.n.; Cañada Larga Road, 0.1 mi n of Valley Road, 30 May 1990, Junak SC-1886; Posa Canyon, ca. 2.3 mi inland from beach, 13 Jun 1965, Philbrick & Benedict B65-1418.

**Rumex salicifolius** Weinm. var. **salicifolius**: Cañada del Puerto, 0.3 mi from Stanton Ranch, 16 Jun 1984, Junak SC-866; near mouth of Christy Canyon, 12 Sep 1989, Junak et al. SC-1417; small canyon half-way between Forney's Cove and Christi Ranch, 23 Jul 1960, Blakley & Muller 3733.

## PORTULACACEAE

**Calandrinia breweri** S.Watson: canyon 1 mi w of Pelican Bay, 24 Mar 1932, Fosberg 7641; burn above Pelican Harbor, 22 Apr 1932, Hoffmann s.n.; lower Willows Canyon, at canyon s of Peak 882, 9 May 1991, Junak SC-2501.

**Calandrinia ciliata** (Ruiz & Pav.) DC.: Prisoners' Harbor, 25 Mar 1932, Fosberg 7656; upper portion of Cañada Christy, ca. 1.4 mi e of Christy Barn, 16 Apr 1991, Junak SC-2242; near mouth of Laguna Canyon, 15 Apr 1991, Junak SC-2236.

**Calandrinia maritima** Nutt.: Fry's, 21 Apr 1932, Hoffmann s.n.; Christy Ranch, 27 Jun 1991, Junak SC-2759; just w of Morse Point, 27 Jun 1991, Junak SC-2766.

**Calyptridium monandrum** Nutt.: Santa Cruz Island, Apr 1888, Brandegee s.n. (UC).

**Claytonia parviflora** Douglas ex Hook. subsp. **parviflora**: ridge 1 mi nw of Picacho Diablo, 13 Feb 1967, Benedict s.n.; 0.3 mi w of top of Portezuela Grade, 12 Apr 1990, Junak SC-1561; Cochies Prietos, 22 Mar 1929, Hoffmann s.n.

**Claytonia perfoliata** subsp. **mexicana** (Rydb.) J.M.Miller & K.L.Chambers: Orizaba Harbor, 28 Feb 1932, Hoffmann s.n.; Christi Pines, 11 May 1969, Benedict s.n.; Coches Prietos Canyon, 1 May 1960, Blakley 3364.

**Claytonia perfoliata** Willd. subsp. **perfoliata**: Eagle Canyon, 11 Apr 1969, Benedict s.n.; ca. 0.3 mi w of The Nature Conservancy cabins at w end of Bosque Mano, 20 Apr 1989, Junak SC-1363.

**Montia fontana** L.: wet ledge, e of Orizaba, 27 Feb 1932, Hoffmann 652.

**Portulaca oleracea** L.: Main Ranch, 14 Jun 1930, Hoffmann s.n.; Stanton Ranch, 16 Oct 1982, Junak SC-597.

## PRIMULACEAE

**Anagallis arvensis** L.: Prisoners' Harbor, 13 Aug 1939, Dunkle 8577 (RSA-POM,SBBG); Rancho del Sur, 9 Apr 1990, Junak et al. SC-1458; road to Fraser Point, near Black Point peninsula, 28 Mar 1989, Young et al. 1862.

**Dodecatheon clevelandii** Greene subsp. **insulare** (Greene) H.J.Thomps.: San Lucas, at edge of Mount Pleasant, 1 Feb 1965, Philbrick et al. B65-126; first canyon s of bunkhouse at Christy Ranch, 22 Mar 1986, Junak SC-1123; Sierra Blanca ridge, in saddle between Peak 1213 and Ragged Mountain, 3 Apr 1991, Junak SC-2157.

**Samolus parviflorus** (Raf.) Hulten: stream above Dick's Harbor, 21 Sep 1930, Hoffmann s.n.; Dick's Canyon, ca. 300 yards from beach, 6 Jun 1981, Junak et al. SC-405; Orizaba Canyon, 23 May 1932, Hoffmann s.n.

## RANUNCULACEAE

**Clematis lasiantha** Nutt.: Portezuela Grade, ca. 2 mi w of Stanton Ranch, 8 Aug 1966, Benedict s.n.; bottom of Portezuelo Grade, 26 Jul 1960, Blakley 3786; Laguna Canyon, 21 Mar 1932, Hoffmann s.n.

**Clematis ligusticifolia** Nutt.: Main Ranch, 15 Aug 1939, Dunkle 8649 (RSA-POM,SBBG); near mouth of Willows Canyon, ca. 100 yards above beach, 13 Sep 1967, Benedict s.n.; Cochies Prietos, 1 Jul 1930, Hoffmann s.n.

**Clematis pauciflora** Nutt.: n branch of Poso Canyon, above head of Alegria Canyon, 5 Jun 1968, Benedict s.n.

**Delphinium parryi** A.Gray subsp. **maritimum** (Davidson) M.J.Warnock: Prisoners' Harbor, 13 Apr 1931, Hoffmann s.n.; first canyon e of winery buildings at Stanton Ranch, 22 Mar 1985, Junak SC-958; ca. 0.25 mi w of Near Point, nw of Morse Point, 18 Apr 1968, Benedict s.n.

**Ranunculus californicus** Benth: ca. 0.25 mi w of canyon on w side of Cavern Point, 28 Mar 1992, Junak SC-3108; upper Cañada Christy at Campo Raton, 21 Apr 1990, Junak SC-1583; Sauces Canyon, 13 Feb 1982, Philbrick & Benedict B82-4.

## RESEDACEAE

**Oligomeris linifolia** (M.Vahl) J.F.Macbr.: spine of e ridge of Coche Point peninsula, 9 Oct 1991, Junak & Young SC-2926; Scorpion Canyon, ca. 1.2 mi inland from Scorpion Anchorage, 23 Apr 1993, Junak SC-3725; just w of Morse Point, 27 Jun 1991, Junak SC-2768.

## RHAMNACEAE

**Ceanothus arboreus** Greene: e of Valdez Anchorage, in e fork of canyon ca. midway between Valdez and Arch Rock, 12 Mar 1981, Junak SC-371; just w of first large canyon w of Rancho Portezuelo ruins, 4 Mar 1992, Junak SC-3013; Cochies Prietos, 22 Mar 1929, Hoffmann s.n.

**Ceanothus megacarpus** subsp. **insularis** (Eastw.) Raven: just s of Peak 1083, on w side of canyon which empties into ocean e of Punta Gorda and Ruby Rock, 7 Nov 1981, Junak & Laughrin SC-453; ca. 1 mi w of Stanton Ranch, 16 Jun 1964, Philbrick et al. B64-46; ridge between Coches Prietos and Canyon del Medio, 30 Apr 1960, Blakley 3329.

**Ceanothus megacarpus** (Eastw.) Raven subsp. **megacarpus**: Cochies Prietos, 22 Mar 1929, Hoffmann s.n.

**Rhamnus californica** Eschsch. subsp. **californica**: upper e fork of Baby's Canyon, 19 Nov 1980, Junak et al. SC-342; 0.25 mi n of Picacho Diablo, 29 Aug 1976, Benedict s.n.; upper e fork of Diablo Canyon, 20 Sep 1990, Junak & Bornstein SC-1996.

**Rhamnus pirifolia** Greene: Twin Harbor, 21 Sep 1930, Hoffmann s.n.; Portezuela Grade, 0.1 mi w of bottom of grade, 7 Oct 1987, Junak SC-1167; Willows Canyon, ca. 1.3 mi inland from beach, 28 Sep 1984, Junak SC-887.

## ROSACEAE

**Adenostoma fasciculatum** Hook. & Arn. var. **fasciculatum**: vicinity of Pelican Bay, 26 Apr 1930, Abrams & Wiggins 68 (RSA-POM,SBBG); ridge s of Canyon del Medio, 2 mi w of Stanton Ranch, 14 Oct 1958, Blakley & Balls B-2592a; ridge e of mouth of Coches Prietos Canyon, 29 Apr 1960, Blakley 3297.

**Adenostoma fasciculatum** var. **prostratum** Dunkle: Limu Pasture, road between Los Pinos del Sur and China Harbor, 1 Feb 1965, Philbrick et al. B65-130; road leading down Sauces Canyon, ca. 400 yards from Ridge Road, 21 Sep 1966, Gankin et al. 855 (DAV,SBBG); pine forest near head of Laguna Canyon, 17 Jun 1964, Philbrick et al. B64-90.

**Aphanes occidentalis** (Nutt.) Rydb.: upper e fork of Orizaba Canyon, 24 Mar 1985, Junak SC-1007; Navy Road just se of Mount Pleasant, on n flanks of Peak 1450, 13 Apr 1992, Junak SC-3121; 0.5 mi w of top of Portezuela Grade, 12 Apr 1985, Junak SC-1023.

**Cercocarpus betuloides** Torr. & A.Gray var. **blancheae** (C.K.Schneider) Little: above w end of Prisoners' Harbor beach, 3 Mar 1992, Junak SC-3006; Arroyo del Medio, 1.1 mi w of Stanton Ranch, 26 Apr 1960, Raven & Smith 15169 (RSA-POM,SBBG); upper e fork of Horqueta Canyon, 3 Jun 1993, Junak SC-3744.

**Heteromeles arbutifolia** (Lindl.) M.Roem.: bluffs just w of Little Scorpion Canyon, 18 Oct 1991, Junak SC-2956; Cascada, ca. 2 mi wnw of Stanton Ranch, 13 Jul 1964, Philbrick B64-295; Willow Creek, ca. 1 mi upstream from ocean, 12 Oct 1958, Blakley & Balls B2540.

**Holodiscus discolor** (Pursh) Maxim: below Cuesta Grade, 20 Mar 1932, Hoffmann s.n.; head of Cañada Cuesta, ca. 0.5 mi w of divide, 9 Jun 1971, Gliessman & Laughrin 616; upper portion of Cañada Christy, in middle e fork of large canyon just s of Embudo Trough, 27 Aug 1980, Junak SC-324.

**Lyonothamnus floribundus** A.Gray subsp. **aspleniifolius** (Greene) Raven: mouth of first canyon e of Dick's Canyon, 12 Mar 1981, Junak SC-373; ca. 0.5 mi ese of Peak 1020, at e edge of Los Pinos del Sur, 18 Dec 1980, Junak & Timbrook SC-358; head of Aguaje Canyon, w of Yellowbanks Anchorage, 25 Mar 1971, Benedict s.n.

**Potentilla anserina** L. subsp. **pacifica** (J.T.Howell) Rousi: wet flat back of beach at mouth of Christy Ranch, 15 Apr 1976, Thorne et al. 47573 (RSA-POM).

**Prunus ilicifolia** (Nutt.) Walp. subsp. **lyonii** (Eastw.) Raven: head of canyon, Pelican Bay, 25 Mar 1941, Moran 754 (RSA-POM,SBBG); Cascada, ca. 2 mi w of Stanton Ranch, 21 Apr 1965, Philbrick & Benedict B65-776; Smugglers' Cove, 14 May 1927, Hoffmann s.n.

**Rosa californica** Cham. & Schltdl.: Cañada del Puerto, 1 mi ne of Stanton Ranch, 11 Oct 1958, Blakley & Balls B-2484; Cañada de la Portezuela, ca. 0.4 mi w of Reparo Viejo, 30 May 1990, Junak SC-1922; Coches Prietos Canyon, 1 mi from mouth, 30 Apr 1960, Blakley 3326.

**Rubus ursinus** Cham. & Schltdl.: canyon draining n from Triangulation Point "Coche Point" to Potato Harbor, 8 May 1968, Philbrick et al. B68-285; pine forest between Embudo and Campo Raton, 18 Jun 1964, Philbrick et al. B64-167; ne extension of Ragged Mountain, 18 Apr 1989, Ayers et al. 667.

## RUBIACEAE

**Galium angustifolium** Nutt. subsp. **foliosum** (Hilend & Howell) Dempster & Stebbins: North Ridge between Picacho Diablo and Lagunitas Secas, 19 Apr 1989, Junak et al. SC-1332; 1.1 mi w of Portezuela Grade summit, 18 Apr 1989, Ayers et al. 653; w side of second large canyon w of Sandstone Point, 18 May 1992, Junak SC-3359.

**Galium aparine** L.: Cañada del Puerto, 0.7 mi inland from Prisoners' Harbor, 6 May 1991, Junak SC-2337; edge of large pine grove above Christy's, 26 Apr 1960, Raven & Smith 15225 (RSA-POM,SBBG); Smugglers' Cove, 14 May 1927, Hoffmann s.n.

**Galium buxifolium** Greene: sea cliff e of Dick's Harbor, 28 Feb 1932, Hoffmann s.n.; Pelican Bay, 3 May 1969, Benedict s.n.; under pines in Buena Vista Pasture, ca. 1 mi e of Centinela, 28 Apr 1968, Benedict s.n.

**Galium californicum** Hook. & Arn. subsp. **flaccidum** (Greene) Dempster & Stebbins: pine woodland ne of Red Mountain, 3 May 1969, Benedict s.n.; Pelican Bay, 14 Aug 1939, Dunkle 8607 (RSA-POM,SBBG); upper Cañada Christy, in lower w fork of Embudo Canyon, 10 May 1990, Junak SC-1813.

**Galium nuttallii** A.Gray subsp. **insulare** Ferris: Scorpion Harbor, 17 Mar 1929, Hoffmann s.n.; first canyon e of winery buildings at Stanton Ranch, 22 Mar 1985, Junak SC-942; ca. 1.3 mi ene of Navy Base, on road to Los Pinos del Sur, 28 Jan 1984, Junak SC-809.

**Galium porrigens** Dempster var. **porrigens**: Cañada del Puerto, 0.25 mi from Stanton Ranch, 15 May 1962, Piehl 62-158; bluff above beach at mouth of Water Canyon, 2 May 1969, Benedict s.n.; lower Buena Vista Grade, 13 Apr 1968, Benedict s.n.

## SALICACEAE

**Populus balsamifera** L. subsp. **trichocarpa** (Torr. & A.Gray ex Hook) Brayshaw: Cañada del Puerto, ca. half way between Stanton Ranch and Prisoners' Harbor, 13 Oct 1958, Blakley & Balls B-2579; w of Red Peak and ca. 2 mi s of Platts Harbor, 5 May 1966, Philbrick B66-193; Alameda, between Stanton Ranch and Bowen Point, 31 Mar 1969, Benedict s.n.

**Populus fremontii** S.Watson subsp. **fremontii**: interior of Santa Cruz Island, 15 Jun 1930, Hoffmann s.n.; foot of Portezuela Grade, 24 Mar 1932, Hoffmann 664; canyon on n side of Central Valley, s of Platts Harbor and ca. 0.25 mi w of intersection of Central Valley and Ridge roads, 21 Apr 1965, Philbrick et al. B65-788.

**Salix bonplandiana** Kunth: upper portion of Cañada del Puerto, 25 Mar 1985, Junak SC-1011; Cascada, ca. 2 mi w of Stanton Ranch, 21 Apr 1965, Philbrick & Benedict B65-779; Hooker Canyon, below upper forks, 4 Nov 1984, Junak SC-941.

**Salix exigua** Nutt.: Cañada del Puerto, 0.5 mi from Stanton Ranch, 30 Sep 1984, Junak SC-904; ca. 0.5 mi w of University of California Field Station, 2 Jul 1968, Benedict s.n.; West Aguaje Spring, n of Sandstone Point, 24 Mar 1971, Philbrick et al. B71-28.

**Salix lasiolepis** Benth.: Cañada del Puerto, 14 Mar 1961, Blakley 3983; lower end of dump at Stanton Ranch airstrip, 9 Apr 1990, Junak et al. SC-1471; East Aguaje Spring, w of Yellowbanks Anchorage, 24 Mar 1971, Philbrick et al. B71-13.

**Salix lucida** Muhlenb. subsp. **lasiandra** (Benth.) E.Murray: Valdez, 9 Nov 1930, Hoffmann s.n.

## SAURURACEAE

**Anemopsis californica** (Nutt.) Hook. & Arn.: Prisoners', 15 Jun 1930, Rowntree s.n.; Prisoners' Harbor, 13 Oct 1958, Blakley & Balls B-2578; Prisoners' Harbor, 21 Sep 1990, Junak & Bornstein SC-2002.

## SAXIFRAGACEAE

**Heuchera maxima** Greene: Pelican Bay, 14 Aug 1939, Dunkle 8618 (RSA-POM,SBBG); small canyon s of Stanton Ranch, 14 Mar 1961, Blakley 4003; s fork of Smugglers' Canyon, ca. 0.3 mi e of Peak 511, 7 Nov 1992, Junak SC-3527.

**Jepsonia malvifolia** (Greene) Small: e side of Tinkers Canyon, 3 Mar 1992, Junak SC-3004; n side of Central Valley, along Valley Peak Road, 16 Mar 1992, Junak SC-3049; ridge between Coches Prietos and Alberts Anchorage, 30 Apr 1960, Blakley 3357.

**Lithophragma cymbalaria** Torr. & A.Gray: Scorpion Canyon, ca. 1 mi inland from beach, 19 Mar 1989, Junak SC-1223; first canyon e of winery buildings at Stanton Ranch, 22 Mar 1985, Junak SC-956; n face of Peak 1808, se of High Mount, 29 Mar 1969, Benedict s.n.

**Saxifraga californica** Greene: lower Dick's Canyon, 10 Apr 1990, Junak SC-1481; Scorpion Harbor, 17 Mar 1929, Hoffmann s.n.; Central Valley, along road from Stanton Ranch to South Ridge, 14 Feb 1982, Junak SC-462.

## SCROPHULARIACEAE

**Antirrhinum kelloggii** Greene: burn above Pelican Harbor, 22 Apr 1932, Hoffmann s.n.; just upstream from Cascada, 10 May 1981, Junak SC-379; e side of San Justiniano Canyon, 30 May 1991, Junak SC-2720.

**Antirrhinum multiflorum** Pennell: North Ridge at saddle just w of Picacho Diablo, 19 Apr 1989, Junak et al. SC-1342; ridgetop above e fork of Ram Canyon, 20 Sep 1990, Junak & Bornstein SC-1990; Cañada Cascada, on e-facing slope directly above dam, 15 Jun 1969, Benedict s.n.

**Antirrhinum nuttallianum** Benth. subsp. **subsessile** (A.Gray) D.M.Thomps.: first major canyon e of Profile Point, 12 Jun 1965, Philbrick & Benedict B65-1358; creek bed between Cuesta Grade and Christy Barn, 7 Jul 1961, Smith 6455; ridge between Coches Prietos Canyon and Alberts Anchorage, 29 Apr 1960, Blakley 3299.

**Castilleja affinis** Hook. & Arn. subsp. **affinis**: near mouth of Dick's Canyon, 10 Apr 1990, Junak SC-1482; Cañada de la Portezuela, just e of Reparo Viejo, 21 Apr 1990, Junak SC-1589; Aguaje Canyon, n of Middle Anchorage, 24 Mar 1971, Philbrick et al. B71-22.

**Castilleja attenuata** (A.Gray) T.I.Chuang & Heckard: just e of Pelican Bay, 18 Apr 1986, Junak SC-1134; just w of Prisoners' Harbor, 14 Apr 1967, Philbrick B67-53; Islay Canyon Road, ca. 0.9 mi from University of California Field Station, 21 Apr 1986, Junak SC-1157

**Castilleja densiflora** (Benth.) T.I.Chuang & Heckard: mouth of Orizaba Canyon, 26 Mar 1932, Wolf 2892 (RSA-POM,SBBG); w end of Prisoners' Harbor, 22 May 1983, Junak SC-680; hills s of Smugglers' Valley, 9 Mar 1985, Unser s.n.

**Castilleja exserta** (A.Heller) T.I.Chuang & Heckard: near Christy's, 21 Mar 1932, Hoffmann s.n.; 1.1 mi w of Portezuela Grade summit, 18 Apr 1989, Ayers et al. 651; Near Point, 17 May 1967, Benedict s.n.

**Castilleja lanata** A.Gray subsp. **hololeuca** (Greene) T.I.Chuang & Heckard: Well Canyon, ca. 1 mi w of Campo del Norte, 21 Mar 1967, Benedict s.n.; Cañada de la Portezuela, 0.6 mi e of top of Centinela Grade, 11 May 1985, Junak SC-1085; e side of large canyon sw of Los Pinos del Sur, 2 Jun 1993, Junak SC-3735.

**Keckiella cordifolia** (Benth.) Straw: e fork of lower canyon above Lady's Harbor, 20 Aug 1967, Philbrick s.n.; 1 mi e of Stanton Ranch, 17 May 1962, Breedlove 2877 (CAS-DS,SBBG); n side of Las Torres ridge, 1 mi w of intersection of Coches Prietos and San Justiniano roads, 12 Apr 1990, Junak SC-1549.

**Linaria canadensis** (L.) Dum.Cours.: bluffs between Prisoners' Harbor and Pelican Bay, 15 May 1967, Benedict s.n.; Islay Canyon Road, 1.7 mi from intersection with South Ridge Road, 15 Apr 1991, Junak SC-2199; Smugglers' Canyon, ca. 0.25 mi inland from beach, 9 Mar 1985, Unser s.n.

**Mimulus brandegei** Pennell: stony summit of third ridge w of Christy's Harbor, 21 Mar 1932, Wolf 2800 [label data is incorrect; see taxonomic treatment for this taxon].
**Mimulus cardinalis** Douglas ex Benth.: Valdez, 13 Jun 1930, Hoffmann s.n.; first canyon w of Reparo Viejo Canyon, 30 May 1990, Junak SC-1899; Horqueta Canyon, ca. 0.7 mi inland from beach, 3 Jun 1993, Junak SC-3758.
**Mimulus flemingii** Munz: pine grove above Chinese Harbor, 15-18 May 1962, Piehl 62-137; head of Islay Canyon, 1.9 mi from University of California Field Station, 5 Jun 1983, Junak SC-748; Alberts Anchorage, 6 Mar 1967, Benedict s.n.
**Mimulus guttatus** Fisher ex DC.: upper w fork of Scorpion Canyon, ca. 1.6 mi inland from Scorpion Anchorage, 23 Apr 1993, Junak SC-3720; Arroyo del Medio, 1.1 mi w of Stanton Ranch, 26 Apr 1960, Raven & Smith 15179 (RSA-POM,SBBG); upper middle drainage of Smugglers' Canyon, 25 Mar 1971, Philbrick & Ricker B71-35.
**Mimulus longiflorus** (Nutt.) A.L.Grant: rock slide area between Pelican Bay and Prisoners' Harbor, 22 May 1983, Junak SC-698; upper portion of Cañada Christy, in large canyon just s of Embudo Trough, 27 Aug 1980, Junak SC-326; Willows Canyon, 29 Mar 1989, Young et al. 1876.
**Verbascum thapsus** L.: Prisoners' Harbor, 12 Aug 1939, Dunkle 8555 (RSA-POM,SBBG); Buena Vista, 11 Oct 1958, Blakley & Balls B-2545; just inland from mouth of Smugglers' Canyon, 13 Sep 1991, Junak & Meadows SC-2830.
**Veronica anagallis-aquatica** L.: Prisoners' Harbor, 30 Apr 1982, Junak SC-510; La Cascada, just downstream from swimming hole, 12 Apr 1990, Junak SC-1566; Coches Prietos Road, at creek crossing below Alameda de los Coches, 9 May 1994, Junak SC-3934.

## SOLANACEAE

**Datura wrightii** Regel: Prisoners' Harbor, 13 Aug 1939, Dunkle 8570 (RSA-POM,SBBG); Scorpion Anchorage, just behind beach, 19 Oct 1991, Junak SC-2978; lower Laguna Canyon, ca. 0.25 mi n of beach, 17 Oct 1990, Junak SC-2076.
**Lycium californicum** Nutt.: extreme tip of San Pedro Point, 29 Apr 1971, Philbrick et al. B71-96; near tip of San Pedro Point, 11 Mar 1994, Junak SC-3820; sea cliff, above and e of extreme s end of Christi Beach, 27 Mar 1969, Benedict s.n.
**Lycopersicon esculentum** L.: w end of beach at Lady's Harbor, 25 Aug 1988, Junak SC-1208; Prisoners' Harbor, 18 Jun 1984, Junak SC-884; mouth of Smugglers' Canyon, 13 Sep 1991, Junak & Meadows SC-2832.
**Nicotiana clevelandii** A.Gray: cliffs near Forney's, 30 Jun 1930, Hoffmann s.n.; near e end of Willows Beach, 16 Mar 1984, Junak SC-844; slope above beach, Valley Anchorage, 24 Mar 1929, Hoffmann s.n.
**Nicotiana glauca** Graham: w end of beach at China Harbor, in small canyon behind fishing shacks, 18 Oct 1990, Junak SC-2104; "No Man's Land" on w side of Montañon ridge, along trail to High Mount, 9 Oct 1991, Junak & Young SC-2923; s side of mouth of canyon, Smugglers' Cove, 25 Mar 1971, Philbrick & Ricker B71-43.
**Solanum clokeyi** Munz: bluffs, w portion of China Harbor, 1 May 1982, Junak SC-522; ca. 0.3 mi w of site of Rancho Portezuelo, 2 Apr 1991, Junak & Bornstein SC-2120; upper Sauces Canyon, 4 Mar 1967, Benedict s.n.
**Solanum douglasii** Dunal in DC.: Valdez, 8 Nov 1930, Hoffmann s.n.; stream bed near Christi Ranch, 16 May 1962, Piehl 62-219; Laguna Canyon, ca. 0.1 mi inland from beach, 17 Oct 1990, Junak SC-2078.
**Solanum elaeagnifolium** Cav.: corrals near airstrip, Christi Ranch, 6 Oct 1977, Laughrin s.n.; stockpen at Christy Barn, 10 Jul 1979, Junak & Philbrick SC-128.

## TAMARICACEAE

**Tamarix ramosissima** Ledeb.: Prisoners' Harbor, 20 Aug 1977, Daily s.n.; Christy Canyon, below Christy Ranch, 22 Jun 1991, Junak SC-2733; Pozo Canyon, ca. 0.3 mi inland from beach, 22 May 1994, Junak SC-3999.

## URTICACEAE

**Hesperocnide tenella** Torr.: n-facing slope of High Mount, 7 May 1968, Philbrick & McPherson B68-262; shaded canyon ne of High Mount, Montañon ridge, 30 Mar 1969, Benedict s.n.; below Main Ranch, 12 Apr 1931, Hoffmann 220.
**Parietaria hespera** B.D.Hinton var. **californica** B.D.Hinton: e fork of first canyon w of Profile Point, 12 Jun 1965, Philbrick & Benedict B65-1348a; first major canyon w of Prisoners' Harbor, 15 May 1967, Benedict s.n.; s side of Potato Harbor, 22 Apr 1993, Junak SC-3702.

**Urtica dioica** L. var. **holosericea** (Nutt.) C.L.Hitchc.: upper part of first canyon w of pine forest above China Harbor, 2 Aug 1991, Junak SC-2808; Cañada de la Portezuela, between ruins of Rancho Portezuelo and Reparo Viejo, 9 May 1990, Junak SC-1759; near mouth of Laguna Canyon, 17 Jun 1964, Philbrick et al. B64-70.
**Urtica urens** L.: Scorpion Canyon, ca. 1.2 mi inland from Scorpion Anchorage, 22 Apr 1993, Junak SC-3710; Stanton Ranch, 7 Feb 1986, Junak SC-1117; Cochies Prietos, 22 Mar 1929, Hoffmann s.n.

### VALERIANACEAE

**Centranthus ruber** (L.) DC.: 0.25 mi w of El Reparo Viejo, 14 Oct 1958, Blakley & Balls B-2594; Central Valley Road ca. 1 mi e of divide, 14 Jul 1964, Philbrick B64-298; Cañada de la Portezuela, in first gully w of Reparo Viejo overlook, 8 May 1990, Junak SC-1671a.

### VERBENACEAE

**Verbena lasiostachys** Link. var. **scabrida** Moldenke: Prisoners' Harbor, 12 Aug 1939, Dunkle 8548 (RSA-POM,SBBG); Cañada Islay, ca. 1 mi w of Stanton Ranch, 16 Jun 1964, Philbrick et al. B64-32; small dry gully running ne from Forney's Cove, 4 Nov 1974, Philbrick B74-153.

### VIOLACEAE

**Viola pedunculata** Torr. & A.Gray: ne flanks of Picacho Diablo, 19 Apr 1989, Ayers et al. 693; ridge on s side of Central Valley, 200 yards w of El Sobranti, 13 Mar 1961, Blakley 3974; hills sw of Smugglers' Cove, 11 Mar 1985, Unser s.n.

## Monocotyledonous Angiosperms

### AGAVACEAE

**Agave americana** L.: point on e side of Pelican Bay, 26 Nov 1977, Philbrick & Junak B77-58; tip of peninsula on e side of Pelican Bay, 14 Sep 1991, Junak SC-2874.

### ALLIACEAE

**Allium lacunosum** S.Watson var. **lacunosum**: road to Picacho Diablo and Lagunitas Secas, 0.3 mi n of Centinela gate, 30 May 1979, Junak et al. SC-104; lower part of Horqueta Canyon, ca. 0.2 mi inland from beach, 3 Jun 1993, Junak SC-3763.
**Allium praecox** Brandegee: coastal flats just w of Prisoners' Harbor, 20 Apr 1989, Junak et al. SC-1386; Central Valley e of road to Lagunitas Secas, ca. 0.2 mi n of Valley Road, 9 May 1990, Junak SC-1750; ridge between 2 forks of Rancho Nuevo Canyon, 18 Apr 1993, Junak SC-3635.
**Bloomeria crocea** (Torr.) Coville: Orizaba Flats, w of Olsen's Cave, 20 Apr 1989, Junak & Ayers SC-1379; Cañada de la Portezuela, ca. 0.1 mi w of Rancho Portezuelo ruins, 9 May 1990, Junak SC-1727; upper e fork of Horqueta Canyon, 3 Jun 1993, Junak SC-3747.
**Brodiaea jolonensis** Eastw.: road to Lagunitas Secas, 1.5 mi n of Valley Road, 19 Apr 1989, Ayers et al. 686; ca. 0.25 mi w of Portezuela Grade, 28 Apr 1960, Smith 6121; just nw of Stanton Ranch, 21 Apr 1989, Junak SC-1390.
**Dichelostemma capitatum** (Benth.) A.W.Wood: e side of Prisoners' Harbor, 13 Mar 1961, Blakley 3942; road to South Ranch, 24 Mar 1932, Hoffmann s.n.; ridge between Coches Prietos Canyon and Alberts Anchorage, 29 Apr 1960, Blakley 3287.
**Triteleia hyacinthina** (Lindl.) Greene: ca. 0.2 mi w of The Nature Conservancy cabins at w end of Bosque Mano, 16 Apr 1992, Junak SC-3213.

### AMARYLLIDACEAE

**Narcissus tazetta** L.: Christy Ranch, 3 Feb 1995, Junak SC-4077; Islay Canyon Road, ca. 0.2 mi s of Central Valley Road, 16 Feb 1979, Junak & Hochberg SC-9; Islay Canyon Road, ca. 0.9 mi from University of California Field Station, 28 Jan 1980, Junak & Hochberg SC-226.

### CYPERACEAE

**Carex barbarae** Dewey: 1 mi below Main Ranch, just below Ravenswood, 26 Mar 1932, Hoffmann s.n.; stream to Prisoners' Harbor, ca. 0.5 mi n of Stanton Ranch, 26 Mar 1983, Ferren-UCSB Botanical Society 2253 (SBBG,UCSB).

**Carex globosa** Boott: under pines near Pelican Bay, 14 Jun 1930, Hoffmann s.n.; first canyon e of winery buildings at Stanton Ranch, 22 Mar 1985, Junak SC-944; Willows Canyon, ca. 1.8 mi inland from beach, 29 Mar 1989, Junak SC-1252.
**Carex harfordii** Mack.: above Fry's Harbor, 10 Sep 1931, Hoffmann 597; 1 mi inland from Dick's Harbor, 7 Dec 1930, Hoffmann 113; Pelican Bay, 1 Sep 1928, Hoffmann s.n.
**Carex senta** Boott: Dick's Canyon, ca. 400 yards s of beach, 6 Jun 1981, Junak et al. SC-406; Canyon de la Siesta, 26 Mar 1932, Wolf 2891 (RSA-POM,SBBG); ca. 0.25 mi upstream from La Cascada, 19 Mar 1984, Junak SC-860.
**Carex subbracteata** Mack.: Scorpion Harbor, 17 Mar 1929, Hoffmann s.n.; upper w fork of first major s branch of Scorpion Canyon, 29 Apr 1971, Philbrick B71-83.
**Carex triquetra** Boott: Santa Cruz Island, 28 Mar 1925, Hoffmann s.n.
**Carex tumulicola** Mack.: s of China Harbor, ne of Navy Compound, 10 May 1973, Philbrick & Philbrick B73-320; South Ridge Road, 1.2 mi e of Christy Barn and airstrip, 30 May 1979, Junak et al. SC-113; ridge n of Ragged Mountain, 2.5 mi n of Punta Arena, 11 Jun 1965, Philbrick & Benedict B65-1318.
**Cyperus involucratus** Rottb.: Valdez, 14 Jun 1930, Hoffmann s.n.
**Scirpus californicus** (C.A.Mey.) Steud.: Prisoners' Harbor, 13 Aug 1939, Dunkle 8581 (RSA-POM,SBBG); Prisoners' Harbor, 13 Jul 1964, Philbrick B64-289; lower portion of Cañada del Agua, 11 Sep 1989, Junak et al. SC-1398.
**Scirpus maritimus** L.: just inland from mouth of Cañada de los Sauces, 8 Oct 1991, Junak SC-2917.
**Scirpus pungens** Vahl: sandy depression adjacent to creek at mouth of Poso Canyon, 18 Apr 1968, Benedict s.n.; small depression just behind beach at mouth of Pozo Canyon, 16 Oct 1990, Junak SC-2051.

## IRIDACEAE

**Sisyrinchium bellum** S.Watson: hill s of old winery building at Stanton Ranch, 14 Mar 1961, Blakley 3995; lower portion of Cañada de los Sauces, 16 Apr 1991, Junak SC-2259; ridge above airstrip near Christi Ranch, 16 May 1962, Piehl 62-208.

## JUNCACEAE

**Juncus bufonius** L.: first major canyon w of Prisoners' Harbor, 22 May 1983, Junak SC-685; Cañada de la Portezuela, along Cañada Larga Road at first creek crossing n of Valley Road, 7 May 1990, Junak SC-1649; small canyon half way between Forney's Cove and Christi Ranch, 22 Jul 1960, Blakley & Muller 3734.
**Juncus mexicanus** Willd.: Corral de Siesta, 17 Oct 1958, Blakley B-2640; w of Rancho del Medio, 27 Jun 1981, Hochberg & Hill 704.
**Juncus patens** E.Meyer: coastal flats w of second major canyon w of Pelican Bay, 8 Oct 1987, Junak SC-1182; lower part of Cañada Larga, 24 Jun 1969, Benedict s.n.; Coches Prietos Road at ne end of Las Torres ridge, 27 Mar 1983, Junak & Turner SC-650.
**Juncus xiphioides** E.Meyer: Cañada del Puerto, 0.3 mi downstream from Stanton Ranch, 10 Sep 1982, Junak SC-570; lower Cañada del Agua, 3 Nov 1984, Junak SC-925; lower part of Cañada Larga, 24 Jun 1969, Benedict s.n.
**Luzula comosa** E.Meyer: headland on e side of canyon on e side of Cueva Valdez, 13 Sep 1989, Junak et al. SC-1429; upper Cañada Christy at Campo Raton, 29 May 1990, Junak SC-1863; Sierra Blanca ridge, just n of Peak 1213, 18 Apr 1989, Junak et al. SC-1317.

## LILIACEAE

**Calochortus albus** Douglas ex Benth.: West Twin Canyon, just above beach, 20 Apr 1989, Ayers et al. 709; Central Valley, 3 mi w of top of Portezuela Grade, 4 May 1966, Philbrick B66-181; first canyon s of Sauces Canyon, ca. 0.25 mi inland, 20 May 1967, Benedict s.n.
**Calochortus catalinae** S.Watson: Cañada de la Portezuela, just w of Reparo Viejo, 21 Apr 1990, Junak SC-1587; ridge between 2 forks of Rancho Nuevo Canyon, 18 Apr 1993, Junak SC-3636; above beach at Coches Prietos, 17 Apr 1989, Ayers et al. 633.
**Calochortus luteus** Douglas ex Lindl.: Portezuela Grade, ca. 3 mi w of Stanton Ranch, 18 Jun 1964, Philbrick et al. B64-185; near Stanton Ranch airfield, 7 Jun 1981, Junak & McIntire SC-418; road to Coches Prietos, 0.2 mi n of Alameda de los Coches Prietos, 30 May 1991, Junak SC-2705.
**Lilium humboldtii** Roezel & Leichtlin subsp. **ocellatum** (Kellogg) Thorne: Valdez, 13 Jun 1930, Hoffmann s.n.; Cañada del Puerto, 2.2 mi inland from Prisoners' Harbor, 31 May 1990, Junak SC-1969; upper Cañada Christy, 0.1 mi e of Campo Raton, 29 May 1990, Junak SC-1861.

**Zigadenus fremontii** (Torr.) S.Watson: Islay Canyon Road, just w of University of California Field Station, 18 Apr 1989, Ayers et al. 680; road to Fraser Point, near Black Point peninsula, 28 Mar 1989, Young et al. 1864; Valley Anchorage, 24 Mar 1929, Hoffmann s.n.

## ORCHIDACEAE

**Epipactis gigantea** Douglas ex Hook.: lower Water Canyon, 20 Nov 1980, Junak & Hochberg SC-350; head of small stream which heads on sw side of Picacho Diablo and flows into Puerto Zuelo Ranch, 25 Mar 1932, Wolf 2881 (RSA-POM); Malva Real Canyon, 0.1 mi inland from mouth, 17 Sep 1992, Junak SC-3477.

**Piperia elongata** Rydb.: middle portion of Pelican Canyon, 23 Jun 1980, Philbrick & Leaberry B80-5; pine forest above China Harbor, 9 May 1979, Junak et al. SC-62; near upper portion of Islay Canyon Road, 9 Jul 1982, Junak SC-564.

**Piperia michaelii** (Greene) Rydb.: bluff, extreme s end of Christi Beach, 6 Jun 1968, Benedict s.n.

## POACEAE

**Achnatherum diegoense** (Swallen) Barkworth: coastal flats just w of Prisoners' Harbor, 20 Apr 1989, Junak et al. SC-1387; Islay Canyon Road, ca. 0.75 mi from University of California Field Station, 26 Mar 1983, Junak SC-620; w side of second canyon w of Sandstone Point, 18 May 1992, Junak SC-3358.

**Aegilops cylindrica** Host: ca. 0.3 mi w of The Nature Conservancy cabins at w end of Bosque Mano, 4 Jun 1993, Junak SC-3771; lower Cottonwood Canyon, ca. 0.2 mi w of The Nature Conservancy cabins at w end of Bosque Mano, 29 May 1992, Junak SC-3421; hillside pasture se of Channel Island Field Station, 2 Jun 1973, Barbe 1566 (RSA-POM).

**Agrostis exarata** Trin.: second cove e of Fry's Harbor, 26 Sep 1979, Junak et al. SC-200; Orizaba Canyon, 23 May 1932, Hoffmann 733.

**Agrostis pallens** Trin.: second cove e of Fry's Harbor, 26 Sep 1979, Junak et al. SC-199; Camino de la Casa, 0.1 mi from intersection with South Ridge Road, 30 May 1991, Junak SC-2726; upper portion of Cañada de los Sauces, 2.4 mi from intersection with road to Pozo Canyon, 1 Aug 1991, Junak SC-2794.

**Agrostis viridis** Gouan: Valdez, 15 Jun 1930, Hoffmann s.n.; Main Ranch, 16 Aug 1939, Dunkle 8651 (RSA-POM,SBBG); canyon draining into Coches Prietos from w, 14 Jun 1967, Benedict s.n.

**Andropogon glomeratus** (Walter) Britton, Sterns, & Poggenb. var. **scabriglumis** C.S.Campb.: Cañada del Puerto, 2.5 mi inland from Prisoners' Harbor, 21 Sep 1990, Junak & Bornstein SC-2017; Cañada del Puerto, at first creek crossing n of Stanton Ranch, 8 Nov 1991, Junak SC-2980; second canyon w of The Nature Conservancy cabins at w end of Bosque Mano, 19 Jul 1985, Junak & Bornstein SC-1099.

**Aristida adscensionis** L.: Cañada del Puerto, ca. 1.6 mi inland from Prisoners' Harbor, 3 Mar 1992, Junak SC-2986; Cañada de la Portezuela, in first canyon w of Reparo Viejo Canyon, 30 May 1990, Junak SC-1901; road to Coches Prietos beach, ca. 0.25 mi inland from coastline, 17 Mar 1984, Junak SC-854.

**Aristida ternipes** Cav. var. **hamulosa** (Henrard) J.S.Trent: just w of large Eucalyptus grove w of Stanton Ranch, 6 May 1963, Smith 8143; ridge s of Islay Pasture, 0.75 mi w of Stanton Ranch, 11 Mar 1961, Blakley 3877; road to Coches Prietos beach, 0.4 mi s of intersection with South Ridge Road, 31 Jul 1991, Junak SC-2774.

**Arundo donax** L.: mouth of Cañada Christy, 2 Oct 1992, Junak SC-3493; mouth of Cañada Christy, just behind beach, 7 Aug 1993, Junak SC-3794.

**Avena barbata** Link: Santa Cruz Island, 28 Mar 1925, Hoffmann s.n.; ridge s of Islay Pasture, 0.75 mi w of Stanton Ranch, 11 Mar 1961, Blakley 3882; ridge between Coches Prietos Canyon and Albert Anchorage, 29 Apr 1960, Blakley 3291.

**Avena fatua** L.: Prisoners', 14 Jun 1930, Hoffmann s.n.; Navy Compound, 1 Feb 1965, Philbrick et al. B65-141.

**Avena sativa** L.: University of California Field Station building, 15 Apr 1967, Philbrick B67-70.

**Brachypodium distachyon** (L.) P.Beauv.: lower part of road on e side of Scorpion Canyon, 18 Oct 1991, Junak SC-2946; South Ridge Road opposite Picacho Diablo, ca. 1.1 mi e of intersection with connector road to Centinela, 28 May 1991, Junak et al. SC-2625; just inland from mouth of Smugglers' Canyon, 13 Sep 1991, Junak & Meadows SC-2831.

**Bromus arizonicus** (Shear) Stebbins: Santa Cruz Island, 1888, Brandegee s.n. (UC).

**Bromus berterianus** Colla: burn above Pelican Harbor, 22 Apr 1932, Hoffmann s.n.

**Bromus carinatus** Hook. & Arn. var. **carinatus**: w end of China Harbor, 16 Apr 1983, Junak & Turner SC-674; Cañada de la Portezuela, 1.8 mi w of top of Portezuela Grade, 18 Apr 1989, Junak et al. SC-1303; s side of Alameda de los Coches Prietos, 17 Apr 1989, Junak et al. SC-1267.

**Bromus carinatus** var. **maritimus** (Piper) C.L.Hitchc.: Santa Cruz Island, Aug 1893, Yates s.n.; sea cliff, Cochies Prietos, 1 Jul 1930, Hoffmann s.n.

**Bromus catharticus** Vahl: Christy Ranch, 4 Apr 1987, Thomas s.n.
**Bromus diandrus** Roth: Santa Cruz Island, 14 May 1929, Hoffmann s.n.; Prisoners' Harbor, 12 Aug 1939, Dunkle 8564 (RSA-POM,SBBG); hill above Cochie Canyon, 12 Apr 1931, Hoffmann s.n.
**Bromus hordeaceus** L.: Prisoners' Harbor, 12 Aug 1939, Dunkle 8566 (RSA-POM,SBBG); Main Ranch, 15 Jun 1930, Hoffmann s.n.; hill on w side of Coches Prietos Anchorage, 1 May 1960, Blakley 3385.
**Bromus laevipes** Shear: pines above Christy's, 29 Jun 1930, Hoffmann s.n.
**Bromus madritensis** L. subsp. **madritensis**: near third small cove w of Pelican Bay, 20 Apr 1989, Junak et al. SC-1365; 0.3 mi e of Buena Vista, 10 May 1990, Junak SC-1786; ridge se of Stanton Ranch, 8 May 1963, Smith 8162.
**Bromus madritensis** subsp. **rubens** (L.) Husn.: China Harbor, 10 Mar 1928, Hoffmann s.n.; Christy's Harbor, 22 Mar 1932, Hoffmann s.n.; upper middle drainage of Smugglers' Canyon, 25 Mar 1971, Philbrick & Ricker B71-32.
**Bromus stamineus** Desv.: Stanton Ranch, 28 Apr 1960, Raven & Smith 15290 (RSA-POM,SBBG); Stanton Ranch, 22 Apr 1990, Junak SC-1626.
**Calamagrostis rubescens** Buckley: pine forest on ridge between n and s forks of Canyon Cervada, 22 Jul 1960, Blakley & Muller 3726 (CAS-DS,SBBG).
**Cortaderia selloana** (Schult.) Asch. & Graebn.: Main Ranch, 26 Mar 1932, Hoffmann s.n.; n of road to Fraser Point, in w fork of canyon ca. 0.5 mi e of Rancho Nuevo ruins, 11 Oct 1991, Junak SC-2942; small canyon on e side of Laguna Canyon, ca. 1 mi n of Peak 1031, Jan 1983, Laughrin s.n.
**Cynodon dactylon** (L.) Pers.: Pelican Bay, 14 Jun 1930, Hoffmann s.n.; Stanton Ranch, 19 Jun 1964, Philbrick B64-200; road to Coches Prietos beach, 0.8 mi s of intersection with South Ridge Road, 12 Apr 1990, Junak et al. SC-1538.
**Distichlis spicata** (L.) Greene: Prisoners' Harbor, 15-18 May 1962, Piehl 62-258; Cañada Cebada, 0.2 mi e of Christy Barn, 12 Sep 1989, Junak et al. SC-1402; Coches Prietos beach, 17 Apr 1989, Young et al. 1908.
**Elymus glaucus** Buckley subsp. **glaucus**: small e-w oriented canyon which drains into w side of Potato Harbor, 15 Jun 1981, Junak SC-433; upper Cañada Christy, between Campo Raton and Embudo Trough, 10 May 1990, Junak SC-1807; Smugglers' Cove, 14 May 1927, Hoffmann s.n.
**Festuca arundinacea** Schreb.: University of California Field Station, 21 Apr 1989, Junak SC-1389.
**Gastridium ventricosum** (Gouan) Schinz & Thell.: rocky ridge near Pelican, 14 Jun 1930, Hoffmann s.n.; Cañada de la Portezuela, ca. 0.2 mi w of Rancho Portezuelo ruins, 8 Oct 1991, Junak SC-2899; upper e fork of Horqueta Canyon, 3 Jun 1993, Junak SC-3749.
**Hainardia cylindrica** (Willd.) Greuter: head of Wagon Gulch, Christi Beach, 13 May 1967, Benedict s.n.; cove ca. 5.5 mi w of Christi Ranch, 12 Jun 1965, Philbrick & Benedict B65-1344; lower Cañada de los Sauces, ca. 0.3 mi inland from beach, 5 Aug 1994, Junak SC-4054.
**Hordeum brachyantherum** Nevski subsp. **californicum** (Covas & Stebbins) Bothmer, N.Jacobsen, & Seberg: bluffs just w of Santa Cruz Island West Latitude Station "717", 13 Apr 1985, Junak SC-1038; University of California Field Station, 11 May 1990, Junak SC-1833; mouth of Cañada de los Sauces, 8 May 1990, Junak SC-1688.
**Hordeum intercedens** Nevski: field near junction of Cañada de Christi and Cañada Cebada, ca. 1 mi e of Christi Ranch, 22 Apr 1965, Philbrick et al. B65-821; Fraser Point, 29 Apr 1960, Raven & Smith 15305 (RSA-POM,SBBG); sw of Ragged Mountain, 16 Apr 1967, Philbrick B67-146.
**Hordeum marinum** Hudson subsp. **gussoneanum** (Parl.) Thell.: head of ridge between two canyons draining w toward Coche Point, sw of Triangulation Point "Coche Point", 8 May 1968, Philbrick et al. B68-277; Central Valley, 0.2 mi w of University of California Field Station, 10 May 1990, Junak SC-1772.
**Hordeum murinum** L. subsp. **glaucum** (Steud.) Tzelev: shore of small lake on ridgetop 5.2 mi e of Stanton Ranch, 25 Apr 1960, Raven & Smith 15146 (RSA-POM,SBBG); upper beach at Christy Ranch, 23 Apr 1965, Smith 8667; Main Ranch, 15 Jun 1930, Hoffmann s.n.
**Hordeum murinum** subsp. **leporinum** (Link) Arcang.: Cañada del Puerto, 0.25 mi n of Stanton Ranch, 14 Mar 1961, Blakley 3994; Alberts Anchorage, 6 Mar 1967, Benedict s.n.
**Hordeum vulgare** L.: edges of field, just e of Stanton Ranch, 28 Apr 1960, Raven & Smith 15287 (RSA-POM,SBBG).
**Koeleria macrantha** (Ledeb.) Schult.: hills s of Christy Ranch, on second ridge s of bunkhouse, 18 Apr 1993, Junak SC-3631; s slope of Central Valley, 1 mi e of Stanton Ranch, 17 May 1962, Breedlove 2878 (CAS-DS,SBBG); pines s of lower Sauces Canyon, 15 Apr 1967, Philbrick B67-101.
**Koeleria phleoides** (Vill.) Pers.: ridgetop just above Valdez Canyon, 30 May 1979, Junak et al. SC-110.

**Lamarckia aurea** (L.) Moench: Scorpion Harbor, 17 Mar 1929, Hoffmann s.n.; Central Valley, at w end of Bosque Mano, 9 May 1990, Junak SC-1710; ridge between Coches Prietos and Alberts Anchorage, 30 Apr 1960, Blakley 3343.

**Leymus condensatus** (C.Presl) A.Love: bluffs just w of mouth of Scorpion Canyon, 14 Sep 1991, Junak & Meadows SC-2852; upper Cañada Christy, on upper part of Centinela Grade, 11 Oct 1991, Junak SC-2934; upper portion of Pozo Canyon Road, 27 Jun 1991, Junak SC-2762.

**Leymus triticoides** (Buckley) Pilg.: Prisoners', 14 Jun 1930, Hoffmann s.n.; Main Ranch, 1 Jul 1930, Hoffmann s.n.; lower Cañada de los Sauces, 0.4 mi upstream from intersection with road to Pozo Canyon, 1 Aug 1991, Junak SC-2797.

**Lolium multiflorum** Lam.: shore of small lake on ridgetop 5.2 mi e of Stanton Ranch, 25 Apr 1960, Raven & Smith 15145 (RSA-POM,SBBG); field near junction of Cañada de Christi and Cañada Cebada, ca. 1 mi e of Christi Ranch, 22 Apr 1965, Philbrick et al. B65-818; road from Christi Ranch to lower Sauces Canyon, near crossing of Sauces Creek, 13 Jun 1965, Philbrick & Benedict B65-1389.

**Lolium perenne** L.: Main Ranch, 15 August 1939, Dunkle 8634 (RSA-POM,SBBG).

**Lolium temulentum** L.: Prisoners', 15 Jun 1930, Hoffmann s.n.; stream at Christi Ranch, between bunkhouse and beach, 22 Apr 1965, Philbrick et al. B65-287.

**Melica imperfecta** Trin.: upper w fork of Scorpion Canyon, ca. 1.6 mi inland from Scorpion Anchorage, 23 Apr 1993, Junak SC-3722; rocky buttes at e end of ridge between Central Valley and Cascada Canyon, w of La Cascada, 5 Mar 1992, Junak SC-3037; road to Fraser Point, ca. 0.5 mi e of Rancho Nuevo ruins, 11 Apr 1990, Junak SC-1524.

**Muhlenbergia microsperma** (DC.) Kunth: mouth of Dick's Canyon, ca. 100 ft from beach, 6 Jun 1981, Junak & Daily SC-409; just w of foot of Portezuela Grade, 28 Mar 1989, Junak et al. SC-1237; extreme e end of South Ridge Road, e of Peak 777, 24 Feb 1984, Junak SC-831.

**Nassella cernua** (Stebbins & Love) Barkworth: w fork of Scorpion Canyon, ca. 1.5 mi inland from Scorpion Anchorage, 23 Apr 1993, Junak SC-3713; road to Lagunitas Secas, at summit of first ridge s of North Ridge, 19 Apr 1989, Junak et al. SC-1326; mouth of Laguna Canyon, 17 Apr 1993, Junak SC-3618.

**Nassella lepida** (Hitchc.) Barkworth: road to Navy Compound, ca. 1 mi e of Prisoners' Harbor, 20 Apr 1965, Philbrick et al. B65-711; road to Lagunitas Secas, at summit of first ridge s of North Ridge, 19 Apr 1989, Junak et al. SC-1327; near top of Laguna Canyon Road, 0.7 mi s of intersection with South Ridge Road, 17 Apr 1993, Junak SC-3607.

**Nassella pulchra** (Hitchc.) Barkworth: ca. 0.9 mi e of West Point, 13 Apr 1985, Junak SC-1046; road to Valley Peak, 0.3 mi n of intersection with Valley Road, 31 May 1990, Junak SC-1944; Sierra Blanca ridge, on n flanks of Ragged Mountain, 3 Apr 1991, Junak SC-2162.

**Parapholis incurva** (L.) C.E.Hubb.: stream at Christi Ranch, between bunkhouse and beach, 22 Apr 1965, Philbrick et al. B65-825; Fraser Point, 29 Apr 1960, Raven & Smith 15305a (RSA-POM,SBBG).

**Paspalum distichum** L.: Prisoners' Harbor, 26 Jul 1960, Blakley 3784.

**Pennisetum clandestinum** Hochst. ex Chiov.: Prisoners' Harbor, 20 May 1967, Benedict s.n.; canyon just n of Smugglers' Field airstrip, 29 Apr 1971, Philbrick et al. B71-102; e Central Valley, in small gully near Ranch airstrip, 20 Nov 1975, Cox s.n.

**Phalaris aquatica** L.: w of Campo Grande airfield, at head of n fork of Scorpion Canyon, 19 Oct 1991, Junak SC-2976; 0.3 mi w of top of Portezuela Grade, 29 May 1990, Junak SC-1850; South Ridge Road, just w of intersection with road to Alameda de los Coches Prietos, 30 Jul 1982, Junak SC-568.

**Phalaris caroliniana** Walter: road e of Fraser Point peninsula, at intersection with road to West Point ridge, 13 Apr 1985, Junak SC-1028.

**Phalaris minor** Retz.: field near ne coast, half way between Scorpion Rocks and San Pedro Point, 29 Apr 1971, Philbrick et al. B71-93; Valley Anhorage, 24 Mar 1929, Hoffmann s.n.; w end, 23 Jun 1927, Hoffmann s.n.

**Phalaris paradoxa** L.: vernal wet area, ca. 1 mi w of Stanton Ranch, 3 May 1966, Philbrick & Clarke B66-159.

**Piptatherum miliaceum** (L.) Cosson: Prisoners' Harbor, 7 Oct 1991, Junak SC-2886; near mouth of Scorpion Canyon, 14 Sep 1991, Junak SC-2869; Central Valley between Centinela and top of Portezuela Grade, 16 Apr 1967, Philbrick B67-127.

**Poa annua** L.: Santa Cruz Island, 1888, Brandegee s.n. (UC); Scorpion Ranch, 28 Mar 1992, Junak SC-3097; Willows Canyon, 26 Mar 1983, Junak SC-629.

**Poa howellii** Vasey & Scribn.: wooded slope above Christy's, 29 Jun 1930, Hoffmann s.n.; base of pine forest, w end, 12 Apr 1931, Hoffmann s.n.

**Poa secunda** J.Presl subsp. **secunda**: mouth of Dick's Harbor, 10 Apr 1990, Junak SC-1474; just w of The Nature Conservancy buildings at w end of Bosque Mano, 23 Mar 1985, Junak SC-962; lower portion of Laguna Canyon, ca. 0.1 mi inland from beach, 17 Mar 1992, Junak Sc-3060.
**Polypogon interruptus** Kunth: Santa Cruz Island, 28 Mar 1927, Hoffmann s.n.; Prisoners' Harbor, 12 Aug 1939, Dunkle 8551 (RSA-POM,SBBG); Valley Anchorage, 24 Mar 1929, Hoffmann s.n.
**Polypogon monspeliensis** (L.) Desf.: Santa Cruz Island, Aug 1893, Yates s.n.; first major canyon e of Profile Point, 12 Jun 1965, Philbrick & Benedict B65-1364a; Christy's, 28 Jun 1930, Hoffmann s.n.
**Schismus arabicus** Nees: Central Valley, at La Cascada creek crossing, 9 May 1990, Junak SC-1711.
**Schismus barbatus** (L.) Thell.: Cañada del Puerto, Ravenswood, 22 Aug 1977, Philbrick & Cummings B77-40; ridge w of Picacho Diablo, ca. 0.2 mi e of trailhead from Lagunitas Secas Road, 4 Apr 1991, Junak SC-2184; bluffs just w of Coches Prietos beach, 19 Mar 1992, Junak SC-3093.
**Vulpia bromoides** (L.) A.Gray: mouth of Fry's Harbor, 10 Nov 1930, Hoffmann s.n.; s of Pelican Canyon, 22 Apr 1932, Hoffmann 682; Cochies Prietos, 22 Mar 1929, Hoffmann s.n.
**Vulpia microstachys** (Nutt.) Benth. var. **pauciflora** (Beal) Lonard & Gould: North Ridge between Picacho Diablo and Lagunitas Secas, 19 Apr 1989, Junak et al. SC-1339; n of Red Peak, 23 Apr 1979, Thorne et al. 52490 (RSA-POM).
**Vulpia myuros** (L.) C.C.Gmel. var. **hirsuta** (Hack.) Asch. & Graebn.: Santa Cruz Island, 28 Mar 1925, Hoffmann s.n.; Prisoners' Harbor, 25 Apr 1979, Thorne et al. 52613 (RSA-POM); Main Ranch, 15 Aug 1939, Dunkle 8640 (RSA-POM,SBBG).
**Vulpia octoflora** (Walter) Rydb. var. **hirtella** (Piper) Henrard: Pelican Bay, 25 Apr 1930, Abrams & Wiggins 46 (CAS-DS); ridgetop immediately s of bunkhouse at Christy Ranch, just e of buildings, 17 Apr 1993, Junak SC-3596; coastal flats ca. 0.1 mi e of Near Point, 3 Apr 1993, Junak SC-3583.

## POTAMOGETONACEAE

**Potamogeton pectinatus** L.: Prisoners' Harbor, May 1930, Clokey 4854 (RSA-POM,LA); Prisoners', 15 Jun 1930, Hoffmann s.n.
**Ruppia maritima** L.: Prisoners' Harbor, 2 Nov 1969, Benedict s.n.; Rancho Nuevo Canyon, 18 Apr 1993, Junak SC-3641.

## TYPHACEAE

**Typha domingensis** Pers.: canyon s of Cueva Valdez, 13 Sep 1989, Junak et al. SC-1428; Central Valley at La Cascada, 11 Sep 1982, Junak SC-577; lower portion of Cañada de los Sauces, 8 Oct 1991, Junak SC-2918.
**Typha latifolia** L.: Prisoners' Harbor, 30 Aug 1992, Philbrick s.n.

## ZANNICHELLIACEAE

**Zannichellia palustris** L.: streambed, 0.25 mi below Main Ranch, 23 Sep 1932, Wolf 4165 (RSA-POM).

## ZOSTERACEAE

**Phyllospadix scouleri** Hook.: Potato Harbor, 15 Jun 1981, Philbrick s.n.; Forney's Cove, 2 May 1982, Junak & Timbrook SC-532; Coches Prietos, 10 May 1973, Philbrick & Philbrick B73-311.
**Phyllospadix torreyi** S.Watson: Prisoners' Harbor, 12 Aug 1939, Dunkle 8567 (RSA-POM,SBBG); Forney's, 30 Jun 1930, Hoffmann s.n.; Smugglers', 15 Jun 1930, Hoffmann s.n.
**Zostera marina** L.: Christi Beach, 9 May 1973, Philbrick & Philbrick B73-302.
**Zostera pacifica** L.: Chinese Bay, 27 Dec 1928, Johnstone s.n. (RSA-POM); Pozo Beach, 18 Apr 1989, Young et al. 1932.

# APPENDIX III. Doubtful and Excluded Taxa.

The taxa listed here have been reported for Santa Cruz Island but we were either unable to locate herbarium specimens that document their occurrence or have found that specimens cited by other workers were misidentified. Some authors (e.g., Eastwood) apparently did not voucher all published reports with specimens. Reports by some authors are apparently based on unsubstantiated or misidentified specimens in earlier literature. We also list a few taxa that do not appear to be naturalized but have been planted or occur as waifs in areas away from buildings. This list is not intended to be complete, because most unsubstantiated reports for Santa Cruz Island were included in Appendix II of Wallace (1985). Taxa are listed by currently accepted names (Hickman 1993). Names used in the original citation are included if they differ from current ones. Standard abbreviations (Holmgren et al. 1990) are used for herbaria, except for the Santa Cruz Island Reserve, which is abbreviated SCIR.

## Pteridophytes

### PTERIDACEAE

**Cheilanthes newberryi** (D.Eaton) Domin Reported as *Notholaena newberryi* D.Eaton by Eastwood (1941); no specimen found.

**Notholaena californica** D.Eaton Reported as *N. candida* Hook. by Greene (1887); no specimen found.

## Dicotyledonous Angiosperms

### APIACEAE

**Sanicula laciniata** Hook. & Arn. Reported by Greene (1887), Brandegee (1890), and Yates (1890a); no specimen found.

### ASTERACEAE

**Agoseris heterophylla** (Nutt.) Greene Reported by Munz (1935), Eastwood (1941), Dunkle (1950), Foreman (1967), Munz (1974), Smith (1976), and Wallace (1985), who cited a specimen at RSA-POM; no specimen found.

**Ambrosia psilostachya** DC. Reported by Greene (1887); no specimen found.

**Conyza coulteri** A.Gray Reported by Greene (1887), Brandegee (1890), Yates (1890a), Hall (1907), Eastwood (1941), Smith (1952, 1976), Munz (1974), and Wallace (1985), who cited a misidentified specimen; Dunkle 8579 (SBBG, NY) is *Baccharis plummerae* subsp. *plummerae*.

**Filago arizonica** A.Gray Reported by Brandegee (1890); no specimen found.

**Hazardia squarrosa** (Hook. & Arn.) Greene var. **obtusa** (Greene) Jeps. Reported by Clark (1979), who cited a single misidentified specimen in unpublished list of exsiccata examined during the study; Wolf 4150 (RSA-POM) is *H. squarrosa* var. *grindelioides* (DC.) W.D.Clark.

**Hesperevax caulescens** (Benth.) A.Gray Reported as *Evax caulescens* var. *humilis* Jeps. by Eastwood (1941); no specimen found.

**Hesperevax sparsiflora** (A.Gray) Greene var. **brevifolia** (A.Gray) Morefield. Reported as *Evax sparsiflora* var. *brevifolia* (A.Gray) Jeps. by Dunkle (1950); no specimen found.

**Lasthenia glabrata** Lindl. subsp. **coulteri** (A.Gray) Ornduff Reported as *Lasthenia glabrata* Lindl. by Eastwood (1941); no specimen found.

**Layia glandulosa** (Hook.) Hook. & Arn. Reported by Wallace (1985), who cited a specimen at RSA-POM; no specimen found.

**Malacothrix californica** DC. Reported by Munz (1935), Munz and Keck (1963), Eastwood (1941), Munz (1974), and Smith (1976); no specimen found.

**Senecio lyonii** A.Gray Reported by Eastwood (1941), Gentry (1949), and Dunkle (1950); no specimen found.

### BORAGINACEAE

**Amsinckia douglasiana** A.DC. Reported by Clokey (1931); no specimen found.

**Amsinckia lycopsoides** Lehm. Reported by Greene (1887), Brandegee (1890), and Yates (1890a); no specimen found.

**Cryptantha intermedia** (A.Gray) Greene Reported as *Krynitzkia intermedia* A.Gray by Brandegee (1890); no specimen found.
**Cryptantha leiocarpa** (Fisch. & C.A.Mey.) Greene Reported as *Krynitzkia leiocarpa* Fisch. & C.A.Mey. by Greene (1887), Brandegee (1890), and Yates (1890a); no specimen found.

## BRASSICACEAE

**Cakile edentula** (Bigelow) Hook. Reported as *Cakile edentula* var. *californica* (A.Heller) Hulten by Smith (1976); no specimen found.
**Caulanthus inflatus** S.Watson Reported by Gentry (1949), who cited a single specimen of questionable origin, and by Smith (1976); we were not able to examine Elmore 453.
**Lepidium virginicum** L. var. **menziesii** (DC.) C.L.Hitchc. Reported as *Lepidium menziesii* DC. by Greene (1887), Brandegee (1890), and Yates (1890a); no specimen found.

## CANNABACEAE

**Cannabis sativa** L. Known from planted populations near Pelican Bay and Alberts Anchorage; not naturalized.

## CAPPARACEAE

**Isomeris arborea** Nutt. Reported as *Isomeris arborea* var. *globosa* Cov. by Eastwood (1941); no specimen found.

## CARYOPHYLLACEAE

**Arenaria serpyllifolia** L. Reported by Munz (1974) and Wallace (1985), who cited a misidentified specimen; Williams 100 (RSA-POM) is *Stellaria media* (L.) Vill.
**Spergula arvensis** L. Reported by Hoffmann (1932a), who cited a collection by Leroy Abrams; no specimen found.

## CHENOPODIACEAE

**Chenopodium album** L. Reported by Greene (1887), Brandegee (1890), Yates (1890a), Eastwood (1941), and Wallace (1985), who cited two misidentified specimens; Clokey 4917 and Howell 6237 are *C. berlandieri* Moq.
**Suaeda moquinii** (Torr.) Greene Reported as *S. torreyana* S.Watson by Greene (1887), Brandegee (1890), Yates (1890a), Eastwood (1941), and Dunkle (1950); no specimen found.

## CONVOLVULACEAE

**Dichondra donnelliana** Tharp & M.C.Johnst. Reported by Tharp and Johnston (1961), who cited a single misidentified specimen, and Smith (1976); Wolf 2841 (RSA-POM) is *D. occidentalis* House.

## CUSCUTACEAE

**Cuscuta californica** Hook. & Arn. Reported by Smith (1976); no specimen found.

## EUPHORBIACEAE

**Ricinus communis** L. Reported by Greene (1887), Brandegee (1890), and Yates (1890a); no specimen found.

## FABACEAE

**Lotus argophyllus** (A.Gray) Greene var. **ornithopus** (Greene) Ottley Reported by Dunkle (1950) and Foreman (1967); no specimen found. Also reported as *Syrmatium ornithopum* (Greene) Greene by Eastwood (1941).
**Lotus humistratus** Greene Reported by Wallace (1985), who cited a single misidentified specimen; Pierson 11084 (RSA-POM) is *L. wrangelianus* Fisch. & C.A.Mey.
**Lupinus arboreus** Sims Reported by Clokey (1931), Eastwood (1941), and Smith (1976). A yellow-flowered specimen, collected at Pelican Bay in 1930 (UC), was examined by R. Philbrick (SBBG records) but we were unable to find it.
**Lupinus nanus** Benth. Reported by Greene (1887), Brandegee (1890), Yates (1890), and Eastwood (1941a); no specimen found.
**Trifolium microdon** Hook. & Arn. Reported by Greene (1887), Brandegee (1890), and Dunkle (1950); no specimen found. Also reported as *T. m.* var. *pilosum* Eastw. by Hoffmann (1932b), Munz (1935), Eastwood (1941), Smith (1952), Munz and Keck (1963), Foreman (1967), Munz (1974), and Smith (1976); Hoffmann 211 and Hoffmann 673 are *T. microcephalum* Pursh.

## GENTIANACEAE

**Centaurium exaltatum** (Griseb.) Piper Reported by Eastwood (1941) and Wallace (1985), who cited a misidentified specimen; Greene s.n. (ND-G 38921) is *C. davyi* (Jeps.) Abrams. Also reported as *Erythraea douglasii* A.Gray by Greene (1887), Brandegee (1888, 1890), and Yates (1890a); no specimen found.
**Centaurium muhlenbergii** (Griseb.) Piper Reported by Jepson (1909-1943), who cited a single specimen, Eastwood (1941), and Smith (1976); we were unable to locate Jepson 12089.
**Centaurium venustum** (A.Gray) B.L.Rob. Reported by Clokey (1931), Gentry (1949), Smith (1952), and Smith (1976); no specimen found.

## HYDROPHYLLACEAE

**Phacelia grandiflora** (Benth.) A.Gray Reported by Hoffmann (1932b), Eastwood (1941), and Smith (1976); all specimens from Santa Cruz Island labeled *P. grandiflora* have been determined as *P. viscida* (Benth.) Torr. Gillett (1955) discussed the taxonomic confusion between these two species.
**Phacelia ramosissima** Lehm. var. **latifolia** (Torr.) Cronquist Reported as *P. suffrutescens* Parry by Greene (1887), Yates (1890a), and Eastwood (1941); no specimen found.
**Phacelia tanacetifolia** Benth. Reported by Munz (1974) and Smith (1976); no specimen found.

## LAMIACEAE

**Salvia apiana** Jeps. Reported by Eastwood (1941); no specimen found.
**Stachys ajugoides** Benth. var. **ajugoides** Reported as *S. ajugoides* Benth. by Jepson (1909-1943); no specimen found.

## OLEACEAE

**Fraxinus** sp. Known from a single collection in Cañada del Puerto; not naturalized.

## OXALIDACEAE

**Oxalis albicans** Kunth subsp. **pilosa** (Nutt.) G.Eiten Reported by Smith (1976) and Wallace (1985), who cited a specimen at SBM (now at SBBG); no specimen found.

## POLEMONIACEAE

**Allophyllum gilioides** (Benth.) A.D.Grant & V.E.Grant Reported as *Gilia gilioides* Greene by Dunkle (1950); no specimen found.
**Gilia achilleifolia** Benth. subsp. **achilleifolia** Reported by Munz (1974) and Smith (1976); no specimen found.
**Linanthus bicolor** (Nutt.) Greene Reported as *Gilia bicolor* (Nutt.) Brand by Eastwood (1941); no specimen found.

## POLYGONACEAE

**Rumex maritimus** L. Reported by Greene (1887), Brandegee (1890), Yates (1890a), and Eastwood (1941); no specimen found. Also reported as *R. fueginus* Phil. by Smith (1976).

## ROSACEAE

**Prunus** sp. Known from several scattered locations as escapes from cultivation; apparently not naturalized.

## RUTACEAE

**Ruta chalepensis** L. Reported by Wallace (1985), who cited a single specimen; Daily 623 (SCIR) was collected from plants cultivated at Stanton Ranch; they are not naturalized.

## SALICACEAE

**Salix gooddingii** C.R.Ball Reported by Eastwood (1941); no specimen found.

## SAXIFRAGACEAE

**Heuchera micrantha** Lindl. Reported by Clokey (1931) and Eastwood (1941), apparently on the basis of a misidentified specimen; Clokey 4952 is *H. maxima* Greene.
**Heuchera pilosissima** Fisch. & C.A.Mey. Reported by Brandegee (1888, 1890); no specimen found.

### SCROPHULARIACEAE

**Castilleja foliolosa** Hook. & Arn. Reported by Eastwood (1941) and Dunkle (1950); no specimen found.
**Mimulus brevipes** Benth. Reported by Munz (1974) and Smith (1976); no specimen found.
**Mimulus floribundus** Lindl. Reported by Greene (1887), Brandegee (1890), Yates (1890a), Eastwood (1941), and Smith (1976); no specimen found.

## Monocotyledonous Angiosperms

### AGAVACEAE

**Yucca** sp. Known from small planted population at Fry's Harbor; not naturalized.

### ALLIACEAE

**Allium amplectens** Torr. Reported by Eastwood (1941); no specimen found.
**Allium hyalinum** Curran Reported by Brandegee (1888, 1890) and Yates (1890a); no specimen found.
**Ipheion uniflorum** Raf. Known from single population in Islay Canyon; not naturalized.

### CYPERACEAE

**Carex abrupta** Mack. Reported by Clokey (1931), Eastwood (1941), Munz (1974), and Smith (1976); no specimen found.
**Carex angustata** Boott Reported by Greene (1887), Brandegee (1890), Yates (1890a), and Eastwood (1941); no specimen found.
**Carex douglasii** Boott Reported by Eastwood (1941); no specimen found.

### IRIDACEAE

**Iris** sp. Known from plants growing in debris pile along Islay Canyon Road; not naturalized.

### JUNCACEAE

**Juncus balticus** Willd. Reported by Greene (1887), Brandegee (1890), Yates (1890a), Eastwood (1941), Smith (1952), Smith (1976), and Wallace (1985), who cited a single specimen at SBM (now at SBBG); no specimen found.
**Juncus effusus** L. Reported by Greene (1887), Brandegee (1890), Yates (1890a), and Eastwood (1941); no specimen found.
**Juncus effusus** L. var. **brunneus** Engelm. Reported by Wallace (1985), who cited a specimen at SBBG; no specimen found.
**Juncus effusus** L. var. **pacificus** Fernald & Wiegand Reported by Wallace (1985), who cited a specimen at RSA-POM; no specimen found.

### LILIACEAE

**Calochortus venustus** Benth. Reported by Brandegee (1888) and Yates (1890a); no specimen found.

### ORCHIDACEAE

**Piperia unalascensis** (Spreng.) Rydb. Reported as *Habenaria unalascensis* (Spreng.) S.Watson by Wallace (1985), who cited a specimen at RSA-POM; no specimen found.

### POACEAE

**Bromus ciliatus** L. Reported by Brandegee (1888, 1890) and Yates (1890a); no specimen found.
**Bromus vulgaris** (Hook.) Shear Reported by Jepson (1909-1943), who cited a Brandegee collection, Munz (1935), Eastwood (1941), Dunkle (1950), Munz (1968), and Smith (1976); no specimen found.
**Dissanthelium californicum** (Nutt.) Hook. Reported by Eastwood (1941); no specimen found.
**Elytrigia repens** (L.) Nevski Reported as *Agropyron repens* (L.) Beauv. by Greene (1887); no specimen found.
**Hordeum depressum** (Scribn. & Small) Rydb. Reported by Wallace (1985), who cited a single specimen; we were not able to examine Abrams and Wiggins 125.
**Monanthochloe littoralis** Engelm. Reported by Wallace (1985), who cited a misidentified specimen; Hoffmann s. n. (SBM 5510) is *Distichlis spicata* (L.) Greene.
**Phalaris canariensis** L. Reported by Greene (1887) and Brandegee (1890); no specimen found.

## TYPHACEAE

**Typha angustifolia** L. Reported by Eastwood (1941), Dunkle (1950), Smith (1952), and Foreman (1967); no specimen found.

# GLOSSARY OF TERMS

Compiled by Julie Broughton

**abundant**. Distribution on Santa Cruz Island for which a taxon is exceptionally widespread and known from more than 30 localities.
**acaulescent**. Apparently without a stem; leaves and inflorescences usually basal; sometimes with a subterranean stem or a stem protruding only slightly aboveground.
**accrescent**. Growing and enlarging with age.
**achene**. A dry, indehiscent fruit with 1 locule and 1 seed.
**acuminate**. Tapered (often abruptly) to a point, the margins concave.
**acute**. Tapered to an often sharp point; margins straight or somewhat convex but converging at less than a right angle.
**adnate**. Fusion of unlike parts, such as stamens fused to the corolla.
**alternate**. Either referring to leaf arrangement in which each node bears 1 leaf or to arrangement of flower parts, such as stamens located between petals rather than in front of petals.
**androgynous**. Spikelet in Cyperaceae, with pistillate flowers borne below the staminate flowers (see gynacandrous).
**annual**. Completing a life cycle (from germination to reproduction and death) in one year or growing season.
**anther**. Pollen-producing part of a stamen.
**anthesis**. Flowering period; the time during which a flower is reproductively functional.
**apex**. The tip or distal end of a structure.
**apiculate**. Terminated abruptly with a minute point.
**appressed**. Pressed against or closely applied to the surface of a structure.
**aristate**. Tipped with an awn or bristle.
**attenuate**. Tapered gradually to a point, the margins _ straight.
**auriculate**. Usually referring to leaf bases with ear-shaped appendages.
**ascending**. Curving or angling upward from base.
**awn**. A bristle-like appendage, either as an appendage of a larger structure, as in the glumes and lemmas of some Poaceae, or as a separate part, as in the pappus of some Asteraceae.
**axil**. Angle formed at a node by a leaf or branch of the stem (adjective: axillary).
**axile**. Usually referring to placentation of an ovary with 2 or more locules, in which the placentas occur at the center of the ovary and at the junction of the internal septa.
**barbed**. With sharp, downward- or backward-pointing projections; usually used to describe trichomes or similar structures.
**beak**. With a prolonged, usually narrow, and thick apex.
**biennial**. Completing a life cycle (from germination to reproduction and death) in two years or growing seasons.
**bilabiate**. Bilateral flower with 2 (upper and lower) lips.
**bilateral**. A form of symmetry in which the structure, usually a flower, is divisible into 2 halves (mirror images).
**bisexual**. Flower with both fertile stamens and fertile pistils.
**blade**. The flattened, expanded part of a leaf or petal.
**bract**. A reduced, often modified, leaf-like structure. (adjective: bracteate).
**caducous**. Falling off very early or prematurely, as in some sepals or petals.
**callus**. A thick, sharp to blunt extension of the lemma base in Poaceae.
**calyx**. Collective term for the outer series of the perianth or sepals, which are usually green and enclose the flower in bud.
**campanulate**. Bell-shaped.
**canescent**. Covered with fine, usually grayish to white trichomes.
**capitate**. Head-shaped or subglobose.
**capsule**. A dry fruit developing from a compound ovary (of one or more carpels), dehiscent from the apex by one or more lines of dehiscence.
**carinate**. A longitudinal, often sharp ridge, like the keel of a boat.
**caryopsis**. The fruit of Poaceae, characterized by the fusion of the seed coat to the ovary wall.
**catkin**. A spike composed of unisexual flowers with inconspicuous perianths, sometimes pendent, and often with conspicuous bracts.
**caudex**. Woody base of a herbaceous perennial.
**caulescent**. With an obvious, leafy stem.
**cauline**. Borne on a stem; not basal.
**ciliate**. Leaves, petals, or sepals with marginal trichomes (diminutive: ciliolate).
**circumscissile**. Dehiscing by means of a transverse line which circles the fruit or anther.
**cleistogamous**. Inconspicuous flowers, usually remaining closed and self-pollinating.
**coma**. A tuft of hairs, usually borne on a seed (adjective: comose).
**common**. Distribution on Santa Cruz Island for which a taxon is widespread and conspicuous within one or more habitats and is known from 21-30 localities.
**conic**. Shaped like a cone, with the widest end at the base.

**connivent**. Structures that converge but which are not fused together.
**cordate**. Heart-shaped.
**coriaceous**. With a leathery texture.
**corolla**. Collective term for the inner series of the perianth or petals, which are usually colorful.
**corona**. A crown-like or ring-like structure composed of free to fused appendages located between the corolla and stamens or inserted on the corolla.
**corymb**. A flat-topped or rounded inflorescence in which the pedicels are of varying length (adjective: corymbose).
**crenate**. Margins with shallow, rounded teeth. (diminutive: crenulate).
**cuneate**. Wedge-shaped or triangular, with narrow end at point of attachment.
**cyathium**. Flower-like inflorescence in Euphorbiaceae, composed of 1 stalked, naked pistillate flower surrounded by few to many, naked staminate flowers, together subtended by an involucre of bracts.
**cyme**. A branched inflorescence in which the terminal flower(s) open(s) before the lateral ones on any axis (adjective: cymose).
**deciduous**. Falling at the end of one season of growth or life; not evergreen.
**decumbent**. Stems whose bases lie or rest on the ground, but whose distal portions ascend or curve upward.
**decussate**. Usually applied to leaf arrangement in which pairs of opposite leaves alternate at right angles with those above and below.
**deflexed**. Abruptly bent downward.
**dehiscent**. A fruit or anther that opens by means of sutures, lids, pores, or teeth.
**deltate**. Usually equilaterally triangular.
**dentate**. Margins bearing teeth (often sharp) that project at right angles to the leaf axis (diminutive: denticulate).
**dichotomous**. Branching repeatedly in pairs or forks, as in some stems or leaf venation.
**didynamous**. With 4 stamens, 2 of which are shorter than the other pair.
**dioecious**. Plants unisexual, the staminate and pistillate flowers on separate plants of the same species.
**disjunct**. A geographical distribution in which a significantly wide gap separates the distribution of plants within a species, as in some California taxa that also occur in Chile.
**dorsal**. The back or outer side (facing away from the axis) of a structure, such as leaves, bracts, petals, or fruits (opposite of ventral).
**drupe**. A fleshy or pulpy, indehiscent, fruit with one seed enclosed by a hard, often stone-like wall.
**ellipsoid**. A three-dimensional structure with the shape of an ellipse.
**elliptic**. A two-dimensional outline, shaped like an ellipse, with the length about twice as long as the width.
**equitant**. Leaves folded along midrib, arranged in 2 overlapping rows along the stem.
**erose**. Irregularly and minutely toothed.
**evergreen**. Remaining green during the dormant season; usually applied to plants that retain their leaves throughout the year.
**exserted**. Extending beyond or out of, as with stamens that protrude beyond the throat of the corolla.
**farinose**. Covered with a meal-like powder.
**fascicle**. A dense cluster of flowers or leaves.
**filiform**. Thread-like.
**fimbriate**. Fringed.
**flexuous**. Having a more or less zigzag form; often applied to stems or stem-like structures.
**foliaceous**. Leaf-like; applied to sepals or bracts that resemble leaves in texture or appearance.
**follicle**. Dry, many-seeded fruit dehiscent along one side; usually derived from a simple pistil.
**foveolate**. Pitted; often applied to fruits or seeds with uniformly pitted surfaces.
**funnelform**. Funnel-shaped, usually referring to a corolla, with the tube gradually widening toward the throat.
**galea**. The hood-like, upper lip of a bilabiate corolla (adjective: galeate)
**glabrate**. Becoming glabrous with maturity or almost glabrous.
**glabrous**. Surfaces without hairs.
**glandular**. With gland-like trichomes or secretions.
**glaucous**. Covered with whitish to grayish, waxy or powdery film that is often rubbed off easily.
**globose**. A three-dimensional structure that is spherical or round in outline.
**gynacandrous**. Spikelet in Cyperaceae, with staminate flowers borne below the pistillate flowers (see androgynous).
**hastate**. With the shape of an arrowhead, but with the basal lobes spreading almost at right angles.
**heterostylous**. Having flowers with different kinds of style (and stamen) lengths.
**hirsute**. Covered with coarse, stiff trichomes (diminutive: hirsutulous).
**hispid**. Covered with stiff or bristly trichomes (diminutive: hispidulous).
**hyaline**. Translucent to almost transparent.
**hypanthium**. Structure derived from the fusion of sepals, petals, and stamens, often forming a tube.
**imbricate**. Overlapping, as in the scales of a cone or the bracts of some involucres.
**incurved**. Bending inwards.
**indehiscent**. A fruit or anther that does not open by means of sutures, lids, pores, or teeth.

**indusium**. In ferns, the leaf appendage that covers or encloses the sorus or sori.
**inflorescence**. The arrangement of flowers on the rachis or in a cluster.
**intergrade**. To merge gradually from one form to another through a more or less continuous series of intermediates.
**internode**. The portion of the stem between two adjacent nodes.
**involucre**. One or more whorls of bracts, usually subtending 1 or more flowers.
**involute**. Margins that are rolled toward the upper surface.
**keeled**. With a ridge or crease, usually along the dorsal side of a folded structure, as in the glumes or lemmas of certain Poaceae.
**laciniate**. Margins that are deeply and often irregularly cut.
**lanate**. Covered with loosely entangled, long, woolly trichomes.
**lanceolate**. Shaped like a lance; narrow, tapered at both ends and widest at a point between the middle and the petiole.
**lenticular**. Shaped like a lens or discus, with both sides convex.
**ligule**. In Asteraceae, the strap-shaped part of a ray corolla; in Poaceae, the appendage located at the junction of the blade and the sheath.
**limb**. The expanded, often flat, apical portion of a structure (e.g., corolla).
**linear**. Shape that is long, narrow, and with a uniform width.
**lobe**. A segment or division of a structure, usually with sinuses that extend up to one-half the distance to the midrib or center of a leaf blade, petal, or ovary.
**locule**. The chamber(s) within an ovary, anther or fruit.
**margin**. The edge of a leaf blade or perianth part.
**membranous**. With the consistency of a membrane; usually thin and soft.
**megasporangium**. Organ in which the larger of two spores (megaspore) is produced in some vascular plants (e.g., Selaginellaceae).
**mericarp**. One segment of a schizocarp (indehiscent fruit) that falls from the fruit when the fruit splits lengthwise along the septum; the seeds remain enclosed in the original locule.
**microsporangium**. Organ in which the smaller of two spores (microspore) is produced in some vascular plants (e.g., Selaginellaceae).
**monoecious**. Plants bisexual, but with both kinds of unisexual flowers (staminate and pistillate) on the same plant.
**mucronate**. With a sharp terminal point or spiny tip, as at the apex of a leaf or bract.
**naturalized**. Populations of alien (non-native) plant taxa that reproduce (usually by seed) from one generation to the next without human intervention.
**nectary**. Gland-like structure that secretes nectar.
**node**. The joint of a stem at which leaves are attached and axillary buds are produced.
**nutlet**. Small, dry, indehiscent fruit with hard walls.
**ob-**. Prefix used to describe a shape with attachment at other end (e.g., obconic: cone-shaped, but with point of attachment at narrow end; oblanceolate: a lanceolate-shaped structure, but with point of attachment at narrowest end; obovate: egg-shaped, but with point of attachment at narrow end).
**oblong**. Shape that is longer than broad, but with parallel sides.
**occasional**. Distribution on Santa Cruz Island for which a taxon is fairly widespread within one or more habitats but with a scattered distribution and known from 11-20 localities.
**opposite**. Either referring to leaf arrangement in which each node bears 2 leaves or to arrangement of flower parts, such as stamens located in front of petals.
**ovary**. The part of the pistil that encloses the ovules and that develops into a fruit after pollination.
**ovate**. Outline in the shape of an egg, with attachment at widest end.
**ovoid**. A three dimensional structure with the shape of an egg.
**ovule**. Structure inside ovary that develops into a seed.
**palea**. A chaff-like or scale-like bract. In Asteraceae, a scale-like pappus; in Poaceae, the upper and smaller of two bracts subtending a flower.
**paleaceous**; chaff- or scale-like.
**palmate**. Radiating from one common point, like the fingers of an open hand; usually used to describe the venation or lobing of a leaf or other structure.
**panicle**. A compound inflorescence in which the branches are racemose (adjective: paniculate).
**papilionaceous**. A bilateral corolla in the Fabaceae, composed of a banner, 2 wings, and 2 additional petals usually fused into a keel.
**papilla**. Swollen or rounded projections.
**papillate**. Covered with papillae (diminutive: papillose).
**pappus**. The highly modified calyx of Asteraceae, composed of awns, bristles, capillary (hair-like) segments, or scales.
**pedicel**. The stalk that subtends a single flower (adjective: pedicellate).
**peduncle**. The stalk that subtends an inflorescence (adjective: pedunculate).
**peltate**. A flat structure subtended by a stalk that is attached to its lower surface instead of the margin.

**perennial**. Living for more that two years or growing seasons and flowering and fruiting repeatedly throughout the life of the plant.
**perianth**. Collective term for the calyx and corolla.
**petal**. One segment or lobe of the corolla.
**petaloid**. Petal-like; usually used to describe a structure (often a sepal) that resembles a petal in color, shape, or position.
**petiole**. A stalk that subtends a leaf blade and connects it to the stem.
**pilose**. Covered with soft, straight trichomes.
**pinnate**. Radiating in 2 rows from a linear axis, like the parts of a feather; used to describe the veins or lobes of a leaf or to describe the arrangement of leaflets in a pinnately compound leaf.
**pinnatifid**. A deeply lobed or dissected leaf with pinnate venation.
**pistil**. Female reproductive structure of a flower, composed of an ovary, style, and stigma.
**pistillate**. A unisexual flower bearing only pistils.
**placentation**. The arrangement or orientation of the placentas within the ovary, and to which the ovules are attached.
**plumose**. Plume-like; usually used to describe a structure with lateral, hair-like processes or appendages, such as the pappus of Asteraceae.
**pollinium**. The pollen mass derived from anthers in either the Asclepiadaceae or Orchidaceae.
**polygamous**. With both unisexual and bisexual flowers on the same plant.
**prostrate**. Usually used to describe stems that are lying flat on the ground.
**puberulent**. The diminutive form for pubescent, in which trichomes are scarcely visible to the unaided eye.
**pubescent**. Usually used to describe a surface with short, soft trichomes.
**punctate**. A surface characterized by translucent to opaque, often glandular dots, pits, or depressions.
**pustulate**. Trichomes with swollen bases, like a blister (diminutive: pustulose).
**raceme**. A simple inflorescence (without branches) composed of a single rachis bearing pedicellate flowers (adjective: racemose).
**rachis**. The central axis of an inflorescence or of a pinnately compound leaf.
**radial**. A form of floral symmetry in which the perianth parts radiate from the center, like the spokes of a wheel.
**rare**. Distribution on Santa Cruz Island for which a taxon is known from only one to three, often very restricted localities.
**receptacle**. The part of the flower on which the calyx, corolla, stamens, and pistils are attached; in Asteraceae the flat surface of the head, on which flowers are attached.
**recurved**. Curved gradually, either downward or backward.
**reflexed**. Bent downward.
**reniform**. Kidney-shaped.
**reticulate**. Net-like; with a network of veins or minute ridges.
**retrorse**. Usually used to describe trichomes that are bent abruptly downward or backward.
**revolute**. Margins that are rolled toward the lower surface.
**rhizome**. A horizontal, underground stem, with scale-like leaves and often producing leafy shoots (adjective: rhizomatous).
**rhombic**. Outline in the shape of a diamond, angular, often with 4 straight edges, and with the widest point midway between the base and the apex.
**rosette**. An arrangement of leaves in which leaves are crowded and basal.
**rotate**. A corolla or calyx with a very short tube and an abruptly spreading, flat, often circular limb.
**rugose**. A surface with a wrinkled appearance (diminutive: rugulose).
**saccate**. Sack-like or bag-like.
**sagittate**. Usually shaped like an arrow-head, with the basal lobes pointed downward or backward.
**salverform**. A corolla with a long, slender tube and an abruptly spreading, flat limb.
**samara**. An indehiscent fruit with wing-like appendages, as in Aceraceae.
**scaberulous**. Diminutive form of scabrous.
**scabrous**. Surface that feels rough to the touch, like sand-paper; covered with short, stiff trichomes.
**scape**. An inflorescence and peduncle usually arising directly from a basal rosette of leaves (adjective: scapose).
**scarce**. Distribution on Santa Cruz Island for which a taxon is locally common but known only from 4-10 localities.
**scarious**. Usually used to describe leaf or bract margins that are thin, dry, tan to brownish, and often translucent.
**schizocarp**. A dry indehiscent fruit that splits at maturity into 1-few-seeded, closed segments (mericarps).
**scurfy**. A surface covered with minute, flat scale-like trichomes.
**sepaloid**. A sepal-like structure (often a bract) that resembles a sepal in color, shape, or position.
**septum**. Wall that separates locules in an ovary or fruit.
**sericeous**. Usually used to describe long, slender, very soft (silky) trichomes that are often appressed.
**serrate**. Margins bearing teeth (often sharp) that project forward or toward the leaf apex (diminutive: serrulate).
**sepal**. One segment or lobe of the calyx.
**sessile**. Attached directly by the base, such as leaves without petioles and flowers without pedicels.

**setaceous**. With a bristle-like structure or bristle-like apex.
**setose**. A surface covered with bristle-like trichomes.
**sinuate**. A margin that appears wavy or undulate, with waves that are parallel to the plane of the blade.
**sinus**. The indentation or cleft between the lobes of a margin.
**sorus**. The part of a fern leaf or leaflet that bears sporangia (plural: sori).
**spatulate**. Shaped like a spatula, with a rounded apex and tapered gradually to the base.
**spike**. A simple inflorescence (without branches) composed of a single rachis bearing sessile flowers (adjective: spikate).
**spine**. Stiff, sharp-pointed structure; usually arising from the epidermis (adjective: spinose; diminutive: spinulose).
**spinescent**. Spine-like
**sporangium**. The sac or case-like structure enclosing spores in plants not producing seeds.
**spore**. A simple, reproductive, single-celled structure, usually haploid and capable of developing into a new individual.
**sporophyll**. A leaf that produces sporangia and spores; usually one of several that comprise the cone in gymnosperms or the cone-like structures in some fern allies.
**spur**. A slender, tubular or sac-like projection from either a petal or sepal and usually containing a nectar gland.
**squarrose**. With tips spreading rigidly at right angles or reflexed; usually used to describe phyllaries in the Asteraceae.
**stamen**. Male reproductive structure of a flower, usually composed of a stalk-like filament and a terminal, pollen-producing anther.
**staminate**. A unisexual flower bearing only stamens.
**staminode**. A sterile stamen, often highly modified in appearance, and sometimes petal-like
**stellate**. Star-shaped; often referring to a trichome with three or more branches radiating from a common point.
**stigma**. The part of the pistil on which pollen is normally deposited and germinates (adjective: stigmatic).
**stipe**. Usually used to describe the stalk subtending an ovary in the genus *Astragalus* and the family Euphorbiaceae (adjective: stipitate).
**stipules**. A pair of appendages at the base of the petiole, variable in form but often leaf-like or bract-like (e.g., Rubiaceae).
**stolon**. A horizontal, aerial stem, often with long internodes producing leafy shoots (adjective: stoloniferous).
**strigose**. A surface covered with stiff, appressed, straight hairs (diminutive: strigulose).
**style**. The narrow segment of a pistil that separates the ovary from the stigma.
**sub-**. A prefix, usually meaning "almost", "slightly", or "somewhat".
**subulate**. Triangular in shape and sharp-pointed, like an awl.
**subtend**. Located below or beneath a structure, often in close proximity.
**succulent**. Fleshy (with a high water content) and often thickened.
**taproot**. The primary root of an annual or biennial plant, usually with lateral secondary roots.
**tendril**. A slender, coiled structure by which a climbing plant becomes attached to its support and derived either from leaves, leaflets, stipules, or stems.
**terete**. Cylindrical and usually round in cross section.
**terminal**. At the apex or top of a structure.
**ternate**. Usually used to describe a compound leaf divisible into 3 leaflets.
**tomentose**. Covered with densely interwoven, usually matted trichomes (diminutive: tomentulose).
**trichome**. A hair-like, epidermal structure, composed of 1 or more cells.
**trifoliate**. With 3 leaves.
**trifoliolate**. A compound leaf with 3 leaflets.
**truncate**. Usually used to describe an apex or base that appears cut off at right angles to the axis.
**tube**. A hollow structure, such as the fused portion of a corolla.
**tuberculate**. Covered with tubercles, which are swollen, wart-like or tuber-like projections on a surface.
**turbinate**. The opposite of conic or obconic; top-shaped.
**umbel**. A simple inflorescence with pedicels of equal length and arising from a common point (adjective: umbellate).
**undulate**. A margin that appears wavy with waves perpendicular to the plane of the blade.
**unisexual**. Flower with either fertile stamens or fertile pistils, but not both.
**urceolate**. Urn-shaped; a corolla that is wide at the bottom and contracted at the throat; characteristic of some species of Ericaceae.
**utricle**. An indehiscent, bladder-like fruit, usually with 1 seed.
**vascular**. With conductive tissue (e.g., xylem and phloem).
**ventral**. The front or inner side (facing the axis) of a structure, such as leaves, bracts, petals, or fruits (opposite of dorsal).
**villous**. With long, soft, unmatted trichomes (diminutive: villosulous).
**viscid**. Sticky.
**whorled**. Either referring to leaf arrangement in which each node bears 3 or more leaves or to the arrangement of flower parts, in which many parts radiate from the center.

# INDEX

**Families in Bold,** COMMON NAMES IN CAPS

Note: only the taxonomic treatments are indexed; index includes both common and scientific plant names.

NOTES